W0107049

Science of
Hard Materials

Science of Hard Materials

Edited by
R. K. Viswanadham
Reed Rockbit Company
Houston, Texas

D. J. Rowcliffe
SRI International
Menlo Park, California

and
J. Gurland
Brown University
Providence, Rhode Island

Plenum Press • New York and London

Library of Congress Cataloging in Publication Data

International Conference on the Science of Hard Materials (1981: Jackson, Wyo.)
  Science of hard materials.

  "Proceedings of the International Conference on the Science of Hard Materials,
held August 23–28, 1981, in Jackson, Wyoming"–T.p. verso.
  Includes bibliographical references and index.
  1. Wear resistant materials. I. Viswanadham, R. K. II. Rowcliffe, David J. III.
Gurland, J. (Joseph) IV. Title.
TA418.26.I55 1981   620.1′1   82-22450
ISBN 978-1-4684-4321-9

ISBN 978-1-4684-4321-9   ISBN 978-1-4684-4319-6 (eBook)
DOI 10.1007/978-1-4684-4319-6

Proceedings of the International Conference on the Science of Hard Materials,
held August 23–28, 1981, in Jackson, Wyoming

©1983 Plenum Press, New York
Softcover reprint of the hardcover 1st edition 1983

A Division of Plenum Publishing Corporation
233 Spring Street, New York, N.Y. 10013

All rights reserved

No part of this book may be reproduced, stored in a retrieval system, or transmitted
in any form or by any means, electronic, mechanical, photocopying, microfilming,
recording, or otherwise, without written permission from the Publisher

PREFACE

This volume contains the proceedings of the first International Conference on the Science of Hard Materials held in Moran, Wyoming, Aug. 23-28, 1981. The objective of the conference was to review and advance the state of knowledge of the basic physical and chemical properties of hard materials and show how these properties influence performance in a variety of applications. To this end, the 49 contributed papers and the four keynote papers by Prof. Fischmeister and Drs. Hintermann, Exner and Almond, present an excellent overview of the state of the art in the "science" of hard materials. The contents of these proceedings also reflect the fact that hard metal technology is now well matured and several aspects of the behavior of these materials are well understood and firmly established.

Structure-property relationships in this class of materials are currently well known. Pitfalls in some of the traditional test methods have been recognized and new test methods are being developed which discriminate between intrinsic material properties and flaw content and distribution. Application of fracture mechanics, although a late comer to the hard materials area (as compared to other structural materials), is rapidly gaining acceptance and new fracture toughness test methods are being developed. Application of modern analysis and analytical techniques to these materials has begun and entirely new and unexpected information has been obtained.

For a variety of reasons, "hard metals" have dominated the research and development scene of "hard materials". To a certain extent this is reflected in these proceedings, and fundamental questions about hard material behavior remain unanswered. Hopefully, the development of new test methods and improved understanding of materials behavior will rekindle interest in other systems that were once prematurely abandoned. Fundamental questions regarding hard materials can only be answered through increased dialogue between hard materials researchers and other scientists. A second conference on the Science of Hard Materials is being planned and will, hopefully, go further in promoting and nurturing this dialogue.

    The editors gratefully acknowledge the financial assistance
provided for this conference by the National Science Foundation,
Reed Rock Bit Co., GTE Precision Metals Group, Cemented Carbide
Producers Association, Kennametal, Teledyne/Firth-Sterling and
TRW Carbide Tools Division. The editors also wish to acknowledge the
colorful and insightful concluding remarks provided by Prof.
Fischmeister which are included in these proceedings. The discussions
at the conference were very lively and an attempt has been made to
keep editorial changes to a minimum to preserve their "human" flavor.

Reed Rock Bit Co., TX   R.K. Viswanadham

SRI International, CA   D.J. Rowcliffe

Brown University, RI   J. Gurland

July, 1983

CONTENTS

CONTENTS

SURFACE TREATMENTS

MECHANICAL BEHAVIOR

NEW HORIZONS

CONFERENCE KEY NOTE PAPER

DEVELOPMENT AND PRESENT STATUS OF THE SCIENCE AND TECHNOLOGY OF

HARD MATERIALS

Hellmut F. Fischmeister

Max-Planck-Institut für Metallforschung
Institut für Werkstoffwissenschaften
Stuttgart, FR Germany

INTRODUCTION

Unlike most other fields of science and engineering, hard materials have not yet become the subject of perennial conferences. Therefore it seems appropriate at the start of this meeting to take stock of our situation, asking ourselves:

- where do we stand today?

We must consider not only the present state of our art but also the changing scenery in which we practice it, and in which the limitation of natural resources now has become an important aspect. We should also ask what we did not know - or even think possible - some time ago. Looking back over the way we came will give courage to us when we return to the toil of slow progress and it may remind us of useful shortcuts.

- what brought us here?

In a field like hard materials, which is cultivated both by scientists and by technologists, we may particularly want to scrutinize the interplay (or confrontation?) of our disciplines. What have we achieved by intuition, practical experience, and trial-and-error, and where has scientific reasoning decided our success? - More important, how should we combine the two approaches in the future?

- <u>where do we go from here?</u>

     In most fields of scientific endeavour, one can see paths in
the distance which appear ready to be walked - if only we knew how
to force the first stretch. A conference such as this seems a good
occasion to identify key problems. Areas of research which have been
abandoned by earlier workers may suddenly assume new promise, thanks
to a new concept or method. Do we see such areas, and such key
ideas?

     To answer all these questions in detail would surpass the scope
of this paper and the capability of this author - it is, in fact,
a fitting goal for a whole conference. I shall try to select some
relevant examples, but the choice will be necessarily subjective.

     The second purpose of this keynote lecture is to review the
state of the art in the field of hard metals (cemented carbides).
Since my choice of topics is dictated by their illustration value,
the review will remain sketchy. Systematic reviews are found in
the literature[1-4].

INVENTIONS AND BREAKTHROUGHS - TECHNOLOGY VS. SCIENCE

     The history of man-made hard materials begins with Moissan's
synthesis of WC in 1900. It might have stopped right there had not
the breakthrough of the incandescent tungsten lamp created a large
demand for ultrahard wire drawing dies, offering immediate rewards
to anyone who could replace diamond by a synthetic material. The
same industry had demonstrated the promise of the new technique of
powder metallurgy, to which the tungsten filament owed its very existance as a commercial product. This made the setting for Karl
Schröter's invention of WC-Co composites in 1923.

     Others before him had realized that WC was a candidate for the
replacement of diamond, and that the brittleness of cast carbides
could be overcome by pulverization and hot reconsolidation[5]. The
principle of liquid phase sintering had not yet been consciously
grasped, but it was known that sintered materials, whose porosity
could no longer be reduced by available pressing and sintering techniques, could be greatly improved by infiltrating their pores with a
liquid metal. It was also known that the densification of tungsten
could be facilitated by additions of nickel, which had to be distill-
ed off subsequently by vacuum annealing[5,6]. When applied to sintered
WC compacts, this techique had, however, given poor quality because
of imperfect penetration of the melt.

     Schröter mixed WC and Co powders and sintered the compacts
"slightly below the melting point of the cobalt". Since the phase
diagram WC-Co was not yet available, he could not know that he was

actually forming a eutectic melt. In all probability, he was also unaware that he was using a principle by which the Incas had solidified platinum dust, using a gold-silver binder[7]. The sintered WC-Co material surpassed its hot-pressed or infiltrated predecessors, and the generic term "Sinterhartmetall" (sintered hard metal) was as much a mark of this success as the tradename under which the product entered the market - Widia, " wie Diamant".

In the immediate sequence (1929 - 1931), a driving force comes into play which has triggered many technical developments: patent circumvention. This gave rise to alloys like TaC-Ni (Fansteel's "Ramet") or Mo_2C-TiC-Ni (DEW's "Titanit", "Cutanit").

The great potential of TiC for the high-speed cutting of steel, which is due to its superior hot hardness and oxidation resistance, was recognized early. Alloying of tungsten carbide with TiC, Mo_2C and TaC was introduced by Schwarzkopf, Comstock and McKenna in the period from 1929 to 1935. Until the late 1950's, the patent situation continued to dictate the development of alloyed cutting grades: Widia X, a WC-TiC grade, Titanit U (TiC-Mo_2C-WC-Co, under Schwarzkopf's patents), and the WC-TaC-Ni-Co grades of Carboloy and Vascoloy-Ramet are famous examples.

During the 1950's, there was still intense competition on the basis of grade chemistry, although the present pattern of cutting grades in the K, M and P categories was beginning to take shape. Understanding of the mechanisms of chip formation and tool wear was developing[8-10]. More and more mechanical engineering expertise was brought into the hard metal field in connection with intense testing of cutting performance and the establishment of entire tooling systems by the more powerful firms in the sixties. The sixties brought the clamped tool insert (throwaway tip) which revolutionized chip cutting productivity. For a time it was feared that hard metal sales would drop because one insert now offered eight cutting edges with less hard metal than the single edge of a traditional soldered tool, but the fears were refuted by the rapid overall growth of industrial and consumer goods production during that decade of affluence. - For a long time, the brunt of cutting tool development shifted to the mechanical field (optimization of edge and chip breaker geometry). Powder metallurgical progress during that period mostly concerned improved production control, giving narrower property and dimensional tolerances, and the scaling-up and automation of production. Attritor milling [11] shortened comminution times, and spray-drying of the mill charge with lubricant added to the milling medium brought improved compactability[12] and better dimensional control.

Wear resisting coatings on hard metal substrates were the breakthrough of the 1970's in the field of cutting tools[13-16]. Fig. 1 shows the spectacular increase of typical cutting speeds

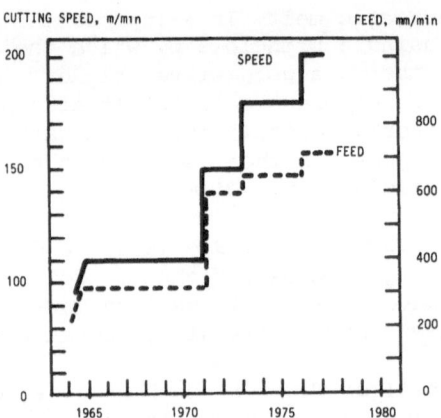

Fig. 1.   *Development of feeds and speeds in typical metal*
          cutting operations. Coatings introduced 1969/70
          (Friberg and Aronsson, Ref.[17])

and feeds[17] which resulted. - For the first time since Schroeter,
we see the introduction of an entirely new concept and process, and
again it is brought about by the marriage of techniques from hitherto unconnected areas. This time, the role of science is much more
obvious to the present-day observer: chemical vapour deposition[18]
could not have evolved without a great deal of basic research into
the thermochemistry of transition metal halides. In fairness, however, let us concede that Moisson's synthesis of WC was as "basic"
a piece of research in 1900 as was halide thermochemistry in the
fifties.

    Important developments were the avoidance of eta phase embrit-
tlement at the coating/substrate interface by carbon adjustment at
the surface; double ($TiC-Al_2O_3$) and multiple coatings (TiC-TiN).
Very likely, the technical developments in the near future will be
channelled by the ownership conditions for the main patents - a
situation reminiscent of bulk alloy development in the 1930's. To
this observer, controlled utilization of epitaxial growth relations
and expansion characteristics of the various layers in coatings
still seem to offer a wide field for research. The wear properties
of various candidates for surface materials (carbides, nitrides etc)
seem to be well enough known. Rather less is known about the mechanical interplay of substrate and coating under actual cutting conditions, and this would appear another rewarding area for research.
An interesting example was described recently by Nemeth[19] who increased the Co content locally beneath the rake face to improve the
toughness of the substrate without loosing wear resistance on the
flank. - Microcracks seem to be unavoidable in coatings, but they
might be put to use following the example of ceramics, which can
actually be made less brittle by introducing finely dispersed
cracks[20].

In contrast to CVD coatings, the metallurgical development of the various carbide grades in the period from 1930 to 1960 was mainly based on inventors' intuition. The wave of systematic research into the crystal chemistry, phase relations, and physical properties of hard substances which emanated from the schools of Hägg in Uppsala, Novotny, Benesovsky and Kieffer in Austria and Samsonov in Kiev provided a basis for understanding the success of the intuitive developments, but the greater part of this worldwide effort remained without impact on hard metal technology, except for providing a systematic map of the scenery which helped hard metal developers to orient themselves in the jungle of possible systems. Great practical value must be ascribed to some of the early phase relation studies which clarified the high temperature reactions between Co, W and C[21,22], W, Ti and C and other industrially important systems[23,24], and of which the majority emanated from the group Kieffer - Benesovsky - Nowotny. They too, however, brought mainly insight after the fact.

Science emerged from backstage to take the lead in the case of one particular group of alloys: the TiC-Ni-Mo grades launched by Ford in 1959. Alloys belonging to this composition family were commercial as early as 1930, but they succumbed to the WC-containing grades because of poor toughness. They entered the scene in another role in the 1950's, when research efforts unprecedented in the field of powder metallurgy were devoted, simultaneously in many countries, to the search for high temperature composites for gas turbine applications - the 'cermets'. The cermet era was perhaps the greatest flop of government-induced research in the history of powder metallurgy in that it brought forth no large scale industrial product. However, it produced spin-offs, some of which are even now bringing late rewards. An early spin-off was a greatly improved understanding of the wetting of solid materials by liquid metals[25-27]. In this context, Parikh and Humenik demonstrated how Mo added to TiC-Ni forms a well-wettable shell of Mo-rich carbide around the poorly wettable TiC. The improved wetting was put to use for optimizing the microstructure of cutting grades. It was understood by then that the brittleness of TiC-Ni grades derives from the paucity of slip planes and the large size of the TiC crystals, combined with their tendency to form networks joined by carbide-carbide boundaries. Improved wetting allowed the production of a fine grained structure with much reduced carbide contiguity, and this brought the TiC-Ni-Mo grades close to the conventional WC-TiC-TaC-Co cutting grades in toughness[28], while preserving the higher wear resistance of TiC.

The next step in the development of the TiC-Ni-Mo grades was initiated in 1974 by Rudy[29] who showed that the miscibility gap in the (Ti,Mo)(C,N) system could be used to effect spinodal decomposition of alloys with certain levels of Mo and N, producing an extremely fine microstructure in which each grain of TiN-rich (α')

solid solution is surrounded by a shell of the MoC-rich α"-phase
which is easy to wet. Good wetting is achieved without introducing
too much Mo (which lowers the wear resistance of TiC) by the strategic deployment of a small amount of molybdenum at the surface of
the carbide grains. Both the shell structure and the very fine
grain size are consequences of the spinodal decomposition reaction;
they are achieved by a heat treatment which would certainly not
have suggested itself to conventional grain size control thinking.
To the author's mind, this is the most interesting application of
materials science principles in hard alloy development, combining,
as it does, Rudy's own research on phase relations in the Ti-Mo-C-N
system with a piece of phase transformation theory that had been
perfected only a few years earlier[30] and with the art of 'microstructure design', to effect an order-of-magnitude improvement
over the empirical state of the art. Commercial success has not
yet materialized; the advent of coatings has reduced the need for
highly wear resistant materials. This work has found an interesting
sequel in Viswanadham's development of (Ti,V,Mo)C-NiMo alloys whose
toughness seems to bring them near candidacy for mining grades. In
these experimental materials, toughness and strength are optimized
by controlled partitioning of Mo between the binder and the carbide
[31], and the plasticity of the binder seems to be improved by the
structure of the carbide/binder phase boundary which enables it to
act as a dislocation source[32].

In the area of mining tools and wear parts, the major breakthrough in the recent past has been the possibility to virtually
eliminate porosity by hot isostatic pressing (HIP) in the early
1970's [33-35]. As will be discussed in a later section, the rupture
strength of hard metals is critically affected by pores which act
as internal notches. HIP treatment increases the average rupture
strength by heating the really poor individuals, which also narrows
the variance within a batch of given size[36,37], cf. Fig. 2.

The importance of pore removal can be visualized when it is
recalled that a single pore revealed during final polishing of a
drawing die or of a Sendzimir roll may cause the loss of an expensive piece of hard metal plus finishing costs already invested. For
tools that are reconditioned by grinding and polishing, pores necessitate as much material removal as actual wear[36].

HIP treatment does not improve intrinsic strength (hardness,
yield strength, fracture toughness). In fact a slight loss of hardness may result from the coarsening of the structure (and the con-
comitant growth of Co layer thickness) during the thermal cycle.
However, the low incidence of internal flaws allows parts of higher
hardness to be used at a required level of rupture strength, making
for a net gain in wear resistance. WC of 0.7 μm grain size with
only 3 w/o of Co can be given enough strength by HIP to be used
for drawing dies at a hardness level of 2000 [34]. Other flaw-sen-

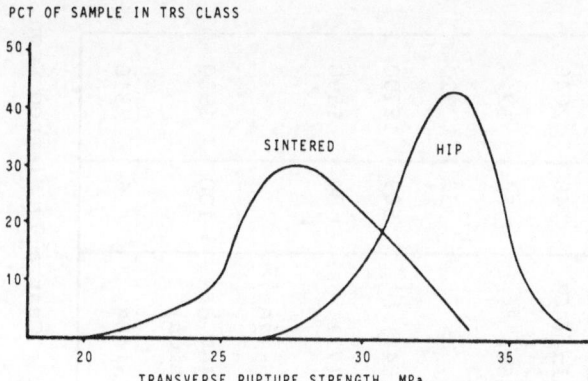

PCT OF SAMPLE IN TRS CLASS

Fig. 2: *Influence of hot isostatic repressing on the trans-*
verse rupture strength of a typical WC-Co rock
drilling grade (Lardner, Ref. 36)

sitive properties, like compressive strength and fatigue strength, are improved in a manner analogous to TRS[37].

HIP is hardly applied to cutting tools, whose performance is less sensitive to pores - with one important exception: the submicron carbide alloys which have to be sintered at temperatures low enough to prevent excessive grain growth. Such grades are used for wear parts of high compressive strength, and these are given low Co contents as in the example above. However, submicron WC grades with about 10 w/o Co are finding increasing application in cutting applications requiring high toughness, which used to be reserved for high speed steels[3,38,39].

THE RAW MATERIALS SCENE

Although the pessimistic view of "Limits to Growth" has been tempered by subsequent analysis, the development of supply and prices in the 1970's has kindled anxiety about the raw materials base of the hard metal industry anew. Within a few years, prices have risen five- to tenfold. Tungsten and cobalt have come to appear critical, and this has engendered a great deal of substitution research. The most developed approaches seem to be the dilution of W by Mo in the hexagonal carbide (stabilized by small amounts of N or B plus Ti, Ta or Hf)[40-42], the partial replacement of Co by Ni or Fe[42], and the substitution of WC-Co by (Ti,V,Mo)C-NiMo alloys[42]. While W has recently become cheaper than Mo per unit weight, Mo is still cheaper per unit volume.

Relevant data are compiled in Table I. The "static resource life" is the ratio of presently known, proven and indicated reserves to present annual consumption. Since both quantities change

Table 1

Metal	Abundance in earth's crust,[43] ppm	Reserves proven+indicated Mt	Geographical distribution (*) WIC %	DC %	CPE %	Static resource life years	Annual consumption kt	Growth Rate %	Country of largest reserves	Price (Aug. 81) $/lb	$/litre
W	1.2	2.05[44]	24.9	12.0	63.1[44]	49[44]–131[44]	42[44]	3.8[44]	P.R.China(46%)	48	2040
Mo	1	8–11[45]	47	35	18[45]	80[45]–105	105[46]	5[45]	USA (46%)[45]	54	1210
Ti	5000	500[43]	39	32	29[47]	16000[43]	30[43]		————	32	320
Ta	2	0.1[48,43]	1.9	98.1[48] not incl.	not incl.	67	1.5[43]		Zaire(64%)	509	18700
Nb	100	11[48]	6.3	97.3[48] not incl.	not incl.	10000	8[43]	5–6[49]	Brazil(72%)	84	1590
V	100	10[43]	97.6	2.4[50] not incl.	not incl.	700–1000[50]	10[43]		S.Africa(76%)[50]		
Co	15	3.7[51]	10	63.	27[51]	118[51**])	30	3.3[52]	Zaire(46% of production)[53]	100	1930
Ni	50	64–75[54,55]	23	62	15[54]	64[54]–105[55] ***)	930[54]	3.8[54]	Oceania(45%)	16	310

*)   WIC = Western industrialized countries: DC = Developing Countries; CPE = Centrally Planned Econ.

**)   with sea nodule potential: $> 10^4$ [51,54]

***) with sea nodule potential: $> 10^3$ [51,55]

with time, the static lives must not be taken literally, but they
provide a scale of relative indicators. "Semi-dynamic life times"
take into account the expected increase of consumption but not the
probable discovery of additional reserves. They are the real doomsday
figures: at an annual growth rate of 4 %, the semi-dynamic life
of tungsten is only 27 years[44] - and tungsten is the main element
in 98 % of present hard metal grades[41].

Fortunately, experience allows us to estimate the rate at
which our knowledge of reserves will be expanded by continued prospection. In many cases, the past growth rate of reserves has been
faster or at least equal to that of consumption. Tungsten is a
good example: from 1956 to 1978 the static resource life of this
metal has increased from 27 to 49 years[44]. Nickel is another case
in point: known land-based reserves have increased from 14 Mt in
1950 to 28 Mt in 1964 and to 75 Mt in 1976[55] - a 'growth' rate
which has outflanked consumption by a factor two. If the reserve
potential of sea nodules is taken into account, the depletion
dates for both Ni and Co move into the very far future. The nodules contain between .23 and .79 % Co and between .38 and 1.28 % Ni.
Five projects which were under way in 1979 would together cover about
15 % of the present Co consumption[54]; another prediction is that
10 deep sea mining operations each harvesting 3 Mt nodules per
year each could cover all of the present Co consumption[51]. However,
there seems to be consensus that nodule harvesting will not become
operational before the 1990's, and when it does, it will be limited
by the market's capacity for the main product, manganese[52]. Mo is
abundant in sea water (1 g/l) but not enough to justify production[56].
Returning to cobalt, we note that in absolute terms it is not a
scarce metal at all. The recent problems were entirely due to political unrest in the regions where cobalt is produced. Similarly,
the main issue with tungsten is its concentration in areas where
supply does not react to demand in the same way as in the free
market economies. Substitution of W by Mo and of Co by Ni would
have a rather limited effect on global husbandry, but it would
create safety margins for the industry in situations of political
disruptions of supply. In a situation of desperate scarcity, nickel
could be more easily made available for hard metals than cobalt,
simply because its yearly tonnage is ten times larger, giving greater scope for savings in less-essential uses. For Mo vs. W, the
tonnage ratio is only 2,5 : 1.

A real scarcity is threatening in the case of tantalum, where
as much as 40 % of the very limited world production went into hard
metals in 1978. The production of tantalum is tied to tin, of which
it is a byproduct[57]. Current production is below demand, and existing inventories are expected to be depleted in the 1990's [92]. Consequently the price of tantalum has increased almost tenfold in
the last five years[48,57]. This has led to a decrease of Ta levels
in cutting grades[58], to dilution with Nb (up to 50 % in Europe,

less in Japan and in the USA), and to attempts to replace TaC by (Nb,Hf)C, so far only in the laboratory[59]. In WC grades, TaC serves as a grain growth inhibitor; conceivably, tantalum could be saved by low temperature sintering and hot pressing.

Although serious scarcity of W and Co, Mo and Ni does not seem imminent, there is bound to be an ultimate limit. The 'growth' of known reserves does not create resources - it only brings us closer to the real bottom. In the long perspective, the technology of hard materials must shift to the really plentiful materials such as Ti. For the next two decades, the economic incitament for this will in all probability be insufficient for serious industrial engagement; here lies perhaps the greatest challenge for basic , foundation- or government-sponsored research in the hard materials field.

In this context, the cutting grades are not the most critical: already now they contain a good deal of the more plentiful metals, and coatings will further alleviate the situation[17]. The rapid development of ceramic[60] and superhard cutting materials will remove another part of the load on W, Ta, and Co. Promising developments are sintered diamond and cubic boron nitride[1,61], transformation-toughened alumina[62,63] and - perhaps - the sialons which have the great advantage that they can be sintered without high pressure. All of these are made from plentiful materials. Diamonds and CBN require large amounts of energy, mainly for the recycling of high pressure armature parts.

The real problem lies in the rock drilling grades where no fully satisfactory replacement has yet been found for the classical WC-Co compositions. The closest approach so far to the toughness-hardness combination of WC-Co is offered by Viswanadham's experimental (Ti,V,Mo)C-NiMo grades[31]. The only alternative based on really plentiful raw material is sintered synthetic diamond[64,65] (with the reservations expressed above); if more widespread application of this material should become technically and economically viable, it would also engender drastic changes of drilling technique.

Meanwhile, recycling is the obvious way of resource conservation, and it is rapidly growing in volume. The Coldstream[66] process has been practiced for a long time, and the quality of the product allows large amounts to be added to virgin powders. More recently, leaching of hard metal binder by Zn to free the carbide[67] has come to fruition, and plant capacity for this process is increasing. Neither of these processes has good capacity for chemical refining, flotation being the only practicable method. By roasting and alkaline leaching, material could be brought back into the chemical cycle for virginal material, but at great cost. The chlorination process developed by the Axel-Johnsson company of Sweden[68] offered excellent promise for purification, but has not survived economically. Increasing price levels might make it a candidate for consideration again. Let us note in passing that increasing dilution of W by Mo would create serious recycling problems with existing processes. Further research on recycling processes must certainly remain a first-order priority.

In concluding this chapter, let us throw a quick glance at the effects of outside technology changes. Machining is steadily moving towards higher cutting speeds and more stable machines, which would favour ceramic grades and diamond tools. Grinding, electrochemical and laser machining will make inroads on conventional cutting and milling, and so will design changes which substitute machined parts by precision castings, stampings and drawn parts. This process has gone on for a long time, yet its effect on the consumption of hard metal tools seems to have been offset by volume growth and an increase of difficult machining operations (e.g., superalloys). Coatings have lowered the growth rate of cutting tools less than was originally expected (however, the present level of 25 % coated tools is still far below the expected saturation level[58] which some experts put as high as 80 %). Some inroads on WC-Co cutting grades will also be made by sintered and coated[15] high speed steels.

Mining tools have seen a still more spectacular growth, and this will certainly continue. One reason is the acute need for energy prospection (oil, gas, geothermal drilling), another is the shift from fluid to solid mining in the energy field. Coal for envisaged large gasification and liquefaction plants must be cut, reamed and crushed, and gangue must be moved, calling for large volumes of drilling and cutting bits and wear parts. In mining for minerals and metals, there will be an inescapable movement towards lower grade ores and deeper levels, leading to an exponential increase of rock drilling and cutting[69]. Since there will always be considerable dissipation of hard metal in mining, great demands will be placed on the raw materials base for the tools.

MICROSTRUCTURE - THE KEY TO PROPERTIES (?)

Following the pioneering work of Gurland[70-72] the microstructure of cemented carbides is customarily described in terms of four parameters which can be measured by lineal analysis[73-77]

- f, the volume fraction of binder (= $1-f_c$, the volume fraction of carbide)
- λ, the mean free path in the binder phase or 'binder layer thickness' (measured along a randomly oriented straight line)
- d, the linear intercept size of the carbide grains
- G, the 'contiguity' of the carbide grains, i.e. the fraction of the total surface of carbide grains which is shared with other carbide grains[73,74,78]

The four are related to each other by[79]

$$\frac{\lambda}{d} = \frac{f}{(1-f)(1-G)} \tag{1}$$

This relation allows the direct measurement of the cobalt layer
thickness (which is frought with image resolution problems) to be
replaced by more reliable determinations: G only requires the
counting of carbide and cobalt layers grains along the test line,
d is usually large enough to be measured satisfactorily, and f is
known from chemistry.

The concept of mean free path seems to have been used as early
as 1942 by Gensamer, Pearsall, Pellini and Low[80] to describe the
strength of pearlitic steels. It became widely popular owing to the
intuitive appeal of Orowan's 1947 model for particle strengthening
[81], which - although intended and valid only for a special group
of materials under special circumstances - inspired model building
almost throughout the field of physical metallurgy. More appropriate
to hard metals, and soon equally popular, was the Hall-Petch model
of 1951-1953 for the flow of polycrystals[82,83]. This and the beginning interest in the quantitative description of microstructures
[84,85] set the scene for the first major success in the area of
microstructure - property relations for particle composites. The
cemented carbides have remained the model case in this area[86,87]
ever since.

The mean free path approach proved especially useful for the
description of hardness[88-90] and coercive force[79], and also -
at least so it seemed at the time - for rupture strength[88], in
that it allowed a unified description of these properties mainly
in terms of one important variable (Fig. 3). - Beside the mean
free path, also the carbide grain size and its distribution[79] and/
or the volume fraction of binder phase showed some influence but
these were less important. Contiguity was found to correleate strongly with the rupture strength of WC-Co alloys'[1], cf. Fig. 4. In
alloys containing TiC or other γ-carbide formers, the grains of
γ-phase develop long contiguous chains, and the contiguity of the
γ-phase has been found to have a strong influence on the hardness
[90,91] and rupture strength[91] of these alloys.

The concept of contiguity took some of the fuel out of an old
issue which had been pursued with great acerbity and experimental
ingenuity, but no decisive results, by earlier researchers: do the
carbide grains form a skeleton, or are they all separated by thin
layers of binder? Contiguity provided a measure of the degree of
carbide/carbide contact formation, and the idea that this could vary
in wide limits replaced the polarity "contacts throughout" vs. "none
at all", helping to explain such observations as 'the considerable
residual strength of carbide specimens from which all cobalt had
been leached out[92]. The underlying problem remained: is there a very

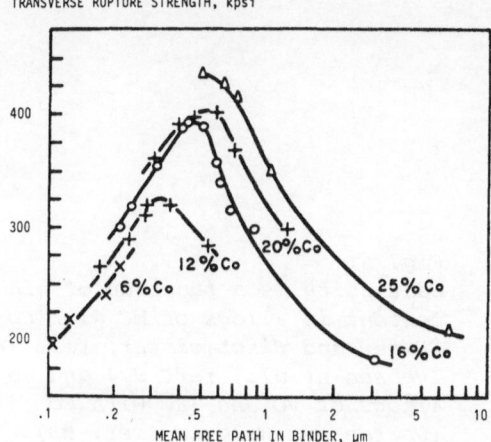

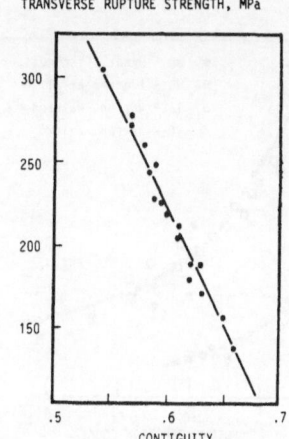

Fig. 3: Transverse rupture strength as a function of mean free path in the binder phase Gurland 1955 (Ref. 88)

Fig. 4: Influence of contiguity on rupture strength. Gurland 1963 (Ref. 71)

thin cobalt layer - non-leachable, one would have to imagine - between carbide grains apparently in contact? The observation that finite dihedral angles form at the joints of such boundaries with the binder phase suggests that cobalt is not present (at least in its normal state) in the boundaries. Recent STEM-measurements by SHARMA et al[93] appear to this reviewer compatible with the assumption that carbide-carbide boundaries are formed, in which Co is enriched, following the general tendency that elements which are sparingly soluble in the lattice segregate to the grain boundaries [94]; the authors themselves seem to prefer the interpretation of a very thin Co layer precluding the formation of a continuous carbide skeleton. Still higher resolution (both spatial and chemical) is attainable by field ion microscopy (FIM) combined with an atom probe, and this technique has recently been applied to the problem [95], indicating that some Co seems to be always present in WC/WC boundaries. However, preparing a FIM specimen so that a grain boundary is present in the narrow tip which can be imaged is difficult, and the technique does not really invite statistical study of many boundaries. In addition, the detection of some Co between WC grains does not distinguish a WC-WC boundary enriched with segregated Co from a sandwich WC-Co-WC where the Co is present with the properties of the bulk metal. The technique of lattice fringe imaging in high resolution electron microscopy[96] could make that distinction, and such studies will no doubt be performed in the near future.

Contiguity changes slowly during sintering[70,76]. For practical purposes, it can be assumed to be a function only of cobalt content in alloys produced (by different producers) according to normal practice (Fig. 5).

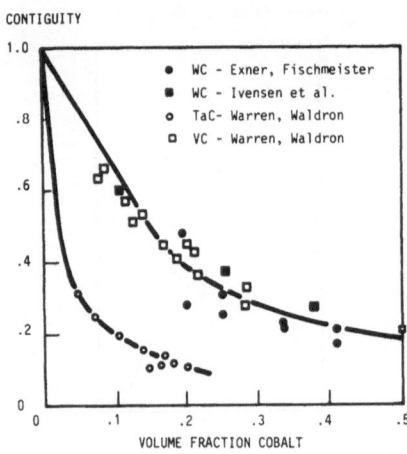

CONTIGUITY

- ● WC - Exner, Fischmeister
- ■ WC - Ivensen et al.
- ○ TaC- Warren, Waldron
- □ VC - Warren, Waldron

VOLUME FRACTION COBALT

Fig. 5:
Contiguity is a function of binder
content in alloys of WC with Co
(Exner and Fischmeister, ref. 76,
Ivensen et al., ref. 97) and in
alloys of VC and TaC with Co
(Warren and Waldron, ref. 98)

The usefulness of the structural parameters λ, d, f and G for the empirical quantification of structure - property relations was verified by many subsequent studies, and has been extended to such properties as the 'work of crack initiation' and fracture toughness, as will be discussed below. The evidence available at the end of the 1960's is summarized in a review by Exner and Gurland[99]. However, the relations formulated at that time were essentially empirical, backed by rather sketchy theoretical concepts. Nevertheless, they served as guidelines in the development of improved grades, e. g., when the beginning industrial synthesis of diamonds called for superhard grades, and the answer was found in finegrain carbides with a very low binder content. It may be argued that this was just an extrapolation of practical experience; as an example of such a view, we quote from a review by Lardner in 1977 [3]: "At the end of the 1960's, metallurgical developments had become almost stagnant, and little had been achieved in explaining the unusual properties of the alloys". The present author does not fully agree to this view, but it does contain an important truth: that scientific models were slow to be accepted by industry at that time, partly, as Lardner points out, because the industry comprised many small companies who did not have the resources to keep in touch with scientific results or to carry on strong research efforts of their own. This situation changed drastically during the following decade in which most small companies were absorbed by those big enough to bear the increasing cost of research.

During the 1970's there followed a period of critical reappraisal and, later, of refinement of the the theory of structure - property relations. The criticism had two independent causes. The first was the discovery, by Suzuki[100,101] and others [102-104], that the properties of the binder phase were strongly affected by its content of dissolved tungsten and carbon, which can vary from 2 to 26 % for W and from 0.04 to 0.12 % for C, according to a constant solubility product[4]. This means that small

changes of the carbon content, which will result from varying carbon activity of the sintering atmosphere, will produce great changes in the tungsten content of the binder. Similar effects occur in
the Ni binder of TiC-Ni-Mo alloys[105,106]. Tungsten affects the binder properties either by solid solution strengthening, which is
strong because of the large difference in atomic radii, or by precipitation of intermetallic phases. There may also be an indirect
effect due to an influence of tungsten content on the grain size of
the binder phase[107]. The mechanical properties[37,104] and the precipitation behaviour[108] of synthetic binder phase alloys have been
studied. Attempts to improve the properties by intentional precipitation treatments have brought modest improvement in hardness
at the expense of toughness[101,102], but in time the understanding
gained led to closer control of the residual gas composition in
vacuum sintering. Some results concerning the relation of sintering
atmosphere to properties have been published recently[108,109] and a
solid basis for processing has been established by the redetermination of the ternary system W-Co-C with complete carbon isoactivity
lines[110,111]. - Until these phenomena had been fully elucidated,
the binder composition remained an uncontrolled degree of freedom,
and properties influenced by it were subject to scatter which made
the property - structure relations suspect.

A second point which discredited the earlier measurements was
the observation by Andersson[112] and Suzuki[113], that rupture of cemented carbide specimens is normally induced by internal flaws
which can be detected by scrutiny of the fracture surface. By careful avoidance of porosity-creating impurities and foreign inclusions,
Andersson succeeded in increasing the average rupture strength of
WC-Co grades by ca. 20 %, simultaneously narrowing the spread between individual specimens. For a time then, all ideas about structure-property relations were rejected as based on spurious property
values. Only recently has a more balanced view resulted from the
application of fracture mechanical thinking. While the initiation
of fracture is due to defects, their propagation is determined by
the microstructure. Inasfar as common production practice at a
given state of the art creates a roughly constant defect population
(and there seems to be surprising similarity in the defect populations of carbides from different producers), comparable rupture
strengths result. They reflect the product of the defect distribution *and* the material's ability to "disarm" cracks by blunting
or plastic energy dissipation. The latter is, of course, a function
of the material's plastic flow behaviour and therefore of its
microstructure. This will be discussed in more detail in the section on toughness.

The hardness of cemented carbides, and such other properties
as thermal conductivity and elastic moduli, are not influenced
by sporadic defects. Their explanation must still be sought in the
microstructure and the properties of the constituents. Exact model-

ling of the properties of particle composites is a very difficult
task. The soundest results had been obtained, at the end of the
1960's, for the elastic moduli[114,115], using an upper- und lower
bound approach which produced a fairly narrow prediction band that
agreed well with experimental values. The models used were dispersions of mono—size cubes or ellipsoids in a matrix. A different
approach was tried in the author's group: to calculate by the finite element method (FEM) the elastic response of an actual microstructure (reduced to the two dimensions of a plane section, but
within that restriction, with realistic shapes and distributions
of the phases), cf. Fig. 6. The results were surprisingly good and
encouraged the use of the method also for forays into the elastic-
plastic behaviour of WC-Co composites where it proved possible to
model the mechanical hysteresis which occurs on loading and unloading of such materials[116-119].

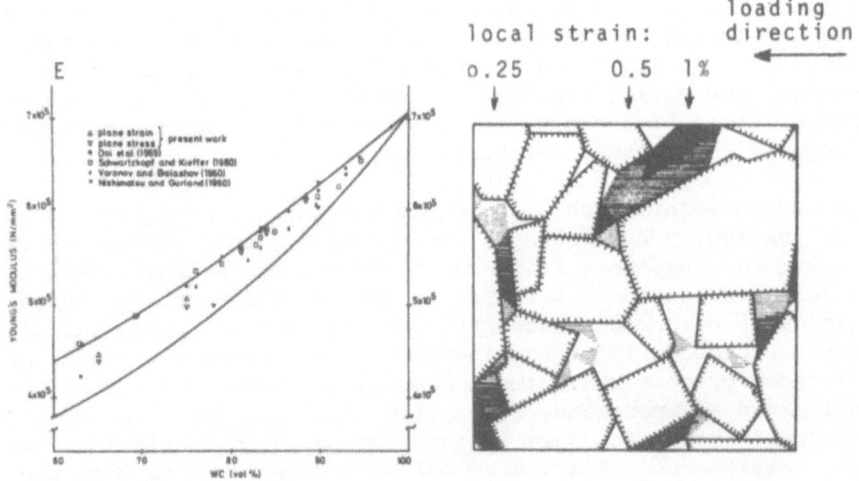

*Fig. 6: Elastic modulus of WC-Co alloys, calculated by
FEM and strain distribution in the binder phase
from FEM calculations   Jaensson and Sundström
(Refs. 116, 117)*

Modelling *plastic* properties of two-phase microstructures is
much more difficult, because there is a change in the load distribution between the phases when one of them begins to flow plastically.
Yet, the past decade has brought significant advances. A rather extensive study of the deformation behaviour of two-phase materials
with varying phase properties, volume fractions, and mean free path
lengths (including the systems WC-Co, α-iron-martensite, α-iron-
pearlite, Cu-CuAl martensite and -pearlite) dealt with the way in
which the individual phases deform in the composite[87]. In particular, the partitioning of stress and strain between the phases was
studied experimentally and by FEM modelling. It was found that the

behaviour of the phases could be described by a simple mixture rule originally proposed by Tamura[120] which states

$$\sigma = f_\alpha \sigma_\alpha + f_\beta \sigma_\beta \qquad (2)$$

$$\varepsilon = f_\alpha \varepsilon_\alpha + f_\beta \varepsilon_\beta \qquad (3)$$

$$(\sigma_\beta - \sigma_\alpha) = (\varepsilon_\beta - \varepsilon_\alpha) \cdot q \qquad (4)$$

The coefficient q in equation (4) is a function of the strain which the system has already sustained; it increases with increasing strain, but at present it cannot be calculated apriori. For a laminated composite, q = 0 would signify series loading (equal stress), and the other extreme q = ∞ would signify parallel loading (equal strain). Increasing deformation seems to move all particle composite systems towards the parallel loading situation[87], but this occurs only at strains much beyond what can be encountered in hard metals.

At small deformations, the strain in the WC phase will depend on the carbide contiguity. If all WC particles were completely surrounded by a contiguous Co phase, the material would deform by flow of the binder between the rigid carbide grains. In the opposite extreme, if the carbide grains formed completely continuous chains throughout the specimen, large scale yielding could not occur without deformation of the carbide phase. It stands to reason that the strain in the carbide phase would be greatest at carbide/carbide grain boundaries constituting narrow links of the chains, and this is borne out by slip line observations by Bartolucci-Luyckx [121] and by Lee and Gurland[122]. The latter authors demonstrated that long WC chains must exist in all WC-Co alloys containing up to 50 % binder, and similar conclusions were drawn by Warren and Gratwohl[123] for TaC-Co alloys. The length of the chains cannot be determined from metallographic sections, but it can be replaced by a parameter called the contiguous volume fraction of the carbide phase (Lee and Gurland[122,124]):

$$f_{cont} = f_c \cdot G \qquad (5)$$

Applying plastic limit analysis to the yielding of an idealized structure of continuous carbide chains interspersed with (isolated) binder regions, Lee and Gurland[122] arrived at identical expressions for the lower and upper bounds:

$$\sigma = \sigma_c f_c G + \sigma_b (1 - f_c G) \qquad (6)$$

In terms of the general mixture rule[87,120] the model implies that at the yield point of the composite, both phases are subject to equal strain (q = ∞), which means that most of the load is carried by the continuous carbide phase. In view of observations on strain and stress partitioning in other systems, the equipartition of

strain is probably a crude approximation, but it seems to be good
enough in practice as demonstrated by the good fit of equation (6)
with experiment (fig. 7). (The figure relates to hardness rather
than yielding; in plastically homogenous materials, hardness reflects the flow stress at about 7 % strain[125].)

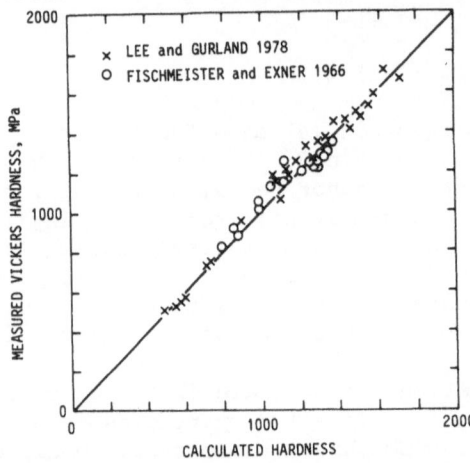

*Fig. 7: Calculated and measured hardness values for WC-Co
alloys according to the model of Lee and Gurland
(Ref. 122)*

In applying equation (6), the 'in situ' yield stress of each
phase has to be inserted. This is different from the yield stress
of bulk single phase specimens because of the limited free path
for dislocation movement. Hall-Petch type relations have been found
both for the binder layers in WC-Co[126,127] and for binder-free
WC[122]:

$$H_\beta \;=\; 304 \;+ 12{,}7 \; \lambda^{-1/2} \; Kg/mm^2 \qquad\qquad (7)$$

$$H_\alpha \;=\; 1382 \;+ 23{,}1 \; d^{-1/2} Kg/mm^2 \qquad\qquad (8)$$

Thus, a satisfactory understanding of the elastic and plastic properties of cemented carbides seems to be within reach. It is more
sophisticated than the early mean-free-path approach, but should
be viewed as a refined confirmation of that approach. Its most
important application is likely to be in the field of fracture
toughness.

TOUGHNESS - A PUZZLE AND A CHALLENGE

Rupture strength

The scatter of rupture strength in cemented carbides was for
a long time considered due to problems of measurement. In 1962
Gücer and Gurland proposed a stochastic theory for the strength of

hard metals[128,129] which also predicted distributions of fracture
strength. Their approach is still of interest: they applied weakest
link statistics to two models of a cemented carbide structure, one
with completely connected and one with completely dispersed carbide
grains, and arrived at the conclusion that the completely dispersed structure should be about 7 times stronger. Gurland's ideas
stimulated several Swedish workers to apply Weibull analysis to the
strength distribution of hard metals[79,130,131] – mainly because it
promised a way of describing and predicting the *volume dependence*
of fracture probability. Weibull's analysis[132] is based on the
concept of intrinsic points of weakness randomly distributed
throughout the material. The great interest and effort which at that
time was devoted to residual thermal stresses in hard metals*)[133,
134] turned speculation about the nature of these points of weakness at first towards sites of high residual stress[79], which was
not a fruitful direction.

   At any rate, there was a growing awareness in the early 1960's
that the scatter of transverse rupture strength was not merely a
nuisance but that it would some day reveal important insights. At
the author's laboratory at Söderfors, statistical analysis of TRS
values was used routinely in a survey of carbides from a wide range
of commercial sources and of experimental alloys intended originally
as a stringent test of the structure - property relations discussed
above. In the course of this study, Anderson discovered that most
materials were made up of two different populations, yielding bimodal strength distributions. In all specimens of the low-strength
populations, fracture could be traced to pores or inclusions, even
though all metallographic pore ratings were better than A1/B1 - considered safe and harmless at that time. The fracture-initiating defects were found to be much larger than A1/B1; they had escaped
metallographic detection by virtue of their scarcity, but would be
brought out by the fracture process. In the high-strength specimens,
the fracture origin could not be associated with visible defects,
and Anderson concluded that these specimens represented the intrinsic
strength of the material. Ways were sought to produce defect-free
specimens. The pores were thought due to impurities which developed
a high gas pressure when enclosed by liquid binder at the sintering
temperature, and therefore a high temperature vacuum treatment was
used to drive off all volatile material from the carbide powder.
Hygiene in powder handling was also improved in order to eliminate
non-metallic inclusions. A striking improvement resulted.

   A short description of these results was published three years
later[112]. It caused a good deal of speculation about the way in
which the pores had been eliminated; this was described in 1976
[136]. Anderson's work focussed the attention of others on fracto-

---

*) For more recent results on residual thermal stresses in hard
    metals see[4,135]

graphic studies of internal flaws. Bartolucci-Luyckx found that
cavities on fracture surfaces contained remnants of Si, Al, Mg, Fe,
Ca, S and Ti even when apparently empty[137,138], and she observed
that the WC grains around inclusions exhibited slip lines, which
did not occur elsewhere on the fracture surface: indication of
plastic deformation in connection with the stress release at
fracture. Suzuki[113] verified the importance of pores and established an empirical correlation between the stress (at fracture)
at the site of the responsible flaw, and its size[139-144]. The role
of pores and exogenic inclusions was further verified by Amberg and
Doxner, who analyzed various causes of porosity and gave a detailed
description of hygiene measures to avoid such defects in the manufacture of critical parts[142]. The thermochemical basis for the removal of pore-forming impurities by high-temperature treatment is
beginning to be established[143].

During that period, the concepts of fracture mechanics were
spreading from the specialists to the science community at large,
and the first fracture toughness data for hard metals became available[144,148]. Exner, Walter and Pabst[144] used the fundamental fracture mechanical relation

$$a_c = Q \cdot (K_{Ic}/\sigma_r)^2 \qquad\qquad (9)$$

(a_c = size of the defect initiating flaw, Q = a constant reflecting
the geometry of the flaw; K_{Ic} = fracture toughness; σ_r = rupture
strength of an individual specimen) to derive hypothetical flaw
size distributions for sintered alumina and for hard metals. Similar
ideas were developed independently in the field of high speed steels
by Johnson[149] and Johansson[150]. In using equation (9) to derive
flaw sizes, one has to know the value of Q which is appropriate to
the shape of the defect. Exner chose the theoretical value for a
penny-shaped internal or half-circular surface crack. Almond and
Roebuck[157], Johansson[150] and the present author[152] measured the
sizes of enough flaws to derive an empirical value of Q from a plot
after the manner of fig. 8, where Q can be found from the slope of
the regression line through the origin when K_{Ic} is known.

In Almond's material, the strongest among the specimens no
longer showed visible flaws at the fracture origin; the fracture
seemed to originate from large WC crystals or aggregates. Such
fractures occurred at constant stress, indicating that the "limiting
strength" of the material had been reached.

In lower quality material studied by Fischmeister et al.,sever-
al types of defects could be distinguished, each having its own Q
value in accordance with its notch severity. After hot isostatic
pressing, the voids disappeared, but the other defect types survived
as would be expected. The intrinsic level was not reached.

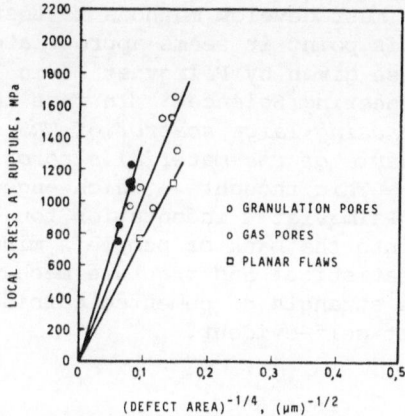

*Fig. 8: Relation between rupture strength and flaw size in
a number of hard metals containing different sorts
of flaws (Fischmeister, Diblik, Kompek, Ref. 152)*

Fracture mechanics, and the fractographic study of flaws, has
brought us a new understanding of rupture stress: the distribution
of TRS values reveals the distribution of defects in the material,
and it does so much more efficiently than any other method known
to us. Fracture toughness, on the other hand, is goverened by the
microstructure. It controls the sensitivity of the material to a
given size and shape of defect - because it reflects the material's
ability to prevent the growth of cracks by plastic blunting and
energy dissipation at the crack tip. With a "constant" distribution
of pores and inclusions (which will result from constant manufacturing practice), TRS will be determined by the resistance to crack
propagation. The probability of pore formation during sintering depends systematically on Co content and carbide grain size. These
two facts together explain the approximate agreement of various
authors' results on the "structure dependence" of rupture strength.
Lindau[153] has proposed an explanation along partly similar lines,
assuming constant flaw size but recalling that the fracture toughness increases with λ, which would account for the ascending branch
of Gurland's plot. The descending part is ascribed simply to the decrease of flaw sizes in a regime where the preexisting flaws are
too small to cause fracture.

Thus, rupture strength now assumes the character of a sensitive
(albeit destructive) monitoring method for internal defects[154,155].
Recognizing the volume dependence of rupture strength which results
from the stochastic nature of crack nucleation, we should use the
standard rupture test bar for large cutting tools, wear parts and
mining tools where in operation the stressed volume is of similar
magnitude to that of the test bar. For assessing the toughness of

metal cutting edges, we must develop methods to test much smaller
stressed volumes. At this point it seems appropriate to quote the
following from an address given by Palmqvist[156] in 1954 (!) to the
Swedish Academy of Engineering Sciences: *"In fact the fracture seeks
out internal defects, causing large scatter of TRS values and making
these values more a measure of the material's porosity than of its
intrinsic properties"*. - This thought - which engendered the subsequent development of Palmqvist's indentation toughness test -
seems to have receded into the back of people's minds for a very
long time, until the statistical and fracture mechanical understanding of the rupture strength of cemented carbides brought it
back as something almost self-evident.

Palmqvist's surface crack resistance

Palmqvist's concept of measuring the cracking resistance of
a hard and brittle material by evaluating the cracks formed around
indentations was originally an attempt to complement the rupture
test by something aiming at a different, but no less important
aspect of toughness. Its fate through the years since its introduction in 1957[157,158] would be a case study for the tortuous progress of ideas as interesting as the story of porosity and rupture strength. Here we can only record its main turning points.
Initial doubts about the reproducibility and relevance of the
measurements were resolved when the influence of residual stresses
due to grinding was clarified[118,159-161]. It is now acknowledged
that with proper precautions[159], surface preparation is no longer
a cause of undue scatter[162]. While the *onset* of cracking at the
corners of a Vickers pyramid indentation is hard to observe and
to reproduce, the *extension* of the cracks per unit load increment

$$W = \frac{dL}{dP} = \frac{L}{P} \tag{10}$$

can be measured with good accuracy, and it is now fairly agreed
that this quantity constitutes a material property, called the crack
resistance[159] or the surface toughness.

Despite some criticism[163] there seems to be a growing con-
viction that the crack resistance is a reflection of practical
toughness behaviour, especially with respect to those aspects which
determine the performance of mining tools [164-166] . Confidence
in the method has been strengthened by the development of theoretical understanding of cracking at indentations[167-174]. Three different types of crack may form at indentations[167,173], depending on
the degree of brittleness: Hertzian cone cracks, lateral vents
(planar cracks emanating radially from the indentation) and median
vents (e.g., penny-shaped or half-penny-shaped cracks centered on
the impression and propagating in the direction of the indentation
force). The Palmqvist cracks belong to the second group. Because
the shape and depth of the crack front and the stress field around

the indentation are complicated, the relation of Palmqvist crack
resistance to bulk fracture toughness K_{Ic} (or critical strain
energy release rate G_{Ic}) is also complicated, but available analyses
indicate that a relation should exist[169,172,174,175]. The theoretical expressions vary according to the treatment of the stress field
and the assumptions made for the crack contour, and their predictions
are only in qualitative agreement with experiment. Practically useful, semi-empirical approximations have been developed[164,165,168,175].

    Assuming that G_{Ic} and hardness are both determined primarily
by the mean free path in the binder, Viswanadham and Venables[164]
have proposed the following simple form for the analysis of crack
resistance data

$$\frac{1}{W} = A(H - H_o) \tag{11}$$

where the slope constant A should be fairly insensitive to the
microstructure and chemistry of the alloy, while H_o reflects the
in situ hardness of the binder. Experimental values conform to
equation (11) as shown by fig. 9. In this plot, each alloy system
(meaning a given combination of binder and carbide, but with varying carbide content and grain size and thus with varying free path)
is represented by a single straight line. The toughness potential
of the alloy system is the greater, the farther to the right this
line is situated. Relying on the constancy of the slope constant
A, the toughness to be expected at various hardness levels can thus
be predicted on the basis of Palmqvist-type measurements on a single
alloy. This method of toughness characterization seems to be gaining
acceptance, at least for screening purposes.

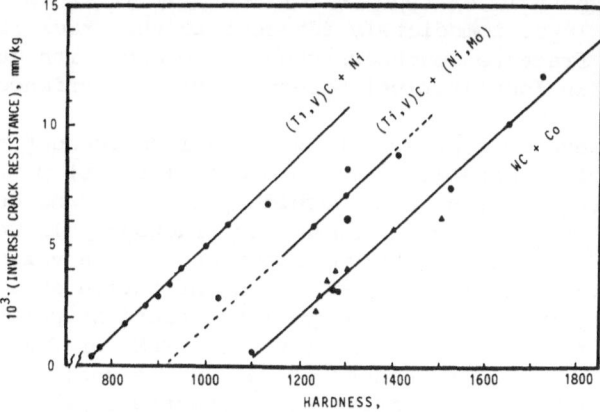

*Fig. 9: A plot of crack resistance vs. hardness shows the
'toughness potential' of different alloy systems
(Viswanadham and Venables, Ref. 32)*

In a very thorough study of the influence of experimental
parameters in the Palmqvist test, Almond and Roebuck[150] have cri-
ticized that the cracks propagate through heavily deformed and
work-hardened material. This is true, but it may actually make the
measurement more applicable to operating performance since mining
tools suffer heavy deformation in both the carbide and the binder
phase.

Fracture toughness

Fracture toughness (K_{IC}) and the related critical strain
energy release rate G_{IC}, reflect a material's ability to impede
crack propagation by energy dissipation. In principle, the following
types of energy sink may contribute to this:

   (i) Surface energy of the crack surface, which may consist of
       four types:
       "c" - transgranular in the carbide
       "cc" - intergranular in the carbide
       "cb" - carbide/binder phase boundary separation
       "b" - transgranular through binder
  (ii) plastic deformation of the binder
 (iii) plastic deformation of the carbide
  (iv) microcracking of carbide grains in the vicinity of the
       crack[176]
   (v) crack branching (which has been observed in hard metals
       under compression[177])

Many attempts have been made to construct physical models for the
fracture toughness of cemented carbides and its dependence on
microstructural parameters[153,178-184]. All of these recognize the
role of plastic deformation in the binder, seeing that a large part
of the fracture path is of type "b" or "cb". If it is assumed that
only the binder layer immediately adjacent to the crack is plasti-
cized, then the fracture toughness should decrease with decreasing
mean free path, in (qualitative) agreement with experience.

The next stage of sophistication takes into account the fact
that the ratio of carbide (c + cc) to binder (cb + b) parts of
the fracture path increases with carbide grain size and decreasing
binder content. Despite a great amount of fractographic work[16,71,
72,88,92,177,184-198], the relative importance of the various fracture modes is still a matter of debate. Quantitative analyses have
been attempted by metallography of polished sections perpendicular
to the crack plane, by evaluation of replica-TEM or SEM fractographs,
by X-ray fluorescence measurements of Co and W signals from the fracture surfaces, and by quantitative Auger spectroscopy.

The metallographic method has insufficient resolution to safely
distinguish types cc, cb, and b. With SEM fractography, reliable

distinction of cb and cc paths is problematic, and small patches
of b and c type fracture may be overlooked. There is also a stereological problem involved in the assessment of true area fractions
from projected images of fracture surfaces when the angular orientation of the fracture facets in the two phases is dissimilar[199,
200]. Determinations by X-ray fluorescence are falsified by contributions from the bulk below the fracture surface. This error is excluded in the case of Auger spectroscopy[184,198], which is a truly
surface sensitive method. However, segregation of Co to cc-bounda-
ries as well as remnants of Co on the carbide side of cb fractures
will give rise to excessive values for binder-type fractures (cb+b).
Lea and Roebuck[198] have shown that ion etching of the surface
reduces the Co signal from WC-Co fracture surfaces. A saturation
level achieved by prolongued ion etching should reveal the true
ratio of binder and carbide fracture paths (with reservations for
details of the sputter process). - These Auger studies have been
made with spectrometers of limited lateral resolution, averaging
over relatively large areas of the fracture. New high resolution
spectrometers would allow a more direct assessment of the role of
binder and carbide fracture paths. No doubt such measurements will
be made in the near future.

    The present situation of fracture toughness modelling is one
of reasonable qualitative understanding, but without a truly conclusive model. Hard metals offer too narrow a range with respect
to both fracture toughness and microstructural parameters to allow
clear differentiation between models on the basis of agreement between experimental and calculated data - especially in view of the
scatter of the measurements. To make matters worse, recent results
on the influence of precracking techniques cast serious doubt on
the validity of many measurements hitherto available[201]. The
development of precracking methods has been the subject of considerable effort[144,145,147,180,182,202-204]; the most consistent
results beeing obtained with a short round bar with a chevron notch
[202].

    The practical value of fracture toughness as a property of
hard metals has often been questioned, because no dramatic differences occur between grades of widely different performance. We
should recall, however, that fracture toughness decides the sensitivity of all hard metal products to preexisting flaws. In the
special case of coated inserts the relevant flaws may be cracks in
the coating[205], making the fracture toughness of the substrate
the decisive property for allowable coating thickness. An understanding of fracture toughness based on a true physical model could
certainly be an essential element in the philosophy of hard alloy
design.

'Toughness' of metal cutting tips

    The term 'toughness' is often used by tool engineers for a
rather fuzzy conglomerate of tool properties. When hard pressed,
they will retrench to two essential symptoms, edge crumbling or
'microchipping', and premature wear in intermittent cutting. There
is some connection between the two, in that early chipping can change
the effective tool geometry so as to produce severe wear later on.
However, intermittent cutting performance should also be strongly
influenced by the tool material's resistance to thermal and mechanical fatigue, as indicated by the characteristic crack patterns
which appear on milling edges.

    All the ideas discussed above pertain to large pieces of hard
metal, such as rolls, drawing dies, of drill shafts - cases of
'macrotoughness'. For the performance of cutting edges, an entirely
different philosophy of 'microtoughness' is needed, because we are
dealing with stressed volumes two or three orders of magnitude
below those in a conventional rupture test bar[155].

    The stressed volume will normally be too small to contain
coarse flaws of the type discussed above. In addition, one would
expect the composite to become tougher at high temperatures, making
the critical flaw size still larger (note, however, one recent and
surprising indication that the fracture toughness of hard metals
declines with temperature[206].) Suzuki[207] and Ueda et al.[208]
have shown that above 700 °C, fracture in hard metals is not initiated by the type of defects which are responsible for room temperature rupture; in normal bend specimens fracture seems to ema-
nate from grain boundaries of the binder phase[107]. In the small
volume of a cutting edge, this is again a very unlikely mechanism.

    Possible but purely speculative crack initiators in the chip-
ping of cutting edges could be large carbide grains or boundaries
between such grains. Another, equally speculative mechanism would
be the propagation of cracks (statically or by fatigue) from fis-
sures created at carbide/binder boundaries by thermal expansion
stresses or by mechanical impact, both acting repeatedly. A loosen-
ing of individual carbide grains in their binder settings has been
identified as part of the wear mechanism of rotary-percussive rock
drills under conditions of strong frictional heating[209].

    To advance our understanding of cutting edge 'toughness',
research should be directed at the probability of encountering
flaws in small volumes, for instance, by extremal statistics analysis
of the volume dependence of rupture stress[210]. Test methods for very
small stressed volumes would have to be developed; approaches
might be evolved from existing indentation[211] or 'edge crumbling
tests' based on mutual indentation of tool edges, or on the loading of tips close to their edges by hardness indentors[212], or on

controlled impact of a cutting edge against a piece of work material[213]. Such tests would have to be extended to high temperatures and loading rates. In addition, high temperature measurements of fracture toughness and of fatigue crack propagation should be undertaken. With the exception of hot hardness, and some little work on hot rupture [191,208,214,215], the mechanical properties of hard metals at high temperatures are virtually a white spot on our map, urgently in need of exploration.

WEAR

Metals cutting tools

The wear processes involved in metal cutting have been the subject of much experimental study since the 1950's, and various reviews are available[8,9,10,216-218].

The following processes are believed to contribute, with weights depending on the operating conditions:

- adhesive wear (welding and rupture)
- abrasive wear (plowing, e.g., by hard inclusions of the work material)
- microfracture (edge crumbling)
- fatigue (thermal and mechanical)
- diffusion between tool and work (which may be enhanced by thermoelectric effects in the tool-work couple)
- oxidation
- built-up edge formation.

The demarcations between these mechanisms are not always clear-cut. Neither can the temperature - load - regions of their predominance be stated with exactitude, although 'machining charts' have been constructed[10,217] which show the main causes of wear for a given tool-work combination in terms of feed and cutting speed. The flank and the rake face are usually subject to different wear mechanisms, and even within different areas on one face, different mechanisms can be active. For instance, oxidation acts mainly at the edges of the wear region, where it can cause characteristic furrows.

The tool temperatures are often in excess of 1000 °C. Their exact determination is difficult; an interesting approach has been used by Naerheim who showed that local temperatures right underneath the wear land could be derived by electron-microanalysis of the dissolution of carbide in the binder.

The realization of the laminar character of the flow of work material next to the rake face - with a stationary, strongly ad-

herent layer right at the interface[217,220] - should bring into
the foreground the creep of the composite under high shear loads,
the plucking-out of carbide grains from the matrix (which must
depend on the high temperature strength of the carbide/binder interface), and diffusional reactions between tool and work material.
Possible mechanisms for edge crumbling or 'microchipping' and research problems germane to that area have been discussed above in
the section on 'toughness of metal cutting tips'.

The advent of coatings has brought about important shifts of
emphasis. Some mechanisms loose their importance altogether for
coated tools. Coatings prevent the diffusion between tool and work
material which in their absence leads to property degradation in
the tool; instead, diffusional reactions may now occur between work
and coating (in the case of carbide coatings [221]). Coatings prevent oxidation, and protect the tool substrate from adhesive wear.
Coatings also seem to reduce friction and lower the tool temperature in a given set of cutting conditions; there are indications
that the 'white layer' normally present at the tool/work interface
does not form on coated tools[222].

It seems to the present reviewer that a large part of the
earlier understanding of tool wear in metal cutting will have to be
reappraised in the light of these developments, and that important
gaps need to be filled quickly. These concern the processes at the
coating/work interface and the adherence of coating to substrate
under heavy shear loads at high temperatures. Adherence may be
influenced by many production parameters, among them the presence
of oxygen at the interface[211]. Break-down of adherence may also
be caused by bulging of the coating when the substrate deforms
plastically in the crater region[223].

With respect to substrate properties of importance for the
performance of coated tools, thermal conductivity[224], creep and
thermal as well as mechanical fatigue must be emphasized. The
thermal shock behaviour of oxide and ceramic cutting tool materials
has been studied in considerable depth by Mai[225], including
measurements of elevated temperature fracture toughness. Mechanical
and thermal fatigue may be responsible for edge crumbling, and are
known to cause cracking in milling edges[226-228]. Creep must
be held responsible for the plastic blunting of cutting edges.
There is some relation between the damage mechanisms which act in
creep and in high temperature fatigue; in the case of high temperature alloys, this has been studied extensively[229,230] and a transfer
of ideas from that field to hard metals should prove fruitful.-Far
too little experimental work has been done on high temperature
fatigue and on creep of hard metals. This makes a recent extensive
study by Osterstock[231] all the more valuable. Creep deformation
of hard metals seems to proceed by sliding of carbide grains in
the binder while the carbide grains themselves experience little

deformation, except at high temperatures and for high volume fractions of carbide. Much further research seems required to fully elucidate the mechanism of creep deformation in cemented carbides. Again, understanding based on a physical model should prove highly valuable in alloy development.

Rock drilling tools

Mechanistic studies of the wear of rock drilling tools began about a decade later than for metal-cutting tools [232-234], and to this day, their number has remained rather limited. An excellent recent review is available [166]. The difficulties were greater than in the case of metal cutting: inhomogeneity of the 'work material', the combination of chemical attack with wear, the superposition of sliding and impact wear, and others. Also it was more difficult to devise representative laboratory experiments. Conventional laboratory wear tests in which hard metal test pieces are brought into contact with pebbles, sand, or slurries of the mineral component [235-239] give little basis for the understanding of wear under the much more severe temperature and load conditions of rock drilling.

Models of the crushing and chipping action of the tool bit on the rock in rotary and percussive drilling were described by Larsen-Basse [240] in 1974. Important components of the wear process are

   (i) preferential abrasion of the binder (erosion)
  (ii) extrusion of the binder to the wear surface[241]
 (iii) uprooting of carbide grains from their setting in the
       binder[241,242]
  (iv) cracking (fragmentation of carbide grains)
   (v) attritional abrasion of the carbide (in soft rocks[243])
  (vi) spalling of the composite under point impact
 (vii) chipping or flaking of the composite after crack growth
       by mechanical and/or thermal fatigue.

Abrasive wear[241] predominates on the gauge or side faces of the drill bit; its front is subject to abrasive and impact wear, with the proportion of impact damage increasing with rock hardness and percussive load. Thermal fatigue becomes apparent by a pattern of crazing cracks which, when rounded out by abrasion, assumes the characteristic appearance of 'snake-skin' [228,244-246]. This phenomenon is most pronounced in soft rocks of low abrasive power, especially in pure magnetite [246]. However, thermal fatigue cracks have been observed in both rotary[240,247] and percussive[248] applications.

Thermal fatigue cracks propagate solely along intergranular paths [245,246,249,250] and often show prolific branching [244,246]

Their propagation direction is into the bulk at right angles to the surface. Spalling then occurs by lateral crack propagation. Speculatively, this could be explained by the action of mechanical fatigue. In rotary-percussive drilling, the shear forces will set up tensile stresses at 45° to the wear surface, and fatigue cracks would propagate at right angles to these. When spalling occurs, the thermal fatigue cracks often have penetrated far deeper into the material than the thickness of the spall, leaving a new surface with the same damage pattern[243]. - It has been claimed that thermal fatigue produces shallow craters in soft, and deep ones in hard rock because the impact forces are different. The present reviewer believes the difference to be due to the different rates of abrasion: hard rocks will wear down the surface so that it stays close behind the advancing thermal fatigue cracks.

Larsen-Basse[240] showed that in a series of experiments with varying Co content the effects of abrasive and impact wear could be separated. Abrasive wear was found related to hardness and Co content, with a monotonous increase with the volume fraction of binder; impact wear seemed to be related to compressive or transverse rupture strength, showing a minimum where these quantities maximize (i.e., at 5 - 10 pct Co).

In interpreting abrasive wear tests[241], one has to keep in mind the universal 'transitional behaviour' of mineral-metal couples [251]: as the hardness ratio of bit to abrasive falls to unity, the wear rate increases very rapidly, and for ratios below unity, wear becomes independent of the hardness of the abrasive. The importance of Co removal for abrasive wear of rock drilling tools was verified by Blomberry, Perrott and Robinson[252] who observed that chemical etching of the binder to a depth as small as 0,2 μm drastically increased the wear rate in a subsequent abrasive test. They also proved that tool surface temperatures above 800 °C must occur in the rotary drilling of sandstone, because glassy compounds of Co and silica were found on the gauge faces and on the trailing edges of the tool tips. Preferential erosion of Co was verified by all subsequent researchers. Areas with local enrichment of Co were specially prone to pitting (which emphasizes the importance of the milling and mixing stage in production!).

CONCLUSION AND OUTLOOK

Some important areas in which recent progress may be noted can only be briefly mentioned. - The understanding of *bonding* in carbides, nitrides etc. has made considerable progress. New spectroscopic methods (e.g., XPS) have made the experimental determination of valence band structure and charge transfer in hard phases much easier, and experimental and theoretical density of state functions seem to be converging into better-than-semi-quantitative agreement[256-263]. It is to be hoped that an extension of such work into the domain of bonding between carbide and binder may prove possible; there are indications[264,265] that the work of adhesion is a function of the electronic structure of the two components. Remembering the tasks that confront us with respect to materials substitiution, this is another area in which understanding based on a physical model would be of immediate value in alloy development.

Wetting measurements were popular in the 1960s but seem to have reached a stage of disillusionment. It is to be hoped that potential progress in the theoretical understanding of wetting on the one hand and the availability of ultra high vacuum technology and surface sensitive monitoring methods (e.g.,AES) on the other may rejuvenate this field of research. There is convincing evidence that traces of oxygen have a strong influence on binder/carbide adhesion in the system TiC/Ni [266] and one may speculate whether some of the brittleness of TiC-based hard metals could be removed by extreme hygiene with regard to oxygen and nitrogen. Striking improvements by small additions of carbide-forming elements to the liquid have been found for the wetting of diamonds by Cu[267,268]

Renewed interest in the thermodynamic basis of WC-Co hard metal production has brought a detailed elucidation of the eutectic reactions which occur during solidification of the binder, using the technique of directional solidification [269,270], and the sintering process of WC-Co alloys has been re-studied with great accuracy [271-273]. Unfortunately, our knowledge about the sintering of TiC- and TaC-containing hard metals is not as well developed as for the classic WC-Co system.

Looking back over the development of the science of hard metals through the last two and a half decades, there are two interesting lists of 'problems in urgent need of research' by Palmqvist[156] (1954) and Duwez[274] (1957). Many problems which then loomed large have now been solved or reduced to insignificance:

- accurate chemical analysis for C, Co, Ta and Nb has become trivial;

- particle size analysis and powder characterization in general has moved into the background - not so much because better methods are available, but because powders are produced on a much larger scale in modern plant with rigid process control;

- improved control of chemistry and sintering atmospheres have largely eliminated the problem of erratic eta phase formation;

- HIP and high temperature carburization have greatly reduced the threat of porosity;

- the processes which occur in the binder during cooling from sintering are understood and under control;

- the nature of toughness and the causes of its variation are now
  fairly well understood, and considerable improvement has been
  achieved;

- the nature, magnitude and effects of internal stresses are now
  well elucidated;

- the 'sintering activity' and 'the grain size memory' of powders -
  highly mystifying and nebulous ideas not so long ago - are beginning to be understood, at least on a phenomenological basis. Grain
  size control and the kinetics of solid solution formation in WC-
  TiC-TaC grades are well in hand;

- the systematic study of carbides, nitrides etc. which appears as
  one of the desiderata on Duwez' list has been accomplished. The
  present author is probably at variance with most hard metal scientists in thinking that this line of study has only recently
  borne fruit (in alloy development in the face of raw material
  shortness);

- single crystal property studies were advocated by Duwez. We now
  have a large stock of data on properties such as hot hardness of
  both stoichiometric and non-stoichiometric single crystals of
  carbides. Transmission electron microscopy - made possible by the
  development of a suitable thinning technique by Lehtinen[275] -
  allows crystallographic studies on microcrystalline specimens,
  and the nature of the slip systems in carbides is well established.

- Duwez strongly emphasizes the need of methods for the analysis of
  structural geometry. Quantitative characterization of cemented
  carbide microstructures is now a matter of routine

On the other hand, there are a number of problems where little progress may be recorded:

- the problem of 'skeleton formation' has remained a perennial
  teaser;

- understanding of the nature of the bonding forces which give
  carbides their outstanding properties is not sufficiently developed, or at least it has not come down to the community of
  hard metal technologists;

- the reason why 'cobalt and tungsten carbide have formed the best
  marriage ever' among all potential composite components is still
  not understood - and we may find ourselves short of these two
  unique metals in a foreseeable future;

- nondestructive test methods for carbide compacts or sintered
  pieces are still largely lacking. Their necessity has been reduced in comparison to twenty-five years ago by better production
  routines, but the problem is renewing itself by increasing demands
  on quality and size of carbide pieces.

The present review has pinpointed a few problems which were not (or hardly) thought of at the time of Palmqvist's and Duwez' lists:

- the partitioning of strain between binder and carbide: in cases where virtually all of the strain is taken by the binder, leaving none for the carbide, studies of dislocation processes in the carbide are irrelevant. Instead, attention should be given to dislocation processes in the binder and at binder/carbide boundaries [32]. As far as dislocations in the carbide are concerned, the slip systems of interest are those operating at high temperatures;

- the true nature of rupture strength distributions: are they really of Weibull type [210,276] - and what are the physical reasons behind this form of distribution? - In particular, the volume dependence [210] of rupture strength should be elucidated with greater certainty;

- the exact nature of the energy-dissipating processes at crack tips, and the true nature of fracture paths under different conditions of propagation: building blocks for a true physical understanding of fracture toughness;

- the application of transformation toughening principles [277-279] to cemented carbides;

- the microproperties involved in the 'toughness' of metal cutting edges;

- thermal fatigue and high temperature plasticity of cemented carbides, the mechanisms involved in deformation and damage accumulation, and their dependence on microstructure;

- improved understanding of the factors which determine the adhesion of coatings to hard metal substrates;

- understanding of the processes which contribute to the wear of coatings in metal cutting;

- improved understanding of wear in rock drilling and cutting.

Obviously, we are not running out of problems. Let us start working on them.

REFERENCES

1. O. Ruediger and H. E. Exner, Powder Metall. Intern. 8:7 (1976).
2. O. Ruediger, Techn. Mitteil. Krupp 70:599 (1977).
3. E. Lardner, Powder Metall. 21:65 (1978).
4. H. E. Exner, Int. Metall. Reviews 24:149 (1979).
5. R. Kieffer and F. Benesovsky, in: "Harmetalle", Springer, Wien (1965).
6. R. Kieffer and W. Hotop, in: Pulvermetallurgie Und Sinterwerk stoffe," Springer, Berlin (1943).
7. C. G. Goetzel, in: "Treatise on Powder Metallurgy" Interscience, Vol. I, New York (1949).
8. E. M. Trent, Machinist 95:1651 (1951).
9 H. Optiz, G. Ostermann, and M. Gappisch, "Beobachtungen ueber den Verschleiss von Hartmetallwerkzeugen, Forschungsbericht des Wirtschafts- und Verkehrsministeriums Nordreheim-Westfalen NR 668 (1958).
10. E. M. Trent, Production Engineer 38:105 (1959).
11. W. Mader and K. F. Mueller, Radex-Rundschau 535 (1971).
12. W. Mader and F. Zeman, Machinenmarkt 81:29 (1975).
13. U. Koenig, D. Dreyer, N. Reiter, J. Kolaska, and H. Grewe, "10th Plansee Seminar," Reutte/Tyrol (1981).
14. B. Lux and H. Schachner, "9th Plansee Seminar," Reutte/Tyrol Preprint No. 34 (1977).
15. W. Schintlemeister, O. Pacher, W. Wallgram, and J. Kanz, Metall 34:905 (1980).
16. B. Nidikom, J. T. Davies, Planseeber. Pulvermet 18:29 (1980).
17. J. Friberg and B. Aronsson, in:"Tungsten," (Proc. 1st Intern. Tungsten Symp. Stockholm 1979) Mining Journ. Books, London 85 (1979).
18. R. K. Yee, Intern. Metals Rev. 1:19 (1978).
19. B. J. Nemeth, A. T. Santhanam, and G. P. Grab, "Proc. 10th Plansee Seminar" Reutte/Tyrol 1:613 (1981).
20. N. Claussen, J. Am. Ceram. Soc. 59:49 (1976).
21. P. Rautala and J. T. Norton, Trans. AIME 194:1045 (1952).
22. J. Gurland, Trans. AIME 200:185 (1954).
23. H. Nowotny, R. Kieffer and O. Knotek, Berg-und Huettenmaenn. Monatsh. 96:6 (1951).
24. R. Kieffer and F. Benesovsky, "Hartstoffe", Springer, Wien (1963).
25. J. R. Tinklepaugh and W. B. Crandall, "Cermets," Reinhold Publ. Corp., New York (1960).
26. M. Humenik and N. M. Parikh, J. Am. Ceram. Soc. 39:60 (1956).
27. N. M. Parikh and M. Humenik, J. Am. Ceram. Soc. 40:315 (1957).
28. D. Moskowitz and M. Humenik, "Mod. Dev. Powd. Met.," Plenum New York 3:83(1966).
29. E. Rudy, S. Worcester, and W. Elkington, 8th Plansee Seminar, Reutte/Tyrol, Preprint No. 30 (1974).
30. J. C. Cahn, Trans. AIME 242:166 (1968);

31. R. K. Viswanadham, B. Sprissler, W. Precht, and J. D. Venables, Met. Trans. 10A:599 (1979).
32. R. K. Viswanadham, Met. Trans. 10A:1631 (1979).
33. E. S. Hodge, Powder Metall. 7:168 (1964).
34. E. Lardner and D. J. Bettle, Metals and Mater 7:540 (1973).
35. H. F. Fischmeister and D. J. Bettle, Powder Metall. 21:119 (1978).
36. E. Lardner, Powder Metall. 18:277 (1975).
37. H. Grewe and J. Kolaska, Metall. 32:989 (1978).
38. D. J. Bettle and J. D. Murray, "Conf. on Recent Advances in Hard Metals Production," Loughborough Univ. Paper 35 (1979).
39. E. Lardner and S. Iggstroem, "Proc. 10th Plansee SEM," Reutte/Tyrol 1:549 (1981).
40. R. Kieffer, P. Ettmayer, and B. Lux, "Conf. on Recent Advances in Hard Metals Production", Loughborough Univ. Paper 33 (1979).
41. H. Holleck, Metall. 33:1064 (1979).
42. H. Holleck, L. Prakash, and F. Thummler, "Conf. on Recent Advances in Hard Metals Production," Loughborough Univ, Paper 25 (1979).
43. R. Meyer, "Colloquium on Material and Energy Conservation in Powder Metallurgy," Soc. Franc. Met. Paris, 25 (1977).
44. F. Bender, "Tungsten," (Proc. 1st Intern. Tungsten Symp." Stockholm 1979) Mining Journ. Books, London, 2 (1979).
45. M. Ruehle, Metall. 33:1311 (1979).
46. A. Sutulov, Metall. 33:1310 (1979).
47. T. W. Farthing and R. E. Goosey, "Tungsten," (Proc. 1st Intern. Tungsten Symp. Stockholm 1979) Mining Journ. Books, London, 125 (1979).
48. W. Aschenbrenner and R. Palme, Metall. 35:262 (1981).
49. W. Gocht, Metall. 33:774 (1979).
50. A. M. Sage, in: "Future Metal Strategy", J. Nutting, ed., The Metals Soc., London (1980).
51. W. Krajewski, Metall. 35:81 (1981).
52. W. Krajewski, Metall. 33:1299 (1979).
53. M. Johnstone, "Conf. On Recent Advances in Hard Metals Production," Loughborough Univ. Paper 3 (1979).
54. W. O. Gluschke, Metall 33:1305 (1979).
55. R. B. Nicholson and P. G. Cranfield, in: "Future Metal Strategy," J. Nutting, ed. The Metals Soc., London (1980).
56. A. A. Nijkern, in: "Future Metal Strategy," J. Nutting, ed., The Metals Soc., London (1980).
57. I. R. Friedman, "Conf. on Recent Advances in Hard Metals Production," Loughborough Univ. Paper 1 (1979).
58. S. Ekemar, "1st International Symposium on Tantalum," Tantalum Producer International Study Centre Brussels, 41 (1978).
59. R. Kieffer, G. Trabesinger, and N. Reiter, Planseeber. Pulvermet. 17:25 (1969).
60. E. D. Whitney, Powder Metall. Intern. 10:16 (1978).
61. R. H. Wentorf, R. C. De Vries, and F. P. Bundy, Science 108:873 (1980).

62. N. Claussen and D. P. Hasselman, in: "Thermal Stresses in Severe Environments," Plenum Press, New York (1980).

63. R. Stevens, Trans. Brit. Ceramic Soc. 80:3/81 (1981).

64. L. E. Hibbs, Jr., and R. M. Wentorf, Jr., High Temperatures-High Pressures 6:409 (1974).

65. E. Hibbs, Jr. and M. Lee, in: "Intern. Conf. Wear of Mater," W. Glaeser et al., eds., New York (1977).

66. J. Walraedt, Powder Metall. Int. 2:77 (1970).

67. P. G. Barnard, A. G. Starliper and H. Kenworthy, U.S. Pat. 3.595.484 (1971).

68. L. Ramqvist, in: "Modern Developments in Powder Metetallurgy," H. Hausner, Ed., Plenum Press, New York (1971).

69. W. O. Alexander, in: "Future Metal Strategy, J. Nutting, ed., The Metals Soc., London (1980).

70. J. Gurland, Trans. AIME 215:601 (1959).

71. J. Gurland, Trans. AIME 277:1146 (1963).

72. C. Nishimatsu and J. Gurland, Trans. ASM 52:469 (1960).

73. J. Gurland, in: "Proc. 4th Plansee Seminar," F. Benesovsky, Ed., Reutte/Tyrol (1962).

74. J. Gurland, "Quantiative Microscopy," R. T. Dehoff, ed., McGraw-Hill, New York (1968).

75. H. E. Exner, and H. F. Fischmeister, Prakt. Metallogr. 3:18 (1966).

76. H. E. Exner and H. F. Fischmeister, Archiv Eisenhuettenw. 37:417 (1966).

77. H. E. Exner, Powder Metall. 13:429 (1976).

78. J. Gurland, Trans. AIME 212:452 (1958).

79. H. Fischmeister and H. E. Exner, Archiv Eisenhuettenw. 37:499 (1966).

80. N. Gensamer, E. E. Pearsall, W. S. Pellini, and J. R. Low, Trans. ASM 30:983 (1942).

81. E. Orowan, "Symposium on Internal Stresses," Institute of Metals, London (1947).

82. E. O Hall, Proc. Phys. Soc 64B:747 (1951).

83. N. J. Petch, J. Iron Steel Inst. 173:25 (1953).

84. C. S. Smith and L. Guttman, Trans. AIME 197:81 (1953).

85. E. E. Underwood, Metals Eng. Quart 1:70 (1961).

86. H. F. Fischmeister, J. Microscopy 95:119 (1972).

87. H. Fischmeister and B. Karlsson, Z. Metallk. 68:311 (1977).

88. J. Gurland and P. Bardzil, Trans. AIME 203:311 (1977).

89. H. Fischmeister, Tekn. Tidskr 761 (1965).

90. H. Fischmeister and J. Drott, unpublished work (1964).

91. H. G. Stjernberg, Powder Metall. 13:1 (1970).

92. J. Hinnueber and W. Kinna, Stahl und Eisen 82:31 [1962].

93. N. K. Sharma, I. D. Ward, H. L. Fraser, W. S. Williams, J. Am. Ceram. Soc. 63:195 (1980).

94. M. P. Seah and E. D. Hondros, Proc. Roy. Soc. A335:191 (1973).

95. A. Henkered, M. Hellsing, H. Norden, and H. O. Andren, these proceedings (1982).

96. D. Van Dyck, "Diffraction and Imaging Techniques in Material Science", S. Amelinckx et al, ed, North Holland 1"355 (1978).
97. V. A. Ivensen, O. N. Eiduk, and V. A. Chistakova, Poroshk. Metall 5:84 (1974).
98. R. Warrne and M. B. Waldron, Powder Metall. 15:166 (1972).
99. H. E. Exner and J. Gurland, Powder Metall. 13:13 (1970).
100. H. Suzuki, Trans. JIM 7:112 (1966)
101. H. Suzuki and H. Kubota, Planseeber. Pulvermet. 14:96 (1966).
102. H. Johnsson, Powder Metall. 15:29 (1972).
103. H. Johnsson, Planseeber. Pulvermet. 21:187 (1973).
104. O. Ruedigerer, D. Hirschfeld, A. Hoffmann, J. Kolaska, G. O. Ostermann, and J. Willbrand, Techn. Mitt Krupp 19:1 (1971).
105. D. Moskowitz and M. Humenik, Int. J. Powder Metall. Technol. 14:1 (1978).
106. P. O. Snell, Planseeber. Pulvermet. 22:91 (1974).
107. H. Suzuki, K. Hayashi, and Y. Taniguchi, Planseeber. Pulvermet. 27:215 (1979).
108. H. Jonsson and B. Aronsson, J. Inst. Met. 97:281 (1969).
109. O. Pacher, W. Schintlmeister, and T. Raine, Powder Metall. 23 :189 (1980).
110. T. Johansson and B. Uhrenius, Met. Sci. 12:83 (1978).
111. L. Akesson, Jernkontorets Forskining D223 Stockholm, (1978).
112. P. B. Anderson Planseeber. Pulvermet. 15:180 (1976).
113. H. Suzuki, K. Hayashi, J. Japan Soc. Powd. Met. 15:369 (1968).
114. B. Paul, Trans. AIME 218:36 (1960).
115. Z Hashin and S. Shtrikman, J. Mech. Phys. Solids 11:127 (1963).
116. B. O. Jaensson, Mat. Sci. Eng. 9:339 (1972).
117. B. O. Jaensson and B. O. Sundstroem, Mat. Sci. Eng. 9:217 (1972).
118. B. O. Jaensson, Mat. Sci. Eng 8:41 (1971).
119. B. O. Sunstroem, Mat. Sci. Eng. 12:265 (1973).
120. I. Tamura, Y. Tomota, and H. Ozawa, "Proc. 3rd Int. Conf. Strength of Metals and Alloys," Cambridge 1:611 (1973).
121. S. Bartolucci and H. H. Schloesin, Acta Met. 14:337 (1966).
122. H. C. Lee and J. Gurland, Mat. Sci. Eng. 33:125 (1978).
123. G. Gratwohl and R. Warren, Mat. Sci. Eng. 14:55 (1974).
124. J. Gurland, Mat. Sci. Eng. 40:59 (1979).
125. B. Karlsson and G. Linden, Mat. Sci. Eng. 17:153 (1975).
126. W. Dawihl and B. Frisch, Cobalt 22:1 (1964).
127. L. Lindau, Sand. J. Met. 6:90 (1977).
128. D. E. Guecer and J. Gurland, J. Mech. Phys. Solids 10:365 (1962).
129. D. E. Guecer and J. Gurland, Jernkont. Ann 147:111 (1963).
130. H. E. Exner, Materialpruef 7:375 (1965).
131. S. Amberg, Qutoed in Reference 129.
132. W. Weibull "Ingenjoers-Vetensk.-Akad. Handl.", IVa; Stockholm, 151 and 153 (1939).
133. R. Bernard, Jernkont. Ann. 147:22 (1963).
134. S. Amberg, Jernkont. Ann. 147:218 (1963).

135. H. Hoffmann and H. Blumenauer, Wiss Zeitschr. Tech. Hochsch. Magdeburg 24:119 (1980).
136. P. B. Anderson, Dissertation Tech. Univ. Graz 1976.
137. S. Bartolucci Luyckx, Acta Met. 23:109 (1975).
138. S. Bartolucci Luyckx, Metall. 33:732 (1979).
139. H. Suzuki and K. Hayashi, Planseeber. Pulvermet. 23:24 (1975).
140. H. Suzuki, T. Tanase and K. Hayashi, Planseeber. Pulvermet. 23:121 (1975).
141. H. Suzuki and T. Tanase, Planseeber. Pulvermet 24:271 (1976).
142. S. Amberg and H. Doxner, Powder Metall. 20:1 (1977).
143. J. Qvick, P.-O Snell, B. I. Nolaeng and M. E. Richardson, "Proc. 10th Plansee Seminar" Reutte/Tyrol 1:717 (1981).
144. H. E. Exner, A. Walter and R. Pabst, Mat. Sci. Eng. 16:231 (1974).
145. M. Ingelstroem and H. Nordberg, Report Swedish Inst. Metal Research IM-948; Stockholm (1975), see also Eng. Fract. Mech. 6:597(1974)
146. J. L. Chermant, A. Deschanvres, A. Yost and R. Meyer, Mater. Res. Bull. 8:925 (1973).
147. P. Kenny, Powder Metall. 14:22 (1971).
148. R. C. Lueth,"Fracture Mechanics of Ceramics," R. C. Bradt, D. P. Hasselmann, and F. F. Lange, eds., Plenum Press, New York (1974).
149. A. R. Johnson,Metall. Trans. 8A:891 (1977).
150. H. Johansson and R. Sandstroem, Mat. Sci. Eng. 36:175 (1978)
151. E. A. Almond and B. Roebuck, Met. Sci. 11:458 (1977).
152. H. Fischmeister, J. Diblik and W. Kompek, to be published.
153. L. Lindau, Fracture 1977", 4th Int. Conf. On Fracture, Waterloo 2A:215 (1977).
154. C. Peters, "Conf. on Recent Advances in Hard Metals Production," Loughborough Univ. Paper 27 (1979).
155. H. F. Fischmeister and L. R. Olsson, "Cutting Tool Materials,"ASM, Metals Park, Ohio 111 (1981).
156. S. Palmqvist, "Pulverteknik", FKO-Meddelande (Swedish Academy of Engineering Science, Stockholm), NR 16:53(1954).
157. S. Palmqvist, Jernkont. Ann, 141:300 (1957).
158. S. Palmqvist,Arch. Eisenhuettenw. 33:629 (1962).
159. H. E. Exner, Trans. AIME 245:677 (1969).
160. A. Hara, M. Megata, and S. Yazu, Powder Metall. Int. 2:43 (1970).
161. P. O. Snell and E. Paernamaa, Planseeber. Pulvermet. 21:271 (1973).
162. C. T. Peters, J. Mater. Sci. 14:1619 (1979).
163. E. A. Almond and B. Roebuck, "Conf. On Recent Advances in Hard Metals Production"Loughborough Univ., Paper 31 (1979).
164. R. F. Viswanadham and J. D. Venables, "Advances in Hard Material Tool Technology," R. Komanduri, ed., Carnegie Press, Pittsburgh, (1976).

165. E. L. Exner, J. R. Pickens, and J. Gurland, Met. Trans. 9A:736 (1978).
166. C. M. Perrott, Ann. Rev. Mater.Sci. 9:23 (1979).
167. I. M. Ogilvy, C. M. Perrott, and J. W. Suiter, Wear 43:239 (1977).
168. C. M. Perrott, Wear 47:81 (1978).
169. B. R. Lawn and E. R. Fuller, J. Mater. Sci. 10:2016 (1976).
170. C. M. Perrott, Wear 45:293 (1977).
171. A. G. Evans and T. R. Wilshaw, Act Met. 24;939 (1976).
172. B. R. Lawn and R. Wilshaw, J. Mater. Sci. 10:1049 (1975).
173. R. Warren and H. Matzke, in these proceedings (1981).
174. B. R. Lawn and R. Wilshaw, "Fracture of Brittle Solids", Cambridge Univ. Press, Cambridge (1975).
175. B. R. Lawn and M. V. Swain, J. Mater. Sci. 10:113 (1975).
176. G. Gille, "Proc. 6th Intern. Conf. Powder Metall," Dresden Paper No. 9 (1977).
177. B. Roebuck and E. A. Almond, "Conf.on Recent advances in Hard Metals Production, Loughborough Univ. Paper 28 (1979).
178. T. Johannesson, "4th Europ. Symp. Powder Met." Grenoble, Paper No. 5-11 (1975).
179. M. J. Murray and C. M. Perrott, "Proc. 1976 Int. Congr. Hard Materials Tool Technol" R. Komanduri, ed, Carnegie-Mellon Univ., Pittsburgh (1976).
180. J. L. Chermant and F. Osterstock, J. Mater. Sci. 11:1939 (1976).
181. C. Chatfield, PM 78-SEMP (5th Europ. Symp. Powder Metall., Stockholm 2:57 (1978).
182. J. R. Pickens, J. Gurland, Mat. Sci. Eng. 33:135 (1978).
183. M. Nakamura and J. Gurland, Met. Trans. 11A:141 (1980).
184. R. K. Viswanadham, T. S. Sun, E. F. Drake, and J. A. Peck, J. Mater. Sci.16:1029 (1981).
185. N. M. Parikh J. Am. Ceram. Soc. 40:335 (1957).
186. G. S. Kreimer and N. A. Alexeyeva, Phys. Metals Metallogr. 13:117(1962).
187. T. Fukatsu and T. Sasahara, J. Jap. Soc. Powder Metall. 10:30 (1963).
188. S. Bartolucci-Luyckx, Acta Met. 16:535 (1968).
189. A. Mason and P. Kenny, Metallurgia 81:205 (1970).
190. A. Hara, T. Nishikawa, and S. Yazu, Planseeber. Pulvermet. 18:28 (1970).
191. M. J. Murray and D. C. Smith, J. Mater. Sci 8:1706 (1973).
192. J. L. Chermant, M. Coster, G. Hautier, and P. Schaufelberger, Powder Metall. 17:85 (1974).
193. J. L. Chermant, M. Coster, and F. Osterstock Metallography 9:503(1976).
194. S. Bartolucci-Luyckx, "Fracture 1977," 4 th Int. Conf. on Fracture, Waterloo, 2:223 (1977).
195. M. J. Murray, Proc. Roy. Soc. A356:483 (1977).
196. J. L. Chermant, A. Deschanvres, and F. Osterstock, Powder Metall 20:63 (1977).
197. B. Roebuck and E. A. Almond, Met. Powder Rept. 1:28 (1979).

198. C. Lea and B. Roebuck, Met. Sci. 15:262 (1981).
199. B. Karlsson and K. Wright, "Microstructural Science," G. Petzow,
     D. Albrecht, J. McCall, eds., Elsevier, New York 9:3059
     (1981).
200. B. Karlsson and K. Wright, "Quantitative Analysis of Multiphase
     Fracture Surfaces Using Profile Measurements," Practicaul
     Metallogr. 19 (1982), to be published.
201 H. Bretfeld, F. W. Kleinlein, D. Munz, R. F. Pabst and H.
     Richter, Z. Werkstofftechn 12:167 (1981).
202. L. M. Barker, Eng. Fracture Mechan. 9:361 (1977).
203. H. Huebner and U. Engel, Z. Werkstofftechn. 9:128 (1978).
204. E. A. Almond and B. Roebuck, Met. Technol. 5:92 (1978).
205. K. G. Stjernberg, Met. Sci. 14:189 (1980).
206. A. L. Maystrenko, N. K. Konovalenko, and G. Gille, "7th Intern.
     Powder Metall. Conf. DDR", Dresden, Paper 20, (1981).
207. H. Suzuki and Y. Tanaguchi, Planseeber. Pulvermet. 25:23
     (1977).
208. F. Ueda, H. Doi, Y. Fujiwara, and H. Masatomi, Powder Metall.
     Intern. 9:32 (1977).
209. J. Larsen-Basse, C. M. Perrot, and P. M. Robinson, Mat Sci. Eng.
     13:83 (1974).
210. E. A. Almond, Met Sci. 12:587 (1978).
211. E. A. Almond, R. S. Irani, and B. Roebuck, Mat. Sci. Eng. 44:173
     (1980).
212. E. A. Almond, Communication at seminar on toughness tests for
     tool materials, IVF Institute for Production Technology
     Research, Gothenburg (1973).
213. M. Hirao, R. Murata, U. Kasuya, H. Takeyama, Ann. CIRP 28:29
     (1979).
214. G. S. Kreimer, "Strength of Hard Alloys," Consultants Bureau,
     New York (1968).
215. W. Dawihl and M. K. Mal, Cobalt 26:25 (1965).
216. M. Opitz and M. Gappisch, Int. J. Mach. Tool Des. Res. 2:43
     (1962).
217. E. Trent, "Metal Cutting," Butterworths, London (1977).
218. E. M. Trent, Treatise Mater. Sci. Techol. 13:443 (1979).
219. Y. Naerheim, Power Metall. Intern. 11:31 (1979).
220. E. M. Trent ISI Special Report No. 94:11 (1967).
221. Ch. Lesniak and E. Bryjak, Planseeber. Pulvermetall. 25:112
     (1977).
222. H. Jonsson, Wear 32:151 (1975).
223. V. C. Venkatesh, A. S. Raju, and K. Srinivasan, Ann. CIRP 25:5
     (1977).
224. H. Tanaka, "Cutting Tool Materials, ASM, Metals Park (1981).
225. Y. W. Mai, in: "Proc. Int. Conf. Fract. Mechan. Technol., G. C.
     Sih, ed., Nordhoff, Alphen (1977).
226. S. M. Bathia, P. C. Pandey, and H. S. Shan, Wear 51:201 (1978).
227. S. M. Bhatia, P. C. Pandey, and H. S. Shan, Precision
     Engineering 1:148 (1979).

228. G. E. Spriggs and D. J. Bettle, Powder Metall. 18:53 (1975).
229. H. Zenner, Z. Werkstofftechn 8:271 (1977).
230. M. Speidel and A. Pineau, in: "High Temperature Alloys for Gas Turbines," D. Coutsouradis et al., eds., Appl. Sci. Publ., London (1978).
231. F. Osterstock, Doctoral Thesis, University de Caen (1980).
232. H. Koto, T. Watanabe, and T. Nakamura, J. Jap. Mater. Soc. 21:429 (1957).
233. A. Latin, Metallurgia 63:211and 267(1961).
234. L. Pons, J. Chevillon, and P. Steff, Compt. Rend. 255:2100 (1962).
235. K. Wellinger and M. Uetz, Aufbereitungstechnik 8:319 (1963).
236. H. Wahl, Aufbereitungstechnik 10:305 (1969).
237. Cemented Carbides Producers Assoc., Standard CCPA P-112.
238. H. Feld, and P. Walter, Z. Werkstofftechn. 7:300 (1976).
239. H. Feld, Z. Werkstofftechn. 9:172 (1978).
240. J. Larsen-Basse, Powder Metall 16:1 (1973).
241. J. Larsen-Basse, Trans. ASMe/J. Lubric. Technol. 101:208 (1979).
242. J. Larsen-Basse, Wear Mater. 2:453 (1979).
243. C. M. Perrott and P. M. Robinson, J. Austral. Inst. Met. 19:22 (1974).
244. S. G. Bailey and C. M. Perrott, Wear 29:117 (1974).
245. K. G. Stjernberg, U. Fischer, and N. I. Hugoson, Powder Metall 18:89 (1975).
246. H. Johnsson, Planseeber. Pulvermet. 24 :108 (1976).
247. G. Schumacher and G. Ostermann, Cobalt 4:471 (1969).
248. H. J. Osborn, Powder Metall. 12;471 (1969).
249. L. Pons and J. Chevillon, and PH. Steff, Rev. Met. 60:325 (1963).
250. M. Lagerquist, Powder Metall. 18:71 (1975).
251. M. M. Kruschov, Wear 28:69 (1974).
252. R. I. Bloombery, C. M. Perrott, P. M. Robinson, Wear 27:383 (1974).
253. A. Hara, S. Yazu, "Proc. 10th Plansee Sem." Reutte/Tyrol I:581 (1981).
254. R. I. Blombery, C. M. Perrott, P. M. Robinson, Mat. Sci. Eng. 13:93 (1974).
255. R. S. Montgomery, Wear 12:309 (1968).
256. H. Ihara, M. Hirabayashi, H. Nakagawa, Phys. Rev. B14:1707 (1976).
257. A. L. Hagstroem, L. I. Johansson, B. E. Jacobsson, S.B.M. Hagstroem, Sol. State Comm. 19:647. (1976).
258. K. Schwarz, J. Phys. C. Sol. St. Phys. 10:195 (1977).
259. M. Gupta, V. A. Gubanov, and D. E. Ellis, J. Phys. Chem. Solids 38:499 (1977).
260. L. P. Mokhracheva, V. A. Tskhai, and P. V. Geld, Phys. Stat. Sol B87:49 (1978).
261. P. Weinberger, Ber. Bunsengesellsch. 81:804 (1977).
262. V. A. Gubanov, A. L. Ivanovskii, G. P. Shveikin, and J. Weber, Sol. State Comm. 29:743 (1979).

263. P. Weinberger, K. Podloucky, C. P. Mallett, and A. Neckel, J.
     Phys. C. Sol. St. Phys. 12:801 (1979).
264. L. Ramqvist, Jernkont. Ann. 153:159 (1969).
265. H. Goretzki and W. Scheuermann, "7th Plansee Seminar"
     Reutte/Tyroll 4:50 (1971).
266. K. Taehtinen and M. H. Tikkanen, Powder Metall. Intern. 11:80
     (1979).
267. Yu. B. Naidich and G. A. Kolesnichenko, Poroshk. Metall. 3:23
     (1964). Henry Brutcher Transl. 6331.
268. Yu. B. Naidich and G. A. Kolesnichenko "Surface Phenomena in
     Melts and Powder Metallurgical Processes," Kiev 158 (1963).
269. L. Westin and T. Franzen, Scand. J. Metall. 8:105 (1979).
270. D. Jaffrey, J. W. Lee, and J. D. Browne, Powder Metall. 23:140
     (1980).
271. B. Meredith and D. R. Millner, Powder Metall. 19:38 (1976).
272. L. Akesson, Thermochim. Acta 29:327 (1979).
273. J. W. Lee, D. Jaffrey, and J. D. Browne, Powder Metall. 23:57
     (1980).
274. P. Duwez, J. Metals 150 (1957).
275. B. Lehtinen, J. Phys. E. J. Sci. Instr. II 1:673 (1968).
276. K. Trustrum and A. De S. Jayatilaka, J. Mater. Sci. 14:1080
     (1979).
277. V. F. Zackay, E. R. Parker, D. Fahr, and R. Busch, Trans. ASM
     60:252 (1967).
278. W. W. Gerberich, P. L. Hemmings, and V. F. Zackay, Met. Trans.
     2:2243 (1977).
279. N. Claussen and G. Petzow, "Proc. 4th Cimtec (Int. Meet. Mod.
     Ceramics Technol.)"S. Vincenzini, ed., Nat. Sci. Res.
     Council, Faenza, 680 (1979).

DISCUSSION

W. Williams:

A comment about the analysis of grain boundaries containing cobalt.
That measurement was done with a dedicated STEM, having a spatial
resolution of a few angstroms, although because of x-ray fluorescence
there is some spreading of the effective volume. But in the twenty
angstrom grain boundaries shown in the slide from Sharma's thesis it
is our belief that there is, without doubt, cobalt in that grain
boundary, and I think that it does not take very many atoms to
exhibit bulk properties of a solid. Therefore my feeling is that
the cobalt in the grain boundary is really metallic cobalt, although
containing some carbon and tungsten in solid solution as you would
expect for that thermal history.

H. Fischmeister:

Thank you for supplying the details. This is really a question of
religion...because it carries with it many corollaries. One is, how
do the properties of very thin layers of cobalt metal change with the
layer thickness. It used to be said that once you get down to mean
free paths of the order of a few angstroms, you have practically
dislocation free regions and these should have theoretical strengths.
In which case, it seems to me that we should no longer consider these
regions as regions of cobalt as we know cobalt in the technological
sense. Something has given me pause in this belief which I have held
for many years. That is the observation by Viswanadham of grain
boundary ledges acting as dislocation sources. I don't think you'll
find the layers were quite as thin as the ones we're discussing now.
If I understood that paper correctly, it shows that there is structure to the boundaries between carbide and the binder, and this
structure may involve misfit dislocations and these grain boundary
ledges could act as sources of dislocations even in layers which
would be thin enough to be highly defect free. On the other hand, I
won't be convinced that this is really cobalt until I see a diffraction pattern of it. There are so many examples of the extremely
strong segregation of elements which fit poorly into the bulk to
grain boundaries. It is very difficult for me to think that the
situation in WC-Co should be different from what we know about, let's
say, ferrite-phosphorus, ferrite-arsenic. In all these cases we have
strong segregation creating monolayers or double monolayers of the
impurity and then profiles which peter out in something like five
angstroms from the boundary. The dimensions are very similar to what
we seem to have or what we might have in the STEM analysis shown
here. I hope that there will be more discussion about this during
the conference.

No Name Given:

Would you comment on the role of binders in silicon carbides and
nitrides of various types? Would aspects you used in considering
cobalt and tungsten carbide apply?

H. Fischmeister:

Yes, you are talking about silicon carbide, silicon nitride materials
which have a thin layer of amorphous material between the crystalline
regions. This layer is usually thought to be detrimental because it
lowers the high temperature strength of the materials. I cannot
really comment very much on this. I share the view that this amorphous layer is probably detrimental to the high temperature properties. On the other hand, I'm not sure that we are reaching the
temperatures where it becomes really detrimental in a cutting application. This is based not on figures but simply on the talk there
is about successful applications of SiAlONs in cutting.

No Name Given:

I'm more concerned about wear than cutting. At lower temperatures
where the silicon supposedly helps the toughness aspects.

H. Fischmeister:

I wouldn't be surprised if it turned out that these interlayers
improve the toughness. They would have an action on crack propagation, and if it's not a question of high temperature application,
they might very well have a positive effect.

A. Krawitz:

One ceramic system you did not mention is the alumina-zirconia
system that has been receiving attention recently. I wonder if you
have any general comments on these.

H. Fischmeister:

I may have been misunderstood because I did show a slide with
zirconia particles embedded in alumina. I did mention transforma-
tion-toughening. Again, the picture is not clear enough to me to
go forth and say that this is a material of the future. I think it
is a material of immense promise.

a) Because it's made of plentiful materials (the materials will
remain cheap from what we can see) and b) the process is not too
expensive. It does involve hot pressing but it's not a very complicated operation. The principle, of course, is so appealing to a

scientist that it goes against my grain to believe that they should not be good materials.

Let me say something I forgot to mention in the lecture. I think we should consider the principle of transformation-toughness very seriously. It doesn't work only in ceramics. Although it's effects in ceramics have been perhaps the most dramatic because we all think of ceramics as inherently brittle materials. We still do. However, transformation toughening works in metals. I would like to remind you of the work of Gareth Thomas which shows that thin films of austenite which can undergo stress-induced transformation can increase the toughness of steels by about a factor of one hundred percent at a given strength level. You get a doubling of K_{IC}. And I would like to just mention the idea that one could look for metallic systems to serve as binders in cemented carbides which undergo stress-induced transformation. I'm leaving this as a puzzle--something to think about.

A. Krawitz:

Along those lines, even in the WC-Co system, it has become known in recent years that cobalt itself undergoes a martensitic transformation. The work of Sarin and work that we will present on Friday documents this as a function of fatigue, as well as monotonic loading.

H. Fischmeister:

I am aware of the potential role of the cobalt transformation. But it seems to me that there are binder systems which exhibit a greater volume effect connected with the transformation. What you are really looking for to make optimum use of transformation toughening is something that has a similarly strong volume effect as we see in zirconia.

H. Holleck:

Perhaps I can make a short comment concerning the gap between hard metals and high speed steels. I feel that this gap will be closed with materials based on the titanium carbide (ex: ferro-TiC) and other carbides.

PHASE EQUILIBRIA AND CRYSTAL STRUCTURES

OF TRANSITION METAL NITRIDES

Peter Ettmayer and Alfred Vendl

Technical University of Vienna
A-1060 Vienna, Austria

ABSTRACT

Binary and ternary phase diagrams of the transition metals of the 4A, 5A and 6A group of the periodic system with nitrogen have been investigated and in parts reevaluated. Nitrogen pressures up to 1000 bar have been used to further the knowledge about nitride phase diagrams. Revised versions of the systems Ta-N and Mo-N are presented. Ternary nitride phases T^5T^6N (T^5=Nb,Ta) (T^6=Cr,Mo) are observed in the systems Nb-Cr-N, Ta-Cr-N, Nb-Mo-N and Ta-Mo-N. These phases belong to the filled-up Θ-CuTi-type with five-fold coordinated nitrogen atoms. The phase Mo(Mo,Ta)$_2$N is isotypic with a number of uranium-platinum metal-carbides and belongs to the filled-up MoSi$_2$-type. Ternary nitride formation seems to be much more frequent than in the corresponding carbide systems.

INTRODUCTION

Nitrides of the transition metals have many properties in common with the corresponding carbides. Both are "interstitial compounds", they are hard, high melting materials with excellent electrical and thermal conductivities. Carbides and nitrides are often isostructural and usually form solid solutions with each other. There are, however, important differences, which might be bound up with a different bonding character of the nitrogen atom compared with that of

47

the carbon atom. While carbides nearly exclusively form
compounds which may be simply described as close packed
- or nearly close packed - arrangements of metal atoms
with small nonmetal atoms in the interstitial sites,
deviations from this simple concept are encountered
much more frequently in nitrides. A further difference
is noted in the fact that nitrogen is gaseous and
therefore nitride systems are pressure dependent,whereas
in carbide systems the influence of pressure can be
neglected. While the carbide systems are reasonably well
known, many of the binary nitride phase diagrams are
still inexactly known and there exists a great deal of
controversy about proposed diagrams.

One particular problem confounding phase diagram
studies of nitrides is the sensitivity of phase
stabilities to methods of preparation: low temperatures
give rise to the formation of metastable nitride phases,
whereas at high temperatures the dissociation pressure
of the nitrides may well exceed normal pressure. Nitriding
with flowing ammonia has proved to be very effective to
prepare nitrides with a high nitrogen content, but
interpretation of phase equilibria in terms of phase
diagrams is often extremely difficult because of the
slow equilibration rates and the ill defined nitrogen
potentials encountered during nitridation. From the
standpoint of phase diagrams and thermodynamics, the use
of nitrogen gas is much preferred. Investigations at
elevated nitrogen pressure have made it possible to
reevaluate, at least partially, the system Ta-N and Mo-N
and to collect some additional data about other nitride
systems.

EXPERIMENTAL

The experiments have mainly been performed in two
internally heated autoclaves in which pressures up to
1000 and 300 bar can be attained. Autoclave I consists
of a tungsten tube that is resistance heated and
mounted in an appropriate water-cooled pressure vessel.
In autoclave II, shown in fig.1, rods or tubes are
mounted between water cooled electrodes. In this autoclave
melting can be observed directly through a viewing port
with a quartz window. The temperature is observed with
an optical pyrometer, the absorption by the window and
the high-pressure gas atmosphere was taken into
consideration and corrected according to previous
calibration tests. Further details of the construction
of the autoclaves are given elsewhere (1,2).

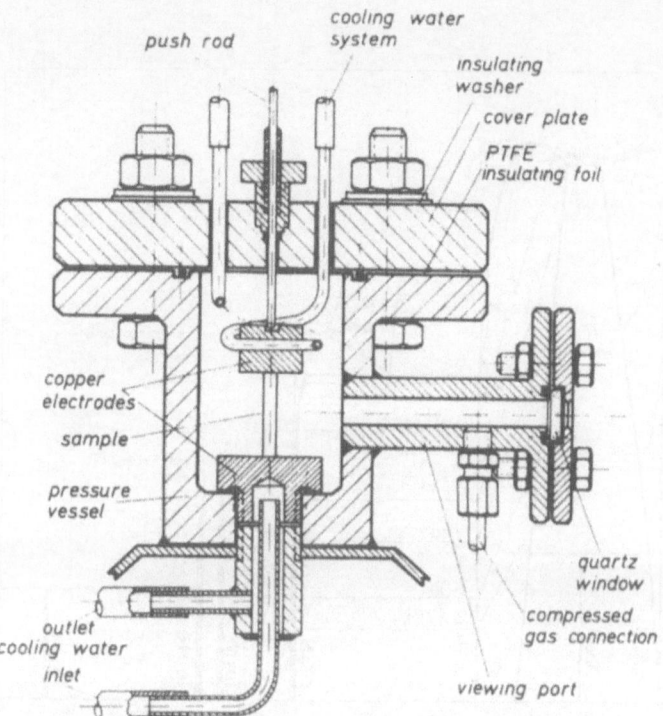

Fig.1. Autoclave II, designed for the determination
       of melting points at pressures up to 300 bar

THE SYSTEM Ta-N

     The phase diagram for the system Ta-N, proposed in
fig.2, is based upon information from the literature
(3,4,5) and own work (6). The solubility of nitrogen in
tantalum has been investigated extensively and thermodynamic data have been derived both for the solid
solution and the phase Ta_2N at the nitrogen poor boundary
(7). Between the saturation boundary and the phase field
of Ta_2N some more or less controversial superstructure
phases have been found, the exact nature and equilibrium
conditions of which are not known. Therefore these
phases have not been included in the phase diagram.

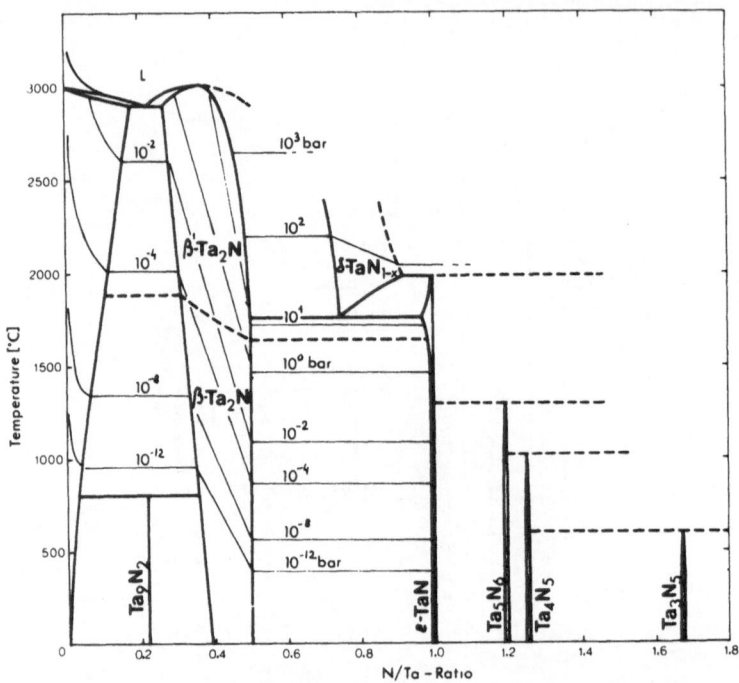

Fig.2. Proposed phase diagram of the system Ta-N

        The phase Ta$_2$N apparently undergoes an order -
disorder transition in the nitrogen sublattice somewhere
between 1500 and 2000°C, the exact transition temperature
has not yet been established. The ordered phase belongs
to the ε-Fe$_2$N type (8) and is distinguished from the
disordered β'TaN phase (L'3 type) by a different axial
ratio a/c of the hexagonal host lattice. The melting
point of Ta$_2$N has been determined both by own work (2) as
well as by Booker et al (9) to be at 2950± 50°C. It is
to be suspected, that the often cited melting point of
TaN in reality has to be attributed to Ta$_2$N. According
to our own experiments TaN undergoes decomposition to
yield Ta$_2$N at temperatures greater than 1500°C and
nitrogen pressure under 1 bar.
    ε-TaN undergoes a phase transition at approximately
1950°C and 40-100 bar nitrogen to yield a slightly
understoichiometric δ-TaN with B1 type of structure.

The exact nature of the phase transition is not yet known, possibly it is a congruent phase transition of the syntectic type to give stoichiometric or even slightly hyperstoichiometric δ-TaN$_{1\pm x}$ as encountered in the Nb-N system.

The melting point of TaN has not yet been determined, but taking into account that the dominant binding forces have to be attributed to the strong Ta-Ta interactions and only secondarily to the relatively weak Ta-N interactions, melting of δ-TaN is expected to take place at about 2800-3000°C and at a nitrogen equilibrium pressure of about 1000-3000 bar.

The phase θ-TaN reported by Schönberg (10) and frequently observed in reaction products between Ta-powder and nitrogen or ammonia at temperatures below 1400°C is believed by Brauer et al (11) to be a high pressure phase, formed at pressure levels in the order of over 20 kbar. Hence, θ-TaN is to be considered as a metastable phase which forms only transiently during nitridation.

Two tantalum nitrides Ta$_5$N$_6$ and Ta$_4$N$_5$ with nitrogen contents higher than 50 at% have been prepared and structurally characterized by Gilles (12) by controlled thermal decomposition of Ta$_3$N$_5$. Ta$_5$N$_6$ may also be obtained by reaction of high pressure nitrogen (350 bar at 1200°C) with tantalum powder (B), while higher tantalum nitrides such as Ta$_4$N$_5$ and Ta$_3$N$_5$ could not yet be synthesized by this method. They have been obtained as a result of an ammonolytical reaction between TaCl$_5$ or (NH$_4$)$_2$TaF$_7$ with ammonia (14,15) under closely controlled temperature conditions.

The crystal structures of the tantalum nitride phases are characterized by a transition from the octahedral structural elements Ta$_6$N in Ta$_2$N and δ-TaN$_{1-x}$ to the trigonal prismatic structural element in θ-TaN and further on to structural elements not observed in transition metal carbide phases. According to a re-determination of the structure of ε-TaN by neutron diffraction (16), the Ta-atoms are sixfold coordinated by equidistant nitrogen atoms in form of a trigonal prism, whereas the nitrogen atoms are situated in the center of a foursided pyramid of Ta atoms (Fig.3). The same structural element is encountered in Ta$_5$N$_6$, (fig.4) where some of the positions of the hexagonal arrangement of metal atoms are left unoccupied. In Ta$_4$N$_5$ the tantalum atoms are still sixfold coordinated by nitrogen atoms,

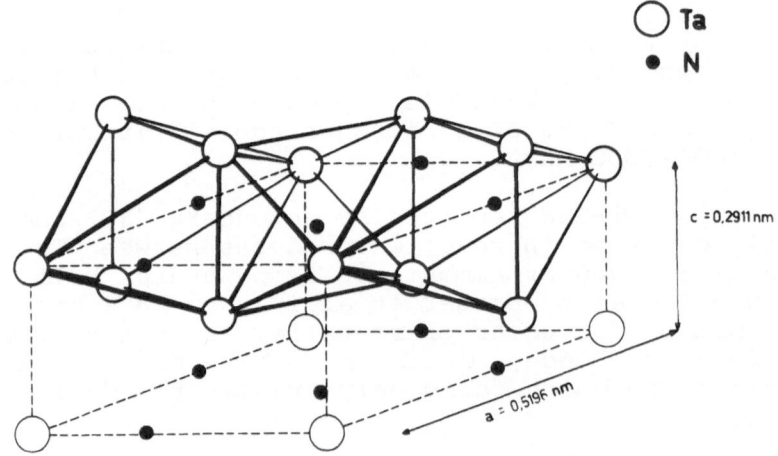

Fig.3. Arrangement of atoms in the lattice of ε-TaN

but 1 nitrogen atom has only 4 next Ta-neighbours in a
planar arrangement (fig.5). In Ta_3N_5 the nitrogen atoms
according to a structure determination of Strähle (17)
are situated in the center of tetrahedra (fig.6),
whereas the tantalum atoms are still sixfold coordinated
by nitrogen atoms in form of slightly distorted octahedra
(fig.7). In this context it is interesting to note, that
in a ternary phase of the composition $(Nb,Ta)_8N_9$ the
structural elements of both θ-TaN-trigonal prisms T_6N -
and ε-TaN - foursided pyramids T_5N (T=Ta,Nb) - are
combined (18).

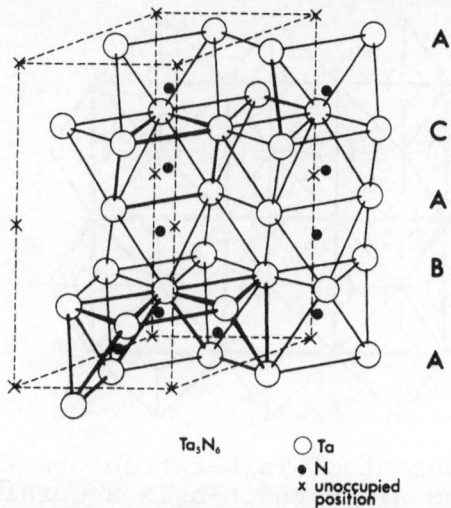

Fig.4.Arrangement of atoms in the lattice of Ta_5N_6

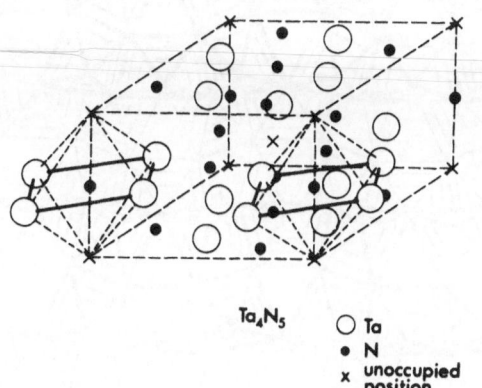

Fig.5. Arrangement of atoms in the lattice of Ta_4N_5

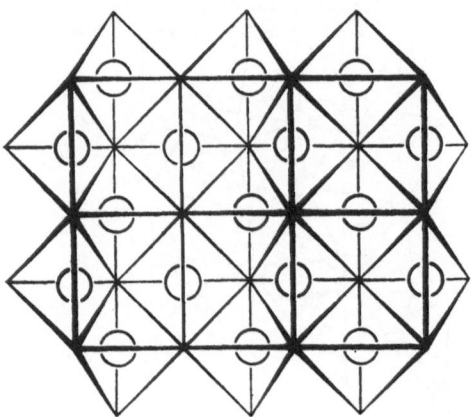

Fig. 6. Arrangement of Ta_4N-tetrahedra in Ta_3N_5
        sighted along the b-axis according
        to Strähle (ref. 17)

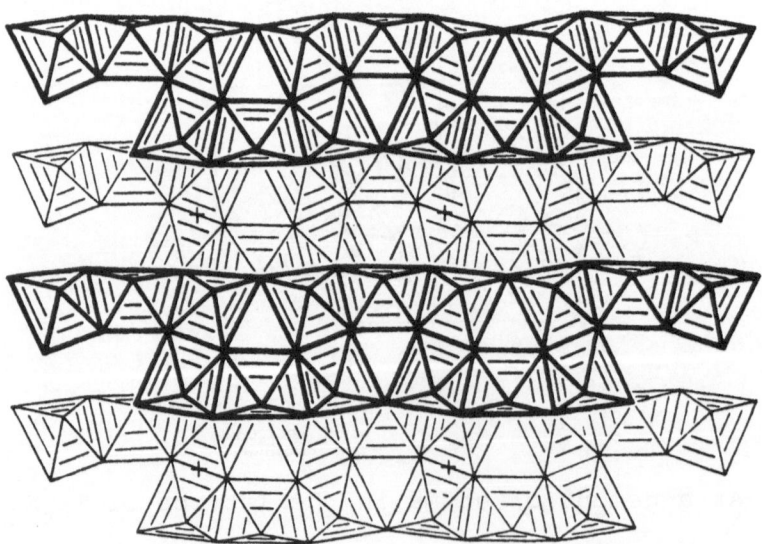

Fig. 7. Arrangement of TaN_6-octahedra in Ta_3N_5
        sighted along the a-axis according
        to Shrähle (ref. 17)

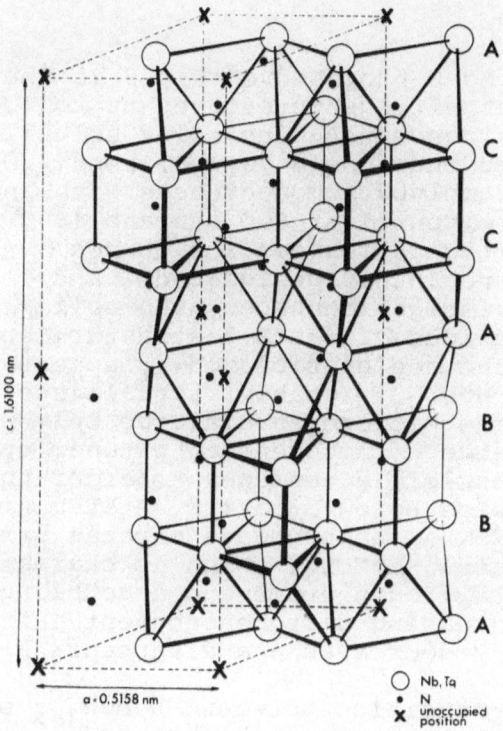

Fig.8.   Atomic arrangement in the lattice of $(Nb,Ta)_8N_9$

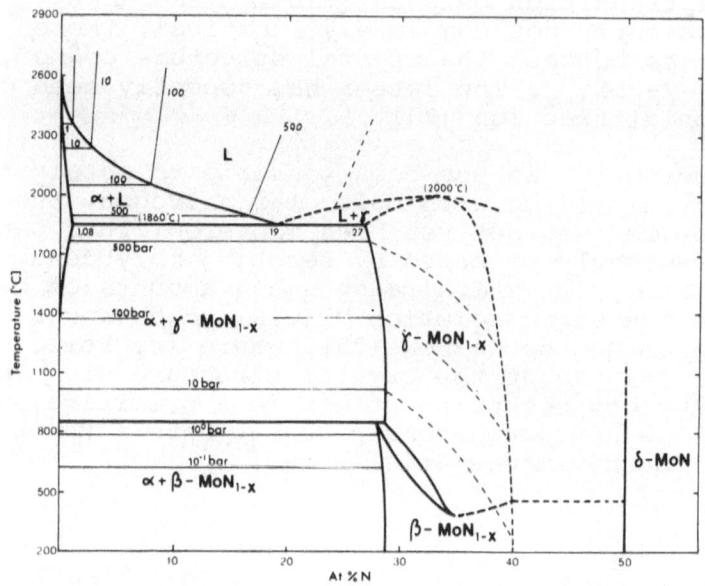

Fig.9.   Proposed phase diagram for the system Mo-N

THE SYSTEM Mo-N

The system Mo-N proposed in Fig.9 is the result of
investigations at nitrogen pressures up to 1000 bar (19).
Previous results concerning the solid solubility of
nitrogen in molybdenum could be confirmed (20). The
melting point of molybdenum decreases with increasing
nitrogen pressure, until at 800 bar and 1860°C the
nonvariant eutectic is formed. The phases β-MoN$_{1-x}$ and
γ-MoN$_{1-x}$ were previously designated β and γ-Mo$_2$N, but
according to investigations about the solid solubility
of MoN$_{1-x}$ in nitrides of the B-1 structure such as CrN
and VN (21,22) the designation MoN$_{1-x}$ appears to be more
aptly chosen. MoN$_{1-x}$ probably crystallizes in the B1
structure with the nitrogen sublattice only partially
occupied. The range of homogeneity extends up to about
40 at% N with possibly a tendency to ordering in the
nitrogen sublattice below 33 at% N. At temperatures
below 860°C γ-MoN$_{1-x}$ undergoes a crystal structure change
to the tetragonal β-MoN$_{1-x}$, which is characterized by an
ordering of the nitrogen atoms and a doubling of the
c-axis. With increasing nitrogen content in β-MoN$_{1-x}$
the tetragonality decreases and disappears at 33,3 at% N.

The phase transition between β-MoN$_{1-x}$ and γ-MoN$_{1-x}$
seems to be of the same type as observed in the subcarbide
region of many carbide systems and possibly similar to
the phase transition between γ-NbN$_{1-x}$ and δ-NbN$_{1-x}$.
While apparently not completely identical, there are many
similarities between the crystal structure of β-MoN$_{1-x}$
(23) and γ-NbN$_{1-x}$. The latter has recently been determined
by neutron diffraction (24).

Hitherto it has not been possible to obtain the phase
δ-MoN by nitriding with molecular nitrogen; the pressure
level required has not yet been mastered. This phase is
more conveniently prepared by reacting molybdenum powder
with streaming ammonia. Already small amounts of carbon
stabilize the solid solution Mo(C,N) which apparently
extends from MoC to δ-MoN (25). There are still
controversies about the crystal structure of δ-MoN.
The metal atoms occupy positions in a primitive, hexagonal
lattice, the nitrogen atoms probably have a tendency to
ordering in the interstitial voids.

TERNARY NITRIDE PHASE DIAGRAMS

The miscibility of MoN_{1-x} with the B-1 phases
CrN (21) and VN (22) indicate that MoN_{1-x} probably has a
defect B1 type structure with only partially occupied N
positions. The existence of an extended range of a
hexagonal solid solution $(Cr,Mo)_2N$ leads to the conclusion
that a hexagonal phase Mo_2N is only slightly less stable
than the cubic phase MoN_{1-x} (figs.10 and 11).

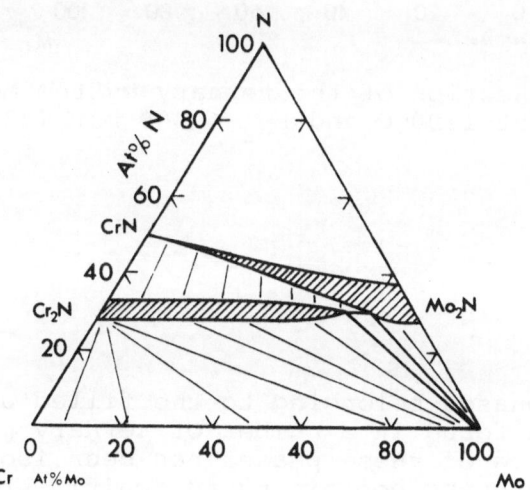

Fig.10. Section of the ternary system Mo-Cr-N
at 1000°C and P_{N_2} ≼ 300 bar (21)

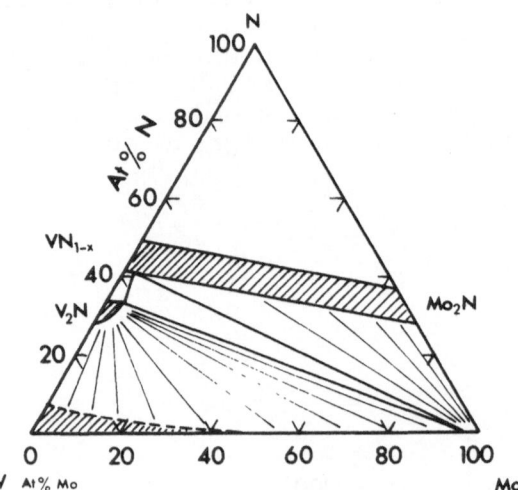

Fig. 11. Section of the ternary system Mo-V-N
at 1100°C and $P_{N_2} \leqslant 300$ bar (22)

      Ternary phases belonging to the filled up θ-TiCu
type have been found in a number of ternary phase diagrams.
A representative of these phases has been isolated from
niobium and chromium containing austenitic steels and was
designated Z-phase (26). The structure has been determined
(27,28). This ternary phase is characterized by alternating
double layers of V A metal (Nb,Ta) and VI A metal (Cr,Mo)
atoms. The nitrogen atoms are coordinated by five
equidistant V A metal atoms in form of a four sided
pyramid (fig.12). The members of the Z-phase family are
listed in table 1.

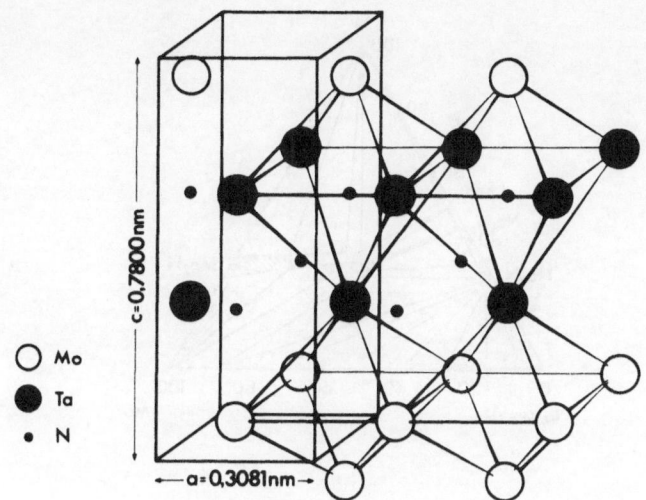

Fig.12. Atomic arrangement of TaMoN

Table 1: Lattice parameters of Z phases
Space group P4/nmm - D_{4h}^{17}

Phase	Lattice parameters a	c	Ratio a/c	Lit.ref.
θ-CuTi	0,3118	0,5921	0,5266	
NbCrN	0,3028	0,7360	0,4082	27,28
$Ta_{1-x}Cr_{1+x}N$	0,3004	0,7334	0,4096	(28)
$NbMoN_{1-x}$	0,3095	0,7799	0,3968	29
TaMoN	0,3081	0,7800	0,3950	30
CaGaN	0,3570	0,7558	0,4723	31

Isothermal sections of the systems Nb-Mo-N and Ta-Mo-N
have been investigated by Vendl (32,33) see fig.13.

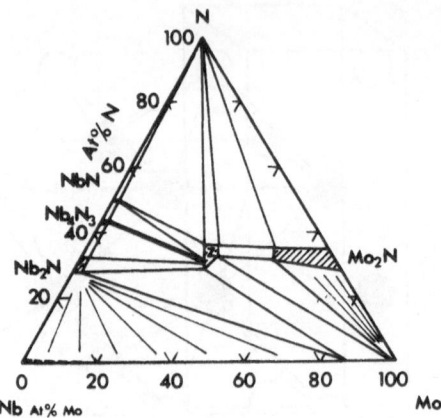

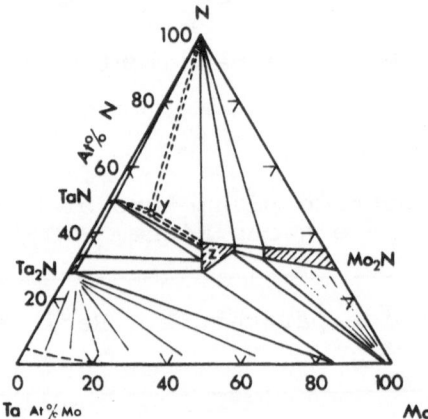

Fig.13. Isothermal sections of the systems Nb-Mo-N
        and Ta-Mo-N at 1100°C, p_{N_2} 300 bar

     Besides the Z-phase MoTaN a further ternary nitride
phase with the composition $Mo(Mo,Ta)_2N_2$ has been observed
(34). That phase is formed at 1600°C and at a nitrogen
pressure of 360 bar. The phase crystallizes with a tetragonal lattice in the space group $P4/nmm - D_{4h}^{17}$, the metal
atoms occupy the positions of Mo and Si in $MoSi_2$. The
nitrogen atoms are situated similar to MoTaN in the interstitial sites, thus forming again four sided pyramids
with nitrogen atoms in the centers.

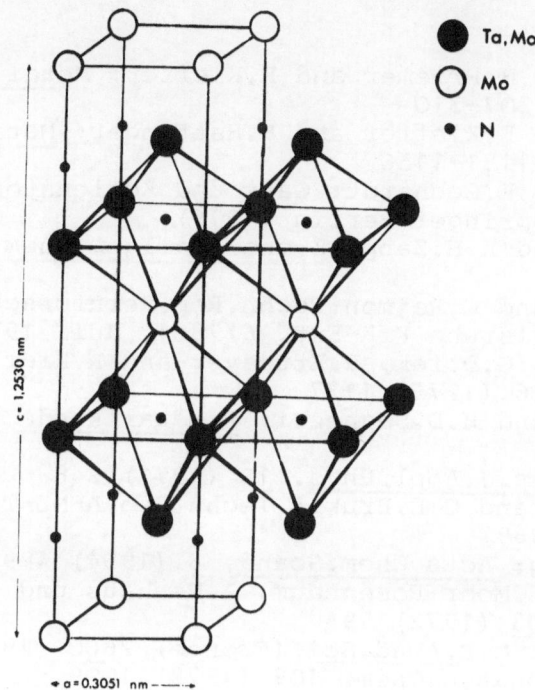

Fig.14. Atomic arrangement in the phase $Mo(Ta,Mo)_2N_2$

The phase $Mo(Ta,Mo)_2N_2$ belongs to a family of ternary carbides and nitrides with filled up $MoSi_2$-type of structure. Some representatives of this familiy are listed in table 2.

Table 2: Lattice parameters of phases with the filled up $MoSi_2$ type structure

Space group I4/mmm

Phase	Lattice parameters (nm)		Ratio c/a	Ref.
	a	c		
$Mo(Ta,Mo)_2N_2$	0,3051	1,253	4,107	34
U_2RuC_2	$0,344_9$	1,257	3,644	35,36
U_2OsC_2	0,346	1,258	3,635	35
U_2RhC_2	0,3464	$1,251_5$	3,613	35,36
U_2IrC_2	0,3478	$1,247_5$	3,587	35,36,37
U_2PtC_2	$0,352_8$	$1,256_7$	3,562	35,36

REFERENCES

1. P.Ettmayer, H.Priemer and R.Kieffer: Metall,
    Berlin, 23 307-310
2. P.Ettmayer, R.Kieffer and F.Hattinger: Metall,
    Berlin, 28 1151-1156
3. E.Fromm and E.Gebhardt: Gase und Kohlenstoff in
    Metallen, Springer-Berlin (1976)
4. G.Brauer and K.H.Zapp: Z.anorg.allgem.Chem. 277
    (1954) 129
5. C.Politis and G.Rejmon:Techn.Rep.Kernforschungs-
    zentrum Karlsruhe KfK-Ext. 6/78-1, Juli 1978
6. J.Gatterer, G.Dufek, P.Ettmayer and R.Kieffer:
    Mh.Chem. 106 (1975) 1137
7. E.Gebhart and H.D.Seghezzi: Z.Metallkunde 52
    (1961) 464
8. N.Terao: Jap.J.Appl.Phys. 10 (1971) 248
9. P.H.Booker and C.E.Brukl: Techn.Rep.AFML-TR-69-117
    Part VI (1969)
10. N.Schönberg: Acta Chem.Scand. 8 (1954) 199
11. G.Brauer, E.Mohr-Rosenbaum, A.Neuhaus und A.Skokan:
    Mh.Chem. 103 (1972) 794
12. J.C.Gilles: C.R.Acad.Sci.(France) 266C (1968) 546
13. A.Vendl: Monatsh.Chem. 109 (1978) 1009
14. G.Brauer, J.R.Weidlein and J.Strähle:
    Z.anorg.allg.Chem. 348 (1966) 298
15. G.Brauer, J.R.Weidlein: Angew.Chem. 77 (1965) 218
16. A.N.Christensen: Acta Cryst. B34 (1978) 261-263
17. J.Strähle: Z.anorg.allg.Chem. 402 (1973) 47-57
18. P.Ettmayer and A.Vendl: J.Less Common Metals
    72 (1980) 209-217
19. H.Jehn and P.Ettmayer: J.Less Common Metals
    58 (1978) 85-98
20. H.Jehn and P.Ettmayer: High Temp-High Press.
    8 (1976) 83-94
21. P.Ettmayer, A.Vendl and R.Kieffer:
    High Temp.-High Press., 10 (1978) 699-702
22. A.Vendl: Monatsh.Chem. 110 (1979) 685-91
23. J.H.Evans and K.H.Jack: Acta Crystallogr.
    10 833-834
24. G.Heger and O.Baumgartner: J.Phys.Chem: Solid
    State Physics 13 (1980) 5833-41
25. P.Ettmayer:Monatsh.Chem.101 (1970) 1720--30
26. H.Hughes: J. Iron Steel Inst. 200 (1962) 1011
27. D.H.Jack and K.H.Jack:J.Iron Steel Inst. 210
    (1972) 790-92
28. P.Ettmayer: Monatsh.Chem. 102 (1971) 858-863
29. A.Vendl: Monatsh.Chem. 110 (1979) 103-108
30. A.Vendl: Monatsh.Chem. 109 (1978) 1001-1004

31. P.Verdier, P.L'Haridon, M.Maunaye and R.Marchand:
    <u>Acta Cryst.</u> B30 (1974) 226-228
32. A.Vendl: <u>Monatsh.Chem.</u> 110 (1979) 1099-1107
33. A.Vendl: <u>Monatsh.Chem.</u> 110 (1979) 879-885
34. P.Ettmayer and A.Vendl: <u>Monatsh.Chem.</u> 111
    (1980) 547-550
35. H.Holleck: <u>J.Nucl.Mat.</u> 28 (1968) 339
36. H.R.Haines and P.E.Potter:
    <u>Nature</u> 1969, 221 (5187) 1238-39
37. A.L.Bowman, G.P.Arnold and N.H.Krikorian:
    <u>Acta Cryst.</u> Sect.B 27 (1971) (Pt 5) 1067-68

PHASE EQUILIBRIA AND STRUCTURAL CHEMISTRY WITHIN TERNARY SYSTEMS:

ACTINIDE METAL-BORON-CARBON

Peter Rogl

Institute of Physical Chemistry
University of Vienna
A-1090 Wien, Austria

EXTENDED ABSTRACT

Early interests in the constitution and thermodynamic equilibria of the high melting actinidemetal boride and carbide phases were stimulated by their possible use as nuclear fuel materials. Thus the existence of several ternary borocarbide compounds has been reported in the literature however only few were characterized then.

Meanwhile most of the phase equilibria and compounds have been confirmed in our laboratory and in particular their crystal structures have been established from X-ray single crystal counter data (ThBC, $Th_3B_2C_3$, ThB_2C).

A detailed reinvestigation of melting behavior as well as of thermodynamic equilibria within both systems Th-B-C, U-B-C revealed the formation of several new compounds, which are observed from isothermal sections at $1000^{\circ}C$, $1600^{\circ}C$, $1800^{\circ}C$: $Th_3B_2C_3$ (unique structure type); U_2BC_2 and $\ell,h-UB_2C$ (ThB_2C-type, related to AlB_2-type with complex B-C ordering; phase transition $\sim 1675^{\circ}C$).

From melting point analysis (Pirani-technique, Ar) congruent melting behavior with relatively high melting temperatures was determined for the ternary borocarbides: ThBC ($2101 \pm 22^{\circ}C$), ThB_2C ($2040 \pm 14^{\circ}C$), UBC ($2144 \pm 23^{\circ}C$), UB_2C ($2282 \pm 28^{\circ}C$).

Using the clear-cross principle in studying possible phase reactions,the thermodynamic stabilities (Gibbs free energy of formation) have been estimated in case of the compounds UBC (- $\Delta G_f \geqslant$ 157 kJ/mole) and UB_2C (- $\Delta G_f \geqslant$ 201 kJ/mole).

Owing to this relative high stability of the ternary actinidemetalborocarbides as compared to their binary boride and carbide systems, ternary phase equilibria are dominated by the (Th,U)BC- and (Th,U)B_2C-phases. Subsequently no phase equilibria between boron or "B_4C" on the one hand and actinide metal carbides on the other hand were found to exist.

According to the classification scheme of ternary metal borides, zig-zag boron-boron chains (B-B $\sim$ 1.76-1.82 Å) with adjacent carbon atoms (B-C $\sim$ 1.55 Å) appear to be the basic structural unit in borocarbides of composition MBC as has been shown for the structure types of ThBC, UBC and YBC. From geometrical as well as structural chemical arguments YBC can be regarded as a link between the structure types of AlB_2 and α-ThSi$_2$ based upon simple shift operations. Similarly the ThBC-type is generated by a double shift (vector:1/2,0,1/2) out of the UBC-type structure.

From the viewpoint of crystal symmetry the correspondence of the metal-boron sublattice in the structure types of ThBC (P4$_1$22), αMoB(I4$_1$/a 2/m 2/d) and similarly for the pair UBC, CrB (both C2/m 2/c 2$_1$/m) can be expressed in a theoretical crystallographic group-subgroup relationship. Thus a hypothetical structure (I4$_1$22; ThBC-type with idealized point positions) is found intermediate between αMoB (low temperature form) and ThBC. In this respect the shift operation ThBC-UBC among actinide borocarbides is an interesting analogue of the αMoB-CrB correlation for transition metal monoborides.

Considering the boron-boron aggregation in borocarbides MBC a gradual replacement of boron chains by boron pair formation i.e. C-B-B-C groups is observed with increasing radius ratio R_M/R_B (M= U,Y,Th). The developed structural regularities are confirmed from the crystal structure of $Th_3B_2C_3$ (P2/m). Again the stability of this borocarbide can be understood by a topochemical combination of structural units: ThC + 2ThBC = $Th_3B_2C_3$.

DISCUSSION

T. Lundstrom:

I would like to ask about the boron carbide. Do you get the same
range of homogeneity for the different temperatures or does it
change?

P. Rogl:

It decreases slightly at lower temperatures. The range is fourteen
atomic percent at 1600°C, from 9 to 21 atomic percent.

H. Holleck:

My question to you concerns the boron carbide region. You know that
boron carbide is a very interesting material for hard metal technology. There is some new literature from Russia which show that very
small amounts of metals in boron carbide raise the hardness from
4000 up to 5500 Vickers. One can not explain this exactly. It can
be a solution effect or a precipitation effect. Have you examined
this region in detail?

P. Rogl:

We have looked for solubility of U in B_4C and we didn't observe any
change in the lattice parameter. I guess there is very limited
solubility--maybe less than one atomic percent. And on the grain
boundaries you already notice uranium carbide coming out if you
prepare samples with one atomic percent. So the solubility limit
should be less than one atomic percent.

H. Holleck:

How does boron influence the transformation temperatures for the
binary uranium carbide?

P. Rogl:

Boron hardly influences the transformation temperature. We didn't
really investigate very closely this region because we made isothermal sections at 1000, 1600 and 1800 degrees. But boron does not
have any influence on the kinetics of the transformation.

H. E. Exner:

I would like to address this question together with the comment that
was made at the start of the Conference that we don't understand

why hardness increases or decreases and we have no appreciation for
that. I think the only way is to ask the physicist how the electronic states of bonding change. Swedish work and work done at our
laboratories where we could show that the bonding stage very closely
corresponds to properties like hardness and wetting behavior. This
may be a wrong approach. But I am not a physicist but a physical
metallurgist. So we use wetting to say something about bonding
stage. The opposite should be done. The binding stage should be
investigated and then, I think, hardness behavior and wetting behavior could be predicted from that.

H. Holleck:

It is only true if you have a small amount in solution. From the
crystal chemistry standpoint, it's hard to understand that you can
put thorium in solution in boron carbide.

H. E. Exner:

I think that the states of binding are shifted so much by very small
additions. For example, in the WC system, you have such a small
range of homogenity that changes are very dramatic from one side to
the other of the composition range. While titanium carbide is easier
to investigate and there you have all these nice correlations. So if
you have systems, which have very small solubility of elements which
give dramatic changes in the electronic states, then you may expect
dramatic property changes there.

P. Rogl:

For instance, in β -boron, (and probably Professor Lundstrom can tell
you more about this) small amounts of dissolved metals can change
the hardness value.

H. Holleck:

There are other examples. For instance, in vanadium carbide, very
small amounts in solution, can raise the hardness by 2000 but you
cannot reproduce this in most cases. But as you say, this is a
problem of very small amounts in solution perhaps. Can you say
anything about the role of the valency of the actinide elements on
the stability of these structures?

P. Rogl:

Actually we do not observe any valency changes whether in uranium
or thorium. Thorium probably has a very pronounced four plus state.
We do not observe any magnetic properties on the uranium borocarbide
as well as on the thorium borocarbide. You probably are familiar

with the Hill criterion for actinide metals and magnetic properties
of actinides. We do not have any atomic distances in our structures
which would favor paramagnetic ordering in uranium compounds due to
a very large proper distance for the Hill criterion. So I think
uranium will be 3+.

H. Holleck:

It would be interesting to check this with cerium. Cerium behaves
simiarly to thorium and plutonium.

P. Rogl:

There are some investigations on the cerium-boron-carbon system
which date back to the early sixties. Only one ternary compound
was identified and very likely this compound was shifted to a
different position, probably the 1:1:1 composition. They've indicated a smaller cerium content. But very likely the cerium has
disappeared during the arc melting procedure because it has a higher
vapor pressure. But as far as I know, nobody has done any detailed
investigation on the cerium borocarbide system in the meantime. So
I know there is a $Ce_5B_2C_3$ type structure which is isotypical to the
lanthanum compound. But no magnetic properties have been studied
and I would guess cerium behaves like a three plus/four plus mixture
like that which is adopted in ternary borides and borocarbides.

AN EXPERIMENTAL AND THERMODYNAMIC STUDY OF THE Co-W-C SYSTEM
IN THE TEMPERATURE RANGE 1470-1700 K

Leif Åkesson

Coromant Research Center
Sandvik AB
Box 42056, S-126 12 STOCKHOLM, Sweden

INTRODUCTION

In spite of its great practical importance in cemented carbides applications, the Co-W-C system has not been extensively investigated.

Takeda[1] carried out an early investigation, mainly of equilibria involving liquid and carbides.

Rautala and Norton[2] performed the most extensive experimental work known up to now. They determined the liquidus surfaces by means of thermal analysis. Carbides were identified and characterized by X-ray analysis; however, they reported the existence of only one η-type carbide, whose composition consequently was said to vary in both carbon and tungsten contents.

Pollock and Stadelmaier[3] commented on the existence of two carbides, M_6C and $M_{12}C$, and described the equilibria of these with other phases. They presented the results in two tentative isothermal sections, at 1273 and 1673 K, respectively.

The first more extensive thermodynamic study of the Co-W-C
system was performed by Johansson and Uhrenius[4] at 1423 K. They
evaluated some thermodynamic parameters for the FCC-Co (β phase)
and the carbides $M_{12}C$, M_6C and WC. Uhrenius et al[5] and Johansson[6]
have also studied the liquidus temperatures in the Co-W-C system,
and have presented tentative isothermal sections at 1533 and 1713 K,
respectively.

Petrdlik and Tuma[7] have very recently studied the solubility
of carbon and tungsten in the cobalt binder phase and its dependence
on the carbon activity in WC-Co cemented carbides. The investigation
was restricted to only one temperature, namely 1473 K.

It is evident that there is a considerable lack of information
regarding the thermodynamic properties of the Co-W-C system, especially at temperatures above 1473 K. The purpose of the present
work was to supply this thermodynamic information, in order to increase the understanding of the Co-W-C system at temperatures that
are of interest for sintering of cemented carbides.

Due to the large extent of this investigation a comprehensive
presentation of the results will be published elsewhere.

EXPERIMENTAL

Annealing experiments were performed in the temperature range
1473-1698 K. Series of 15-20 small specimens of binary Co-W alloys
with different Co/W-ratios were carburized and simultaneously equi-
librated to reach a constant carbon activity using a flowing gas
mixture of CO and CO_2. The specimens were quenched and analysed
for carbon. The different phases were identified using X-ray diffraction analysis and analysed for cobalt and tungsten in an electron
microprobe. The carbon activity for each experiment was determined
from a knowledge of the thermodynamics of the Fe-C system and the

carbon content in a pure iron sample inserted among the other alloys.

THERMODYNAMIC MODELS

For the thermodynamic description of the β phase (FCC-Co) and
the liquid phase a regular solution model suggested by Hillert and
Staffansson[8] was used. This model has also recently been used by
Johansson and Uhrenius[4]. However, in the present work an extension
of the model was used. To be able to describe the maximum in the
liquidus curve in the cobalt-rich side of the binary Co-W system[9,10],
it was necessary to introduce a concentration dependence, according
to a suggestion by Guggenheim[11], of some of the interaction parameters.

The free energy of the M_6C carbide, where M stands for Co and W,
was expressed by an extension of a regular solution model previously
used by Richardson[12] and by Hillert et al[13]. The M_6C carbide was
then assumed to be stoichiometric with respect to the carbon to metal
ratio. The other carbides, MC and $M_{12}C$, were treated as stoichiometric phases with respect to both their carbon and metal contents
in accordance with the results in this work, i.e. they were considered as WC and Co_6W_6C, respectively.

RESULTS AND DISCUSSION

In this paper the most important results regarding the different phases and phase relations in the Co-W-C system will be given.
Also the methods of evaluation of the thermodynamic parameters and
the calculation of phase equilibria will be described briefly.

From the experimental information the following important results and conclusion can be listed:

- The two η carbides, M_6C and $M_{12}C$, were both found to be in equilibrium with the cobalt binder phase, solid or liquid at all temperatures investigated. The $M_{12}C$ carbide was present at low carbon

activities and M_6C at somewhat higher carbon activity values.

- Only the M_6C carbide was found to be simultaneously in equilibrium
with binder phase and WC, i.e. the three-phase equilibrium: binder
phase + M_6C + WC existed at all the temperatures studied. This is
in contrast to the conditions at lower temperatures reported earlier[4,14], where the three-phase equilibrium β phase + $M_{12}C$ + WC was
reported to exist at 1423 K and 1273 K, respectively. As a consequence the authors did not find the equilibrium between binder
phase and M_6C.

- This is related to the fact that the stability range of the M_6C
carbide is wider at the high temperatures studied in this work. This
is demonstrated in Fig. 1. M_6C was found to be stable in the composition range $Co_{3.1}W_{2.9}C$ ($Y_W = 0.48$) to $Co_{2.2}W_{3.8}C$ ($Y_W = 0.64$),
where the limits correspond to equilibrium with β phase and BCC-W,
respectively. At 1423 K Johansson and Uhrenius[4] reported that M_6C
was only stable in a narrow composition range, viz. $Co_{2.2}W_{3.8}C$ to
$Co_{2.0}W_{4.0}C$. Pollock and Stadelmaier[3] have also found that the composition range of the M_6C carbide increases with temperature in
accordance with the results given above.

- The temperature of formation of the stable four-phase equilibrium:
solid and liquid binder phase, WC and graphite was found to be
1548 $\pm$ 2 K. This is the lowest temperature at which liquid phase
is stable in the Co-W-C system, i.e. the lowest stable eutectic
temperature. In the literature[1,5], the existence of liquid phase
is reported at lower temperatures, but in these cases the liquid
was supercooled and then solidified through a metastable eutectic
reaction including η carbide (mostly M_6C). Furthermore, the carbon
content was found to have a strong influence on the eutectic temperature. A lower carbon content (lower carbon activity) results
in a higher eutectic temperature.

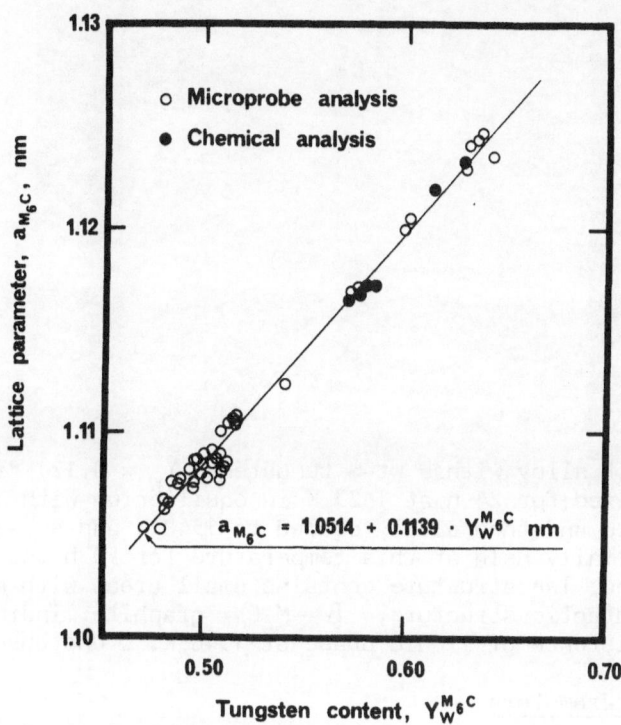

Fig. 1. The dependence of the lattice parameter of $(Co,W)_6C$
on the tungsten content in the carbide.

It is worth mentioning that the abovementioned stable eutectic
temperature was verfied by two independent types of experiments:
using differential thermal analysis (DTA) of a WC-Co sample with
an excess of carbon and using double isothermal heat treatments.
Using the latter technique traces of liquid phase could be detected
in samples isothermally heat-treated at temperatures down to 1548 K,
see Fig. 2.

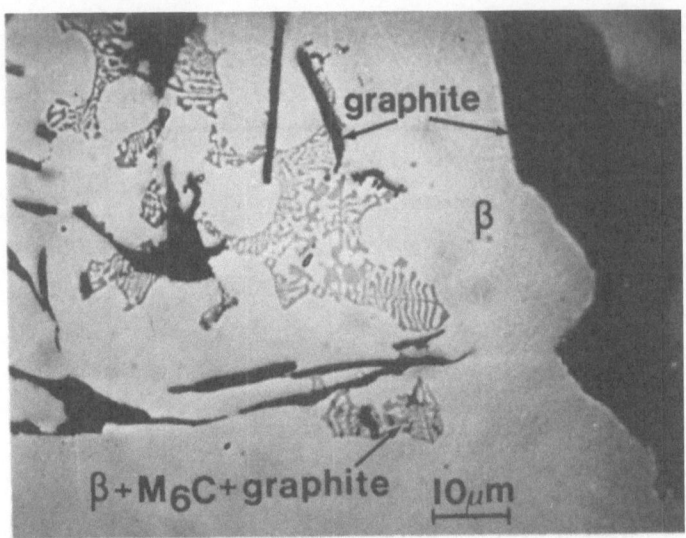

Fig. 2. A cobalt alloy with 30 wt-% tungsten (Y_W = 0.12) first
        carburized for 24 h at 1623 K in equilibrium with
        graphite and then slowly cooled to 1548 K and subsequently
        isothermally held at this temperature for 50 h and then
        quenched. The structure contains small areas with a very
        fine eutectic structure, β + M_6C + graphite, indicating
        the existence of liquid phase at 1548 K.   Unetched.

Thermodynamic Parameters

The evaluation of the parameters was performed by computer-
ized linear regression analysis of the experimental results for
one phase at a time. The experimental data was obtained from about
25 experiments at different temperatures and carbon activities.
Also, in some cases, experimental data found in the literature
was used.

The parameters for the binary β phase, (Co,C), were evaluated
using data on the solubility of carbon in cobalt at different carbon activities. In the evaluation data from the present work was

used together with data from some earlier work[15-17]. The parameter
for the ternary β phase, (Co,W,C), was evaluated using the carbon
content in single-phase alloys with different tungsten contents and
at different carbon activities, i.e. evaluating the slope of the
carbon isoactivity lines in the β phase.

The parameters for the liquid phase were evaluated in the
same way as for the β phase. In the evaluation some experimental
data from other investigations[18-21] was also used.

The free energy of formation from the β phase for the carbides Co_6W_6C and WC was evaluated using experimental data on the
solubility of the carbides in the β phase at equilibrium at different carbon activities. The parameters for the M_6C carbide were
evaluated using experimental data, i.e. solubility, composition
and carbon activity, from three different equilibria, namely $M_6C/β$,
M_6C/WC and $M_6C/M_{12}C$.

Using the evaluated thermodynamic parameters isothermal sections can be calculated at arbitrary temperatures. Sections at
1698 K are given; the solubility lines for the carbides and isoactivity lines for carbon are shown in Figs. 3 and 4. Furthermore,
it is possible to calculate all existing three-phase and four-phase
equilibria involving the phases for which thermodynamic parameters
have been evaluated. As an example, the four-phase equilibrium
β + liquid + M_6C + WC was calculated to exist at 1628 K and the
carbon activity = 0.32. This equilibrium indicates the lowest temperature at which M_6C can be formed in contact with liquid phase
and WC. This reaction is of interest in the manufacture of cemented
carbides, due to the coarse-grained M_6C that then can be formed.

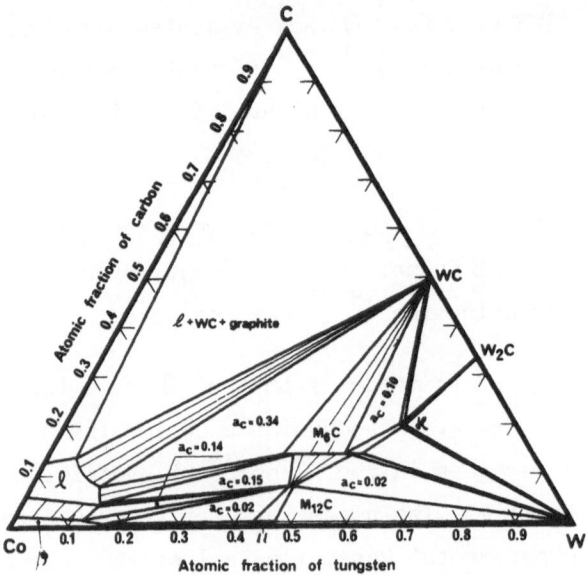

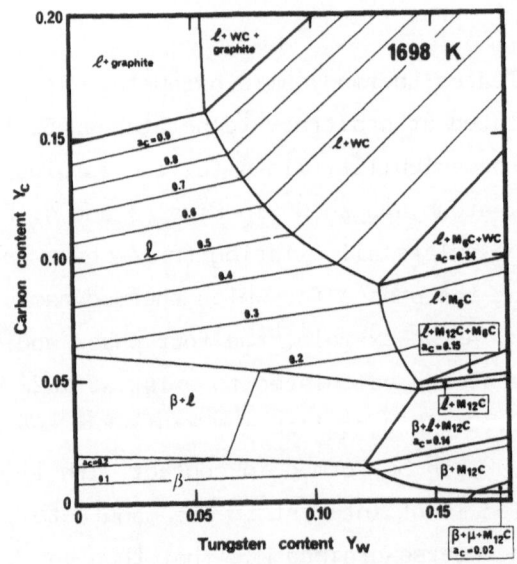

Figs. 3 and 4.   Calculated isothermal sections of the Co-W-C
system at 1698 K.

SUMMARY AND CONCLUSIONS

The experimental technique used in this study was designed to allow carburisation of binary Co-W specimens to carbon equilibrium at temperature up to about 1700 K.

Some of the experimental results can be summarized:

- The formation temperature of the stable ternary eutectic consisting of β phase, WC and graphite was determined to be 1548 K. At lower carbon contents the eutectic temperature increases and the eutectic consists only of β phase and WC.

- The two η carbides, M_6C and $M_{12}C$, were both found in equilibrium with binder phase, solid or liquid at all temperatures studied.

- Only the M_6C carbide was found in equlibrium with binder phase and WC simultaneously. As a consequence the three-phase equilibrium, binder phase + $M_{12}C$ + WC, is not stable at the temperatures studied.

- The composition range of the M_6C carbide was found to be $Co_{3.1}W_{2.9}C$ to $Co_{2.2}W_{3.8}C$ with the corresponding lattice parameters 1.106 and 1.125 nm, respectively. The compositions range of M_6C was almost temperature-independent in the studied interval, 1470-1700 K.

By means of chemical and electron microprobe analyses of carburized Co-C and Co-W-C specimens augmented with data from other investigations, thermodynamic parameters have been determined according to an extended regular solution model for the β phase, the liquid phase, the carbides $M_{12}C$, M_6C and WC. Using the thermodynamic parameters phase equilibria, solubility lines and isothermal sections can be calculated in the Co-W-C system which is of great value for the understanding and development of processes and new grades in the cemented carbide manufacturing industry.

ACKNOWLEDGEMENTS

This work was performed at the Coromant Research Center, Sandvik AB, Stockholm. It was supported financially by the Swedish Board for Technical Development (STU).

The author is grateful to Sandvik AB for permission to publish this work.

REFERENCES

1. S. Takeda, A metallographic study of the action of the cementing materials for cemented tungsten carbide, Sci. Rep. Tōhoku Univ., (Sendai), Honda Anniversary Volume: 864 (1936).

2. P. Rautala and J. T. Norton, Tungsten-cobalt-carbon system, Trans. AIME, 194:1045 (1952).

3. C. B. Pollock and H. H. Stadelmaier, The eta carbides in the Fe-W-C and Co-W-C systems, Met. Trans., 1:767 (1970).

4. T. Johansson and B. Uhrenius, Phase equilibria, isothermal reactions, and a thermodynamic study in the Co-W-C system at 1150°C, Metal Science, 12:83 (1978).

5. B. Uhrenius, B. Carlsson and T. Franzén, A study of the Co-W-C system at liquidus temperatures, Scand. J. Metallurgy, 5:49 (1976).

6. T. Johansson, Solidification studies in the Co-W-C system, Thesis, Univ of Uppsala, Inst. of Chemistry, UUIC-B 18-67 (1975).

7. M. Petrdlik and H. Tuma, Bestimmung des Kohlenstoff- und Wolframgehalt in der Kobalt-verbindungsphase bei WC-Co-Hartmetallen bei 1200°C an Hand der Kohlenstoffaktivität und des Gleichgewichtsdiagramms, in: "Proceedings of the 10th Plansee-Seminar, Vol. 2, H. M. Ortner, ed., Metallwerk Plansee, Reutte (1981).

8. M. Hillert and L.-I. Staffansson, The regular solution model for stoichiometric phases and ion melts, Acta Chem. Scand., 24:3618 (1970).

9. M. Hansen, "Constitution of binary alloys",
    McGraw-Hill, New York, (1958).

10. ASM Metals Handbook, Volume 8, 8th edition,
    Metals Park, Ohio, (1973).

11. E. A. Guggenheim, The theoretical basis of Raoult's law,
    Trans. Faraday Soc., 33:151 (1937).

12. F. D. Richardson, The thermodynamics of metallurgical carbides
    and of carbon in iron, J. Iron Steel Inst., 175:33 (1953).

13. M. Hillert, T. Wada and H. Wada, The α - γ equilibrium in
    Fe-Mn, Fe-Mo, Fe-Ni, Fe-Sb, Fe-Sn and Fe-W systems,
    J. Iron Steel Inst., 205:539 (1967).

14. R. Otterberg, Phase equilibria in the Co-rich part of the
    Co-W-C system at 1000°C, Internal Report No. 2293,
    Sandvik AB, Stockholm (1975).

15. R. P. Smith, The activity of carbon in iron-cobalt alloys
    at 1000°C, Trans. AIME, 233:397 (1965).

16. K. K. Rao and M.E. Nicholson, The solubility of carbon in
    cobalt-nickel alloys at 1000°C, Trans. AIME, 227:1029
    (1963).

17. J. C. Swartz, Solubility of graphite in cobalt and nickel,
    Met. Trans., 2:2318 (1971).

18. H. Schenck, M. G. Frohberg and E. Steinmetz, Untersuchungen
    über wechselseitige Aktivitätseinflüsse in homogenen
    metallischen Mehrstofflösungen, Teil I. Experimentelle
    Untersuchungen der Systeme Mn-C-X, Co-C-X, Ni-C-X im
    flüssigen Zustand, Arch. Eisenhüttenwes., 34:37 (1963).

19. E. T. Turkdogan, R. A. Hancock and S. I. Herlitz, The solu-
    bility of graphite in manganese, cobalt and nickel,
    J. Iron Steel Inst., 182:274 (1956).

20. Y. Kojima and K. Sano, The solubility of graphite in nickel
    and cobalt and effects of alloying elements, Tetsu to
    Hagane, 47:897 (1961).

21. Ya. V. Pozhidaev, B. P. Burylev, J. E. Dobrovinskiy and
    V. D. Ivanova, Izv. VUZ/Chern. Met., 12:11 (1975).

DISCUSSION

H. E. Exner:

I would agree that very good phase diagrams are necessary for understanding the behavior and the microstructure of cemented carbides.
I have two questions. People at Krupp have pointed out that if you
anneal WC–Co you get a series of supersaturation decays into $M_{12}C$,
M_6C and finally to Co_3W. Would this show up in your phase diagram
that you get after long annealing, a stable situation where you have
cobalt with Co_3W precipitates instead of eta phase. The second
question is, if you don't cool down rapidly, you may end up with
fine precipitates of eta phase in a structure where you have insufficient carbon to avoid eta phase at sintering temperature. So what
I'm saying is: if you anneal long, you get Co_3W precipitation in
the binder phase and if you cool slowly, you may get eta phase in
this structure in spite of the fact that you have enough carbon for
sintering.

L. Akesson:

Isn't the annealing for getting Co_3W performed at lower temperatures
than I have studied in this work? I have only studied conditions at
temperatures between 1200 and 1425°C. In this temperature range, you
don't get these compounds. In fact, if you have very low carbon
activity values, you get this mu phase Co_7W_6 instead, at these high
temperatures. But this is at very low carbon activity values. There
is almost no carbon in the alloy then. So I am not able to comment
about annealing at very low temperatures and I haven't performed any.

H. E. Exner:

If you draw a cross-section of this diagram, would you find eta phase
in any case? Or, would you find a cross-section where you have a
more or less pseudobinary section?

L. Akesson:

The only binary section is a quasibinary between cobalt and tungsten
carbide if you are at the proper carbon content. They are all very
close to this boundary for eta carbides. But I haven't done any
calculations at very low temperatures. My thermodynamic parameters
are not so good as to extrapolate to very low temperatures.

NEUTRON SCATTERING STUDIES OF THE DEFECT STRUCTURES IN TiC_{1-x} AND NbC_{1-x}

V. Moisy-Maurice, C.H. de Novion, N. Lorenzelli

SESI, Bât. 31, CEN, BP 6,
92260 Fontenay-aux-Roses, France

A.N. Christensen

Chemistry Division, Aarhus University, Aarhus C
Denmark

ABSTRACT

Single crystals of TiC_{1-x} and NbC_{1-x} were studied by elastic neutron diffuse scattering at room temperature in the $(1\bar{1}0)$ reciprocal lattice plane ; the spectra of $TiC_{0.76}$, $TiC_{0.79}$ and $NbC_{0.73}$ were analysed by the Sparks and Borie method, which allowed to determine the first Cowley-Warren short-range order coefficients and a shortening (0.03 Å) of the average first neighbor metal-carbon distances. The order-disorder transformation in TiC_{1-x} ($0.52 \leqslant 1-x \leqslant 0.67$) was studied by high temperature powder neutron diffraction ; the transition temperatures (≈ 760-790°C) and critical coefficients β were determined. The results are discussed in terms of interatomic pair potentials.

INTRODUCTION

Many of the interesting physical properties of refractory transition metal carbides with rocksalt structure, in particular the transport and plastic properties, are controlled by a departure from the stoichiometric composition MC, due to a large content of intrinsic carbon vacancies. The origin of the stability of these non-stoichiometric compounds, and the nature of the terms

responsible for ordering of vacancies are still largely
unknown. Recent theoretical calculations suggest that the
stabilization of TiC_{1-x} (which exists in the range
$TiC_{0.50-0.97}$) has mainly an entropic origin [1].

Several superlattice structures corresponding to
long-range ordering of carbon vacancies were found in
these compounds [2], in particular V_8C_7 [3] and V_6C_5 [4] in the
VC_{1-x} carbide, a "Ti_2C" superstructure [5] in the TiC_{1-x}
carbide, and Nb_6C_5 [6,7] in NbC_{1-x}.

The "Ti_2C" superstructure, characterized by (1/2 1/2
1/2) type superlattice reflexions, which was first observed using powder neutron diffraction by Goretzki [5], is
still unresolved ; indeed, Parthé and Yvon [8] showed that
two structures corresponding to the Fd3m (cubic superlattice, with lattice parameter $b = 2a_{fcc}$, and described
as sequences of (111) carbon planes alternately 1/4 and
3/4 full), and R3m (Cu-Pt type, consisting of alternately
empty and full (111) carbon planes) space groups gave
exactly the same powder diffraction patterns in the absence of rhomboedral distortion for the latter. This superstructure was observed at compositions up to $TiC_{0.71}$ after
long anneals below 600°C [9]. Bell and Lewis [10], on the
behalf of electron microscopy observations, assigned to
these compounds the Fd3m structure ; but recent observation by Russian workers [11] of weak superlattice spots
(001, 110,...) led them to propose a new (rhomboedral)
superstructure, of formula Ti_8C_5, and which may be deduced
from Ti_2C Fd3m by filling with carbon atoms the (111)
planes originally 3/4 full.

The Nb_6C_5 superlattice was observed by electron diffraction on samples slowly cooled below 1000°C [6,7] ; it
was assigned an hexagonal structure (space group $P3_1$ and/
or $P3_2$) isomorphous to the V_6C_5 form discovered by
Venables et al. [4], with $a_{hex} \simeq (\sqrt{6}/2)a_{fcc}$ and
$c_{hex} \simeq 2\sqrt{3}\ a_{fcc}$; in this structure, carbon vacancies are
found third neighbours on the metalloid f.c.c. sublattice.

Short-range ordering of carbon vacancies was also
observed in the $TiC_{0.5-0.7}$ range (in rapidly cooled samples) and in all the NbC_{1-x} carbides, by the existence of
diffuse streaks in the electron diffraction patterns [12].
A quantitative interpretation of this short-range ordering
was only made for a $VC_{0.75}$ single crystal from electron [13]
and neutron [14] diffuse scattering data, and for NbC_{1-x}
powders studied by neutron scattering [15] ; but in all
these studies, the atomic displacements were not taken
into account.

Here, we present elastic diffuse neutron scattering studies of the non-stoichiometric carbides TiC_{1-x} and NbC_{1-x}. It allowed to obtain for the samples which do not show long-range ordering, the short-range order parameters of the carbon sublattice and the static atomic displacements due to the size effect of vacancies. On the other hand, high temperature powder neutron diffraction allowed to study the order-disorder transition and, in a preliminary way, the kinetics of ordering in $TiC_{0.52-0.67}$. Up to now, only the V_8C_7 and V_6C_5 order-disorder transitions, which are thermodynamically of first order, had been studied by electrical resistivity and DTA [16].

EXPERIMENTAL

Samples

Single crystals TiC_{1-x} and NbC_{1-x} were grown from sintered specimens (with nominal purities 99.9 %) by a zone melting procedure described elsewhere [17]. The crystals were grown on seeds parallel to the <1$\bar{1}$0> axis. The NbC_{1-x} (1-x = 0.73 and 0.80) samples were slowly cooled (65 hours) from the melting point T_F down to 400°C. The titanium carbide samples had different thermal treatments: $TiC_{0.64}$ and $TiC_{0.79}$ were slowly cooled (respectively 55 and 65 hours) from T_F to 400°C ; but $TiC_{0.76}$ was annealed 4 days at 600°C, before a more rapid cooling down to room temperature. The samples were subsequently cut in cylindrical form ($\simeq$ 1 cm^3) using spark erosion, and their orientation checked by Laue back reflexion method. Their carbon content was determined chemically.

The TiC_{1-x} powder samples (1-x = 0.52, 0.58, 0.63, 0.67) were prepared by sintering at 1600°C under vacuum cold-pressed mixtures of titanium hydride TiH_2 and graphite. They contained about 500 ppm weight metallic impurities. Carbon, oxygen and nitrogen contents were determined by chemical analysis ; the measured O/Ti and N/Ti atomic ratio were about 0.5 10^{-2}. The $TiC_{0.52}$ sample contained ~ 2 % free titanium.

Neutron scattering

The diffuse neutron scattering experiments were performed on the two axis D7 spectrometer of the HFR (ILL, Grenoble)[18], at room temperature and under vacuum. The incident wavelength was 3.14 Å, and time-of-flight analysis allowed to select only the elastically scattered neutrons. The data were obtained for $0.2 \leqslant q = 4\pi\sin\Theta/\lambda \leqslant 4$ Å^{-1}

and calibrated to absolute units by comparison with a
vanadium standard, after correction for background and
absorption.

The high temperature neutron diffraction experiments
were performed on the spectrometer D1B of the HFR (ILL,
Grenoble) [18] ; it incorporates a 400 cell multidetector,
which allowed the simultaneous measurement of scattered
neutrons within the angular range $2\theta = 14$ to $94°$, with
steps of $0.2°$. The incident wavelength was 2.524 Å. The
samples, enclosed in niobium containers, were placed in a
furnace with niobium resistor and screens. The temperature
was regulated with a stability of $1°C$. The thermal gradient on the samples (9 mm diameter, 35 mm length) was
measured to be about $2°C$.

DIFFUSE NEUTRON SCATTERING EXPERIMENTS

Results

On figures 1, 2 and 3 are shown the elastic scatte-
ring differential cross-sections $d\sigma/d\Omega$ in the $(1\bar{1}0)$ reciprocal lattice plane for $TiC_{0.64}$, $TiC_{0.76}$, $TiC_{0.79}$,
$NbC_{0.73}$ and $NbC_{0.80}$. The data are given in mbarn per metal
atom and are not corrected for incoherent scattering
$d\sigma/d\Omega)_{inc}$: the incoherent scattering of Nb and C are weak
(21 and 1 mbarn respectively [19]), but the value for titanium (≈ 210 mbarn [19]) corresponds to about half of the
scattering measured for $TiC_{0.76}$ and $TiC_{0.79}$. The diffuse
scattering figures are symmetric relative to the <001> and
<110> reciprocal axis, as predicted for statistically
cubic crystals. But the intensities are far from periodic
in the reciprocal space, indicating that atomic displacements are important.

The samples may be classified into two categories :

a) The $TiC_{0.64}$ (Fig. 1) and $NbC_{0.80}$ (Fig. 3b) samples show large diffuse scattering maxima.

In the case of $NbC_{0.80}$, these are found in 1/2 1/2
1/2, 1/2 1/2 3/2,..., and in 2/3 2/3 0, 1/3 1/3 1,...,
positions which are characteristic of the hexagonal superstructure Nb_6C_5 described in the introduction [6,7]. The
observed scattering corresponds to the superposition of
diffraction patterns of four types of domains, the c axis
of each of which is parallel to one of the four cube
diagonals <111>.

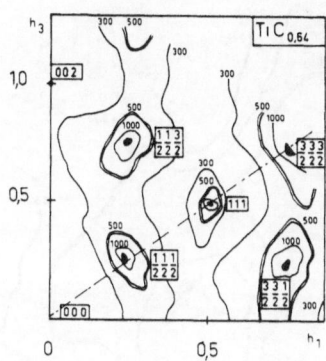

Fig. 1.   Elastic diffuse neutron scattering cross-section
dσ/dΩ (in mbarn per metal atom) measured at 300 K
in the (1$\bar{1}$0) reciprocal lattice plane of TiC$_{0.64}$.

In the case of TiC$_{0.64}$, the intensity maxima are in
1/2 1/2 1/2, 1/2 1/2 3/2 and 3/2 3/2 1/2, characteristic
of the "Ti$_2$C" superstructures described above. We are not
able to choose between the cubic Fd3m structure, or a mixture of the four axial domains of the rhomboedral R$\bar{3}$m
(Cu-Pt type) structure.

In both samples, bands of diffuse scattering, roughly parallel to the <001> direction, and similar to those
observed by Billingham et al.[12] by electron diffraction
on similar samples, are observed. A preliminary analysis
of the TiC$_{0.64}$ spectrum by the Sparks and Borie method[20],
showed that the bands are partly due to short range ordering between carbon vacancies (J.P. Landesman and
V. Moisy-Maurice, unpublished) ; the short-range order
contribution to dσ/dΩ could not be fitted by a single set
of Cowley-Warren α coefficients ; therefore we suggest a
presence of ordered precipitates of "Ti$_2$C$_{1+y}$" in a short-
range ordered matrix. Approximate sizes L of ordered
domains were deduced from the integral width of the

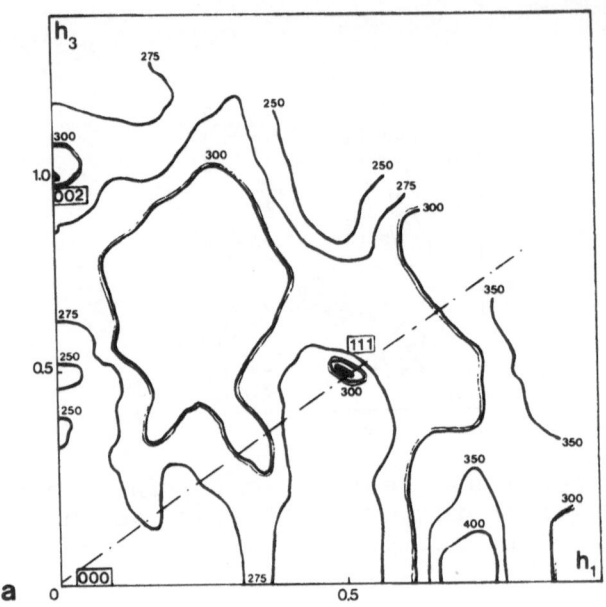

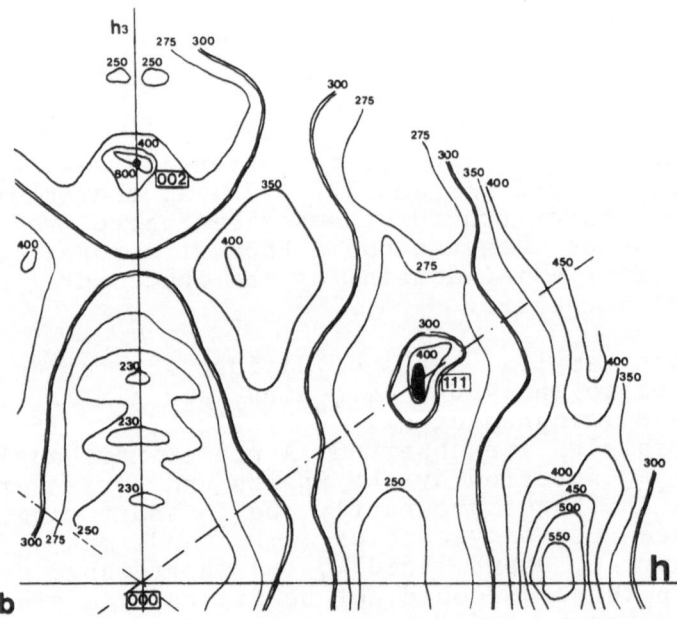

Fig. 2.   Elastic diffuse neutron scattering cross-section
          $d\sigma/d\Omega$ (mbarn per metal atom) measured at 300 K in
          the $(1\bar{1}0)$ reciprocal lattice plane.
          a : TiC$_{0.76}$ b : TiC$_{0.79}$.

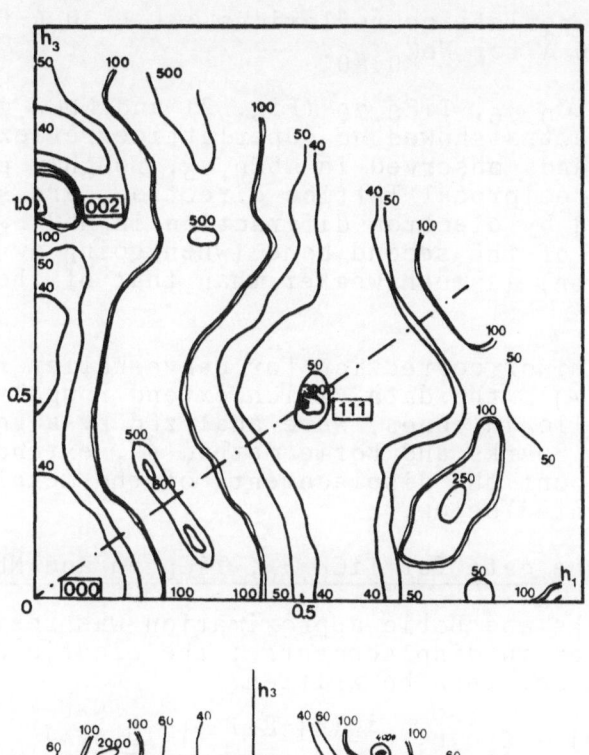

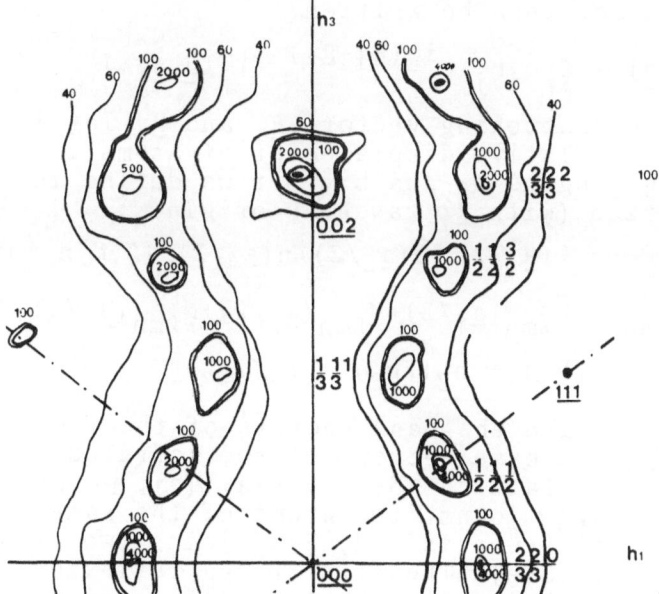

Fig. 3. Elastic diffuse neutron scattering cross-section $d\sigma/d\Omega$ (mbarn per metal atom) measured at 300 K in the $(1\bar{1}0)$ reciprocal lattice plane.
a : $NbC_{0.73}$ b : $NbC_{0.80}$.

1/2 1/2 1/2 superlattice reflexions : L $\simeq$ 30 Å for $TiC_{0.64}$ and 80 Å for $NbC_{0.80}$.

b) The $TiC_{0.76}$, $TiC_{0.79}$ (Fig. 2) and $NbC_{0.73}$ (Fig. 3a) sample spectra showed no superlattice reflexions. The diffuse bands observed in $NbC_{0.73}$, roughly parallel to the <001> reciprocal lattice direction, are similar to those observed by electron diffraction in NbC_{1-x} [12], but the intensity of the second band (when going away in the <110> direction) is much weaker than that of the first one.

After a minor correction for Debye-Waller factors ($B_M = B_C = 0.25$ Å^2), the data, which extend roughly in the two first Brillouin zones, were analyzed by a generalization of the Sparks and Borie method [20], extended to take into account the displacements of the metal atoms. This is presented below.

Analysis of the data for $TiC_{0.76}$, $TiC_{0.79}$ and $NbC_{0.73}$

The Sparks and Borie approximation was restricted to the first order in displacements ; the elastic diffuse cross-section may then be written :

$$d\sigma/d\Omega \ (\underline{q}) = \sum_{ij} b_i b_j \ e^{i\underline{q}(\underline{R}_i - \underline{R}_j)} \ (1 + i\underline{q}.\underline{\delta}_{ij})$$

where $\underline{q}$ is the scattering vector, $\underline{R}_i$ and $\underline{R}_j$ lattice sites, $\underline{\delta}_{ij} = \underline{\delta}_i - \underline{\delta}_j$ the relative displacement of atoms i and j of scattering lengths b_i and b_j. Let us define in the rocksalt lattice (with $\underline{R}_j$ taken as origin) :

$$\underline{R}_i = \underline{R}_{\ell mn} = \ell (\underline{a}_1/2) + m(\underline{a}_2/2) + n(\underline{a}_3/2) \quad (\ell,m,n \ integers)$$

$$\underline{\delta}_i = \underline{\delta}_{\ell mn} = L_{\ell mn}(\underline{a}_1/2) + M_{\ell mn}(\underline{a}_2/2) + N_{\ell mn}(\underline{a}_3/2)$$

$$\underline{q} = 2\pi[h_1(2\underline{b}_1) + h_2(2\underline{b}_2) + h_3(2\underline{b}_3)]$$

Here, $\underline{a}_1$, $\underline{a}_2$, $\underline{a}_3$ are the base vectors of the f.c.c. lattice, $\underline{b}_1$, $\underline{b}_2$, $\underline{b}_3$ the base vectors of the corresponding b.c.c. reciprocal lattice ($\underline{b}_1 = (\underline{a}_2 \times \underline{a}_3)/(\underline{a}_1, \underline{a}_2, \underline{a}_3)$). If $\ell + m + n$ is even, $R_{\ell mn}$ joins two atoms of the same nature (C-C or M-M) ; if $\ell + m + n$ is odd, $R_{\ell mn}$ joins two atoms of different nature (C-M). The first Bragg reflexion (111) corresponds to $h_1 = h_2 = h_3 = 1/2$. In the (1$\bar{1}$0) reciprocal lattice plane, $h_2 = h_1$, and $f(h_1, h_2, h_3)$ may be written $f(h_1, h_3)$.

One obtains : $d\sigma/d\Omega = d\sigma/d\Omega)_{SRO} + d\sigma/d\Omega)_{AD}$, with :

$$\left.\frac{d\sigma}{d\Omega}\right)_{SRO} = Nx(1-x)\ b_C^2 \sum_{\substack{\ell+m+n\\ \text{even}}} \left[\alpha_{\ell mn}\ \cos(2\pi h_1\ell)\ \cos(2\pi h_1 m) \right.$$
$$\left. \cos(2\pi h_3 n)\right]$$

(short-range order cross-section, independent of displacements).

$$\left.\frac{d\sigma}{d\Omega}\right)_{AD} = N \sum_{\ell mn} (h_1\ B_{\ell mn}^L + h_1\ B_{\ell mn}^M + h_3\ B_{\ell mn}^N)$$
$$\times\ \sin 2\pi(h_1\ell + h_1 m + h_3 n)$$
$$= h_1 Q(h_1,h_3) + h_3 R(h_1,h_3)\ \text{(displacement cross-section).}$$

For $\ell+m+n$ even : $B_{\ell mn}^L = 2\pi x(1-x)b_C^2\ (\frac{1-x}{x}+\alpha_{\ell mn})\ L_{\ell mn}^{C-C}$

$\ell+m+n$ odd : $B_{\ell mn}^L = 4\pi(1-x)\ b_C\ b_M\ L_{\ell mn}^{C-M}$

(1-x is the carbon content ; the average interatomic metal displacements are zero : $\delta_{\ell mn}^{M-M} = 0$ for $\ell+m+n$ even).

$d\sigma/d\Omega)_{SRO}$, Q and R are periodic functions of h_1 and h_3, and may be obtained by linear combination of the cross-sections measured in different cells of the $(1\bar{1}0)$ reciprocal plane. Their Fourier transforms, $\mathcal{A}_{ijk}$, ρ_{ijk} and χ_{ijk} are respectively linear combinations of the short-range order coefficients $\alpha_{\ell mn}$ and of the relative displacement components $L_{\ell mn}^{C-C}$ and $L_{\ell mn}^{C-M}$. The detailed formalism used and the complete set of equations are presented elsewhere [21]. The α coefficients were limited to eight shells of neighbours, and the displacements to four shells of neighbours (two C-M and two C-C), in which particular case $\delta_{\ell mn}$ on the average must be parallel to the interatomic vector to maintain statistically the cubic symmetry.

The α coefficients obtained are given in Table 1 for the three studied samples. The relative average radial displacements $\delta_{\ell mn}$ are given (in angströms) in Table 2.

Table 1 - Cowley-Warren short-range order coefficients between carbon atoms and vacancies deduced for $TiC_{0.76}$, $TiC_{0.79}$ and $NbC_{0.73}$.

$\alpha_{\ell mn}$	$TiC_{0.76}$	$TiC_{0.79}$	$NbC_{0.73}$
$\alpha_1 = \alpha_{110}$	-0.005(0.005)	+0.010(0.005)	-0.095(0.010)
$\alpha_2 = \alpha_{200}$	-0.080(0.005)	-0.175(0.005)	-0.275(0.010)
$\alpha_3 = \alpha_{211}$	+0.013(0.005)	+0.010(0.005)	+0.051(0.005)
$\alpha_4 = \alpha_{220}$	+0.006	+0.025	+0.072
$\alpha_5 = \alpha_{310}$	-0.003	+0.015	+0.044
$\alpha_6 = \alpha_{222}$	+0.025	+0.070	-0.030
$\alpha_7 = \alpha_{321}$	-0.007	-0.015	-0.020
$\alpha_8 = \alpha_{400}$	+0.003	+0.005	+0.030

Table 2 - Relative radial average atomic displacements (in angströms) between C-M and C-C near neighbours in $TiC_{0.76}$, $TiC_{0.79}$ and $NbC_{0.73}$.

$\delta_{\ell mn}$ (Å)	$TiC_{0.76}$	$TiC_{0.79}$	$NbC_{0.73}$
δ_{100}^{C-M}	-0.030(0.010)	-0.030(0.010)	-0.033(0.010)
δ_{111}^{C-M}	+0.006(0.004)	+0.006(0.004)	+0.011(0.006)
δ_{110}^{C-C}	+0.011(0.004)	+0.010(0.004)	+0.006(0.004)
δ_{200}^{C-C}	+0.008(0.004)	+0.007(0.004)	+0.004(0.004)

Discussion

The above results show that the degree of short-range ordering is much larger in $NbC_{0.73}$ than in $TiC_{0.76}$ and $TiC_{0.79}$, but that the atomic displacements are very similar. From the obtained parameters, the $d\sigma/d\Omega)_{AD}$ and $d\sigma/d\Omega)_{SRO}$ cross-sections were reconstructed and compared to the experimental values [21] : for $TiC_{0.76}$ and $TiC_{0.79}$, the agreement is rather good, especially for $d\sigma/d\Omega)_{AD}$;

for $NbC_{0.73}$, the calculated $d\sigma/d\Omega)_{SRO}$ does not show such a sharp maximum as the measured, which clearly is due to the fact that correlations extend further than eight shells of neighbours.

Although the short-range ordering is found much weaker in $TiC_{0.76}$ and $TiC_{0.79}$ than in $NbC_{0.73}$, two common conclusions may be deduced for the three samples from the observation of Tables 1 and 2 : the negative values of α_2 (meaning that vacancies avoid to put themselves as second neighbours), and the shortening (0.03 Å) of the average metal-carbon first neighbour distances. Both facts suggest, as will be shown below, a strong value of the second neighbour pair interaction potential between a vacancy and a carbon, via the metal atom which is at half distance. The fact that this metal atom moves away from the vacancy and goes nearer the carbon is also found in the Ti_2C [5] and V_8C_7 [3] superlattices. A strengthening of the metal-carbon bonding due to a redistribution of the metal d electrons originating from the dangling bonds may be suggested.

On the other hand, important differences are observed between the three samples : α_1 is found strongly negative and α_3 positive in $NbC_{0.73}$, showing that in this compound vacancies prefer the third neighbour positions as in the long range ordered superstructure Nb_6C_5 (for which $\alpha_1 = \alpha_2 = -0.2$, $\alpha_3 = +0.2$) (qualitatively similar results were obtained from the study of NbC_{1-x} powders [15]). On the contrary, in $TiC_{0.76}$ and $TiC_{0.79}$, $\alpha_1 \simeq \alpha_3 \simeq 0$ (in Ti_2C : $\alpha_1 = \alpha_3 = 0$, $\alpha_2 = -1$).

A surprising observation comes from the fact that the α coefficients are found about twice larger in $TiC_{0.79}$ than in $TiC_{0.76}$ (see Table 2), although the compositions are very near, and that ordering in TiC_{1-x} seems to appear more easily at low carbon concentration (see $TiC_{0.64}$, Fig. 1). In our opinion, this has to be attributed to the different thermal treatments of the samples : $TiC_{0.76}$ was rapidly cooled after a long anneal at 600°C, when $TiC_{0.79}$ was more slowly cooled down to room temperature ; the larger α coefficients in $TiC_{0.79}$ suggest that this sample is representative of a lower temperature. On the other hand, the fact that the displacements were found practically equal in the two samples, and therefore less dependent of thermal treatments, is a good proof of the validity of our separation of $d\sigma/d\Omega)_{SRO}$ and $d\sigma/d\Omega)_{AD}$.

As only the $TiC_{0.76}$ spectrum seems to correspond to a well defined thermal equilibrium (at 600°C), and as the short-range ordering in this sample is weak, we applied

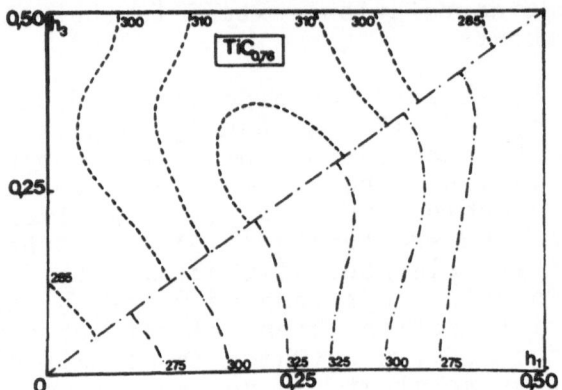

Fig. 4.   Calculated short-range order diffuse cross-
          section $d\sigma/d\Omega)_{SRO}$ in the $(1\bar{1}0)$ reciprocal
          plane of $TiC_{0.76}$:

          —·—· : reconstructed from the $\alpha_{\ell mn}$ values given
          in Table 1.
          ------ : reconstructed from the first and second
          neighbour pair interaction potentials
          V_1 = 3.2 and V_2 = 7.1 meV.

the mean-field theory of Clapp and Moss [22] to deduce
interatomic pair potentials from the $d\sigma/d\Omega)_{SRO}$ contribution. Indeed, according to these authors :

$$\alpha(\underline{q}) = C^*/[1+2x(1-x) \ V(\underline{q})/kT]$$

where $\alpha(\underline{q})$ and $V(\underline{q})$ are the Fourier transforms of the
$\alpha_{\ell mn}$ and interatomic pair potentials

$$V_{\ell mn} = \frac{1}{2} (V_{\ell mn}^{C-C} + V_{\ell mn}^{\square-\square}) - V_{\ell mn}^{C-\square}$$

($\square$: vacancy, C : carbon atom) ; C^* is a normalization
constant and T the absolute temperature (873 K in the
present case). The fit was made with only two potentials
between first and second neighbours of the metalloid
sublattice, V_1 and V_2. We found : $V_1 \simeq 3.2 \ 10^{-3}$ eV and

$V_2 \simeq 7.1 \ 10^{-3}$ eV for $TiC_{0.76}$. On Fig. 4 are compared the reconstructed $d\sigma/d\Omega)_{SRO}$ calculated with the eight $\alpha_{\ell mn}$ values given in Table 1 and with the two above values of pair potentials. The agreement is reasonable. The positive values of V_1 and V_2 with $V_2/V_1 > 0.5$ let us predict, according to Clapp and Moss [22], that long-range ordering at low temperature will appear with 1/2 1/2 1/2 type superlattice reflexions ; this is in agreement with the superstructures observed at lower carbon content.

HIGH TEMPERATURE NEUTRON DIFFRACTION EXPERIMENTS

Results : the order - disorder transition in TiC_{1-x}

At room temperature, the powder neutron diffraction spectra of the four sintered TiC_{1-x} samples (1-x = 0.52, 0.58, 0.63 and 0.67) showed three broad and weak diffuse maxima in 1/2 1/2 1/2, 1/2 1/2 3/2 and 3/2 3/2 1/2. From the integral width of these lines, one could deduce ordered domain sizes of approximately 50 Å.

The samples were heated and held at a temperature (typically 650°C) where one could observe with the multi-detector D1B a progressive narrowing and increase of the integrated intensities of these lines, corresponding to a simultaneous increase of the ordered domain size and of the degree of order within the domains. The kinetics of the process decreased considerably with increasing carbon content. After the average size of domains reached $\sim$ 150 Å, the integrated intensities of the lines remained constant (although narrowing of the linewidth continued to occur, due to domain growth). Therefore, one considered that the long-range order parameter within a domain had attained its maximum value at the annealing temperature.

The samples were then heated by steps of 10°C : the superlattice reflexions were found to decrease, and disappeared at a critical temperature T_C. The values of T_C were found to be : 775±10°C for $TiC_{0.52}$, 765±5°C for $TiC_{0.58}$, 785±5°C for $TiC_{0.63}$ and 770±5°C for $TiC_{0.67}$. (For $TiC_{0.52}$, which contained some free titanium, the transition was less well defined). Above T_C, weak diffuse maxima were still observed on the diffraction spectra, even up to 900°C.

On figure 5, one has reported the thermal dependence of the 1/2 1/2 1/2 superlattice reflexion intensity for $TiC_{0.58}$ and $TiC_{0.63}$. (The behaviour of the 1/2 1/2 3/2 and

V. MOISY-MAURICE ET AL.

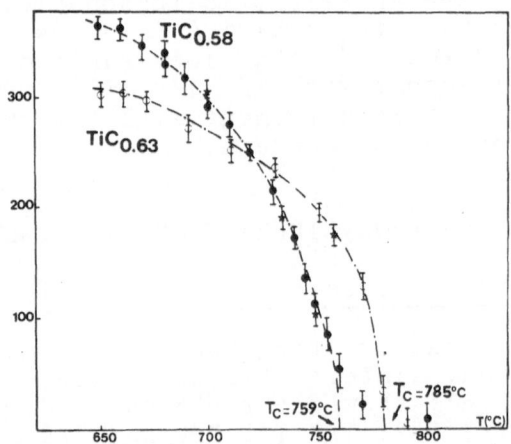

Fig. 5. Thermal dependence of the superlattice reflexion
         intensities $I_{1/2\ 1/2\ 1/2}$ of $TiC_{0.58}$ and $TiC_{0.63}$,
         measured during heating (circles) and cooling
         (stars).

         $.\_.\_.\_$: Fit with a function $I = A(T_C-T)^{2\beta}$.

3/2 3/2 1/2 reflexions was found quite similar). The curve
for $TiC_{0.58}$ is typical of a second order type transition,
with no apparent discontinuity of the long-range order
parameter S. But in the case of $TiC_{0.63}$, the decrease
around T_C is more rapid, and a small discontinuity of S
cannot be excluded. For both samples, the intensities
measured during the cooling and heating procedures were
found to be placed on the same curves (see Fig. 5) : the
transitions showed no hysteresis.

    Assuming a second order type transition, we fitted
the intensity of the 1/2 1/2 1/2 reflexion by a formula
$I = A(T_C-T)^{2\beta}$, I being proportional to the square of the
long-range order parameter S. The fit is shown on Fig. 5.
The β values were : 0.24±0.02 for $TiC_{0.52}$, 0.25±0.02 for
$TiC_{0.58}$ and 0.18±0.02 for $TiC_{0.63}$. For this latter sample,

the possible existence of a small discontinuity of S at T_C might have reduced the apparent value of β. For $TiC_{0.67}$ the kinetics were too slow to ensure that thermal equili-brium was attained, and the fit was not made.

Discussion

From the theoretical point of view, the second order type transition found for $TiC_{0.52}$ and $TiC_{0.58}$ is in agree-ment with the Landau-Lifshitz criteria [23]. Indeed, the latter allow the $Ti_2C \leftrightarrow TiC_{1-x}$ transition to be of second order type, as :

- the proposed superstructure space groups ($Fd3m$ or $R\bar{3}m$) are subgroups of the rocksalt structure space group ($Fm3m$) ;

- the wave vectors for ordering are of the $1/2$ $1/2$ $1/2$ type ;

- the combination of three superlattice wave vectors does not give a vector of the reciprocal lattice of TiC_{1-x} or Ti_2C.

The measured critical coefficient value $\beta = 0.25$ is in reasonable agreement with the value determined theore-tically ($\beta = 0.31$) for the 3 dimensions f.c.c. rigid lattice Ising model, with only short-range order pair interactions [24]. This suggests that the interactions res-ponsible for ordering in TiC_{1-x} are rather short-range, and justifies the analysis of diffuse scattering made above in terms of near neighbour pair potentials. We also remind that Kanamori [25] has shown that the stability (at 0 K) of the superstructures V_8C_7 and V_6C_5 of the VC_{1-x} rocksalt structure compound, may be explained on the basis of pair interaction potentials up to five neighbour shells. One would therefore infer that the "chemical" interaction energy between the electronic structure perturbations induced by two vacancies predominates on the long range "elastic" interaction energy between the displacement fields around these two vacancies (which varies as r^{-3}). Indeed , in the Pd-H system, where the phase diagram may be described on the basis of elastic interactions between H interstitials, the critical coefficients are found near the mean field theory values ($\beta = 0.5$) [26].

Nevertheless, a difficulty still occurs : we applied the Cowley [27] mean field theory to calculate the thermal variation of the long-range order parameter S in the Ti_2C_{1+y} ($Fd3m$ or $R\bar{3}m$) type structures, with the excess carbon occupying randomly the vacancy sublattice. Limiting the pair interaction potentials to the third neighbours in the metalloid f.c.c. sublattice, one finds :

$$\ln \left[\left(\frac{1-x}{x} - S^2\right) \left(\frac{x}{1-x} - S^2\right) / (1+S^2)^2 \right] = -12 \frac{V_2}{kT} S^2$$

(the first and third neighbour pair interaction potentials V_1 and V_3 do not contribute to the S value). The order-disorder transition is calculated to be of second order type, and from the value $T_C = 1038$ K for $TiC_{0.58}$, one obtains $V_2 \simeq 28 \ 10^{-3}$ eV, i.e. four times more than the V_2 value deduced from the $TiC_{0.76}$ $d\sigma/d\Omega)_{SRO}$ diffuse scattering data. This large discrepancy cannot be reasonably attributed to a change of the pair interaction potentials with composition, or to the neglect of V_4. It is also much larger than the error usually attributed to mean field theories (which generally overestimate T_C by about 20 %)[23]. More fundamentally, the applicability of the pair interaction approximation (neglecting many-body interactions) to transition metal carbides has to be questionned ; the role of the elastic interactions on the values of the pair potentials, of the critical temperatures and coefficients has also to be evaluated.

ACKNOWLEDGEMENTS

We wish to thank Drs.W. Just and P. Convert for their help during the neutron scattering experiments performed on the HFR-ILL (Grenoble), and Prof. F. Gautier for many usefull discussions. Carlsbergfonden is acknowledged for the spark erosion machine. The department of Geology at Aarhus University is acknowledged for the carbon analysis of the single crystals.

REFERENCES

1. L. M. Huisman, A. E. Carlson, C. D. Gelat and
       H. Ehrenreich, Phys. Rev. B 22:991 (1980).
2. C. H. de Novion and V. Maurice, J. Phys. Colloq.
       38:C7-211 (1977).
3. Y. Guérin and C. H. de Novion, Rev. Int. Hautes Temp.
       et Refract. 8:314 (1971).
4. J. D. Venables, D. Kahn and R. Lye, Phil. Mag.
       18:177 (1968).
5. H. Goretzki, Phys. Stat. Sol. 20:K141 (1967).

6. M. H. Lewis, J. Billingham and P. S. Bell, in :
   "Electron Microscopy and Structure of Materials",
   E. Thomas, R.M. Fulrath and R.M. Fisher, ed.,
   University of California Press (1972), p. 1084.
7. J. D. Venables and M. H. Meyerhoff, in : "Solid
   State Chemistry", N.B.S. Special Publication 364
   (1972), p. 583.
8. E. Parthé and K. Yvon, Acta Cryst. B 26:153 (1970)
9. V. T. Em, I. Karimov, V.F. Petrunin, I. Khidirov,
   I. S. Latergaus, A. G. Merzhanov,
   I. P. Borovinskaya and V. K. Protudina, Sov. Phys.
   Crystallogr. 20:198 (1975).
10. P. S. Bell and M. H. Lewis, Phil. Mag. 24:1247 (1971).
11. B. V. Khaenko, S. Y. Galab and M. P. Arbuzov,
    Kristallog. 25:112 (1980).
12. J. Billingham, P. S. Bell and M. H. Lewis, Acta
    Cryst. A 28:602 (1972).
13. M. Sauvage and E. Parthé, Acta Cryst. A 28:607 (1972).
14. M. Sauvage, E. Parthé and W. B. Yelon, Acta Cryst. A
    30:597 (1974).
15. B. E. F. Fender, in : "Chemical Applications of Ther-
    mal Neutron Scattering", B.T.M. Willis, ed.,
    Oxford University Press, Oxford (1973), p. 250.
16. W. S. Williams, Prog. Solid State Chem. 6:57 (1971).
17. A. N. Christensen, J. Crystal Growth 33:99 (1976).
18. "I.L.L. Neutron beam facilities available for users",
    Institut Laue-Langevin, Grenoble (1978).
19. G. E. Bacon, "Neutron Diffraction", Clarendon Press,
    Oxford (1975).
20. C. J. Sparks and B. Borie, in : "Local Atomic Arran-
    gements studied by X-Ray Diffraction", Gordon and
    Breach, New-York (1965), p. 5.
21. V. Moisy-Maurice, Ph D Thesis, Université de
    Strasbourg, France (1981), Report CEA-R-5127 (1981).
22. P. C. Clapp and S. C. Moss, Phys. Rev. 142:418 (1966)
    and 171:754 (1968).
23. D. de Fontaine, Solid State Phys. 34:73 (1979).
24. J. Als-Nielsen, in : "Phase Transitions and Critical
    Phenomena" , Vol. 5A, Academic Press, London
    (1976), p. 87.
25. J. Kanamori, in : "Modulated Structures 1979",
    American Institute of Physics, New-York (1979),
    p. 117.
26. J. P. Burger, in "Solid State Phase Transformations
    in Metals and Alloys", Les Editions de Physique,
    Orsay, France (1980), p. 211.
27. J. M. Cowley, Phys. Rev. 77:669 (1950).

DISCUSSION

W. Williams:

I wanted to remark on the splendid nature of this research. I think
this is absolutely beautiful work and I'm very, very impressed by it.
A comment on the last remark that you made, namely the suitability
of pairpotentials and mean field theories for describing order/dis-
order transformations in this system. I grant that's a very risky
business. However, in the case of vanadium carbide, we have had
some success in treating the V_6C_5 and V_8C_7 first order transitions
using Bragg-Williams mean field theory and getting a fairly decent
agreement between the measured latent heat of the transformation and
the value that would be estimated from the mean field theory using
the measured temperature of the transformation. I wonder if you
would comment on that situation?

C. H. de Novion:

We have some results in VC which show second neighbor potentials
stronger than the first. So, it would be well worth to re-do the
calculations using the two potentials. But you also have lattice
deformation. If you have a carbon vacancy, the metal first neighbors
go away from them. Then if you put another vacancy, its energy will
depend on the site. So lattice deformation plays a role in the
ordering. What I deduced from the diffuse scattering data was
deduced only from the short range order pure contribution and the
ordering temperature takes into account everything and also the
distortion. So perhaps this is a source of my discrepancy between
the two pairpotentials deduced from the two aspects.

G. Gruzarsky:

Did you say that your diffuse scattering measurements were qualitatively different for $TiC_{0.76}$ and $NbC_{0.73}$?

C. H. de Novion:

A qualitative difference? Yes. Because in $NbC_{0.73}$, the diffuse
scattering in electron diffraction is like spaghetties. In $TiC_{0.76}$,
the diffuse scattering is like hills and indents. But I must say,
these were slowly cooled samples and perhaps if they were rapidly
cooled, you would have more resemblance. Slowly cooling the samples
favors ordering in the niobium carbide. So I would say qualitative
difference, yes.

G. Gruzarsky:

Was there any evidence for nearest neighbor carbon vacancies in the titanium carbide?

C. H. de Novion:

No. For the first neighbors in titanium, we found about the same as in the random distribution. It's for the second neighbors that we find vacancy-carbon association. In the niobium carbide we find effect on both first and second neighbors for vacancy-carbon association.

A. Krawitz:

I have a brief comment and a question. The Borie-Sparks method is ideally suited for neutron diffraction. Most measurements using it have involved x-rays, but because of the constancy of the neutron scattering factors it is ideally suited. I am concerned somewhat about the fact that you used the first order Borie-Sparks. Are you not concerned that you might have some thermal diffuse scattering contamination in your separation?

C. H. de Novion:

The thermal diffuse scattering was experimentally eliminated because we used time of flight analyzers which for neutrons are quite efficient. But still there is a second order static displacement term which should be used to refine the Sparks and Borie approximation. But for this you have to go further away in the reciprocal space and use another spectrometer. As we find other important displacements of the atoms compared to the sites, it will be necessary to say that the first order Sparks and Borie approximation is perhaps not enough. But the thermal diffuse scattering was eliminated.

ELECTRON RADIATION DEFECTS IN TaC_{1-x} AND $TiC_{0.97}$

J. Morillo, C.H. de Novion, J. Dural

SESI, Bât. 31, CEN, BP 6
92260 Fontenay-aux-Roses, France

ABSTRACT

The electrical resistivity changes of $TaC_{0.99}$ and $TaC_{0.80}$ have been measured at 21 K during irradiation with electrons of incident energies ranging from 2.5 to 0.25 MeV : a non-zero production rate is observed, even at the lowest energies. The recovery of defects was followed up to 400 K for $TaC_{0.99}$ and $TiC_{0.97}$ irradiated with 2.25 MeV electrons and up to 160 K for $TaC_{0.80}$ irradiated with 0.75 MeV electrons. The results are compared to fast neutron radiation damage data. For $TiC_{0.97}$ and $TaC_{0.99}$, the contributions of the different defects to the production rates and recovery spectra are tentatively separated, and a rough estimate of Frenkel pair resistivities is given.

INTRODUCTION

Low temperature radiation damage studies in compounds such as transition metal carbides MC_{1-x} are very interesting for several reasons :
1) Fundamental information about point defects in crystals where the bonding is altogether metallic, covalent and ionic.
2) Because of the very different masses and electronic structures of M and C atoms, one may hope to get large differences in the threshold energies and the recovery stages, and therefore to separate the contribution of different types of radiation-induced point defects.
3) It is possible to vary externally the concentration of

103

one of the point defects : the carbon vacancy.
4) Several refractory carbides such as UC and PuC, are
potential nuclear reactor fuels ; the fundamental aspects
of radiation damage in these compounds has not been much
studied yet. [1]

Up to now, radiation effect studies in transition
metal carbides have been mainly restricted to neutron
irradiation at 100°C [2,3,4]. Under these conditions, some
of the induced defects (probably interstitials) are mobile ; so these experiments mainly concern agglomerates
rather than isolated point defects. Nevertheless, Iseki
et al [3] have observed in $TiC_{0.62}$ irradiated with fast neutrons a large annealing stage around 600°C ; from the measured migration energy (4.6 eV), this stage was assigned
to free migration of carbon vacancies. Neutron irradiation was also observed to have little influence on the
superconducting transition of NbC, compared to the case
of the A-15 compounds [4]. At least, we want to mention two
irradiation experiments performed in an electron microscope :
- the disordering of the V_6C_5 superstructure [5] (where C
atoms and vacancies form ordered arrays on the f.c.c.
metalloid sublattice).
- the observation of voids in previously deformed single
crystals of $TiC_{0.93}$, irradiated at room temperature with
100 keV electrons and annealed around 800-900°C [6].
Both experiments show that the transfer of a carbon atom
into a neighbour vacant site of the metalloid sublattice
in a non-stoichiometric carbide MC_{1-x} needs a low energy ;
the value found by Venables and Lye [5] (5.4 eV), is near
the carbon self-diffusion (migration) energy.

In the present paper, we report low temperature electrical resistivity measurements obtained on titanium monocarbide $TiC_{0.97}$ and tantalum monocarbides $TaC_{0.99}$ and
$TaC_{0.80}$, which were irradiated with electrons at 21 K, and
which were annealed after irradiation up to 400 K. The
results will be compared to fast neutron radiation damage
experiments [7].

EXPERIMENTAL

Samples

The polycrystalline samples were approximately 0.2 to
0.3 mm thick, 2 mm wide, and from 5 to 25 mm long. They
were prepared by M. Lequeux [8] by direct carburation at
high temperature (2000°C for TaC) of metal foils embedded

in graphite. $TaC_{0.80}$ was obtained by annealing under vacuum (5 days at 1730°C) partly carburated tantalum foils.
The compositions were deduced from lattice parameter and
electrical resistivity measurements, with an absolute
accuracy of ±0.01 in x. For $TaC_{0.99}$, the measured superconducting temperature T_c = 9.7±0.2 K corresponds to
crystals very close to the stoichiometric composition [9].
The total metallic impurity content was typically 200 ppm
weight, and the N and O contents about 20 and 60 ppm
weight respectively.

Irradiations and damage recovery

The electron irradiation experiments were performed
at 21 K in the VINKAC [10] low temperature irradiation facility on the Van de Graaff electron accelerator at CEN
Fontenay-aux-Roses. Doses and other experimental conditions are given in Table 1. The electron flux was measured directly on a Faraday cage placed behind the sample.
The electron energy was corrected for the losses in hydrogen, in the window of the cryostat and in the sample :
the average loss was approximately 250 keV for all incident energies (between 500 keV and 2.75 MeV). The spread
of particle energy in the samples around the mean value
was ±100 keV. The electrical resistivities of the samples
were measured during the irradiation, indium providing
extremely good ohmic contacts.

After the electron irradiations, it was possible,
without taking the samples out from the hydrogen bath, to
carry out 10 minutes isochronal recovery steps in a furnace at increasing temperatures. The sample was heated
for 10 minutes at a temperature T_r (with a stability of
1 K) in the furnace, then cooled to 21 K for the electrical resistivity measurements.

RESULTS

Damage rates (2.25 MeV electrons)

For $TiC_{0.97}$ and $TaC_{0.99}$ irradiated with 2.25 MeV
electrons, the resistivity versus dose dependences are
not exactly linear, but show slight negative curvatures
(as in the case of neutron irradiation at 21 K [7]). For
both samples, the irradiation temperatures, total doses
$\Delta\phi_t$, total resistivity increases $\Delta\rho_t$ and average damage
rates $\Delta\rho_t/\Delta\phi_t$ are summarized in Table 1, where fast neutron irradiation data [7] are given for comparison.

Table 1.   Resistivity variations induced by fast neutron
           (f.n.) and 2.25 MeV electron irradiation in
           $TiC_{0.97}$ and $TaC_{0.99}$.
           (In ref. 7, the electron incident energy pre-
           sented in Table 1 is the nominal value, 2.5 MeV;
           it had not been corrected for the 0.25 MeV
           losses which have been taken into account in
           the present article).

Compound	Irradiation particle	T irradiation (K)	$\Delta\phi_t$ (particle /cm^2)	$\Delta\rho_t$ ($\mu\Omega$-cm)	$\Delta\rho_t/\Delta\phi_t$ $\times 10^{18}$
$TiC_{0.97}$	f.n.	21	$1.3\ 10^{18}$	333	25
	electrons	21	$8.4\ 10^{18}$	10	1.2
$TaC_{0.99}$	f.n.	~40	$1.8\ 10^{18}$	5.0	2.8
	electrons	21	$5.6\ 10^{18}$	0.8	0.14

     For $TaC_{0.80}$, the $\Delta\rho$ versus $\Delta\phi$ curve showed a large
negative curvature and depended strongly on the electron
beam intensity ; this is due to the fact that the irradiation at 21 K occurs within an annealing stage (see
Fig. 2) ; therefore, no significance can be attributed
to this curve.

     The most significant feature in the results reported
in Table 1 is the difference in the damage rates $\Delta\rho_t/\Delta\phi_t$
between TiC and TaC. $\Delta\rho$ increases 10 times faster in TiC
than in TaC in both the neutron and the electron irradiation.

Electron energy dependence of the damage rate in TaC_{1-x}

     The damage rate due to electron irradiation at 21 K
was measured in $TaC_{0.99}$ and $TaC_{0.80}$ for different electron
incident energies, ranging from 0.25 to 2.50 MeV (after
correction for energy loss). Because of the ±100 keV width
of the particle energy spectrum in the sample, and of the
uncertainties on the measurement of the electron flux, the
results must be considered only as semi-quantitative,
especially for low energy electrons. Nevertheless, these

results, presented on Fig. 1 for $TaC_{0.99}$ in the form of a $\Delta\rho/\Delta\phi$ versus E curve, show that the damage rate at 0.25 MeV is not much lower than at 2.50 MeV ; no clear threshold energy appears from this curve.

The results for $TaC_{0.80}$ are similar, but with larger uncertainties, due to the presence of an annealing stage at the irradiation temperature 21 K.

Recovery of defects

On Fig. 2 are shown several recovery curves for electron irradiated tantalum carbides TaC_{1-x} :
- for $TaC_{0.99}$ irradiated with 2.25 MeV electrons (total dose 5.6 10^{18} electrons/cm^2) ;

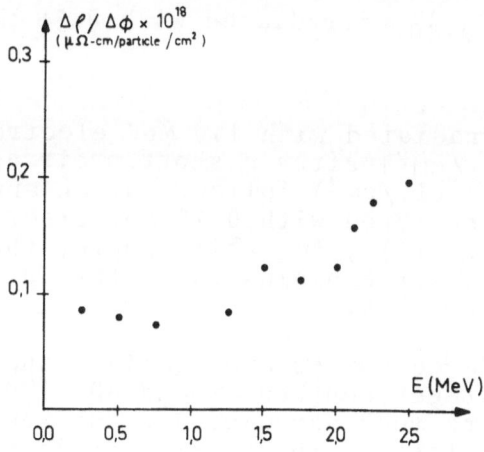

Fig. 1.   Damage rate $\Delta\rho/\Delta\phi$ measured in $TaC_{0.99}$ irradiated with electrons of energy E.
(The value for E = 2.25 MeV is somewhat higher than that given in Table 1 : this is due to the slight negative curvature of the $\Delta\rho$ versus $\Delta\phi$ curve ; here, the data correspond to the initial slope of these curves).

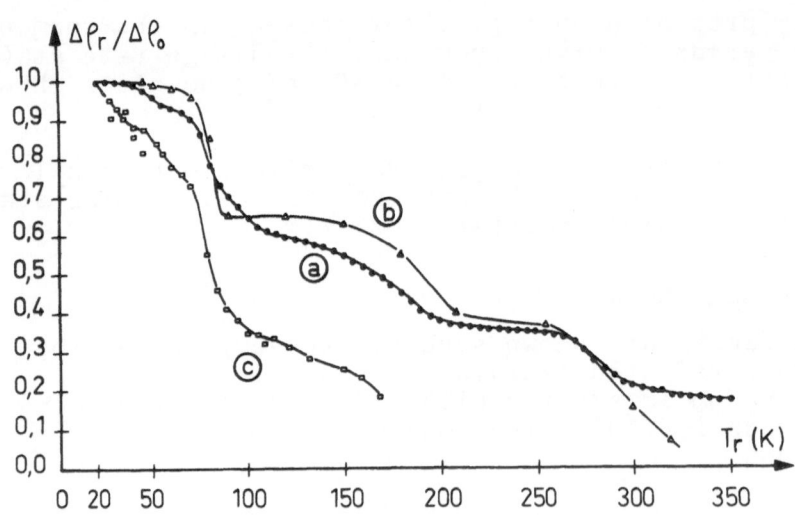

Fig. 2.   Variations of the electrical resistivities $\Delta\rho_r$ after recovery at a temperature T_r, normalized to the resistivity increase after electron irradiation $\Delta\rho_0$ at 20 K.
a : $TaC_{0.99}$ irradiated with 2.25 MeV electrons.
b : $TaC_{0.99}$ irradiated with 1.0 MeV electrons (after a short preirradiation at 2.5 MeV followed by an anneal at 400 K).
c : $TaC_{0.80}$ irradiated with 0.75 MeV electrons.

- for $TaC_{0.99}$ irradiated with 1.0 MeV electrons (total dose $4.7 \ 10^{18}$ el./cm^2) after a short preirradiation at 2.5 MeV ($0.5 \ 10^{18}$ el./cm^2) followed by an anneal at 400 K;
- for $TaC_{0.80}$ irradiated with 0.75 MeV electrons (total dose $3.2 \ 10^{18}$ el./cm^2) ; for this sample, the rupture of a resistivity contact explains the interruption of the recovery curve at 160 K.

The two $TaC_{0.99}$ curves are similar, showing three well separated stages centred around 80, 170 and 280 K. But the stages are much narrower and the recovery at 320 K nearly complete in the sample irradiated at 1.0 MeV. The $TaC_{0.80}$ curve shows a continuous recovery starting at the irradiation temperature (21 K) and a large stage around 80 K. Such a continuous recovery is typical of disordered alloys.

Fig. 3 shows the results for titanium carbide $TiC_{0.97}$ irradiated at 2.25 MeV, i.e. :

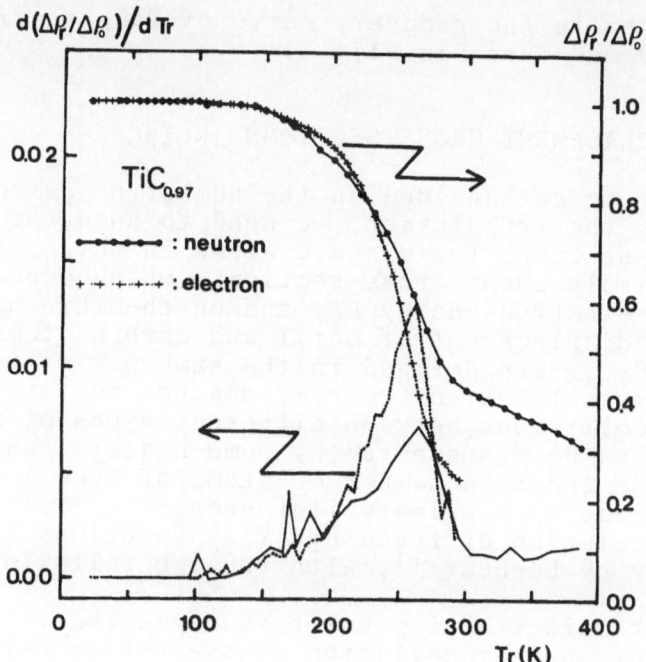

Fig. 3.   Variations of the residual resistivities $\Delta\rho_r$
in TiC$_{0.97}$ after recovery at a temperature T_r,
normalized to the resistivity increase after
irradiation $\Delta\rho_o$, along with the derivatives
of these curves.

- the variations of the residual resistivity $\Delta\rho_r$ after
recovery at a temperature T_r, normalized to the resistivity increase after the irradiation, $\Delta\rho_o$;
- the derivatives of the previous curves for neutron and
electron irradiations.
These recovery spectra divide into two parts : many, not
well defined, small stages between 100 and 230 K, and a
large stage centred at 267 K. The main difference between
the electron and neutron irradiations is in the intensity
of this stage centred at 267 K, the recovery being much
more important after electron irradiation.

Comparing the recovery curves for TaC$_{0.99}$ and TiC$_{0.97}$
(Figures 2 and 3), one sees that both compounds present
an important stage between 200 and 300 K, but that the
low temperature stage present in TaC$_{0.99}$ is absent in
TiC$_{0.97}$. We also remark that the first substage observed
around 50 K in TaC$_{0.99}$ irradiated at 2.25 MeV, is much

less pronounced in the recovery curve of $TaC_{0.99}$ irradiated at 1.0 MeV (see Fig. 3).

ELECTRON DISPLACEMENT CROSS-SECTIONS IN TaC

In order to get an idea on the number of defects introduced by the irradiation, we need to know the displacement cross-sections for M and C atoms in MC_{1-x}, σ_{dM} and σ_{dC}. These displacement cross-sections, which depend on the incident electron energy E_e, and on the threshold energies for displacement of metal and carbon atoms (E_d^M and E_d^C) in MC_{1-x}, are defined in the same way as for a monoatomic crystal. But of course, one has to take into account the collisions between different types of atoms. This effect can be represented by some $n_{ij}(T_i)$ functions which describe the mean number of atoms of type j displaced in a cascade by a primary i of energy T_i (i,j = M and/or C). A calculation of these $n_{ij}(T_i)$ functions has been made recently by Lesueur [11], with the two following hypothesis :
 i) the electronic stopping power is neglected ;
ii) the law of nuclear collision cross-section between ions is of the type $T_1^{-p} T_2^{1-p}$ (T_1 is the incident ion energy, T_2 the transferred energy, which depends only on the type of the incident ion).

We shall restrict here to the case of electron irradiation in stoichiometric TaC (M = Ta, x = 0). In this case, the maximum transferred energies with 2.5 MeV electrons are about T_m^M = 100 eV for Ta and T_m^C = 1600 eV for C ; with E_e = 0.5 MeV, it is about 10 eV for Ta and 140 eV for C. Therefore the hypothesis i) is entirely justified [12]. For the hypothesis ii), the best values for p in the concerned energy range are p = 0.055 for Ta and 0.3 for C [13].

So in our case, a displacement cross-section σ_{di}^e (i = C, M) will be written (for example in the case i = C) as a generalization of the formula of Oen [14] :

$$\sigma_{dC}^e = \int_0^{T_m^C} \sigma_C^e (\nu_C + n_{CC}) dT_C + \int_0^{T_m^M} \sigma_M^e \, n_{MC} \, dT_M$$

where σ_i^e is Mott's differential cross-section for a collision between the electron of energy E_e and the atom i, with an energy T_i transferred to i ; ν_i is the probability for a stable displacement of atom i, with, in a one

step model : $\nu_i = 0$ for $T_i < E_d^i$ and 1 for $T_i > E_d^i$. In the above formula, the dependence of the σ_i^e and n_{ij} with T_M, T_C and E_e have not been explicitly written. In the displacement cross-section of carbon atoms, σ_{dC}^e, there are therefore three contributions :
1) The primary displacement term (ν_C) which is the same as in a monoatomic target.
2) The cascades originated by carbon atoms, which can be considered as the equivalent of the modified Kinchin and Pease contribution (n_{CC}) [15].
3) The cascades originated by metal atoms, which is a completely new term (n_{MC}) compared with monoatomic targets.

   In the present case ($i = C$ in TaC), the influence of these collisions between different types of atoms is important ; this can be seen on Fig. 4 which represents the displacement cross-sections of C in TaC with and without taking into account the collisions between diffe-

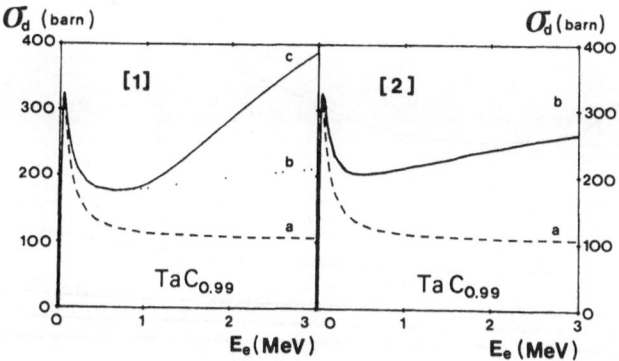

Fig. 4.   Calculated displacement cross-section for C
           in TaC as a function of electron energy :
           [1]: Taking into account the Ta-C collisions
           with the three contributions defined in the text.
           a = contribution 1 ; b = contributions 1) +2) ;
           c = contributions 1) +2) +3).
           [2] : Without taking into account the Ta-C colli-
           sions : a = primary cross-section ; b = total
           cross-section (primary + cascade).

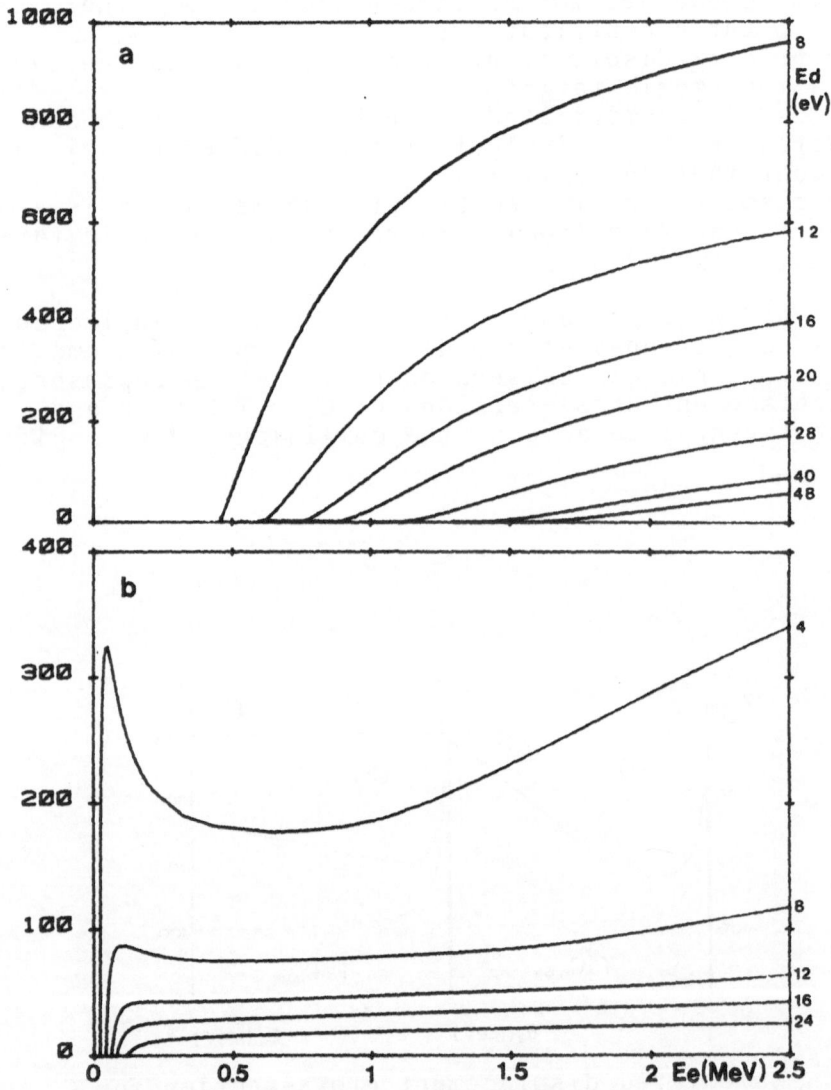

Fig. 5.   Total calculated displacement cross-sections
          of C and Ta atoms by electrons in $TaC_{0.99}$ for
          different threshold energies.
          a : σ^{e}_{dTa} b : σ^{e}_{dC}

rent types of atoms, as a function of the electron energy for a threshold energy E_d^C = 4 eV.

The displacement cross-section for the metal atom, σ_{dM}^e, may be written similarly to σ_{dC}^e ; but in the case of σ_{dTa}^e in TaC, contributions 2 and 3 are negligible.

On Fig. 5 are shown the calculated total displacement cross-sections σ_{dC}^e and σ_{dTa}^e in TaC, for different threshold energies of carbon and metal atoms (the threshold energy E_d^i of one type of atom i, does not influence the cross-section σ_{dj}^e for the other type of atom j, even if one takes into account the collisions between i and j atoms [11]). One may see on Fig. 5b that σ_{dC}^e shows a minimum around 0.5 MeV if E_d^C is smaller than 10 eV ; this minimum is qualitatively observed on Fig. 1 which represents the defect production rate in TaC for different incident electron energies. On the other hand, no abrupt change of slope is observed on Fig. 1, which would correspond to the tantalum displacement threshold energy : this suggests, when comparing to Fig. 5a, that E_d^{Ta} is larger than 30 eV.

From these displacement cross-section calculations, using the threshold energies suggested above ($E_d^C \simeq 5$ eV, $E_d^M \simeq 40$ eV), the final atomic concentrations of defects of each atom at the end of the electron irradiations reported in Table 1 (with E_e = 2.25 eV), are found of the order of 10^{-3} to 10^{-4}.

DISCUSSION

The general behaviour of $TiC_{0.97}$ and TaC_{1-x} during irradiation as well as during recovery is very similar to that of a monoatomic metal [16]. This, together with a consideration of the high conduction electron densities in these compounds justifies the hypothesis used for the calculation of the number of defects made above : mainly that they are produced as in metals only through nuclear collisions and not via electronic excitations as in ionic crystals. So it seems reasonable to consider (in matter of irradiation) carbides as diatomic metals. However, a greater number of recovery stages may be expected since there are two kinds of vacancies and interstitials, which will give rise to their own recovery stages. Moreover, their interaction will introduce new phenomena in the same way as impurities do in metals.

Obviously, the present results are not sufficient to make any precise determination of the Frenkel pair resistivities, or to attribute definitely any of the recovery stages to a specific defect. However, they allow to state the following points :

1) The stages observed in our recovery experiments are very probably related to the migration of interstitials.
This is because, from the available self-diffusion data in transition metal monocarbides [17,18], the migration energies for carbon and metal vacancies are estimated to be about 4 to 5 eV ; this gives for vacancies migration stages in a recovery spectrum at temperatures much higher than those studied here.

2) The defects created by electron irradiation in TaC_{1-x} ($E_e \leqslant 2.5$ MeV), and at least the first large recovery stage observed in Fig. 2 around 80 K, are mainly due to carbon defects.
This was already suggested from the comparison made above of Fig. 1 and 5, and from the displacement energy of carbon atoms (5.4 eV) measured by Venables and Lye [5] in V_6C_5. Indeed, the non-zero production rate observed at electron energies as low as 0.25 MeV cannot be due to tantalum defects : it would mean an $E_d^{Ta} < 5$ eV, which seems unreasonable ($E_d^M \simeq 30$ eV in the isomorphous compound SmS [19], and $E_d^{Ta} = 32$ eV in pure tantalum metal [16]). We attribute tentatively the large stage observed around 80 K to the free migration of carbon interstitials.

3) It is difficult to give an answer about the presence of Ta defects in TaC irradiated with high energy electrons.
The increase of production rate with incident electron energy above 1.5 MeV (see Fig. 1) can be as well attributed to the carbon displacement cross-section variation (see Fig. 5a) as to the appearance of displaced tantalum with a high threshold energy ($E_d^{Ta} \simeq 40$ to 60 eV, see Fig. 5b). The similarity of the recovery curves of $TaC_{0.99}$ irradiated with 2.25 and 1.0 MeV suggest that Ta defects are negligible ; but the 1.0 MeV curve should be considered with caution because of the preliminary short irradiation at 2.5 MeV, even if followed by an anneal at 400 K.

4) On the contrary, in $TiC_{0.97}$, $\Delta\rho$ (induced by fast neutrons or 2.25 MeV electrons) is probably mainly due to titanium defects, carbon interstitials being probably mobile at 20 K, and carbon vacancies already present

in the proportion of 3 %.
This is deduced from the absence of low temperature
(T < 150 K) recovery stages such as in TaC_{1-x}, and from
the similarity of the $TiC_{0.97}$ recovery curves with those
of isomorphous SmS and YS [7,19] ; indeed the latter both
show only high temperature (160-250 K) recovery stages.
In the case of SmS, threshold energy determinations at
20 K show that the sulfur atom is mobile at 20 K, which
means that the observed stages are due to metal defects[19].
Therefore, we suggest that in $TiC_{0.97}$ the recovery
stage centred around 270 K is due to the migration of
titanium interstitials ; the migration of titanium vacancies occurs probably in the high temperature (900 °C)
stage observed by Iseki et al [3] from the recovery of
lattice parameter in neutron irradiated specimens.

In principle, the points discussed above lead to an
estimation of the titanium Frenkel pair resistivity ρ_F^{Ti}
in $TiC_{0.97}$ and of the carbon Frenkel pair resistivity in
$TaC_{0.99}$ (in the latter case from the production rate
in the range 0.5-1.0 MeV). We obtain from the relation
$\Delta\rho = \sigma_{di}^{e} \rho_F^{i}$ the following values per atomic percent
Frenkel pair : $100 < \rho_F^{Ti} < 500$ μΩ-cm in $TiC_{0.97}$ and
$1 < \rho_F^{C} < 10$ μΩ-cm in $TaC_{0.99}$. The value of ρ_F^{C} in TaC is
consistent with the specific resistivity for a 1 % atomic
concentration of carbon vacancies (6 μΩ-cm [20]) and is of
the same order of magnitude as the specific Frenkel pair
resistivities in monoatomic metals with similar conduction electron densities [16]. The Frenkel pair resistivity
in $TiC_{0.97}$ is much larger than in $TaC_{0.99}$; a similar
difference is observed for the specific resistivities of
carbon vacancies associated to non-stoichiometry [20] ;
this could be explained by the difference in the conduction electron density which is 30 times smaller in TiC
than in TaC, TiC behaving somewhat as a semi-metal [20].

The experiments presented in this paper have brought
two puzzling results :
- The non-zero production rate of defects in $TaC_{0.80}$ at
an energy (0.75 MeV) where only carbon displacements are
thought to occur. These defects anneal largely at low
temperature (T < 100 K) and therefore are thought to be
carbon interstitials ; the stability of these interstitials in the presence of 20 % carbon vacancies is surprising.
- The different behaviour of carbon interstitials in
$TiC_{0.97}$ and TaC_{1-x}, suggesting mobility in the irradiation conditions (21 K) in the first case but not in the
second.

It is only a systematic study of production rates
and recovery stages for different compositions MC_{1-x} and
incident electron energies (including experiments below
21 K) that will allow to understand these behaviours.
Nevertheless, the explanations remain probably in the
existence of complex defects (not simple metal and carbon
Frenkel pairs) ; the interactions between metal and carbon
defects play probably an important role. Because of the
lighter mass of Ti compared to Ta, the number of metal
defects is much larger in $TiC_{0.97}$ than in $TaC_{0.99}$ irradiated with electrons of energies $E_e \leqslant 2.25$ MeV, and this
might affect differently the mobility of the carbon interstitial.

ACKNOWLEDGEMENTS

We wish to thank Drs D. Lesueur and L. Zuppirolli
for many usefull discussions, and Dr M. Lequeux (O.M.,
ONERA, France) who furnished the samples.

REFERENCES

1. B. G. Childs, J. C. Ruckman and K. Buxton, in :
     "Carbides in Nuclear Energy", Vol. 2, Macmillan
     and Co Ltd, London (1964), p. 849.
2. M. S. Koval'chenko and V. V. Ogorodnikov, Poroshkovaya
     Metallurgiya 9:48 (1966).
3. M. Iseki, S. Ushijima and T. Kirihara, Kakuyugo
     Kenkyu 43:127 (1981) (in Japanese).
4. D. Dew-Hughes and R. Jones, Appl. Phys. Lett.
     19:565 (1969).
5. J. D. Venables and R. G. Lye, Phil. Mag. 19:565
     (1969).
6. D. K. Chatterjee and H.A. Lipsitt, J. Less Common
     Met. 70:111 (1980).
7. J. Morillo, C. H. de Novion and J. Dural, Rad. Effects
     55:67 (1981).
8. M. Lequeux, Thèse de 3ème cycle, Orsay, France (1972).
9. A. L. Giorgi, E. G. Szklarz, E. K. Storms,
     A. L. Bowman and B. T. Matthias, Phys. Rev.
     125:837 (1962).
10. J. Dural, Thèse d'Université, Poitiers, France (1980).
11. D. Lesueur, to be published in Phil. Mag. (1981).
12. J. Linhard, M. Scharf and H.E. Schiott, Dansk. Vid.
     Selsk., Mat. Fijs. Medd. 33:1 (1963).

13. K. B. Winterbon, S. Sigmund and J. B. Sanders,
     Dansk. Vid. Selsk., Mat. Fijs. Medd. 37:14
     (1970).
14. O. S. Oen, Report ORNL-4897, Oak Ridge, USA (1973).
15. M. T. Robinson, Phil. Mag. 12:741 (1965) and
     17:639 (1969).
16. "Vacancies and Interstitials in Metals", A. Seeger,
     D. Schumacher, W. Schilling and J. Diehl, ed.,
     North-Holland Publishing Company, Amsterdam
     (1970).
17. W. F. Brizes and E.I. Salkowitz, Scripta Met., 3:659
     (1969).
18. S. Sarian, J. Appl. Phys. 39:3305 (1968) and
     40:3515 (1969).
19. J. Morillo, to be published.
20. W. S. Williams, Prog. Solid. State Chem., 6:57
     (1971).
19. J. Morillo, C.H. de Novion and J.P. Senateur, J.
     Phys. Colloq. 40:C5-348 (1979).
     J. Morillo, to be published.
20. W.S. Williams, Prog. Solid. State Chem., 6:57
     (1971).

DISCUSSION

R. Sinclair:

Again I would preface my question by congratulating the authors on a
very interesting piece of work. My question concerns the nature of
the carbon defects which were produced. The carbon is already in the
octahedral interstices in the metal sublattice. So if they're displaced to another interstitial site then they're just in their regular sublattice. Do you believe, then, that they go to tetrahedral
sites to create the defect?

C. H. de Novion:

What calculation gives--simple calculations--is that you would have
a pair of atoms in the octahedral sites. We take a C-C pair oriented
in the <111> direction. For example, in uranium carbide, at high
temperature, you can make large solid solutions, and it will keep the
cubic metal sublattice and in the octahedral sites you have got a
percentage of isolated carbons in the middle of the site and carbon
pairs. Russian workers have also made calculations for this type of
radiation defect. This is the most probable with no experimental
evidence.

R. Sinclair:

So you're suggesting then that you put two carbons into an octahedral site.

C. H. de Novion:

Yes. Split interstitial, a little like in the FCC metals.

H. Matzke:

In the uranium-carbon system, I think the only way to explain thermal mobility is to invoke that interstitials are migrating even in the substoichiometric region. Would you be able to see something like this with your techniques? To really detect the presence of a small amount of split interstitials in a substoichiometric carbide?

C. H. de Novion:

Concentrations of defects created here are a few 10^{-4} typically-- that's not much. It's at low temperature, but you see them. The distance they have to migrate to go back to a vacancy is about ten interatomic distances. That's why we see recovery stages at low temperatures. We did not measure the migration energy of the defects involved in the stages found here. Possibly we make more complex defects and if the metal atoms can be displaced, for example, we could make carbon-metal interstitial pairs or more complicated defects than we thought at first.

H. Matzke:

Just one suggestion. A means of seeing what you create is, of course, high voltage electromicroscopy in transmission where the damage is created while you look at the sample and you can see whether you form groups of metal atoms or voids or whatever. In this way, we can determine the displacement energies of the different complexes.

C. H. de Novion:

Yes.

W. Williams:

The very large contribution to the resistivity which you measure does suggest, very particularly for TiC, that the defect is a very complicated one which has an enormous elastic strain associated with it. That is, if I remember your slide for titanium carbide, you found an incremental resistivity of something like 500 $\mu\Omega$cm something like several hundred $\mu\Omega$cm per atomic percent defects. Now the number

that we obtained from resistivity measurements for substoichiometric
materials in 1964 was fifteen$\mu\Omega$cm per atomic percent vacancies.
The number you find is very much larger than that, which suggests
to me that you're looking at a defect, as you've already suggested
in your talk, that is some complex. Perhaps the dumbbell split-
interstitial will do it. But it's something with a very large
elastic strain associated with it. Wouldn't you agree with that?

C. H. de Novion:

Yes, we think that we displace the titanium in this compound and of
course a titanium interstitial has a big perturbation on the system.

W. Williams:

Would you also agree that the value obtained by Venables in electron
microscopy for the vanadium carbide case refers to simple displacement of a carbon atom from one lattice site to an adjoining vacancy
and hence is more likely to represent a diffusional migration energy.
In the radiation damage case you're looking at just the strain
energy required to stuff a carbon atom into some peculiar interstitial site, which would be the first half of the energy barrier seen
by a diffusing particle at high temperatures, or a carbon atom that's
moved from one equilibrium lattice site to another. I'm raising the
question of whether these are in fact the same mechanism or rather
they're totally different even though the numbers appear to be
similar?

C. H. de Novion:

In principle, they are different. I agree with you, that in the
case of Venables, it's from an octahedral lattice site to another
one and here it should be, for example, from a split-interstitial
octahedral site. So it's not the same barrier energy, although the
order of magnitude is the same.

DEFECT STRUCTURE OF WC DEFORMED AT ROOM AND HIGH TEMPERATURES

M. K. Hibbs*, R. Sinclair*, and D. J. Rowcliffe**

* Department of Materials Science and Engineering
  Stanford University
  Stanford, California 94305

** SRI International
   Menlo Park, Ca. 94025

ABSTRACT

Single crystals of WC, deformed by micro-indentation at room temperature and at 1000°C, are examined by transmission electron microscopy in order to determine the mechanism of slip. The plastic deformation induced by indentation occurs by the motion of partial dislocations with Burgers vectors $1/6 <11\bar{2}3>$. These partial dislocations combine in pairs to form extended dislocations with Burgers vectors $1/3 <11\bar{2}3>$. Deformation in samples indented at 1000°C takes place by the same mechanism.

Reactions occur between the leading partial dislocations of faults on intersecting slip planes. The new partial dislocation formed at one of these intersections is shown to have Burgers vector $1/6 [1\bar{2}10]$. This reaction causes the defect configuration to became sessile and may be the first step in a crack nucleation mechanism similar to that proposed by Cottrell for bcc metals. The observation of defect pile-ups at tips of microcracks formed near room temperature, high load (500 gm) indentations, supports this suggestion. A high density of defects found near the cracked edge of a specimen indicates that plastic deformation may accompany crack propagation under some circumstances.

INTRODUCTION

While there is great interest in improving the toughness, hard-

121

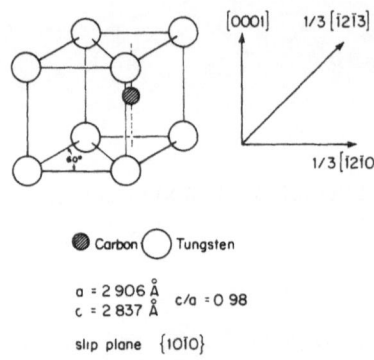

Fig. 1. Unit cell of WC. a corresponds to the 1/3 <$1\bar{2}10$> directions, the cell edges and the cell face diagonal in the horizontal
plane. c is the [0001] direction, the vertical cell edge. The
{$10\bar{1}0$} slip planes are the vertical cell faces and vertical diagonal
plane through the unit cell.

ness, and wear resistance of cemented carbide cutting tools by altering their composition and microstructure, little is known of the
basic deformation and wear processes of the consituent phases. This
study addresses this problem by examining plastic deformation and
fracture induced by micro-indentation of WC, one of the carbides used
in these composite materials. The defects resulting from room temperature and high temperature (1000°C) indentation of single crystals
of WC are characterized by transmission electron microscopy (TEM).
The interaction of deformation and fracture is also examined at a
microscopic level using this technique.

Background

     The crystal structure of WC is hexagonal with W atoms in the
0,0,0 positions and carbon atoms in the 1/3,2/3,1/2 positions (Fig. 1).
The unit cell dimensions are virtually identical, ie. c/a≈1. a corresponds to the 1/3 <$11\bar{2}0$> directions while c lies along [0001]. The
{$1\bar{1}00$} planes are the vertical cell faces and the vertical diagonal
plane through the unit cell.

     Optical slip step analyses of indented WC single crystals per-
formed by Takahashi and Friese[1] and Luyckx[2] indicate that slip occurs
on {$1\bar{1}00$} planes in <$11\bar{2}3$> directions. However, most of the subsequent transmission electron microscopy studies have been performed on
as-sintered WC-Co composites and not on deformed materials.
Johannesson and Lehtinen[3,4] and Bolton and Redington[5] observed networks consisting of undissociated dislocations with Burgers vectors

[0001] and $1/3<11\bar{2}0>$ intersecting at extended nodes. They believe that the nodes are dissociated dislocations with total Burgers vector $1/3<11\bar{2}3>$ split into partial dislocations of the type $1/6<20\bar{2}3>$. Hagege, et. al.[6] also observed networks of dislocations with Burgers vectors [0001] and $1/3<11\bar{2}0>$, but demonstrated that the nodes were dislocations with total Burgers vectors $1/3<11\bar{2}3>$ dissociated into partial dislocations of the type $1/6<11\bar{2}3>$. They also found stacking faults extending across entire carbide grains. Some of these faults were found to lie on $\{1\bar{1}00\}$ planes with displacement vector $\underline{R}=1/6<11\bar{2}3>$, while others existed on $\{11\bar{2}3\}$ planes with $\underline{R}=1/3<1\bar{1}00>$. Hibbs and Sinclair[7] characterized defects resulting from micro-indentation of single crystals. They found stacking faults bounded by two identical partial dislocations with Burgers vectors $1/6<11\bar{2}3>$ resulting in $1/3<11\bar{2}3>$ total Burgers vectors. This work is summarized later in this paper.

EXPERIMENTAL PROCEDURES

Bulk specimens were prepared from mixtures of WC and Co powders melted in an argon atmosphere at $1650^{\circ}C$ and then cooled at rates down to $2.5^{\circ}C/hr$. WC single crystals were precipitated in a Co matrix by this process. The Co was then etched away with warm HCl. The best crystals were in the form of triangular plates about 3mm on a side and 1mm thick. The large faces correspond to the basal plane while the remaining three sides are $\{1\bar{1}00\}$ planes.[8]

Samples were deformed by micro-indentation on the basal plane at room temperature with low loads (20-50 gm), at room temperature with high loads (500 gm), and at high temperature ($1000^{\circ}C$) with high loads (11 kg). The samples deformed at room temperature were first ground to a thickness of 100 μm and polished on both sides. Fifty to seventy-five indents were places in a square array 100-200 μm apart on one polished surface using a Vickers micro-hardness tester. No cracking was observed around indentations in the samples deformed with 20-50 gm loads. The 500 gm load was used to initiate macroscopic cracks around indentations. Radial and ring cracks were observed optically. The samples deformed at $1000^{\circ}C$ were first indented on a polished basal plane surface using a 3.2mm diameter sapphire ball.[8] The sample was then ground down to 100 μm and polished on the undeformed side.

All samples were thinned to electron transparency using a micro-ion mill. The bulk of the milling was done from the non-indented side in order to preserve the defects resulting from the deformation, but each sample was also milled from the indented side for one or two hours to remove the deformation produced by grinding and polishing. The foils were observed in a Philips EM400 operated at 120kV. The position of indents in the TEM thin foils were determined by comparing low magnification electron micrographs with optical micrographs of the prepared samples.

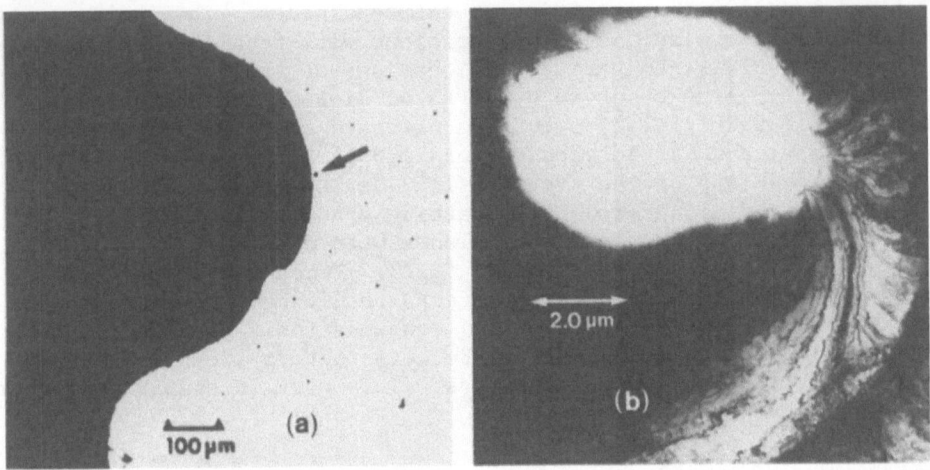

Fig. 2.   (a) Optical micrograph of a deformed TEM specimen of WC.
Indentations have been placed in a square array. The arrow points
to the location of an indentation which appears as a large hole in
the low magnification electron micrograph in (b).

RESULTS

Deformation at Room Temperature

    Figures 2a and 2b contain an optical micrograph, and low magni-
fication electron micrograph, respectively, of a sample indented at
room temperature. The arrow in 2a points to an indentation which was
correlated as described above to the hole in 2b. From the information that this hole is an indentation position, and from the knowledge that the grown-in dislocation density is quite low,[7] it is concluded that the defects observed in this region are the result of
plastic deformation.

    A higher magnification electron micrograph of an area near this
indentation is shown in Figure 3. The defect structure consists of
a high density of stacking faults with symmetric fringe contrast
in bright field, indicating that the displacement vector is one half
a lattice translation vector. The streaks in the corresponding diffraction pattern indicate that these faults lie on $\{1\bar{1}00\}$ planes.
Contrast analyses[9] were performed on these faults and gave the result
that bounding partial dislocations have identical Burgers vectors of
the type $1/6<11\bar{2}3>$ resulting in total Burgers vectors for the defects
of the type $1/3<11\bar{2}3>$. Projections of a perfect crystal and a faulted
crystal onto a $(1\bar{1}00)$ slip plane are shown in Figure 4. The faulted
crystal is created by moving the top layer of tungsten atoms across
the layer of carbon atoms with displacement vector $R=1/6[11\bar{2}3]$. This
fault vector preserves the trigonal prismatic coordination of carbon
atoms by tungsten atoms. In fact, the faulted structure appears to

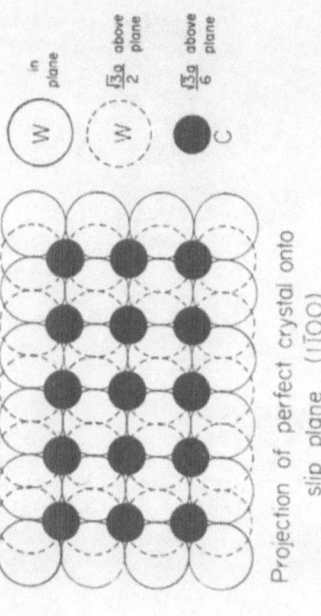

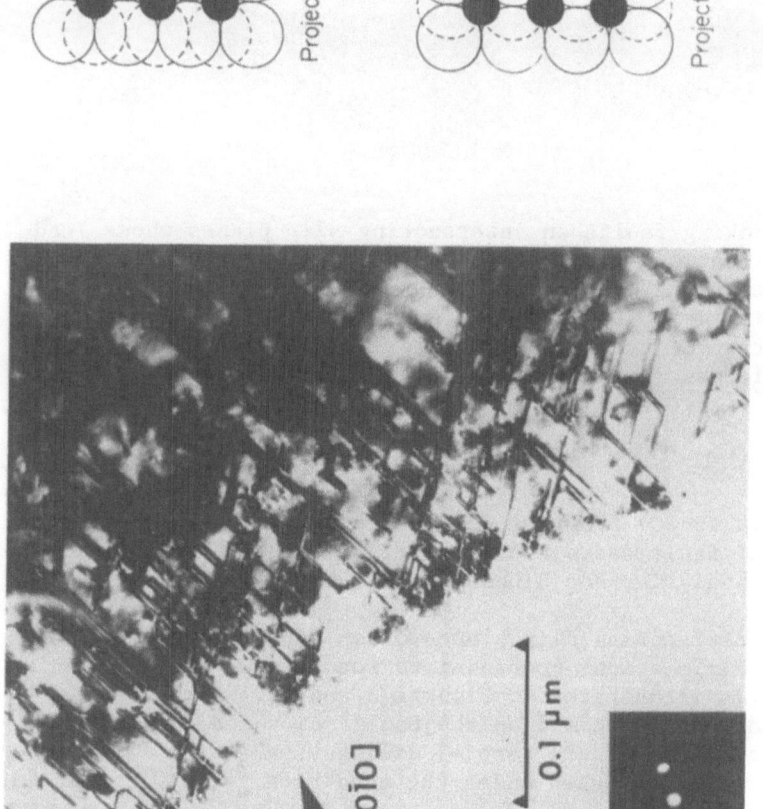

Fig. 3

Fig. 3. A high magnification electron micrograph taken near the hole shown in Fig. 2(b). A high density of stacking faults may be seen lying on {1$\bar{1}$00} planes. The streaks in the diffraction pattern along the <1$\bar{1}$00> directions confirm that the faults lie on {1$\bar{1}$00} planes.

Fig. 4. Projection of the perfect and faulted structures of WC onto the (1$\bar{1}$00) slip plane. The fault was created by moving the top layer of W atoms across the C atom layer with the displacement vector R=1/6[11$\bar{2}$3].

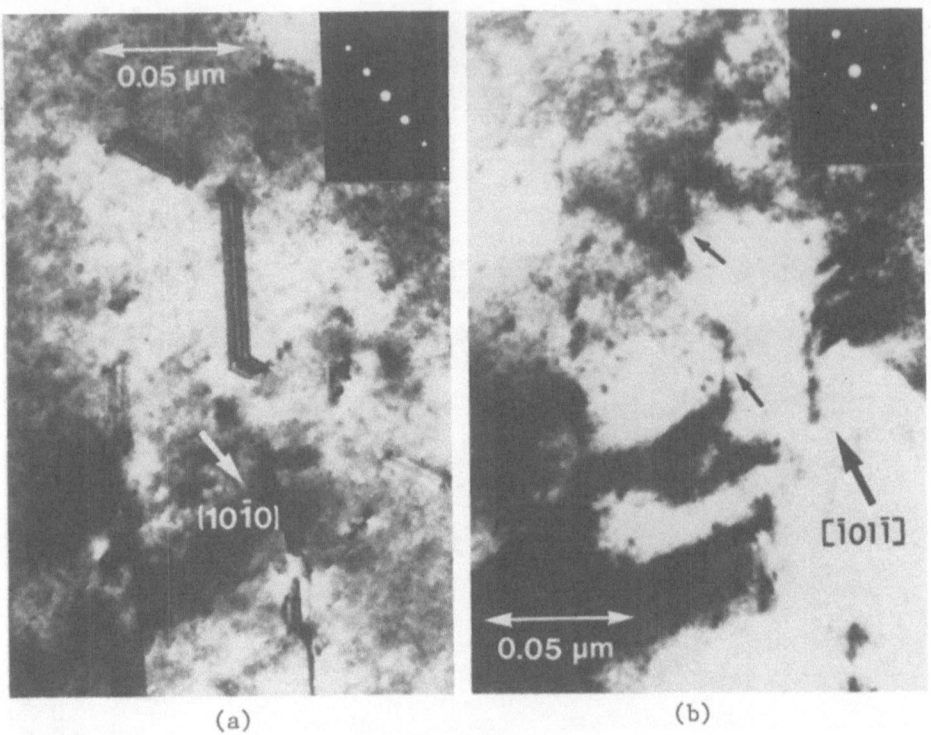

(a)   (b)

Fig. 5. Two stacking faults on intersecting slip planes whose leading partial dislocations have reacted to form a new partial dislocation. (a) The fringe contrast of both faults is visible for diffracting condition g=[10$\bar{1}$0]. (b) The common partial dislocation and
the fringe contrast of the two faults are invisible while the trailing partial dislocations are both visible. Diffracting condition
is described by g= [$\bar{1}$01$\bar{1}$].

be merely rotated by 90° from the perfect structure. This suggests
that this type of fault has a relatively low energy. For further
details of this analysis, see Hibbs and Sinclair [7].

Faults on intersecting slip planes often appear to react, the
leading partial dislocations combining to form a new partial dislocation. This is illustrated in Figures 3 and 5a. A contrast analysis experiment was performed on the faults pictured in Figure 5
to determine the nature of the partial dislocation at the fault intersection. Figure 5a was taken under the condition g=[10$\bar{1}$0], that is,

Table 1. Summary of the contrast experiment performed on the defect configuration in Figure 5.

reflection	visibility				
	fault 1	fault 2	b_1	b_2	b_3
$10\bar{1}0$	V	V	–	–	–
$11\bar{2}0$	V	I	–	V	–
$01\bar{1}0$	I	V	I	–	–
$2\bar{1}\bar{1}0$	I	V	V	–	–
$01\bar{1}1$	V	I	–	V	–
$\bar{1}01\bar{1}$	I	I	V	V	I
$\bar{1}100$	V	I	–	I	–
$1\bar{1}0\bar{1}$	I	V	I	–	–

b_2

fault 2

b_3 b_1

fault 1

$$\underline{b}_1 = \tfrac{1}{6}[2\bar{1}\bar{1}3] \qquad \underline{b}_2 = \tfrac{1}{6}[\bar{1}\bar{1}2\bar{3}]$$

$$\underline{b}_3 = \tfrac{1}{6}[1\bar{2}10]$$

V: visible, I: invisible, –: fringe contrast of a fault interferes with the observation of the partial dislocation.

the $(10\bar{1}0)$ planes are at the Bragg diffraction position, and fringe contrast from both faults is visible. The diffraction condition for 5b was $\underline{g}=[\bar{1}01\bar{1}]$. The fringe contrast for both faults and the partial dislocation at the intersection are all invisible. The small arrows point to the trailing partial dislocations of the two faults, which are still visible. Table 1 summarizes the analysis performed on this defect configuration. Normally, at least two disappearance criteria are necessary to determine the Burgers vector of a dislocation. (Dislocation contrast disappears when $\underline{g}\cdot\underline{b}=0$.) However, this was not possible in the present case since the stacking fault fringe contrast obscured the disappearance of the dislocation in question. Because of the limitations of the specimen holder, it was impossible to tilt to two different reflecting conditions for which the fringe contrast of both faults and the intersecting dislocation disappeared. Therefore, to carry out the analysis, one additional assumption was made which was shown to be valid by the work described above. That is, that the total Burgers vectors of the faults are of the type $1/3 <11\bar{2}3>$ and that the leading and trailing partials of any fault are identical. In this case the trailing partial of one fault is found to have Burgers vector $\underline{b}_1=\pm 1/6[2\bar{1}\bar{1}3]$, while the trailing partial of the other has Burgers vector $\underline{b}_2=\pm1/6[\bar{1}\bar{1}2\bar{3}]$. Combinations of these two partials would result in a reaction to form a partial dislocation with either a $\pm1/6[1\bar{2}10]$ or a $\pm1/2[10\bar{1}2]$ Burgers

$\frac{1}{6}[2\bar{1}\bar{1}3] + \frac{1}{6}[\bar{1}\bar{1}2\bar{3}] = \frac{1}{6}[1\bar{2}10]$

$\frac{1}{6}[2\bar{1}\bar{1}3] + \frac{1}{6}[\bar{1}\bar{1}2\bar{3}] = \frac{1}{6}[1\bar{2}16]$

$\frac{1}{6}[2\bar{1}\bar{1}3] + \frac{1}{6}[11\bar{2}\bar{3}] = \frac{1}{2}[10\bar{1}0]$

$\frac{1}{6}[2\bar{1}\bar{1}3] + \frac{1}{6}[11\bar{2}3] = \frac{1}{2}[10\bar{1}2]$

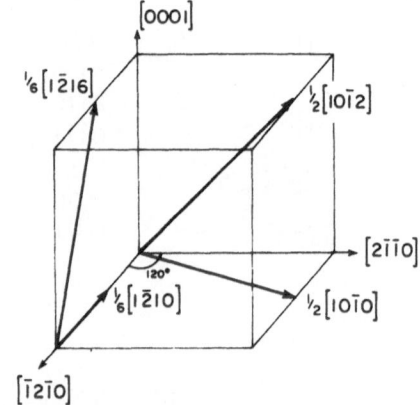

$\underline{b}$	$\frac{1}{6}[1\bar{2}10]$	$\frac{1}{6}[1\bar{2}16]$	$\frac{1}{2}[10\bar{1}0]$	$\frac{1}{2}[10\bar{1}2]$
$\lvert b\rvert^2$	4.22Å²	10.16 Å²	10.56Å²	18.61Å²

$2\lvert b\rvert^2 = 8.25\text{Å}^2$ for $\underline{b} = \frac{1}{6}\langle 2\bar{1}\bar{1}3\rangle$

Fig. 6. The four possibilities for reactions between partial dislocations with Burgers vectors of type $1/6\langle11\bar{2}3\rangle$. The directions of the Burgers vectors of the resulting partial dislocations are shown within the unit cell, along with values of $\lvert b\rvert^2$ for each of the four possible Burgers vectors.

vector. The invisibility of this dislocation for $g=[\bar{1}01\bar{1}]$ is consistent with $\underline{b}_3=\pm1/6[1\bar{2}10]$ and not with $\underline{b}_3=\pm1/2[10\bar{1}2]$. We conclude that a dislocation reaction has occurred to form a new partial dislocation with Burgers vector $\underline{b}_3=\pm1/6[1\bar{2}10]$.

Four possibilities exist for the Burgers vector of a new partial dislocation resulting from the reaction of partial dislocations of type $1/6\langle11\bar{2}3\rangle$. These are listed in Figure 6, along with the values of $\lvert b\rvert^2$ for each and $2\lvert b_p\rvert^2$ for $\underline{b}_p=1/6\langle11\bar{2}3\rangle$. A diagram of these four directions is also given in Figure 6. Since $\lvert b\rvert^2<2\lvert b_p\rvert^2$ for $\underline{b}=1/6\langle11\bar{2}0\rangle$ whereas $\lvert b\rvert^2>2\lvert b_p\rvert^2$ for any of the other three possibilities for $\underline{b}$, it seems likely that only a reaction forming a new $1/6\langle11\bar{2}0\rangle$ partial would be stable. In any case, none of the four possible Burgers vectors of the new dislocation lies on the slip plane of either of the original extended dislocations. Therefore, this reaction should be a mechanism which hinders plastic flow, since the new configuration is expected to be sessile. The reaction may also have implications for fracture.

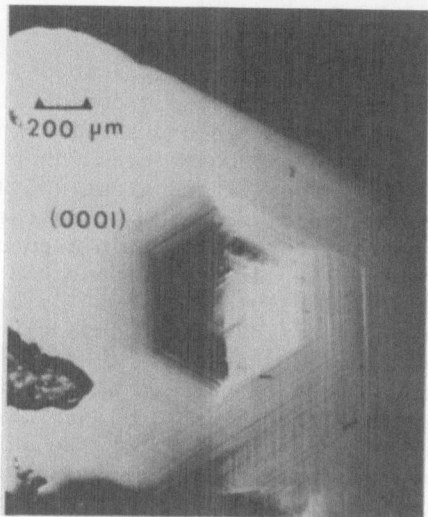

Fig. 7.   An optical micrograph of WC deformed at 1000°C on the
basal plane with a 3.2mm diameter sphere. The slip traces are
consistent with {1$\bar{1}$00} slip planes.

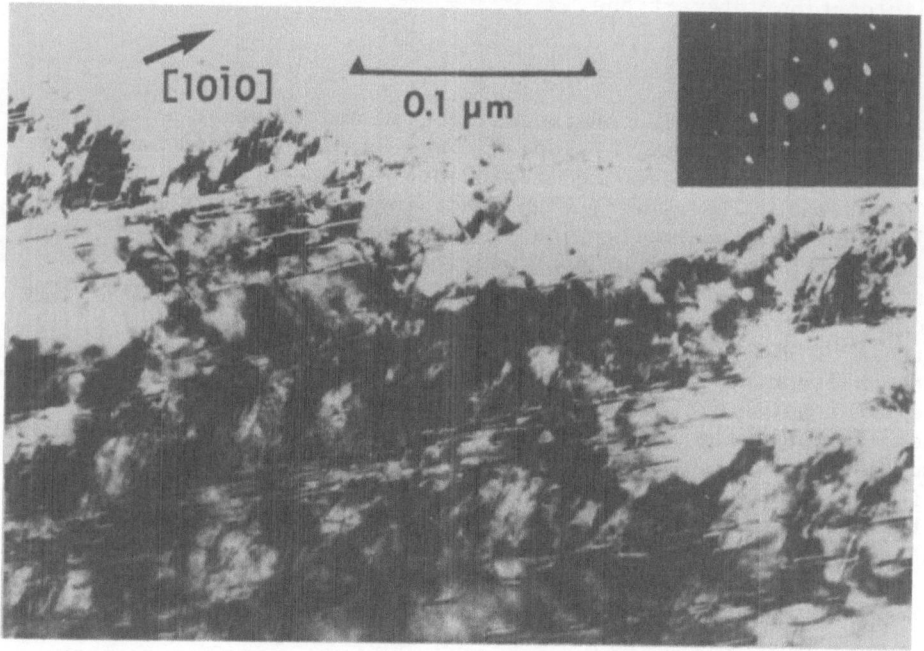

Fig. 8.   Stacking faults lying on {1$\bar{1}$00} planes near an indentation created at 1000°C. The bounding partial dislocations of a
fault have identical Burgers vectors 1/6<11$\bar{2}$3> forming a total
extended dislocation with Burgers vector 1/3 <11$\bar{2}$3>.

Deformation at 1000°C

An optical micrograph of WC deformed at 1000°C on the basal plane
with a 3.2mm diameter sapphire sphere is shown in Figure 7. The slip
traces resulting from the indentation are similar to those resulting
from room temperature indentation, being consistent with {10$\bar{1}$0} slip
planes. However, the slip traces extend about 400 μm from the center
of the indentation as compared to about 20 μm in the room temperature
case. An electron micrograph taken from thin area near the indentation is shown in Figure 8. The defects responsible for the plastic
deformation are stacking faults lying on {10$\bar{1}$0} planes having the
symmetric fringe contrast in bright field similar to that of faults
found near room temperature indentations. A contrast analysis experiment confirmed that these defects have total Burgers vectors of the
type 1/3<11$\bar{2}$3> and have split into identical partial dislocations
with Burgers vectors of the type 1/6<11$\bar{2}$3>. The defects also show
the same propensity for interaction and formation of sessile configurations as in the room temperature deformation case. Therefore, it
seems that the extension of plastic deformation to greater distances
from the center of the indentation at higher temperatures must be
due to a lower threshold energy for nucleation and propagation of
dislocations, rather than any change in slip system or the elimination
of dislocation interactions.

Fracture at Room Temperature

Preliminary studies of the nucleation and propagation of cracks
in WC are reported here. Figure 9 contains a series of cracks which
were observed in the deformed region near an indentation created with
a 500 gm load. The general appearance indicates the cracks lie on
a (1$\bar{2}$10) plane, though they jog in other directions. The center of
the largest crack is about 40 Å wide. There appear to be greater
densities of defects at the crack tips. This may be evidence for a
crack nucleation mechanism in WC similar to that proposed by
Cottrell[10]. If the Cottrell type of mechanism does operate, the
reaction described above, 1/6[2$\bar{1}\bar{1}$3]+1/6[$\bar{1}\bar{1}$2$\bar{3}$]=1/6[1$\bar{2}$10], could provide the sessile nucleus against which other dislocations in the
same slip band pile up under the action of an applied stress. The
resulting crack plane would be (1$\bar{2}$10). This process is illustrated
in Figure 10. The nucleation of cracks on {1$\bar{2}$10} planes at intersecting slip traces has been observed macroscopically by Pattanaik
and Rowcliffe[11].

Another aspect of fracture behavior is illustrated in Figure 11.
Two different areas along the broken edge of a specimen are shown.
Macroscopically, the edge was very straight and along one (1$\bar{1}$00)
plane. Microscopically, the crack is more jagged but the edges are
still along {1$\bar{1}$00} planes. Defects are found at distances less than
1 μm from the foil edge (Figure 11a) and consist of both stacking
faults and very straight undissociated dislocations. Since this

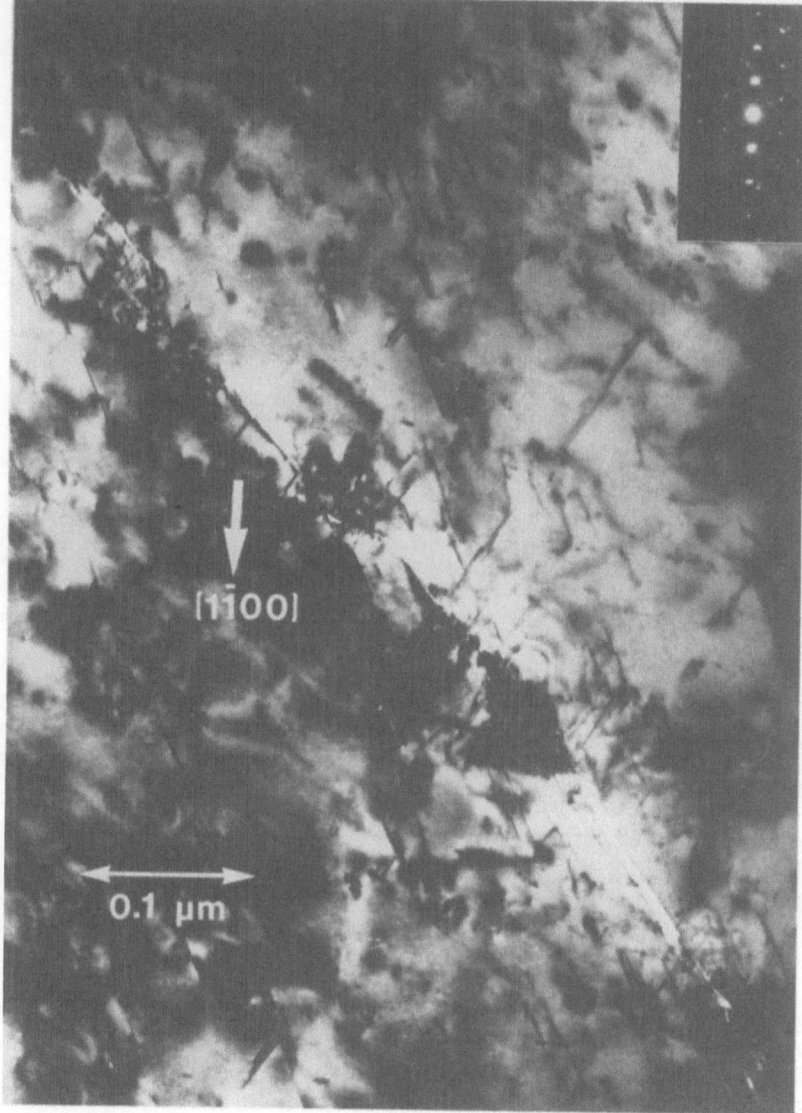

Fig. 9. Microcracks observed near an indentation formed at room
temperature with a 500gm load. Dislocation pile-ups occur at the
crack tips.

broken edge was not near any indentations and all the surface deformation was removed by ion milling, it seems likely that this deformation accompanied the propagation of a crack. Contrast analyses were
performed on undissociated and extended dislocations and both were

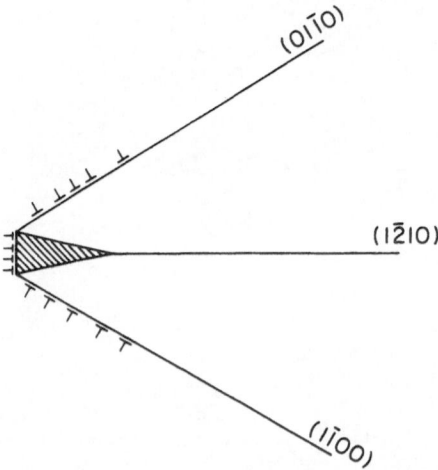

Fig. 10. Illustration of a possible crack nucleation mechanism in WC, similar to that proposed by Cottrell for bcc materials. The leading partial dislocations of extended defects moving on (01$\bar{1}$0) and (1$\bar{1}$00) planes react to form a new partial dislocation with Burgers vector 1/6[1$\bar{2}$10]. Other dislocations pile-up against this sessile configuration under the action of an applied stress, forming a nucleus for cracking on the (1$\bar{2}$10) plane.

found to have Burgers vectors of the type 1/3<11$\bar{2}$3>. Long straight lines may be seen in Figure 11b across which thickness fringes are discontinuous, indicating that thickness or orientation changes abruptly. This may be evidnece of other cracks extending into the foil.

It is not true, however, that all cracks observed in this material are associated with deformation such as this. Some cracks are free of defects along their length, while others have dislocations only at the tip. Crack tip dislocations may be due to deformation or may only be accommodations of lattice mismatch associated with partial closure and healing at the interface[12]. Clearly, much work needs to be done to fully understand the fracture processes in this material. However, deformation does appear to interact with crack nucleation and propagation.

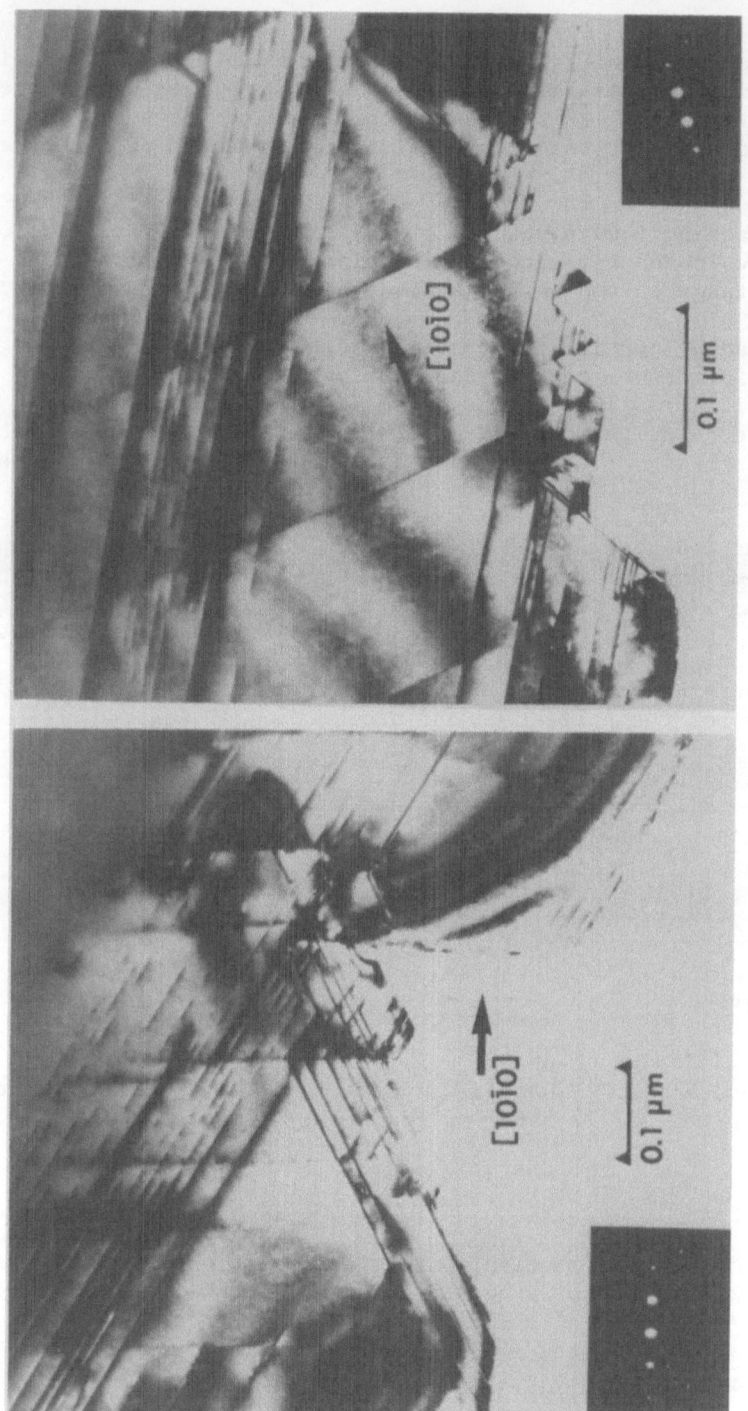

Fig. 11. Two different areas along the broken edge of a specimen. The jagged edges lie along {1$\bar{1}$00} planes. In (b) the long straight lines across which thickness fringes are discontinuous may indicate additional cracks extending into the sample.

CONCLUSIONS

1. Deformation resulting from room temperature micro-indentation occurs by motion of partial dislocations on $\{1\bar{1}00\}$ planes with Burgers vectors of the type $1/6<11\bar{2}3>$. These partial dislocations combine to form extended dislocations with total Burgers vectors of the type $1/3<11\bar{2}3>$.

2. The stacking faults resulting from these partial dislocations are likely to be low energy since the angular relationships and distances between tungsten and carbon atoms are preserved at the fault.

3. The leading partial dislocations of the extended defects often react to form a new partial dislocation. The reaction $1/6[2\bar{1}\bar{1}3]+1/6[\bar{1}\bar{1}2\bar{3}]=1/6[1\bar{2}10]$ is expected to be a stable reaction, according to the b^2 criterion, and has been observed. The new configuration, consisting of two stacking faults on intersecting planes sharing a common partial dislocation, is expected to be sessile.

4. Samples indented at 1000°C deform in the same way as those indented at room temperature, by $1/6<11\bar{2}3>$ partial dislocations which combine to form $1/3<11\bar{2}3>$ extended defects. They also exhibit the same propensity for reactions between faults on intersecting planes. Therefore, the propagation of defects to greater distances from the center of an indentation must be the result of a decreased threshold energy required for the nucleation and propagation of defects.

5. Some evidence is observed for the nucleation of cracks on $\{11\bar{2}0\}$ planes by dislocation reactions and subsequent pile-ups on intersecting slip planes.

6. Defects along a cracked edge may indicate that plastic deformation is associated with some crack propagation in WC.

ACKNOWLEDGEMENTS

The authors wish to thank Mr. S.L. Shinde and Mr. V. Jayaram for helpful discussions. Financial support from the National Science Foundation Grant No. DMR 77-22647 is gratefully acknowledged.

REFERENCES

1. T. Takahashi and E. J. Freise, Determination of the Slip Systems in Single Crystals of Tungsten Monocarbide, _Phil_. _Mag_. ser.8, 12:1 (1965).
2. S. B. Luyckx, Slip System of Tungsten Carbide Crystals at Room Temperature, _Acta_ _Metall_. 18:233 (1970).

3. T. Johannesson and B. Lehtinen, The Analysis of Dislocation Structures in Tungsten Carbide by Electron Microscopy, Phil. Mag. 24:1079 (1971).

4. T. Johannesson and B. Lehtinen, On the Plasticity of Tungsten Carbide, Phys. Stat. Sol. (a) 16:615 (1973).

5. J. D. Bolton and M. Redington, Plastic Deformation Mechanisms in Tungsten Carbide, J. of Mat. Sci. 15:3150 (1980).

6. S. Hagege, J. Vicens, G. Nouet, and P. Delavignette, Analysis of Structure Defects in Tungsten Carbide, Phys. Stat. Sol. (a) 61:675 (1980).

7. M. K. Hibbs and R. Sinclair, Room-Temperature Deformation Mechanisms and the Defect Structure of Tungsten Carbide, Acta Metall. in press (1981).

8. D. J. Rowcliffe, Indentation Deformation of Tungsten Carbide and Tungsten-Titanium Carbide, these proceedings.

9. P. B. Hirsch, A. Howie, R. B. Nicholson, D. W. Pashley, and M. J. Whelan, "Electron Microscopy of Thin Crystals", Butterworth, London (1965).

10. A. H. Cottrell, Theory of Brittle Fracture in Steel and Similar Materials, Trans. AIME 212:192 (1958).

11. S. Pattanaik and D. J. Rowcliffe, to be published (1982).

12. B. R. Lawn, B. J. Hockey, and S. M. Wiederhorn, Atomically Sharp Cracks in Brittle Solids: An Electron Microscopy Study, J. of Mat. Sci. 15:1207 (1980).

MICROSTRUCTURAL CHARACTERIZATION OF DEFORMATION AND

PRECIPITATION IN (W,Ti)C

S. L. Shinde, V. Jayaram and R. Sinclair

Department of Materials Science and Engineering
Stanford University
Stanford, California 94305

ABSTRACT

Transmission electron microscopy has been used to study the deformation produced by Vickers micro-indentations in rock-salt structure carbides with approximate compositions $(W,Ti)C_{0.8}$ and $(W,Ti)C_{0.6}$. In the more severely deformed regions, the dislocation density increases from the annealed value of 10^8 cm^{-2} to around 10^{13} cm^{-2}. The defects were found to lie on all the three major low index planes, $\{111\}$, $\{110\}$ and $\{100\}$ with Burgers' vectors that are consistent with $a/2 <110>$. The operation of these slip planes is consistent with observations of hardness anisotropy in other transition metal carbides, particularly TiC. Indentations at high loads have revealed interactions between slip bands. These sometimes result in the formation of wedge shaped cracks and in the abrupt displacement of slip traces at the intersection of different bands. The latter may reflect the sequential activation of different slip planes during the loading process.

Preliminary precipitation studies on $(W,Ti)C_{0.6}$ are reported, revealing a crystallographic relationship between the bcc (W,Ti) phase and the fcc carbide matrix.

INTRODUCTION

Additions of transition metal carbides, such as TiC, are known to improve the performance of WC-Co alloys used for machining metals[1]. There have been a number of studies on the deformation behavior of cubic carbides, both at room temperature and at temperatures above $800°C$ [2-8]. These have been of two kinds. Above the brittle-ductile transition, which occurs around $800°C$ for TiC,

137

deformation can be introduced in the bulk by compression or bending, and the defects can be studied by examining thin foils in the transmission electron microscope (TEM) [2-4]. At lower temperatures, with the exception of a TEM study on NbC, [9] indirect techniques, such as hardness anisotropy, have been necessary to deduce the operation of particular slip systems [5-8].

The present study was designed to provide direct evidence of the nature of room temperature deformation mechanisms in (W,Ti)C. In addition, insight was generated on the manner in which plastic flow spreads around an indentation at room temperature, which may benefit our understanding of this process microscopically. This work was also concerned with the effects of precipitating a soft metal phase within the carbide matrix, which may be expected to influence the mechanical properties. The bulk of the paper deals with the deformation produced around Vickers indentations in (W,Ti)C, while reporting some preliminary results on the precipitation process.

BACKGROUND

(W,Ti)C is isostructural with TiC up to almost 60% substitution of Ti by W and has the rock-salt crystal structure. Hardness anisotropy measurements using Knoop indenters have suggested that at room temperature, the preferred slip plane in stoichiometric TiC is {110} [8]. With increasing temperature[10] and deviations from stoichiometry[5], the appearance of additional hardness minima along <110> on a cube face have been interpreted as a sign of the onset of {111} slip, although it should be noted that the data are equally consistent with {100} slip. Hardness minima have recently been found in stoichiometric TiC [11].

The arguments for a transition from {110} to {111} slip are supported by studies of bulk deformation, where the brittle-ductile transition is lower [2,3] in $TiC_{0.8}$ compared to TiC and from TEM studies of which, the slip system at $T \geq 800°C$ in TiC has been shown to be {111} <1$\bar{1}$0> [3].

EXPERIMENTAL PROCEDURES

1. Deformation

Both arc-melted, polycrystalline specimens* of previously hot pressed, sintered powders and single crystals** grown by induction floating zone melting were used. The atomic compositions are as follows:

	Ti	W	C	
Polycrystal	33	22	45	$MC_{0.8}$
Single Crystal A	33	22	45	
Single Crystal B	37	25	38	$MC_{0.6}$

* The arc melting was carried out at SRI International, USA
** Single crystals were obtained from MARTIN MARIETTA CORP.

Carbon analyses were made by combustion with V_2O_5, while the Ti : W ratios were determined in a microprobe, using a wavelength dispersive X-ray spectrometer. The accuracy of the composition is estimated to be $\pm$ 1%. Though crystal B was nominally in the two phase region of the ternary equilibrium diagram[12], no trace of the (W,Ti) phase could be detected by X-ray diffraction.

Slices from the bulk were prepared by spark machining or diamond sawing. Pieces suitable for viewing in the TEM were cut and ground to a final thickness of 100–150μm and a 1μm finish with diamond paste. An array of Vickers indents was then placed on one side using loads of between 25 g and 300g and the specimen was ion-beam milled from the reverse side until the hole produced contained indentations in the electron transparent region. A comparison of optical and low magnification electron micrographs made it possible to correlate deformation observed in the TEM with the position of indentations.

2. (W,Ti) Precipitation

Samples of crystal B were heat treated in a tungsten mesh furnace for up to 50 hours at 1600°C in a vacuum of 10^{-5} torr. At this temperature the equilibrium phases expected are (W,Ti),, which is bcc, and (W,Ti)C. No evidence of bulk precipitation was detected by Laue back reflection using Cu-radiation. The specimens were then decarburized by heating in contact with Ti powder at 1600°C. With the composition now further into the two phase field, precititation of the metallic phase occurred. TEM specimens were then made in the same way as for the single phase materials (apart from indentation).

RESULTS

1. Undeformed carbide

Some specimens were examined undeformed, in order to character-ize the defects that had been produced during solidification and cooling. The dislocations appear in networks or singly as in Fig. 1. Using two or more g.b=0 criteria their Burgers vector was determined to be of the type 1/2 <110>. Weak beam dark field images such as in Fig. 2, show that the dislocations are undissociated down to a resolution of 30Å[13].

2. Deformed carbides

In studying the deformation around indentations in the TEM, it is necessary to ensure that the foil is reasonably thick, i.e., >1000Å. The importance of this is illustrated in Fig. 3. In Fig. 3(a) the dislocation loops generated by the indentation, which is the dark central region, are confined to a small region of 3μm around

S. L. SHINDE ET AL.

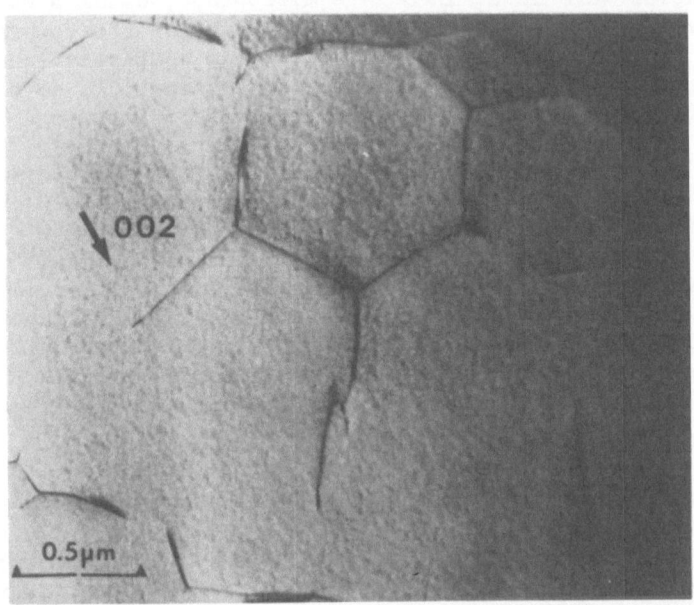

Fig. 1   Dislocation networks in undeformed, anneal-
          ed (W,Ti)C.

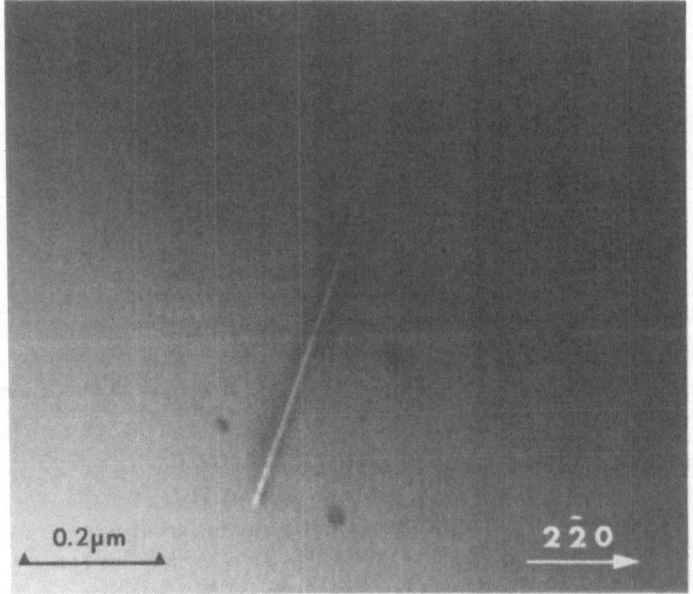

Fig. 2   Weak beam image of dislocation showing the
          absence of dissociation.   Image width is 30Å

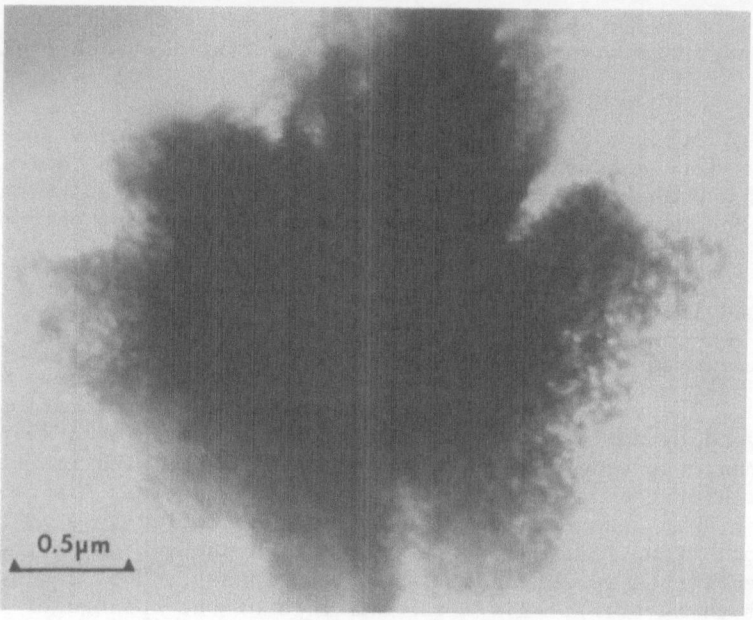

Fig. 3(a)   Deformation is preserved around the indenta-
            tion in the foil shown here where the thick-
            ness is > 1000Å

Fig. 3(b)   Glide of dislocations around the indent has
            occurred in the more extensively thinned foil
            (c.f., fig. 3(a)) through spontaneous stress
            relief.

the center. A large density of defects is visible extending outwards
from the indentation center along low index crystallographic directions.

In Fig. 3(b), the indentation was created by the same load but
the foil is more extensively thinned in that region. The defect
density is now markedly lower and suggests that the stresses have
been spontaneously relieved by glide of the dislocations out of the
specimen.

2.1 Low Load Indentation in $(W,Ti)_{0.8}$

Deformation was introduced in the polycrystalline sample with
a 25g load. The slip planes of the various sets of dislocations
were analyzed by noting that when the projected width of a slip band
is a minimum, the band must then be parallel to the electron beam.
This is illustrated in Figs. 4 and 5 where the specimen has been
tilted such that the group of dislocations identified by A is now
seen end-on. Comparison with the diffraction pattern shows the
$[1\bar{1}\bar{1}]$ direction to be perpendicular to the traces, thus showing the
slip plane to be $(1\bar{1}\bar{1})$. A similar analysis of the triangular pattern
of slip traces shown by B, C, and D in Fig. 4 shows that B and C
are {111} bands, while D is a {110} band. Deformation produced by
indenting on a cube face of the single crystal is illustrated in
Fig. 6. The four-fold symmetry of the dislocation loops (indicated
by X) is apparent. These were determined to lie on {111} . Also
present are additional sets of loops which project at 45° to the
others and correspond to slip on {110}.

2.2 High Load Indentations in $(W,Ti)C_{0.6}$

2.2.1 Introduction Observations. Fig. 7 shows a low magnification
electron micrograph of a 300g indentation on a face close to (012).
The approximate shape of the original indent is indicated by the half
diamond outline and the numbers 1-4 indicate the regions which reveal
glide bands.

A closer look at region 4 reveals three sets of slip traces
which are shown in Figs. 8 and 9. In Fig. 8 the electron beam
direction is close to [011] and edge-on bands corresponding to
$(1\bar{1}\bar{1})$ and (100) planes are visible, shown by A and B respectively.
In Fig. 9 the projection is along $[00\bar{1}]$ and traces corresponding to
(100) and (010) are visible.

In region 1, slip is noticed on (110), indicated in Fig. 10
by the trace C. However, in addition there are other traces that
lie at angles of 60° to [100] in the [001] projection. These cannot be due to slip on any of the three major low index planes,
{100} {110}, or {111}. Traces similar to these have been seen

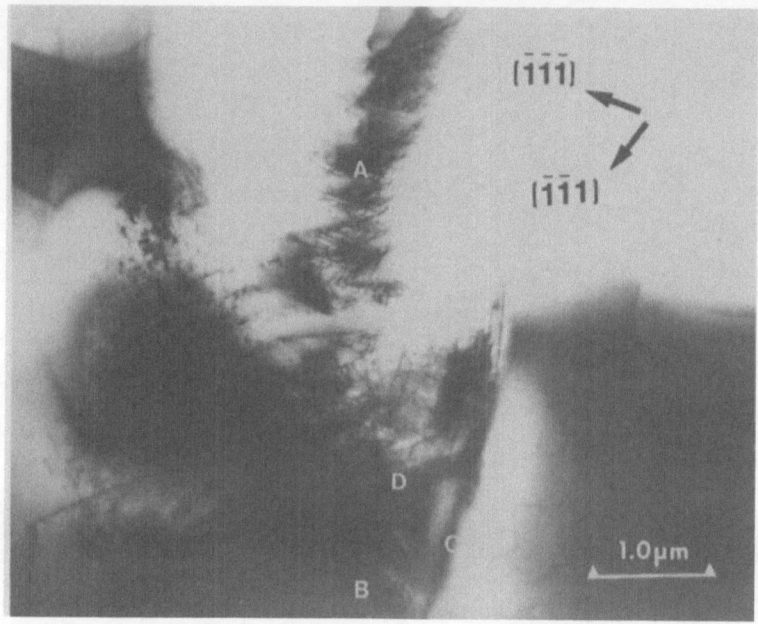

Fig. 4   Deformation around a 25g indent in polycrys-
                talline (W,Ti)C$_{0.8}$. B and C are {111} slip
                bands while D lies on {110}.

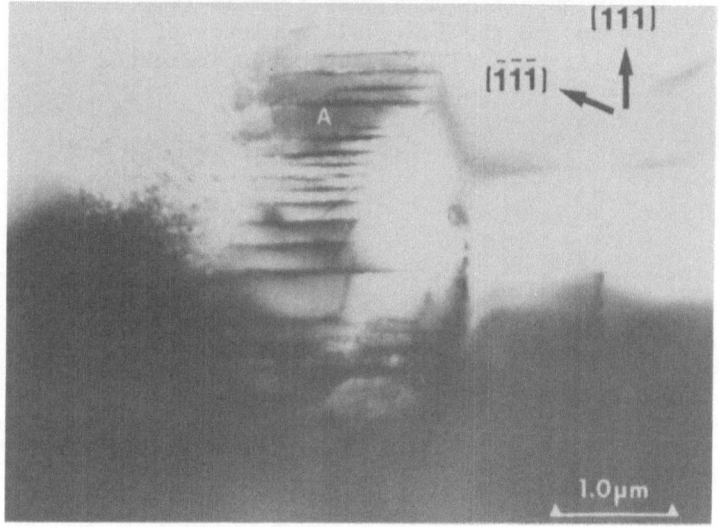

Fig. 5   Slip bands A, which were inclined in fig. 4,
                are now exactly parallel to the electron beam
                and are shown to lie on ($\bar{1}$11).

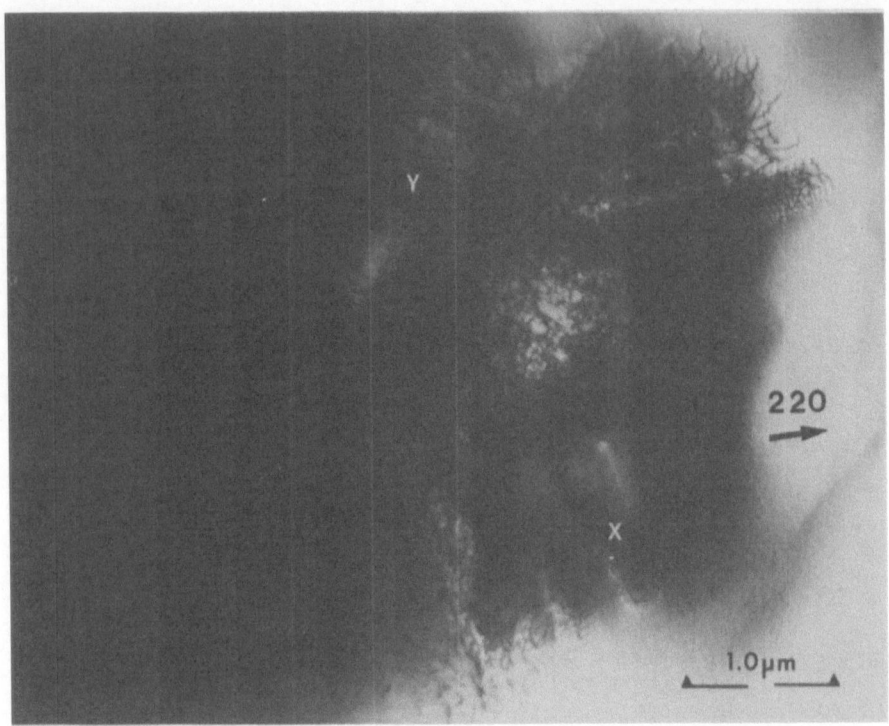

Fig. 6   Indentation on a cube face in $(W,Ti)C_{0.8}$
              showing the four-fold symmetry of the
              {111} and {110} glide bands indicated by
              x and y respectively.

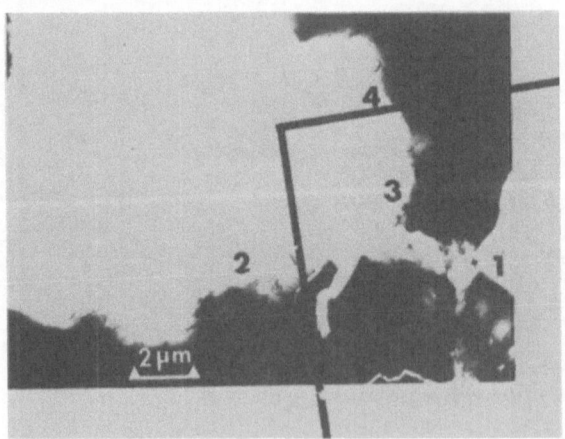

Fig. 7   Low magnification view of a 300g indent in
              $(W,Ti)C_{0.6}$. Deformation in regions 1, 2, 3
              and 4 is shown in more detail later.

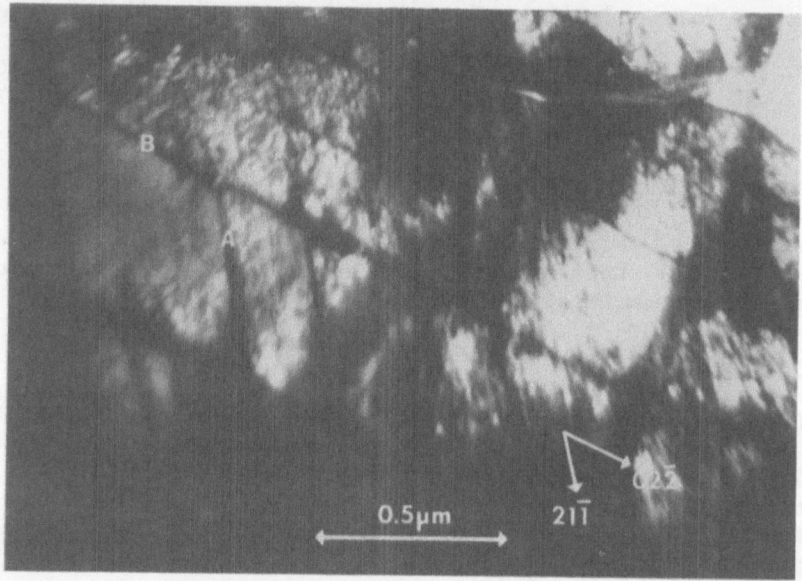

Fig. 8   Slip bands on (1$\bar{1}$1) and (100) indicated by
                A and B respectively are seen edge on in this
                magnified view of region 1 of the indentation
                (W,Ti)$C_{0.6}$. The beam direction is [011].

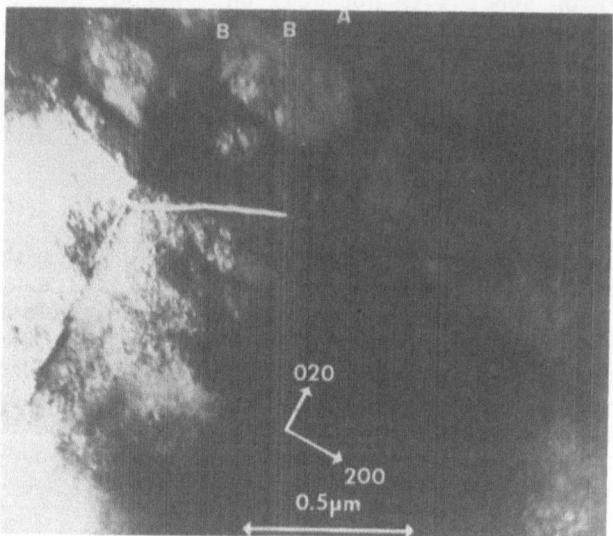

Fig. 9   On tilting the foil about [100] to [001], slip
                bands B continue to be edge on while additional
                traces at right angles, corresponding to slip
                on (010) are now visible parallel to the elec-
                tron beam.

macroscopically around ball indentations made with an 11 Kg. load
in $(W,Ti)C_{0.8}$ at 1000°C [14]. We note that {111} and {100} slip
occur simultaneously in the same region while slip occurs on {110}
where neither of the other planes operates. Thus the mode of deformation varies with position around the Vickers indentation.

2.2.2 Slip Band Interactions. (a) At the intersections of
certain {100} slip bands, wedge shaped cracks were found. These
lie on both {100}, which is the cleavage plane in this material and
on {110}. Fig. 11 shows a crack which has nucleated on {110} and
then appears to branch out along {100}.

This phenomenon of secondary cleavage has been observed in MgO
at slip band intersections[15]. These cracks appear to be distinct
from the macroscopic radial cracks that are observed in hardness
indentations. At this stage it is not possible to establish the
role played by these micro-cracks in determining the fracture behavior of $(W,Ti)C$.

(b) At other intersections a slip trace is frequently found
to be displaced, suggesting that deformation on the two planes was
sequential and not simultaneous. This is shown in Fig. 12 where the
(100) band seen on edge, suffers a displacement of about 500Å where
it is intersected by the inclined slip band. Since the exact nature
of the indentation process is not well understood, we can only speculate that the stress state in a given region must vary significantly
during the loading period. It is also possible that the concentration of the deformation in these slip bands might lead to work hardening of the material thereby causing the stress to rise and initiate
slip on other planes.

2.3 Precipitation

The precipitation was inhomogeneous. While large parts of the
crystal (several hundred microns in size) were still single phase,
other regions showed extensive precipitation. In order to discover
any preferential orientation of the precipitates with respect to the
matrix, selected area diffraction patterns were taken along low index
matrix zone axes. Two of these taken along $[112]_{fcc}$ and $[111]_{fcc}$
are shown in Fig. 13. They reveal a clustering of the bcc precipitate 110 and 200 reflections suggesting an orientation relationship of the type

$$(110)_{bcc} \quad / / \quad (111)_{fcc}$$

$$[1\bar{1}0]_{bcc} \quad / / \quad [11\bar{2}]_{fcc}$$

Fig. 14 shows a bright field micrograph of the precipitate particles,

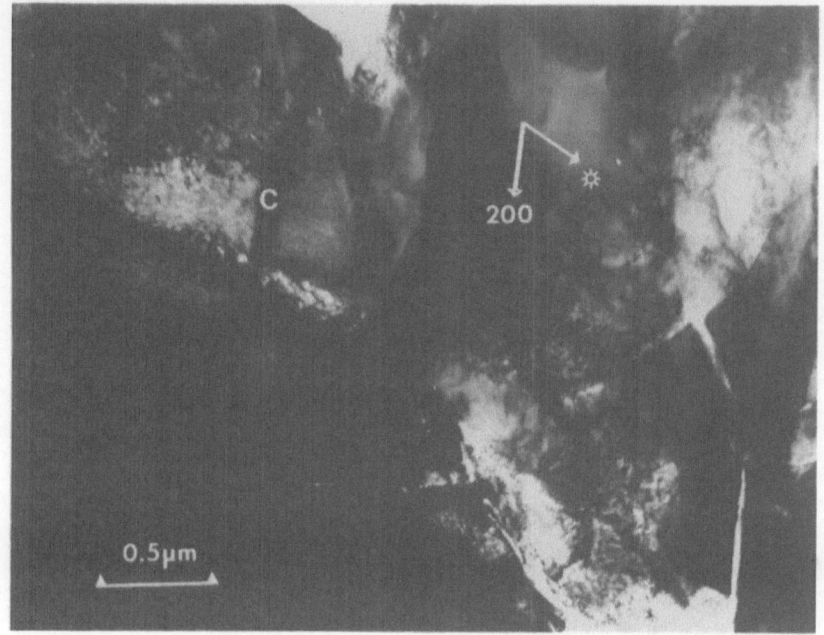

Fig. 10   Slip is visible here on (011), indicated by C
           and on planes not corresponding to any of the
           3 low index planes. One of the latter is in-
           dicated by *. The beam direction is [001].

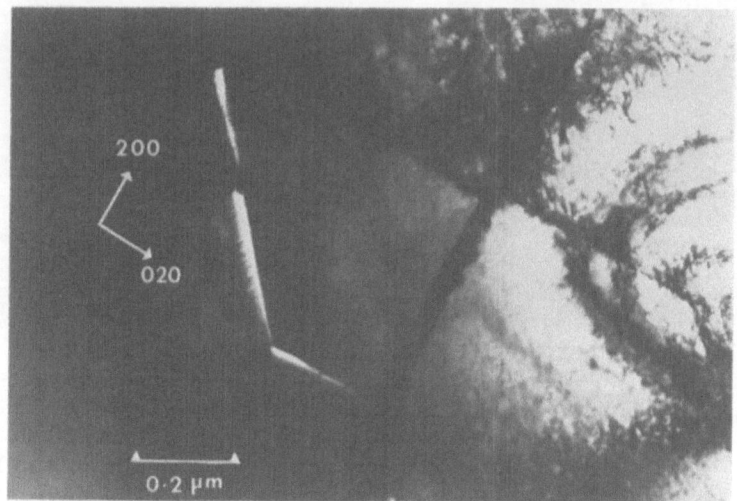

Fig. 11   A wedge shaped cleavage crack, on (110) is
           seen originating from the intersection of two
           {100} slip bands. Subsequently the crack
           appears to branch out along (100).

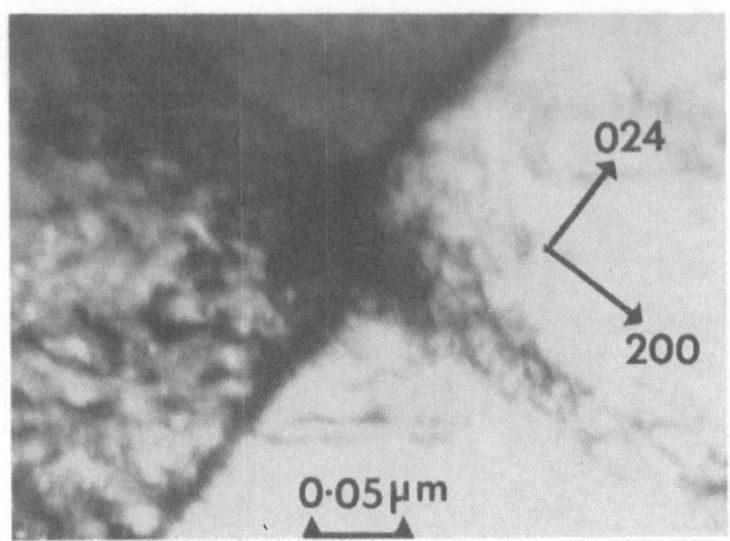

Fig. 12   The (100) slip trace seen on edge is displaced
             by ∿ 500Å where it is intersected by the ortho-
             gonal slip band suggesting that deformation
             initially spread on (100).

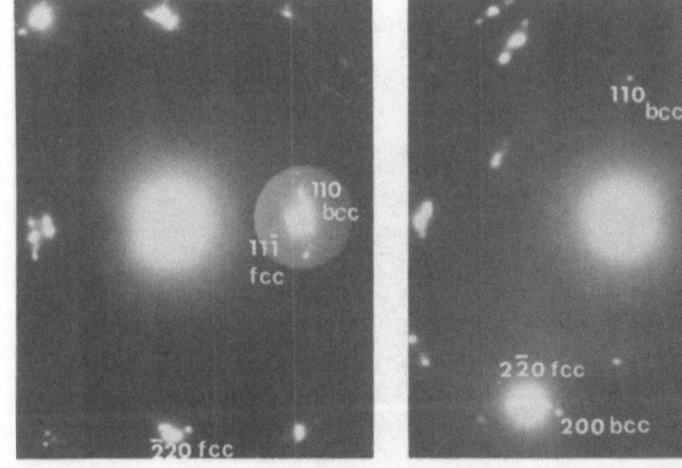

Fig. 13   Selected area diffraction patterns from the
             two-phase material. The electron beam direct-
             ion is [111] fcc on the left and [112] fcc on
             the right. The bcc precipitate 110 and 200
             Debye rings are concentrated near matrix single
             crystal spots suggesting an orientation relation-
             ship of the type

$$(110)_{bcc} \quad // \quad (111)_{fcc}$$
$$[1\bar{1}0]_{bcc} \quad // \quad [11\bar{2}]_{fcc}$$

showing moire fringes and interfacial dislocations.

Further work is in progress to refine the heat treatment and to investigate the deformation behavior of the two phase material.

3. Discussion

It has been shown from this study that the operative slip plane varies with position near the indentation and that slip can occur on all the three major low index planes. These facts are now discussed in terms of previous observations on the relationship between the structure and the deformation behavior in the rock-salt structure transition metal carbides.

In the group IV transition metal carbides, the metal atoms are stacked as in the fcc structure and the carbon atoms occupy the octahedral interstices. The bonding has been estimated to be a mixture of metallic and covalent types. Strong covalent bonds would tend to inhibit slip on {111} and favour slip on {110}, as in ionic crystals. With increasing metallic bonding, the plastic behavior is expected to be similar to that of fcc metals, i.e., slip would be expected on {111}.

The determination of the operative slip planes in the group IV transitional metal carbides has until this study, relied on the hardness anisotropy displayed on the cube faces of single crystals. In this method [16] Knoop hardness indents are made on an {001} face with the long axis of the diamond aligned along different crystallographic directions. When the preferred slip plane is {110} the hardness is a minimum when the long axis is along <100> while minima along <110> indicate the slip plane to be either {100} or {111}. This interpretation has been justified by observations on crystals, whose slip systems were previously known from other techniques [16].

For the stoichiometric carbides, such as TiC, and ZrC, Hannink et al have shown that a preference for {110} slip exists at room temperature [8]. Recent work by Breval has suggested that both {100} and {111}/{100} slip may occur. However, for the substoichiometric carbide, there is agreement from studies on $VC_{0.84}$ [7,8] and $TiC_{0.8}$ [7] that slip must occur on {110} and one or both of {111} and {100}. This is borne out by this TEM study which shows that slip can indeed occur on all three planes at room temperature. It should be noted that {100} slip has not generally been inferred from anisotropy measurements, although it is consistent with them. The reason for this lies in the fact that the high temperature slip plane is {111} and it has been felt that the transistion in hardness, which also occurs with rising temperatures, must therefore reflect the operation of {111} slip planes [7].

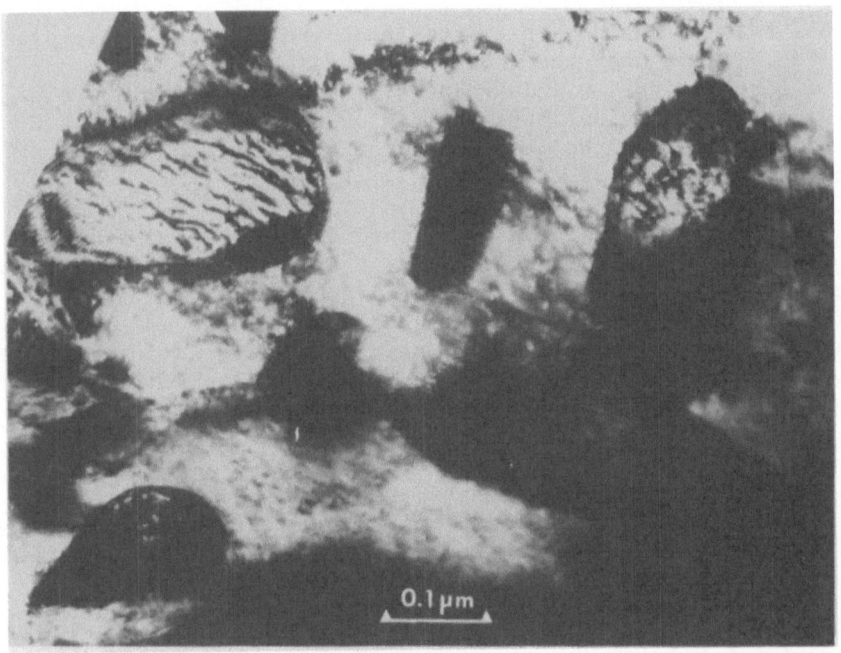

0.1 μm

Fig. 14   Bright field image of W-rich precipitates
           showing moire fringes and interfacial dis-
           locations.

The presence of slip on all three major low index slip planes and in addition on a fourth set of planes suggests that correlations between bonding and slip are more complex than believed. It should be noted that in the absence of ionic bonding, electrostatic arguments against slip on {100} are no longer valid. ZrC [17] has been induced to slip on {100} at 1400°C at critical resolved shear stresses that are only 1.5 x the values for slip on {111} and {110}. Other rock-salt structure compounds, where slip occurs on {100}, are PbS and PbTe [18], both of which exhibit a relatively low degree of ionic bonding. At this time there are no systematic studies on the energies required to nucleate and move a dislocation on various slip planes in this structure when the bonding is primarily a mixture of covalent and metallic types.

CONCLUSIONS

(1) Using transmission electron microscopy, slip has been observed around Vickers indentations in (W,Ti)C. For a load of 25g (W,Ti)$C_{0.8}$ exhibits slip on {111} and {110}, while for 300g (W,Ti)$C_{0.6}$ shows deformation on {100}, {110}, {111} and a fourth set of planes that is, as yet, unidentified.

(2) The observation of slip on all the three major low index planes is consistent with earlier inferences drawn from hardness anisotropy data, on other transition metal carbides having the rock-salt structure and deviation from stoichiometry e.g., $TiC_{0.8}$ $VC_{0.84}$.

(3) Indentations with loads of 300g produce more extensive plastic deformation which results in interactions between slip bands. Cleavage cracks have been found on both the primary {100} cleavage plane and on {110} at the intersection of {100} slip bands.

At other intersections slip bands suffer an abrupt displacement, suggesting that slip on different planes occurs at different times during the loading process.

ACKNOWLEDGEMENTS

The authors would like to thank D. J. Rowcliffe of SRI International for valuable discussions and Walter Precht of Martin Marietta for the single crystals of (W,Ti)C. Financial support from an NSF grant No. DMR-77-22647 is also gratefully acknowledged.

REFERENCES

1. H. E. Exner, "Physical and Chemical Nature of Cemented Carbides," International Metals Reviews, No. 4, 149 (1979).
2. W. S. Williams, J. Appl. Phys. 35:1329 (1964).
3. G. E. Hollox and R. E. Smallman, J. Appl. Phys. 37:818 (1966).

4. G. E. Hollox, Mater. Sci. Eng. 3:121 (1968-69).
5. D. J. Rowcliffe and G. E. Hollox, J. Mater. Sci. 6:1261 (1971).
6. D. J. Rowcliffe and W. J. Warren, J. Mater. Sci. 5:345 (1970).
7. D. J. Rowcliffe and C. E. Hollox, J. Mater. Sci. 6:1270 (1971).
8. R. H. J. Hannink, D. C. Kohlstedt, and M. Murray, Proc. Roy Soc.
    A326:409 (1972).
9. G. Morgan and M. H. Lewis, J. Mater. Sci. 9:349 (1974).
10. D. L. Kohlstedt, J. Mater. Sci. 8:777 (1973).
11. E. Breval, Ph.D. Thesis, Technical University of Denmark, (1980).
12. E. Rudy, J. of the L. Common Metals 33:245 (1973).
13. S. L. Shinde, Proceedings EMSA, 1980.
14. D. J. Rowcliffe, these proceedings.
15. R. J. Stokes, T. L. Johnson, and C. H. Li, Phil. Mag. 4:920
    (1959).
16. C. A. Brookes, J. B. O'Neill, and B. W. Redfern, Proc. Roy. Soc.
    A322:73 (1971).
17. D. W. Lee and J. S. Haggerty, J. Amer. Ceram. Soc. 52:641 (1969).
18. J. J. Gilman, Acta Met. 7:608 (1959).

DISCUSSION

C. A. Brookes:

Two questions, if I may. Did you control the orientation of the
indenter when you made your indentations? And secondly, how do you
define the criteria for 'extensive' nature of slip with regard to
anisotropy?

V. Jayaram:

To answer the second question, in terms of 'extensive', (I don't
know if this is justified) I was referring to the extent of minima
that have been predicted for knoop indentations on the cube face
along the <110> and the <100> directions. If the extent of the
minima reflect the tendency for a slip plane to operate, then one
really observes, particularly in the second case that we studied,
the 300 gram indent, the slip planes are of the <111> and <100> type
and very little of the <110> type. On the first question, the
second crystal was indented on a face that was not a well-defined
crystallographic face. The first crystal, the single crystal, was
indented on a cube face but I'm afraid we don't know the orientation
of the indenter.

C. A. Brookes:

As far as I know, the models which have been used to explain anisotropy don't attempt to explain the degree of anisotropy--only the
nature. And in fact, those materials where you get a very small
amount of plastic deformation are generally those where you get the
greatest degree of anisotropy.

C. H. de Novion:

We have recently observed that in $TiC_{0.6}$, if you anneal it a few
weeks at 700°C, it becomes rhombohedral with a distortion of about
one degree. There are four types of ordered domains in such a
compound. Do you think this will modify the mechanical properties?

V. Jayaram:

Have you observed domain structure within the material?

C. H. de Novion

We did not look in detail at the size of the domains. It's a few
hundred angstroms.

V. Jayaram:

Probably it would make the material more brittle or restrict the
amount of plastic flow due to the difficulty in accommodating the
strains of the various domains.

R. Sinclair:

I think in all the pictures that I've seen, there are no manifesta-
tions of the reaction which you describe. It would be difficult to
see with electron diffraction but would be very noticeable in transmission micrographs because the domain interfaces would show characteristic fringe contrast which is very easy to see. And as I said,
we've seen no manifestation of that in our materials.

C. H. de Novion:

Yes, but you have to anneal it one month.

H. Fischmeister:

My question may be premature because it would call for experimentation that perhaps nobody is prepared to do yet. But it strikes me
that really we would be interested in knowing how high loading rates
would affect the phenomena that we've heard about. So I'm trying to
find out if anyone is prepared to speculate or perhaps even has data
on what high loading rates do to dislocation movements in tungsten
carbide.

D. Rowcliffe:

We are planning to do experiments along these lines.

INDENTATION DAMAGE IN TUNGSTEN CARBIDE AND TUNGSTEN-TITANIUM CARBIDE

David J. Rowcliffe
Materials Research Laboratory
SRI International
Menlo Park California 94025

INTRODUCTION

Tungsten carbide (WC) is the main constituent of cemented carbides used for metal machining and rock drilling. Titanium carbide (TiC) is added to some grades of WC-Co to improve wear resistance in cutting metals. These grades contain both hexagonal WC and cubic (W,Ti)C solid solutions as the principal hard phases. In both applications high stresses and high temperatures are generated locally at the tool surface by contact with asperities or with rock fragments. Under normal rock-drilling conditions the cutting surfaces can reach temperatures up to 400°C [1] and temperatures as high as 800°C might be reached with dull bits or under unusually high loads. [2] Tool-tip temperatures up to 1000°C [3] have been recorded in metal cutting operations.

The deformation characteristics of pure WC and (W,Ti)C, particularly under conditions of contact loading, are not well known. Indentation effects have been studied in WC single crystals only at room temperature [4-7] and no studies have been reported on (W,Ti)C. In WC limited plastic flow can occur at room temperature and the primary slip system is $\{10\bar{1}0\}$ $\langle11\bar{2}3\rangle$. [5,6] Indentation can also cause microfracture and these effects have been described for single crystal WC deformed at room temperature on various planes with sharp and blunt indenters. [7] This work is extended in the present study which describes indentation effects in WC and (W,Ti)C single crystals up to 1100°C.

EXPERIMENTAL PROCEDURES

Single crystals of WC up to 5 mm in size were grown from solution in cobalt as described previously.[7] The crystals were generally idiomorphic plates with basal plane habit. Test samples of (W,Ti)C were cut from a single crystal 1 cm long x 1 cm diameter grown by floating-zone melting (Martin Marietta, Baltimore, Md.). The W/Ti atom ratio was approximately 2/3 and the carbon/metal ratio was 0.8.[8]

Some indentation tests were performed using a standard microhardness testing machine at room temperature. Others were made in the system shown schematically in Figure 1. The system consisted of an indenter and a furnace mounted on the cross-head of a mechanical testing machine. The blunt indenter was a 3 mm sapphire sphere cemented to an alumina rod. The sharp indenter was a Vickers diamond pyramid attached to a molybdenum rod. The crystals were mounted on molybdenum blocks using an intermediate layer of nickel powder and heating to 1450°C under argon. All specimens were polished before testing and set in an anvil attached to the load cell. Heating and indentation were made in an atmosphere of deoxydized argon and temperatures were measured within 2 mm of the surface of the specimen with an estimated accuracy of ± 5°C. To make an indentation, the specimen was heated to the test temperature and the cross-head moved down until the desired load was reached. The cross-head was then stopped for 25 seconds and reversed. This was therefore a constant displacement test in which load relaxation could occur, in contrast to the constant load conditions used in conventional hardness testing.

RESULTS

Tungsten Carbide

Figure 2 summarizes indentation effects at room temperature on the basal plane of WC. Plastic impressions with no visible cracks can be produced using low loads with both types of indenter. With the sharp indenter slip bands extend well beyond the contact zone and the slip traces are consistent with the established $\{10\bar{1}0\}$ slip plane. Cracks were first observed at 5 N, but not every crystal cracked at this load, as can be seen from Figures 2a and 2b. At higher loads, lateral cracks formed under the indentation can intersect the surface as shown in Figure 2b. In addition, radially oriented cracks on $\{11\bar{2}0\}$ planes can arise at the intersection of prominent slip bands. In the absence of lateral cracks close to the surface, the slip-induced cracks are relatively short. When lateral cracks

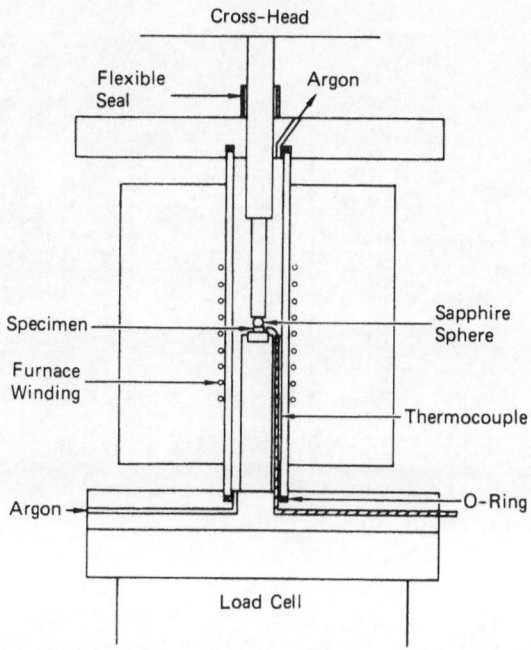

Fig. 1.   Schematic diagram of the system for indentation at high
          temperatures.

intersect the surface then the associated stress relaxation causes
the slip-induced cracks to extend to the periphery of the lateral
crack as can be seen in Figure 2b. A mechanism has been proposed for
the formation of the slip induced cracks[7] and details of the
dislocation interactions are presented elsewhere in these
proceedings.[9]

    With a spherical indenter the slip bands are concentrated within
the contact zone (Figure 2c), in contrast to the extensive surface
slip generated with a sharp indenter. This difference arises because
the stress field under the sphere is mainly compressive whereas a

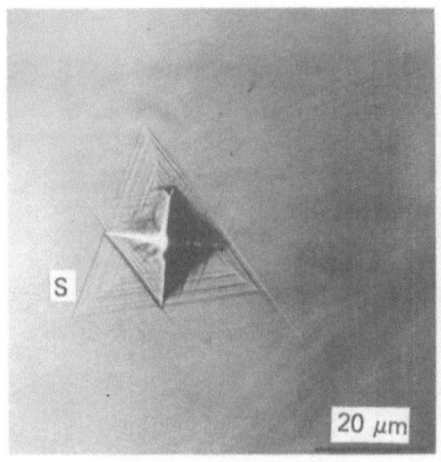

(a) Diamond Pyramid 5 N

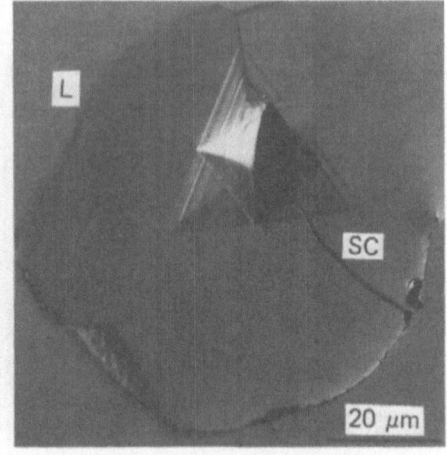

(b) Diamond Pyramid 5 N

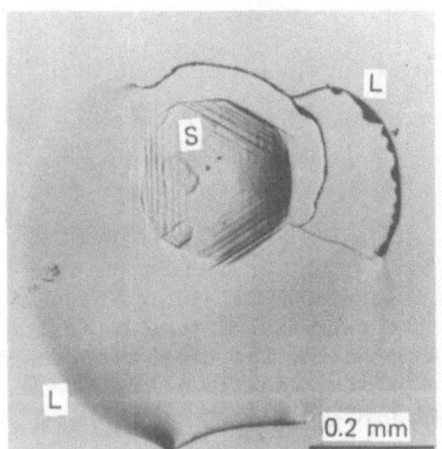

(c) Sapphire Sphere 450 N

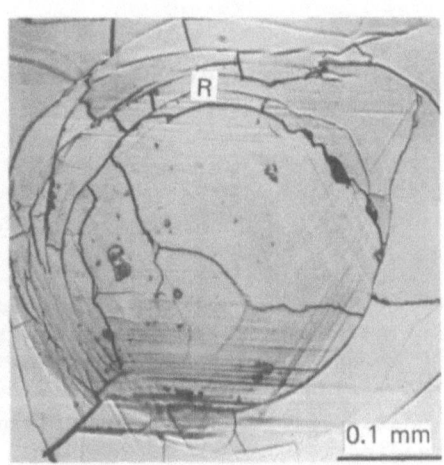

(d) Sapphire Sphere 450 N

Fig. 2. Indentations in WC at room temperature.
S-slipbands; L-lateral crack; SC-slip-induced crack;
R-ring crack.

large shear component is generated off the surfaces of the Vickers
pyramid. Slip-induced cracks are likewise absent at spherical
impressions. Loads around 400 N cause either ring cracks or lateral
cracks, as can be seen by comparing Figures 2c and 2d. Although both
types of crack can form at the same load, they are not found at the
same impression. This effect might be related to the local strength
of the crystal. Work hardening on the loading cycle could cause the
elastic fracture strength to be exceeded to give ring cracks and a
substantial relief of stress. However, if the crystal is able to
support the stress by plastic flow alone then the resulting high
residual stress is sufficient to nucleate lateral cracks on the
unloading cycle.Loads above 450 N produce slip, ring cracks, and
radial cracks.The radial cracks are not associated with slip band
intersections, but appear to be analogous with radial cracks seen at
spherical indentations in polycrystalline ceramics.[10]

The contribution of plastic flow to the indentation process
increases as the temperature of indentation is raised. Figures 3-5
show spherical indentations made up to 1000°C with loads of up to
450 N. An important observation is that the slip trace pattern is
unchanged between room temperature and 1000°C. This suggests that
slip occurs only on $\{10\bar{1}0\}$ planes over the whole temperature range.
In contrast to the deformation behavior at room temperature, extensive slip occurs outside the contact zone above 400°C, as illustrated
in Figure 3. At 1000°C slip extends a distance as much as three
times the diameter of the indentation. Another prominent feature is
the difference in shape between indentations made at room temperature
and those made at higher temperatures. At room temperature the
impressions are circular whereas at higher temperatures they are
strikingly hexagonal with little indication that the indenter was a
sphere. The hexagonal shape could arise from stress relaxation.
Once the cross-head has stopped the indenter cannot penetrate further
but the stress can be relieved by extending the slip-bands in a
direction tangential to the contact surface of the sphere, which
leads to the hexagonal shape. The corresponding relaxation of load
was seen on the load displacement chart of the testing machine. At
room temperature the load relaxation is approximately 5% of the
maximum load, whereas at 1000°C between 30 and 50% of the load is
relaxed during 25 seconds.

The formation of cracks depends on both temperature and load.
The temperature effect predominates because of increased plasticity
and the most sensitive range is below 400°C. Light loads or high
temperatures lead to plastic flow as the main mode of stress relief,
whereas at low temperatures or with high loads some cracking also
occurs. Crack-free impressions were consistently produced at 1000°C
for all loads up to 450 N. Lateral cracks are not observed at this
temperature probably because the stress is relieved by plastic flow
over a relatively large volume of crystal. The yield strength of WC

is apparently sufficiently low at 1000°C to prevent the build-up of
the large residual stress required for lateral fracture.

At 400°C, for loads up to 450 N, some indentations are crack-
free but others show a few small segments of ring cracks as well as
occasional lateral cracks, as indicated in Figure 3a. At 250°C, low
loads produce the crack-free hexagonal-shaped impressions, (Figure
4a) characteristic of high temperature indentation. Higher loads
result in lateral and ring cracking as can be seen in Figures 4b and
5a, but the incidence of ring cracking is small at this temperature. A comparison of Figures 5a and 5b indicates the temperature
sensitivity of crack formation at high loads. Radial cracks are
absent at 250°C but are well-developed at 150°C. Ring cracking is
also more prominent at 150°C but the ring cracks appear as small
segments in contrast to the more complete circular cracks seen at
room temperature.

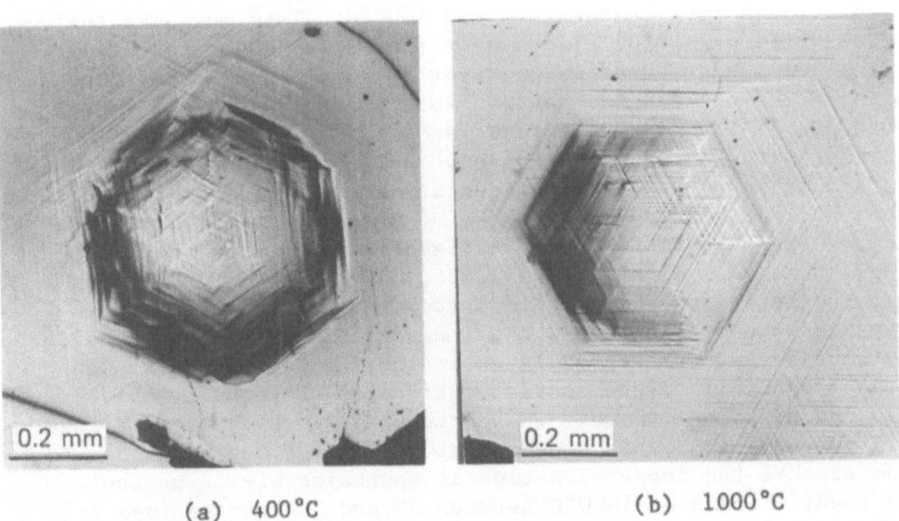

(a) 400°C   (b) 1000°C

Fig. 3.   Spherical indentations in WC with a load of 225 N at two
          temperatures.

Tungsten-Titanium Carbide

In Figure 6, the hardness of (W,Ti)C at room temperature is
plotted as a function of load, and Figure 7 shows examples of
corresponding indentations. The tests were made on {100} planes with
the indenter diagonals oriented in <010> directions. In contrast to
WC, extensive plastic flow does not occur in (W,Ti)C indented at room

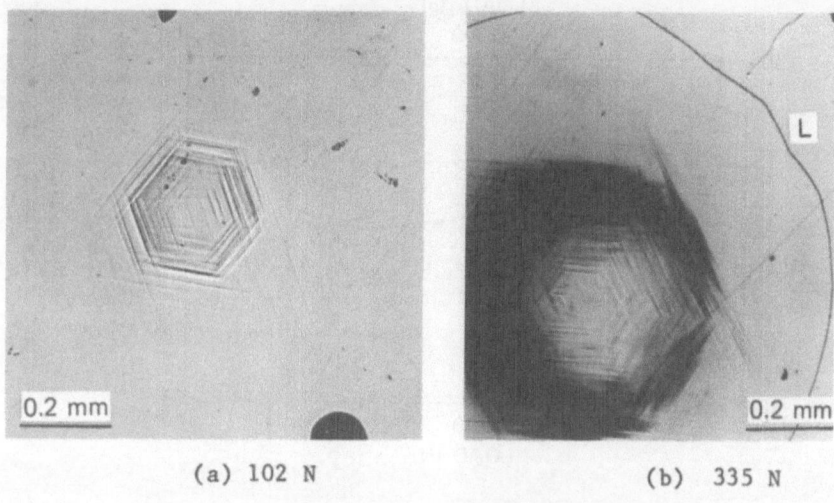

(a) 102 N   (b) 335 N

Fig. 4.   Spherical indentations in WC at 250°C at two loads.
          L-lateral crack.

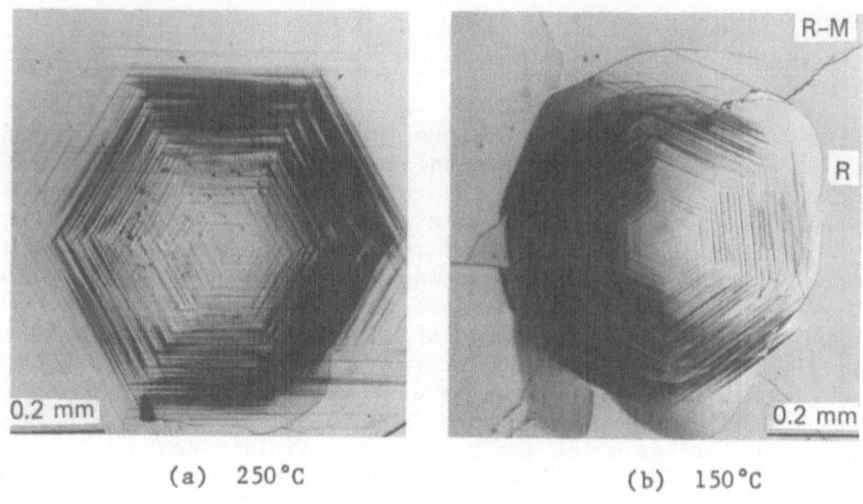

(a)   250°C   (b)   150°C

Fig. 5.   Spherical indentations in WC with a load of 450 N at two
          temperatures.
          R-ring crack; R-M-radial crack.

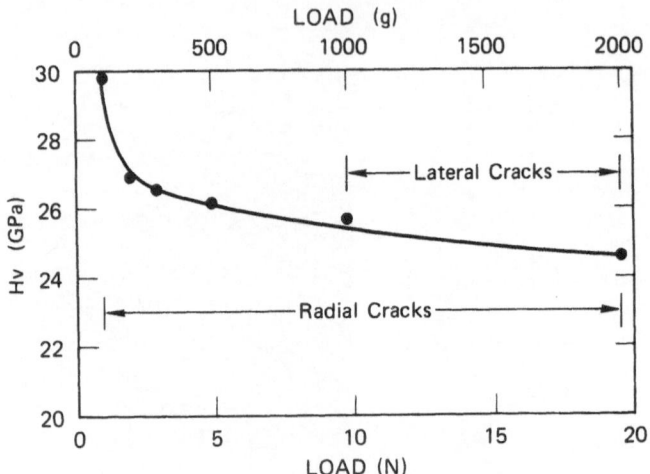

Fig. 6.   Vickers hardness of (W,Ti)C and the types of cracks formed,
as a function of load on the {100} planes.

temperature. With the Vickers diamond, radial cracks form at loads
at least as low as 1 N.   These cracks lie on {010} planes, the usual
cleavage plane for carbides with the rock salt structure.   Lateral
cracks begin to emerge at the surface at loads of 10 N, as shown in
Figure 7.   At higher load, pieces of the crystal become detached as
the lateral cracks intersect the surface and adjacent pairs of radial
cracks.   Above 20 N the damage was too severe for further hardness
measurements to be made.   The hardness is very sensitive to loads up
to about 10 N, the load at which lateral cracks first appear at the
surface.   Lateral crack formation is associated with the residual
stresses induced by the plastic zone immediately under the
indenter.   The load at which lateral cracks initiate probably
corresponds to the maximum plastic deformation that the crystal can
withstand.   Thus the hardness is not expected to decrease further
with increasing load, once lateral cracks begin to form.

Vickers indentations were also made above 600°C on {100} planes
of (W,Ti)C with a load of 20 N.   Radial cracks formed at all temperatures up to 1100°C, but lateral cracks were only visible at
impressions made at room temperature.   The absence of lateral cracks

at 600°C and above, again suggests that the yield stress is
relatively low, as indicated in Figure 8, and a much larger volume of
the crystal is being deformed plastically than at room temperature.

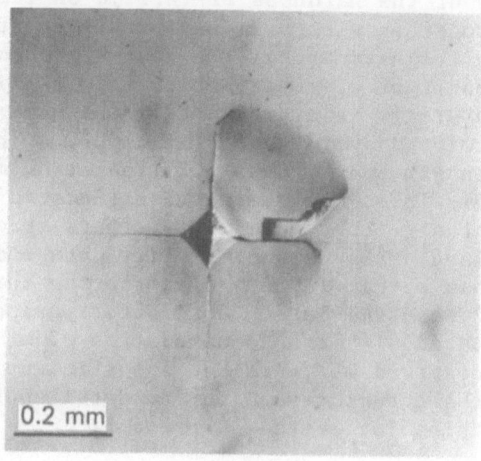

(a) 10 N

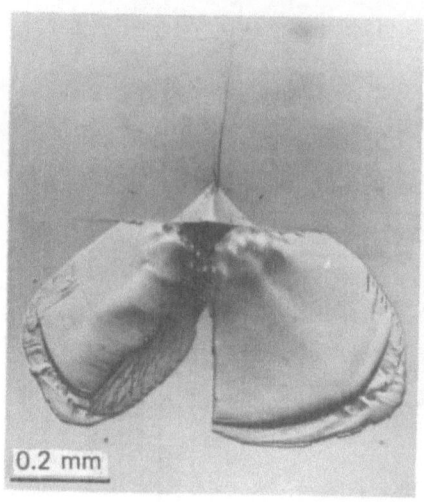

(b) 20 N

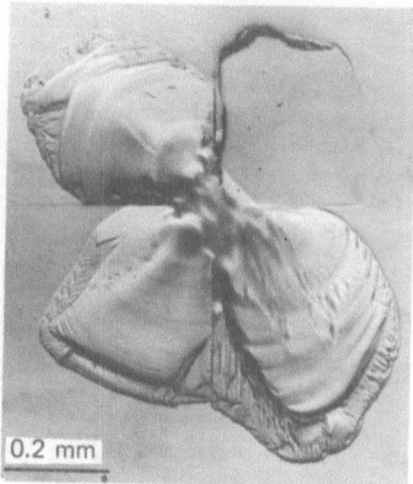

(c) 30 N

Fig. 7. Vickers indentations on the {100} plane at room temperature
         at various loads.

However, there was very little other visual evidence of plastic
flow. Curved traces were seen on the crystal surface immediately
adjacent to Vickers indentations made at 1000°C and 1100°C but their
origin has not yet been identified.

The dependence of the hardness of (W,Ti)C on temperature is
shown in Figure 8 together with data for single crystal $TiC_{0.96}$.[11]
At room temperature the hardness of the near-stoichiometric TiC is
somewhat greater than that of substoichiometric (W,Ti)C. The
hardness of both materials is approximately the same at 600°C but the
solid solution is twice as hard as TiC at 1100°C. The effect of
alloying on the strength and hardness of transition metal carbides is
not well understood. In some systems the hardness of the alloyed
carbide exceeds that of either parent; in others it does not.[12] In
TiC-VC, certain alloys[13] show a much smaller change of yield strength
with temperature than either TiC or VC, mirroring the hardness
behavior of (W,Ti)C reported here. Both substitution of metal atoms
and ordering associated with the accommodation of stoichiometry could
contribute to the observed hardness but insufficient information is
available both on the structure and on the details of slip in (W,Ti)C
to allow the data to be intepreted at present.

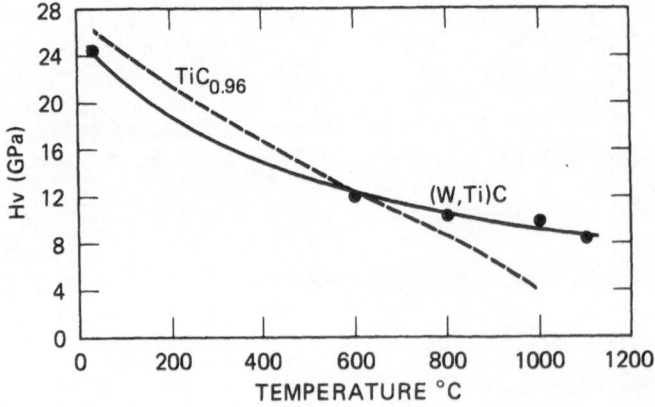

Fig. 8. Vickers hardness of (W,Ti)C as a function of temperature.

Permanent impressions could not be made in (W,Ti)C with a
spherical indenter at any temperature up to 1000°C. Ring cracking
occasionally accompanied by cleavage on {100} occurred up to 400°C
for loads up to 450 N. In the example shown in Figure 9, the
indentations were made on {100} and the effects of cleavage in
modifying the expected circular crack shape are evident. Higher
loads and higher temperatures inevitably resulted in failure of the
sapphire ball before a permanent impression could be produced.
However at 1000°C, some faint traces of what appeared to be slip
lines were seen in several specimens. These traces did not coincide

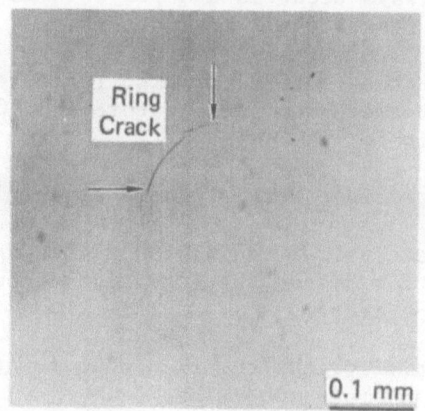

(a)   270 N

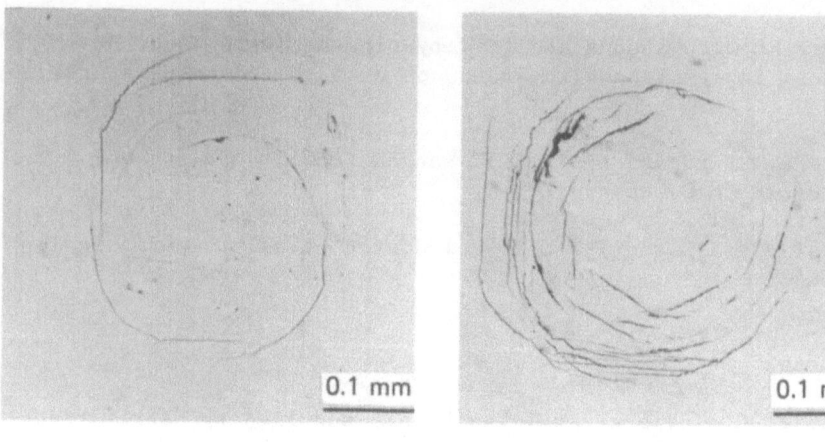

(b)   335 N   (c)   450 N

Fig. 9.   Spherical indentations on   {100} in (W,Ti)C at 250°C at
          various loads.

with those of a single family of low index planes, but a transmission electron microscopy study[8] has indicated that slip can occur on {100}, {110}, and {111} planes in (W,TiC).

CONCLUSIONS

1. Indentation damage was studied on the basal plane of WC single crystals between room temperature and 1000°C with loads up to 450 N.

2. Plastic flow always occurred and played a major role in accommodating the stress for all test conditions.

3. Slip-induced cracks, lateral cracks, radial cracks and ring cracks formed between room temperature and 400°C depending on the loading conditions.

4. Crack-free impressions were produced at all loads above 400°C.

5. Slip occurred exclusively on {10$\bar{1}$0} planes up to 1000°C and plastic relaxation caused the formation of hexagonal impressions above room temperature.

6. In (W,Ti)C, radial cracks occurred down to at least 1 N using a sharp indenter on the {100} plane.

7. Above 10 N lateral cracks emerged at the surface and interacted with radial cracks to cause surface spalling.

8. Ring cracks with shapes modified by cleavage were found at spherical impressions with loads up to 450 N and temperatures up to 400°C.

9. The Vickers hardness of (W,Ti)C ranged from 24 GPa at room temperature to 8 GPa at 1100°C.

10. Slip is very limited in (W,Ti)C indented at all temperatures up to 1100°C.

ACKNOWLEDGMENTS

    The (W,Ti)C single crystal was kindly supplied by Walter Precht of Martin Marietta. This research was supported by the National Science Foundation on NSF Grant No. DMR 77-22647.

REFERENCES

1. C. M. Perrott, Tool Materials for Drilling and Mining, Ann. Rev.
    Mater. Sci. 9:23 (1979).
2. C. M. Perrott, The Effect of Temperature on Performance of Hard
    Metal Rock Bits, Corros. Ind. 653:16 (1980).
3. E. M. Trent, Wear Processes which Control the Life of Cemented
    Carbide Tools, Iron and Steel Institute, Publ. 126, London
    (1970).
4. T. Takahashi and E. J. Friese, Determination of the Slip Systems
    in Single Crystals of Tungsten Monocarbide, Phil. Mag. 12:1
    (1965).
5. S. Luyckx, Slip System of Tungsten Carbide Single Crystals at
    Room Temperature, Acta Met. 18:233 (1970).
6. M. K. Hibbs and R. Sinclair, Room-Temperature Deformation
    Mechanisms and the Defect Structure of Tungsten Carbide, Acta
    Met. 29:1645 (1981).
7. S. Pattanaik and D. J. Rowcliffe, Indentation Deformation and
    Fracture in Tungsten Carbide, to be published.
8. S. L. Shinde, V. Jayaram, and R. Sinclair, Microstructural
    Characterization of Deformation and Precipitation in (W,Ti)C,
    these proceedings.
9. M. K. Hibbs, R. Sinclair, and D. J. Rowcliffe, Defect Structure
    of WC Deformed at Room and High Temperatures, these
    proceedings.
10. A. G. Evans and T. R. Wilshaw, Quasi-Static Solid Particle
    Damage in Brittle Solids-I. Observations, Analysis and
    Implications, Acta Met.24:939 (1976).
11. D. L. Kolhstedt, The Temperature Dependence of Microhardness of
    the Transition-Metal Carbides, J. Mater. Sci. 8:777 (1973).
12. G. Jangg, R. Kieffer, and L. Usner, Gewinnung von Mischkarbiden
    aus dem Hilfsmetallbad, J. Less Common Metals 14:269 (1968).
13. G. E. Hollox, Microstructure and Mechanical Behavior of
    Carbides, Mat. Sci.Eng 3:121 (1968/69).

DISCUSSION

C. A. Brookes:

If you indent silicon at room temperature and heat the indentation to
about 300°C you get slip lines developed on the surface. Have you
observed this sort of thing with the WC crystals? That's the first
question. And the second question is: The hardness of sapphire is
of the same order as tungsten carbide over the entire temperature
range. Did you observe any deformation of the spherical indenter?

D. Rowcliffe:

In answer to the second question: We never observed any plastic
deformation of the indenter. Sometimes with loads up to about 450
Newtons, the indenter would simply fail. But there was no sign of
plastic deformation even at 1000°C. Now to answer the other point--
slip line pattern is developed in WC at all temperatures, even
ambient.

W. Williams:

If titanium carbide single crystals are indented at room temperature
and heated to about 800°C, one then sees dislocation rosette patterns
on etching which are very similar to silicon and other materials. I
was going to comment on the 25°C impression which I think C.A.
Brookes was also interested in, which appeared to be larger in
diameter than the 150°C impression. But I guess that's because of a
ring crack which then begins to develop some local slip around it
due to stress relaxation, so that the total area appears to be
larger. Is that your interpretation?

D. Rowcliffe:

Yes, that's exactly what does happen. And if you remember, on one
of the slides, it shows an indentation at room temperature where
there is just a lateral crack and no ring cracks and another one
which is heavily ring cracked, and they're at the same load and the
one with the ring cracks is very much larger in diameter.

CRYSTAL CHEMISTRY AND ELECTRON ENERGY LOSS SPECTROSCOPY OF

TITANIUM CARBIDE PRECIPITATES IN TiB$_2$

W. S. Williams,* C. Allison† and P. Mochel

Materials Research Laboratory
University of Illinois at Urbana-Champaign
Urbana, IL, U.S.A.

INTRODUCTION

Titanium combines readily with boron, carbon, nitrogen and oxygen to form a variety of refractory compounds. Several have been used in cutting tool inserts--TiC, TiN and TiB$_2$. Although these compounds are generally treated as single-phase materials, they may contain second-phase precipitates which affect their mechanical properties at the high temperatures reached in service.

A well-known example is TiC with TiB$_2$ precipitates. The precipitate was first noticed by Williams, who reported it to be lamellar in morphology and lying on {111} planes in the NaCl-structure TiC.[1] He associated the precipitate with boron and argued from thermodynamic data and crystallographic considerations that it must be TiB$_2$. This identification was confirmed by Venables[2] using autoradiography, electron diffraction and transmission electron microscopy.

In this paper we discuss the converse case: TiB$_2$ crystals containing TiC precipitates. The identification of the precipitate in this case was particularly difficult as electron diffraction gave an ambiguous result. However, with the use of electron energy loss spectroscopy, it was possible to identify which light element was present in the precipitate.

We describe first the TiB$_2$ crystals used and optical studies of precipitates; then we give the electron microscopy and diffraction results, describe electron energy loss spectrscopy and its appli-

*Also with the Departments of Physics and of Ceramic Engineering
†Also with the Department of Physics

cation to the present problem, mention the crystallography of the matrix/precipitate relationship, review the thermochemistry of the system, and discuss the implications for cutting tools. A preliminare account of this work has been published.[3]

OPTICAL MICROSCOPY OF PRECIPITATES IN TiB_2 CRYSTALS

The TiB_2 crystals studied were grown by the Linde Division of Union Carbide Corporation by the arc-Verneuil technique in an argon atmosphere. They were known to contain lamellar precipitates from previous examination by optical microscopy.[4,5] Figure 1 shows the linear traces made by the lamellar precipitates as they intesect the surface of a TiB_2 crystal. The surface is parallel to the basal plane of this hexagonal-structure material. The precipitates are seen to be crystallographically oriented, making angles of 60° with respect to each other.

Thin foils of this crystal were prepared for electron diffraction and electron energy loss spectroscopy by diamond sawing, grinding and polishing, and finally, by ion thinning. Specimen preparation was a major part of this project, as TiB_2 is exceptionally refractory.

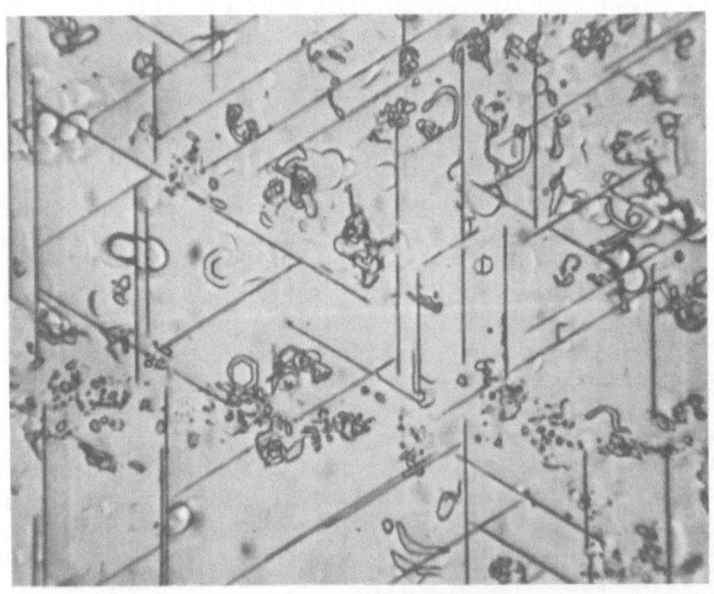

Fig. 1. Precipitates in TiB_2 crystal. Optical micrograph, 1500 X.

TRANSMISSION ELECTRON MICROSCOPY AND ELECTRON DIFFRACTION
OF PRECIPITATES

Thin foil specimens of TiB_2 crystals containing precipitates
were studied in a transmission electron microscope (TEM)(JEOL 200).
Figure 2 is an electron micrograph showing images of some interesting
features. The close correspondence between the electron image and
the optical micrograph suggests that the TEM features are images of
the same precipitates. Analysis of the images confirms that they do
represent precipitates. Electron diffraction showed an fcc pattern
superimposed on an hexagonal pattern (Fig. 3).

The hexagonal electron diffraction pattern corresponded well
with that known for TiB_2, while the fcc pattern could represent a
precipitate consisting of TiC, TiB, TiN or TiO. All have the same
structure and similar lattice constants. To resolve this question,
we employ electron energy loss spectroscopy, which is particularly
sensitive to light elements and has the spatial resolution to examine
a single precipitate.

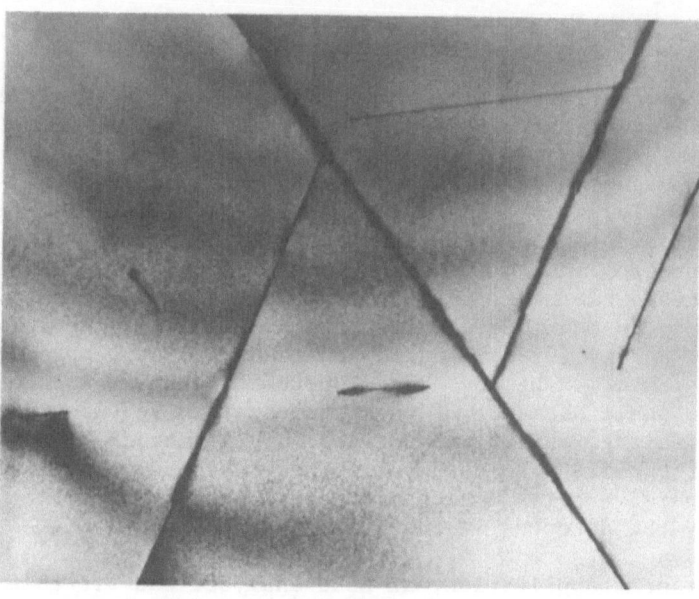

Fig. 2.   TEM micrograph of TiB_2 crystal containing precipitates.

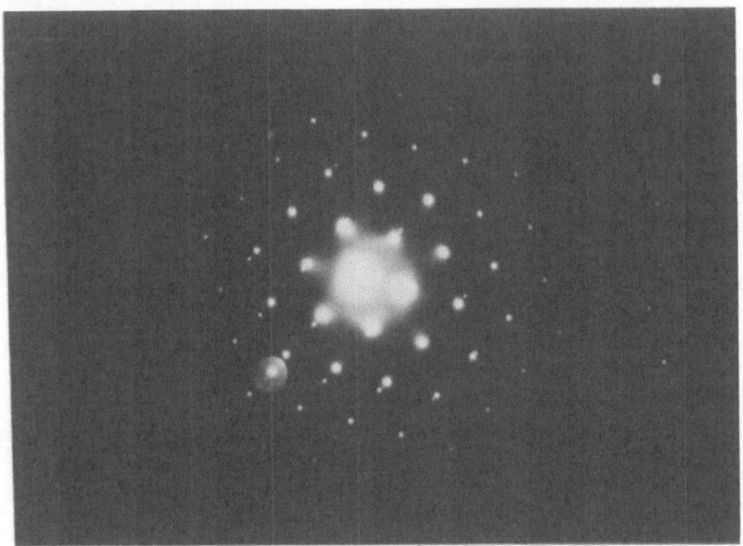

Fig. 3.   Electron diffraction pattern from precipitate and matrix.

ELECTRON ENERGY LOSS SPECTROSCOPY OF PRECIPITATES IN TiB_2

Electron energy loss spectroscopy (EELS) is an electron beam
technique which can perform elemental analyses of solids in a few
cubic Angstroms. It does this by determining how much energy has
been lost by a high-energy electron beam after passing through a
thin specimen foil. A spectrum of the number of electrons which have
lost a given amount of energy is displayed and reveals the presence
of "edges" corresponding to the energy required to excite characteristic X radiation--the basis for the elemental analysis. Thus EELS
is complementary to the more usual approach of identifying atomic
species by the energy (or wavelength) of the X radiation produced by
electron bombardment. The advantages of EELS include the opportunity
to exploit the small diameter of the focal spot in an electron microscope without enlargement of the excited volume through X-ray fluorescence, and the ability to detect light elements normally invisible
to energy-dispersive X-ray detectors because of the absorption of the
soft X rays by the detector window.

In the present application, an EELS analyzer was fitted to a
dedicated scanning-transmission electron microscope (STEM), (Vacuum
Generators HB-5). The beam energy was 100 kV, and the focal spot

was between 5 and 10Å.

With the very thin foil specimens prepared, (several hundred Å), only single scattering events occur--an important requirement for sharp energy-loss edges. Figure 4 shows the energy loss spectra obtained from the TiB_2 matrix and from the unknown precipitate. Only Ti and B edges appear in the spectrum from the matrix, as expected for TiB_2, whereas only Ti and C edges appear for the precipitate (Fig. 4). Since the diffraction pattern is fcc, the precipitate must be rock-salt structure TiC. The other candidate compounds are eliminated by the EELS elemental analysis.

DISCUSSION

The occurrence of precipitates of one titanium compound in another appears not to be an isolated or special circumstance. Fine-scale precipitates of TiB_2 in TiC single crystals, in commercial powders, and in hot-pressed bodies bave been observed.[1] A similar situation holds for the converse case:precipitates of TiC in TiB_2 bodies

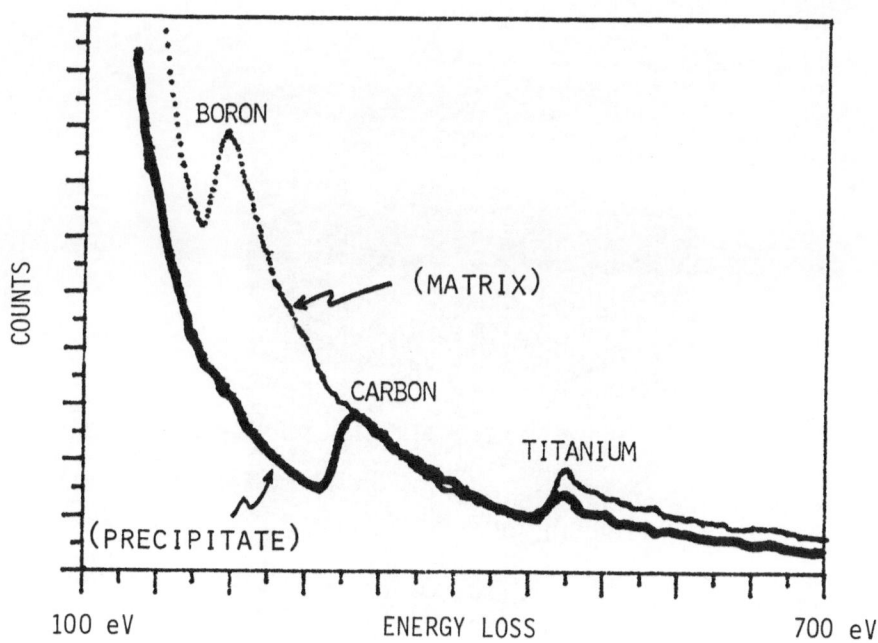

Fig. 4. Electron energy loss spectra for TiB_2 matrix and for precipitate. With inclusion of diffraction information, precipitate identified as TiC.

of various shapes and origins have been observed in the present work.
Single crystals of TiB_2, grown by the arc-Verneuil technique, were
the principal objects of study and displayed the lamellar precipitates characterized above. In addition, although the electron microscopy has not been completed, optical microscopy reveals the presence
of precipitates in arc-melted buttons of TiB_2 and in hot-pressed TiB_2
cylinders.

If such precipitation is a general occurrence in these systems,
several questions come to mind: 1) where do the precipitates come
from? 2) are they stable? 3) how do they affect the properties of
the matrix?

In the case of the boride precipitates in the carbide, the
source of the boride is contamination from some processing operation
that involved boron or the deliberate addition of boron through solid
state diffusion or vapor transport (Fig. 5). The boron that diffuses
in reacts with "excess" titanium in the nonstoichiometric carbide to
produce the compound TiB_2 and increase the carbon-to-metal ratio, x,
in TiC_x.[1]

Fig. 5.   Precipitates of TiB_2 in TiC crystal plastically-deformed at
          1500°C and exposed to boron source. Optical micrograph.

For the converse case of carbide precipitates in borides, the source of carbon is probably contamination from hydrocarbons in the furnaces used to prepare the objects from which specimens were taken.

In either case, the amount of second phase is minute: as little as a tenth of a percent can cause readily-detected precipitation. Indeed, this small amount of precipitation accounts in part for the difficulty in identifying the chemical and structural nature of the second phase, as X-ray diffraction of the bulk does not detect it.

Although the formation of titanium-containing second phase precipitates in the nonstoichiometric compound TiC_x is easy to visualize, it is less clear how they form TiB_2, which is regarded as a stoichiometric, "line," compound. There must be some "excess" titanium present in this case also to combine with carbon impurity to produce the observed TiC precipitates. Otherwise the carbon would be stable in the boride matrix, according to the ternary phase diagram for the system Ti-C-B[6] (Figure 6). However, in view of the minute amount of

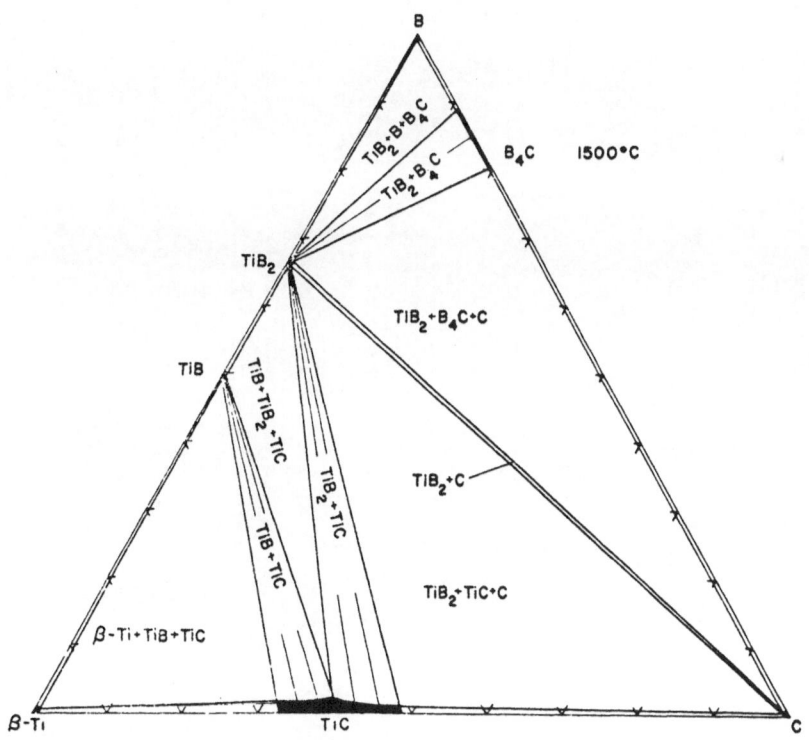

Fig. 6. Ternary phase diagram for the system Ti-C-B.[6]

excess titanium required, as indicated above, localized defects--e.g., stacking faults--might be sufficient.

Indeed, in the case of TiB$_2$ precipitates in TiC, Venables pointed out the similarity of the TEM image of the precipitate to a stacking fault in the matrix and showed that the precipitate is bounded by Shockley partial dislocations.[2] The titanium atoms are close packed in the {11.1} plane of NaCl-structure TiC and in the basal plane of hexagonal TiB$_2$. If the next layer is carbon, it forms another close-packed {111} plane appropriate for TiC; however, if the next layer is boron, it forms an hexagonal net appropriate for TiB$_2$. Thus, in this system, the precipitates lie on {111} planes in the TiC matrix.[1,2]

In the case of TiC precipitates in TiB$_2$ studied here, precipitation was in the prism plane of the matrix. The crystallographic similarity between the prism plane of hexagonal TiB$_2$ and the cube plane of fcc TiC, rotated 45°, makes this orientation plausible.[3,4]

A second kind of precipitate, lying in the basal plane, might also be expected, in view of the above discussion of the case of TiB$_2$

Fig. 7.   Precipitates lying on basal plane of TiB$_2$ crystal. Optical micrograph.

precipitates in TiC. Indeed, another type of precipitate appears to be present in specimens of TiB_2 studied by optical microscopy (Fig.7). In the isomorphic system ZrB_2, precipitates lying in both planes have been reported and justified crystallographically by Leombruno, Haggerty and O'Brian.[7] These precipitates were not identifiable with techniques available at that time, but in view of the present results, it is likely that they were ZrC.

Are these various kinds of precipitates stable after prolonged heat treatment at high temperatures? Although systematic experiments have not been carried out, it follows from thermodynamic evidence of the equilibrium co-existence of boride and carbide phases (Fig. 6) that the precipitates cannot be removed by annealing, though their distribution might be affected. The initial distribution and concentration of them depends on the initial dislocation density and the amount of carbon or boron impurity present.

So far the discussion of borides has dealt only with carbide precipitates. However, as the ternary phase diagram indicates, TiB_2 can coexist with boron if there is excess boron and no carbon, and with B_4C if there is excess boron and carbon. With rapid quenching, a cubic TiB can also be produced. Furthermore, if the stoichiometry is correct for TiB_2, and if there is excess carbon, then graphite flakes could in principle form. Thus there are several other possibilities for precipitation in the TiB_2 system. A precipitate in similar TiB_2 crystals was once reported as $TiB_{2.03}$ by Mersol, Lynch and Vahldiek.[5] However, in view of more recent analyses this identification is open to question.

What impact will precipitates have on mechanical properties? In the case of a TiC matrix, which suffers a rapid loss of yield stress with increasing temperature, the addition of a few tenths of a percent boron to produce TiB_2 precipitates increases the yield stress at 1600°C by a factor of six.[1] In the case of a TiB_2 matrix, a comparison study has not yet been possible because of the presence of precipitates in all crystals examined. However, compression testing of such specimens has failed to produce any gross plastic deformation at loads up to 40 kg/mm^2 and temperatures of up to 1500°C. This result is consistent with observations reported by Haggerty and Lee on the system ZrB_2 containing what we suggest were ZrC precipitates.[8]

SUMMARY AND CONCLUSIONS

The precipitation of transition metal carbides in borides and of borides in carbides has been reviewed. These precipitates form as a consequence of the thermodynamic stability of the two phases in contact and the crystallographic similarities of specific planes in each phase. Identification of the carbide precipitates has been acomplished using electron energy loss spectroscopy, which is particularly useful for light elements and has spatial resolution of a few Angstroms. Practical implications of these precipitates for resisting plastic deformation of the matrix at high temperatues have been recognized for the case of TiB_2 in TiC.[9] However, the converse case has yet to be fully explored.

ACKNOWLEDGMENTS

This work was supported in part by the U. S. Department of Energy under Contract DE-AC02-76ER01198 with the University of Illinois Materials Research Laboratory.

REFERENCES

1. W. S. Williams, Dispersion Hardening of Titanium Carbide by Boron Doping, Trans. Metal. Soc. AIME, 236:211(1966).
2. J. D. Venables, The Nature of Precipitates in Boron-Doped TiC, Philos. Mag., 16:873(1967).
3. P. Mochel, C. Allison and W. S. Williams, Study of Titanium Carbide Precipitates in Titanium Diboride by Electron Energy Loss Spectroscopy, J. Am. Ceram. Soc., 64:185 (1981).
4. W. S. Williams and J. D. Ruggiero, A Preliminary Study of Dislocations and Other Substructure in TiB_2 Single Crystals; unpublished work.
5. S. A. Mersol, C. T. Lynch and F. W. Vahldiek in "Anisotropy in Single-Crystal Refractory Compounds," Vol. 2, 41, F. W. Vahldiek and S. A. Mersol, eds., Plenum Press, New York (1968).
6. E. Rudy, Ternary Phase Equilibria in Transition Metal-Boron-Carbon Silicon Systems, Part V. Compendium of Phase Diagram Data, Air Force Materials Laboratory, Metals and Ceramics Div., Wright-Patterson AFB, Ohio, 606(1969).
7. W. J. Leombruno, J. S. Haggerty and J. L. O'Brian, Structure and Orientation of the Second Phase in ZrB_2 Crystals, Mater. Res. Bull., 3:361(1968).
8. J. S. Haggerty and D. W. Lee, Plastic Deformation of ZrB_2 Single Crystals, J. Am. Ceram. Soc. 54:572(1971).
9. U. S. Patent #3,497,368, W. S. Williams, assignor to Union Carbide Corporation (Feb. 24, 1970).

DISCUSSION

C. Politis:

Do you think it will be possible to make in this system spherical precipitaties about 200Å in size?

W. Williams:

I think spherical precipitates are very unlikely because of the low energy interface available between these two materials which have very similar crystal structures. It's essentially a coherent precipitate.

C. Esnouf:

Do you have any information about the possible influence of these precipitates on the resistance to corrosion by liquid aluminum?

W. Williams:

Not yet. But that is an area of considerable interest.

D. Moskowitz:

You mentioned you had some electron diffraction patterns of the carbide precipitates in TiB_2. I'm curious if the lattice parameter was anywhere near what it should be for pure TiC?

W. Williams:

It's close but not exactly correct. I do have the numbers.

HARDNESS MEASUREMENTS IN THE EVALUATION OF HARD MATERIALS

C.A. Brookes

Department of Engineering Science
University of Exeter
Exeter, UK

INTRODUCTION

The measurement of hardness is probably the most widely applied
method of evaluating the mechanical properties of solids. Its usage
ranges from techniques of quality control to such fundamental studies
as the identification of slip systems. Nevertheless, there are still
many aspects which have been neither thoroughly investigated nor
properly explained. Consequently, the full potential of hardness
measurement, in terms of our understanding of the behaviour of
materials and their properties, is far from having been realised.
In this paper we shall identify some of the aspects of hardness which
are especially relevant to the hardest materials and which would
benefit from further research. In particular, we shall be concerned
with the implications of anisotropy for our understanding of the
effects of varying normal load, temperature, and indentation creep.

EXPERIMENTAL CONDITIONS INFLUENCING HARDNESS MEASUREMENT

Crystallographic orientation – anisotropy

The crystallographic nature of plastic flow and the subsequent
pile-up of material, even in the hardest of single crystals deformed
at room temperature, may distort impressions made by spherical and
conical indenters. Furthermore, such indenters tend to develop
circumferential tensile stresses, around the region of contact,
which encourage brittle fracture and the resultant cracking is
detrimental to the accurate measurement of these materials. Consequently, one of the common pyramidal indenters, i.e. the Vickers,

Knoop or Berkovich, will be normally preferred. However, each of
these indenters will be liable to yield anisotropic results which can
be of two kinds. First, there is the observed variation of hardness
when the indented surface is changed from one crystal plane to
another for a given material. Secondly, there is the variation of
hardness observed when the orientation of the indenter is changed
with respect to a specific crystallographic orientation on a given
crystal plane. In the interests of accuracy, and as the best basis
for comparing the basic properties of crystals, it is clearly advisable to specify both the plane of indentation and the orientation of
the indenter when reporting hardness values.

Whilst all the common pyramidal indenters give anisotropic
results on single crystals, most work on indentation hardness anisotropy has been carried out using the Knoop indenter. This is
probably because the hardness value is determined on the basis of the
length of the long diagonal only and the alignment of this diagonal
with a specific crystallographic direction is self evident and
unambiguous. Also, the Knoop indenter is more likely to give a crack
free impression in those of the hardest solids which combine the
greatest tendency towards brittle fracture with the highest degree of
anisotropy. Reviews of the literature have demonstrated that, with
the exception of some anomalies due to indentation creep considered
later, the nature of anisotropy is determined by the slip systems
which control the indentation process[1,2,3]. Knoop hardness measurements for a selection of cubic crystals, each one chosen to represent
a common set of slip systems in these crystals, are given in Table 1
and it can be seen that the nature of anisotropy for the rocksalt
type crystals is the converse of the rest. It has also been
established that hexagonal crystals can similarly be placed in two
broad categories - one where the anisotropy is determined by
{0001} <1120> and the other by {1100} <1120> slip systems as in the
case of Al_2O_3 and SiC respectively.

The natural and consistent identification of the orientation of
the Knoop indenter, whatever the plane of indentation, is not shared
by the other pyramidal indenters. Thus, symmetrical constraints
imposed by the shape of the indenter will conflict with those due to
the crystal and confusing results will emerge. For this reason, it
would seem a wise precaution to confine the use of the Berkovich
triangular based indenter to the (111) planes and the Vickers square
based indenter to the (001)[4]. Furthermore, and for reasons which are
apparent from a consideration of the various models for explaining
anisotropy, it is advisable to identify the orientation of these
indenters with respect to one of their facets rather than a
diagonal[1→6]. Then, the nature of anisotropy for different indenter
may be directly compared as illustrated in Table 2. (Note that it
has been shown that the Vickers indenter will tend to give somewhat
higher values than the Knoop indenter for the same material[4].)

Table 1.   Knoop Hardness Anisotropy on (001) Planes of a
           Selection of Cubic Crystals, at Room Temperature,
           Representing Various Slip Systems

Crystal	Primary System	Knoop Hardness (GN m^{-2})		Other Examples
		[100]	[110]	
Copper (face-centred cubic)	{111} <1$\bar{1}$0>	0.47	0.34	Al, Ni
Diamond – Type II (diamond cubic)	{111} <1$\bar{1}$0>	103.00	91.00	Si, Ge
Calcium fluoride (fluorspar)	{100} <011>	1.78	1.57	UO_2
Molybdenum (body-centred cubic)	{110} or {112} or {123}, <111>	2.10	1.65	V, Nb, W, Fe(3%Si)
Magnesium oxide (rocksalt)	{110} <1$\bar{1}$0>	4.41	7.35	TiC, VC, HfC, ZrC, NbC

Table 2.   A Comparison of the Nature of Anisotropy for Knoop (H_k)
and Vickers (H_v) Hardness Values (in GN m^{-2}) on (001)
Planes of Cubic Crystals

Crystal	Temp.	Active Slip Systems	H_k [100]	[110]	H_v [100]	[110]	Ref
Titanium carbide	25°C	{110} <1$\bar{1}$0>	20.01	26.95	25.92	28.62	(7)
Titanium carbide	25°C	{110} <1$\bar{1}$0>	–	–	25.48	34.30	(8)
Titanium carbide	25°C	{110} <1$\bar{1}$0>	19.31	23.03	–	–	(2)
Vanadium carbide	350°C	{111} <1$\bar{1}$0>	13.72	10.78	18.03	16.61	(7)
Nickel oxide	25°C	{110} <1$\bar{1}$0>	4.12	5.25	5.18	6.25	(4)

Studies on the anisotropy in Knoop hardness, over the temperature range of -196 to 616°C, form an impressive demonstration of the
application of hardness measurements to the determination of active
slip systems in the transition metal carbides[7,9,10]. These materials
have the rocksalt type crystal structure with bonding that changes
from predominantly covalent to metallic at the higher temperatures.
Their anisotropic indentation hardness behaviour reflects this change
in bonding and can be separated into three different temperature
regimes. At the lowest temperatures, the anisotropy is consistent
with slip on {110} <1$\bar{1}$0> slip systems – i.e. the minimum hardness is
in [100] and maximum in [110] on the (001) planes as indicated in
Table 1. At the higher temperatures, the anisotropy is completely
reversed and can be explained by slip either on {111} <1$\bar{1}$0> or on
{001} <110> systems. At intermediate temperatures, it appears that
slip occurs on either one or the other of these slip systems depending on the orientation of the indenter. Observations on the slip
steps around the indentation, etch pits and individual dislocations,
using electron microscopy techniques, have confirmed that the slip
system in the high temperature regime is {111} <1$\bar{1}$0>.

Most of the models which consistently predict the measured
anisotropy in the Knoop hardness of crystals are based on the calculation of an effective resolved shear stress developed in the
bulk of the crystal and determined by the orientation of the
indenter facets with respect to the operative slip systems. For
example, Brookes et al.[1] suggested the following equation:

$$\tau_e = \sigma \cos \phi \cos \lambda \tfrac{1}{2}(\cos \psi + \sin \gamma) \tag{1}$$

where τ_e is the mean effective resolved shear stress developed by
the four facets; σ is the stress applied by the indenter which need
not be specified for considerations of the nature of anisotropy;

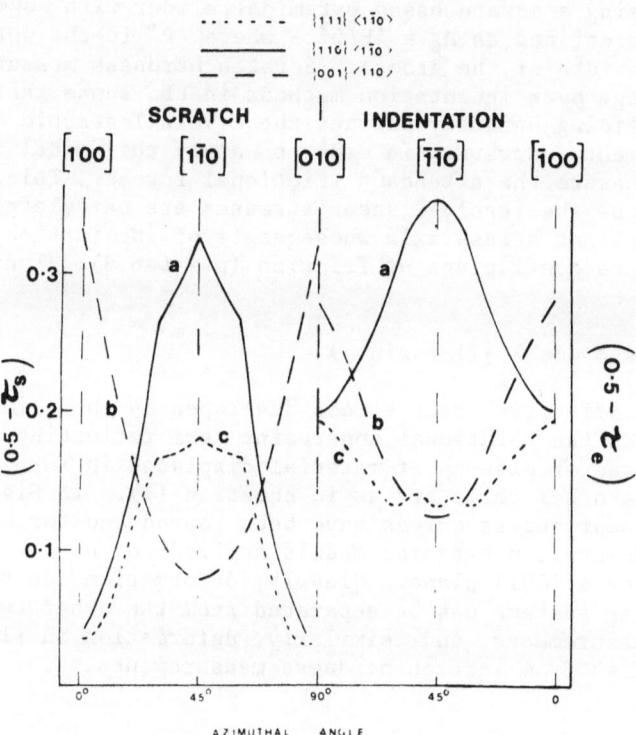

Fig. 1.   A comparison of the resolved shear stress curves for the
          scratch and indentation hardness models calculated for:
               (a)   {110} <1$\bar{1}$0>
               (b)   {001} <1$\bar{1}$0>
               (c)   {111} <1$\bar{1}$0>

cos ϕ cos λ is the Schmid;Boas factor; $\frac{1}{2}(\cos \psi + \sin \lambda)$ is a rotational constraint term reflecting the ease of displacement of material on to the surface and from beneath the indenter. Thus, the higher the value of the mean effective resolved shear stress, for a given normal load, the greater the amount of plastic flow and the lower the hardness. A similar model has now been developed for explaining and predicting anisotropy in the scratch hardness of cubic crystals under experimental conditions where a smooth parallel sided groove can be formed and the hardness expressed as a mean pressure[11]. For example, using a square based pyramidal slider with edge leading, the hardness is defined as $H_s = 4P/w^2$ – where 'P' is the normal load and 'w' is the width of the groove. Scratch hardness measurements have an advantage over indentation methods in the sense that the direction of sliding uniquely defines the crystallographic direction of the measurement. However, in order to apply this model it is necessary to measure the attendant frictional forces. This is necessary because the resolved shear stresses are calculated on the basis of a resultant stress axis whose angle of inclination (θ) is determined by the coefficient of friction ($\mu = \tan \theta$). The relevant equation is:

$$\tau_s = \sigma \cos \phi \cos \lambda \; \tfrac{1}{2}(1 + \sin \delta) \tag{2}$$

where τ_e is the effective shear stress developed by the slider; $\frac{1}{2}(1 + \sin \delta)$ is the rotational constraint term reflecting the nature and degree of pile-up of material displaced in forming the groove; and the other terms are as in equation (1). In Fig. 1 the resolved shear stress curves have been reproduced for both scratch and indentation hardness models applied to the three common slip systems for a (001) plane. Clearly, deformation due to {110} <1$\bar{1}$0> slip systems can be separated from the other two by indentation measurements, and, similarly, deformation on {100} <011> can be identified from scratch hardness measurements.

Normal load

The law of geometrical similarity implies that, for all pyramidal indenters, the hardness of a given material should be independent of the normal load. In general, this law holds for indentations made at normal loads above 2kgf. but at loads below this level, i.e. in the commonly accepted microhardness range[12], there are many exceptions[2]. The major part of the more reliable data shows that the hardness tends to increase with decreasing load (P) such that:

$$P = Ad^n \tag{3}$$

where 'd' is the diagonal length of the indentation; 'A' is a constant; and 'n' is less than 2.00. Deviations from the law of geometrical similarity are relatively small for metal crystals in that values of 'n' usually lie between 1.92 and 2.00[2]. However,

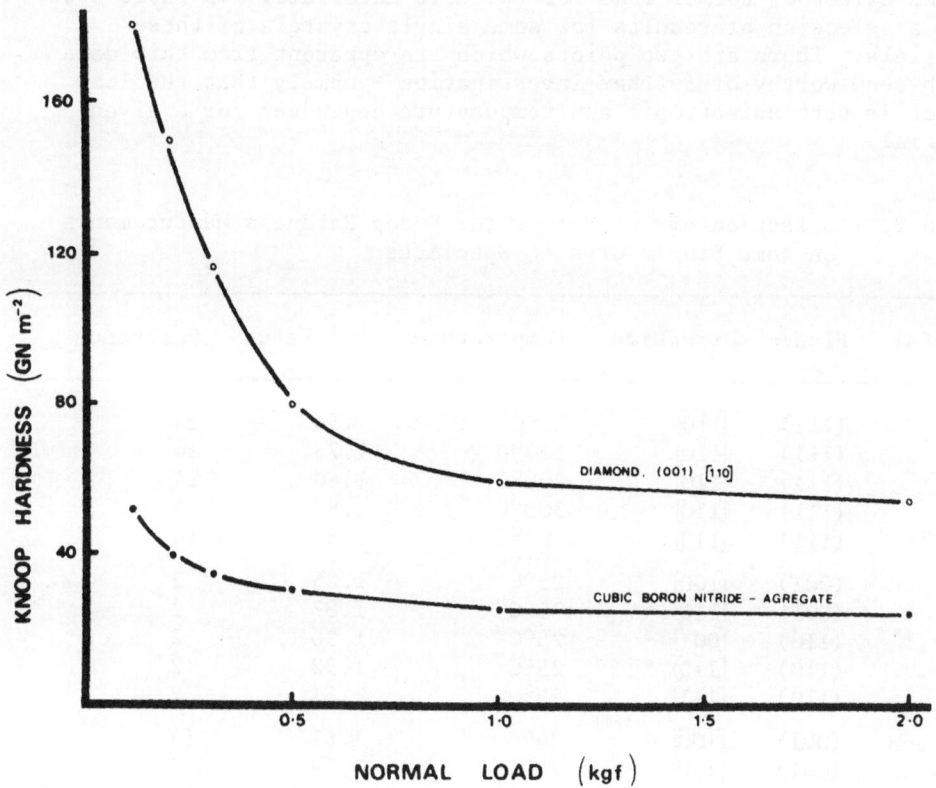

Fig. 2. The effect of normal load on the measured Knoop hardness of a diamond crystal and cubic boron nitride aggregate material.

this effect is more pronounced in the harder crystals and its importance is increased because the normal load has to be limited to less
than 2 kgf. in order to minimise the extent of cracking in these
solids. Figure 2 illustrates this effect for indentations on the
(111) plane of diamond using a Knoop indenter aligned in the $[1\bar{1}0]$
direction[13].

There is a limited amount of reliable published data, i.e. where
the orientation of the indenter has been controlled and reported,
on the effect of normal load for the hard materials. In Table 3 we
give a selection of results for some single crystals of these
materials. There are two points which are apparent from this data
which seem worthy of further investigation – namely that the load
effect is both anisotropic and temperature dependent for a given
material.

Table 3.   Selection of 'n' Values for Knoop Hardness Measurements
            on some Single Crystal Specimens

Crystal	Plane	Direction	Temperature	'n' Value	Reference
Si	(111)	$[1\bar{1}0]$	25°C	1.54	14
Si	(111)	$[1\bar{1}0]$	100°C	1.73	14
Si	(111)	$[1\bar{1}0]$	200°C	1.80	14
Si	(111)	$[1\bar{1}0]$	300°C	1.82	14
Si	(111)	$[1\bar{1}0]$	400°C	1.83	14
MgO	(001)	$[100]$	25°C	1.85	2
MgO	(001)	$[110]$	25°C	1.87	2
MgO	(110)	$[00\bar{1}]$	25°C	1.96	2
MgO	(110)	$[1\bar{1}0]$	25°C	1.92	2
MgO	(110)	$[11\bar{1}]$	25°C	1.81	2
Diamond	(001)	$[100]$	25°C	1.61	13
Diamond	(001)	$[110]$	25°C	1.49	13
TiB$_2$	$(10\bar{1}0)$	$[1\bar{2}10]$	25°C	1.69	cited in 2
Al$_2$O$_3$	(0001)	$[10\bar{1}0]$	25°C	1.80	14
Al$_2$O$_3$	$(11\bar{2}0)$	$[000\bar{1}]$	25°C	1.88	14
Al$_2$O$_3$	$(1\bar{1}00)$	$[000\bar{1}]$	25°C	1.87	14
Al$_2$O$_3$	$(1\bar{1}00)$	$[11\bar{2}0]$	25°C	1.72	14

The load effect is not only observed in single crystals but also in polycrystalline and aggregate materials. Relevant data for polycrystalline specimens include values of 'n' = 1.70 for SiC[16]; 1·92 for a bearing steel (EN 31)[14]; and 1·88 for a tungsten carbide:cobalt tool material.[14] A new aggregate material based on cubic boron nitride is probably worthy of special mention in this respect due to its unusual microstructure (see Fig.3). Note that this specimen had been leached in aqua regia to remove the metallic matrix phase and reveal the nature of interparticle contact and bonding. There are two particularly important differences between this type of material and the more common aggregates – such as the tungsten carbide cermets. First, the particle size is larger than the 1 to 5 microns commonly used in the cermets; the average size of the original cubic boron nitride particles was approximately 10 microns. Secondly, the method of manufacture is such that the primary nitride particles are joined by cubic boron nitride bridges grown as a result of a secondary synthesis during the aggregation process. Thus, the formation of the indentation cannot be wholly accommodated by the movement of rigid particles in a softer deformable matrix. Inevitably, the continuous skeleton of the hard phase must be deformed or fractured during the hardness test. The effect of normal load on the Knoop hardness of this material is shown alongside the results for the diamond crystal, in Fig.2, and the value of 'n' was 1.70.[15]

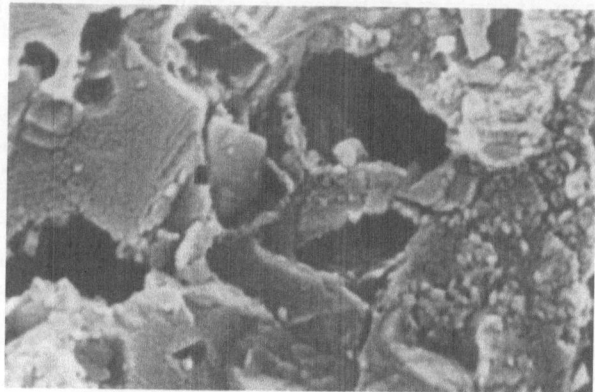

Fig. 3.   The microstructure of cubic boron nitride aggregate material showing the continuous skeleton of the hard phase after the removal, by leaching in aqua regia, of the matrix.

Varying the 'dwell' time

A typical loading cycle, in making a hardness measurement, involves the actual application of a given normal load over a period of about 18 seconds with a further 12 seconds during which the full load is applied. The latter period is often referred to as the 'dwell' time. In some cases, increasing the dwell time leads to an increase in the size of the indentation for a given normal load and hence an apparent reduction in the hardness. This effect is called indentation creep and may be identified with one of three types. First, there is conventional creep which is observed in all crystalline solids when they are deformed at temperatures above $0.4 \ T_m$, on the homologous temperature scale, and which increases with increastemperature. Secondly, there is the type which is mainly observed in ionic and covalent solids at room temperature (generally at low homologous temperatures and light loads) and which decreases with increasing temperature. This is often called anomalous indentation creep[17],[18]. Thirdly, there is that which is observed in some crystalline solids at comparatively low homologous temperatures, up to about $0.4 \ T_m$, but which increases with increasing temperatures.

Morgan[19] has investigated this latter type of indentation creep for a range of metallic, ionic and covalent solids. His work has shown that the effect cannot be predicted or anticipated on the basis of the criteria which govern conventional creep mechanisms. For example, at room temperature ($0.15 \ T_m$) measurable indentation creep was observed for Knoop indentation hardness measurements in the [110] directions but not in the [100] on the (001) plane of MgO[20]. Furthermore, in Figure 4, it can be seen that, whilst there is indentation creep in both directions at 730°C ($0.33 \ T_m$), the rate of creep is still anisotropic. This is a general feature of the rock-salt type crystal structures which slip on {110} <1$\bar{1}$0> and has been used to explain the apparent anomaly in the nature of anisotropy at room temperature of the lower melting point alkali halides. Recently, Kumashiro et al[8] have reported anisotropic rates of indentation creep for Vickers hardness measurements in titanium carbide crystals deformed at 900°C ($0.27 \ T_m$). Figure 5 shows the results obtained by Morgan[19] for silicon crystals deformed at 300°C ($0.34 \ T_m$) and 500°C ($0.34 \ T_m$), indicating a typical increase in the rate of creep, and for a crystal of aluminium oxide at 300°C ($0.25 \ T_m$) reflecting a surprisingly high rate of creep for a comparatively low homologous temperature. Preliminary results[15] for an aggregate of cubic boron nitride are also shown in Figure 5 and again a surprisingly high rate of creep is observed at a temperature of 500°C ($\sim 0.21 \ T_m$).

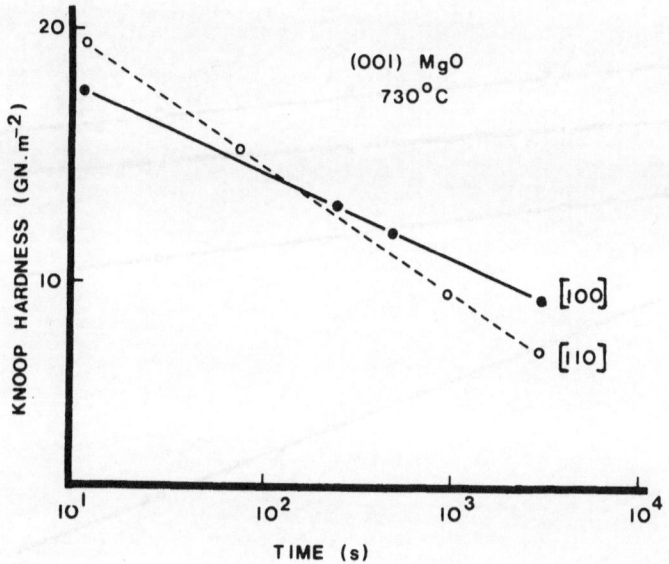

Fig. 4. Anisotropic rates of indentation creep on a (001) plane
of MgO at 730°C (0.33 T_m).

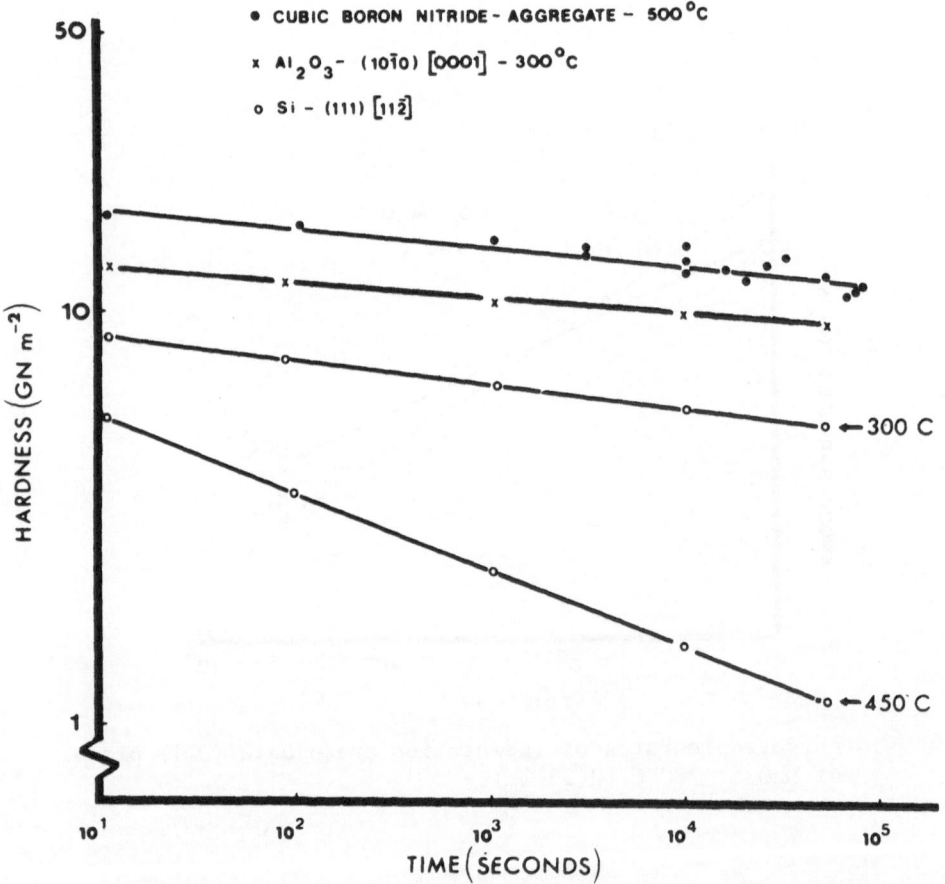

Fig. 5. Indentation creep in cubic boron nitride aggregate material
and single crystals of silicon and aluminium oxide.

Effect of indentation temperature

Some aspects of the effect of temperature have already been described earlier. There appears to be three main consequences for the anisotropic hardness of crystals as a result of increasing the experimental temperature. Two of these have been mentioned; both may actually change the nature of anisotropy. The first reverses the anisotropy for standard (12s) indentations and is due to a change in the active slip systems. This is a common effect in transition metal carbides and is illustrated by the results of

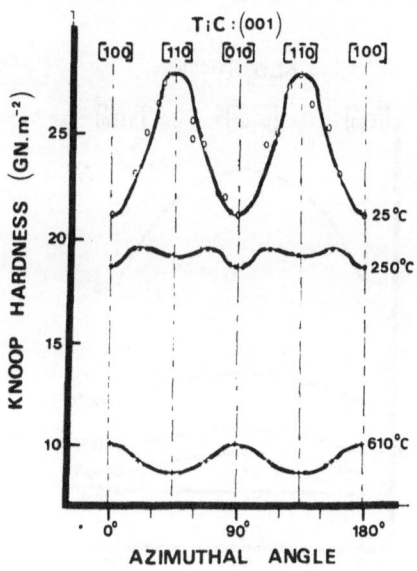

Fig. 6. Showing the reversal in the nature of hardness anisotropy, reflecting the change in the operative slip systems, for titanium carbide crystals.[7]

Hannink et al[7] for titanium carbide crystals and reproduced in
Fig. 6.   Secondly, there is the possibility of a reversal in anisotropy as the result of the indentation creep phenomenon - as in the
case of the rocksalt type crystals and indicated by the results
shown in Fig. 4.   Finally, there is the more common occurrence
where the nature of anisotropy is maintained but the degree is reduced.   This type of behaviour is illustrated in the results for
Al_2O_3, shown in Fig. 7, over the temperature range of $25^\circ C$ to
$1050^\circ C$ (0.13 to 0.57 T_m).[2]

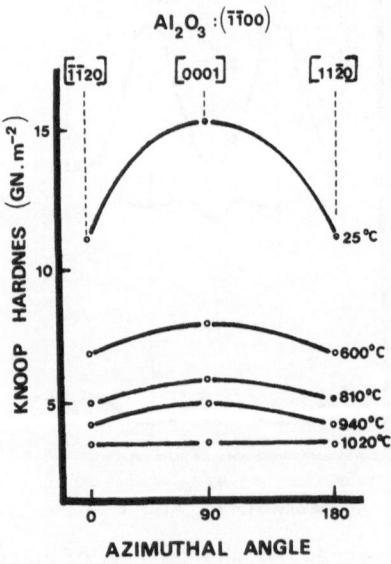

Fig. 7.   Decreasing degree of anisotropy in hardness with increasing
          temperature for Al_2O_3 crystal.[2]

SOME ASPECTS OF DEFORMATION

Studies on the fracture mechanics of the indentation process
constitute an expanding and interesting area of current research but
they are outside the scope of this paper. Nevertheless, it is worth
noting that the extent and nature of crack formation, even though
such cracks do not interfere with the hardness measurement, is often
anisotropic. Figure 8 demonstrates this effect for Berkovich
indentations on the (111) planes of (a) calcium fluouride with
{001} <110> slip systems and (b) cubic boron nitride whose slip
systems are not known. The orientation of the facets on this indenter, with respect to the crystal surface, is the same for both
indentations on a given specimen. However, their orientations are
different in terms of the resolved shear stresses developed on the
active slip systems in the bulk of the crystal. In the case of
calcium flouride, one indentation leads to the formation of radial
cracks whilst there are no cracks formed for the other (harder)
orientation. Also, whilst both orientations give radial cracking in
cubic boron nitride it is apparent that one of the indentations has
also produced some subsurface fracture. In general, we have observed
that, in those solids with a tendency to brittle behaviour, indentations in the soft directions on a given crystal plane are more
inclined to produce cracks than those in the hard direction. However,
it should be emphasised that there is no evidence to support the view
that such cracks contribute to the anisotropy in hardness effect.

Fairly extensive studies have been carried out in this laboratory
on the zone of deformation, as revealed by dislocation etch pit techniques, produced by indentation. We have observed that the load
effect is accompanied by a distinct change in the geometry of dislocations produced in a given crystal. This is illustrated in Figure
9 for Vickers indentation at loads of 0.1 kgf. and 0.5 kgf. in a
[110] direction on a (001) plane of MgO[2]. It is apparent that the
mobility of edge dislocations decreases with respect to screw
dislocations at decreasing normal loads. Also, we have observed that
the depth of the deformation (or dislocated) zone is virtually
independent of orientation and shape of indenter for a given crystal
and normal load. Indeed, provided that its hardness is above a
certain threshold, a softer - and therefore blunt - indenter or slider
will produce the same depth of dislocated material as a rigid indenter. It has been shown that cumulative deformation by the small scale
multiplication and movement of dislocations, induced by repeated
loading of softer solids against hard ones, may workharden materials
which would normally fracture[21]. The degree of workhardening obtainable by such means has been of the order of 30% in such diverse
materials as single crystals of MgO, EN 31 bearing steels and a
tungsten carbide:cobalt tool tip[14].

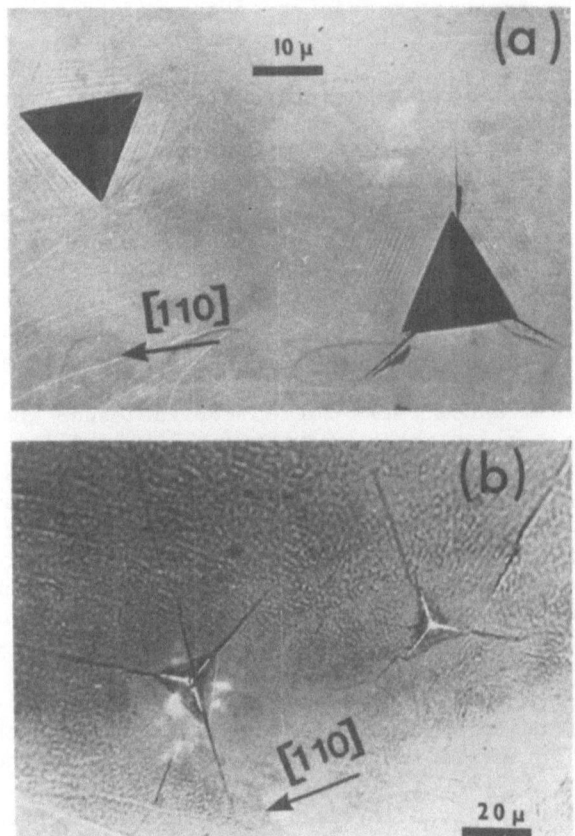

Fig. 8.   Showing the effect of Berkovich indenter orientation on
the formation of cracks in (111) planes of
(a) calcium fluoride and (b) cubic boron nitride.

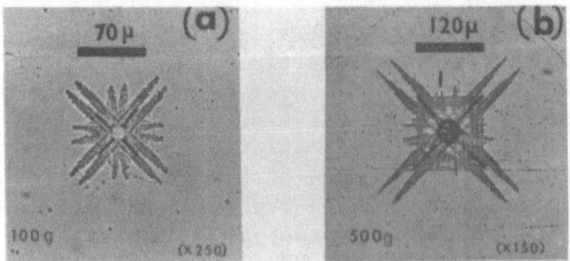

Fig. 9.   Vickers indentations, with facets aligned in [110] on a
(001) MgO plane, etched to reveal dislocation pattern
with normal loads of (a) 100g and (b) 500g.

SUMMARY

In this paper we have attempted to demonstrate that anisotropy
in hardness is an important characteristic to consider when evaluating
the mechanical properties of hard crystalline solids. This phenomenon
has implications for other effects due to variations in the normal
load, indentation dwell time and temperature. It is apparent that our
understanding of the indentation process will be furthered and the
technological application of hardness measurements will be advanced
when these effects have been more thoroughly researched.

ACKNOWLEDGEMENTS

The author would like to acknowledge the help of his colleagues
Dr R.M. Hooper, J.D. Ross and W.A. Lambert and thanks De Beers
Industrial Diamond Division for a grant to the laboratory.

REFERENCES

1. C. A. Brookes, J. B. O'Neill and B.A.W. Redfern Proc. Roy. Soc
   A322, (1971).
2. R. P. Burnand, Ph.D. dissertation, University of Exeter, (1974).
3. B. Moxley, Ph.D. dissertation, University of Exeter, (1974).
4. C. A. Brookes and B. Moxley, J. Phys. E 8 (1975).
5. F. W. Daniels and C. G. Dunn Trans. Am. Soc Metals 41:419 (1949).
6. J. Pospiech and J. Gryziecki, Archiwum Hutnictwa XV, No. 3:267
   (1970).
7. R. J. Hannink, D. Kohlstedt, and M. Murray, Proc. Roy. Soc.
   A326:409 (1972).
8. Y. Kumashiro, A. Itoh, T. Kinoshita, and M. Sobajima J. Mater.
   Sci. 12:595 (1977).
9. G. E. Hollox and D. J. Rowcliffe, J. Mater. Sci. 6:1261 (1971).
10. G. Morgan and M. H. Lewis, J. Mater. Sci. 9:349 (1974).
11. C. A. Brookes and P. Green, Proc. Roy. Soc. A368:37 (1979).
12. B. W. Mott, "Micro-Indentation Hardness Testing," Butterworths,
    London, (1956).
13. C. A. Brookes, "Indentation Hardness of Diamond, Properties of
    Diamond," J. E. Field, ed., Academic Press, London (1979).
14. C. A. Brookes and J. Ross, unpublished work.
15. C. A. Brookes and W. A. Lambert, to be published.
16. M.G.S. Naylor and T. F. Page, "Proc. 5th Int. Conf. on Erosion by
    Solid and Liquid Impact," Cambridge (1979).
17. J. H. Westbrook and P. J. Jorgensen, Trans. AIME. 230:613 (1965).
18. A. R. C. Westood, D. L. Goldheim, and R. C. Lye, Phil Mag. 16:505
    (1967).

19. J. E. Morgan, Ph.D. dissertation, University of Exeter, (1976).
20. C. A. Brookes, R. P. Burnand and J. E. Morgan, J. Mater. Sci.
    10:2171 (1975).
21. C. A. Brookes and M. P. Shaw, Nature 263, No. 5580:760 (1976).

DISCUSSION

R. Rice:

My question is on the load dependence of hardness, which I also
believe is a real effect. But other extraneous factors can contribute to it. Am I correct in assuming that you have already subtracted
the elastic correction and also a correction due to experimental
error in determining the length of the diagonal from your data, or
were you looking at a composite of the data?

C. A. Brookes:

No, we've used it in the way that most people would use it for hardness measurement and we've assumed that these results you see are for
the knoop indentor. We assumed that the long diagonal doesn't change
in length. I know that is open to some question. But, in fact,
we've looked at the length of the indentation with the load on, in a
transparent crystal like magnesium oxide, and with the load off, and
we've not been able to measure a difference in the length of the
diagonal.

R. Rice:

The other question I had concerns the nature of the surface finish.
Paul Becker in my laboratory a number of years ago showed that you
could get a rather significant apparent load dependence of Vickers
hardness primarily, although I believe he also observed it in knoop
hardness, of titanium dioxide and aluminum oxide as a result of very
limited amounts of plastic flow in the surface from the polishing
processes. He was able to demonstrate this by comparing the hardness
as a function of load on single crystals that had been chemically
polished versus those that had been mechanically polished and therefore this could also be an important contribution to the load
dependence. So I'm curious as to what your surface preparation
techniques were and whether this might be at least a contributing
factor to your load dependence as well.

C. A. Brookes:

We did a fairly extensive investigation of these effects for magnesium oxide in which we compared surfaces which were cleaved in air,
surfaces which were freshly cleaved in toluene and surfaces which
were chemically and mechanically polished. Although the absolute
hardness measurements did vary somewhat, the effect was fairly
consistent.

C. Yust:

Can you elaborate on the behavior of the Berkovitch indentation in
calcium fluoride? It is cubic and I don't really understand why the
inversion of the indentor would make any difference.

C. A. Brookes:

Yes, it's a cubic crystal. It's a (111) plane that we're indenting,
and as far as the surface is concerned, those two orientations should
give the same effects. But the indentation process is determined not
by the surface, but by resolved shear stresses in the bulk of the
crystal. Then the two senses of the same direction will, in fact,
give different resolved shear stresses.

MECHANICAL BEHAVIOR AND ELECTRON MICROSCOPY ANALYSIS OF W_2C

J. Dubois, T. Epicier, C. Esnouf, and G. Fantozzi
Groupe d'Etudes de Metallurgie Physique et de Physique des
Materiaux - E.R.A. 463 - I.N.S.A. Batiment 502 - 69621
Villeurbanne Cedex - France

INTRODUCTION

Up to now, a relatively large number of studies has been made on the mechanical behavior of transition metal carbides with the face centered cubic structure (NaCl type). A summary of studies on plasticity until 1970, has been made by Toth.[1] Since then, only tantalum carbide and titanium carbide have been the subject of new works, by Martin[2] and Chermant.[3] In contrast, plastic deformation of hexagonal carbides has been investigated very little. Some studies on the mechanical behavior or microstructure of WC and Mo_2C have been reported.[4-11] Therefore it appears to us of major interest to undertake a study of plastic deformation of W_2C at high temperatures (between 800° and 2200°C). In the present study, electron microscope observations (100 KV and 1 MeV) and results of mechanical deformation tests are associated so as to determine the process responsible for the plastic deformation of W_2C.

Bending experiments have been performed on thin plates of carbide (25 x 4 x 0.3 mm) obtained by carbon diffusion into tungsten sheets.[12] These specimens are polycrystalline with large grains (average grain diameter 0.1 mm) and can contain some very small regions of WC because the carbon content is near stoichiometry. The chemical analysis is given elsewhere[13] and the crystal structure of W_2C is shown in Fig. 1.

The carbide samples have been plastically deformed to strains (ε) of 10^{-2} on a four point bending micromachine described previously.[2-14] This machine is sufficiently rigid (rigidity coefficient is of the same order as the elastic modulus at the temperature of the test) and of high sensitivity to enable us to make all the mechanical

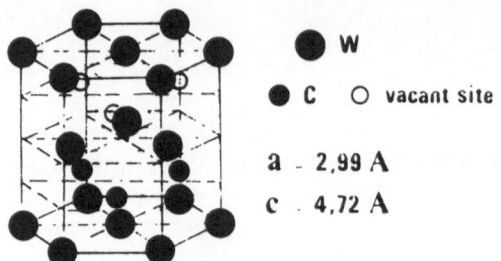

Fig. 1 : Crystal Structure of W_2C

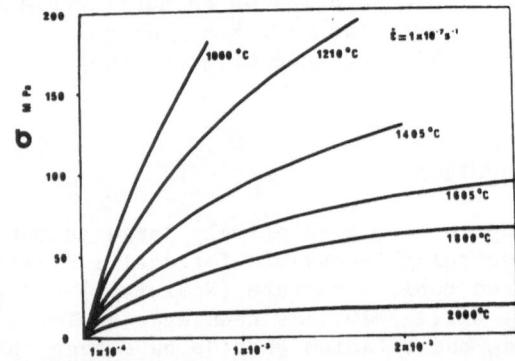

Fig. 2 : Typical Stress-
Strain Curves

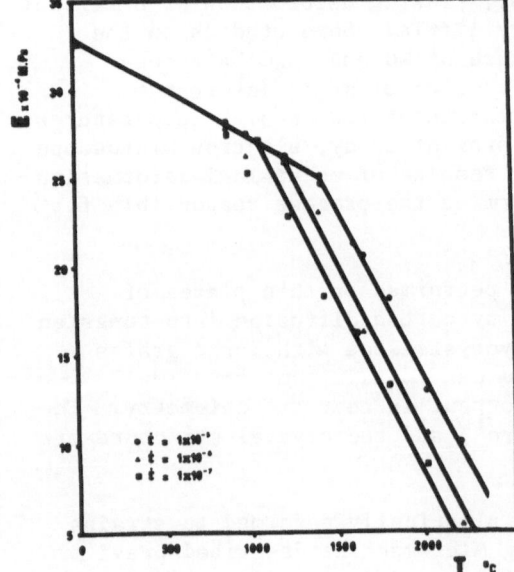

Fig. 3 : Variation of the
Young modulus as a function
of temperature and strain-rate.

tests necessary for a thorough study of the first stages of plastic deformation namely: $\sigma - \varepsilon$ curves, stress relaxation, strain rate (between (10^{-7} s^{-1} to 10^{-4} s^{-1}) and temperature changes.

Thin foils for T.E.M. were prepared either by ion-thinning or by electropolishing with a basic solution cooled to about 0°C (NaOH 5%, NaNO$_3$ 5 %, H$_2$O 40% and ethylene glycol monoethylether 50% by weight).

RESULTS OF MECHANICAL DEFORMATION TESTS

The deformation curves (σ, ε) are shown in Fig. 2, for pre-deformed samples. The great brittleness of our samples has made it necessary to make a predeformation: $\varepsilon = 10^{-3}$ at 2100°C. This procedure offers the advantage of creating a relatively homogeneous dislocation structure. The material shows a high hardening rate, the curve being steepest at the lowest temperature, a behavior which differs from that of cubic carbides.[1-4] The elastic modulus is determined from the slope, $d\sigma/d\varepsilon$ at the origin. The elastic modulus, E, and the yield stress, σ, have been measured as a function of parameters such as: carbon content, temperature, and strain rate. The results of the tests for different strain rates (from 10^{-7} s^{-1} up to 10^{-5} s^{-1}) are shown in Figs. 3 and 4. Below 800°C, fracture occurs without any detectable plastic flow. At ambient temperature E varies between $22 \pm 2 \times 10^4$ MPa at 2.60% C to $33 \pm 2 \times 10^4$ MPa at 3.16% C.

Activation parameters (volumes and energies) were determined using the data of the stress-strain curves. The activation volume was determined from the relation

$$V = kT \left(\frac{\partial \ln \dot{\varepsilon}}{\partial \sigma} \right)_T$$

The values are plotted in Fig. 5a as a function of applied stress, σ. These values agree with those obtained during the relaxation tests or the strain rate changes, Fig. 5b, where each curve represents one temperature. It is worth noting that the activation volume is a little below that found for cubic carbides.[2,3]

The values of internal stress, σ_i, were determined from the relaxation test data, Fig. 5b, or dip-test. Modulus-reduced internal stress is plotted against temperature in Fig. 6; it is worth nothing that the internal stress varies more quickly than the elastic modulus. This behavior can be explained by a recovery effect that becomes more important as the temperature is raised and the strain rate is lowered. Figure 7 shows this effect, at 2000°C, as a function of strain rate (10^{-4} s^{-1} to 10^{-6} s^{-1}). After loading until yielding was indicated, the specimen was unloaded then immediately reloaded.

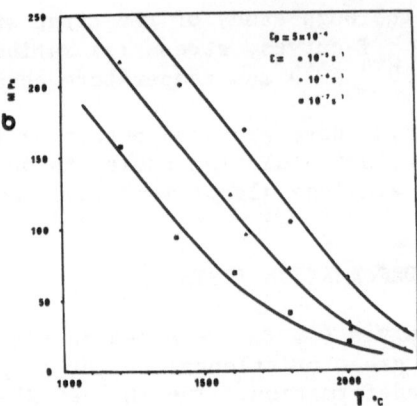

Fig. 4 : Variation of the Yield stress, at 0.05 % plastic
         strain, as a function of temperature and strain
         rate.

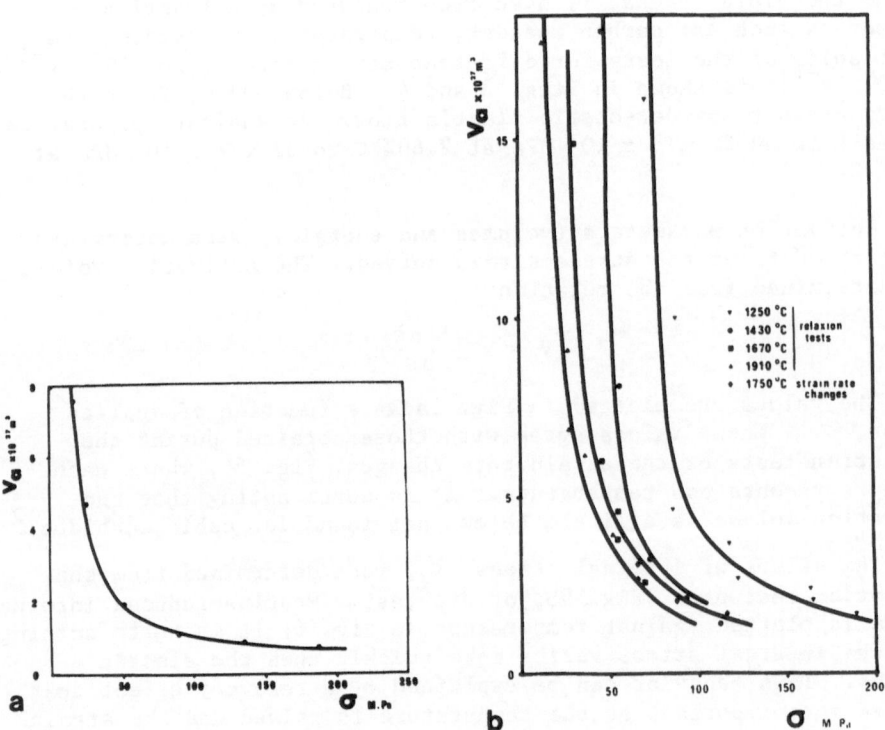

Fig. 5a-b : Activation volume as a function of applied stress.

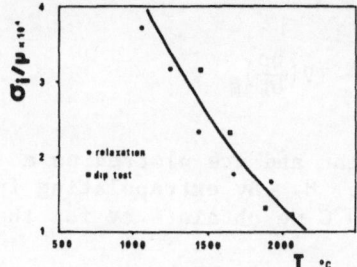

Fig. 6 : Modulus reduced internal stress as a function of temperature.

Fig. 7 : Typical force-relative displacement of knifes curves at various strain-rate.

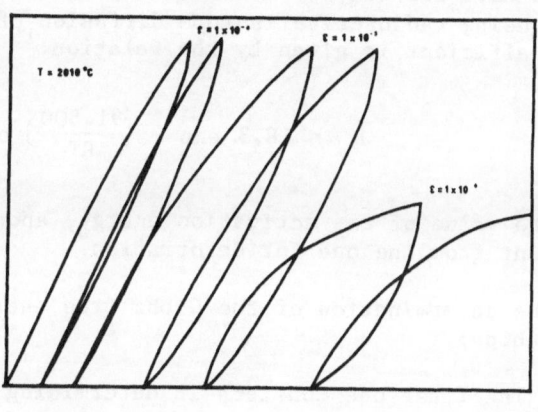

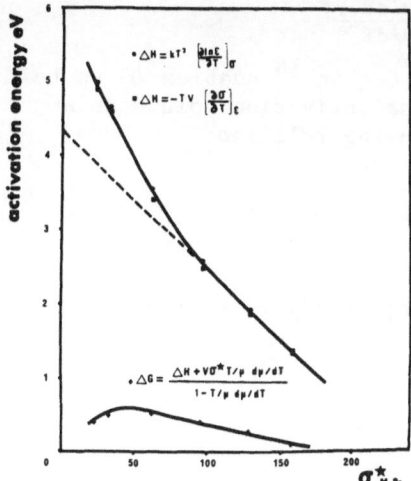

Fig. 8 : Stress dependence of enthalpy and the gibbs free energy.

The apparent activation enthalpy can be directly determined according to the (σ, ε) curves by using the following relations:

$$\Delta H = kT^2 \left(\frac{\partial \ln \dot{\varepsilon}}{\partial T}\right)_\sigma = - TV \left(\frac{\partial \sigma}{\partial T}\right)_{\dot{\varepsilon}}$$

The results thus obtained are similar and are plotted as a function of effective stress, σ^+, in Fig. 8. By extrapolating from those values for temperatures below 1750°C we obtain 4 eV for the activation enthalpy.

We have studied the bulk diffusion of carbon in W_2C by a direct method using radioactive isotope diffusion of ^{14}C. The bulk diffusion coefficient is given by the relation:[15]

$$D_v = 18.3 \ \exp - \left(\frac{91.500}{RT}\right) \ cm^2 \ s^{-1}$$

The value of the activation energy, about 4 eV, is not very different from the one for deformation.

The determination of the Gibbs free energy ΔG, can be made by two methods:

The first one consists in determining ΔH and V at a given temperature for the matching value of the stress, then calculating ΔG using the relation:

$$\Delta G = \frac{\Delta H + T/\mu \, d\mu/dT \ V\sigma^+}{- T/\mu \, d\mu/dT}$$

The second method, described by Cagnon,[16] enables ΔG to be deduced from the variation of the activation volume as a function of σ by using the following relation:

$$\tau = \frac{\mu_o}{\mu} \sigma$$

then to integrate the cuve $V(\tau)$ graphically to obtain ΔG as follows:

$$\Delta G = \frac{\mu}{\mu_o} \int_\tau^{\tau_o} V(\tau) \ d\tau$$

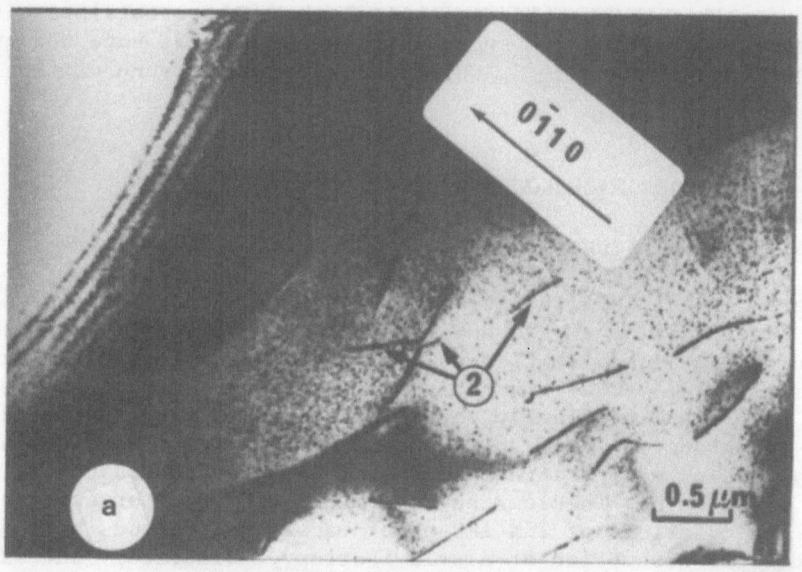

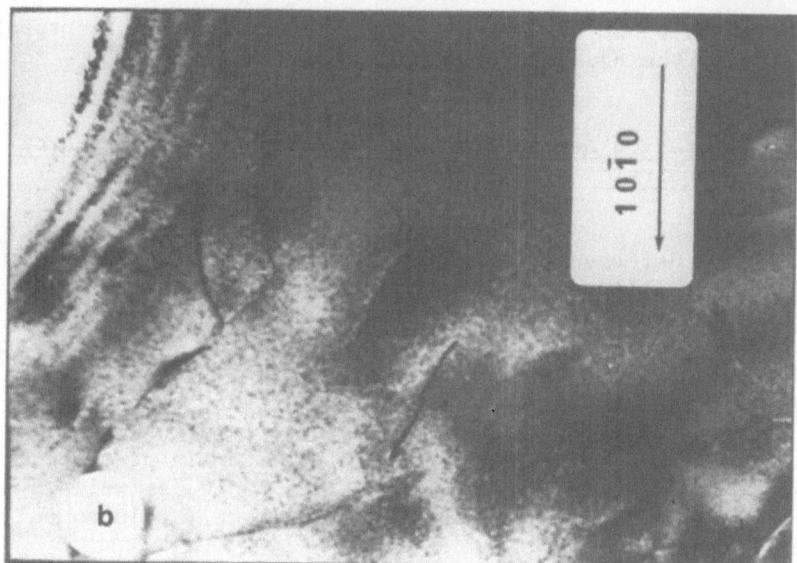

Fig. 9 : Determination of basal screw dislocations

    a) Dislocation 2 are in contrast

    b) Invisibility of dislocation 2 : $\vec{b_2} = \frac{1}{3}\left[1\bar{2}10\right]$

The results are practically equivalent to those obtained with the first method, which are shown in Fig. 8. We most note the great difference between ΔG and ΔH, which shows the importance of the entropy term.

RESULTS OF ELECTRON MICROSCOPE OBSERVATIONS

Slip System in the Range Temperature 1300° to 2200°C

The classical invisibility criterion $\vec{g} \cdot \vec{b} = 0$ has been used for the determination of the Burgers vector of dislocations. Determination of the line direction of dislocations shows that the basal slip system $\langle 11\bar{2}0 \rangle$ (0001) is activated in the temperature range 1300° to 2200°C. We had previously determined the same slip system at ambient temperature.[17] In spite of the anisotropy of W_2C, which probably induces modifications in the dislocation contrast, extinction of dislocations is clear enough to identify the Burgers vectors, Fig. 9. Calculations of TEM images of basal dislocations have been made under the assumption of elstic isotropy. Simulated contrasts are in good agreement with the observed images. This shows that anisotropy is not very important.

Although no other system has been seen, cross-slip is probable, as suggested by the presence of dipoles (Fig. 13).

General Configuration of the Structure Defects as a Function of Temperature and Strain

This study has been performed using a high voltage microscope (1 Mev), on plates deformed between $\varepsilon = 10^{-2}$ to 10^{-3}. Figures 10 to 15 show the configuration of dislocations in the temperature range 1300° to 2200°C. We have observed mixed dislocations but also pure screw or edge dislocations. Generally we can divide the temperature range in two domains. First, below 1750°C, Figs. 10-11, where dislocations are often isolated and random. At the higher test temperatures, however, there is a trend towards a more orderly arrangement. Second, above 1750°C, Figs. 12 to 15, in this domain of temperature we can see a few dislocation networks, Figs. 12-13, as well as dipole trails, Fig. 14 or loops, Fig. 15, resulting from dipole annealing by volume diffusion.

Core Structure of Dislocations

Figure 16 shows a characteristic dissociation observed in W_2C deformed at 1950°C; the perfect dislocation is lying in the basal plane. This dissociation has been previously studied and seems to

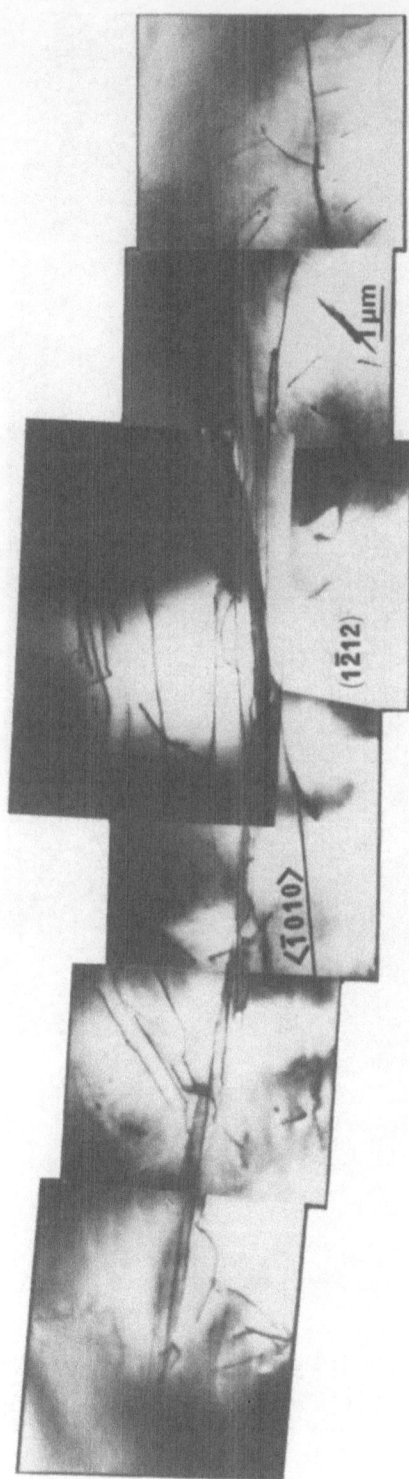

Fig. 10 : Structure of dislocations after deformation test,
$\varepsilon = 1.10^{-3}$ at 1300°C

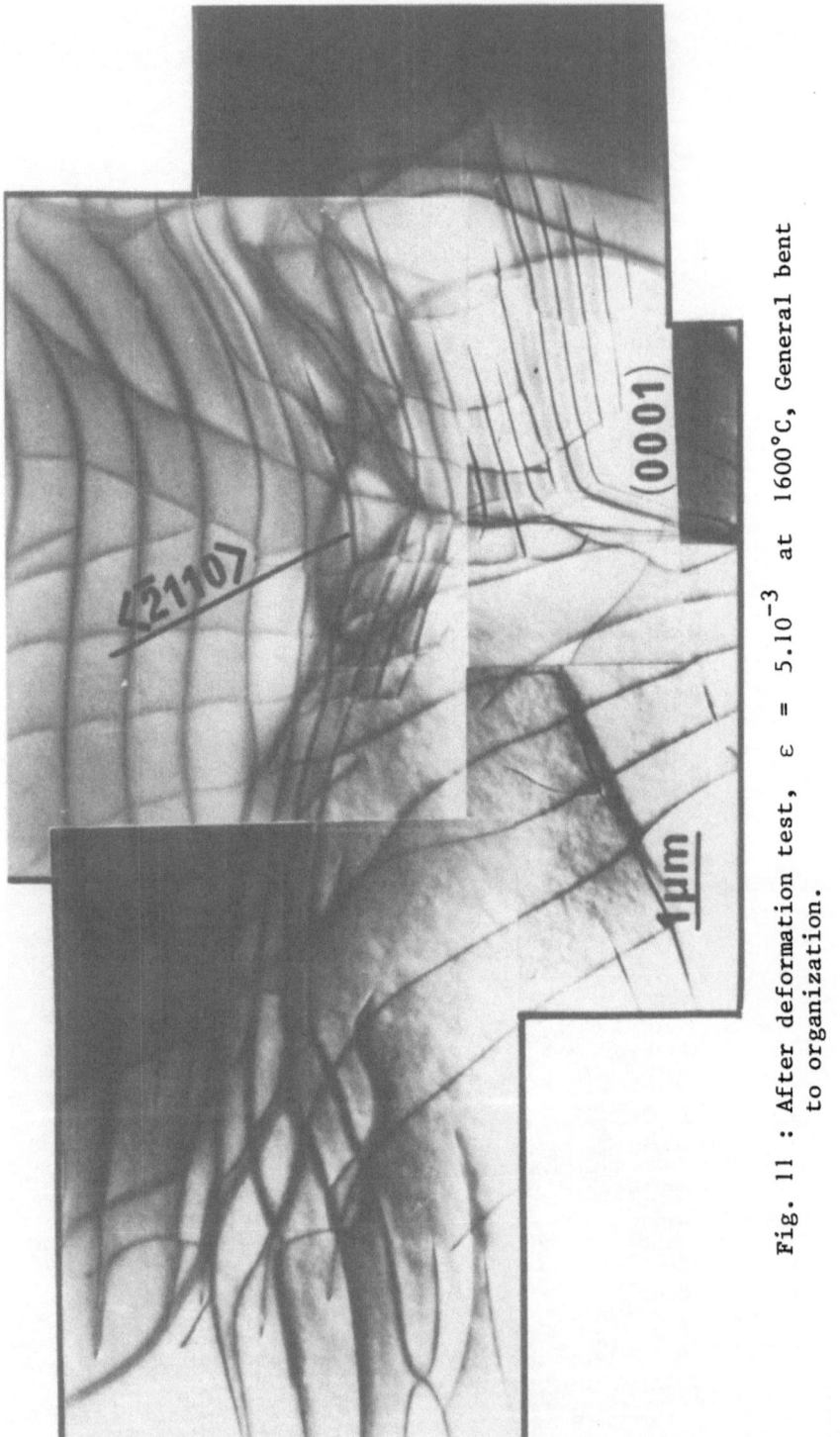

Fig. 11 : After deformation test, $\varepsilon = 5.10^{-3}$ at 1600°C, General bent to organization.

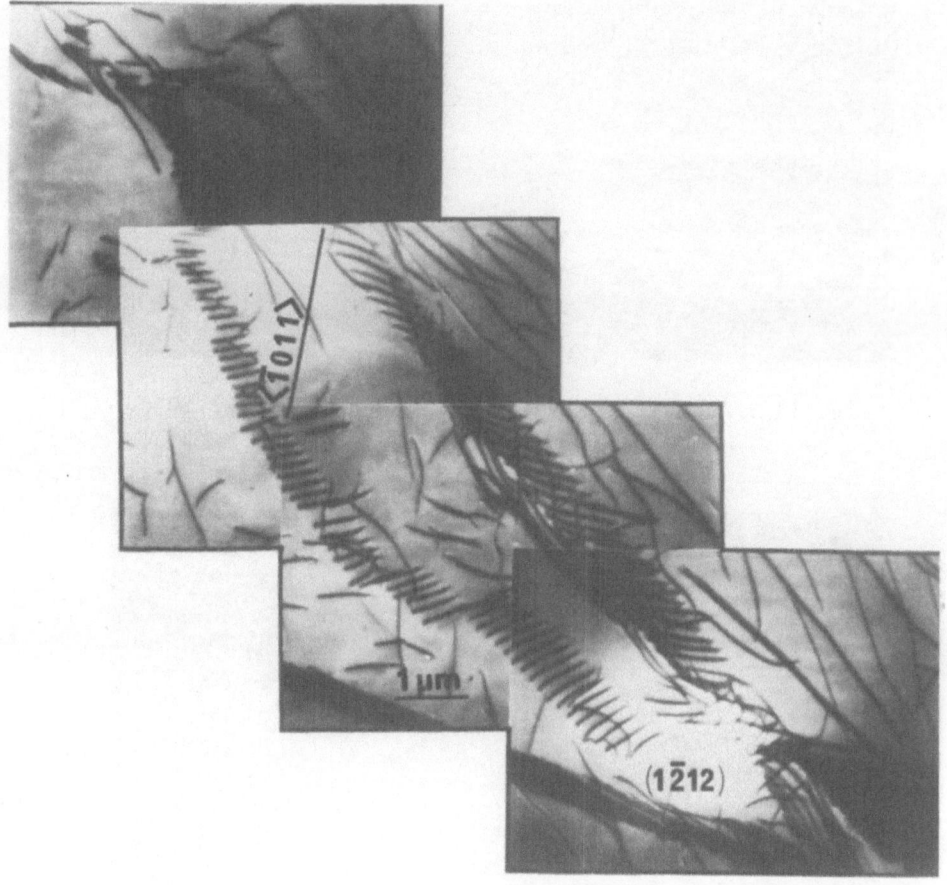

Fig. 12 : A cell, $\varepsilon = 10^{-2}$ at 1750°C.

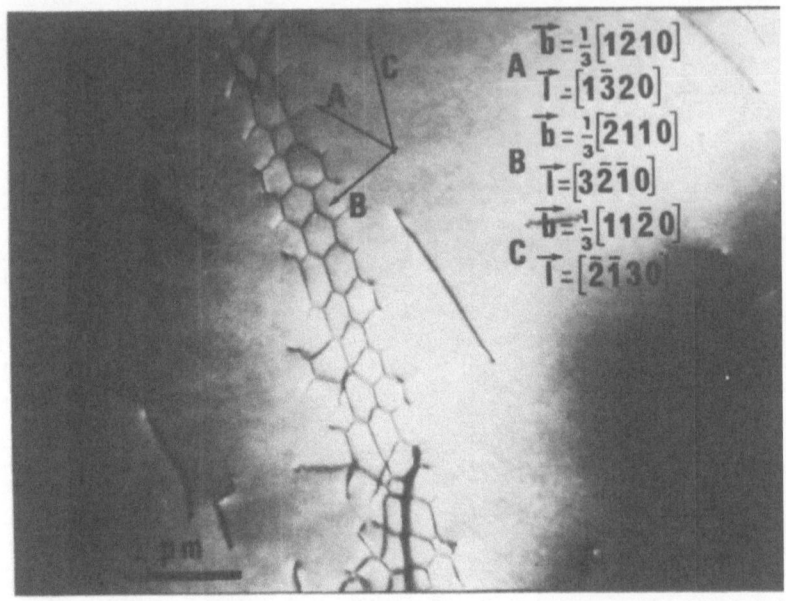

Fig. 13 : An hexagonal network, $\epsilon = 10^{-2}$ at 1750°C.

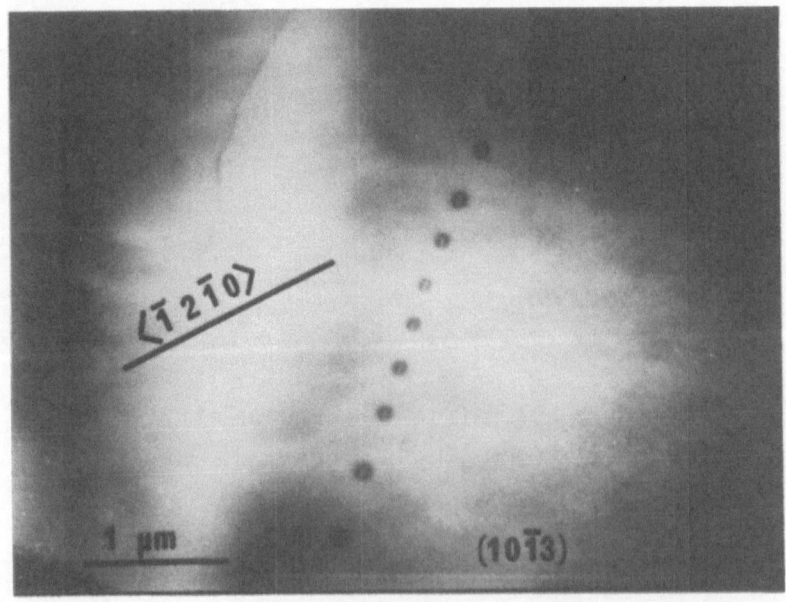

Fig. 15 : A row of loops, $\epsilon = 1.10^{-3}$ at 2100°C

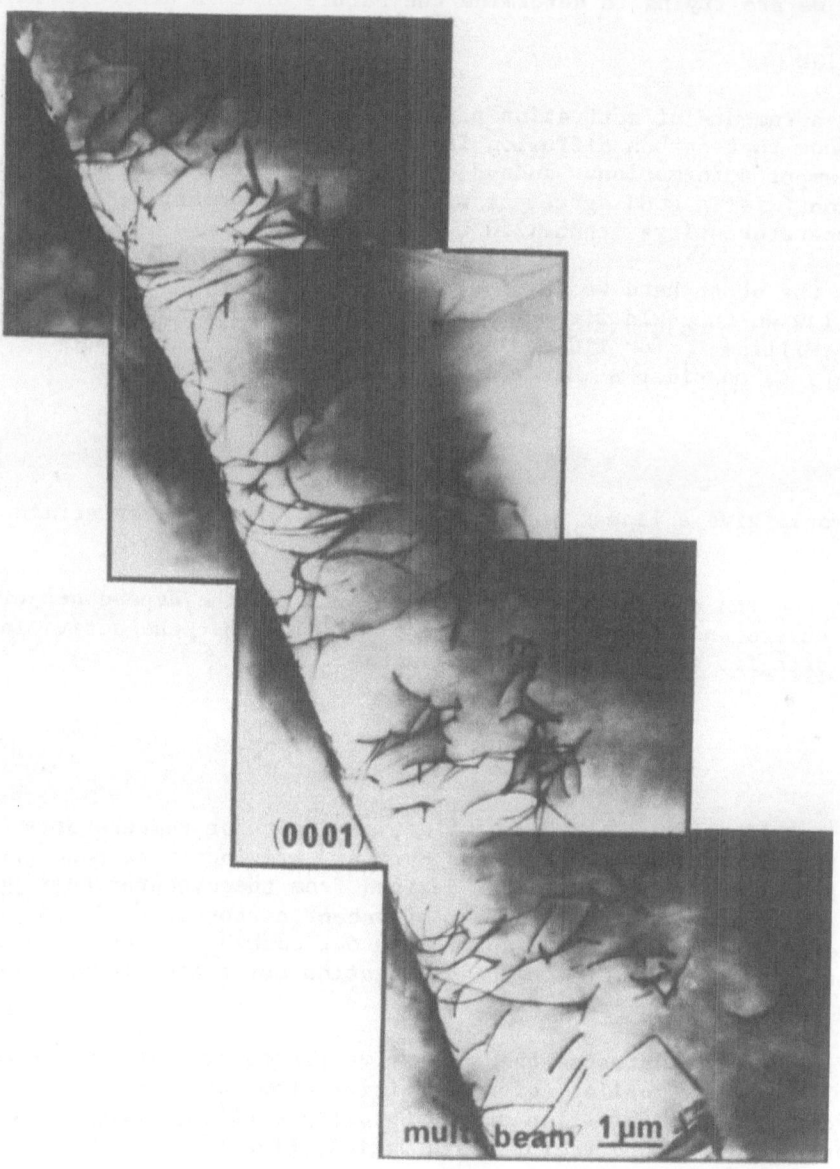

Fig. 14 : A few dipole-trails, $\varepsilon = 1,25 \ 10^{-3}$ at 2000°C.

occur in the basal plane.[18]   Figure 17 shows another dissociation
that occurs on samples deformed at 2100°C.   In this case the two
partial dislocations are out of the basal and prismatic planes.   At
present we are trying to determine the nature of this dissociation.

DISCUSSION

Measurements of activation parameters made during mechanical
tests show that carbon diffusion is the main point controlling the
mechanism of deformation.   Indeed the value of activation energy of
diffusion[15] is in good agreement with that of deformation, at least
for temperatures lower than 1750°C.

On the other hand we have been looking for a type of constitutive law which could fit our results.   A power law such as that
used by Williams[19] for TiC could not be verified.   Thus it was
necessary to consider a more complex relations:

$$\dot{\varepsilon} = K \; \sigma^{\, n} \; \exp - \left(\frac{\Delta Ho - Vo \; \sigma^+}{kt} \right)$$

which would give a linear dependance of σ on T for a given strain
rate.

Indeed the constitutive law is deduced from the dependance of Va
on the stress and temperature.   Figure 18 shows that the activation
volume data fit well to the relationship:

$$Va = kT \left(\frac{\partial \ell n \dot{\varepsilon}}{\partial \sigma} \right) = Vo + n \frac{kT}{\sigma^+}$$

with n = 2.10 and V = 0.74×10^{-27} m^3, at least for temperatures
lower than 2000°C. Rowcliffe[4] and Martin[3] previously proposed such
a relation coherent with a model derived from the synchro-shear process[20] which requires the presence of vacant carbon sites in the
lattice.[2] This model can be applied in our case since the value of
activation volume shows that the dislocation cores are almost free of
carbon atoms.

Figure 19 illustrates the successive phases of this mechanism
controlled by the mobility of interstitial atoms of carbon, and which
is favoured by the formation of two Shockley partial dislocations
similar to these proposed for dissociation, Fig. 16.

Above 1750°C measurements of activation parameters show an
enhancement of activation energy, that is consistent with the
recovery shown previously and with the results of TEM.   Thus the
presence of dislocation out of the basal plane, the appearance of

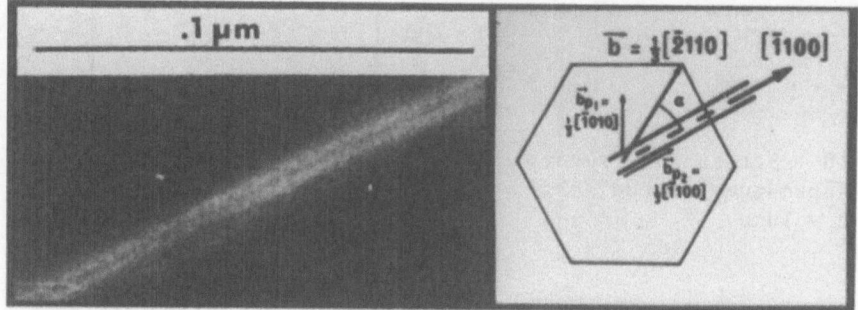

Fig. 16 : Basal dislocation splitted into partials : weak beam dark field shows that the width of the dissociation d $\simeq$ 30 A°.

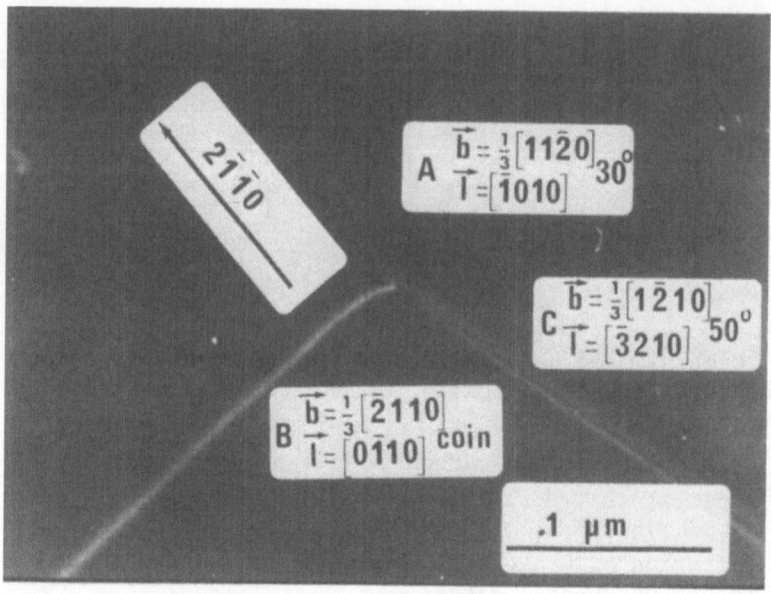

Fig. 17 : Weak beam dark field shows a dissociation, with a width d $\simeq$ 70Å

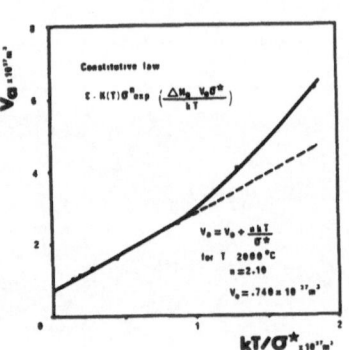

Fig. 18 : Stress and temperature dependence of the activation volume.

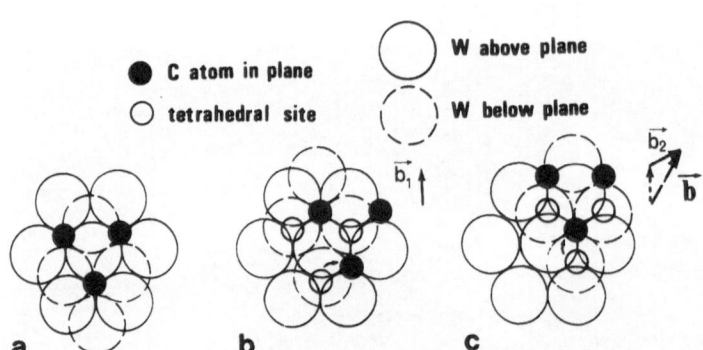

Fig. 19 : Mechanism of shear-diffusion in the basal plane of W_2C
(the carbon layer is schematically represented filled with inters-
ticial atoms).

a) Stacking of tungsten and carbon planes perpendicular to the
   $\vec{C}$ - Axis.

b) Shear b_1 = 1/3 $[\bar{1}010]$ of the low tungsten layer and simultaneous diffusion of C-atoms from previous sites (become tetrahedral) to new octahedral sites.

c) Shear $\vec{b}_2$ = 1/3 $[\bar{1}100]$ and new similar diffusion of carbon
   atoms; passing of the dislocation $\vec{b}$ = 1/3 $[\bar{2}110]$

dipoles and the formation of loops are the result of two mechanisms, cross-slip and climb.

On the one hand, the formation of dipoles is certainly the result of cross slip. Indeed, only the basal slip system is activated and the formation of dipoles by a mechanism proceeding from jogged dislocations is not probable, intersections of dislocations being very limited. Nevertheless the formation of dipoles can be the result of interaction of two dislocations gliding in the basal planes separated by a short distance, as pointed out by Tetelman.[21] In this mechanism, the dislocations can lower their energy by reorientating part of their lengths in parallel positions.

The interaction between dislocations situated in parallel slip planes is shown in Fig. 11. Then, the cross slip segment will be annihilated leaving a dislocation dipole as shown by Fig. 14 and therefore the cross slip mechanism operates during plastic deformation.

On the other hand, for plastic deformation at high temperatures, loops trails are observed, Fig. 15. The same observations have been made in cubic carbides.[22] This formation of loops trails can be explained by a climb process with self or pipe diffusion of vacancies as shown by Hollox and Smallmam.[22] In the same way we observe that the spacing between loops is about three times the loop diameter.

ACKNOWLEDGEMENTS

The authors are grateful to J. Pelissier, C.E.N.G. Grenoble France, for using the 1 MeV electron microscope.

REFERENCES

1. L. E. Toth, Transition Metal Carbides and Nitrides, Academic Press, New York, 1972.
2. J. L. Martin, These Universite de Paris Nord, 1972.
3. J. L. Chermant, Rapport Scientifique A.T.P., in "Rupture et Fluage," contract ISO4, n° 1791, Octobre 1977.
4. D. J. Rowcliffe, Ph.D. Thesis, Cambridge (1965).
5. R.H.J. Hannink, D. L. Kohlstedt, M. J. Murray, Proc. Roy. Soc. A 326:409 (1972).
6. S. R. Shenck, R. J. Gotschall, W. S. Williams, Mater. Sci. Eng. 32:229 (1978).
7. S. B. Luyckx, Acta Met. 24:233 (1970).
8. T. Johannesson, B. Lehtinen, Phys. Stat. Sol. (a), 16:615 (1973).
9. V. K. Sarin, Phil. Mag. 35:139 (1977).

10. S. Hagege, J. Vicens, G. Nouet, P. Delavignette, Phys. Stat. Sol (a), 61:675 (1980).
11. J. Dubois, G. Orange, C. Mai, G. Fantozzi, C.R.Acad. Sci. Paris t. 287, Series B.:53 (1978).
12. J. Dubois, G. Fantozzi, P. F. Gobin, Rapport Scientifique A. T.P. "Rupture et Fluage," contrat ISO4 n° 1791 (1977).
13. J. Dubois, G. Orange, G. Fantozzi, Scripta Met. 14:107 (1980).
14. G. Orange, J. Dubois, G. Fantozzi, P. F. Gobin, Journees d'Automne, Octobre 1979.
15. D Treheux, J. Dubois, G. Fantozzi, Ceramurgia Inter. to be published.
16. M. Cagnon, Phil. Mag. 24:1465 (1971).
17. G. Orange, J. Dubois, G. Fantozzi, P. F. Gobin, Mat. Sci. Eng., 34:291 (1978).
18. T. Epicier, C. Esnouf, J. Dubois, G. Fantozzi, submitted to Scripta Met.
19. W. S. Williams, J. Ap. Phys., 35:1329 (1964).
20. M. L. Kronberg, Acta Met. 5:507 (1957).
21. D. Hull, Introduction to Dislocations Pergamon Press, London, 1968.
22. G. E. Hollox, R. E. Smallman, J. Ap. Phys. 37:818 (1966).

POINT DEFECTS AND MICROHARDNESS IN TRANSITION METAL
BORIDES AND β-RHOMBOHEDRAL BORON

Torsten Lundström

Institute of Chemistry
University of Uppsala
Box 531
S-751 21 Uppsala, Sweden

INTRODUCTION

In early structural research on borides it was not possible to determine with high accuracy the atomic positions and their fractional occupancies. The boron atoms were frequently placed in the structure from space filling considerations. In recent years, however, several highly accurate structure determinations have been carried out by single crystal diffractometry in most cases using X-rays, but in some cases using neutrons. In the present paper we shall discuss the influence of these determinations on our present understanding of the structures, compositions and microhardness of some binary borides and some solid solutions of transition elements in β--rhombohedral boron.

CRYSTAL STRUCTURE AND POINT DEFECTS

Inorganic non-molecular compounds inherently contain point defects (substitutional and interstitial atoms, vacancies) that frequently are associated with a variable composition. Although this is a common situation the composition of the stoichiometric compound is not always easy to specify. For ionic and covalent crystals it is customary to designate a phase by a formula which corresponds to some simple combining ratio, for instance NaCl or GaP. The homogeneity range of the phase is in such crystals fairly small. The transition metal chalcogenides

invariably display large homogeneity ranges, which in
some cases, for instance $Fe_{1-x}O$, do not include the "stoichiometric" composition. For intermetallic compounds,
e.g. the sigma phases, it is in general impossible to
associate a specific formula with a given phase, at least
in the lack of a detailed knowledge of the crystal structure of the phase. Even so the inherent disorder (often
substitutional) is frequently so large that it is insignificant if one formula or another is preferred. The situation is much the same for binary compounds between
transition elements and p elements (H, B, C, N) although
the substitutional disorder is relatively much less important. In the present paper the ideal formula (or composition) of a phase will be based on a crystal structure determination and the defects established denoted as
far as possible.

Crystal Structures and Defects in Some Binary Borides

It has recently been demonstrated[1] that the orthorhombic Mn_4B structure[2] does not contain the earlier assumed half-occupied boron position. A full occupancy was
found in a crystal from a two-phase sample that also contained α-Mn, which leads to the formula o-Mn_2B. The range
of homogeneity of the phase is very small. A small homogeneity range is also displayed by Pd_2B, which crystallizes
in the anti-$CaCl_2$ type structure and thereby is iso-struc-
tural[3] with Co_2C and Co_2N. Empty holes are distributed in
an ordered manner in the Pd_2B structure.

A binary boride of the composition W_2B_5 was originally reported in the W-B system[4] and later in the Ru-B system[5]. The structure of this phase can conveniently be
described using close-packed metal layers (A, B), puckered boron layers of type K, and planar boron layers of type
H as shown in Fig. 1. Kiessling's original structure determination[4] can be given as the stacking sequence AHAKB
HBKAHAK..., although the structure of the boron layers was
not known in detail.

Recently, however, a crystal of the composition $WB_{2.0}$,
which represents the boron-poor side of the homogeneity
range, was investigated[6] by single-crystal diffractometry.
As a result of this investigation the structure can be
given as an AHAK´BHBK´AHAK´... stacking, indicating the
occurrence of vacancies in the K layers (see Fig. 1).
The phase is still homogeneous after an increase in the
boron content of approximately 1.5 atomic per cent[4]. The
mechanism of the varying composition very probably involves a partial filling of the $2(a)$ position in the K

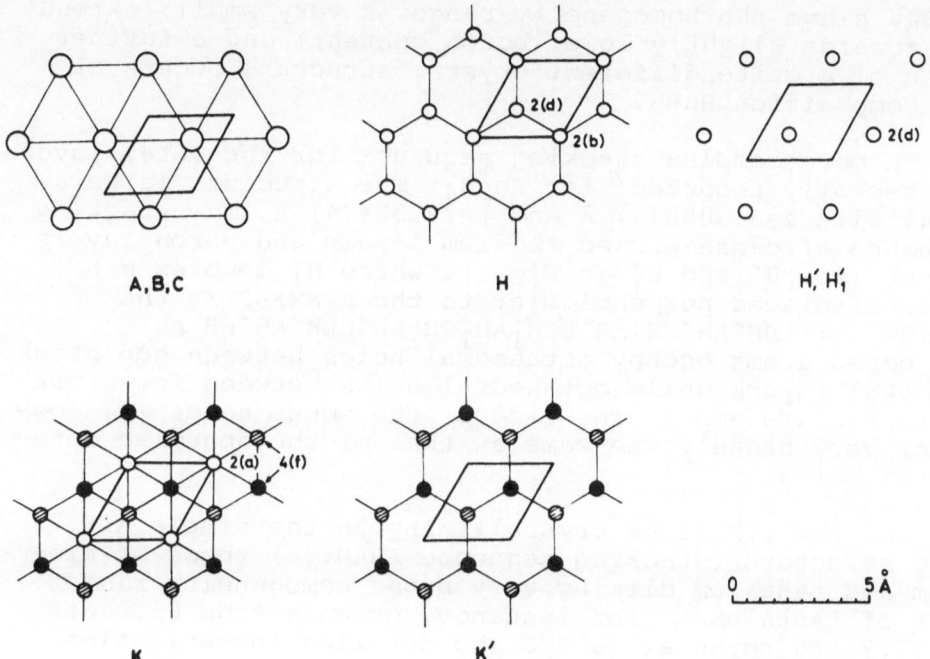

Fig. 1. Layers of boron (H, H⁻, H⁻₁, K, K⁻) and layers of
metal (A, B, C) in Ru_2B_{3-x}, W_2B_{5-x} and Rh_5B_4. The
layers K and K⁻ are puckered while the remaining
ones are planar.

layers (see Fig. 1). The formula may be given as W_2B_{5-x}
with $0.7<x<1$ or as WB_{2+x} with $0<x<0.15$. The former nota-
tion better describes the structure, although the composition W_2B_5 does not lie within the homogeneity range.

A single-crystal X-ray investigation[6] has shown that
the structure of the ruthenium compound must be written
AH⁻AK⁻BH⁻BK⁻AH⁻AK⁻..., using defective K layers (K⁻) as
well as defective H layers (H⁻). (Note that the latter
layers in fact consist of isolated boron atoms, situated
2.905 Å from each other.) The structure thus normally
contains voids in the K and H type boron layers. Using
the stacking sequence mentioned above, the ideal formula
should be written Ru_2B_3. The small homogeneity range may
be indicated by writing Ru_2B_{3-x} with $0<x<0.1$. Although
the metal layer sequence is identical with that in the
originally proposed W_2B_5 type structure, it is not possible to consider Ru_2B_{3-x} as a grossly non-stoichiometric

Ru_2B_5, since the homogeneity range is very small (extending towards slightly lower boron content) and a further phase of a quite different crystal structure occurs at the composition RuB_2.

A more complex stacking sequence for the metal layers was recently reported[7] for Rh_5B_4. The structure is hexagonal with $a=3.3058(2)$ Å and $c=20.394(4)$ Å. The stacking sequence of close-packed rhodium layers and boron layers (layer type H´ and H_1^- in Fig. 1, where H_1^- denotes a H´ layer displaced perpendicular to the c axis) is the following: BH´AH´BH´AH´BCH₁´AH₁´CH₁´AH₁´CBH´AH´BH´AH´... The boron atoms occupy octahedral holes between hcp stacked metal layers while octahedral holes between fcc stacked layers are empty. The size of the empty holes is, however, very closely the same as that of the occupied octahedral holes.

Of the diborides crystallizing in the simple AlB_2 type structure (stacking sequence AHAH...) those of niobium and tantalum display very broad homogeneity ranges. That of tantalum[8], for instance, extends from 65.8 at% to 73.2 at% boron at 2000°C. No detailed investigation has been reported of the mechanism for the change in composition within the homogeneity range but it has been postulated that hyperstoichiometric compositions are most likely formed by incorporation of additional boron rather than by creation of metal vacancies, and that (0,0,1/2) is the most probable position for the additional boron[8,9]. However, since the space available at this position is extremely limited and since the mechanism proposed does not explain the observed constancy in unit cell volume over the homogeneity range the mechanism seems improbable. I suggest that a hyperstoichiometric composition implies interstitial boron atoms in the positions (1/3,2/3,1/2) and (2/3,1/3,1/2) and the occurrence of vacancies in the tantalum position. This would easily explain the increase in the c axis, the decrease in the a axis and the constancy of the unit cell volume over the homogeneity range.

The composition of the boron-richest phase in the systems Mo-B and W-B was a question of controversy for several years. It was originally denoted[10] MeB_4, later[11] MeB_{12} and[12] $Me_{2-x}B_9$, and finally[13] $Me_{1-x}B_3$ with x close to 0.20. The last-mentioned composition is based on a complete structure determination[13], which revealed holes and partially filled atomic positions in the structure. This may be described as a stacking of metal layers (A´ and B´ in Fig. 2) and planar hexagonal boron (H) layers. The prevailing stacking sequence A´HB´HA´H... is such that

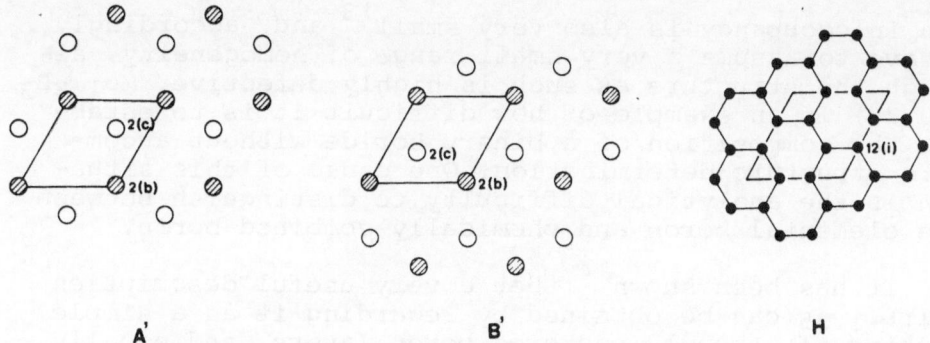

Fig. 2. Defective planar metal layers (A´, B´) and planar
        boron layers (H) in $Mo_{1-x}B_3$. The dumbbell-like
        holes are centered at (2/3,1/3,1/4). Position
        2(b) is partially occupied. *From Ref. 13.*

there are dumbbell-like holes in the structure centered
at (2/3, 1/3, 1/4) and extending in the c direction. These
holes have a convenient size for boron atoms but even in
boron-saturated samples no such atoms were detected in
the holes. Accordingly, there are no boron-boron contacts
in the c direction. The atomic position 2(b) (see Fig. 2)
in the metal layers is occupied only to 60 %. The varia-

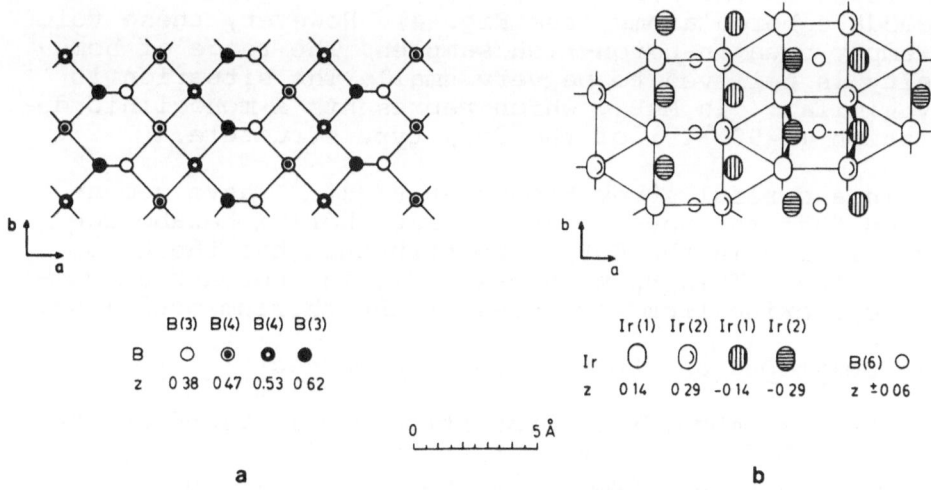

Fig. 3. Puckered boron layers (a) and double layers of
        metal atoms (b) in $IrB_{\sim 1.35}$. Position B(6) is not
        fully occupied. *From Ref. 15.*

tion in occupancy is also very small[13] and, accordingly, we have to assume a very small range of homogeneity, although the structure as such is highly defective. $Mo_{1-x}B_3$ ($x \approx 0.20$) is an example of how difficult it is to establish the composition of a binary boride without a complete structure determination. One cause of this situation is the analytical difficulty to distinguish between free elemental boron and chemically combined boron.

It has been shown[14] that a very useful description of $IrB_{\sim 1.35}$ can be obtained by regarding it as a simple stacking of strongly puckered boron layers, and equally strongly puckered double layers of iridium atoms (Fig. 3). A redetermination of the structure[15] then showed that it contains holes, which are never filled by boron atoms, and in addition an atomic position that is only partially occupied by boron atoms. This implies that there is no three-dimensional boron network in the structure. The phase displays a range of homogeneity and it is fairly safe to assume that the varying composition arises from a varying occupancy in the partially filled boron position in the metal double layers (B(6) in Fig. 3).

The crystal structure of CrB_4 is illustrated[16] in Fig. 4, which shows a projection of the orthorhombic unit cell in the direction of the short axis (2.866 Å). The boron atoms form a three-dimensional network, which contains channels, occupied by the chromium atoms. There are also voids in the structure, which are large enough to accomodate boron atoms (see Fig. 4). However, these voids are empty even in boron-rich samples. The range of homogeneity is believed to be very small. The situation is very similar[17] in MnB_4, which represents a monoclinic deformation ($\beta = 90.31^{\circ}$) of the CrB_4 type structure.

In a careful study Etourneau *et al.*[18] have demonstrated that the non-stoichiometric thorium hexaboride crystallizing in the CaB_6 type structure has the homogeneity range $Th_{1-x}B_6$ with $0 < x < 0.22$. The boron-rich compositions arise from vacancies at the thorium positions.

Solid Solutions and Defects in β-rhombohedral Boron

The β-rhombohedral boron structure is based on the boron icosahedron as the main structural element. It can be described using complex B_{84} units and B_{10} units[19,20]. A B_{84} unit consists of a central boron icosahedron, whose atoms are bonded along the quasi-fivefold axes to the vertex atoms of twelve half-icosahedra. A B_{10} unit is formed by three half-icosahedra, which share one trian-

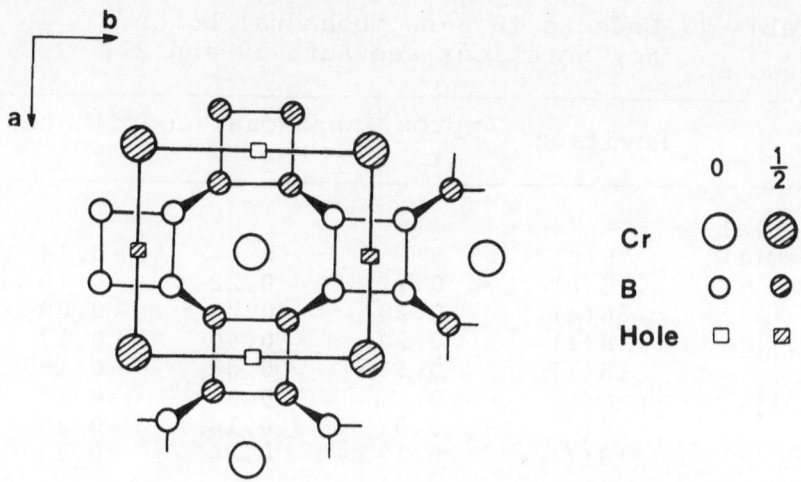

Fig. 4. The crystal structure of CrB_4 as viewed along the short c axis.

gular face in pairs and have the vertex atom in common. The two structural units described are linked together via their half-icosahedra to form new complete icosahedra. The linkage leads to a rigid three-dimensional framework, in which all boron atoms form part of an icosahedron and which can be described using a hexagonal unit cell with $a=10.9276$ Å and $c=23.814$ Å.[21] There are two further non-equivalent boron atoms (B(15) and B(16)) which do not form part of any icosahedron. One of these, B(15), is six-coordinated and situated between two B_{10} units. The other position, B(16), is eight-coordinated. This position often shows a low occupancy.

The icosahedron is a compact arrangement of atoms, the radius of the central hole being 0.90 r_B. The linkage of the boron icosahedra in β-rh. boron, however, leads to a very open structure with only 36 % space filling ratio ($r_B=0.88$ Å). The structure contains several holes[22] which are sufficiently large to accommodate transition metal or p element atoms. In addition to interstitial atoms β-rhombohedral boron also can contain substitutional atoms and a considerable number of vacancies (in the B(13) and B(16) position). The approximate coordinates of these three types of defects have been collected in Table 1.

Table 1. Defects in β-rhombohedral boron.
For notations see Ref. 19 and 22.

Defect	Position	Approx. hexagonal coordinates		
		x	y	z
Interstitial				
A(1) [=M(1)]	6(c)	0	0	0.14
A(2)	18(h)	0.11	0.22	0.10
A(3)[a]	36(i)	0.28	0.25	0.05
D [=M(2)]	18(i)	0.20	0.40	0.17
D'	18(i)	0.19	0.38	0.15
E [=M(3)]	6(c)	0	0	0.23
F(1)[a]	18(i)	0.07	0.14	0.25
F(2)[a]	18(i)	0.12	0.24	0.25
Substitutional				
B(1)[b]	36(i)	0.17	0.17	0.18
B(4)[c]	36(i)	0.24	0.25	0.35
Vacancy				
B(13)	18(h)	0.06	0.12	0.55
B(16)	18(h)	0.06	0.11	0.12

[a] No occupancy found experimentally
[b] B(1) can be replaced by Si or Ge
[c] Two B(4) atoms can be replaced by one Sc

Crystallographic studies of the defects have been carried out on several materials during the last few years. Mostly a single crystal of the solid solution has been studied using an X-ray diffractometer to measure the intensities. The results obtained are generally of high quality and some of them are given in the Tables 2 and 3. The holes are invariably incompletely filled and in some cases the occupancy is very low indeed (see Table 2).

The holes A(1)-A(3) (see Table 1) are situated outside the triangular faces of the central icosahedron in the B_{84} unit. An atom in such a hole would coordinate 12 neighbouring boron atoms. The A(1) hole accommodates most types of atoms with the exception of those with large atomic radii for instance scandium. Silicon can be accommodated in A(2) while so far no evidence exists for the occurrence of any atoms in A(3), F(1) and F(2).

Table 2. Composition for some solid solutions in β rhombohedral boron.

| Structure | Metal content (at%) | | Unit cell volume (A^3) | Ref. |
	From structure refinement	From micro-probe analysis		
β-rh. B	-	-	2462.7	21
$SiB_{\sim 36}$	2.70	2.74±0.14	2509.0	23
$ScB_{\sim 28}$	3.49	3.2 ± 0.5	2508.4	24
$CrB_{\sim 41}$	2.44	2.7 ± 0.3	2482.5	22
$MnB_{\sim 23}$	4.12	3.6 ± 0.4	2508.6	25
$FeB_{\sim 49}$	2.01	-	2473.3	26
$CuB_{\sim 23}$	3.8	4.17±0.02	2500.2	27
$CuB_{\sim 28}$	3.45	2.7 ± 0.4	2489.9	25
$GeB_{\sim 90}$	1.11	-	2481.8	28

Table 3. Occupancy (in percent) of interstitial and substitutional positions and vacancy occurrence
in β-rhombohedral boron.

| Structure | Interstice | | | | Substitution | | Vacancy | |
	A(1)	A(2)	D	E	B(1)	B(4)	B(13)	B(16)
β-rh. B	-	-	-	-	100	100	27	75
$SiB_{\sim 36}$	46.4	4.8	0	0	86.7 B / 13.3 Si	100	26	100
$ScB_{\sim 28}$	0	0	31.4	72.7	100	92.5 B / 5.7 Sc	38.7	100
$CrB_{\sim 41}$	71.9	0	18.0	0	100	100	28.3	100
$MnB_{\sim 23}$	25.6	0	43.1	66.2	100	100	35.0	100
$FeB_{\sim 49}$	50.7	0	18.5	0	100	100	27.4	100
$CuB_{\sim 28}$	6.1	0	43.1[a]	50.5	100	100	38.9	79.1
$CuB_{\sim 23}$	8	0	46	61	100	100	21	87
$GeB_{\sim 90}$	20.9	0.5	2.9[b]	0	3.4 Ge / 96.6 B	100	27	100

[a]Two positions, situated 0.44 Å from each other
[b]Two positions, situated 0.76 Å from each other

The D and E positions have 15 boron neighbours. The
E position can occasionally be occupied by transition
metals. The D hole has been found to be partially occupied in all solutions studied with the exception of that
of silicon. Within this hole, copper or germanium atoms
are distributed on two separate positions, situated 0.4 Å
and 0.8 Å from each other, respectively.

It is interesting to note that scandium substitutes
for two B(4) atoms[24] while a silicon[23] or a germanium[28]

atom can substitute for a B(1) atom. In GeB~90 a low but
significant germanium occupancy (0.42% and 1.15% respectively) was found in two other positions in addition to
those listed in Table 2.

   Note that, although a formula MeB_x is given for a
specific composition in the Tables, it represents only
one composition in the range of solid solubility. Occasionally it represents the maximum solubility of the element in β-rhombohedral boron. The nature of the solid solution of copper in β-rhombohedral boron has been studied
in detail by Lundström and Tergenius[29]. It was shown that
in a series of copper-boron alloys the unit cell volume
increases continuously and nearly linearly with the copper
content. The preferential occupancy of up to four non-
-equivalent crystallographic positions by copper was determined in the same alloys. It was found that copper
initially is accommodated in the D position (Table 1)
and subsequently, as the copper content is increased, in
the E and A(1) positions.

MICROHARDNESS

   Results from the measurement of Vickers microhardness within the homogeneity range of TaB_2 are displayed[30]
in Fig. 5. The results were obtained using a Durimet
microhardness tester and a load of 0.49 N. A minimum of
eight impressions were made on each polycrystalline sample.
The determination of the composition of the samples,
which were prepared from the elements through arc-melting
and subsequently heat treated at 1800°C, was carried out

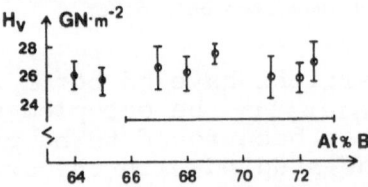

Fig. 5. Microhardness (H_V) versus composition in the
        homogeneity range (indicated) of TaB_2.

using the cell dimensions versus composition curves in
Ref. 12. It appears from Fig. 5 as if the microhardness
does not depend on the composition. Although the composition dependence of the microhardness is much lower
than that in TaC[31] for instance, it is possible that a
weak composition dependence might be masked by the relatively low precision of the measurements. To reveal such
a dependence it would be interesting to collect Knoop
data for single crystals of different compositions using
the most composition-sensitive face of the crystals for
the hardness impressions.

The microhardness of the solid solutions of the 3d
transition metals in β-rhombohedral boron have been measured.[32] The increase in microhardness relative to pure
β-rh. boron (H_V=33.6 MN·m^{-2}) has been plotted versus the
increase in the unit cell volume (a lattice strain parameter) in Fig. 6. Only the solid solutions for which a
single crystal X-ray study is available were incorporated
(see Table 3). It is evident that a hardening effect is
obtained and that manganese and scandium are the most
effective hardeners in spite of the relatively large
difference in atomic size (r_{Mn}=1.31 Å and r_{Sc}=1.60 Å).
The increase in microhardness in saturated solid solutions
of these two elements in β-rh. boron is approximately 25%.

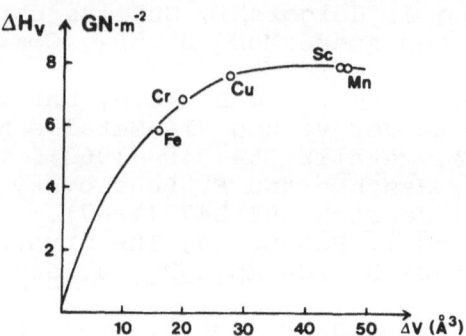

Fig. 6. Hardness increase (ΔH_V) versus unit cell volume
        increase (ΔV) for some solid solutions of tran-
        sition metals in β-rhombohedral boron.

Acknowledgement. Financial support from the Swedish Natural Science Research Council is gratefully acknowledged.

REFERENCES

1. L.-E. Tergenius, Refinement of the Crystal Structure of Orthorhombic Mn_2B, formerly denoted Mn_4B, J. Less--Common Metals, in print.
2. R. Kiessling, The Borides of Manganese, Acta Chem. Scand. 4:146 (1950).
3. E. Hassler, T. Lundström and L.-E. Tergenius, The Crystal Chemistry of Platinum Metal Borides, J. Less--Common Metals 67:567 (1979).
4. R. Kiessling, The Crystal Structures of Molybdenum and Tungsten Borides, Acta Chem. Scand. 1:893 (1947).
5. C.P. Kempter and R.J. Fries, Crystallography of the Ru-B and Os-B Systems, J. Chem. Phys. 34:1994 (1961).
6. T. Lundström, The Structure of Ru_2B_3 and $WB_{2.0}$ as Determined by Single-Crystal Diffractometry, and Some Notes on the W-B System, Arkiv Kemi 30:115 (1968).
7. B.I. Noläng, L.-E. Tergenius and I. Westman, The Crystal Structure of Rh_5B_4, J. Less-Common Metals, in print.
8. E. Rudy, St. Windisch and Y.A. Chang, Ternary Phase Equilibria in Transition Metal-Boron-Carbon-Silicon Systems. Part I. Related Binary Systems. Volume X. Systems V-B, Nb-B and Ta-B. AFML-TR-65-2.
9. R. Kiessling, The Borides of Tantalum, Acta Chem. Scand. 3:603 (1949).
10. A. Chrétien and J. Helgorski, Sur les borures de molybdène et de tungstene MoB_4 et WB_4, Compt. Rend. 252: 742 (1961).
11. E. Rudy, F. Benesovsky and L. Toth, Untersuchung der Dreistoffsysteme der Va und VIa-Metalle mit Bor und Kohlenstoff, Z. Metallk. 54:345 (1963).
12. H. Nowotny, H. Haschke and F. Benesovsky, Bor-reiche Wolframboride, Monatsh. 98:547 (1967).
13. T. Lundström and I. Rosenberg, The Crystal Structure of the Molybdenum Boride $Mo_{1-x}B_3$, J. Solid State Chem. 6:299 (1973).
14. T. Lundström, Preparation and Crystal Chemistry of Some Refractory Borides and Phosphides, Arkiv Kemi 31:227 (1969).
15. T. Lundström and L.-E. Tergenius, Refinement of the Crystal Structure of the Non-stoichiometric Boride $IrB_{\sim 1.35}$, Acta Chem. Scand. 27:3705 (1973).
16. S. Andersson and T. Lundström, The Crystal Structure of CrB_4, Acta Chem. Scand. 22:3103 (1968).
17. S. Andersson and J.-O. Carlsson, The Crystal Structure of MnB_4, Acta Chem. Scand. 24:1791 (1970).
18. J. Etourneau, R. Naslain and S. La Placa, L'hexaborure de thorium non-stoichiometrique $Th_{1-x}B_6$. J. Less--Common Metals 24:183 (1971).

19. J.L. Hoard, D.B. Sullenger, C.H.L. Kennard and R.E. Hughes, The Structure Analysis of β-Rhombohedral Boron, J. Solid State Chem. 1:268 (1970).

20. B. Callmer, An Accurate Refinement of the β-Rhombohedral Boron Structure, Acta Cryst. B33:1951 (1977).

21. A. Jimenez Crespo, L.-E. Tergenius and T. Lundström, The Solid Solution of 4d, 5d and Some p Elements in β-Rhombohedral Boron, J. Less-Common Metals 77:147 (1981).

22. S. Andersson and T. Lundström, The Solubility of Chromium in β-Rhombohedral Boron as Determined in $CrB_{\sim41}$ by Single-Crystal Diffractometry, J. Solid State Chem. 2:603 (1970).

23. M. Vlasse and J.C. Viala, The Boron-Silicon Solid Solution: A Structural Study of the $SiB_{\sim36}$ Composition, J. Solid State Chem. 37:181 (1981).

24. B. Callmer, A Single-Crystal Diffractometry Investigation of Scandium in β-Rhombohedral Boron, J. Solid State Chem. 23:391 (1978).

25. S. Andersson and B. Callmer, The Solubilities of Copper and Manganese in β-Rhombohedral Boron as Determined in $CuB_{\sim28}$ and $MnB_{\sim23}$ by Single-Crystal Diffractometry, J. Solid State Chem. 10:219 (1974).

26. B. Callmer and T. Lundström, A Single-Crystal Diffractometry Investigation of Iron in β-Rhombohedral Boron, J. Solid State Chem. 17:165 (1976).

27. I. Higashi, T. Sakurai and T. Atoda, Crystal Structure of CuB_{23}, J. Less-Common Metals 45:283 (1976).

28. L.-E. Tergenius and T. Lundström, The Crystal Structure of $GeB_{\sim90}$, J. Less-Common Metals, in print.

29. T. Lundström and L.-E. Tergenius, On the Solid Solution of Copper in β-Rhombohedral Boron, J. Less-Common Metals 47:23 (1976).

30. T. Lundström, B. Lönnberg and I. Westman, to be published.

31. D. Santoro, Variation of Some Properties of Tantalum Carbide with Carbon Content, Trans. AIME 227:1361 (1963).

32. J.-O. Carlsson and T. Lundström, The Solution Hardening of β-Rhombohedral Boron, J. Less-Common Metals 22:317 (1970).

DISCUSSION

C. Politis:

Excellent results. Could you tell how the single crystals were produced?

T. Lundstrom:

Well, for crystal structure determinations, the crystals need not be very large. But for more detailed studies of the hardness, one should have large single crystals. All the hardness values I have shown here were obtained from polycrystalline materials. In general we prepared the crystals by arc melting and later heat treatment at high temperature.

C. Politis:

Did you also produce single crystals by vapor transport?

T. Lundstrom:

Yes, we have tried. But it is difficult with boron because of the difficulties in finding a suitable container material.

QUALITATIVE AND QUANTITATIVE INTERPRETATION OF MICROSTRUCTURES

IN CEMENTED CARBIDES

Hans Eckart Exner

Max-Planck-Institut für Metallforschung
Institut für Werkstoffwissenschaften
Stuttgart, FR Germany

INTRODUCTION

Cemented carbides (or hardmetals) can be conveniently defined in terms of microstructure: They consist of single crystals of one or more carbide phases which form a continuous skeleton the free space of which is filled by a continuous metal binder. Due to their exceptional technical performance in cutting and wear applications, the microstructure of these composite materials has been subject to extensive scientific investigations. Cemented carbides have also been model systems for studying the densification and the development of microstructures during liquid phase sintering (see, for example[1-5]) as well as for finding experimental relationships between microstructure and properties of two-phase alloys (see, for example, [6-10]).

The physical and chemical nature of the carbide phases and binder metals used as well as their interaction have been described recently by this author[11]. This paper aims to provide a basis for understanding the importance of the microstructural features for the properties and the application of hardmetals. In attempting to review the present knowledge, emphasis is laid on tungsten carbide cobalt alloys due to the fact that this system has been studied most extensively and also is the most important one from a technical point of view. However, multicomponent cemented carbides and WC-free alloys are also considered whenever pertinent information exists. A short survey of experimental techniques is given which reviews the historical development and the present state of metallographic methods and instruments for quantitative analysis of microstructure in order to provide a convenient source to widespread literature.

METHODS FOR CHARACTERIZING THE MICROSTRUCTURE

Light microscopy is the most important means for quality control purposes (see, for example[12,13]). For coarse grades, light microscopy is also useful for microstructural evaluations in scientific studies, and the preparation of cross sections and contrasting techniques are well developed[14-19]. In most technical grades, pores, impurities, binder phase regions and carbide grains can often not be adequately studied due to insufficient resolution, and electron microscopy must be used. Replica methods have been worked out for transmission electron microscopy allowing high resolution imaging of etched cross sections[17,20-27] and fracture surfaces[28-32]. Stereographs of fracture surfaces using this technique were published as early as 1956[28]. Replica techniques have been nearly completely substituted by scanning electron microscopy using backscattered or secondary electron imaging. Thin foil transmission electron microscopy has been successfully attempted by various authors and used extensively for studying the deformation and fracture behaviour and the nature of precipitates in the binder of WC-Co alloys[17,33-43]. Photo emission microscopy is an excellent means of revealing the structure of multiphase hardmetals[43,44] but is restricted for widespread practical use by the high price of the instrument.

Microprobe analysis was applied early for studying the composition of grain boundaries and interfaces[45,46] as well as the composition of mixed carbides and impurity phases[43,47-52]. Frequently, the answers obtained by this technique are not conclusive due to lack of spatial resolution. More recently, a number of high-resolution analytical electron microscopy and diffraction techniques as well as other modern analytical methods (see, for example, 53-59) have been employed for revealing the structure and composition of phases, grain boundaries, interfaces, and fracture surfaces of cemented carbides. It can be hoped that the new instruments and methods will settle several of the open question discussed further below.

Quantitative light microscopy was introduced to WC-Co alloys by Gurland as early as 1955[60], see also[6,51,52] and approximately 10 years later, Exner and Fischmeister[22,23] followed up this work using replica electron microscopy. Since then, numerous authors have studied the geometry of WC-Co microstructures in relation to powders and processing conditions used during production and heat treatment. Later on, TiC-based cemented carbides and multiphase hardmetals were also extensively studied by quantitative metallography[5,49,63-68]. The availability of accurate data is presumably the major reason why this relatively complicated alloy system is used frequently as an experimental model for studying the kinetics of microstructural development as well as microstructure/property relationships[1-10].

The geometric parameters as well as basic techniques for measuring them have been frequently described (see, for example, [22,23, 25,31,61,63,68-72]) and there is no need to discuss them in full detail here. Progress has been made in instrumentation for image analysis of cemented carbides. Manual lineal analysis was used extensively in early work, and mechanical stages with electronic counting and recording devices have been developed[25,69,70] which still have major merits. In addition, a large number of advanced instruments are now commercially available[72]. Electronic planimeters connected to a microprocessor and a general purpose computer (as, for example, the Kontron MOP, Videoplan and IBAS systems, the Leitz ASM, or similar systems), automatic.image analysers using interactive modules to solve detection problems (as, for example, available with the Cambridge Quantimet systems, the Leitz TAS, and the Kontron IBAS, among others), and fully automatic analysis using a television camera with the light microscope or an epidiascope, or on-line scanning electron microscopy can be used for quantitative analysis of microstructural geometry. While fully automatic analysis has been problematic in all but the most simple cases and tedious detection procedures were employed occasionally (see, for example[73]) full grey level image storages now available with the automatic analysers mentioned above may solve some of the problems in the near future.

The most relevant parameters are the mean grain size of the carbide crystals, the mean free path in the binder phase, and the contiguity of the carbide skeleton. These average quantities can be related without any assumptions on the shape and size distribution to the interface and grain boundary areas per unit volume which, in turn, can readily be measured from metallographic cross sections. Care has to be taken when calculating three-dimensional shape parameters or size distributions for the carbide grains from planar or linear measurements since these are interrelated and no assumption-free calculation procedure is available. In some cases, the assumption of an ideal spherical or cubic shape is a good approximation (as, for example, in the VC-Co and VC-Ni or TaC-Co systems respectively [64-66]). The carbide crystals in WC-Co alloys as well as in TiC-Ni and most mixed carbide systems assume more complex shapes. In these cases, evaluation of three-dimensional size distributions and of spatial shape of the carbide grains is difficult if not impossible, and data measured on the planar cross section (e.g. linear intercept or intersect area distributions and planar shape factors) should be used[74].

Dihedral angles can easily be determined by means of the digitising units mentioned above, and the true dihedral angle can be estimated from the angle occurring with maximum frequency. However, care must be taken here, too: a distribution of angles may exist in the real structure due to anisotropic interfacial energies. In principle, this distribution could be calculated from planar data[75]. In practice, however, this would require excessive efforts in measurement and calculations and has not yet been attempted.

In spite of these difficulties, quantitative image analysis provides valuable information on the microstructural geometry and is a necessary prerequisite for understanding the effects of processing on the properties of cemented carbides.

MICROSTRUCTURAL FEATURES AND THEIR EFFECTS

Discussing the microstructure of cemented carbides, the following features are of interest:

1. Binder phase
   Grain size
   Mean free path
   Crystal structure
   Precipitates
2. Carbide phases
   Crystal size
   Crystal shape
   Homogeneity and complex structures
   Grain boundaries and carbide/carbide interfaces
3. Carbide/binder interface
4. Other phases
   Impurities
   Pores

Although it is difficult to deal with the influence of these microstructural features independently of each other, they will be considered separately in the following discussion.

MICROSTRUCTURE OF BINDER PHASE

Grain (Domain) Size

The geometry of the microstructure of the binder phase develops during liquid phase sintering by filling the space between the carbide grains. In this way it assumes a continuous, complex-shaped structure. Two types of interfaces are formed during sintering and cooling respectively: Interfaces between the binder phase and the carbide phase(s) and grain boundaries in the binder phase. It is well established that very coarse Co grains (domains of identical crystal orientation) develop during freezing after sintering at cooling rates typical in cemented carbide production: Sandford and Trent[76] estimated the cobalt grain size in WC-Co alloys to be in the order of 1 mm. Willbrand and Wieland[77] confirmed this number by an X-ray microbeam technique for slowly cooled specimens while large equi-oriented cobalt regions were not detected in quenched specimens by these authors using X-rays. Metallographic etching techniques also proved unsuccessful[77]. More recently, Viswanadham[78] was able to reveal the grain size of nickel and nickel-molybdenum binders used in

74 % vol. (V, Ti) C cermets by etching and found values of 0.5 to
1 mm. Viswanadham attributed a high significance to this finding
with respect to mechanical properties and wear. Suzuki, Hayashi and
Taniguchi[80] studied the grain size in WC-Co and WC-TiC-TaC-Co alloys.
For WC-Co alloys, they found a systematic dependence upon cobalt content, carbide content, and cooling rate as shown in Fig. 1. These
authors demonstrated a close correlation between domain size and
high-temperature transverse rupture strength. Le Roux[80], using X-
ray analysis, showed that the cubic/hexagonal transformation of the
cobalt phase during heat treatment at approx. 1000 K reduced the
size of the grains by a factor of two from 1 to 0.5 mm, and found
a pronounced influence of this heat treatment on transverse rupture
strength. Considering the fact that the amount of grain boundary
area is negligible compared to other interfaces (carbide/binder and
carbide/carbide) the present author is inclined to attribute this
correlation to the simultaneous variation of other microstructural
parameters as suggested by Suzuki and co-workers[79] and Le Roux[80] as
an alternative explanation. Almond and Roebuck[81] found a grain size
of only 50 to 100 μm in WC-Co alloys with fine carbide grains which
agrees with the lower values shown in Fig. 1.

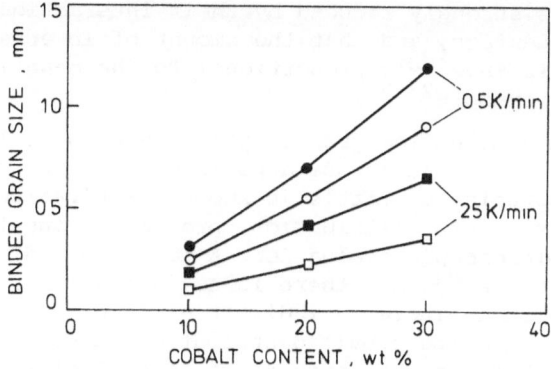

Fig. 1.   The variation of the grain size (size of domains of identi-
          cal crystal orientation) with cobalt content, carbon con-
          tent, and cooling rate, according to Suzuki and Hayashi and
          Taniguchi[79]. The full symbols indicate a low, the open sym-
          bols a high carbon content of the alloys.

Free Path Length

Due to the fact that thousands to millions of carbide crystals
are embedded in a single grain of the binder phase, the mean free
path (defined by the arithmetic mean of the distances from one
carbide/binder interface to the other measured in the binder phase)
is two to four orders of magnitude smaller than the binder grain
size, i.e. less than 1 μm in most hardmetals. It seems obvious that
the mechanical behaviour of the binder and, in turn, of the cemented
carbide is strongly related to this mean free path which corresponds
to the distance a dislocation can move (provided the binder regions
are free of precipitates). Gurland[60-62] was first to point out this
correlation for the transverse rupture strength of WC-Co alloys, and
this parameter has been extensively used to explain the mechanical
behaviour of hardmetals (for earlier references see[6]). The close correlation between hardness and mean free path is quantitatively understood on a theoretical basis (though other microstructural parameters
must also be considered[6,71,82] and hardness can also be described by
a modified rule of mixtures[5]). The increase of transverse rupture
strength with mean free path is not as easily interpreted: it is not
only related to the increasing ductility of the binder but also to
the decreasing probability of the presence of critical pores due
to improved sinterability. The fracture toughness measured by the
critical stress intensity factor is, on the other hand, independent
of such critical flaws and shows a clear dependence on the mean
linear free path (for references see[83]).

A direct and theoretically well-founded linear correlation also
exists between the reciprocal mean free path and the coercive force
of cemented carbides which is only slightly modified for alloys containing different carbides (see, for example[63,82]). This correlation
is based on the fact that the mobility of the Bloch walls during
demagnetization is strongly reduced by their interaction with the
carbide/binder interface, and that the amount of interface per unit
volume of binder is inversely proportional to the mean linear free
path (compare for example[23,72]).

Obviously, the mean free path lengths measured as linear intercepts along a straight line across a polished section are not constant but vary appreciably. Little is known quantitatively at
present about this length distribution, presumably due to resolution
problems during microscopic evaluation. A log-normal distribution
was observed by Freytag[84], but there is no indication if this finding can be generalized. There is indirect evidence based on magnetic
measurements[84,85] that the cobalt distribution in WC-Co alloys depends upon the carbon content which seems to control the redistribution of the cobalt during heating to the sintering temperature. The
milling process exerts a strong influence on the distribution of cobalt. Insufficient mixing during milling causes large cobalt pools
[86,87]. These effects may influence the distribution of linear intercepts and thus the mechanical and physical properties. e.g., cobalt
pools cause porosity[86] and problems in finishing tool surfaces[87].

Crystal Structure

Bulk cobalt is hexagonal at room temperature and cubic at temperatures above approx. 700 K. In WC-Co alloys, the phase transformation is sluggish and is completely suppressed during cooling, and only cubic cobalt exists after normal processing. The stabilisation of cubic cobalt to room temperature is favoured by mechanical constraints developing as a result of differential thermal contraction during cooling. Stressing and/or deformation during mechanical testing or during performance cause partial or complete transformation to occur[88], and a mixture of hexagonal and cubic cobalt may be present. On the other hand, dissolved tungsten raises the equilibrium transformation temperature, according to Giamei[89] to more than 1000 K. Annealing below that temperature results in hexagonal phase. Other authors[84,90,91] suggest that dissolved tungsten stabilizes the cubic phase down to room temperature as do dissolved carbon and nickel[90,91]. Since the properties of hexagonal and cubic cobalt are fairly similar, these observations seem of little interest. However, in addition to the reduction of binder phase grain size discussed above and the precipitation of intermetallic phases discussed below, the deformation induced transformation leads to a high work hardening rate which, as postulated recently by Brabyn, Cooper and Peters[90], may dominate the tensile properties and toughness of WC-Co alloys. This interesting idea is sustained by the observations shown in Fig. 2: Nickel additions to WC-Co alloys increase strength and toughness up to a maximum at a ratio of approximately 2/3 Co and 1/3 Ni where stacking fault energy reaches a minimum and stacking fault width a maximum. This hypothesis may offer some clues upon the erratic response of the mechanical behaviour of WC-Co alloys upon annealing (for references see[11,43]). More work is under way to elucidate the role of phase transformations in yielding and fracture of cemented carbides[91].

In iron-based binders, the structure of the binder phase is much more easily changed by heat treatment. Following earlier work on WC-based alloys with iron-nickel binder by Moscowitz, Ford and Humenik[92], Prakash[93-95] showed in a recent study how the structure and morphology of the iron-based binder phase varies as a function of composition and heat treatment and how these changes influence the properties. This interesting study shows that mixed binders may offer substantial technical advantages due to the various possibilities of martensitic and precipitation hardening well known from steel technology (for detailed references see[93]).

Precipitates

Precipitates which are much smaller than the binder phase regions between the carbide grains are nonmetallic impurities and graphite (which effect the properties very similarly to pores), carbides and intermetallic compounds precipitated during cooling or heat treatment in the solid state. The precipitation kinetics

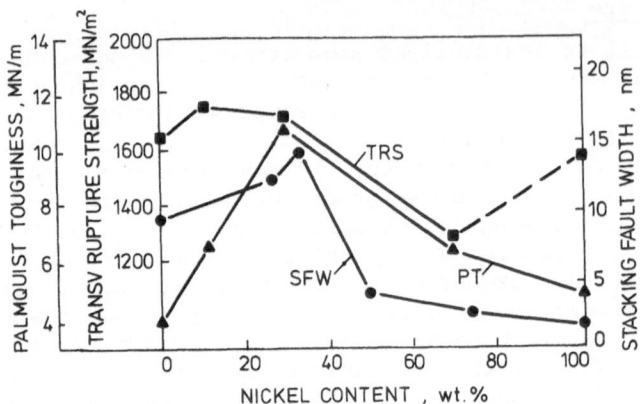

Fig. 2.   Toughness, transverse rupture strength and stacking fault
          width of WC-10wt%(Co,Ni) alloys as a function of nick-
          el content in the binder, according to Brabyn, Cooper and
          Peters[90].

of cobalt alloys containing tungsten and carbide have been studied
in considerable detail by various authors[41,43,88,95-101]. Though the
conditions for precipitation in the binder phase of cemented carbides and in bulk alloys are not identical, the principal features
seem to be the same. From the supersaturated solid solution present
after cooling from sintering temperature, the eta-carbide $(W,Co)_6C$
precipitates, according to Grewe and Kolaska[99] initially as Co_2W_4C
which changes composition to the more cobalt-rich variations Co_3W_3C
and Co_4W_2C which finally decays into Co_3W. From the established
phase diagram W-C-Co (for references see [11]), this intermetallic
phase should not be in equilibrium with WC at any temperature. Local
concentration gradients could arise due to the slow diffusion of
tungsten in cobalt. It seems more likely, however, that the transformation of the cubic into the hexagonal phase is the reason for
the formation of the intermetallic Co_3W due to a sudden decrease of
tungsten solubility in the binder. At any rate, finely dispersed
Co_3W has been proven to be present in heat treated WC-Co alloys
simultaneously with the hexagonal cobalt modification[43,97,100]. Heat
treatment is feasible, and improved compressive strength and cutting performance have been observed after annealing while toughness
is usually reduced[100].

     In addition to a preliminary study on cobalt alloys with addi-
tions of tungsten, titanium and carbon[102], the phase equilibria and

the precipitation behaviour of the Co-Mo-W-C and the Ni-Mo-W-C systems have been thoroughly investigated in a recent study[103]. Carbide phases of eta-type seem to be the only precipitates in the binder phase in these systems. On the other hand, there is no clear indication of the nature of precipitates in iron-based binders investigated by Prakash[93] though the precipitation behaviour of steels of similar composition is well established. A very careful study[104] of the microstructure of a Ni/Al binder in WC-based cemented carbides must be mentioned here which showed that the nature and amount of intermetallic particles precipitated during cooling and annealing corresponds well to the phase diagram of the bulk alloy. Heat treatment in the solid state allowed control of the size and distribution of these precipitates without changing the geometric structure of the carbide phase. This makes this system an ideal model for studying the effect of binder phase properties.

The toughness of TiC-based cemented carbides is closely related to the precipitation behaviour of the nickel-molybdenum binder. In a recent study, Moscowitz and Humenik[105] have reviewed earlier work showing the appearance of Ni_3Ti during heat treatment of TiC-Ni. The presence of this phase is consistent with the binary phase diagram as well as with the ternary phase diagram Ti-C-Ni[106] and has been demonstrated earlier by Snell[107]. Moscowitz and Humenik[105] modified the precipitation behaviour by adding several percent of aluminum to the binder which, according to the Ni-Ti-Al phase diagram[108] and to neutron diffraction studies[55] results in γ' precipitates. These precipitates are thought to be the reason for reduced transverse rupture strength and improved resistance to plastic deformation during cutting[105]. Binder phase strengthening by γ' precipitates was also achieved by Doi and Nishigaki[109]. There seems to be a great potential to optimize the performance of cemented carbides by modifying the microstructure of binder phases through doping and precipitation annealing.

MICROSTRUCTURE OF CARBIDE PHASES

Crystal Size

Besides carbon content, the crystal size of the carbide phases is the most carefully controlled variable in the production of hardmetals. There is no room here to discuss the large number of studies on the influence of the conditions of powder production, milling, and sintering on the carbide grain size in cemented carbides and on the kinetics of grain growth during annealing. (The author's collection contains approximately 100 published papers on these topics). Only a few general findings will be mentioned.

Most impurities in the starting powders seem to have no notable
effect on normal carbide crystal growth in WC-Co alloys or to reduce
growth tendency[110,111]. Pronounced grain growth inhibition is exerted
by other transition metal carbides, especially by those of the 5th
group of the periodic table. (For a comparison of these carbides and
references see[112]). These are used in industry to control grain
growth and allow the production of submicrograin alloys (for references see[43]). Grain refinement has also been observed for Li, Na, K,
Ni, Sn, V[111] and Fe, Mo, Cr, SiO_2, Al_2O_3[110] while Ni,P and C have
been reported to increase tendency for normal grain growth[111,113,114].
Discontinuous grain growth (locally exaggregated growth of selected
crystals) is superimposed upon normal grain growth under certain conditions. These include use of excessive sintering times or active
powders[114-116], and other reasons, such as recarburisation of eta
phase[117] (see below). This discontinuous growth can easily be qualitatively detected by visual inspection (Fig. 3) and can be analysed
quantitatively by separating the normal from the discontiunuous
fraction[114,115]. Impurities containing P seem to be the only chemical activators for discontinuous growth. Little is known about the

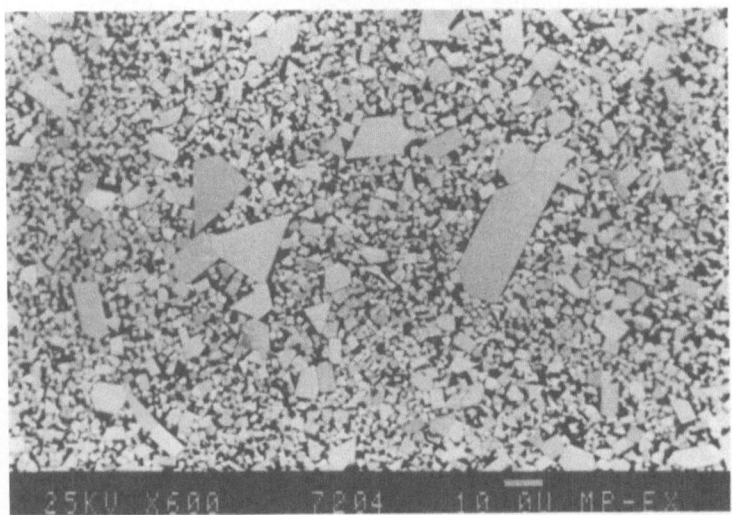

Fig. 3.   Discontinuously grown crystals in a WC-9 wt. % Co alloy
          annealed 25 h at 1400 °C (Scanning electron micrograph,
          backscattered electron image).

mechanisms of grain growth enhancement or inhibition by impurities
or additions, and the basic understanding is limited to some good
guesses. More basic work would be necessary to get a theoretical
basis for these effects of high technical interest.

The width of the carbide crystal size distribution in sintered
WC-Co alloys depends only very little on that of the powder or on
sintering conditions. Usually the linear intercepts in the carbide
grains show a distribution very close to a logarithmic Gaussian

distribution (see, for example,[22,113,114,118]). While a wide size range may be present in the carbide powder (due to the formation of fragments during milling of coarse powders[118] or due to mixing of carbide powders of widely varying mean particle size[22,72,118])giving rise to enhanced growth, the width of the log-normal distribution is nearly constant[22,118]. The size distribution widens and loses its log-normal shape when discontinuous grain growth occurs[22,113, 114]. Large grains are known to reduce toughness and are avoided as far as possible in technical grades. Little is known, however,of what the optimum size distribution should be, a fact which is presumably due to the limited amount of quantitative information on size distribution in technical materials.

Qualitatively, the processes leading to coarsening of the WC crystals at sintering temperature can easily be observed by studying micrographs (see, for example,[22]). Solution and reprecipitation due to size differences (Ostwald ripening), coalescence of carbide crystals by movement of the grain boundaries and filling corners of irregular crystals by precipitation from the molten binder are the principal mechanisms for normal grain growth. The quantitative interpretation of carbide growth is difficult, however, due to the superposition of these processes, due to the irregular crystal shape, and due to the small variation of WC crystal size in the technical temperature and time intervals. Great care must be taken in applying theoretical approaches to growth kinetics even for the most simple two-phase composites, and most of the present conclusions seem at least oversimplified.

The growth of carbide crystals in tungsten-free cemented carbides has also been studied [3,5,64-68,119]. For model alloys (usually with high binder content) simple kinetics were observed. Two observations may be mentioned: High carbon TiC is stable with respect to grain growth which can be activated to a level typical for technical alloys by addition of Ti or TiO_2[64]. This behaviour of TiC is in contrast to that of WC where increasing carbon content activates grain growth. The second observation is interesting from a theoretical point of view: In VC-Ni and VC-Co alloys with 40 vol.% binder, the size distribution of the spherical carbide crystals assumes a stationary shape predicted theoretically for reaction controlled Ostwald ripening[3,64]. In spite of the perfect agreement between theory and experiment in these and other cases, the development of microstructure in cemented carbides lends itself very little as an experimental model for checking theoretical results due to the high volume fraction leading to the superposition of growth mechanisms mentioned above and due to the large variety of crystal shapes discussed in the next section.

This situation is further complicated in systems with complex binders or carbides as for example in the usual compositions for cutting tools, e.g. (W,Ti)C-Co or (Ti,Mo)C-Ni,Mo[49,67,107,120], and almost nothing is known about grain coarsening, the size distributions and the interaction of growth kinetics in multiphase carbide alloys as for example the commercially important WC-(W,Ti)C-Co materials. In these mixed carbides, homogeneity is not achieved during sintering and chemical driving forces rather than surface energy considerations determine the formation of microstructure. This problem is addressed further below.

Crystal Shape

Tungsten carbide has a highly anisotropic structure and therefore develops anisotropic crystal shapes during growth which can be described as flat triangular prisms with truncated edges[11]. In alloys with a binder content typical for technical hardmetals, this shape is usually not fully developed due to impingement with other crystals and coalescence. However, most of the crystal sections observed in a polished cross section can easily be interpreted by this equilibrium configuration[11,36]. This shape is not only typical for WC crystals grown in cobalt but also in other binders (Ni, Fe and mixed binders)[92-95].

TiC is usually rounded or irregular in technical alloys but becomes more and more angular with carbon content deviating more and more from stoichiometry[64,68]. Mixed crystals with TiC as main component behave similarly, and (Ti ,W)C shows an irregularly rounded but most often equiaxed shape in technical alloys[11,49,67,119]. VC and Mo_2C develop nearly spherical crystals typical for isotropic interfacial energy[64,65,121,122]. TaC and NbC, on the other hand, are highly anisotropic and form cubes with more or less rounded edges and corners when sintered with Ni or Co[64,121,122]. If these carbides are added to WC-Co alloys, they are visible as spotted structures in polished and etched cross sections[123]. According to Suzuki and Yamamoto[123] these irregular features precipitate during cooling from the molten binder phase and contain abnormally large amounts of tungsten in solid solution. Their shape depends on the carbon content of the alloy and varies when Ni or Fe are added to the cobalt binder.

Eta phases (i.e. double carbides of W and Co or Fe) can be present as large angular crystals if the carbon content is low enough that the three phase region WC-eta-melt is reached at the sintering temperature. Small eta-phase precipitates showing up as small black regions in etched cross sections, may form during cooling in a critical temperature range from the melt or from the solid binder at small deviations from the stoichiometric carbon content. If cooling is slow enough and carbon potential is increased during cooling, the eta-phase may be recarburized and decay into WC and Co[117]. No harm to mechanical properties is done in this case while lower toughness but pronouncedly increased wear resistance have been observed when small eta-phase precipitates are retained to room temperature by quenching[84,85]. The large eta areas formed during sintering are, however, detrimental to cutting performance and toughness since they are sites of crack nucleation due to their inherent brittleness and due to the cobalt depletion of the surrounding alloy[124,125]. In addition, they act as nuclei for other unwanted features (pores and large carbide grains[117,124]) if recarburized during cooling. By qualitative inspection of the microstructure of fracture surfaces and cross sections the actual and possible action of eta phase as a fracture source can be evaluated.

Homogeneity and Complex Carbides

While tungsten carbide exists only in a small range of carbon content very close to the stoichiometric composition and hardly takes any other elements in solid solution, the cubic carbides of the 4th and 5th group of the periodic table have a wide composition range with respect to carbide content and large solubility potentials for other carbides . Mixed carbide powders, e.g. $(W,Ti)C$, are produced under such conditions that they are not in equilibrium with the binder phase at sintering temperature, and mixtures of single carbide powders as, for example, WC-TiC or TiC-Mo_2C are obviously in a non-equilibrium state at sintering temperature. Equilibrium phases are not formed readily by solid state diffusion since sufficient diffusivities are only obtained at temperatures beyond 1800 °C (for references see[126]): According to May and Krämer[126], the diffusion distance for W in TiC at 1500 °C is less than 0.05 µm per hour. On the other hand, homogenization can also be effected by solution of nonequilibrium phases in the binder and reprecipitation of equilibrium phases. Due to its high chemical force this mechanism is very effective and much faster than the solution-precipitation processes driven by capillary forces. The result is crystals with a cored microstructure, i.e. a nucleus of original composition is surrounded by the equilibrium phase (Fig. 4).

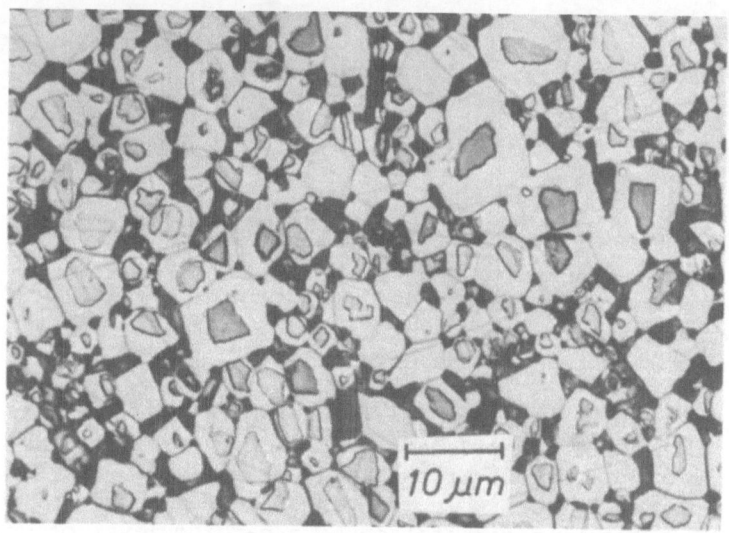

Fig. 4.   Cored carbide crystals in a TiC-20 wt% Mo_2C-10 wt% Ni alloy. From Rüdiger and Exner[43].

The mechanism and the kinetics of coring of (Ti,W)C crystals in W-Ti-C-Co alloys have been studied extensively by May[49,119] and other authors (for references see[43]). The coring phenomenon has also been investigated for WC-TiC-TaC-Co[47,49] and for TiC-Mo$_2$C-Ni[43] alloys and is now well understood. The effect of cored carbide microstructure on turning performance has been found very positive[43] but does not seem to be utilized systematically for technical grades.

In tungsten carbide-cobalt alloys, layers surrounding the carbide crystals were detected by Drott[21] in 1964, using a special etching technique. It is interesting to note that similar layers can be seen in published micrographs taken by replica transmission electron microscopy as early as 1956[28]. Based on microprobe profiling, inhomogeneous composition of tungsten carbide crystals has been reported by Peter, Jung and Kohlhaas[46] who suggested a cobalt-rich layer at the carbide-cobalt interface. More recently, we have observed double boundaries usually on the basal planes of large discontinuously grown tungsten-carbide crystals in scanning electron micrographs of long-time annealed WC-9 wt. % Co alloys[115]. Figure 5 shows an example.

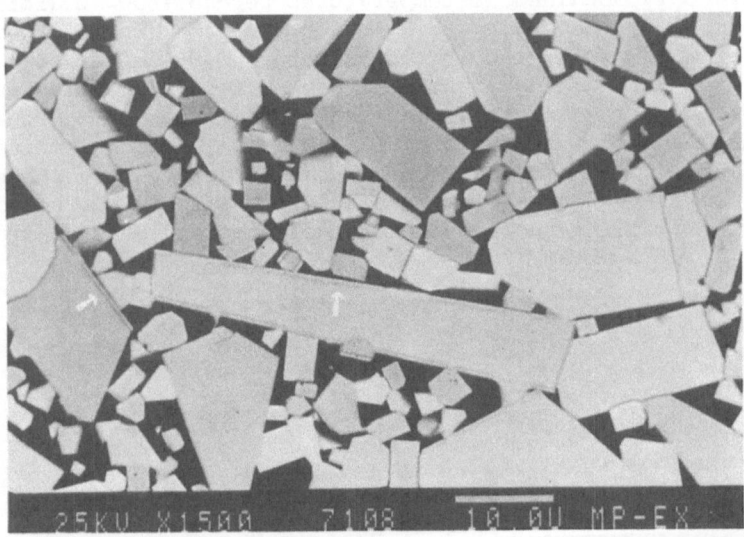

Fig. 5.   Scanning electron micrograph (backscattered electron image) of a WC-9 wt. % Co alloy annealed 25 h at 1400 °C. Double boundaries are visible parallel to the basal planes of large tungsten carbide crystals.

It is still not unequivocally clear if these layers are artefacts or real, and, with a few exceptions[27,39,93], Drott's results have not been discussed any further. If these layers are real, they are presumably not formed by coring in the way discussed above for cubic crystals but rather by precipitation of a carbide layer of deviating composition (e.g. of eta type) during freezing of the tungsten and carbon rich melt. It is certainly not worthwhile at present to spe-

culate on the effect of such layers on the properties of WC-Co
alloys and ways to influence them. It seems definitely worthwhile,
however, to use modern instrumentation to obtain a clear picture of
the surface composition of tungsten carbide crystals in contact with
cobalt.

Complex microstructures may arise when the carbides (or carbo-
nitrides) undergo phase transformations. Alloys with special properties can be produced in this way as, for example, by spinodal decomposition. An interesting material has been developed by Rudy,
Worchester and Elkington[127]: Particles of one phase are surrounded
by layers of another phase deviating slightly in composition after
spinodal decomposition of a titanium/molybdenum carbonitride in a
nickel-molybdenum binder. The resulting microstructure is virtually
insensitive to coarsening during annealing and gives a high wear resistance in cutting applications. There are numerous phase reactions
in the complex carbide systems which may result in interesting microstructures not utilized yet in cemented carbides. Recently, some
developments in this direction have been described by Holleck and
co-workers[95,106,128].

Grain Boundaries and Carbide/Carbide Interfaces

The nature of grain boundaries between tungsten carbide crystals in WC-Co alloys and even their existence has been a matter of
controversial discussion through three decades (for early references
and detailed discussion see[6,22]) and is still more or less vigorous-
ly debated during conferences (as, for example, during the present
one) and in literature. In spite of recent investigations of fracture
surfaces by Auger spectroscopy[54,129] and microbeam X-ray analysis[57]
which demonstrate appreciable amounts of cobalt in the boundaries
between carbide crystals, the present author clings to the view that
the contacts between tungsten carbide grains normally do not contain
cobalt layers with the structure or the properties of the binder
phase. According to this view the contacts visible in light or electron micrographs are real grain boundaries, and it is proposed here
that cobalt is segregated to a degree depending upon the types of
grain boundaries discussed in the following.

Microstructural inspection of well annealed tungsten carbide-
cobalt alloys shows that there are three types of grain boundaries
present (compare Figs. 3, 5 and 6): By far the largest fraction are
straight boundaries (Fig. 6,I). It has shown by electron diffraction
that orientations forming high-coincidence site grain boundaries prevail between adjacent tungsten carbide grains[130] confirming an
earlier suggestion by Gurland (private communication 1967, based
upon earlier studies[117,131] on the angles between the crystal faces
of adjacent carbide grains). According to atom-probe investigations
[53] these straight grain boundaries show no evidence of segregated
cobalt. The second type are faceted grain boundaries which occur
occasionally as for example in Fig. 6 at position II. It is suggested

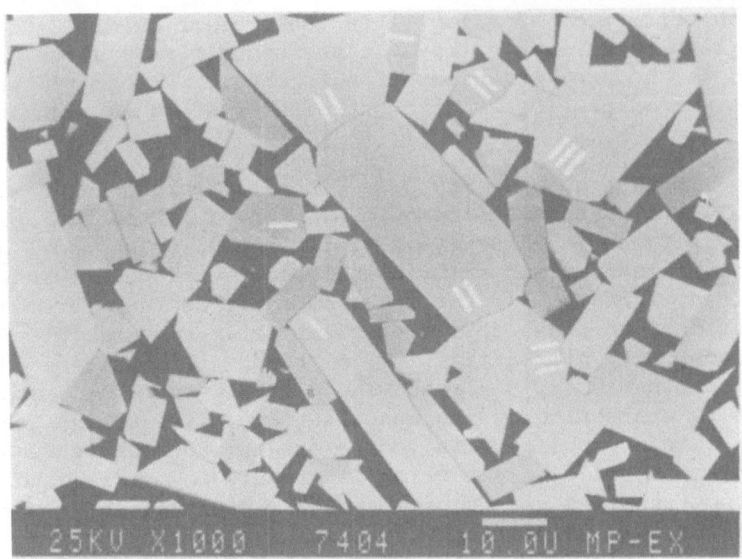

Fig. 6.   Different types of grain boundaries in WC-9 wt.% Co alloys
           annealed 25 h at 1400 °C. I: Straight grain boundaries,
           II: Faceted grain boundaries, III: Curved grain bounda-
           ries (Scanning electron micrograph, backscattered electron
           image).

here that these faceted boundaries form typically between large
grains if the grain boundary does not change its position during
extended times at temperature. The third type are curved grain boundaries (see, for example, Fig. 6, position III) which form between
grains with arbitrary mutual orientation and thus have a high degree
of irregularity[22]. It seems very likely that these grain boundaries
accommodate larger amounts of cobalt which facilitates material transport across the grain boundary and thus coalescence of the adjacent
crystals. During coarsening, the low-energy high-coincidence site
boundaries are statistically favoured, and due to this dynamic selection the fraction of curved boundaries decreases during annealing.
This hypothesis is consistent with most experimental findings reported in literature but needs to be confirmed by more detailed
studies.

     The existence of grain boundaries in cubic carbide based hard-
metals has not been doubted in view of the limited wetting by the
molten binder. The present author does not know of any studies on
the segregation behaviour and the properties of the grain boundaries in cubic carbides and of interfaces developing between different carbide phases in cubic carbide based and multiphase cemented
carbides. More work seems necessary to establish sufficient knowledge in this field since the nature of the grain boundaries and
interfaces has not only interesting theoretical aspects but is also
an important feature for the mechanical behaviour of hardmetals:

Though cobalt layers between all carbide crystals do not seem necessary for explaining plasticity (as suggested for WC-Co alloys in early work, for a discussion and references see[22]), the amount of binder metal segregation and the structure of grain boundaries and interfaces may influence their effectivness as fracture initiation site and as preferred paths for crack formation.

INTERACTIONS BETWEEN CARBIDE AND BINDER

From the effects between carbide and binder phases, some possible reactions occurring between tungsten carbide and cobalt as well as coring of cubic crystals and microstructural coarsening due to solution and precipitation of carbide in and from the binder have been discussed above. Another important microstructural aspect related to carbide-binder interaction is the geometry of the carbide skeleton. A measure of skeleton formation is the contiguity parameter defined by Gurland[131], (see also[22,121]): Contiguity is the ratio of grain boundary density/total surface density of the carbide crystals in the alloy. This parameter is influenced by sintering conditions and strongly dependent on the binder content.For cubic carbides forming rounded grains it has been shown by Warren and Waldron[121] that contiguity is linearly related to the dihedral angle which, in turn, is determined by the ratio of grain boundary energy/and carbide-binder interfacial energy. On the other hand, contiguity seems to be governed by geometrical constraints rather than by equilibrium considerations in those cases where angular crystals predominate as for tungsten carbide based alloys and others where highly anisotropic interfacial energies prevail.

Open questions remain regarding the electronic interchanges taking place at the carbide-binder interface which determine interfacial energy, adhesion and wetting, and thus microstructural geometry. In spite of several ambitious studies (for references see[11]), the effects of small chemical changes as exerted by dopants, impurities and sintering atmospheres are not yet fully understood. Otherwise, the effects of the interaction between carbide and binder phases on the microstructure follow the general rules of physical metallurgy and are relatively easy to interpret. The effect of contiguity on the properties of cemented carbide has been discussed in earlier papers (see for example[6,61]) and more recently by Bartolucci-Luyckx[132] and Gurland[133].

IMPURITIES AND PORES

Modern production methods guarantee a high level of purity and a low amount of residual porosity. Thus, pores and impurities are of rather infrequent occurrence in commercial grades of hardmetals and can hardly be detected by microscopic inspection. Nevertheless, these flaws play a decisive role in fracture initiation and are a main feature in microscopic quality control[12,13,19,134-139].

Additional phases caused by impurities may be visible in polished microsections if larger amounts of ore components are allowed to get through to the final product or if dust or dirt get into sintered parts due to lack of cleanliness during production. Due to the "clinical" conditions introduced by most hardmetal producers during the last decade, the probability of sectioning particles of these phases - if they are present at all - is near zero. In fracture surfaces these particles may show up at sites of crack initiation and are easily detected and analyzed by scanning electron microscopy[19,51].

Pores, on the other hand, can hardly be completely avoided by normal sintering. Lardner and Bettle[136] and Amberg and Doxner[137] have reviewed the causes of residual porosity, and the latter authors have shown that a large area (in the order of 3000 cm²) must be analyzed to get statistically significant information on the pore size distribution. Obviously, pores due to different causes have different appearances, and a full range of shapes and sizes may exist in poorly produced grades. Isostatic hot pressing has been introduced to reduce residual porosity during the seventies and proven successful in those cases where large pores are especially harmful[136,137,142,143].

Finally, the so-called carbon pores should be mentioned. Free carbon is usually removed during polishing and therefore appears as porosity in a metallographic section. Very similarly as described for eta phase above, free carbon may be formed during liquid phase sintering or precipitate during slow cooling. In the first case, large crack-like features are visible in the microstructure, while small black spots occur in the second case[40]. Both types of free carbon can act as stress raisers and are harmful to toughness if not even larger pores caused by other reasons play a major role in fracture initiation[40,144].

CONCLUDING REMARKS

The material reviewed here shows that the microstructure of cemented carbides is an interesting and rewarding field of metallographic research. New techniques have opened a deeper insight than was possible only a few years ago, but a large variety of details is still open for further investigations some of which need sophisticated instrumentation and techniques of research. A large majority of investigations has been carried out on tungsten carbide-cobalt alloys. The most intriguing open questions in this system are the nature of the grain boundaries and carbide-binder interfaces, the kinetics of eta-phase and Co_3W precipitation during cooling and annealing, and the exact correlations between the particular microstructural features and the properties. Other alloys have found less attention though some of them are standard grades and widely used in technical applications. Above all, the microstructure of the three-phase alloy (Ti, W)C-WC-Co should be investigated in detail in order to be able to understand the behaviour of the more complex cemented hardmetals. Also, more quantitative relationships between microstructural parameters and properties relevant for practical use should be established. With the large variety of high-power instruments and useful techniques available, considerable progress should be achieved in the near future.

REFERENCES

1. W.D. Kingery, E. Niki and M.D. Narasimhan, Sintering of Oxide and Carbide-Metal Compositions in Presence of a Liquid Phase, J. Amer. Cer. Soc. 44: 29 (1961).

2. T.J. Whalen and M. Humenik, Mechanisms and Microstructural Aspects of Liquid Phase Sintering, Progr. Powder Met. 18: 85 (1963).

3. H.E. Exner, Ostwald-Reifung von Übergangsmetallkarbiden in flüssigem Nickel und Kobalt, Z. Metallkde. 64:273 (1973).

4. W.J. Huppmann and G. Petzow, Particle Rearrangement during Liquid Phase Sintering of Several Carbide-Metal Combinations, in: "Modern Developments in Powder Metallurgy," Vol. 9, p. 77, Metal Powder Industries Federation, Princeton (1977).

5. M. Coster, Etude du Paramétre de Voisinage, Metallography 9:415 (1976).

6. H.E. Exner and J. Gurland, A Review of Parameters Influencing some Mechanical Properties of Tungsten Carbide Cobalt Alloys, Powder Met. 13: 13 (1970).

7. B.O. Sundström, Elastic-Plastic Behaviour of WC-Co Analysed by Continuum Mechanics, Mat. Sci. Eng. 12: 265 (1973).

8. B. Paul, Prediction of Elastic Constants of Multiphase Materials, Trans. Met. Soc. AIME 218: 36 (1960)

9. J. Gurland, "Some Aspects of the Fracture of Metallic Composites," in: Fundamental Phenomena in the Materials Sciences, Vol. 4, p. 177, Plenum Press, New York (1967)

10. H.C. Lee and J. Gurland, Hardness and Deformation of Cemented Tungsten Carbides, Mat. Sci. Eng. 33: 125 (1978).

11. H.E. Exner, Physical and Chemical Nature of Cemented Carbides, Intern. Met. Rev. 24: 149 (1979).

12. G.A. Wood, Quality Control in the Hardmetal Industry, Powder Met. 13: 338 (1970).

13. E. Lardner and G.A. Wood, Quality Control and Testing Methods for Hardmetals, in: Recent Advances in Hardmetal Production, Vol. 1, Paper 30, Loughborough University and Technology an Metal Powder Report (1979).

14. A. Op Het Veld and P. Borgers, A Preparation Method for Revealing the Structure of Hard Metals, Pract. Metallogr. 4: 235 (1967).

15. W. Peter, E. Kohlhaas and O. Jung, Revealing of Hard Metal Structures by Interference Vapour Deposition, Pract. Metallogr. 4: 284 and 605 (1967).

16. H. Grewe, Structural Investigations on Hard Metals,
       Pract. Metallogr. 6: 411 (1969).
17. R. Arndt and E. Hillnhagen, Special Techniques for the Metallo-
       graphic Examination of Hard Metals, Pract. Metallogr.
       9: 309 (1972).
18. W.U. Kopp and O. Linke, Preparation and Structure of Sintered
       Carbides, Pract. Metallogr. 16: 257 (1979).
19. W. Mader, G. Grassmück, J. Blaha and P. Warbichler,
       Identifizierung von Poren und Inhomogenitäten im Gefüge
       von Sinterhartmetallen, Pract. Metallogr., Special Issue
       10: 155 (1979).
20. J. Hinnüber, O. Rüdiger and W. Kinna, An Electron-Microscope
       and X-Ray Investigation of the Milling of Tungsten
       Carbide/Cobalt Mixtures, Powder Met. 8: 1 (1961).
21. J. Drott, "The Observation of an Intermediate Layer in WC-Co
       Alloys," in: Proc. 3rd European Reg. Conf. Electron Micro-
       scopy, p. 159, Publishing House of Czechoslovak Academy
       of Sciences, Prague (1964).
22. H.E. Exner and H.F. Fischmeister, Gefügeausbildung von ge-
       sinterten Wolframkarbid-Kobalt-Legierungen, Archiv
       Eisenhüttenwesen 37: 417 (1966).
23. H.E. Exner and H.F. Fischmeister, Die Anwendung der Linear-
       analyse zur quantitativen Gefügebeschreibung von Hart-
       metallen, Prakt. Metallogr. 3: 18 (1966).
24. B. Lehtinen, Increasing the Three-Dimensional Impression of
       Replicas, Pract. Metallogr. 6: 500 (1969).
25. E. Hillnhagen and J. Willbrand, Stereometric Electron Metallo-
       graphy on WC-Co Alloys, Pract. Metallogr. 6: 135 (1969).
26. A. Hara, T. Nishikawa and S. Yazu, The Observation of the
       Fracture Path in WC-Co Cemented Carbide Using a Newly
       Developed Replica Method with Electron Microscope,
       Planseeber. Pulvermet. 18: 28 (1970).
27. B.O. Jaensson, Deformation Phenomena of High Strength WC-Co
       Alloys Studied by Means of a New Electron Microscope
       Replica Locating Technique, Pract. Metallogr. 9: 624
       (1972).
28. W. Mader, Über die Anwendung der Elektronenmikroskopie zur
       Untersuchung von Hartmetallen, Radex-Rundschau: 247 (1956)
29. K. Hayashi, H. Suzuki and I. Kawakatsu, Observations on Sur-
       face Cracks and Fracture Surfaces of Cemented Carbides,
       J. Japan Inst. Met. 32: 993 (1968).
30. E. Kohlhaas, D. Jung and A. Fischer, Electron Microscope
       Studies of Fracture Surfaces in Hard Metals, Pract.
       Metallogr. 9: 243 (1972).
31. M. Coster, Application des méthodes quantitatives et stéréo-
       logiques à l'etude de la croissance et des charactér-
       istiques méchaniques d'échantillons polyphasés: cas des
       matériaux carburés de tungstène-cobalt. Ph. D. Thesis,
       University of Caen (1974).

32. A.N. Pilyankevich, V.N. Paderno, V.N. Klimenko and V.A. Maslyuk, An Electron Microscopical Study of Hard Alloys Based on Chromium Oxide, Soviet Powder Met. Met. Ceram. 552 (1977).

33. S. Bartolucci and H. H. Schlossin, Plastic Deformation Preceeding Fracture in WC-Co Alloys, Acta Met. 14: 377 (1966).

34. G. Persson, Thin Foils of WC-Co Hard Metals, Nature 218: 159 (1968).

35. B. Lehtinen, The Thinning of Tungsten Carbide-Cobalt Hard Metals to Electron Transparency by Electropolishing, J. Sci. Inst. (J. Phys. E.)1: 673 (1968).

36. A. Hara, T. Nishikawa and T. Nishimoto, Transmission Electron Microscopy of WC-Co Cemented Carbides, J. Japan Soc. Powders and Powder Met. 16: 310 (1970).

37. T. Johannesson and B. Lehtinen, The Analysis of Dislocation Structure in Tungsten Carbide by Electron Microscopy, Phil. Mag. 24: 1079 (1971).

38. R. Arndt, Plastizität von Hartmetallen auf WC-Co Basis, Z. Metallkde. 63: 274 (1972).

39. H. Johnsson, Studies of the Binder Phase in WC-Co Cemented Carbides Heat Treated at 650 °C, Powder Met. 15: 29 (1972).

40. H. Johnsson, Composition and Microstructure of Binder Phase of Slowly Cooled WC-Co Cemented Carbides, Planseeber. Pulvermet. 21: 187 (1973).

41. H. Le Roux, Microstructure of Reversibly Crystallising Turbostratic Graphite in Heat-Treated WC-Co Hard Metals, Acta Met. 24: 299 (1976).

42. V.K. Sarin and T. Johannesson, On the Deformation of WC-Co Cemented Carbides, Metals Science 9: 472 (1975).

43. O. Rüdiger and H.E. Exner, Application of Basic Research to the Development of Hard Metals, Powder Met. Intern. 8: 7 (1976).

44. H. Gahm, S. Karagöz and G. Kompek, Metallographic Methods for the Characterization of the Microstructure of Cemented Carbides, Pract. Metallogr. 18: 14 (1981).

45. H.F. Fischmeister, Review on Discussions of the Colloquium on Cemented Carbides in Ehrendal, Jernkont. Ann. 147: 200 (1963).

46. W. Peter, O. Jung and E. Kohlhaas, Kobalt in den Karbiden von Hartmetallen, in: Preprints 2nd European Powder Metallurgy Symposium, Vol. 2, Paper 7.4, Stuttgart (1968).

47. W. Peter, E. Kohlhaas and O. Jung, Über die inhomogene Zusammensetzung von Wolframkabid und von Mischkarbiden WC-TiC und WC-TiC-TaC in Hartmetallen, in: Preprints 2nd European Powder Metallurgy Symposium, Vol. 2, Paper 7.6, Stuttgart (1968).

48. W. Mader and K.F. Müller, Identifizierung der in technischen Hartmetallen auftretenden Phasen mit Hilfe der Mikrosonde, Planseeber. Pulvermet. 16: 39 (1968).

49. W. May, Zum Reaktionsverhalten übersättigter (Wolfram, Titan, Tantal)-Karbide in flüssigem Wolfram, Techn. Mitt. Krupp 30: 15 (1972).

50. U. Bäckmann and B. Lehtinen, Quantitative X-Ray Analysis of Tungsten Carbide-Cobalt Cemented Carbides in the Scanning Electron Microscope, J. Sci. Inst. (J. Phys. E.) 4: 955 (1971).

51. S. Bartolucci-Luyckx, Role of Inclusions in the Fracture Initiation Process in WC-Co Alloys, Acta Met. 23: 109 (1975).

52. H. Suzuki, K. Hayashi and T. Yamamoto, The Influence of Anomalous Phases on the Strength of Titanium Base Cermets, Planseeber. Pulvermet. 26: 92 (1978).

53. A. Henjered, M. Hellsing, H. Nordén and H.O. Andrén, Atom-Probe and STEM X-Ray Microanalysis of WC-Co Cemented Carbides, This conference (1981).

54. D.T. Quinto, M.N. Haller and G.J. Wolfe, Low-Z Element Analysis in Hard Materials, This conference (1981).

55. A. Krawitz, E. Drake, R. DeGroot, C. Vasel and W. Yelon, Neutron Diffraction of Cemented Carbide Composites, This conference (1981).

56. C.S. Yust and P.S. Sklad, Characterization of TiB_2-Ni Cermets by Transmission and Analytical Electron Microscopy, This conference (1981).

57. N.K. Sharma and W.S. Williams, Microchemical Analysis of Grain Boundaries in Cemented Carbides, J. Amer. Cer. Soc. 58: 1031 (1979).

58. C. Lea and B. Roebuck, Fracture Topography of WC-Co Hard-metals, Metals Science 15: 262 (1981).

59. R.K. Viswanadham, T.S. Sun, E.F. Drake and J.A. Peck, Quantitative Fractography of WC-Co Cermets by Auger Spectroscopy, J. Mat. Sci. 16: 1029 (1981).

60. J. Gurland and P. Bardzil, Relation of Strength, Composition and Grain Size of Sintered WC-Co Alloys, Trans. AIME 203: 311 (1955).

61. J. Gurland, The Fracture Strength of Sintered Tungsten Carbide-Cobalt Alloys in Relation to Composition and Particle Spacing, Trans. AIME 227: 1146 (1963).

62. J. Gurland, Current View on the Structure and Properties of Cemented Carbides, Jernkont. Ann. 147: 4 (1963).

63. K.G. Stjernberg, Some Relationships between the Structure and the Properties of WC-TiC-Co Alloys, Powder Met. 15:1 (1970).

64. H.E. Exner, E. Santa Marta, and G. Petzow, "Grain Growth in Liquid-Phase Sintering of Carbides," in: Modern Developments in Powder Metallurgy, Vol. 4, p. 315, Plenum Press, New York (1971).

65. R. Warren, Microstructural Development During the Liquid Phase Sintering of VC-Co Alloys, J. Mat. Sci. 7: 1434 (1972).

66. G. Grathwohl and R. Warren, The Effect of Cobalt Content on
     the Microstructure of Liquid-Phase Sintered TaC-Co
     Alloys, Mat. Sci. Eng. 14: 55 (1975).
67. L. Lindau and K.G. Stjernberg, Grain Growth in TiC-Ni-Mo and
     TiC-Ni-W Cemented Carbides, Powder Met. 19: 210 (1976).
68. J.L. Chermant and M. Coster, Quantitative Analysis of
     Parcticle Growth in TiC-Ni Alloys, J. Microscopy 109:
     269 (1977).
69. H.F. Fischmeister, Geräte und Verfahren der quantitativen
     Metallographie, Prakt. Metallogr. 2: 251 (1965).
70. H.F. Fischmeister, Characterization of Porous Structure by
     Stereological Measurements, Powder Met. Intern. 7: 178
     (1975).
71. H.E. Exner, Methods and Significance of Particle- and Grain-
     Size Control in Cemented Carbide Technology, Powder Met.
     13: 429 (1970).
72. H.E. Exner and H.P. Hougardy, Einführung in die quantitative
     Gefügeanalyse (Introduction to Quantitative Analysis of
     Microstructures), Monography edited by Deutsche Gesell-
     schaft für Metallkunde, Oberursel (1982), in print .
73. T. Werlefors and C. Eskilsson, On-Line Computer Analysis of
     WC-Co Structures Imaged in a SEM, Metallography 12: 153
     (1979).
74. H.E. Exner and H.L. Lukas, The Experimental Verification of
     the Stationary Wagner-Lifshitz Distribution of Coarse
     Particles, Metallography 4: 325 (1971).
75. C.A. Stickels and E.E. Hucke, Measurements of Dihedral Angles,
     Trans. Met. Soc. AIME 230: 795 (1964).
76. E.J. Sandford and E.M. Trent, "The Physical Metallurgy of
     Sintered Carbides," in: Spec. Rep. No 38, p. 84, Iron and
     Steel Institute, London (1947).
77. J. Willbrand and U. Wieland, The Size of Coherent Domains in
     the Binder Metal of Cobalt-Bonded Tungsten Carbide,
     Intern. J. Powder Met. 8(2): 89 (1972).
78. R.K. Viswanadham, Binder Grain Size in Cemented Carbides,
     Metallography 12: 333 (1979).
79. H. Suzuki, K. Hayashi and Y. Taniguchi, Effects of Domain
     Size of Binder Phase on High Temperature Strength of WC-
     Co Cemented Carbides, Planseeber. Pulvermet. 27: 215
     (1979).
80. H. Le Roux, "The Martensitic Transformation and the Transverse
     Rupture Stength of Heat Treated WC 25 wt% Co Hard-
     Metals," in: Proceed. 10th Plansee Seminar, Vol. 1, p. 529
     Metallwerk Plansee, Reutte, Austria (1981).
81. E.A. Almond and B. Roebuck, The Origin of WC Substructure and
     the Effect of Processing on the Microstructure of WC-Co
     Hardmetals, in: Proceed. 10th Plansee Seminar, Vol. 1,
     p. 659, Metallwerk Plansee, Reutte, Austria (1981).

82. H.F. Fischmeister and H.E. Exner, Gefügeabhängigkeit der
     Eigenschaften von Wolframkarbid-Kobalt-Hartlegierungen,
     Archiv Eisenhüttenwesen 37: 499 (1966).

83. K.H. Zum Gahr and A. Fischer, Einfluß des Gefüges von WC-Co
     Hartmetallen auf die Bruchzähigkeit und den Abrasiv-
     verschleiß, Metall 35: 38 (1981).

84. J. Freytag, Auswirkung von Zusätzen auf die Bindephase einer
     WC-12 Gew. % Co Hartlegierung, Ph. D. Thesis, Universität
     Stuttgart (1977).

85. H.E. Exner, J. Freytag, G. Petzow and P. Walter, Sinterverlauf
     eines WC-12 Gew.% Co Hartmetalls mit unterschiedlicher
     Bindephasenzusammensetzung, Planseeber. Pulvermet. 26: 90
     (1978).

86. H.F. Fischmeister and H.E. Exner, Beobachtungen über den
     Mahlvorgang bei Hartmetallpulvern, Planseeber. Pulvermet.
     13: 178 (1965).

87. D.G. Evans, "The Relationships between Structure and Perfor-
     mance of Cemented Carbide Tools," in: Proceed. Recent
     Advances in Hardmetal Production, Vol. 2, Paper 39,
     Metal Powder Report, London (1979).

88. H. Suzuki, T. Yamamoto and H. Sakanoue, Binder Phase Trans-
     formation in WC-Co Cemented Carbides, J. Japan Inst.
     Metals 32: 993 (1968).

89. A.F. Giamei, J. Burma, S. Rabin, M. Cheng and E.J. Freise,
     The Role of Allotropic Transformation for Cobalt Alloys,
     Cobalt 40: 140 (1968).

90. S.M. Brabyn, R. Cooper and C.T. Peters, "Effects of Substitution
     of Nickel for Cobalt in WC Based Hardmetal," in: Proceed.
     10th Plansee Seminar, Vol. 2, p. 675, Metallwerk Plansee,
     Reutte, Austria (1981).

91. B. Roebuck and E.A. Almond, A Comparison of the Deformation
     Characteristics of Co and Ni Alloys Containing Small
     Amounts of W and C, in: Proceed. 10th Plansee Seminar,
     Vol. 1. p. 493, Metallwerk Plansee, Reutte, Austria (1981).

92. D. Moscowitz, M.J. Ford and M. Humenik, High Strength Tungsten
     Carbides, Intern. J. Powder Met. 6 (1): 55 (1970).

93. L. Prakash, Weiterentwicklung von Wolframkarbid-Hartmetallen
     unter Verwendung von Eisen-Basis-Bindelegierungen, Ph. D.
     Thesis,Universität Karlsruhe, Kernforschungszentrum
     Karlsruhe, Report KfK 2984 (1981).

94. L. Prakash, H. Holleck, F. Thummler and P. Walter, "The Influence
     of Binder Composition on the Properties of WC-FE/Co/Ni
     Cemented Carbides," in: Modern Developments in Powder Metal-
     lurgy, Vol. 14, p. 255, Metal Powder Industry Federation,
     Princeton (1981).

95. F. Thummler, H. Holleck and L. Prakash, "Ergebnisse zur Weiter-
     entwicklung von Hartstoffen und Hartmetallen," in: Proceed.
     10th Plansee Seminar, Vol. 1, p. 459, Metallwerk Plansee,
     Reutte, Austria (1981).

96. H. Johnsson and B. Aaronson, Microstructure and Hardness of
     Cobalt-Rich Co-W-C Alloys after Ageing in the Temperature
     Range 400-1000 °C, J. Inst. Metals 97: 281 (1969).

97. O. Rudiger, Compostion and Properties of WC-Co Alloys, Int.
     J. Powder. Met. 7: 29 (1971).

98. D.L. Tillwick and I. Joffe, Magnetic Properties of Co-W Alloys
     in Relation to Sintered WC-Co Compacts, Scripta Met.
     5: 479 (1973).

99. A. Hoffmann and R. Mohs, Gleichgewichtsuntersuchungen im ko-
     baltreichen Teil des Systems Co-W-C bei 1250 °C, Metall
     28: 661 (1974).

100. H. Grewe and J. Kolasha, "Gezeite Einstellung von Lösungszu-
     standen in der Bindephase technischer Hartmetalle und
     Folgerungen daraus," Metal 35 (563 (1981), and in: Proceed.
     10th Plansee Seminar, Vol. 1, p. 509, Metallwerk Plansee,
     Reutte, Austria (1981).

101. G. Wirmark and G.L. Dunlop, Phase Transformation in the Binder
     Phase of Co-W-C Cemented Carbides, This conference (1981).

102. O. Rüdiger and A. Hoffmann, Reine Kobaltlegierungen mit Zusätzen
     von Kohlenstoff, Wolfram und Titan, Metall 24: 723 (1970).

103. W.D. Schubert, P. Ettmayer, B. Luxana, W. Ohlsson, "Phasengleich-
     gewichte in den Systemen Co-Mo-W-C und Ni-Mo-W-C," in:
     Proceed. 10th Plansee Seminar, Vol. 2. p. 871, Metall-
     werk Plansee, Reutte, Austria (1981).

104. R.K. Viswanadham, P. Lindquist and J.A. Peck, Preparation and
     Properties of WC-(Ni,Al) Cemented Carbides, This conferen-
     ce (1981).

105. D. Moskowitz and M. Humenik, "Cemented TiC Base Tools with
     Improved Deformation Resistance," in: Modern Developments
     in Powder Metallurgy, Vol. 14, p. 307, Metal Powder
     Industry Federation, Princeton (1981).

106. H. Holleck, and H. Kleykamp, "The Constitution of Cemented Car-
     bide Systems," in: Modern Developments in Powder Metallurgy,
     Vol. 14, p. 234, Metal Powder Industry Federation,
     Princeton (1981).

107. P.O. Snell, The Effect of Carbon Content and Sintering Tempera-
     ture on Structure Formation and Properties of a TiC-24 %
     Mo-15 % Ni Alloy, Planseeber. Pulvermet. 22: 91 (1974).

108. A. Taylor and R.W. Floyd, The Constitution of Nickel-Rich Alloys
     of the Nickel-Titanium-Aluminum System, J. Inst. Metals
     81: 25 (1952/3).

109. H. Doi and K. Nishigaki, "Binder Phase Strengthening through
     Precipitation of Intermetallic Compound in Titanium Car-
     bide Base Cermet with High Binder Concentration," in:
     Modern Developments in Powder Metallurgy, Vol.10, p. 525
     Metal Powder Industries Federation, Princeton (1977).

110. H. Tulhoff, "On the Grain Growth of Tungsten Carbide in Cemented
     Carbides," in: Modern Development in Powder Metallurgy,
     Vol. 14, p. 247, Metal Powder Industries Federation,
     Princeton (1981).

111. M. Schreiner, E. Alizadeh, T. Schmitt, E. Lassner and B. Lux, "Einfluß geringer Konzentrationen von Fremdelementen auf die WC-Co-Hartmetallsinterung und die Produktionseigenschaften," in: Proceed. 10th Plansee Seminar, Vol. 2, p. 811, Metallwerk Plansee, Reutte, Austria (1981).

112. H. Grewe, H.E. Exner and P. Walter, Behinderung des Kornwachstums in Hartmetall-Legierungen vom ISO-K 10-Typ durch Zusatzkarbide, Z. Metallkde. 64: 85 (1973).

113. G.J. Rees and B. Young, A Study of the Factors Controlling Grain Size in Sintered Hard-Metal, Powder Met. 14: 1 (1971).

114. K.M. Friederich, H.E. Exner and H. Tulhoff, "Quantitative Erfassung des diskontinuierlichen Kornwachstums in WC-Co-Hartligierungen," Proceed. 10th Plansee Seminar, Vol. 2, p. 795, Metallwerk Plansee, Reutte, Austria (1981).

115. K.M. Friederich, Quantitative Erfassung der Vergröberungsneigung von Wolframkarbidkristallen in Hartlegierungen, Ph. D. Thesis, Universität Stuttgart (1981).

116. H. Tulhoff, Discussion of K.M. Friederich and H.E. Exner, Quantification of Continuous and Discontinuous Growth in Cemented Cabides, This conference (1981).

117. J. Gurland, A Study of the Effect of Carbon Content on the Structure and Properties of Sintered WC-Co Alloys, Trans. AIME 200: 285 (1954).

118. H.E. Exner, A. Walter, P. Walter and G. Petzow, Auswirkung der Wolframkarbid-Ausgangspulver auf gesinterte Wolframkarbid-Hartmetalle, Metall 32: 443 (1978).

119. M.B. Waldron, Stand und Forschung auf dem Gebiet der Hartmetalle, Neue Hütte 23: 317 (1978).

120. W. May, Phase Decomposition and Grain Growth in (W,Ti)C-Co Alloys, J. Mat. Sci. 6: 1209 (1971).

121. R. Warren and M.B. Waldron, Microstructural Development during the Liquid-Phase Sintering of Cemented Carbides, Powder Met. 15: 166 (1972).

122. R. Warren, Effects of Carbide Composition on the Microstructure of Cemented Binary Carbides, Planseeber. Pulvermet. 20: 299 (1972).

123. H. Suzuki and T. Yamamoto, Spotted Structures in Cemented Carbides Containing Small Amounts of Tantalum and Niobium Carbide, J. Japan Soc. Powd. Met. 16: 235 (1969).

124. S. Bartolucci-Luyckx, "Some Features of the Eta Phase in Substoichiometric WC-Co Alloys," in: Proceed. 10th Plansee Seminar, Vol. 1, p. 629, Metallwerk Plansee, Reutte, Austria (1981).

125. C.T. Peters and R. Cooper, Effects of Eta Phase Precipitation on the Structure and Mechanical Properties of WC-5 % Hardmetals, Planseeber. Pulvermet. 26: 181 (1978).

126. W. May and E. Krämer, Volumendiffusion und ihr Einfluß auf die Mischkristallbildung, Planseeber. Pulvermet. 22: 107 (1974).

127. E. Rudy, S. Worchester and W. Elkington, Modified Spinodal Alloys for Tool and Wear Applications, High Temperatures-High Pressures 6: 497 (1974).

128. H. Holleck, Constitutional Aspects in the Development of New Hard Materials, This conference (1981).

129. C. Lea and B. Roebuck, Fracture Topography of WC-Co Hardmetals, Metals Science 15: 262 (1981).

130. S. Hagége, Structure and Importance of Interfaces in WC-Co Composites, Lecture at Max-Planck-Institut Stuttgart (1981), to be published.

131. J. Gurland, Observations on the Structure and Sintering Mechanism of Cemented Carbides, Trans. Met. Soc. AIME 215: 601 (1959).

132. S. Bartolucci-Luyckx, "Contiguity and Fracture Process of WC-Co Alloys," in: Proceed. 5th International Conference on Fracture, Cannes (1981), to be published.

133. J. Gurland, A Structural Approach to the Yield Strength of Two-Phase Alloys with Coarse Microstructures, Mat. Sci. Engineering 40: 59 (1979).

134. P.B. Anderson, Hartmetalle erhöhter Zähigkeit, Planseeber. Pulvermet. 15: 180 (1967).

135. N.I. Romanova, G.S. Kreimer and V.I. Tumanov, Effect of Residual Porosity on the Fatigue Life of Tungsten-Carbide Cobalt Alloys Subjected to Cyclic Cantilever Bending, Soviet Powder Met. Met. Ceram.: 737 (1979).

136. E. Lardner and D.J. Bettle, Isostatic Hot Pressing of Cemented Carbides, Metals and Materials 7: 540 (1973).

137. S. Amberg and H. Doxner, Porosity in Cemented Carbide, Powder Met. 20: 1 (1977).

138. H. Suzuki and K. Hayashi, Strength of WC-Co Cemented Carbides in Relation to their Fracture Sorces, Planseeber. Pulvermet. 23: 24 (1975).

139. H. Suzuki, K. Hayashi and T. Yamamoto, The Influence of Anomalous Phases on the Strength of Titanium Carbide Phase Cemented Cermets, Planseeber. Pulvermet. 26: 42 (1978).

140. E.A. Almond, Strength of Hardmetals, Metal Science 12: 587 (1978).

141. E.A. Almond and B. Roebuck, The Mechanical Testing of Cemented Carbides, Trans. J. British Ceram. Soc. 79: 53 (1980).

142. H. Grewe and J. Kolaska, Vorgänge und Eigenschaftsveränderungen beim heißisostatischen Nachverdichten von Hartmetallen, Metall 32: 989 (1978).

143. E. Lardner, Isostatic Hot Pressing of Cemented Carbide, Powder Met. 18: 47 (1975).

144. D.N. French and D.A. Thomas, The Nature and Effects of Excess Carbon Defects in Carbide-Cobalt Alloys, Intern. J. Powder Met. 3 (3): 7 (1967).

DISCUSSION

F. Rymas:

Dr. Exner, occasionally during the manufacture of very fine grained
WC-Co alloys very large crystals can form in a matrix of very fine WC
particles. What are the possible causes for this occasional very
exaggerated grain growth?

H. E. Exner:

The next lecture by one of my students who just got his degree is on
this subject. But I would like to point out that exaggerated grain
growth in all materials we know of, seems to be related to one
factor. If you're just at the edge where one grain growth mechanism
goes into another one, for example, precipitation controlled to
grain boundary movement controlled--at this edge you are prone to
get large grains. I think that discontinuous grain growth is one of
the least understood phenomena and I don't have a better explanation.
We just know that you can stop it by adding tantalum carbide which
in fact is done. People in technology know how to prevent it but I
don't think that scientists understand it yet.

G. Dearnaley:

You mentioned that carbon stimulates diffusion in the solid phase
during sintering. Is it known how this happens and what defects
take part in the diffusion mechanism?

H. Exner:

When I talked about diffusion I meant surface diffusion of cobalt
over tungsten carbide. The coercive force increases much earlier if
there is a high cobalt content than if there is a low one. This is
the only indication that the distribution of cobalt is related to
the carbon content and I also think that carbon stimulates all the
thermally activated procsses in WC-Co and I don't think this is
understood yet. Even the effect of TaC addition to the others--I
don't think that the real mechanism is understood. It is also
related to the carbon content of the binder phase. But why? And
how? I don't know.

R. Warren:

If you wish to invoke the chemical-driven coalescence process in
WC-Co, this implies a certain amount of solubility of cobalt in
tungsten carbide, does it not?

H. E. Exner:

No. Let me switch over to titanium carbide again. If you looked at
this structure, you saw these big black flakes which were carbon
that came out of the titanium carbide. So I think the composition
of the carbide itself, with respect to carbon content is different.

R. Warren:

So you mean it could also be brought about by carbon.

H. E. Exner:

Yes, the carbide powder you use is not in equilibrium with the binder
phase, with respect to carbon content. In the WC-Co system, the
range is so narrow it's hard to observe and it's also hard to show.
But in other carbides you can easily show that the titanium carbide
phase which is precipitated is of a different composition than the
one inside, only with respect to carbon content.

R. Warren:

It should perhaps be possible to do a quanititative theoretical
calculation of the driving force if one can take into account this
narrow range in carbon.

H. E. Exner:

That's a very good idea because I don't know anything about the
relation between interfacial energy and chemical driven energy in
this system.

P. Sklad:

You mentioned briefly the formation of Co_3W precipitates in the
cobalt binder. Since there's been so much work done on these types
of precipitates in the superalloy industry, I wonder if you might
comment on the possibility of using that sort of precipitation as a
means of changing the properties of the composites?

H. E. Exner:

These studies were done by a Swedish group with essentially carbide-
free alloys. They claim that the properties can be changed quite
dramatically by changing the binder phase composition and by having
Co_3W precipitaties. As I said earlier, it is not known what the
optimum binder properties are due to the fact that they are related
so much to performance. But I think there is large potential in
heat treating even WC-Co alloys. Moskowitz and Prakash have shown

that if you use other binders, nickel-iron binders, heat treatment
really makes a tremendous difference in properties and wear behavior.
But there is a potential in heat treating and to use these precipitates as hardening agents.

A METHOD FOR QUANTIFICATION OF DISCONTINUOUS GRAIN GROWTH AND

ITS APPLICATION TO CEMENTED CARBIDES

K.M. Friederich and H. E. Exner

Max-Planck-Institut für Metallforschung
Institut für Werkstoffwissenschaften
Stuttgart, FR Germany

ABSTRACT

The discontinuous grain growth in tungsten carbide-cobalt alloys is investigated quantitatively. For this purpose a mathematical procedure was developed to separate the size distribution of continuously and discontinuously grown carbide crystals. It is assumed that the linear intercept distributions of both fractions can be described by two-parametric functions of the same type (e.g. logarithmic normal distribution). With this method variations in tendency for discontinuous growth of tungsten carbide powders can be characterized. The application is shown for cemented carbides with 12 wt-% Co produced from different WC-powders annealed at 1400 °C up to 25 h.

INTRODUCTION

Controlling grain growth during sintering is of major importance in production of technical hard metals due to the fact that carbide grain size has a pronounced effect on the mechanical properties[1-3]. In spite of careful control, tungsten carbide powders show varying tendency for continuous[4] and discontinuous[5] grain growth. The reasons for these variations are not yet fully clear, and there is no exact method available to predict the coarsening behaviour of a particular tungsten carbide powder. In order to find quantitative parameters for growth tendency, a mathematical method has been developed for separating the continuous and the discontinuous grain fractions and for characterizing both fractions separately. This method may be useful for separating superimposed distributions of grains and particles in other materials as well.

EXPERIMENTAL DETERMINATION OF INTERCEPT DISTRIBUTIONS

From metallographic cross sections the spatial distribution
of crystal sizes cannot be obtained directly since only statistical
sections of the crystals are visible. These can be characterized
using planar parameters like intersect area, maximum diameter,
equivalent circle diameter, and other quantities related to the
planar section. If automatic detection cannot be used (as in case
of scanning electron micrographs of cemented carbides) measurement
of linear intercepts is not only more objective and more accurate
but also much faster than planar evaluation. Therefore, linear intercept distribution analysis is advantageously used to characterize
crystal growth.

If two distributions (e.g. those of a continuously and a discontinuously grown crystal fraction) are superimposed, wide size
distributions are observed. As a correlary, small frequencies will
be obtained in the extreme (i.e. the smallest and the largest) size
classes especially if one of the two fractions is represented in
the structure by a small number. In order to fill these classes with
a sufficient number of counts a large number of total counts would
be required. To reduce this effort a method proposed by Fischmeister
et al[6] is adopted here which has been successfully applied for analysing particle size distributions of tungsten carbide powders[6-8].
This method is used here in a slightly modified form for experimental determination of wide intercept length distributions in hard
metals as follows:

Scanning electron micrographs of polished and etched WC-Co
samples are taken at two magnifications adjusted to the range of
crystal sizes present (e.g. 600 and 1500 x, examples are shown in
Figs. 1 and 2). These micrographs are evaluated on a digitizing
tablet. It is convenient to have on-line computer facilities to
store and to process the data. In our laboratory, a Videoplan manufactured by Kontron, Munich, is available in which a high resolution
tablet is equiped with a microprocessor, general purpose computing
facilities, and 64 kbyte storage. Straight parallel lines are drawn
across the micrographs and intercept length of carbide grains are
measured by just touching the end points of each intercept with the
stylus. For classification of intercepts, 30 logarithmically equidistant classes in the range of 0.1 to 100 μm are used corresponding
to a geometric scale with modulus $^{10}\sqrt{10}$.

At the lower magnification (Fig. 1) the coarse fraction
can easily be measured. A large number of intercepts are below
2 μm in this example; these are not considered at this magnification
and only the number of intercepts between 2 μm and maximum length,
l_{max}, are registered. At the higher magnification (see Fig. 2) all
intercepts are measured. In this case, long intercepts (close to

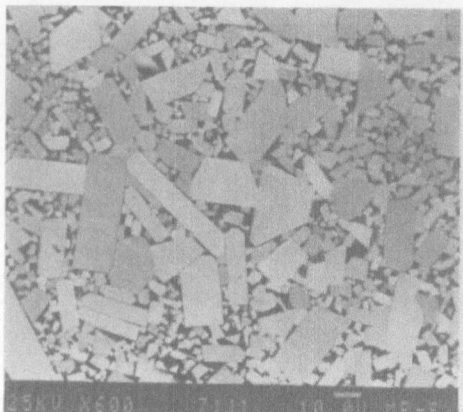

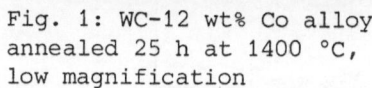

Fig. 1: WC-12 wt% Co alloy   Fig. 2: WC-12 wt% Co alloy
annealed 25 h at 1400 °C,   annealed 25 h at 1400 °C,
low magnification   high magnification

l_{max}) occur at a very low frequency but the classes close to 2 μm
are adequately represented. Summing up for a number of micrographs
taken from the same sample to reduce fluctuations due to inhomogeneity the two data sets are then combined by simply dividing the
frequencies by the total length of lines for the higher and the
lower magnification, respectively.

The superposition procedure is shown in Fig. 3: In the range
where the counts overlap similar frequencies are obtained for both
magnifications and the mean of the two counts is used to represent
frequency in these classes. In the example shown in Fig. 3, these
are three classes ranging from 2.5 to 3.2, 3.2 to 4.0 and 4.0 to
5.0 μm, respectively. The frequencies of smaller classes are taken
from the count made at high magnification and for larger classes
only the count at low magnification is considered. Finally, the distribution is normalized to 100 % total frequency.

This method makes possible to evaluate the frequency of large
crystals with sufficient accuracy at tolerable time expenditure.
If the fine particles are rare, a similar procedure can be used
counting small grains only in micrographs taken at a third (higher)
magnification.

SEPARATION OF SUPERIMPOSED DISTRIBUTIONS

A suitable statistical procedure for separating superimposed
distributions is the maximum-likelihood method which, however, requires a large computing effort. Rather large programming space is
necessary for applying a method used by Howard et al [9]. In this

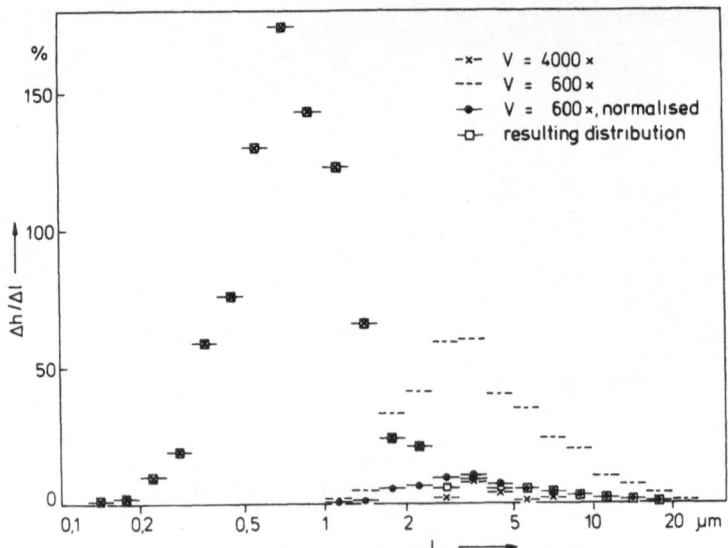

Fig. 3: Determination of intercept
length distribution by superposition
of data measured at 600 and 4000 x,
respectively

method, it is assumed that the individual distributions are of
Gaussian type in a linear or logarithmic scale and the five independent parameters are fitted by iteration. Using this method
with our data confirmed the finding of Howard et al[9] for brain cells
that iteration only converges if the input for the first iteration
step is very close to the final result. Therefore, this method only
works satisfyingly if overlap of two distributions is not pronounced,
i.e. when the experimental distribution exhibits two clearly separated peaks. In order to avoid excessive computing time and to reduce sensitivity for input, a simpler method was developed which
requires moderate storage space and computing capacity and, therefore, is easily implemented for the relatively small Videoplan computer. The only assumption is, in accordance to earlier attempts of
separating bimodal size distributions, that both individual distributions are of the same type of a two parametric distribution
function and that this type (preferentially a Gaussian distribution
with any sensible function of the argument) is known in forehand.
Based on earlier experimental observations which indicate that the
intercept length distribution of tungsten carbide particles in
sintered hard metals follows closely a logarithmic Gaussian (log
normal) distribution, the well known equation of this distribution
[10,11] was used for characterizing the intercept distribution of
each of the two fractions:

$$H(\ln l) = \frac{1}{\ln\sigma \cdot \sqrt{2\pi}} \cdot \int\limits_{0}^{\ln l} \exp\left[-\frac{1}{2} \cdot \left(\frac{\ln l/l_{50}}{\ln\sigma}\right)^2\right] d \ln l \qquad (1)$$

This type of distribution was confirmed to apply to most alloys
investigated here prior to occurence of discontinuous grain growth[12].
We use the usual substitution $x = \ln l$ and $s_g = \ln\sigma$ and get

$$H(x) = \frac{1}{s_g \cdot \sqrt{2\pi}} \int\limits_{-\infty}^{x} \exp\left[-\frac{1}{2}\left(\frac{x-x_{50}}{s_g}\right)^2\right] dx \qquad (2)$$

Figure 4 shows two logarithmic Gaussian distributions A and B
the parameters of which were chosen arbitrarily ($l_{50,A} = 0.8 \ \mu m$,
$x_{50,A} = -0.223$, $s_A = 0.5$; $l_{50,B} = 5 \ \mu m$, $x_{50,B} = 1.609$, $s_B = 0.8$).
These two distributions were superimposed with a numerical con-
centration $c_A = 70 \ \%$, $c_B = 30 \ \%$ which results in the fully drawn
curve in Fig. 4.

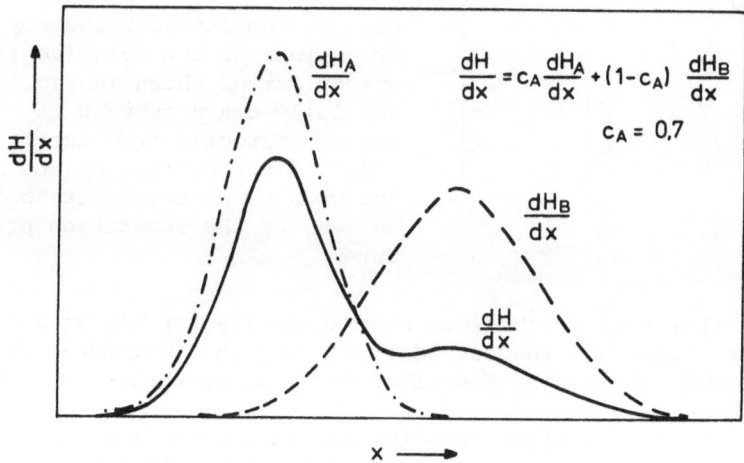

Fig. 4: Superposition of two idealy
logarithmic normal size distributions

In Fig. 5, the corresponding cumulative frequency curves are
displayed in a logarithmic probability paper. In general, the cumulative frequency up to a value of $x = \ln l$ is

$$H(x) = c_A \cdot H_A + (1-c_A) \cdot H_B \qquad (3)$$

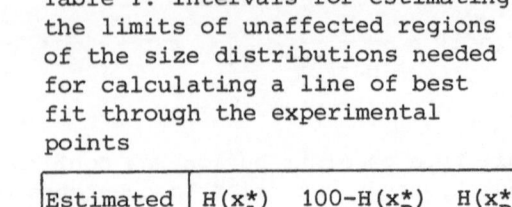

Table 1: Intervals for estimating
the limits of unaffected regions
of the size distributions needed
for calculating a line of best
fit through the experimental
points

Estimated value	$H(x^*_A)$ %	$100-H(x^*_B)$ %	$H(x^*_B)$ %
<0.02	–	95	5
0.02-0.05	1	90	10
0.05-0.10	2	80	20
0.10-0.20	5	60	40
0.20-0.40	10	40	60
0.40-0.60	20	20	80
0.60-0.80	40	10	90
0.80-0.90	60	5	95
0.90-0.95	80	2	98
0.95-0.98	90	1	99
>0.98	95	–	–

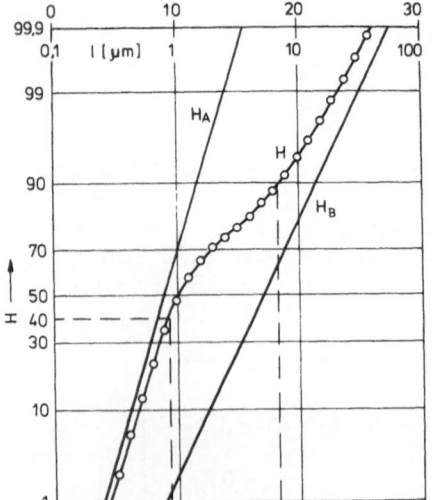

Fig. 5: Cumulative frequency distributions of the relative frequency curves shown in Fig. 4.
The fully drawn curve H is calculated directly from the individual distributions H_A and H_B,
the open circles are data points
derived by the separation procedure

A curve of this type is obtained experimentally for linear intercept distributions, and the method of finding the individual distributions A and B will now be described for this synthetic sample.

In the range of overlapping, the contributions of both fractions must be considered. There is, however, a limiting value x^*_A
below which the contribution of the larger fraction B can be neglected, and a limiting value x^*_B above which the contribution of
the smaller fraction A is negligible. If the median l_{50} of fractions
A and B are pronouncedly different and the geometric standard deviations s_g is in the same order, these limits will be well defined.
In practice these prerequisites do not restrict application of the
method since discontinuously grown crystals are usually pronouncedly coarser than the continuous fraction, while their distribution
width is usually quite similar on a logarithmic scale. On the other
hand, the distributions usually overlap to a degree that two distinct peaks are not discernible due to the fact that statistical

fluctuations of frequencies obscure the picture.

Experience shows that the numerical concentration of the finer fraction A, c_A, lies in a range of $\pm$ 5 % around the turning point of the cumulative frequency curve. In a first step of the separation we take the cumulative frequency at the turning point as a first estimate of c_A which is easily determined by visual inspection of the log normal plot. In the example shown in Fig. 5, the turning point is obviously in the frequency range of 70 to 75 % and we use c_A = 72 %. (In this synthetic case we know the exact value, c_A = 70 %, which we neglect here by purpose for demonstration.) The next step is to choose suitable intervals for defining the limiting values x_A^* and x_B^*. These can be freely chosen , we use integer values of the cumulative frequency equidistantly distributed along the logarithmic scale. The corresponding cumulative frequencies H_A^* and H_B^* can be taken from Table 1. If the estimated value of c_A falls in an interval i, H_A^* is taken as the lower limit of the next lower interval i-1 and H_B^* as the upper limit of the next higher interval i+1. For turning points in the frequency ranges below 2 % and above 98 %, calculation of the distribution parameters for the fraction represented by less than 2 % cannot be estimated to a sufficient degree of accuracy. In these cases further calculations should not be attempted and the corresponding numerical density should be set zero.

In our example, the estimated c_A is 72 % and, skipping the value of 60 %, H_A^* = 40 %. x_A^* then is interpolated using the linear approximation

$$x_A^* = x_{i-1} - \frac{H_{i-1} - H_A^*}{H_{i-1} - H_{i-2}} \cdot (x_{i-1} - x_{i-2}) \qquad (4)$$

where x_{i-1} and x_{i-2} are the upper and lower class limits and H_{i-2} and H_{i-1} the corresponding cumulative frequencies of the experimental interval i-1. From Fig. 5 we get i = 11, H_A^* = 40 %, x_{i-2} = -0.231 and x_{i-1} = 0, H_{i-2} = 35.3 % and H_{i-1} = 48.2 %. Using the equation (4) we find x_A^* = -0.147. x_B^* we obtain in the same way: Starting with c_A = 72 % the cumulative frequency at the upper limit of the next higher interval i+1 is, according to Table 1, 90 %. This value is chosen for H_B^*. x_B^* is determined by interpolation according to equation (4) where H_A^* is substituted by H_B^* and i-2 and i-1 by i+1 and i+2, respectively.

For our example, Fig. 5 clearly shows that the fraction B contributes very little to the superimposed distribution up to the limit x_A^*, namely $(1-c_A/100) \cdot H_B(x_A^*)$ = 0.5 %. Above x_B^*, the contribution of fraction A is $100 - H_A(x_B^*) \cdot c_A/100$ which amounts to less than 0.1 %. It is obvious that neglecting these parts of the distributions will not cause any significant errors.

The next step is to extrapolate the distribution curve for fraction A from the experimental distribution curve in the range $x_O < x < x_A*$ and for fraction B the experimental data in the range x_B* to x_{max} to the full range x_O to x_{max}. (x_O and x_{max} are the lower class limit of the lowest experimental class and the upper class limit of the highest class, respectively). Since a log normal distribution is represented as a straight line in the log normal plot we can find the line of best fit by a regression analysis. We plot $H_{A,j} = H_j/c_A$ versus x_j (where H_j is the experimental cumulative frequency at the upper class limit of class j, x_j) in the range x_O to x_A*, and $H_{B,j} = (H_j-c_A \cdot 100\%)/(1-c_A)$ versus the corresponding x values in the range x_B* to x_{max}.

For the calculation of the lines of best fit, the computer needs fixed values of cumulative frequencies which are chosen as shown in Table 2. From tables for the standard normal distribution [10,11] those multiples y of the standard deviation are taken for which the cumulative frequencies H_{th} correspond to the values given in Table 2. The corresponding x values are found by linear interpolation according to

$$x = x_j - \frac{H_{A,j}-H_{th}}{H_{A,j}-H_{A,j-1}} (x_j - x_{j-1}) \qquad (5)$$

For the regression analysis, y must be chosen as abscissa since y values are fixed while the x values are open to experimental scatter.

Fig. 6 shows the regression lines for the synthetic example. By using the well-known regression formulae the coordinate pairs $x(H_{th})$ and $y(H_{th})$ yield the intercept x_{50} (at y = 0) and the slope s_g of the straight line defined by

$$x(H_{th}) = x_{50,A} + s_{g,A} \cdot y(H_{th}) \qquad (6)$$

Applying the same procedure for fraction B, we get the median $x_{50,B}$ and the geometric standard deviation $s_{g,B}$ for this fraction (Fig. 6). With these results and the estimate c_A we can easily calculate the frequencies of the superimposed distribution by using equation (3).

Following the standard procedure for a χ^2-test, we take the squares of deviations between the experimental relative frequencies $\Delta H_{j(exp)}$ and the relative frequencies $\Delta H_{j(cal)}$ which can be found in good approximation from

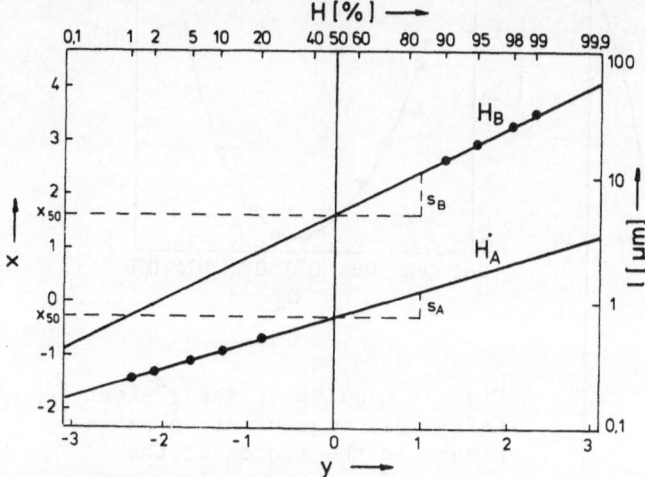

Fig. 6: Determination of distribution parameters by means of line fitting

Table 2. Multiples of standard deviations for specified values of frequencies.

H(x) %	y
1	-2.324
2	-2.053
5	-1.645
10	-1.281
20	-0.841
40	-0.253
50	0
60	0.253
80	0.841
90	1.281
95	1.645
98	2.053
99	2.324

$$\Delta H_{j(cal)} = \frac{1}{s_g \cdot \sqrt{2\pi}} \exp\left[-\frac{1}{2}\left(\frac{x_j - x_{50}}{s_g}\right)^2\right] \cdot (x_j - x_{j-1}) \qquad (7)$$

and obtain χ^2 by summing these deviations for all experimental classes (total number k) according to

$$\chi^2 = \sum_{i=1}^{k} \frac{\left[\Delta H_{j(exp)} - \Delta H_{j(cal)}\right]^2}{\Delta H_{j(cal)}} \qquad (8)$$

The value of χ^2 is determined for a systematic variation of c_A in an interval of ± 0.1 of the initial estimate in steps of $\Delta c_A \doteq 0.01$. The observed minimum value of χ^2 indicates optimum fit. The χ^2 value for a single log normal distribution using the experimental median and the experimental standard deviation is also calculated. For a bimodal distribution this value will be pronouncedly higher than the minimum value indicating that the experimental distribution

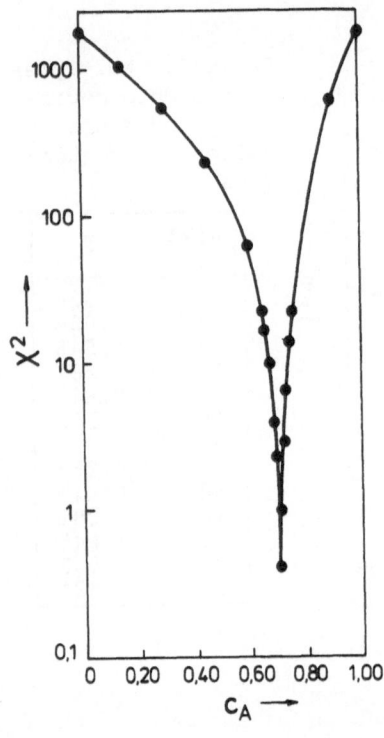

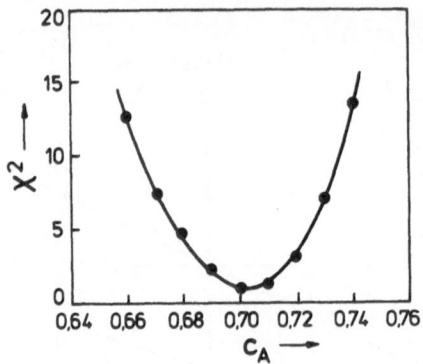

Fig. 7: Results of the χ^2-test. Left: full concentration range, right: in the region of the minimum

can not be described adequately by a single log normal function (i. e. by c_A = 100 % or c_B = 100 %).

   Fig. 7 shows this procedure graphically for the full concentration range (c_A = 0 to 1) and for the range of c_A (estimated) ± 0.1, i.e. between 66 and 74 %. There is a pronounced minimum at c_A = 70 % indicating that the input is well reproduced by separation procedure. Table 3 shows that there is a close agreement between the calculated parameters of the individual distribution and the original values. This close agreement is also demonstrated in Fig. 5 where the data points for the recalculated distribution

Table 3: Comparison between input and calculated size distribution parameters

	fraction A		fraction B		c_A
	l_{50} (µm)	s_g	l_{50} (µm)	s_g	
input	0.8	0.5	5.0	0.8	0.7
calculated	0.789	0.500	5.019	0.796	0.70

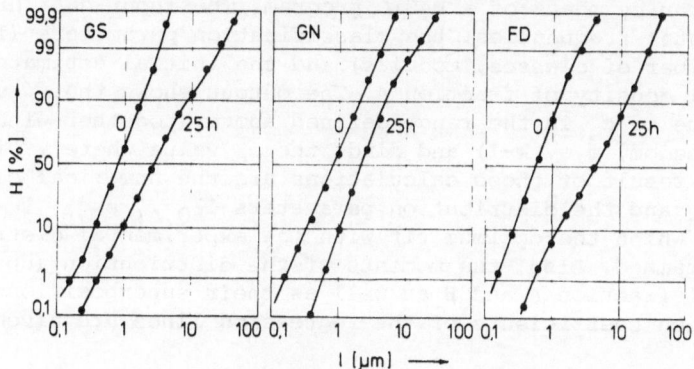

Fig. 8: Experimental distributions of linear intercepts of carbide crystals in the sintered state and after 25 h annealing at 1400 °C

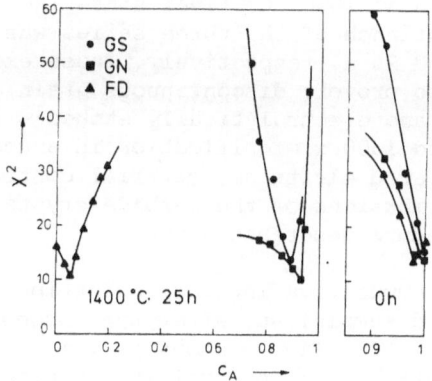

Fig. 9: Results of the χ^2-test for the linear intercept distribution shown in Fig. 8

follow very closely the original distribution curve.

The complete computing procedure is carried out in the Video-plan calculator by means of a BASIC program. The input data are the experimental frequencies, the classification parameters (lowest limit x_o, number of classes, modulus) and the initial estimate for the numerical density of fraction A. The output shows the χ^2 values for each value of c_A in the range defined above (together with the degree of freedom, i.e. k-1) and finds the c_A value where χ^2 is a minimum. The result of these calculations are the numerical concentration c_A and the distribution parameters $l_{50,A}$, $s_{g,A}$, $l_{50,B}$ and $s_{g,B}$ for which the optimum fit with the experimental distribution is obtained. Also, the moments of the distribution (up to the third) of fraction A and B as well as their superpositions and the correlation coefficients of the regression lines are given by the program.

RESULTS FOR WC-Co ALLOYS

Starting with three tungsten carbide powders produced by different technical processes, all steps of industrial hard metal production were followed up at a laboratory scale to produce three series of hard metal samples. An attritor was used for mixing the tungsten carbide with 12 wt% cobalt with all milling parameters (weight of powder, balls and liquid, speed, milling time) kept constant. Green compacts were produced at identical filling volume and compacting pressure, and all samples were sintered simultaneous-ly (presintering 30 min at 1000 °C, final sintering 45 min at 1350 °C). One sample of each of the three series was annealed at 1400 °C for 1, 3, 8 and 25 h, respectively. These excessive heat treatments were used to provoke discontinuous grain growth. All samples were polished and electrolytically etched[13] and photo-graphed at e.g. 600 and 1500 x magnification in a scanning electron microscope (back scattered electrons, material contrast). The linear intercept distributions of the carbide crystals were measured according to the procedure described above.

Fig. 8 shows the cumulative intercept distribution of the initial stage (sintered sample) and after the longest annealing treatment (25 h at 1400 °C). Fig. 9 shows χ^2 values for each of these samples. The sintered samples exhibit a minimum χ^2 value at a numerical concentration close to $c_A = 100$ % while the annealed samples exhibit pronounced minima which differ for the three series.

Compared to the synthetic example the deviations between the calculated and the experimental distribution curves are much larger here. The obvious reason for these deviations lies in the fact that the log normal distribution is not a perfect approximation

of the real size distribution even in those cases where no discontinuous crystal growth has occurred. Thus, agreement with a
log normal distribution is not obtained on a 95 % significance
level for any of the experimental distributions. Considering the
experimental scatter, this is not surprising. Finding the χ^2 values
in the order of the degree of freedom k-1 we think that the goodness
of fit is adequate for practical purposes, and that the procedure
outlined above yields useful results to describe discontinuous
crystal growth. As shown elsewhere[12], the log normal distribution
can easily be substituted by any other two-parametric distribution
in case that the log normal distribution does not agree to the experimental results.

In Fig. 10 the calculated distribution parameters for the
continuously and discontinuously grown crystal fractions and the
numerical concentrations are shown. The three powders were graded
with respect to their tendency to grow discontinuously as follows:

GN: Coarse powder , normal growth tendency
GS: Coarse powder , stable
FD: Fine powder, pronounced tendency for dis-
      continuous growth

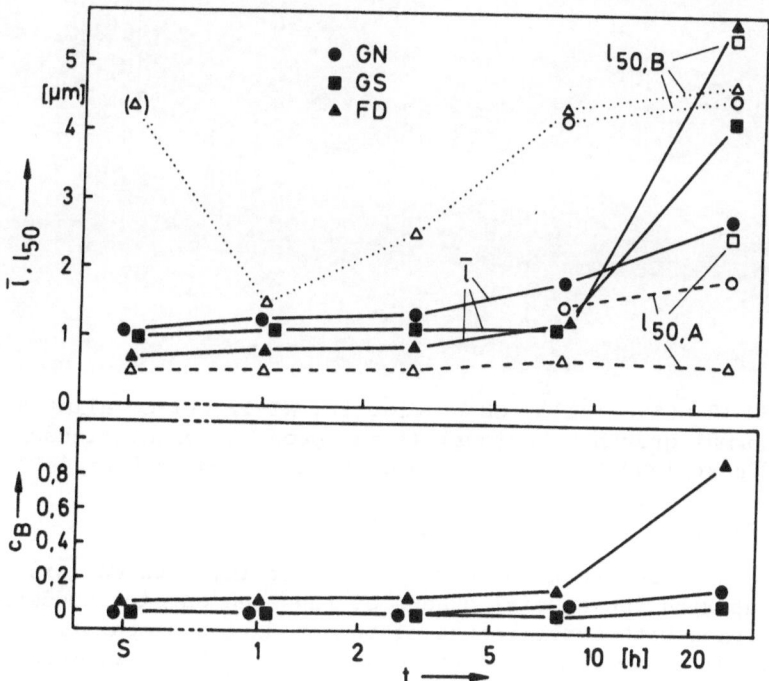

Fig. 10:   Numerical density of the discontinuous
fraction and distribution parameters of three
cemented carbide grades as a function of heat
treatment

At annealing times up to 3 h, very little discontinuous grain
growth is observed. In this range, the mean linear intercept increases in a similar way for the three series. For extended sintering times (8 and 25 h) pronounced coarsening takes place due to
discontinuous crystal growth (Fig. 11). Fig. 12 shows that for
powder FD nearly all crystals belong to the discontinuously grown
fraction. Discontinuous grain growth is observed for powder FD in
the sintered stage ($c_B > 0$). c_B increases during annealing, and
the discontinuous fraction B coarsens while the continuous fraction
A does not coarsen to any noteworthy degree. The pronounced overall
coarsening as indicated by the mean linear intercept is mainly due
to the transition from continuous into discontinuous grain growth
rather than to normal coarsening of the individual fractions. As
shown in Fig. 12, most of the crystals of the alloy FD sintered
for 25 h at 1400 °C have undergone discontinuous growth. These results show that the coarsening of WC-Co alloys should not be analyzed using simple time laws derived for ideal Ostwald ripening[14,15].
A more complete discussion of coarsening behaviour of WC-12 % Co
hard metals produced from different WC powders including higher
annealing temperatures can be found elsewhere[12].

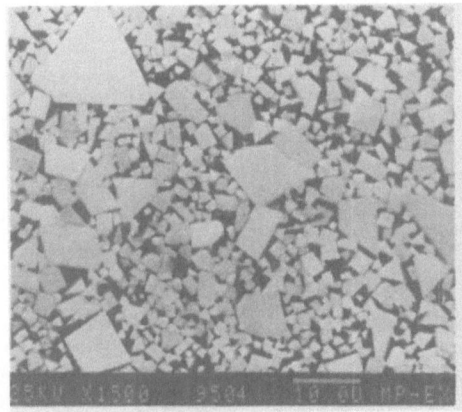 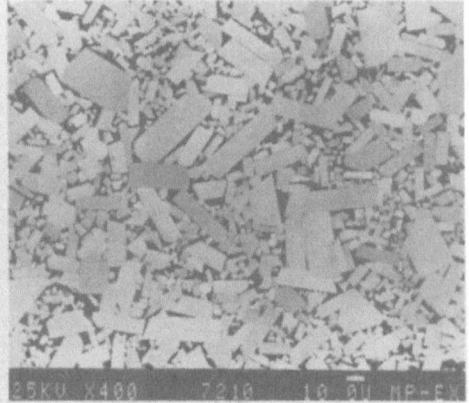

Fig. 11: WC-12 wt% Co alloy GN   Fig. 12: WC-12 wt% Co alloy FD
(coarse, normal growth behaviour)   (fine, prone to discontinuous
annealed 8 h at 1400 °C   growth) annealed 25 h at 1400 °C

ACKNOWLEDGEMENTS

    We are grateful to Hermann C. Starck Berlin, Werk Goslar, for
providing us with the special grades of tungsten carbide powders
and for helping with experimental equipment during preparation of
the hard metal alloys. We also would like to thank Dr. D. Bormann,
Dr. B. Krismer and Dr. H. Tulhoff for enlightening discussions.

Note: A german version of this paper was presented at the 10th
Plansee Seminar, Reutte, 1981 and published in the Proceedings
Vol. 2, pp. 795-810.

REFERENCES

1. E. Lardner, The control of grain size in the manufacture of sintered hard metal, Powder Metall. 13: 394 (1970).
2. H. Fischmeister and H.E. Exner, Gefügeabhängigkeit der Eigenschaften von Wolframkarbid-Kobalt-Hartlegierungen, Archiv Eisenhüttenwesen 37: 499 (1966).
3. H. E. Exner and J. Gurland, A review of parameters influencing some mechanical properties of tungsten carbide-cobalt alloys, Powder Metall. 13: 13 (1970).
4. H. E. Exner, A. Walter, P. Walter and G. Petzow, Auswirkungen der WC-Ausgangspulver auf gesinterte WC-Co-Hartmetalle, Metall 32: 443 (1978).
5. H. Tulhoff, On the grain growth of WC in cemented carbides, in: "Modern Developments in Powder Metallurgy (Proceedings 1980 International Powder Metallurgy Conference, Washington, DC)", MPIF, New Jersey, in press.
6. H. F. Fischmeister, C.A. Blände and S. Palmqvist, A comparative study of methods for particle size analysis in the sub-sieve range, Powder Metall. 7: 82 (1961).
7. H. E. Exner, Struktur und Eigenschaften der Hartlegierung WC 10 % Co, PhD Thesis at Montanistische Hochschule Leoben, Austria (1964).
8. H. F. Fischmeister, H. E. Exner and G. Lindelöf, Particle size analysis in cemented carbide technology, in: "Modern Developments in Powder Metallurgy Vol 1", H. H. Hauser, ed., Plenum Press Publ. Corp., New York (1966).
9. V. Howard, L. Scales and R. Lynch, The numerical densities of alpha and gamma motoneurons in the trigeminal motor nucleus of the rat: a method of determining the separate numerical densities of two mixed populations of anatomically similar cells, in: "Proceedings Fifth International Congress for Stereology", Mikroskopie 37: 229 (1980).
10. L. Sachs, "Angewandte Statistik", Springer Verlag, Berlin (1974).
11. K. Bosch, "Angewandte mathematische Statistik", Rowohlt Taschenbuch Verlag, Reinbek/Hamburg (1976).
12. K. M. Friederich, Quantitative Erfassung der Vergröberungsneigung von Wolframkarbidkristallen in Hartlegierungen, PhD Thesis at Universität Stuttgart, Germany (1981).
13. W. Peter, E. Kohlhaas and A. Fischer, Präparationsmethoden für elektronenmikroskopische Gefügeaufnahmen von Hartmetallen, Prakt. Metallogr. 5: 115 (1968).
14. H. Grewe, H. E. Exner and P. Walter, Behinderung des Kornwachstums in Hartmetall-Legierungen vom ISO-K10-Typ durch Zusatzkarbide, Z. Metallkd. 64: 85 (1973).
15. J. L. Chermant, M. Coster, A. Deschanvres and A. Iost, Étude de la cinétique de croissance de système carbure-métal, J. Less-Common Met. 52: 177 (1977).

DISCUSSION

R. K. Viswanadham:

How did you measure the activity of the powders?

K. M. Freiderich:

Using a method published some years ago by Grewe in Metall. The
solution is an organic mixture of methanol, pyridine and a little
bit of iodine.

R. K. Viswanadham:

If it dissolves faster do you characterize that as more activity?

K. M. Freiderich:

Yes.

E. Engle:

Can you give us a little more information on the sintering temperature and the effect of the furnace atmosphere.

K. Friederich:

1400°C, and for times up to twentyfive hours in good vacuum.

W. Williams:

The discontinuous grain growth which has been discussed here could
occur via the fortuitous placement of particles having similar
crystallographic orientations in a small region in space in the
composite, which due to reprecipitation on their surfaces would lead
to a rather large grain. The grain boundaries could then be swept
out leaving a single crystal particle. This would be a statistical
fluctuation which might occur occasionally and lead to these very
large grains.

K. Friederich:

I think that's the main reason for the discontinuous grain growth.

O. Knotek:

Have you any idea if this change of activity is influenced by traces
of other elements?

K. Friederich:

That is possible.

E. Kimmel:

You implied that a lot of this chemical activity change or the change that you saw was related to the methods used in the processing or manufacturing of the powders. Could you elaborate on this?

K. Friederich:

The differences in the powders were in the reduction step from the oxide to the metal and afterwards the powders were all treated in the same way. You can perhaps ask Dr. Tulhoff from Starcke the producer of the powders.

H. Tulhoff:

The powders used in these tests were produced in our laboratory and the differences were not in the carburization of the powders but in the reduction step. The last slide in the presentation compared powders reduced in hydrogen at 800°C with those reduced in plasma-hydrogen. The powder reduced in plasma-hydrogen has a much higher tendency to exhibit exaggerated grain growth. We feel that this is due to lattice disturbances produced under the plasma reduction conditions. The third powder was reduced by carbon as was done many years ago and no difference in behavior during sintering could be observed.

CARBIDE-MATRIX REACTIONS IN WEAR RESISTANT ALLOYS

O. Knotek, W. Wahl, H. Reimann and P. Lohage

Institut für Werkstoffkunde B
University of Aachen
Aachen (FRG)

INTRODUCTION AND PROBLEM DESCRIPTION

In various coating processes, nickel-based hard alloy powders are mainly applied either in atomized form or mechanically
pulverized. The alloys have not only good corrosion resistance,
due to the nickel solid solution which often contains chromium,
but also high abrasion resistance and a relatively low melting
point. Aside from nickel and chromium, most nickel hard alloys
contain boron and silicon, and often small amounts of additional
carbon and iron . Boron is very slightly soluble in nickel.
The boride Ni_3B forms a low melting eutectic at about 1.000°C
with nickel or nickel solid solution.

The considerable lowering of the melting point compared to
pure nickel not only simplifies the processing of the hard alloy
powder by means of different powder coating procedures, but also
makes possible the direct fusion of powder layers on substrates
in the furnace. Boron and silicon act simultaneously as deoxi-
dizers and improve both the properties of the coating material
in all coating methods and the bond to the substrate (1). The
diffusion into the substrate during coating procedures has been
discussed earlier (2,3).

The present study aims to discuss the behavior of heterogeneous powder with nickel-base hard alloys. For many applications mixtures with tungsten carbides of different carbon content
are used. During coating, reactions take place between the
nickel-chromium matrix and the added carbides.

We know of similar reactions in other carbide-containing coating alloy systems. Nevertheless, the research on the relationship between abrasion resistance and the reactions is still in an early stage. The experimental study of nickel-based alloys is relatively simple.

These heterogeneous or quasi-alloys used for abrasion resistance are metastable and consequently not in thermodynamical equilibrium.

For example there are tubes filled with W_2C/WC for wear resistant coatings. Aside from the carbide distribution in the coating, all other coating parameters play an important part in abrasion resistance.

Abrasion resistance certainly depends on the phases present, although their grain size, grain size distribution and matrix alteration are also of principal importance. The life of cast mixer blades for concrete compulsatory mixers varies, for example, between 3.000 m^2 and 100.000 m^2 depending on structure of the alloys, mixing agents and the manufacturing process. When applying carbide-containing quasi-alloys, exposed to abrasion and corrosion at high temperatures, reactions of the existing phases during use cannot be excluded completely. Considering the holding times of hot sintering screens in metallurgical technology, these can be of relevance.

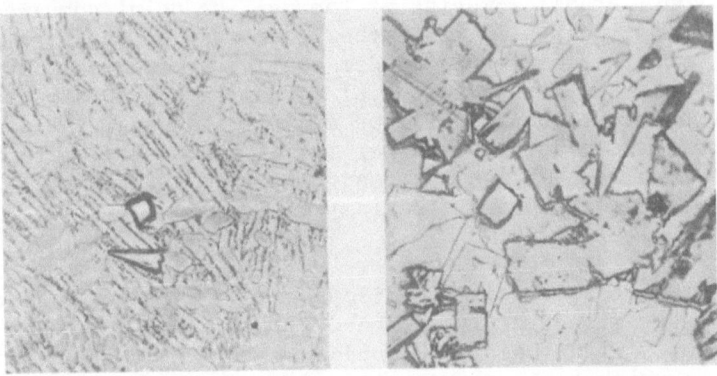

Fig. 1: Metallographic structure of WC-Fe/Ni before (a) and after
(b) heat treatment (1).

Heat treatment causes evident change in structure and subsequent reduction in abrasion resistance, not only in the case of tungsten carbide filled tubes, but also in coatings by sintered WC-Fe/Ni electrodes, as illustrated in Fig. 1.

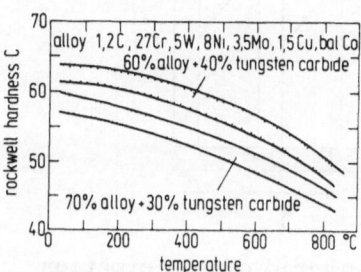

Fig. 2: Hot hardness of sintered WC-CrCoWC - alloy rod coatings

Figure 2 demonstrates the relation of the tungsten carbide content to hot hardness.

It is difficult to follow the reactions which take place during casting or in the fused mass of quasi-alloys or in heterogeneous compounds. For this reason carbide compounds with NiCrBSi powder alloys are most suitable for research, as already mentioned. In these "mechanical" alloys, segregation, relevant in practical application, can be observed due to the extreme density variation between the carbides (WC) and the matrix and furthermore, it can be influenced by employing WC/TiC compound carbides, for example.

At this point we would like to draw attention to a further possibility. In carbon-containing cobalt and iron systems, tungsten monocarbide can be obtained as a stable phase. Not only does the usual cemented carbide, sintered with liquid phase, contain WC, but also in other alloys produced by casting, primary M_6C decomposes to stable WC under certain conditions (4). The following figures show a WC formation-diagram, as well as microstructures of a WC-forming Co-based hard alloy (Figures 3-5).

The stability of M_6C is decisively influenced by the chromium content, for example.

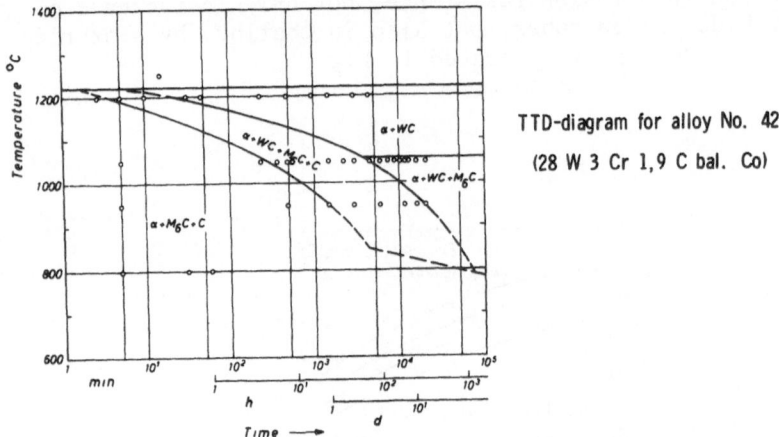

TTD-diagram for alloy No. 42
(28 W 3 Cr 1,9 C bal. Co)

Fig. 3: Diagram of WC-formation

Fig. 4: Alloy in as-cast condition

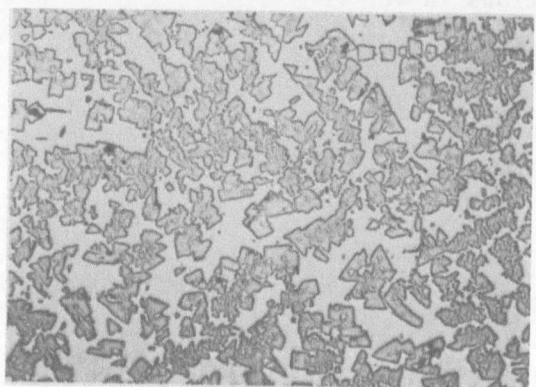

Fig. 5:   Alloy after exposure

It is obvious that, when considering coatings, attention
is mainly drawn to the combination of abrasion resistance and
corrosion resistance with the mechanically strong substrate
material. However, in many cases, fracture strength must be
given priority over other properties needed for higher performance. For this reason, reproducibility in the production of
relatively thick coatings with high carbide content is of great
importance, not only with reference to the reaction in particular
phases, but also from a technological viewpoint. The experiments
described in the following section were conceived with regard to
this aspect.

EXPERIMENTAL PROCEDURES AND DISCUSSION

The following matrix alloys and matrix carbide compounds
were used in the remelting experiments.

Table 1:   Nickel-based hard alloys employed as matrices

Alloy	Cr	Si	B	Fe	C	Ni	T_S	T_L
			wt.-%				°C	
HFA 1	10.0	3.0	2.25	4.25	0.45	bal.	965	1180
HFA 2	13.5	4.25	3.0	4.75	0.75	bal.	965	1035

Table 2:   Heterogeneous hard alloys with added carbides
           (compositions in wt.-%)

```
50 HFA  1 + 50 WC
70 HFA  1 + 30 WC
50 HFA  2 + 50 WC
70 HFA  2 + 30 WC
76 HFA  1 + 24 TiC
88 HFA  1 + 12 TiC
76 HFA  2 + 24 TiC
88 HFA  2 + 12 TiC
50 HFA  1 + 50 (W, Ti) C 50/50
70 HFA  1 + 30 (W, Ti) C 50/50
50 HFA  2 + 50 (W, Ti) C 50/50
70 HFA  2 + 30 (W, Ti) C 50/50
50 HFA  1 + 50 (W, Ti) C 70/30
70 HFA  1 + 30 (W, Ti) C 70/30
50 HFA  2 + 50 (W, Ti) C 70/30
70 HFA  2 + 30 (W, Ti) C 70/30
```

Powders were sprinkled upon AISI 321 steel to serve as a
substrate and processed to a coating in a vacuum furnace
($5 \cdot 10^{-5}$ mbar). Details of these experiments aiming to gain
improved abrasion properties have been published recently (5).
The results are compiled in Figures 6 and 7.

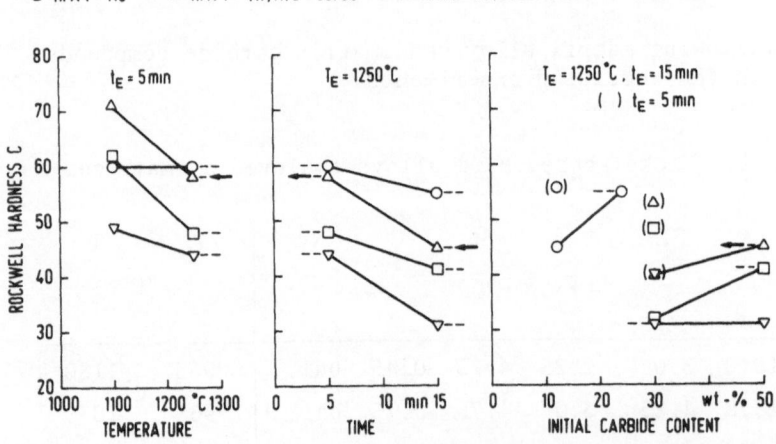

Fig. 6:   Hardness of heterogeneous alloys in relation to time,
          temperature, and initial carbide content based on HFA 1.

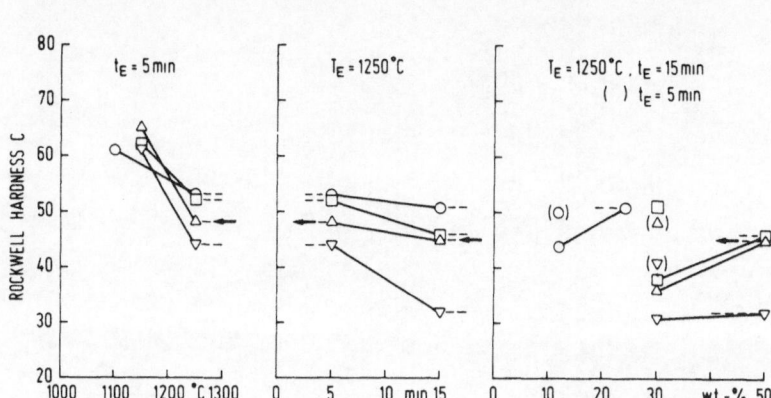

Fig. 7: Hardness of heterogeneous alloys in relation to time,
        temperature, and initial carbide content based on HFA 2.

Figure 6 presents hardness values for all the melted powder
compounds of hard alloy based on one matrix. The figure should
be read from right (initial carbide content) to left (temperature). One example is the change in hardness of the compound
alloy HFA 1 + (W, Ti)C - 50/50. By altering the ratio of the
mixture of HFA 1 and the carbide from 70 : 30 to 50 : 50 weight
percentage, the Rockwell hardness can be raised from 40 to 45
HRC after processing at a temperature of 1250°C and a holding
period of 15 minutes. Another procedure is based upon a weight
ratio of 50 : 50. By shortening the melting period at 1250°C
to 5 mins it is possible to gain a Rockwell hardness of 58 HRC
which is considerably higher than the hardness of 35 to 40 HRC
usually obtained for HFA 1 as a thermal spray coating. A
further hardness enhancement can be gained by lowering the melting temperature to 1100°C while keeping the holding period at
5 mins.

Figure 8 shows the metallographic structures according to
the processing cycles for the composite carbide (W, Ti)C - 50/50
HFA 1. X-ray and analytic tests have shown the following
changes with reference to the phase ratio when raising the
melting temperature and prolonging the processing period: Due
to iron absorption from the base material a large part of the
property-defining CrB decomposes by recrystallization to form
$(Cr, Fe)_5B_3$ and narrow seams of the η-carbides Ni_2W_4C and
probably traces of Fe_3W_3C form on the surface of the added carbides.

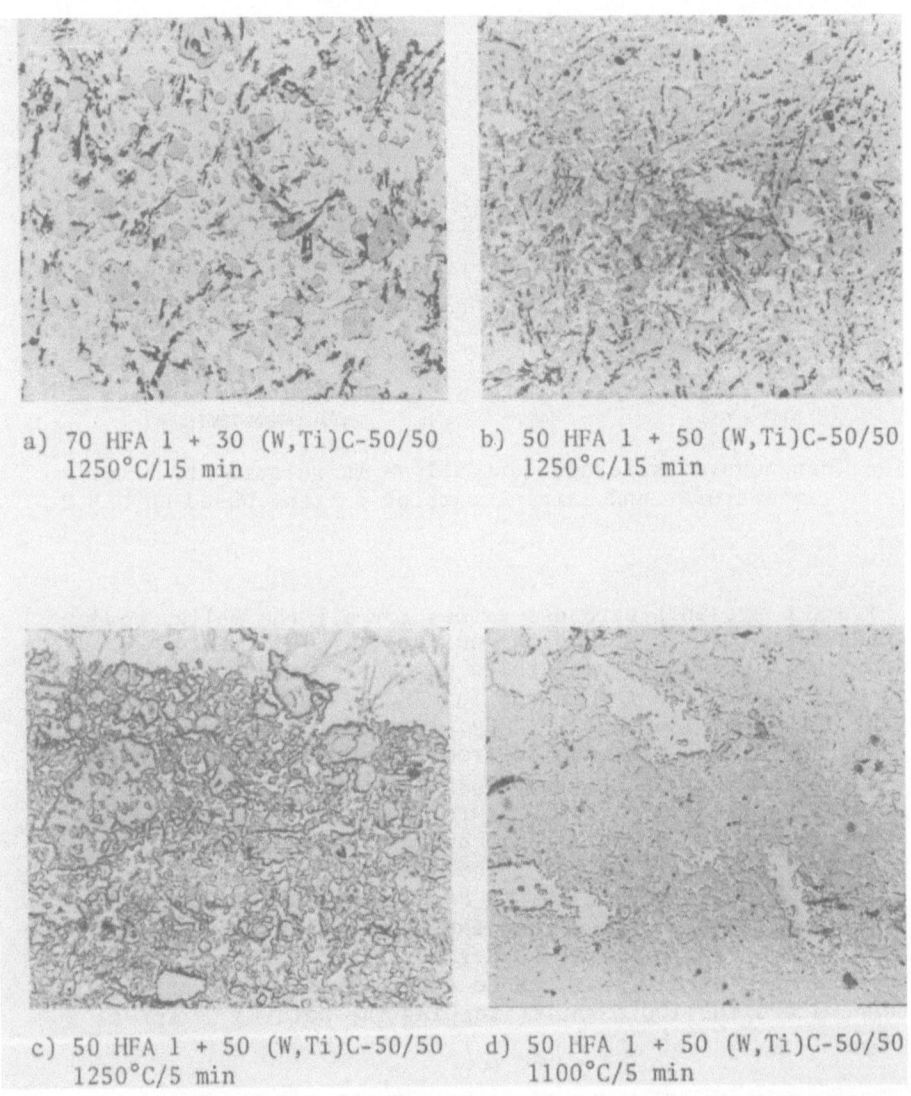

a) 70 HFA 1 + 30 (W,Ti)C-50/50   b) 50 HFA 1 + 50 (W,Ti)C-50/50
   1250°C/15 min   1250°C/15 min

c) 50 HFA 1 + 50 (W,Ti)C-50/50   d) 50 HFA 1 + 50 (W,Ti)C-50/50
   1250°C/5 min   1100°C/5 min

Fig. 8: Metallographic structure of the compound alloys HFA 1
         + (W, Ti) C - 50/50 exposed to different processing
         conditions

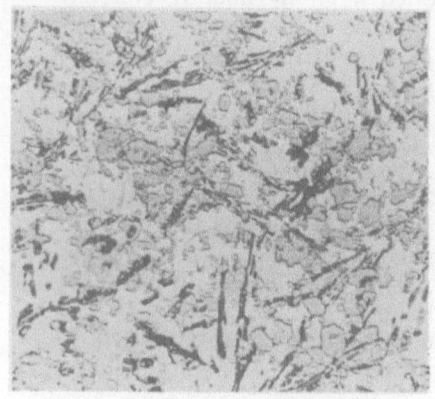 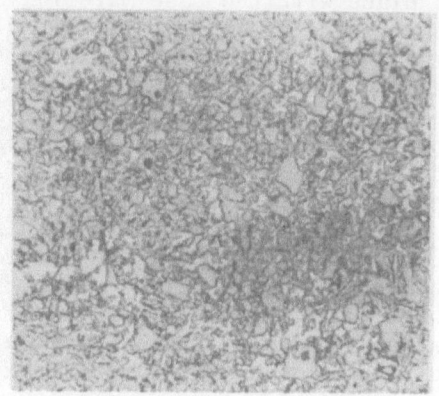

a) 70 HFA 2 + 30 (W,Ti)C-50/50   b) 50 HFA 2 + 50 (W,Ti)C-50/50
   1250°C/15 min   1250°C/15 min

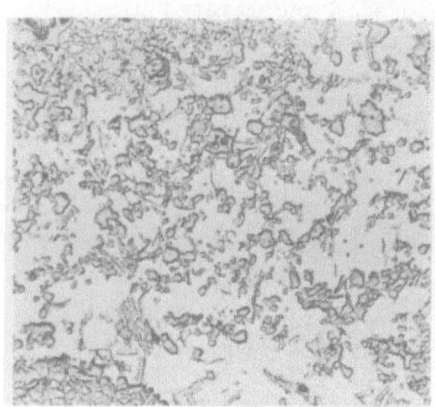 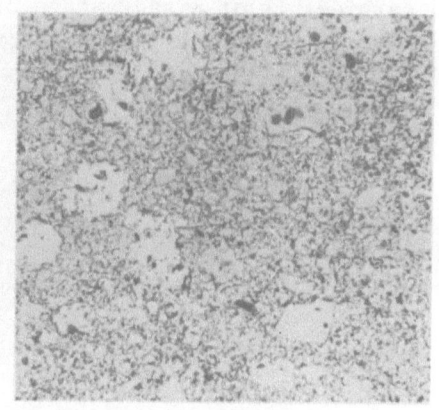

c) 50 HFA 2 + 50 (W,Ti)C-50/50   d) 50 HFA 2 + 50 (W,Ti)C-50/50
   1250°C/5 min   1150°C/5 min

Fig. 9: Metallographic structures of the compound alloys
         HFA 2 + (W, Ti) C - 50/50 exposed to different
         processing conditions

The behaviour of HFA 2 in the compound alloys is shown in Figs. 7 and 9. We have found that high convection diffusion takes place with the base material due to high B- and Si-content during furnace fusion. This caused a higher iron content in the coating than in the HFA 1 compound alloy layers. For this reason a more spontaneous $(Cr, Fe)_5B_3$ crystallization took place and enforced the state of dissolution on the surface of the composite carbides. It is remarkable that in contrast to HFA 1-based alloys, a great reduction in the tungsten content can be gained with low energy input (1150°C/5 mins) in the boundary zones of the composite carbide grains. At higher melting temperatures and longer times the formation of the η-carbides Ni_2W_4C can be accelerated. It appears that the dissolution of composite carbides in such alloys is based on distinctive exponential characteristics. It is also possible to detect iron-containing η-carbides by applying x-ray methods.

According to the results of x-ray and metallographic tests, and also hardness tests, it is possible to conclude that the higher B- and Si-contents in HFA 2 are responsible for the higher iron content originating from the base material, leading to the stabilization of Fe_3W_3C and $(Cr, Fe)_5B_3$. On the other hand, we can state that the higher tungsten content in the added carbides is of disadvantage because of accelerated dissolution and solution leading to the considerable formation of Ni_2W_4C.

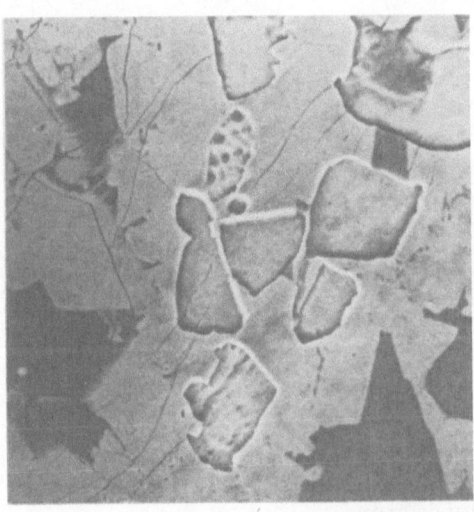

a) 50 HFA 1 + 50 WC; 1250°C/15 min

Fig. 10a: Electron-micrographs of compound alloys after adding
           monocarbides

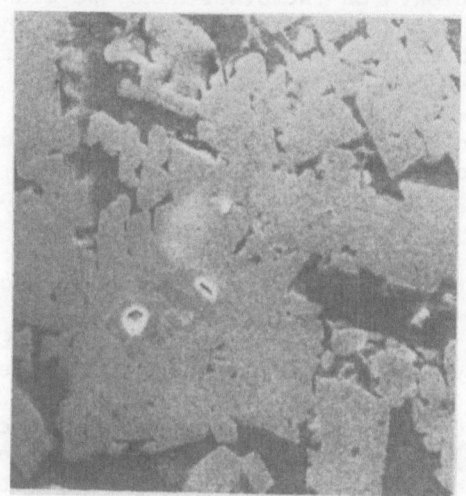

b) 50 HFA 2 + 50 WC; 1250°C/15 min

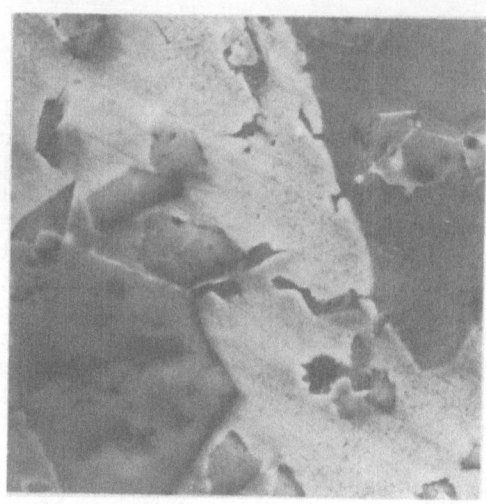

c) 76 HFA 2 + 24 TiC; 1250°C/5 min

Fig. 10b,c: Electron micrographs of compound alloys after
              adding monocarbides

The different influences of the matrix alloys can be explained by Fig. 10. After a high energy input (1250°C/15 mins) in HFA 1 a large amount of WC remains, bordered with η-carbides of different composition. On the other hand, WC dissolved almost completely in HFA 2 under the same thermal condition. A η-carbide matrix with a comparatively small volume of hard phase deposit forms.

TiC shows stable behaviour in HFA 2 (Fig. 10c) even after high processing temperatures. Almost no dissolution is to be observed; the wetting of the particles' surface is complete.

These facts are contrary to the common experience of TiC wetting in powder metallurgy and thermal spray technology.

Several authors note the wetting behaviour of WC, TiC and composite carbides by nickel, where the contact angle of TiC lies at 20° to 38°. Various values between 16° and 21° are given for composite carbides and WC generally has a contact angle of 0° (6,7,8,9). The good wetting properties of TiC found in the present work could be attributed to an improvement of the wetting behaviour of nickel-chromium solid solution as compared to that of pure nickel, as is commonly observed in sintered TiC-hard metals with a Ni-Cr or Ni-Cr-Mo matrix. An influence of Boron is also possible.

The contact angle of 0° in the case of WC indicates alloying between WC and nickel which has been proven in the tests shown. The good wetting behaviour of TiC in the liquid nickel solid solution matrix during furnace fusion allows the presumption that the carbide hard phases of metallic or non-metallic nature, difficult to treat under other procedures, are suitable for this technology. The density of the added hard phases should be adjusted to that of the matrix alloys, as long as these are not in metallurgical interaction with the hard phases, to avoid inhomogeneous layer compositions caused by gravitational segregation. The latter was the case in the processed carbides with pure WC, whereas both composite carbides show only minimal tendency to segregation and the pure TiC tends to float, especially in thick layers (>3 mm).

As a first step of abrasion testing, conditions were selected in which the "low level of wear" is stabilized for the hard phases and the matrix as well.

Under these conditions the bonding of the hard compounds by the matrix can be studied.

In technical applications sometimes one can observe the removal of hard compound crystals or agglomerates by mechanical means. So the optimization of these pseudo coating alloys has to take into account both the forming of new phases and the bonding of the initially added hard compounds.

Abrasion tests were carried out according to the disc/pin principle. First of all, C 45 in a normally annealed condition served as disc material. The stress applied to the cylindrical samples during abrasion measurement amounted to 13 N/cm^2. The disc covered an abrasion distance of 32 km at a relative speed of 1m/s. The most predominant wear mechanisms were determined to be adhesion and abrasion in alternating proportions. Tri-booxidation and surface destruction were barely noticeable mechanisms.

In tests executed up to this time, the heterogeneous alloy 76 HFA 1 + 24 TiC in the fusion cycle 1100°C/5 mins showed an insignificant loss in weight of 1 mg/m^2. Under pure adhesion wear, increased remelting temperature caused a gain in weight of 3 mg/cm^2.

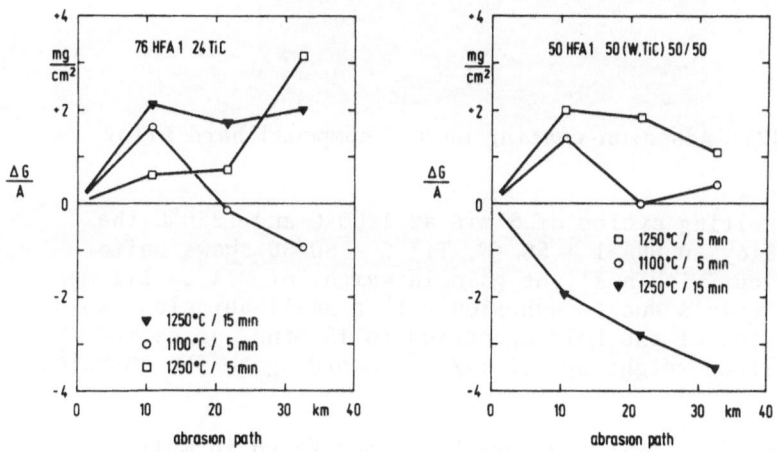

Fig. 11: Wear of compound alloys

A further increase of energy input would result in an abrasion quota of only slight gain in weight.

As it appears, the mechanical bonds between TiC grain and matrix do not attain the actually possible level and the TiC

grain will break out of the matrix under wear. On the other
hand, when the fusion temperature is enhanced, the TiC grain
is soundly anchored in the matrix and shows only slight brittle
fracture behaviour on account of the hardly noticeable diffusion which is the result of a short holding period.

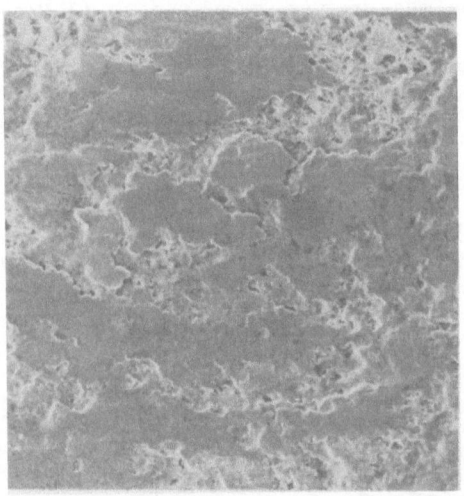

Fig. 12:   Adhesion coating on TiC compound hard alloy

     In remelting cycles of 5 min at 1100°C and 1250°C the
compound alloy 50 HFA 1 + 50 (W, Ti) C - 50/50 shows uniform
wear behaviour with a slight gain in weight of 0.4 to 1.1 mg/
cm^2. The wear is due to adhesion with a small abrasion quota.
A prolongation of the holding period to 15 mins causes reduction in surface weight by 3.5 mg/cm^2 according to the abrasion
mechanism.

     Increased brittle fracture behaviour noted in multi-
component carbides can be considered to be responsible. Further
influential factors to be taken into consideration are the
formation of η-carbide seams around the compound carbide grains
and subsequent brittle fracture susceptibility of the bond
between matrix and compound carbide.

     In the case of all steel coatings, an enhanced wear quota
can be expected with an increase in energy input during processing. The explanation for this lies in the recrystallization

of CrB to $(Cr,Fe)_5B_3$, since CrB is defined by its abrasion resistant character in homogeneous alloys.

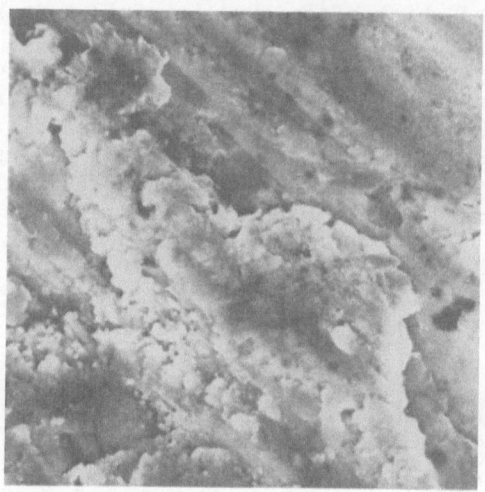

Fig. 13:   Abrasion channels on (W, Ti) C compound hard alloys

Further research is necessary (i.e. C45 H) to enable a more extensive judgment of the wear behaviour of furnace-fused surface coatings, as abrasion cannot be considered as an independent variable, but rather as the superimposed result of wear mechanism of varying causes.

SUMMARY

To enhance wear and corrosion resistance, hard nickel-base alloys are used in powdered form (atomized or crushed) and applied by various hard facing techniques. NiCrBSi hard facing powders are often used either as homogeneous alloys or heterogeneous mixtures with hard materials such as carbides.

Due to the relatively low solidus temperature (1000°C) and the broad solidus-liquidus range of these alloys, most of the sprayed layers can be fused by various means to increase density and improve adhesion to the base metal. Although hardness and wear resistance of the matrix play a significant role in the wear resistance of such a system, a complete description of its wear behaviour requires a knowledge of the specific reactions that occur between the nickel-base matrix and the hard compounds

during fusing. These reactions depend on the specific compositions of the matrix and carbides, their stoichiometry, grain size
and the time and temperature of heat treatment. Under practical
conditions of fusing, the equilibrium state is not achieved.
Wear resistance depends on the specific non-equilibrium structures that are produced which often give improved wear resistance.

   In this investigation a systematic study was made of the
phases produced in nickel-base matrices containing WC and
(W, Ti) C hard materials by thermal cycles typical of those
encountered in practical application. The wear resistance of
the various non-equilibrium structures produced was measured and
correlated to the thermal cycles used in heat treatment.

REFERENCES

1.   M. L. Fiedler and H. H. Stadelmaier, Z. Metallkde, 66: 402
     (1975).
2.   E. R. Stover and J. Wulff, Trans. Met. AIME 215: 127 (1959).
3.   C. B. Pollock and H. H. Stadelmaier, Metallurg. Trans., 767
     (1970).
4.   W. Jellinghaus, Archiv Eisenhüttenw., 40: 843 (1969)
5.   Int. Conf. Metallurgical Coatings, San Francisco (1981).
6.   J. Hinnüber and O. Rudiger, Techn. Mitt. Krupp, Forsch. Ber.
     20: 162 (1962).
7.   M. Humenik and N. M. Parikh, J. Amer. Ceram. Soc., 39: 60
     (1959).
8.   V. F. Funke, V. I. Tumanov and Z. S. Trukhanova, Jzv. Akad.
     nauk SSSR, Metallurg. Toplivo, No. 5: 101 (1961).
9.   L. Ramquist, Int. J. Powder Metallurg. 1, No. 4: 2 (1965).

DISCUSSION

C. A. Brookes:

I'd agree with you that one should be cautious about using hardness
measurements to evaluate resistance to wear but it does worry me a
bit about your Rockwell hardness measurements. Could you explain
how deep the Rockwell hardness indentation goes with respect to the
thickness of your layers?

O. Knotek:

The thickness of the materials was 2 to 3mm. So I think this is
sufficient to avoid substrate effects on the hardness of the coating.
We found the same correlation or noncorrelation with Vickers indentations as well. So I am satisfied that substrate deformation played
no part.

If instead of using solid solution carbides pure WC is used, segregation induced by gravity would be a problem. One improvement of
this coating is that specific gravity of the matrix and the carbide
phase are similar and segregation is minimized.

BINDER-CARBIDE PHASE INTERACTIONS IN TITANIUM

CARBIDE BASE SYSTEMS

D. Moskowitz and H. K. Plummer, Jr.

Engineering and Research Staff
Ford Motor Company

ABSTRACT

Binder-carbide phase interactions have been investigated in titanium carbide base systems using the Energy Dispersive X-ray Analysis (EDXS) capability of the scanning transmission electron microscope (STEM). Concentration gradients of Mo and V within carbide grains are shown after sintering in the presence of a Ni-rich liquid phase. Binder Ti and Mo levels were measured using this technique, which has a major advantage over phase separation analysis due to the ability to locate the analysis position with respect to the microstructural components.

INTRODUCTION

The physical properties of cemented carbide materials are a function of the composition and microstructure of the two constituent phases. This has been well demonstrated for both WC base and for TiC base composites. In the WC-Co system, Kubota, Ishida and Hara,[1] and Suzuki and Kubota[2], among others, have shown that tungsten can be retained in solution in the cobalt binder, and that this affects virtually all of the physical properties. The binder tungsten content, in turn, is an inverse function of the carbon content of the alloy. A third phase, the intermetallic Co_3W, could be produced by precipitation in binders super-saturated with tungsten.

Previous Studies in the TiC-Ni-MoCx System

In 1966, Moskowitz and Humenik[3] published data dealing with the interaction between binder and carbide phases in TiC-Mo-Ni

299

300 D. MOSKOWITZ AND H. K. PLUMMER, Jr.

alloys. They showed how molybdenum diffused substitutionally into
the carbide phase during liquid phase sintering of these alloys,
resulting in a Ni-base binder alloyed with sizeable amounts of
Ti and lesser amounts of Mo. Estimates of Ti content of the
binder in TiC-9Mo-12.5Ni alloys ranged from approximately 3 weight
% to 9 weight %, depending on whether the alloys were high or
low in carbon content, respectively.

Suzuki, Hayashi and Terada[4] studied Ti and Mo levels in the
binder of TiC-Mo-30Ni alloys by electron probe microanalysis.
Table 1 lists the Ti and Mo contents they measured on low and high
carbon alloys of materials containing up to 30% Mo. Their measurements were made on 10 micron-wide regions of the binder.

Table 1: Binder Ti and Mo Contents of TiC-Mo-30Ni Alloys

| | Ti | | Mo | |
Alloy	Low Carbon (%)	High Carbon (%)	Low Carbon (%)	High Carbon (%)
TiC-30Ni	∿9	∿3	–	–
TiC-10Mo-30Ni	∿11	∿5	∿2	∿0
TiC-20Mo-30Ni	∿10	∿4	∿4	∿1
TiC-30-Mo-30Ni	∿9	∿4	∿6	∿3

Snell[5] determined Ti and Mo contents of the binder phase of
TiC-24Mo-15Ni alloys by chemical analysis of the preferentially
dissolved Ni-base phase. His results, which range from ∿3% Ti
and ∿2% Mo for high carbon alloys to ∿10.5% Ti and ∿6.5% Mo for
low carbon alloys, agree reasonably well with Suzuki et al. If
Snell's transverse rupture strength data are plotted as a function
of binder Ti content, there is good agreement with Moskowitz and
Humenik's[6] finding of a strength maximum in the neighborhood of
6-8% Ti (Figure 1). On the other hand, these two sets of data
do not agree as well with respect to tool life vs. binder Ti
content: Snell's tool life is at maximum at considerably lower
Ti contents than the authors had found.

OBJECTIVE

The object of the present investigation was to explore the use of scanning transmission electron microscopy (STEM) methods in analyzing binder and carbide phases in $TiC-MoC_x-Ni$ base materials. Current STEM equipment, equipped with an energy dispersive x-ray detection system, can provide energy dispersive x-ray spectroscopy (EDXS) capability, which can analyze sub-micron-size binder regions, allowing for analysis on a finer scale than electron probe microanalysis. Since the STEM/EDXS method enables one to locate the specific analysis position within the microstructure, it also has a significant advantage over phase separation analysis.

EXPERIMENTAL PROCEDURE

The materials studied consisted of TiC-10Mo-22.5Ni base compositions, designated "7G", which are in use commercially in steel machining operations. A more recent modification[7] called "7GY", containing 7.5% TiN and 10% VC, which is designed for steel milling application, was also examined. Specimens were prepared by diamond sawing to 2.5mm thickness from the bulk material and then lapping to .12mm thickness followed by cutting into 3mm diameter discs using an ultrasonic cutter. They were then further thinned by ion beam thinning using Argon ions at 6 Kv and .4ma current for a period of approximately 50 hours.

INSTRUMENTATION

A Phillips 300 Transmission Electron Microscope, modified by the manufacturer to provide a scanning beam mode of operation, was used in this study. A focused, convergent beam of electrons was directed at a fixed specimen position with a spot diameter of approximately 100 Å at 100 KV accelerating potential. The characteristic X-rays which are excited were detected by a Kevex energy dispersive X-ray detector. A solid-state electron dectector located after the specimen provided STEM images with a resolution of approximately 30 Å from electrons which have passed through the specimen and impinge on the detector. The EDXS detector was followed by a Tracor Northern TN2000 multi-channel analyser system.

Modifications to the instrument were made by adding lead apertures at appropriate positions. In addition, the specimen holder was modified. Both modifications resulted in reduced spurious non-focussed electrons and stray X-rays resulting from interaction of the electron beam with internal microscope parts. In addition, changes were made in the EDXS dectector collimator to increase signal to noise ratio.

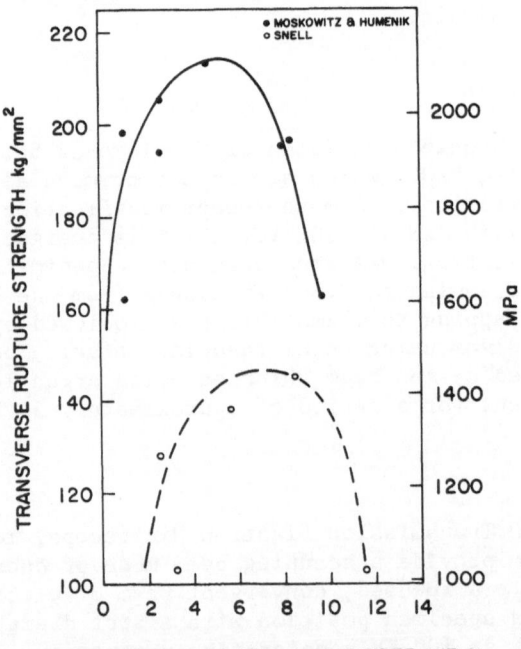

Fig. 1. Transverse Rupture Strength vs. Binder Titanium Content
of TiC-Mo-Ni Materials

Elemental concentrations were determined using a procedure[8] which is based upon the equivalence of the detected intensity of each emitted spectral peak to fixed ratios in known specimens. This yielded an accuracy of $\pm 5\%$ of high concentrations. Concentrations could be determined with reasonable accuracy down to levels of approximately 2%.

RESULTS

Figure 2 shows a photomicrograph at approximately 19,000X magnification plus three sketches of a region of the microstructure of a 7G specimen that was analyzed for Ni, Ti and Mo contents of the binder. The spots dimly visible in the photomicrograph are caused by deposition of carbonaceous material on the sample surface, indicating the beam's position. Since errors caused by analysing material from an adjacent grain or subsurface carbide particle can only decrease the Ni or increase the Ti value, anomalously low Ni values coupled with unusually high Ti values were not averaged into the measurements. Table 2 shows the results of analysing 13 different locations in 3 different binder regions in the 7G material. There is reasonably good agreement between the presently obtained binder Ti values and the earlier one obtained on the same material by phase separation.

Table 2: Ni, Ti and Mo Analyses of 7G Binder

Region	$\underline{Ni}_{AV}$	$\underline{Ti}_{AV}$	$\underline{Mo}_{AV}$
1	75.8	9.0	1.5
2	77.0	7.9	1.0
3	74.0	9.3	2.3
Average	75.6	8.5	1.5
Phase Separation	Bal.	7.4	-

Concentration gradients of Mo and Ti shown in several carbide grains of another region of the 7G sample shown in Fig. 3 illustrates the STEM/EDXS evidence for the diffusion of Mo into TiC. Grain #I in the figure is a typical case of a high concentration of Mo at the grain's outer rim, coupled with little or none of the element at its' core. The opposite effect is seen for Ti, with its highest concentration, corresponding to that of

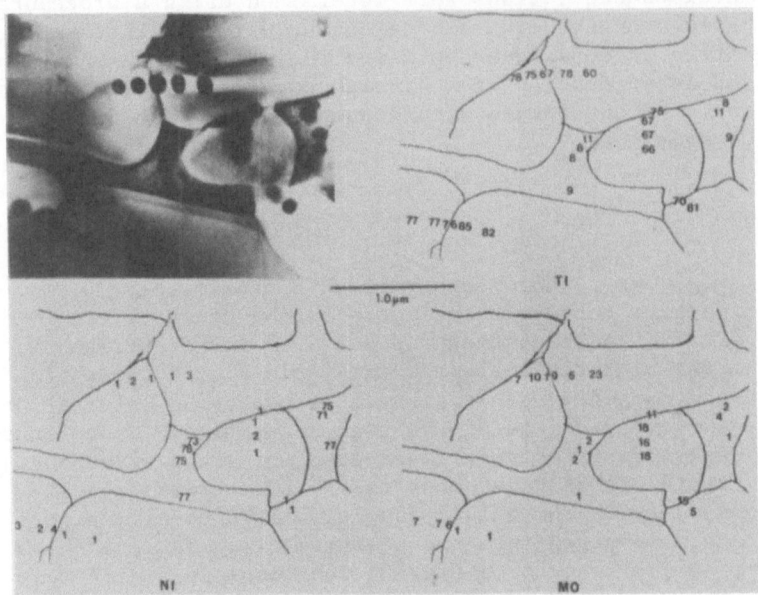

Fig. 2. Photomicrograph and 3 Microstructure Sketches of 7G
Specimen Analyzed for Ni, Ti and Mo Contents

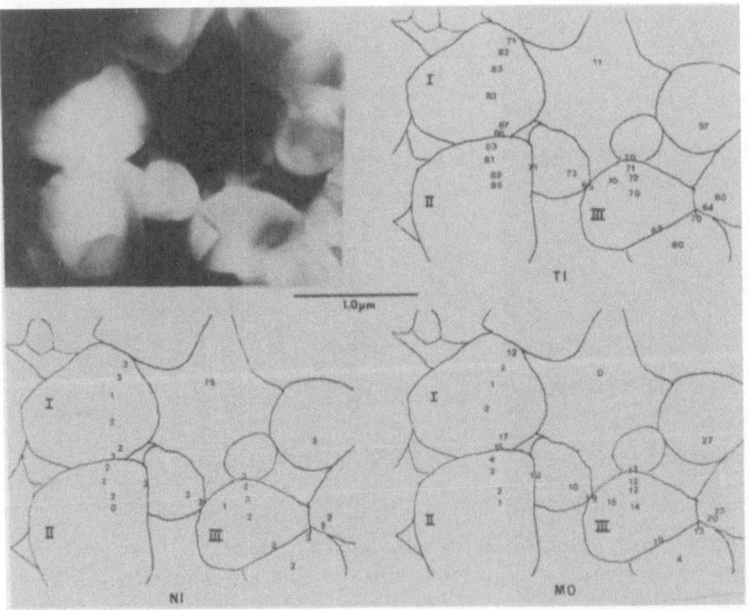

Fig. 3. Concentration of Mo and Ti in Carbide Grains of 7G
Specimen

almost pure TiC, occuring at the grain's core. Note that a similar
effect, although with a lesser concentration gradient, is visible
in grain #II. The relative uniformity in Mo and Ti contents
through Grain III indicates that this represents a cross-section
through a highly Mo-diffused region close to the grain's surface.

Analysis of the 7GY material, in which 17.5% of the TiC was
replaced by 7.5% TiN and 10% VC, shows vanadium to be present in
the binder, in addition to Ti and Mo. This is shown in Table 3:

<u>Table 3: Analyses of 7GY Binder</u>

Region	Ni	Ti	Mo	V
1	76	4.9	3.8	2.6
2	77.5	5.3	1.7	1.9
3	73.1	5.9	5.3	3.0
Average	75.5	5.4	3.6	2.5

The relative amounts of dissolved Mo and V, interestingly,
are roughly in proportion to their nominal content in the overall
alloy.

Figure 4 is a sketch of the microstructure of a typical 7GY
material, showing estimated analyses for Ni, Ti, Mo and V content.
A number of interesting observations can be made from the STEM
analyses shown in this figure. Note that Ti contents tend to be
at a maximum in the central part or core of a carbide grain, and
have lower values in the diffusion zone closer to the grain
boundary. Mo and V contents show the reverse effect, both
elements being most concentrated in the outer zone of a carbide
grain. Grains that contain lesser amounts of Mo in their core region show undetectable concentrations of V. This indicates that
the diffusion rate of Mo into TiC is greater than that of V
into TiC. The diffusion zone of Mo at the carbide grains' outer
periphery, coupled with the almost undiffused TiC core has been
previously observed by electron probe microanalysis, although on
considerably coarser microstructures.

Concentration gradients of Mo and V through what appears to
be the core of a large grain (2-3 microns) in 7GY sample #2
are shown in Grain I of Fig. 5. Both Mo and V are present in

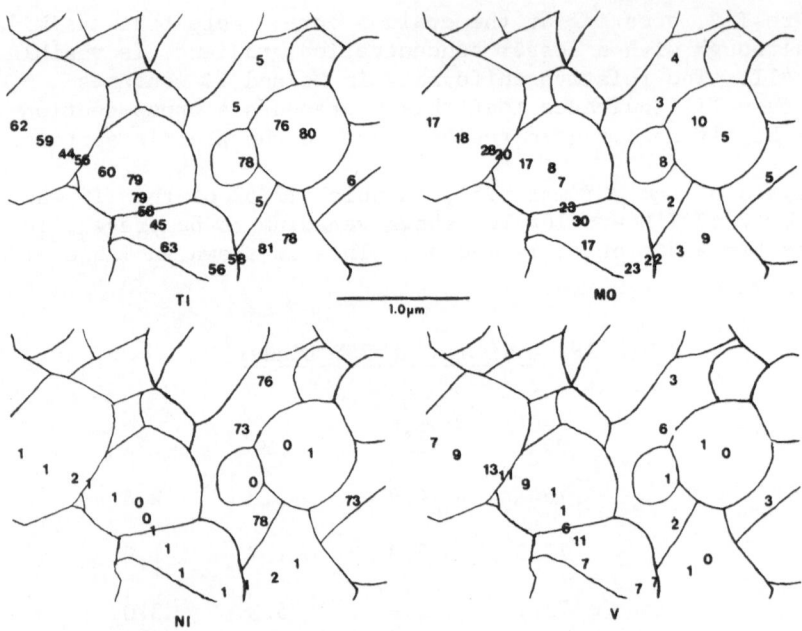

Fig. 4. Microstructural Sketch of 7GY Specimen Showing
Analyses for Ni, Ti, Mo and V Content

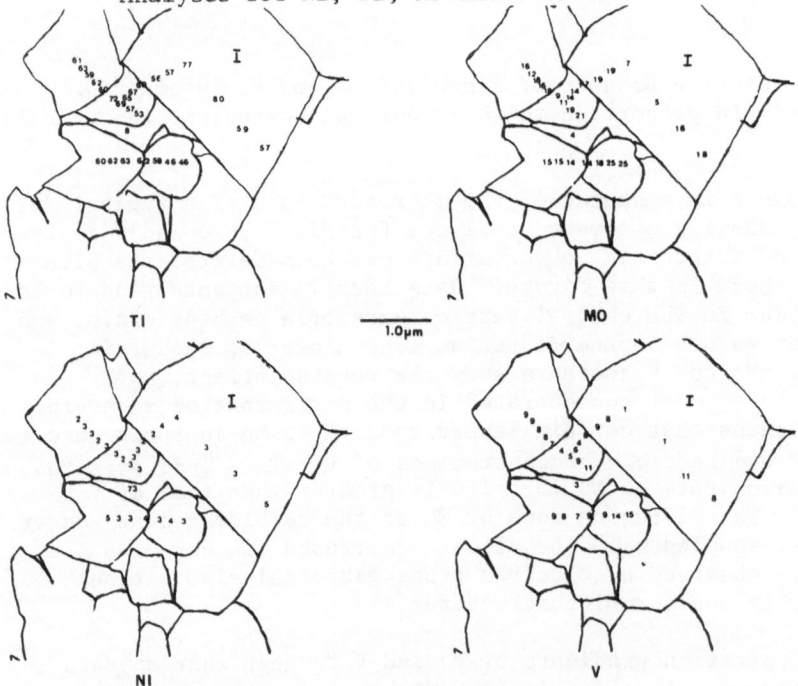

Fig. 5. Concentration Gradients of Mo, V and Ti in 7 GY
Specimen #2

a diffusion zone at the grain's outer perimeter, which corresponds well with a decreased concentration of Ti in that same area. Note that, once again, while some Mo is present in the core, V appears to be essentially absent. The minimal variation in concentration of these elements in the smaller grains in this figure is probably because the cross-section taken could have been adjacent to the outer periphery of these grains. In addition, the diffusion zone's depth may be sufficient to include the entire cross-section of smaller carbide grains.

Several attempts were made during this study to determine whether a Ni film is present at carbide-carbide grain boundaries. Analyses for Ni were run in increments of 0.1 microns, traversing from the interior of one carbide grain across the grain boundary into an adjacent grain. An analysis at a position believed to be on a grain boundry was also included. These results were uniformly negative, indicating that, within our experimental error, a Ni film could not be detected at carbide-carbide grain boundaries.

If we assume that a binder film, if present, would be no more than 20Å wide[9], the STEM/EDXS equipment used in the present study might have difficulty in picking it up. This is due to the fact that the volume of analyzed Ni, under these conditions, is estimated to represent only approximately 5 percent of the total volume excited by the beam, which is close to the actual experimental error.

CONCLUSION

Scanning transmission electron microscopy, combined with the capability for energy dispersive x-ray spectroscopy, is a useful tool in studying phase interactions in cemented carbide systems. Applied to $TiC-MoC_x-Ni$ base materials, the method confirmed the diffusion zone of molybdenum at the outer periphery of carbide grains. Binder Ti values obtained using the STEM/EDXS method compare reasonably well with those obtained by phase separation. Evidence for the diffusion of vanadium into carbide grains in VC-containing compositions is also given. The method has significant advantages over electron probe micro-analysis and phase separation analysis, due respectively, to an order of magnitude greater resolution of the analysis area, and the capability of knowing analysis position with respect to microstructure.

ACKNOWLEDGEMENT

The authors wish to acknowledge the valuable assistance of S. S. Shinozaki in carrying out this study.

REFERENCES

1. H. Kubota, R. Ishida, and A. Hara, Trans. Indian Inst. Metals 17:132 (1964).
2. H. Suzuki and H. Kubota, Planseeber. Pulvermetallurgie 14:96 (1966).
3. D. Moskowitz and M. Humenik, Jr., "Modern Developments in Powder Metallurgy, V. 3," Plenum Press, New York, 83 (1966).
4. H. Suzuki, K. Hayashi, and O. Terada, J. Japan Inst. of Metals 35:146 (1971).
5. P. O. Snell, Planseeber. Pulvermetallurgie 22:91 (1974).
6. D. Moskowitz and M. Humenik, Jr., Int. J. Powder Metallurgy 14:39 (1978).
7. To be published in: "Proceedings of 1980 Int. Powder Metallurgy Conf., June 22-27, 1980," Washington, D.C.
8. G. Cliff and G. W. Lorimer, J. Microscopy V. 103 pt.2:203 (1975).
9. N. K. Sharma, I. D. Ward, H. L. Fraser, and W. S. Williams, J. Am. Ceramic Soc. V. 63, No. 3-4:164 (1980).

DISCUSSION

V. K. Sarin:

The coring effect in these materials, where you have the original titanium carbide grain with a layer of mixed carbide on it, could that be produced by redeposition of dissolved Mo, V etc. rather than diffusion?

D. Moskovitz:

That is a possibility. Nobody has really shown the mechanism. We believe it's diffusion at the sintering temperature into the grain. It could be a precipitation type of effect. I'm calling them diffused zones. I wouldn't rule out redeposition. Maybe the kinetics of it should be looked at.

W. Williams:

I'd like to just carry on the line of discussion having to do with whether the molybdenum was diffused into the particle or might have been reprecipitated. The results on diffusion of titanium in titanium carbide and vanadium in vanadium carbide and other such systems indicate a very high activation energy for diffusion, of the order of 8eV. Because in this case, unlike the case of carbon, it is necessary to form the vacancy as well as to move it. Therefore I wonder whether at the sintering temperature of 1400 or 1500°C it would be possible to activate with any measureable effect a process that requires an activation energy of 8eV.

H. Fischmeister:

Concerning the diffusion question, I'd just like to point out that
the carbide that is dissolved in the binder at the liquid phase
sintering temperature must go somewhere. It can only form a skin.
The point I wanted to make, however, was really another one. I was
intrigued by the fact that in the vanadium-alloyed grades you find a
much lower titanium content in the binder. I found this an interesting thought for speculation because attempts to apply superalloy
principles or strengthening principles (as far as I understand the
situation) have failed so far because of embrittling phases. The
high titanium contents, quite naturally have no superalloy functions.
However, titanium contents in the vanadium-alloyed grades are in the
vicinity of those found in superalloys, (admittedly on the high side)
and maybe this situation would make it worthwhile for another go at
gamma-prime hardening attempts.

PHASE TRANSFORMATIONS IN THE BINDER PHASE

OF Co-W-C CEMENTED CARBIDES

G. Wirmark and G.L. Dunlop

Department of Physics
Chalmers University of Technology
S-412 96 Göteborg, Sweden

ABSTRACT

The role of the binder phase in Co-W-C cemented carbides is discussed and the literature concerning possible phase transformations in the binder is briefly reviewed. Some recent observations made by analytical electron microscopy of precipitation reactions in simulated Co-W-C binder phase alloys are presented.

INTRODUCTION

Cemented carbides were one of the first examples of successful composite materials in which the beneficial properties of two component materials are retained in the final composite. Hardness and wear resistance are provided by the carbide phase while the metallic binder phase contributes strongly to the ductility and toughness of the composite.

The first cemented carbides developed in Germany in the early 1920s were mixtures of WC and Co and alloys based on these two major components still comprise the majority of hard metals manufactured today. Despite the uncertainess of supply and the fluctuating price of Co, this metal is still used for the binder phase of approximately 95% of all cemented carbide production (1). Despite many efforts to find satisfactory alternative binder phase compositions, WC cemented carbides with substitute binder phases have not yet found a prominent place in production. From earlier work it would seem that cobalt has wetting and adhesive properties which are superior to most other metals and it also results in a tougher final product. However, more recent work is casting some doubts on this.

The binder phase of Co-W-C cemented carbides contains significant amounts of W and C which are present in solid solution after cooling from the sintering temperature. The amount of dissolved W and C is less in the solid binder than in the melt due to the diminished solubility of WC in Co with decreasing temperature. This results in the precipitation of WC, usually on existing WC grains, during cooling. Also, as shown by Jonsson and Aronsson (2) prolonged ageing at intermediate temperatures (eg. 600-800°C) can result in a number of other precipitation reactions occurring. Some of these produce fine precipitate dispersions which could strengthen the binder phase. Thus, the prospect of improving the properties of cemented carbides by simple heat treatments is raised. Also, the operating temperatures of cemented carbides are generally rather high (see for example 3) and it may be possible for reactions of this type to occur during prolonged service. This could be expected to alter, for example, the creep strength of the binder and the mechanical properties of the hard metal as a whole.

The present paper examines the various phase transformations which can occur in alloys which are of similar composition to the binder phase of Co-W-C cemented alloys. The literature on the subject is briefly reviewed and some new results which have been obtained by analytical transmission electron microscopy are presented.

THE ROLE OF THE BINDER PHASE

The binder phase of cemented carbides performs two major tasks:

(i)   it enables the composite to be sintered into a dense body;

(ii)   it introduces a ductile component into the microstructure and thus increases the materials toughness.

The sintering behaviour of a carbide/binder metal mixture is to a large extent determined by the wettability of the carbide by the molten binder. Cobalt, and perhaps to a lesser degree Fe and Ni, provides good wettability with WC. This together with the existence of a two phase region (WC + liquid) in the Co-W-C ternary phase diagram (see Fig. 1) ensures that high density cemented carbides which do not contain the brittle η-carbide or graphite phases can be produced by using Co as a binder metal. Similar two phase regions also occur in the Fe-W-C (4) and Ni-W-C (5) systems but as yet Fe and Ni have not been used in many applications. There would still appear to be difficulties in achieving the same good combination of mechanical properties with these metals as can be obtained with Co as the binder.

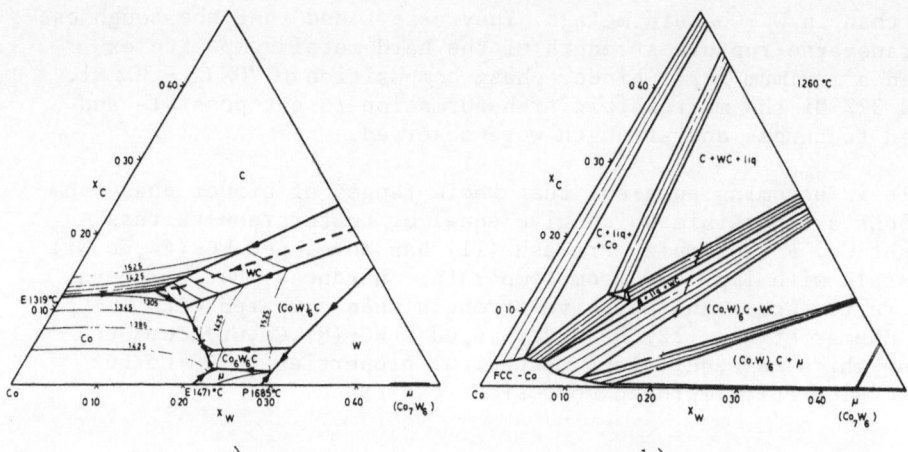

a)   b)

Fig. 1.   The Co-W-C system. (a) Basal projection of liquidus sur-
          faces in the Co-rich corner. (b) Isothermal section at
          1260°C in the Co-rich part. After Uhrenius et.al. (6).

Co has a stronger influence on the toughness of WC hardmetals
than other possible binder phase metals (eg. Fe and Ni). The high
temperature allotropic form of Co (fcc) tends to be stabilized by
dissolved W and C (1) because the M_S temperature is reduced (see
Fig. 2). A contributing factor to the drop in M_S of the binder
phase in hardmetal composites may also be mechanical constraints
to the martensitic α(fcc)$\rightarrow\epsilon$(hcp) transformation which are provided
by the rigid skeleton of WC grains. At room temperature the α(fcc)
form of Co-W-C binder phase solid solutions is metastable with
respect to the low temperature ϵ(hcp) modification and the martensitic transformation $\alpha\rightarrow\epsilon$ can occur with the application of sufficient shear stress (7). Sarin and Johannesson (8) contend that
this stress-induced transformation is the mechanism of plastic deformation of the Co-rich binder phase. If this is the case then
this transformation may provide the reason for why Co-bound WC-
based hardmetals are generally considerably tougher than if other
binder metals are used. As in the case of TRIP steels, it seems
reasonable to assume that considerable mechanical energy may be
absorbed by the martensitic transformation thus providing a good
crack-arrester and giving the composite a relatively high fracture
toughness.

Because of the high price and unstable supply of Co (9) there
has developed a strong interest in the possible complete or partial
replacement of Co by Ni in WC-based hard metals. The alleged superiority of Co has been challenged by a number of workers and
Brabyn et.al. (10) have shown that Co-Ni alloys may in fact be
better than straight Co binders. They found that with high Ni contents the width of the 2 phase WC-liquid field was considerably

wider than in WC-Co hard metals. They also found that the toughness
and transverse rupture strength of the hard metal composite exhibited a maximum for a binder phase composition of 70% Co – 30% Ni.
Beyond 32% Ni the martensitic transformation is not possible and
reduced toughness and strength were observed.

It is becoming apparent that whole ranges of binder phase compositions are possible which give equal or better results than
straight Co. For example, Prakash (11) has developed WC-(Fe,Co,Ni)
hardmetals with improved room temperature hardness, hot hardness,
fracture toughness and abrasive strength when compared with WC-Co.
Also, Ekemar et.al. (12) have developed a WC-(Ni,Co,Cr) cemented
carbide which has equivalent mechanical properties to WC-Co but
with considerably improved corrosion resistance.

EQUILIBRIUM PHASE RELATIONSHIPS

Considerable work has been carried out on phase relationships
in the Co-W-C system in the temperature range 1000-1500°C (for a
review, see Exner (1)). The findings in the W-rich corner of the
system differ substantially between different investigators but
they are all in substantial agreement in the areas of importance
for cemented carbides.

As can be seen from Fig. 1(b) the two phase region WC-liquid
only exists in a narrow range of carbon content which is close to
a W/C atomic rato of 1. Lower carbon contents result in the precipitation of the η-carbide phase and higher carbon contents cause
the precipitation of graphite.

Ternary Co-W-C diagrams have not been established for temperatures below 1000°C. However, the binary Co-W and Co-WC diagrams
shown in Fig. 2 are useful in practice. Pure Co undergoes the allotropic fcc→hcp transformation at 417°C. The rate of transformation
by diffusion is extremely slow and for all practical purposes only
the martensitic transformation is of importance. The characteristics of the martensitic transformation have been described in some
detail by Giamei et.al. (13). The transformation is diffusionless
and occurs by the motion of partial dislocations along the close
packed {111} planes in the f.c.c. lattice. Recent work on the subject has been mainly concerned with the mechanism by which the
partial dislocations can multiply. Horsewell et.al. (16) have shown
how grain boundaries can act as the source of Shockley partials
while Majahan and coworkers (17) have developed a model whereby
grown-in dislocations multiply rapidly and the resulting dislocations separate into Shockley partials. These partials move away
from each other to form overlapping stacking faults. In this way
two a/2 <110> dislocations can give rise to four or six layers of
hcp structure.

According to the binary diagrams in Fig. 2 dissolved W and WC
does not actually bring about a thermodynamic stabilization of the
α(fcc) form of Co. However these elements do cause a substantial
decrease in the M_s temperature. Since the diffusional transformation is generally very slow at intermediate temperatures this reduction in M_s temperature brings about an effective "stabilization"
of the α phase.

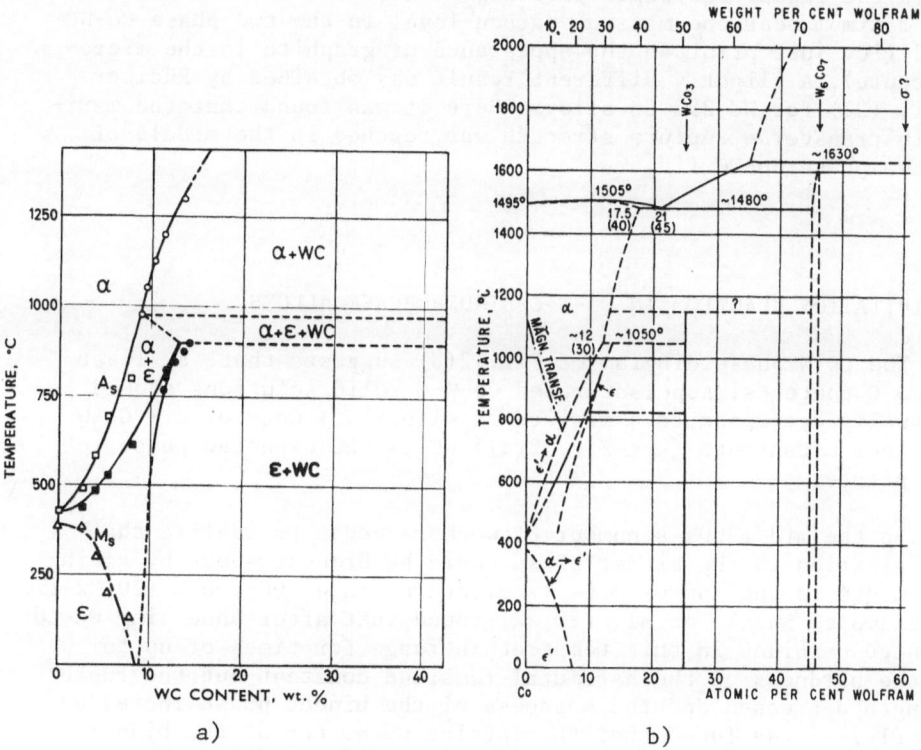

Fig. 2. Co-rich binary phase diagrams. (a) The quasi-binary Co-WC
         system (14). (b) The Co-W system (15).

BINDER PHASE COMPOSITION AND ITS EFFECT ON MECHANICAL PROPERTIES

The amount of WC which can dissolve in the binder phase is
still a matter which is open for discussion. Recent reports indicate that 9.5 wt.% WC can be dissolved at 1250°C (18). According
to Exner (1) there seems to exist a solubility limit of
$[W][C] = 8 \times 10^{-4}$ where the concentrations are given as atomic fractions. When the carbon content is low, up to 26 wt.% W has been
measured in solution (18). Where graphite is present (ie. the
binder phase is saturated with carbon) the solubility of W is only
of the order of 1-3 wt.% (19).

The mechanical properties of cemented carbides are related to
the W/C ratio in the binder phase. The hardness of the binder phase
and also of the composite increases with increasing W/C (20). However
since transverse rupture strength is not only dependent on resistance to plastic deformation but is also dependent upon resistance
to fracture the influence of W/C ratio on this parameter is somewhat different. Suzuki and Kubota (20) noted that for WC-10%
alloys the transverse rupture strength increased with C content
to a maximum near the maximum carbon-level in the two phase Co-WC
field (ie. just prior to the appearance of graphite in the microstructure). A slightly different result was obtained by Rüdiger
et.al. (21) for WC-25% Co alloys. Here it was found that the maximum in transverse rupture strength was reached in the middle of
the two phase Co-WC field.

PRECIPITATION REACTION IN Co-W-C BINDER PHASE ALLOYS

The Co-W phase diagram of Fig. 2(b) suggests that, at least
at low C contents, supersaturated Co-W-C solid solutions should
eventually decompose to a mixture of either α + Co_3W or ϵ + Co_3W.
At higher C contents (see Fig. 2(a)) WC is the expected phase to
precipitate.

In the mid 1960s a number of workers began to realise that
precipitation in the binder phase could be brought about by ageing
treatments in the approximate temperature range 600-800°C (20,22-25).
For example, Suzuki et.al. (20,22) found that after annealing WC-Co
cemented carbides in this temperature range for times of up to
50h the hardness of the hardmetal remained constant but the rupture
strength decreased and the hardness of the binder phase increased
slightly. It was found that the lattice parameter of the binder
decreased as did the specific resistivity of the composite. The
extent of all of these changes increased for alloys having a lower
carbon content i.e. for alloys where the amount of W dissolved in
the binder phase was greatest. From these results it was concluded
that W had been removed from solution in the binder phase during
annealing (22) and that the precipitating phase was probably Co_3W
(20). The kinetics of transformation $\alpha \rightarrow \epsilon$ + Co_3W have been studied
by Toda (26) in a binary Co-25 wt.% W alloy. Metallography indi-
cated that the maximum rate of transformation for this alloy oc-
cured at a temperature between 700 and 800°C and that the reaction
required an incubation period of approximately 10 min.

The most detailed work reported on phase transformations in
Co-W-C alloys was carried out by Jonsson and Aronsson (2). This
work was carried out on Co-W-C alloys which simulated a range of
binder phase compositions with varying W and C contents. Combined

transmission electron microscopy and X-ray diffraction showed that
a whole range of transformation products could arise depending upon
the alloy composition and the ageing temperature. Their findings
can be summarised as follows:

(i) At low ageing temperatures (<400°C) needles of Co_2C
formed in α-Co.

(ii) At higher temperatures in the range 550-750°C fine
coherent precipitates of α' precipitated in α-Co
and ε' precipitated in ε-Co. These precipitates have
ordered crystal structures based on the parent ma-
trices. For example α' has the familiar Cu_3Au (γ')
structure. Dispersions of these precipitates were
found to bring about considerable increases in hardness.

(iii) Ageing at temperatures higher than that for the precipitation of α' or overageing of the α' or ε' dispersions led to intergranular or discontinuous preci-
pitation of Co_3W.

(iv) At the highest ageing temperatures (750-1000°C) coarse
precipitation of the carbide phases η_1, η' and WC
occurred.

The approximate kinetics of these transformations were determined
and presented as temperature-time transformation diagrams. The
wide range of precipitation reactions observed by Jonsson and
Aronsson suggests that it may be possible to tailor binder phase
compositions in order to improve the properties of cemented carbides
either through precipitation during special heat treatments or
through precipitation during service.

Indirect evidence for the precipitation sequence
$\alpha \rightarrow \alpha + \alpha' \rightarrow \alpha + Co_3W$ occurring in the binder phase of WC-25 wt.% Co
hardmetal aged at 700°C has been obtained by Tillwick and Joffe
(27) using magnetic measurements. Jonsson (28,29) has followed up
the previous work on simulated binder phase alloys and showed by
transmission electron microscopy that similar precipitation reactions also occur in the binder phase of aged cemented carbides. It
was found that the precipitation of α' was accompanied by an increase in hardness of the composite and also a decrease in the
transverse rupture strength.

RECENT OBSERVATIONS OF PRECIPITATION IN SIMULATED Co-W-C BINDER
PHASE ALLOYS

Two Co-W-C alloys containing 16.4 wt.% W (5.9 at.%) with in
one case 0.06 wt.% C (0.32 at.%) and in the other 0% C were heat

treated in order to study precipitation reactions at 650°C and
750°C. The heat treatments were carried out by either direct isothermal ageing after solution treatment at 1300°C or by quenching
after solution treatment to room temperature and then ageing. The
specimens were examined by analytical transmission electron microscopy.

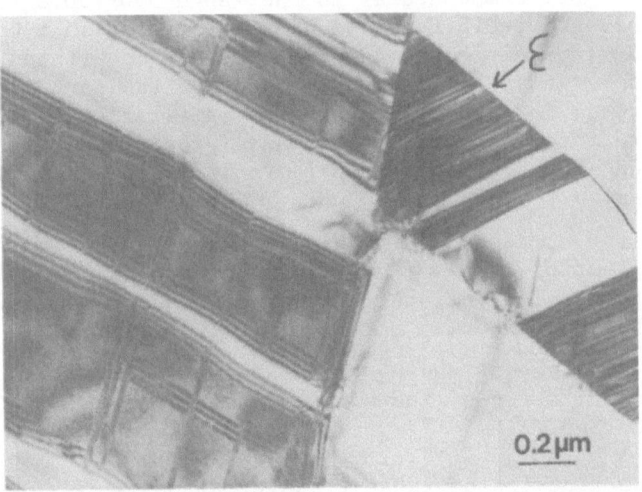

Fig. 3.   The Co-W-0%C alloy in the as-received condition.   An
          area of hcp (ε) is indicated. The fcc (α) phase con-
          tains a number of twins and stacking faults.

Experimental

The alloys, which were supplied by the National Physical
Laboratory England, had been prepared from high purity constituents
by melting, casting and hot extrusion. Details of the manufacturing
procedure are given in (30). Specimens of approximately 1 mm thickness were sealed in evacuated silica capsules and solution treated
for 1 h at 1300°C. The specimens were then aged at 650°C or 750°C
according to one of the following procedures:

(i)   Direct isothermal transformation, viz. direct cooling
      from the solution treatment temperature to the ageing
      temperature.

(ii) Quench ageing, viz. quenching to room temperature from
      the solution treatment temperature followed by heating
      to the ageing temperature.

The ageing time at 650°C was 100 h while times of 25, 50 and 100 h
were employed at 750°C. All ageing treatments were terminated by
quenching into water.

Specimens were examined by X-ray diffractometry using CuKα and CrKα radiation and also in a 200 kV TEM/STEM electron microscope with X-ray analysis facilities.

Results

As-received condition. In the as-received condition the fcc(α) allotropic modification dominated with only a small amount of the h.c.p.(ε) modification present. The amount of ε was somewhat greater in the alloy not containing carbon (see Fig. 3). No other phases were detected by either X-ray diffraction or electron microscopy.

Quench ageing at 650°C. Both alloys had virtually exactly the same microstructure after this heat treatment. The matrix crystal structure was almost totally fcc. Some minor traces of ε-hcp were detected by X-ray diffraction but no ε was found during transmission electron microscopy. The ε detected by X-ray diffraction

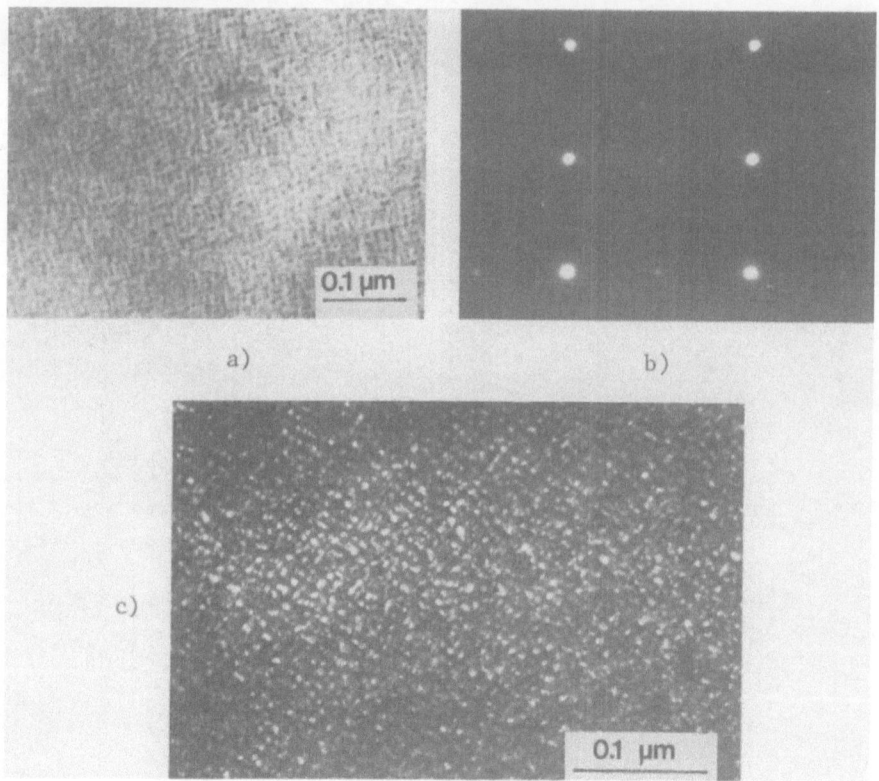

a)   b)

c)

Fig. 4.   Precipitation of fine α' after quench ageing at 650°C.
          (a) bright field. (b) selected area diffraction pattern
          showing superlattice reflections. (c) dark field taken
          using an α' superlattice reflection.

probably arose due to stress induced transformation during mechanical polishing. As can be seen in Fig. 4 a very fine homogeneously dispersed array of α' precipitates was present after the ageing treatment. The average precipitate size after 100 h ageing at 650°C was 50 Å.

Direct isothermal transformation at 650°C. In both alloys the matrix phase was ε. Some small islands of α were observed in the carbon containing alloy. The dominating microstructure was, as shown in Fig. 5, a banded feathery mixture of ε and what is probably the ordered hcp phase ε' which was referred to by Jonsson and Aronsson (2). As yet the structure of this phase has not been unambiguously identified. The high magnification secondary electron image of an electropolished surface in Fig. 6 shows the banded nature of the microstructure more clearly. EDX analysis was carried out in and between the bands and the relevant spectra are shown

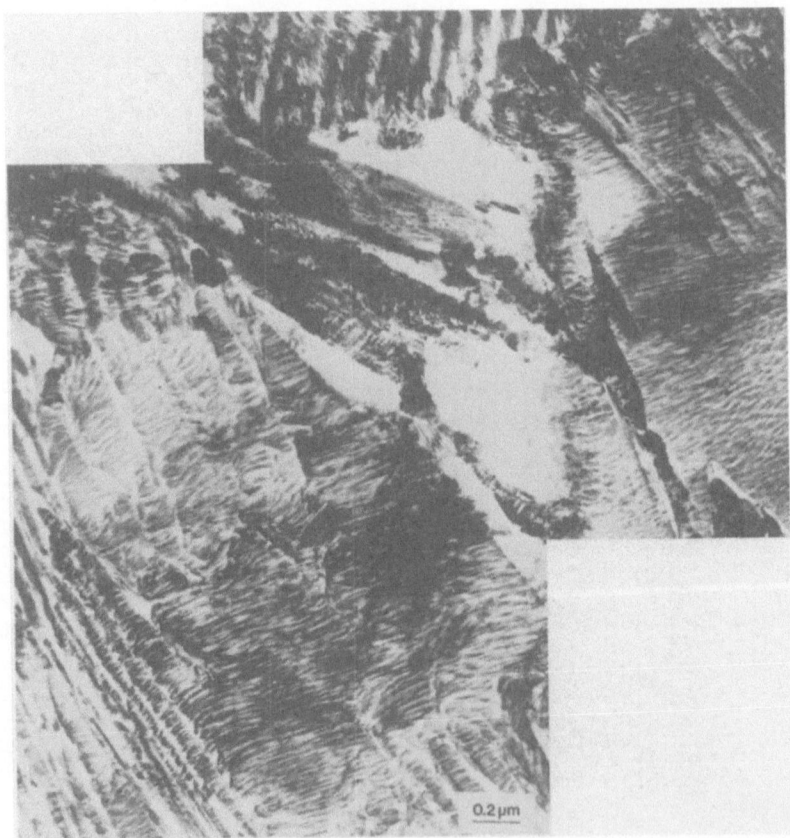

Fig. 5.   The banded feathery microstructure obtained by direct isothermal transformation at 650°C. TEM bright field.

in Fig. 7. Between the bands (ie. in ε at position B in Fig. 6)
the W content was low (below 3%). Although beam spreading in the ε'
containing bands (i.e. at position A in Fig. 6) made it impossible
to perform a fully quantitative analysis it was found that these
areas contained over 20 wt.% W. The possibility cannot be excluded
that the ε' had a composition close to Co_3W.

As mentioned previously the C-containing alloy had small islands of α within the banded ε/ε' microstructure. An example of
this is shown in Fig. 8. These islands contained a very fine dispersion of precipitates of the ordered α' phase.

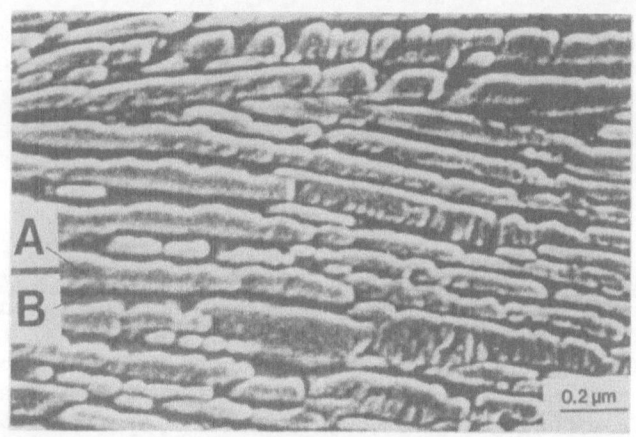

Fig. 6.   Secondary electron image of the banded microstructure
          resulting from direct isothermal transformation at 650°C.

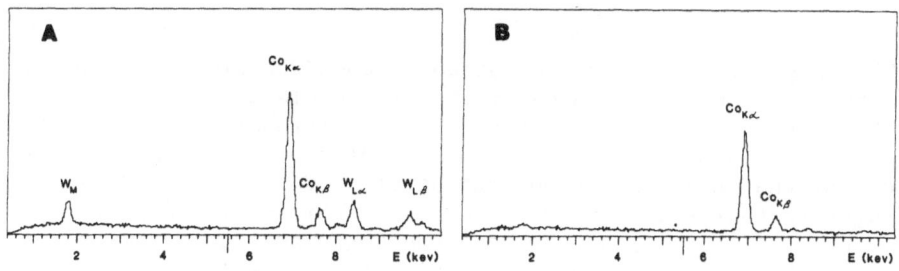

Fig. 7.   EDX spectra of the areas, A and B, indicated in Fig. 6.
          Note the low W content of ε and the higher W content
          in what may be ε'.

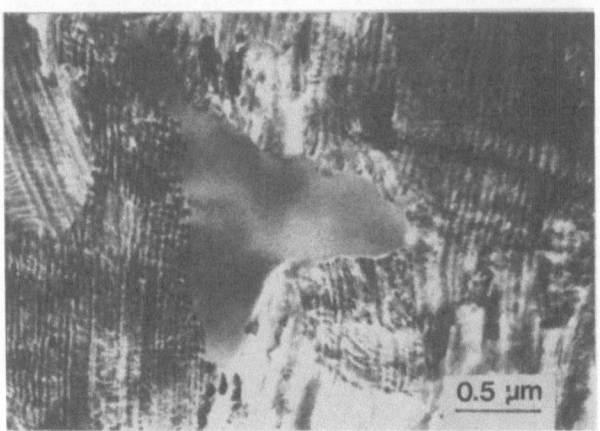

Fig. 8.   An island of α/α' within the banded ε/ε' structure of
the C-containing alloy after direct isothermal transformation at 650°C.

Direct isothermal transformation and quench ageing at 750°C.
At this ageing temperature no differences were noted between the
two alloys. Also, both types of heat treatment resulted in the same
mode of transformation, i.e., the discontinuous precipitation of
Co_3W. The resulting microstructure after complete transformation
and quenching from the ageing temperature is a mixture of lamella
Co_3W in ε (Fig. 9). In accordance with Jonsson and Aronsson (2)
a parallel crystallographic relationship between the two phases
was obtained:

$$\{0001\}_\varepsilon \ // \ \{0001\}_{Co_3W}$$

$$<11\bar{2}0>_\varepsilon \ // \ <11\bar{2}0>_{Co_3W}$$

The incubation time for the discontinuous reaction was between
25 and 50 h and the reaction was virtually complete after 100 h.
Fig. 10 shows the reaction front of partially transformed material
after 50 h at 750°C. Quantitative EDX analysis of the Co_3W lamellae
showed that they had a tungsten content of ∿ 50 wt.% (ie. ∿ 25 at.%)
This suggests that the Co_3W is close to stoichiometric composition
The W content of the ε phase between the Co_3W lamellae was measured
to be 10 wt.% but beam spreading in this case may well have given
a somewhat too high concentration.

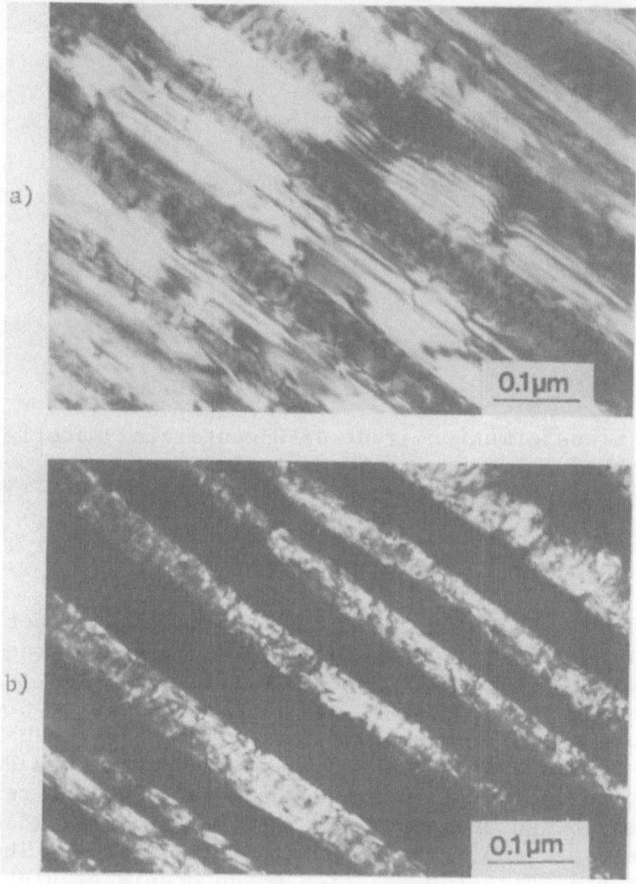

Fig. 9. Lamellar Co$_3$W in ε after transformation at 750°C. (a) bright field. (b) dark field taken using a Co$_3$W precipitate reflection.

DISCUSSION

The observations relating to quench ageing which were described in the previous section are consistent with the results obtained previously by Jonsson and Aronsson (2). However, it should be noted that the present work was on material with much lower carbon contents than that studied by Jonsson and Aronsson. The TTT diagrams obtained by these authors always indicated that carbides precipitated first prior to the appearance of other phases. Jonsson (31) studied alloys with lower carbon contents which are more in keeping with those investigated here. He also found that α' preci-

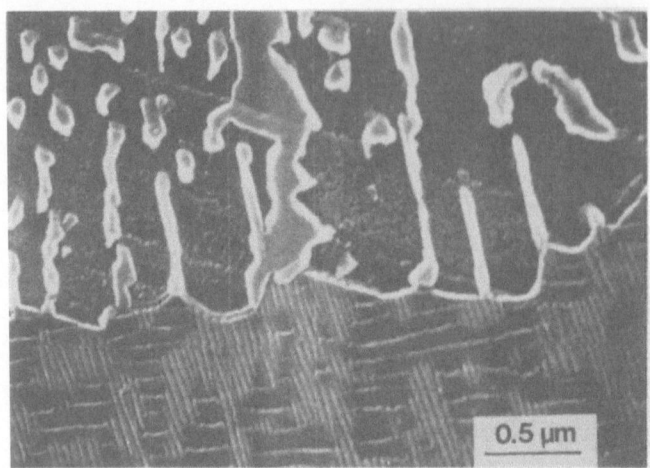

Fig. 10. The transformation front of C-containing material
quenched and aged for 50 h at 750°C.

pitated continuously at the lower ageing temperatures and Co_3W
precipitated discontinuously at higher temperatures.

It was found that the mode of heat treatment (ie. direct isothermal transformation or quench ageing) had no effect on the
transformation product when ageing at 750°C. The resulting microstructure was a mixture of lamellar Co_3W in martensitic ε. Since
750°C is well above the M_s temperature (see Fig. 2) it can only be
concluded that ε formed during quenching of the α which had been
depleted of W during ageing. It can also be noted that the reaction
front was typically bowed bewteen Co_3W lamellae (see Fig. 10) indicating a diffusional rather than a martensitic reaction. Thus,
the discontinuous precipitation reaction which occurs at high temperatures is:

$$\alpha_1 \rightarrow \alpha_2 + Co_3W$$

where α_1 has a higher W content than α_2.

The reaction which has not been reported previously is that
which occurs during direct isothermal transformation at 650°C.
Here the observed transformation product was a banded mixture of
ε and what is probably the ordered ε' phase. The latter phase was
found to have a considerably enhanced W content. Ageing times
shorter than 100 h at this temperature have not yet been investigated but it would seem that this reaction may also be discontinuous. The small fcc islands observed in the carbon-containing alloy
(see Fig. 8) give some support to this contention. Here again the
ageing temperature was well over M_s and therefore it is tempting

to suggest that the reaction at 650°C is:

$$\alpha_1 \rightarrow \alpha_2 + \alpha'$$

where α_1 has a higher W content than α_2 and α' is the ordered pre-cipitate. The decreased W content of α_2 would render it readily transformable to ε on quenching. The question remains: is it also possible for α' to transform simultaneously to ε'? Clearly con-siderably more work is required before these transformations are fully understood.

The transformations discussed here should clearly have a strong influence on the mechanical properties of the binder phase of ce-mented carbides. Under many circumstances the influence on the pro-perties of the composite as a whole can be expected to be rather small (19,28,29), (eg. small increases in hardness and small de-creases in toughness).

The success of Ni-base superalloys strengthened by γ' in high temperature creep resistant applications suggests that uniform dispersions of α' in α such as shown in Fig. 4 should be particu-larly useful in those applications where the common mode of failure of cemented carbides is by creep deformation. The problem then arises of developing a binder phase composition which enables such uniform fine microstructures to be retained at the very high tem-peratures which are involved. The necessary changes in binder phase chemistry should of course not be detrimental to the other properties which are also required of the hardmetal composite.

CONCLUSIONS

1. Phase transformations have been studied in low carbon Co-W-C alloys which simulate the Co-rich binder phase of cemented carbides.

2. Quenching from 1300°C followed by ageing at 650°C re-sulted in copious fine precipitation of the ordered α' phase in fcc α.

3. Direct isothermal transformation at 650°C resulted in a feathery banded microstructure containing hcp and what is probably an ordered hcp phase, ε'.

4. Transformation at 750°C by either quench ageing or direct isothermal transformation resulted in the discontinuous precipitation of lamellar Co_3W.

ACKNOWLEDGEMENTS

The experimental alloys were kindly supplied by Dr E.A.
Almond of the National Physical Laboratory, England and financial
support was received from the Swedish Board for Technical Development. Discussions with B. Aronsson, C. Chatfield, R. Warren and
L. Kjellsson are gratefully acknowledged.

REFERENCES

1. H.E. Exner, Physical and chemical nature of cemented carbides,
    Int. Met. Revs. 243:149 (1979).
2. H. Jonsson and B. Aronsson, Microstructure and hardness of
    cobalt-rich Co-W-C alloys after ageing in the temperature
    range 400-1000°C, J. Inst. Met., 97:281 (1969).
3. Y. Naerheim, A metallurgical method for determining the tem-
    perature in cemented carbide tools, in: Proc. 5th European
    Symposium on Powder Metallurgy, Stockholm, 1978, 99.
4. B. Uhrenius and H. Harvig, A thermodynamic evaluation of car-
    bide solubilities in the Fe-Mo-C, Fe-W-C, and Fe-Mo-W-C-
    systems at 1000°C, Met. Sci., 9:67 (1975).
5. M-L. Fiedler and H.H. Stadelmaier, The ternary system nickel-
    tungsten-cobalt. Z. Metallkde., 66:402 (1975).
6. B. Uhrenius, B. Carlsson and T. Franzén, A study of the Co-W-C
    system at liquidus temperatures, Scand. J. Metall. 5:49
    (1976).
7. H. Suzuki, T. Yamamoto and H. Sakanoue, Binder phase trans-
    formations in WC-Co cemented carbides, J. Jap. Inst. Met.,
    32:993 (1968).
8. V.K. Sarin and T. Johannesson, On the deformation of WC-Co
    cemented carbides, Met. Sci., 9:472 (1975).
9. V.A. Tracey and N.R.V. Hall, Nickel matrices in cemented
    carbides, Powd. Met. Int. 12:132 (1980).
10. S.M. Brabyn, R. Cooper and C.T. Peters, Effects of the substi-
    tution of nickel for cobalt in WC based hardmetal, in:
    Proc. Plansee-Seminar, 1981. 2:675.
11. L. Prakash, Development of tungsten carbide hardmetals using
    iron-based binder alloys, PhD thesis, Institut für Material-
    und Festkörperforschung, Kernforschungszentrum, Karlsruhe
    KFK 2984 (1980).
12. S. Ekemar, L. Lindholm and T. Hartzell, Aspects of nickel as
    a binder metal in WC-based cemented carbides, in: Proc.
    10th Plansee-Seminar, 1981. 1:477.
13. A. Giamei, J. Burman and E.J. Freise, The role of the allo-
    tropic transformation in cobalt-base alloys, (Part I),
    Cobalt, 39:88 (1968).
14. A.F. Giamei, J. Burma, S. Rabin, M. Cheng and E.J. Freise, The
    role of the allotropic transformation in cobalt-base alloys,
    (Part II), Cobalt, 40:140 (1968).

15. M. Hansen and K. Anderko, Constitution of binary alloys, McGraw-Hill, New York (1958).

16. A. Horsewell, B. Ralph and P.R. Howell, An intergranular mechanism for the fcc→hcp martensitic transformation, Phys. Stat. Sol. (a), 29:587 (1975).

17. S. Mahajan, M.L. Green and D. Brasen, A model for the fcc→hcp transformation, its applications, and experimental evidence, Met. Trans. A, 8A:283 (1977).

18. A. Hoffmann and R. Mohs, Gleichwichtsuntersuchingen im Kobalt-reichen Teil des Systems Co-W-C bei 1250°C, Metall, 28:661 (1974).

19. H. Jonsson, Microstructure and hardness of heat-treated Co-W-C alloys with compositions close to those of binder phases of WC-Co cemented carbides, PhD thesis, University of Uppsala, 1980.

20. H. Suzuki and H. Kubota, The influence of binder phase composition on the properties of WC-Co cemented carbides, Planseeb. f. pulvermet., 14:96 (1966).

21. O. Rüdiger, D. Hirschfield, A. Hoffmann, J. Kolaska, G. Oster-mann and J. Willbrand, Composition and properties of the binder metal in cobalt bonded tungsten carbide, Int. J. Powd. Metall. 7:29 (1971).

22. H. Suzuki, M. Sugiyama and T. Umeda, Effect of annealing on properties of sintered WC-Co alloys, Nippon Kinzoku Gakkai-Si, 28:287 (1964).

23. V.I. Tumanov, V.F. Funke, Z.S. Trukhanova, T.A. Novikova and K.F. Kuznetsova, Sov. Powd. Metall., 131 (1964).

24. ibid, Poroshk. Metall., 57 (1964).

25. A.A. Betser and J. Gurland, Some effects of temperature and heat treatment on the strength of sintered WC-Co alloys, Publ. 66-MD-17, ASME (1966).

26. T. Toda, Trans. Jap. Inst. Met., 6:139 (1965).

27. D.L. Tillwick and I. Joffe, Precipitation and magnetic hardening in sintered WC-Co composite materials, J. Phys. D: Appl. Phys. 6:1585 (1973).

28. H. Jonsson, Studies of the binder phase in WC-Co cemented carbides heat-treated at 650°C, Powd. Metall., 15:1 (1972).

29. H. Jonsson, Studies of the binder phase in WC-Co cemented carbides heat-treated at 950°C, Planseeb. f. pulvermetall. 23:37 (1975).

30. B. Roebuck and E.A. Almond, A comparison of the deformation characteristics of Co and Ni alloys containing small amounts of W and C, in: Proc. 10th Plansee-Seminar 1981, 1:493.

31. H. Jonsson, Microstructure and hardness of Co-rich Co-W-C alloys with low carbon contents (0.02-0.05 wt.%). Scand. J. Metall. 5:81 (1976).

DISCUSSION

V. K. Sarin:

Have you observed such a lamellar type of precipitate in WC-Co?

G. Dunlop:

No, we haven't looked yet at any composite materials. We've only
been looking at alloys which are equivalent to the binder phase. I
think H. Jonsson has seen the discontinuous reaction in cemented
carbides.

V. K. Sarin:

Are they in the lamellar form or are they discrete?

G. Dunlop:

I think both the discontinuous lamellar form and the alpha-prime
precipitates have been seen in heat treated cemented carbides.

F. Rymas:

At the beginning of your presentation you mentioned that you were
examining two different alloys. One had .06 carbon and the other
with no carbon. The results that you presented--were they for both
of the alloys or the alloy without carbon?

G. Dunlop:

The behavior of the two alloys is essentially the same. In fact, we
see no distinct difference between those two alloys.

R. Sivan:

Have you tried various heat treatment cycles? Is it possible that
those that you have tried may have been overaged. If it was a very
short cycle, you may get different TRS values?

G. Dunlop:

Excuse me if I misled you. We have not done any mechanical property
measurements on cemented carbides, and I think the work that has
been done is rather limited. I'm sure that you can get a range of
TRS values depending on the type of precipitate which you develop
during heat treatment and how much you age it. And then, of course,
it's not certain that TRS is the property which you want to develop
for many applications.

ION-ETCHING TECHNIQUES FOR MICROSTRUCTURAL CHARACTERIZATION OF CEMENTED CARBIDES AND CERAMICS

Akira Doi, Takeshi Nishikawa and Akio Hara

R & D Group
Sumitomo Electric Industries, Ltd.
Itami, Japan

INTRODUCTION

Prior attempts to delineate the grain boundaries of the binder phase (γ phase) in cemented carbides using thermal and chemical etching methods[1], have not been very successful. These methods, though simple and convenient, often do not produce adequate contrast for detailed microscopic observation and analysis. Other etching methods such as, ion sputter etching or simply ion-etching, are of a more recent origin. They are principally used in the preparation of thin electron-transparent foils from materials where conventional methods of specimen preparation are not applicable. In this report, the feasibility of applying ion-etching methods for microstructural characterization of cemented carbides and ceramics is examined. Ion-etching is found to give improved contrast between binder phase grains in cemented carbides as well as reveal macroscopic heterogenieties in certain ceramics.

EXPERIMENTAL PROCEDURE

The cemented carbide and ceramic samples used in the present study are listed in Table 1. The cemented carbides were produced by conventional sintering methods. The Si_3N_4 sample (C-1) was obtained by sintering a mixture of Si_3N_4 and MgO in a nitrogen atmosphere at 1700°C for 30 mins. The alumina sample (C-2) was sintered in vacuum at 1440°C for 2 hours followed by hot isostatic pressing at 1400°C and 150 MPa for 1 hour. The Al_2O_3 + TiC composite was hot pressed at 1675°C for 30 mins. followed by hot isostatic pressing under conditions similar to those mentioned above.

329

Table 1. Chemical Composition of Cemented Carbides
and Ceramics used in the present study

Sample No.	Chemical Composition in weight %. (Vol %)							Average Grain Size (μm)		
	WC	(TiTaNb)C	TiC	Co	MgO	Si_3N_4 [*]	Al_2O_3 [**]	WC	Si_3N_4	Al_2O_3
CC-1 [†]	85 (76.3)	---	---	15 (23.7)	---	---	---	5~6	---	---
CC-2 [†]	85.4 (78.5)	9.0 (12.6)	---	5.6 (8.9)	---	---	---	5~6	---	---
C-1	---	---	---	---	5	95	---	---	0.5~3	---
C-2	---	---	---	---	0.2	---	bal.	---	---	1~2
C-3	---	---	30	---	0.2	---	bal.	---	---	1~2

  *   Stark, H1 grade
 ** Alcoa, A16 grade
  † Carbides from Sumitomo Electric Ind. Ltd.

Two types of ion-etching devices were used for the present work;
the ion-etching device of the cold cathode type (IE10, Eiko Engineering Co.) used for high speed etching and the dual ion thinning device
of the hollow cathode type (IE20, Ibid) used for precision etching.
These devices are shown schematically in Fig. 1. Procedures used in
the preparation of samples for SEM and TEM examination are shown in
Fig. 2.

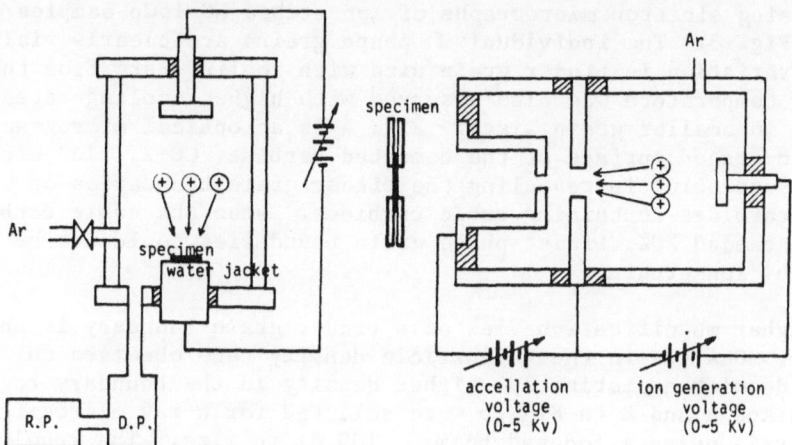

Fig. 1.   Schematic illustration of the ion-etching devices used in
           the present study.

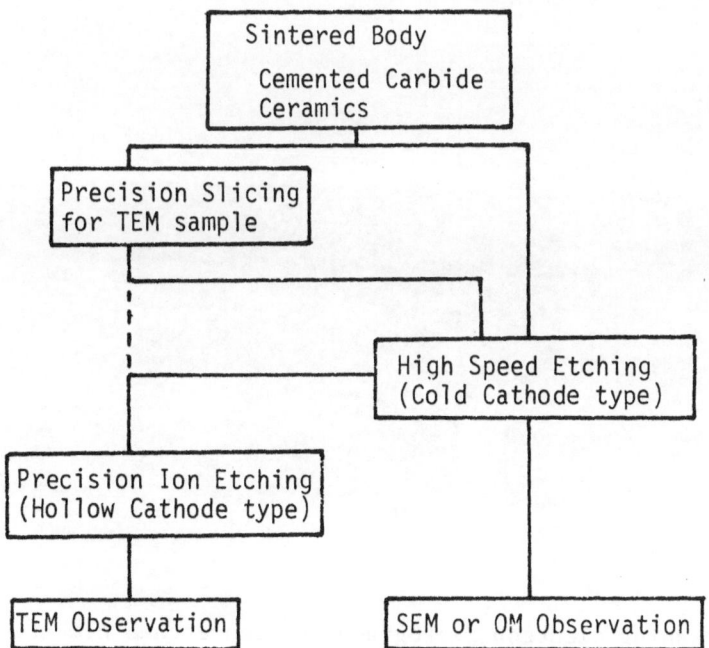

Fig. 2.   Flow chart for preparation of specimens for optical,
           scanning and transmission electron microscopy.

EXPERIMENTAL RESULTS AND DISCUSSION

Scanning electron micrographs of ion-etched WC-15Co samples are
shown in Fig. 3. The individual γ phase grains are clearly visible.
A marked variation in binder grain size with cooling rate from the
sintering temperature was also observed with higher cooling rates
resulting in smaller grain sizes. Fig. 4 is an optical micrograph
of the ion-etched surface of the cemented carbide, CC-2. Ion-etching
was less successful in revealing the binder grain boundaries of
cemented carbides containing cubic carbides. When the cubic carbide
content exceeded 20%, binder phase grain boundaries could not be
revealed by ion-etching.

A higher magnification view of a binder grain boundary is shown
in Fig. 5. Changes in the WC particle density were observed on
either side with a distinctly higher density in the boundary region.
Grains marked A and B in Fig. 5 were selected for x-ray microdiffraction analysis using a focused beam $\sim$ 100 μm in size. The results
of this analysis shown in Fig. 6 indicate strong spatial variations
in the orientation distribution of WC grains from one binder grain
to another.

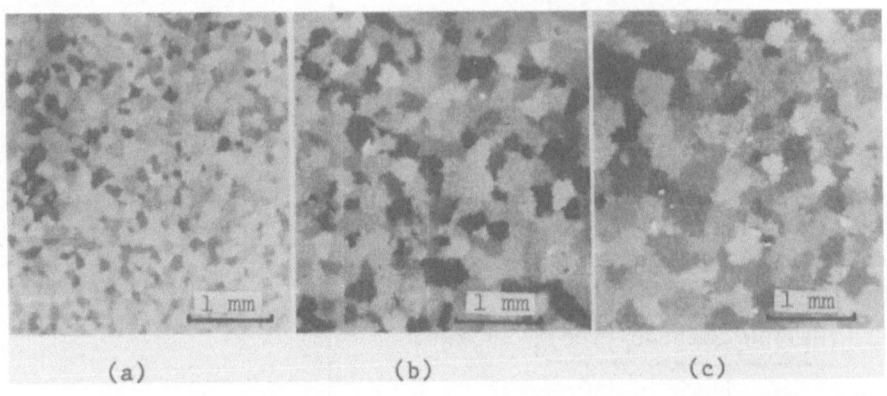

(a)   (b)   (c)

Fig. 3   Scanning electron micrographs of ion-etched cemented
           carbide samples (CC-1) produced with different cooling
           rates from the sintering temperature of 1375°C. The
           average cooling rates were: (a) $\sim$ 40°C/min. (b) $\sim$ 13°C/min.
           (c) $\sim$ 7° C/min.

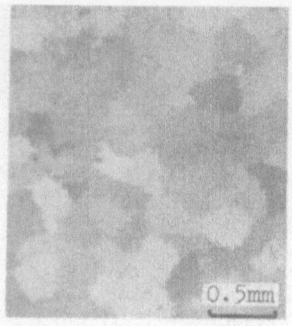

Fig. 4   Optical micrograph of the
          ion-etched surface of a
          carbide sample containing
          cubic carbides (CC-2)

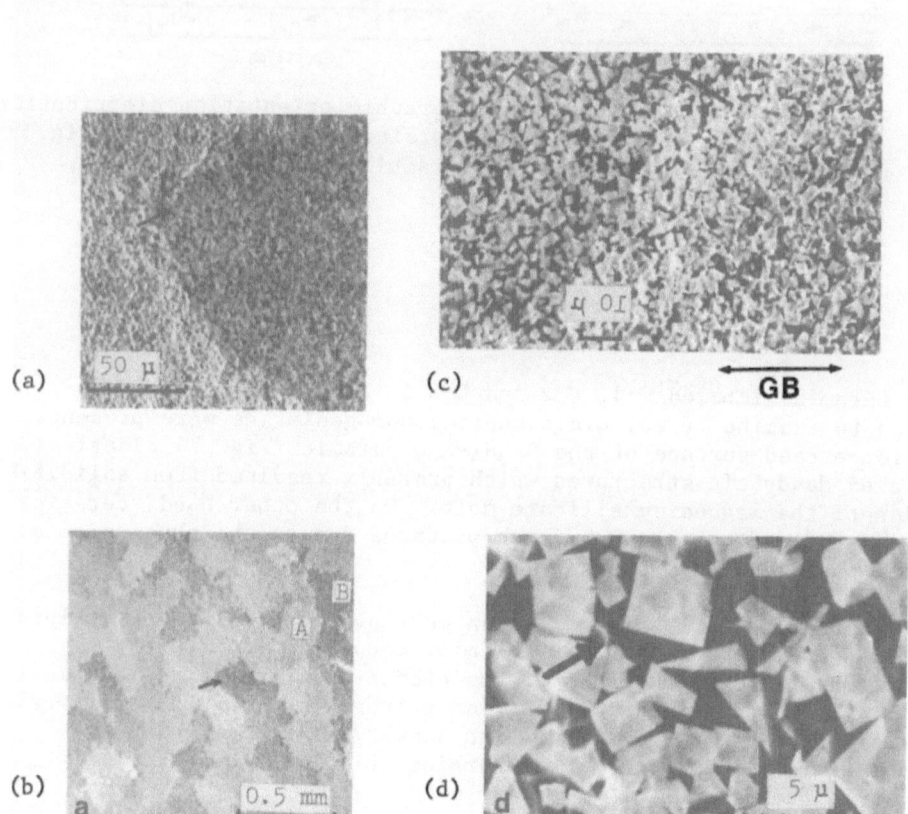

Fig. 5   Scanning electron micrographs of a grain boundary region
          in a WC-15 Co sample at different magnifications. GB
          indicates the approximate position of the grain boundary.
          Binder grains marked A and B were selected for further
          study by x-ray microdiffraction.

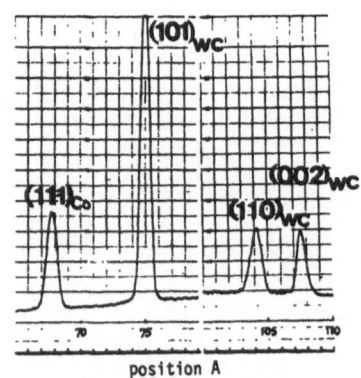 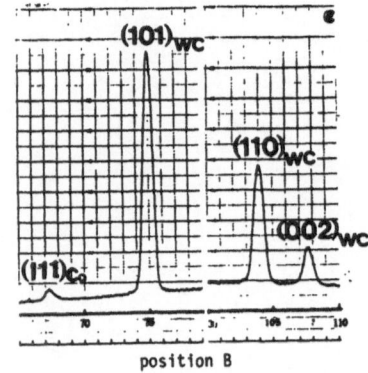

position A   position B

Fig. 6   Variations in the crystallographic orientation distribution
          of WC grains in the binder grains marked A and B in Fig. 5
          obtained by x-ray microdiffraction.

          Ceramics labeled C-1, C-2 and C-3 in Table 1 were also ion-
etched to examine if any microscopic inhomogenieties were present.
The ion-etched surface of the Si_3N_4-MgO ceramic (Fig. 7) clearly
revealed dendritic structures which probably resulted from solidification of the magnesium silicate melt. On the other hand, ceramics
C-2 and C-3 did not show any such patterns due to the absence of a
liquid phase during sintering.

          Thin discs for TEM observation were prepared using conventional
methods[2] and examined in a 200 KV microscope. Figs. 8 and 9
are bright field micrographs of the binder and carbide regions in
CC-1. The selected area diffraction pattern from the binder reveals
several extra spots due to the high density of stacking faults.
Bright field micrographs of the carbide (Fig. 9) show several
contours, the origins of which are unknown.

          Bright field micrographs of the ceramic samples C-1 and C-2 are
shown in Fig. 10 and 11. Micropores in the range of 20 to 100 Å were
observed in sample C-1 and small particles on the grain boundaries of
the Al_2O_3 grains were identified as Ti by energy dispersive analysis.
(see Fig. 12).

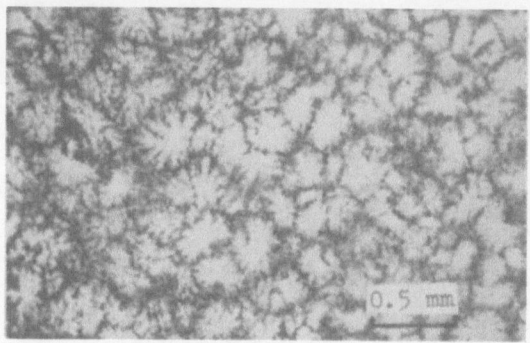

Fig. 7   Scanning electron micrograph of ion-etched Si_3N_4-5% MgO
showing macroscopic heterogeneity produced by dendritic
solidification.

Fig. 8   Bright field micrograph of a binder area in cemented
carbide CC-1 and associated diffraction pattern revealing
extra spots due to stacking faults in the binder.

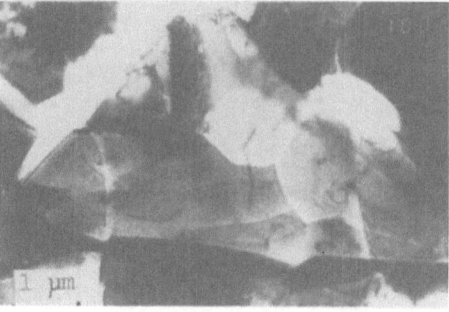

Fig. 9   Bright field view of a carbide grain in cemented carbide
CC-1. Many unexplained contours were observed in the WC
grain.

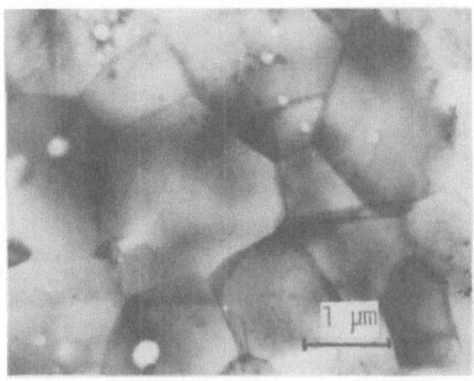

Fig. 10.   Transmission electron micrograph of HIP post-sintered
           Al_2O_3.

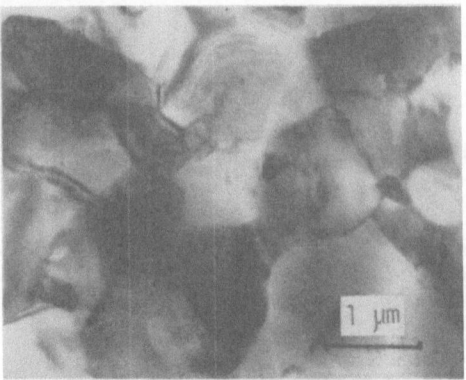

Fig. 11.   Transmission electron micrograph of HIP post-sintered
           Al_2O_3-TiC.

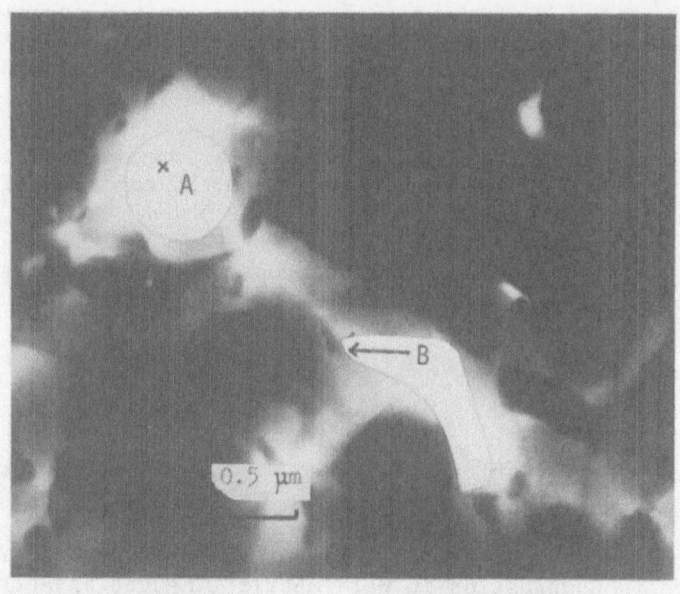

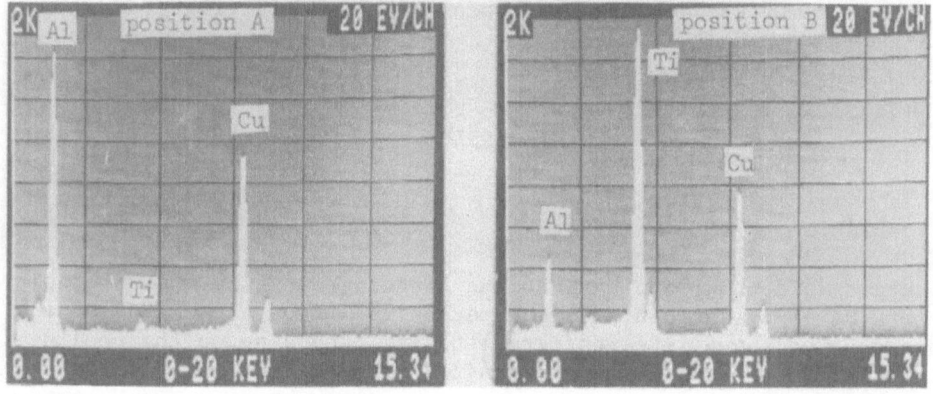

Fig. 12   Energy dispersive x-ray analysis of Al_2O_3-TiC composite.

CONCLUSIONS

1. Ion-etching has been shown to be a useful method for microstructural characterization of both cemented carbides and ceramics. It is uniquely suited for revealing microstructural aspects that conventional etching methods fail to show.

2. Binder phase grains and grain boundaries in cemented carbides can be observed by ion-etching provided the cubic carbide content does not exceed 20%.

3. Ion-etching can also reveal macroscopic heterogenieties resulting from liquid phase sintering of ceramic bodies such as Si_3N_4-MgO.

REFERENCES

1. H. Suzuki, K. Hayashi and Y. Fuke, J. Japan Soc. Powder and Powder Met. 19(8):346 (1978).

2. B. Roebuck and E. A. Almond, "Microstructural Events Preceding Fracture in Compression in Hard Metals", Recent Advances in Hard Metal Production, Loughborough University, 1979.

DISCUSSION

L. Prakash:

Dr. Exner said this morning that the binder grain size most probably doesn't have any influence on the mechanical properties. You have shown that you have different grain sizes, depending on the cooling rate. Have you measured any mechanical properties?

A. Doi:

No.

H. Fischmeister:

Your observation of a preferred orientation relationship between the cobalt and the tungsten carbide grains I find extremely interesting. It brings to mind observations by Suzuki who found that at high temperature the nucleation of rupture in TRS tests occurred at binder phase grain boundaries. And if there is an orientation relationship within each grain, then this becomes immediately understandable.

Maybe I could make a short comment concerning the mysterious contours. Have you considered the possibility that in those places where these contours occurred, you were actually looking at overlapping grains?

A. Doi:

Very possible.

E. Almond:

We often find inclusions in carbide grains and they are cobalt. When they're large they have a hexagonal symmetry. Have you looked at the possibility of these being cobalt? We also find that slip bands go straight through them as though they don't exist because the slip planes are similar. So I think this is one thing you could look at if you're looking for an explanation for the inclusions.

A. Doi:

Thank you.

G. Dearnaley:

I'd like to offer another possible explanation for the mystery of the dislocations traveling across the contours. Perhaps they have been induced by your argon ion bombardment process itself. The reason I suggest this is that I shall show tomorrow that dislocations are produced during ion implantation into cobalt cemented tungsten carbide.

SOME CONSIDERATIONS OF THE EFFECT OF HOT ISOSTATIC PRESSING ON

HARDMETAL STRUCTURE AND PROPERTIES

Roy Cooper

Boart Research Centre
P O Box 1242
Krugersdorp 1740
South Africa

INTRODUCTION

Hot isostatic pressing (HIP) of hardmetal has become an accepted process tool to close the slight residual porosity remaining after sintering. The process involves the heating of sintered hardmetal in an inert gas atmosphere under the simultaneous action of both temperature and pressure. Process conditions for different grades of hardmetal are usually determined empirically.

As hot isostatic pressing can be regarded as a resintering operation under pressure a simple mathematical model has been derived, based on strength changes as a result of changes in density, from which an attempt to isolate the effects of temperature and pressure during HIP has been made, based on experimental results.

If the lower bound of HIP treatment is considered as resintering at the treatment temperature with no applied pressure, and the upper bound of treatment is actual hot isostatic pressing then the effect of temperature and pressure can be found. This could be of commercial significance in helping to tailor HIP conditions between the upper and lower bounds depending on the density increase required to satisfy material specifications. Obviously, significantly more experimental work would be needed to attain this ideal for the total commercial range of hardmetal grades.

1. THEORETICAL CONSIDERATIONS

    Strength improvements on the reduction of porosity in sintered
hardmetal have been reported by Anderson[1], Suzuki and Hayashi[2],
Lardner and Bettle[3] and Rüdiger and Exner[4]. Engel and Hübner[5] have
reported that the strength/flaw size relationship obeys Griffith's
basic strength equation so that the strength is dependent on the
largest microstructural defect such as porosity, inclusions and
coarse grains.

    Amberg, Nylander and Uhrenius[6] have shown that pore closure in
11% cobalt hardmetal treated at 1200°C and 100 MPa occurs by cobalt
extrusion into the pores, and at the higher treatment temperatures
of 1320°C and 1350°C cobalt "lakes" produced by extrusion of cobalt
have also disappeared. They stated that the cobalt "lakes" disappeared due to the cobalt melt wetting the carbide skeleton thereby
sucking the "lakes" into the surrounding structure. It is thought
that this mechanism would lead to the formation of porosity so it
must be concluded that the hot isostatic pressing would augment the
surface tension forces in the component to remove the cobalt "lakes"
by a rearrangement of the carbide grains, as concluded by Ingelström[7].
Ingelström's treatment temperatures imply that grain reorientation
occurs when the binder is solid or just liquid. It is likely that
reorientation would require lower pressures with more liquid binder
present, by treating at a higher temperature. Amberg, Nylander and
Uhrenius[6] have shown thermodynamically that by assuming the carbon
monoxide, formed during sintering and trapped in the pores, disso-
ciates into carbon and oxygen in the cobalt, these enter into solution
in the cobalt during hot isostatic pressing to allow total pore
closure. On the release of the pressure, supersaturation is present
in the cobalt.

    Both Amberg, Nylander and Uhrenius[6] and Ingelström[7] state that
cobalt is extruded into pores at sub-eutectic or eutectic temperatures
at pressures of between 50 MPa and 100 MPa, and Ingelström also states
that grain reorientation occurs at both sub-eutectic and eutectic
temperatures (eutectic is 1325°C from the WC-Co pseudo-binary phase
diagram). The time to remove pores is a function of both pressure
and temperature.

2. MODEL DERIVATION

Assumptions

2.1 Strength decay with increased pore size follows a negative
exponential relationship as proposed by Romanova, Kreimer, and
Tumanov[8].

2.2 Pore size distribution can be replaced by a total pore volume.

2.3 Strength increase is due solely to pore closure.

2.4 The treatment conditions of 1400°C at 160 MPa, which were the maximum used in an ASEA hot isostatic press, are considered adequate to reach what in this paper is termed theoretical density, hence maximising strength.

2.5 Maximisation of strength ignores the deleterious effect of surface defects, inclusions and open porosity which cannot be eliminated by hot isostatic pressing.

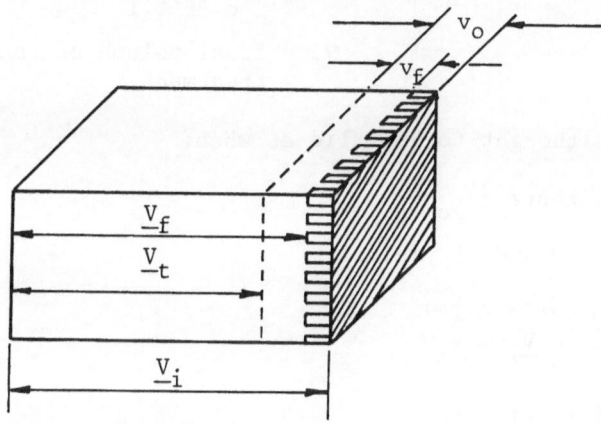

Figure 1: Schematic Diagram of Effect of Hip on Porosity Reduction

Figure 1 shows schematically a volume of porous hardmetal of initial Volume V_i. Hot isostatic pressing at the upper bound conditions reduces the volume to V_t due to the total elimination of porosity. In this condition strength is maximised. The initial total pore volume is considered to be v_o and the final total pore volume after HIP at other than maximum conditions or resintering only is v_f. The final pore volume v_f tends to zero as hot isostatic pressing conditions approach those considered to give theoretical density. Obviously, v_f equals v_o before any HIP treatment.

The mathematical model controlling strength increase by the reduction in porosity, as measured by density increase is considered to be:

$$\sigma = \sigma_o e^{-b\,(P-\delta P)} \quad \dots\dots\dots\dots\dots\dots\dots\dots\dots\dots\dots\dots\dots (1)$$

σ = measured strength

σ_o = theoretical strength

P = ratio of initial pore volume to initial volume of specimen: $v_o/\underline{V}_i$

δP = ratio of difference between initial and final pore volume to initial volume of specimen: $(v_o - v_f)/\underline{V}_i$

$\underline{V}_f$ = final volume of specimen after treatment

Equation (1) is thought to be valid as when:

$v_f \rightarrow 0, \quad \delta P \rightarrow P,$ hence $\sigma \rightarrow \sigma_o$

Substituting for P and δP yields:

$$P - \delta P = \frac{v_o}{\underline{V}_i} - \left\{ \frac{v_o}{\underline{V}_i} - \frac{v_f}{\underline{V}_i} \right\}$$

$$\therefore \quad P - \delta P = \frac{v_f}{\underline{V}_i}$$

but $v_f = \underline{V}_f - \underline{V}_t$

$$\therefore \quad P - \delta P = \frac{\underline{V}_f - \underline{V}_t}{\underline{V}_t}$$

as $V = \frac{M}{\rho}$ and M remains constant

$$\underline{V}_f \rho f = \underline{V}_i \rho_i = \underline{V}_t \rho_t$$

$$\therefore \quad P - \delta P = \frac{M(^1/\rho_f - {}^1/\rho_t)}{M/\rho_i}$$

$$= \rho_i \frac{(\rho_t - \rho_f)}{\rho_t \rho_f}$$

Substituting this in equation (1)

$$\frac{\sigma}{\sigma_o} = e^{-b\rho_i \frac{(\rho_t - \rho_f)}{\rho_t \rho_f}} \dots\dots\dots\dots\dots\dots\dots\dots\dots\dots\dots\dots\dots\dots(2)$$

The magnitude of the function $\rho_i(\rho_t - \rho_f)/\rho_t\rho_f$ is a measure of both the initial porosity level and the effectiveness of hot isostatic pressing; an increase in effectiveness causing a decrease in the magnitude of the function, as then $\rho_f \to \rho_t$, and the function tends to zero. The function also tends to zero with a decrease in initial porosity. As the effectiveness of hot isostatic pressing increases, and if the assumptions made in section 2 are correct the strength of the component increases to approach the theoretical value.

Since $\ln \frac{\sigma}{\sigma_o} = -b\rho_i \frac{(\rho_t - \rho_f)}{\rho_t \rho_f}$

it should now be possible to predict schematically the relationship between σ/σ_o and $\rho_i(\rho_t - \rho_f)/\rho_t\rho_f$.

(a) when $\rho_t = \rho_f$, $\sigma/\sigma_o = 1$, $\ln \sigma/\sigma_o = 0$

(b) when $\rho_i \frac{(\rho_t - \rho_f)}{\rho_t\rho_f} = 0$, $\sigma/\sigma_o = 1$, $\ln \sigma/\sigma_o = 0$

(c) the gradient for the curve of Pressure and Temperature will be steeper than that for temperature only as hot isostatic pressing is more effective in the densification process.

If the inferences (a) to (c) are valid the curve of $\ln \sigma/\sigma_o$ as a function of $\frac{\rho_i(\rho_t - \rho_f)}{\rho_t\rho_f}$ can be respresented as in Figure 2.

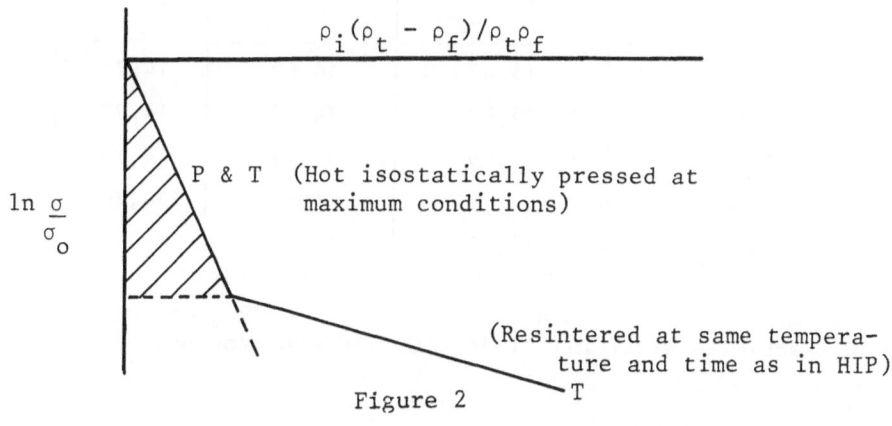

Figure 2

3. EXPERIMENTAL WORK

150 right cylindrical compression test pieces with nominal sintered dimensions 10 mm x 8 mm diameter were produced from an 8%
cobalt, 3-4 µm mean grain size hardmetal, typical of that used for
the percussive drilling of rock. The 150 presintered specimens were
split into 5 groups of 30 specimens and sintered in vacuum at 1325°C,
1350°C, 1375°C, 1400°C and 1425°C respectively to give gradually
reducing amounts of total porosity with increasing sintering temperature. Density, magnetic saturation and coercive force measurements
on each sample were made. Pore size distribution was not measured
as total porosity only is considered in this work.

The 30 specimens sintered at each temperature were split into
three groups of 10 specimens; one sub-group was left in the as-
sintered condition, the second sub-group was hot isostatically
pressed at 1400°C and 160 MPa pressure for one hour, and the third
sub-group was resintered in vacuum at 1400°C for one hour. Density,
magnetic saturation and coercive force were re-measured on each of
the samples which had been hot isostatically pressed or resintered.

All specimens were then fractured in compression, with unlubricated cobalt shims 0,13 mm thick between the test machine platens
and the ground specimen surfaces. The density, compressive strength,
and magnetic property measurements are shown in Tables 1 to 3, and
the derived data for input into the mathematical model are shown in
Table 4. The latter results are shown graphically in Figure 3.

4. RESULTS   Table I

Density Measurements (mean of 10 values)

Initial Sintering Temperature °C	As Sintered g/cm³	Resintered g/cm³	H I P g/cm³
1325	14,67	14,69	14,74
1350	14,68	14,69	14,75
1375	14,69	14,71	14,75
1400	14,71	14,68	14,76
1425	14,70	14,72	14,77

Maximum density reached, taken as ρ_t = 14,79 g/cm³ which
was achieved after HIP at 1400°C, 160 MPa, for one hour.

Table II

Compressive Strength Measurements (mean of 10 values)

Initial Sintering Temperature °C	As Sintered MPa	Resintered MPa	H I P MPa
1325	3907	3893	3989
1350	3900	3824	4051
1375	3838	3907	4072
1400	3852	3900	4127
1425	3796	3838	4030

Maximum compressive strength reached taken as σ_0 = 4170 MPa which was the maximum achieved after HIP at 1400°C, 160 MPa, for one hour.

Table III

Magnetic Properties (mean of 10 values)

Initial Sintering Temperature °C	As Sintered		Resintered		HIP	
	$4\pi\sigma$ emu	H_c Oe	$4\pi\sigma$ emu	H_c Oe	$4\pi\sigma$ emu	H_c Oe
1325	154	103	150	93	153	92
1350	154	99	152	88	152	92
1375	154	93	151	88	153	92
1400	154	95	153	86	153	91
1425	149	95	145	90	148	95

Note: 20 emu ≡ 1% cobalt in the binder

Table IV

$\dfrac{\rho_i(\rho_t - \rho_f)}{\rho_t\rho_f}$ $\times 10^{-3}$	σ MPa	σ/σ_o	$\ln \sigma/\sigma_o$
6,75 ⎤	3893	0,93	-0,07
6,76 ⎥	3824	0,92	-0,08
5,40 ⎬ Resintered	3907	0,94	-0,06
7,43 ⎥	3900	0,94	-0,06
4,73 ⎦	3838	0,92	-0,08
3,37 ⎤	3989	0,96	-0,04
2,69 ⎥	4051	0,97	-0,03
2,69 ⎬ H I P	4072	0,98	-0,02
2,02 ⎥	4127	0,99	-0,01
1,35 ⎦	4031	0,97	-0,03

$\rho_i(\rho_t - \rho_f)/\rho_t\rho_f \times 10^{-3}$

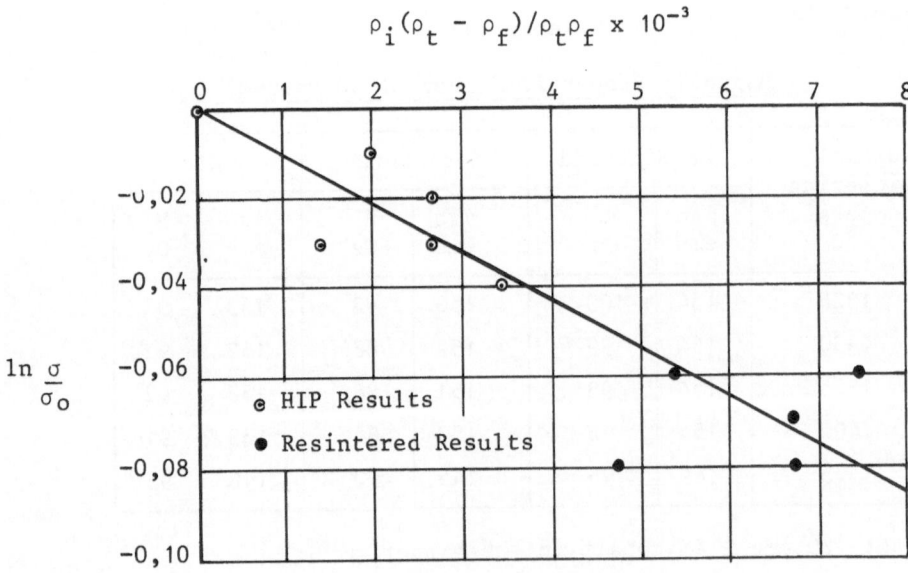

Figure 3

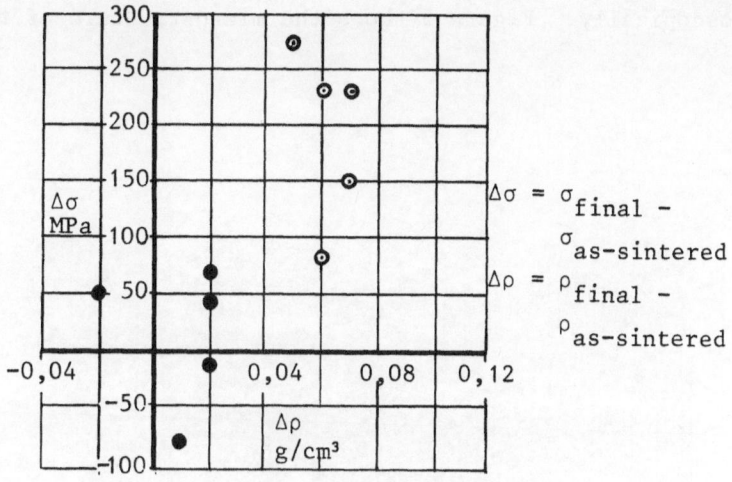

Figure 4

5. DISCUSSION

There is a danger that full mechanical properties are not developed by undersintering the hardmetal used in this investigation to
develop varying porosity levels, as the reactions occurring during
liquid phase sintering may not be completed. The resulting mechanical properties would then be a function of porosity and the reduced
amount of liquid binder produced at sintering temperatures lower
than the commercially used 1425°C. It is considered that apart
from the expected and desired porosity levels the material had
sintered satisfactorily as the magnetic saturation values, which
are a measure of free cobalt, are within the specification
(145-155 emu) for the material, although the coercive force values
for the samples sintered at 1325°C and 1350°C are slightly high
(specification 85-95 Oe) as the sintering reaction would not have
gone to completion at these temperatures due to the paucity of
liquid phase formed. Compressive strength has not been affected by
the high coercive force values, and surprisingly, strength values
have decreased as sintering temperature increased. The density
change with increased sintering temperature has increased as expected (specification 14,65-14,75 g/cm³ as-sintered).

On resintering and HIP the magnetic saturation values are basically unchanged but the coercive force measurements have dropped
slightly, especially for the samples sintered at 1325°C and 1350°C.
The drop for the latter specimens is due to completion of the sintering reaction, and the general slight drop is thought to be due to a
slight degree of grain growth although this could not be detected.

microscopically. Figure 5 shows the microstructure of the hardmetal
used.

Figure 5: 1325°C Hip'd at 1400°C and 160 MPa for 1 Hour

As the microstructure and magnetic properties are satisfactory,
the measured strength changes after resintering and HIP are considered to be a function of changes in the level of porosity only.

Examination of Figures 6 to 8 shows that the reduction in porosity
of the specimens which had been hot isostatically pressed is
significantly greater than that occurring on resintering. Indeed it
is difficult to determine a change in the porosity level after resintering. The influence on pore closure and density increase by
the application of pressure at temperature is, therefore, significantly greater than temperature alone. This confirms that the pressure
augments the surface tension forces occurring in the liquid binder;
the temperature mainly provides liquid phase to allow the pressure
to act on the component to increase densificaton. although limited
densification does occur on resintering.

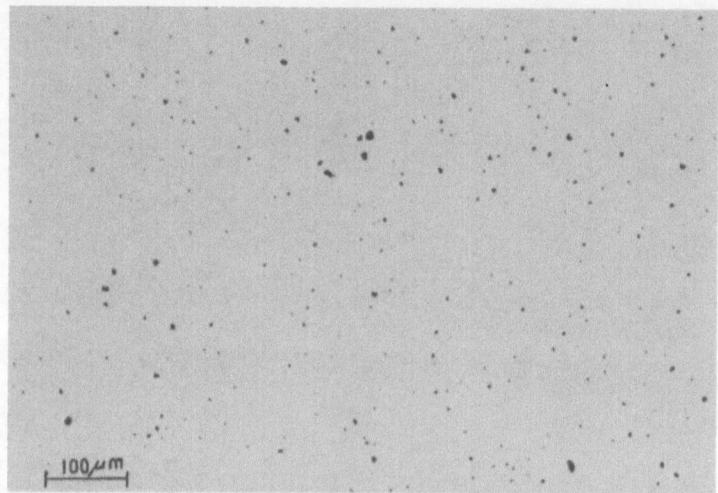

Figure 6: Sintered at 1325°C

Figure 7: Resintered at 1400°C for 1 Hour

Figure 8: 1325°C HIP'd at 1400°C and 160 MPa for 1 Hour

The inferences drawn from the mathematical model have not all
been verified experimentally. Although some scatter of results is
present on Figure 3 the curve is a straight line which confirms
that the strength/density relationship is a negative exponential.
However, the expected inflexion between the HIP and resintered
results as shown schematically on Figure 2 is not present, so the
assumption that the gradient of the HIP portion of the curve should
be greater than that of the resintered portion due to the greater
effectiveness of HIP in densification, is not valid. Although an
inflexion between the HIP and resintered results is not present on
the experimental curve of $\ln \sigma/\sigma_0$ as a function of $\rho_i(\rho_t - \rho_f)/\rho_t\rho_f$
there is evidence that two discrete distributions of data are formed.
The HIP data gives reduced values of $\rho_i(\rho_t - \rho_f)/\rho_t\rho_f$, with
consequently increased strengths relative to the theoretical value.

The distance apart of the two groups of data is a measure of
the improvement in densification and strength increase with pressure
and temperature (HIP) and temperature only (resintering). It can be
seen that the highest value of $\rho_i(\rho_t - \rho_f)/\rho_t\rho_f$ on resintering is
$4,73 \times 10^{-3}$, corresponding to a maximum density of 14,72 g/cm³.
The compressive strength for the material at this value of
$\rho_i(\rho_t - \rho_f)/\rho_t\rho_f$ should have reached 95% of theoretical but this was
not the case. The lowest value for hot isostatic pressing was
$3,37 \times 10^{-3}$ which is equivalent to the minimum density of 14,74 g/cm³
obtainable for the material under the conditions used.

Plotting the results as change in strength as a function of change in density as shown on Figure 4, whilst discriminating between the effect of pressure and temperature and temperature only, does not allow the drawing of the conclusion that both groups of data fall on the same curve as has been shown on Figure 3.

For different combinations of pressure and temperature one would expect Figure 3 to consist of a series of parallel curves with varying positions of the HIP and resintered groups of data on the curves depending on the efficacy of the pressure and temperature conditions chosen. The intercepts of the curves would be lower than that shown in this work as it has been assumed that 1400°C and 160 MPa are the maximum conditions available. For different grades of hardmetal one would expect the same trends to be followed as shown in this work although the densification/strength character-istics would probably be different.

6.   CONCLUSIONS

This work has used a simple mathematical model in an attempt to isolate the effects of the pressure and temperature components of hot isostatic pressing. The data resulting from pressure and temperature and temperature only follow a negative exponential distribution with the two groups of results forming discrete distributions on the curve.

The pressure component has a significantly greater effect on densification and strength increase than the temperature component.

7.   ACKNOWLEDGEMENTS

The author would like to thank Mr S M Brabyn and Mr T A Hack of Boart Research Centre for carrying out the experimental work, and Dr J A Cave of De Beers Diamond Research Laboratory for carrying out the hot isostatic pressing.

REFERENCES

1. Anderson, Planseeberichte fur Pulvermetallurgie 15:180 (1967).
2. Suzuki and Hayashi, ibid 23:24 (1975).
3. Lardner and Bettle, Metals and Materials 7:540 (1973).
4. Rudiger and Exner, Powder Metallurgy International 8:2003 (1978).
6. Ambert, Nylander, Uhrenius, Powder Metallurgy International 6:178 (1974).
7. Ingelstrom, Fifth European Symposium on Powder Metallurgy 3:240, 63 (1979).
8. Romanova, Kreimer, Tumanov, Poroshkova Metallurgya 8:84 (1974).

DISCUSSION

H. E. Exner:

I would like to refer to the potassium effect in tungsten where you
keep pores open by putting a volatile substance in. I think what
you really do when you put the pressure on is to press these pores
which are filled with some gas or some volatile substance. This is
the reason why you can't close them by normal sintering. So if you
now measure the difference between transverse rupture strengths and
the tensile force, the gas pressure in the very small pores adds up
to the tensile stress. And in compression, it effects the opposite
way. I would like to propose the following experiment: After hot
isostatic pressing, to take it up to temperature again without the
stress. I would assume that the porosity occurs again so you would
get a swelling due to the gas pressure included in the pores.

R. Cooper:

I think this has been reported by some of the Japanese workers that
one does get this effect of annealing and the pores do come back.

E. Almond:

I'd just like to make a few comments on the paper. I question the
assumption that pore size doesn't matter. I think it most definitely
does. I think that the well known measurement of density is very
difficult. You only see it in the last figure with a pore content
of .22 volume percent, and if you can solve the problem of measuring
those, I think you can do very well. The third point is the effect
of sintering temperature on compressive strength. What we found is
that if you increase the sintering temperature, you change grain
shape. So you have rounded grains at low sintering temperatures
and these become idiomorphic at high temperatures, and this effects
mechanical properties. We also find that you change the dislocation
density in the carbide so you remove dislocations as you increase
sintering temperature.

R. Cooper:

I wasn't trying to say that pore size distribution isn't important
but just that for simplicity we could substitute it by total pore
volume. The idea behind this process is purely commercial. We
wanted to get rid of holes in hard metal and we need to apply our
normal commercial quality control techniques which are based on
density changes and we think that the density changes as measured
are sufficiently sensitive to pick up what we require. We just
found that all that was happening in the microstructure was that

porosity decreased with increased sintering temperature. We haven't noticed what used to be termed the unrecrystallized structure which is a sort of spherical grain. We've always noticed, even at 1325°C that the grains were angular. Just the size distribution was different.

SURFACE TREATMENTS

Hans Erich Hintermann

Laboratoire Suisse de Recherches Horlogères
2000 Neuchâtel / Switzerland

INTRODUCTION

Coating of bulk materials is one way of producing composites.
The properties of the near-surface region may differ considerably
from those of the bulk substrate material. Thus various, sometimes
opposite properties can be combined in one single workpiece.

Coatings are applied for different reasons: economics, materials conservation, unique properties, esthetics, engineering and design flexibility, etc. Of prime interest in the context of this conference are compound materials which as the predominant property
exert high hardness in the near-surface region. This is especially
true for applications where good friction, wear and abrasion properties are the main objectives. To this end either a coating is deposited onto the substrate, producing an OVERLAY coating by an add-on
process, or the substrate is altered at and beneath its surface by
a diffusion process to produce a DIFFUSION coating. In many cases
the coating is a combination of an OVERLAY and a DIFFUSION coating
with an interface region between the two (Fig. 1). This interface
region may be very thin, sometimes hardly detectable, and depends
strongly on the materials combined, the temperature and time applied.
It determines to a great extent the adhesion and fatigue behavior of
the coating/substrate system.

Several technologies are available to produce hard coatings.
Most of them are exploited industrially in a fairly extensive manner
in high technology applications, i.e.

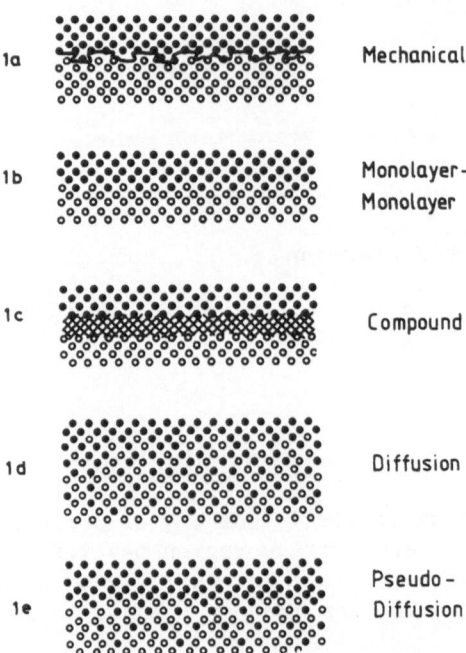

Fig. 1. Coating/substrate interfacial zones (schematic)

- thermal spraying (TS)
- spark hardening (SH)
- arc cladding (AC)
- laser surface hardening (LSH), more particularly
  · laser transformation hardening (LTH)
  · laser glazing (LG)
  · laser cladding (LC), in particular
    laser melt/particle injection (LMPI)
  · surface alloying (SA)
- anodization (A)
- fused salt electrolysis (FSE)
- chemical vapor deposition (CVD) and a special variant of it,
  the powder packing (PP) processes
- thermal diffusion (TD)
- physical vapor deposition (PVD) particularly sputtering (SP),
  ion plating (IP), ion implantation (II) and activated reactive
  evaporation (ARE).

   Many, especially the atomistic deposition processes, have the potential of depositing materials which vary significantly from the conventional metallurgically processed material. The coatings may have high intrinsic stresses, high point defect concentrations, extremely fine grain size, oriented microstructures, metastable phases, macro and micro porosity, a high degree of purity, or may on the contrary contain incorporated impurities. These properties necessarily influence the behavior of these materials under mechanical load, chemical environment, thermal shock or fatigue loading. Metallurgical properties which may be affected include:

- elastic modulus   - diffusion rates
- tensile strength   - friction and wear properties
- fracture toughness   - corrosion resistance
- fatigue strength   - adhesion
- hardness

   In the following we limit our considerations to hard coatings produced by CVD, TD and PVD, and the investigation of properties to adhesion, fatiguing, friction, wear and corrosion.

THE PROCESSES

Chemical vapor deposition (CVD)

   The industrially most advanced hard coatings deposited by CVD are TiC, TiN, Ti(C,N), TiB_2, Al_2O_3, Cr_7C_3, SiC, Si_3N_4, FeB and Fe_2B, and combinations thereof. Mostly metal halide reduction reactions operating at temperatures between 800 and 1100°C are used. At these temperatures and prolonged deposition times, which can go up to several hours, changes in structure and composition of the bulk, the coating and the interface can occur and will influence strongly the mechanical and corrosion behavior of these compound materials. Examples of the type of reactions widely used are:

   For OVERLAY coatings:

$$TiCl_4(g) + CH_4(g) \xrightarrow[H_2]{T,P} TiC(s) + 4HCl(g)$$
   or

$$TiCl_4(g) + \frac{1}{2} N_2(g) + 2H_2(g) \xrightarrow[H_2]{T,P} TiN(s) + 4HCl(g)$$

   Similarly, proceeding from tungsten halides or silanes instead of $TiCl_4$ one obtains the tungsten carbides WC, W_2C and SiC and Si_3N_4 respectively.

For DIFFUSION coatings:

$$BCl_3(g) + \frac{3}{2} H_2(g) \xrightarrow{\quad T \quad} \underline{B(s) + 3HCl(g)}$$
$$\downarrow \text{ substrate Fe}$$

or

$$\underline{FeB(s) + Fe_2B(s)}$$

$$CrCl_2(g) + H_2(g) \xrightarrow[H_2]{\quad T,P \quad} \underline{Cr(s) + 2HCl(g)}$$
$$\downarrow \text{ substrate steel}$$
$$\underline{(Cr_{1-x}Fe_x)_7C \ (s)}$$
$$\text{where } x = 0 - 0.68$$

In a similar way one can obtain silicides, manganides, alumi-
nides, beryllides and other diffusion coatings.

In general the coefficients of thermal expansion, α, of these
hard coating materials differ widely from those of the softer but
tougher substrate materials, as can be seen in Table 1:

Table 1. Thermal Expansion Coefficient α
of Hard Materials (Bulk) at RT

Materials	$10^6 \alpha [K^{-1}]$
FeB	23
Steel	14±4
Stellites	13–15
Cr_7C_3	9.4
TiN	9.35
αAl_2O_3	8.3–9.5
Fe_2B	7.85
TiC	7.4–8.8
VC	7.2
HfC	6.1–6.6
SiC	5.4–6.8
Cemented carbide	5.7 (measured)
Corundum	5.2–6.9
Si_3N_4	2.76

Hence, shear stresses are built up at the coating/substrate interface which can lead to failure by cracking (Fig. 2) or spalling
(Fig. 3) if the adhesion is poor.

Due to the high temperatures generally involved for the CVD of
hard coatings the choice of substrate materials is limited. Ceramics,
cemented carbides, ferrotitanites, steel, some Ni-(Co)-alloys, superalloys and graphite are convenient substrate materials.

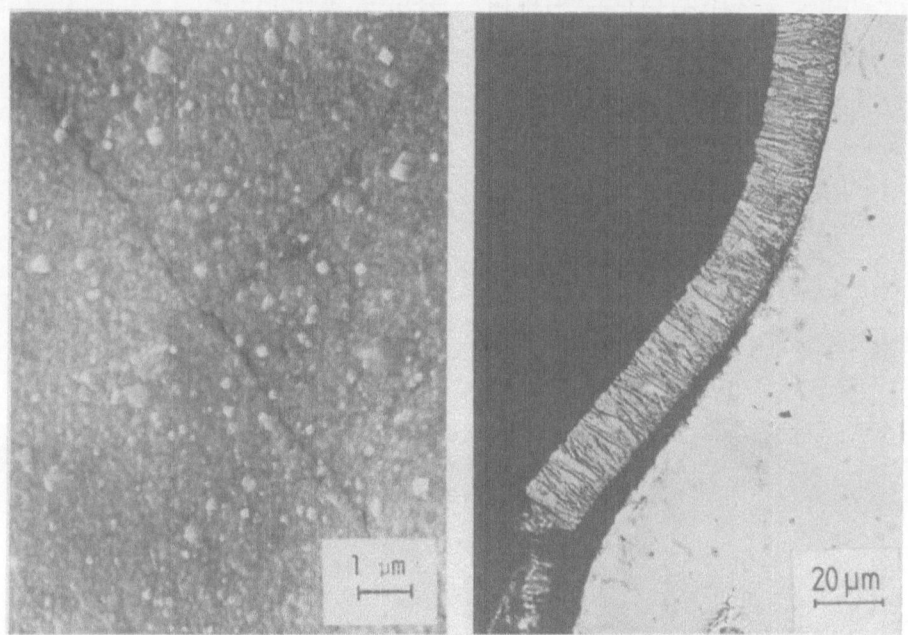

Fig. 2. SEM micrograph of jagged   Fig. 3. Spalling of W coating on
        cracks in a TiC coating   steel
        on cemented carbide
        (fracture strain limit
        $\leqslant 2$ %)

Of them mainly TiC, TiN, Al_2O_3 coated cemented carbide cutting
tools and Cr_7C_3, TiC, TiN coated steel tools of different kinds have
obtained industrial importance. Also boriding of steel and Ni-(Co)-
alloys by CVD or PP are exploited industrially to some degree[1].

When using steel which is economically the most interesting candidate substrate material it is important to consult the temperature-
time-transformation, TTT, diagrams, in order to obtain after coating
and upon cooling the best possible structure and the least deformation
and volume changes (Fig. 4)[2].

CVD has the advantage of high throwing power, uniform and homogeneous coating deposition and high adhesion of these coatings to
the substrate.

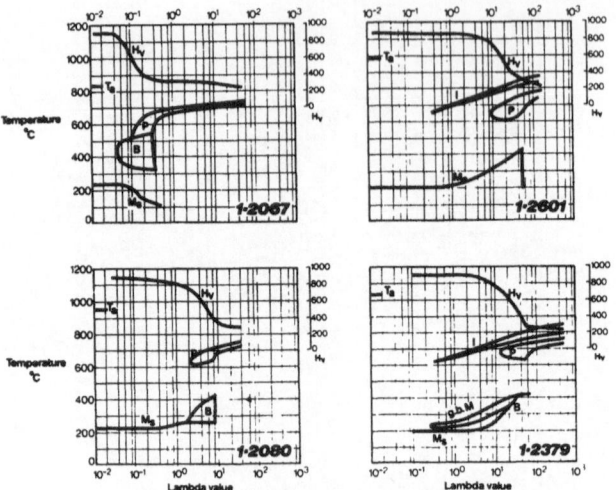

Fig. 4. TTT continuous cooling diagrams of
steels used in CVD with indication of
final hardness values[3].

Thermal Diffusion (TD)

This process applies to steel, cast iron, stellite and cemented
carbides, i.e. very much the same substrate materials as for CVD.
Compact carbide coatings can be formed by merely immersing the workpiece in a molten bath with borax solvent and carbide forming elements, such as ferro alloys of V, Nb, Ti, Cr and B. The carbon contained in the substrate combines with the carbide forming element to
produce the carbide layer - very much according to the same mechanism
as the initial growth of TiC occurs in CVD, where the main carbon
source is the carbon contained in the substrate.

The layers formed are pore-free and strongly adherent. The
throwing power of the process is high, better than that of PVD. The
coating temperature lies typically between 800 and 1200°C, the
dipping time can last from one half to several hours. The deposition
rate depends on temperature, time and carbon content of the substrate
as shown in Fig. 5.

The reaction pot is an open vessel, the environment is non contaminant since no salts vaporize. No protecting gas is needed. Hence,
this process is simple and very economical.

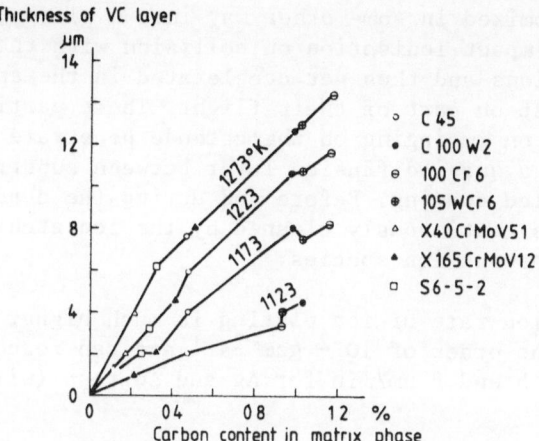

Fig. 5. Effect of carbon content in the
matrix phase and temperature on
VC layer thickness (Arai [4])

The shortcomings of this process are the same as for CVD: deformation of the workpiece and structural changes in the substrate are
possible. Furthermore the surface roughness can be considerable and
the homogeneity low. Hence in this process also, the understanding
of the thermodynamics and kinetics of phase transformations in the
substrate materials are of prime importance, if distortion, volume
changes and alterations of crystalline structure need to be minimized
or avoided.

Physical Vapor Deposition (PVD)

To deposit h a r d materials on an industrial scale, sputtering,
ion plating and in recent years also ion implantation[5] are the most
advanced PVD technologies[6].

Sputtering can be carried out in its simplest version in a diode
configuration. The materials particle current is produced in an electrical discharge in a dilute gas. In a plasma sustained by a potential difference of 1 to several kV between the electrodes the positively
charged gaseous ions are accelerated towards the cathode which under
this ion bombardement pulverizes in an atomic scale and deposits in
the neighborhood of the cathode. The deposition rates are relatively
low, orders of magnitude smaller than for the straight forward evaporation process for which 10^{-4} to 10^{-2} gcm^{-2}s^{-1} are typical deposition
rates.

Ion plating can be considered as a combination of evaporation
and sputtering. The working pressure lies in the same range as that
of sputtering, i.e. 10^{-2} to 10^{-3} mbar. The coating material is

evaporated or atomized in some other way into a plasma, where these species undergo impact ionisation on collision with the positively charged gaseous ions and thus get accelerated in the applied electrical field at least on part of their flight. These particles of the coating material on impinging on the cathode penetrate to some depth into it and form a quasi-diffusion layer between substrate and subsequently deposited coating. Before and during the deposition the substrate surface is continuously cleaned by the ion etching action of the bombarding gaseous ion species.

The deposition rate of ion plating is much higher than that of sputtering, of the order of 10^{-2} $gcm^{-2}s^{-1}$ and can reach 2-3 μm/min for Al and up to 5 and 6 μm/min for Ag and Au resp (with planar magnetron).

Both methods have the advantage to deposit coatings at lower temperatures (<500°C) than CVD and thermal diffusion, thus avoiding deformation and structural changes of the substrate materials. For the same reason hardly any limit exists in the choice of substrate materials. As recent developments have shown also adhesion can be very high, approaching in certain applications those of CVD coatings (TiN coated high speed steel drills[7]). On the other hand, the throwing power of the process is lower than that of CVD.

THE BONDING

With CVD, thermal diffusion, sputtering, ion plating, ion implantation, generally very high bond strength to the substrate can be achieved. In the case of CVD and thermal diffusion the coating and substrate material form transition zones by interdiffusion, as can be seen from Fig. 6.

In the case of ion plating a quasi-diffusion zone beneath the substrate surface can be produced, enhancing adhesion. These interfacial zones can be as thick as a few μm or as thin as only some atomic layers, depending on the materials combination and the production parameters applied.

Since no appropriate and readily available means to measure adhesion and correlate it to an industrially interesting property such as flaking was available on a laboratory scale, and even less so for production control purposes, a special measuring device had to be developed[9]. It is based on a model of Benjamin and Weaver[10] and consists in scratching the coated surface with a diamond pin of defined curvature at a given speed and progressively increased load, until the coating is seizing and is eventually removed locally. The lowest charge applied, L_c, at which the coating begins to seize is then taken

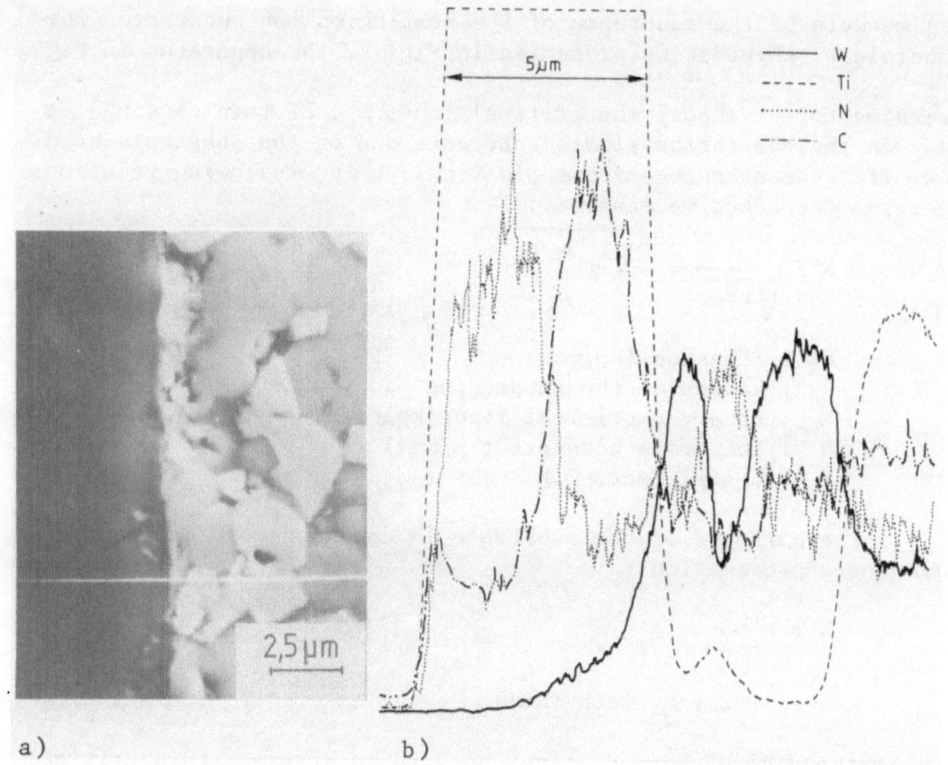

Fig. 6. Scanning electron micrograph (a) and microprobe analysis (b) of a double layer coating TiC/TiN obtained by CVD of TiN onto cemented carbide (68.5 % WC, 12 % TiC, 10 % Ta(Nb)C, 9.5 % Co)[8].

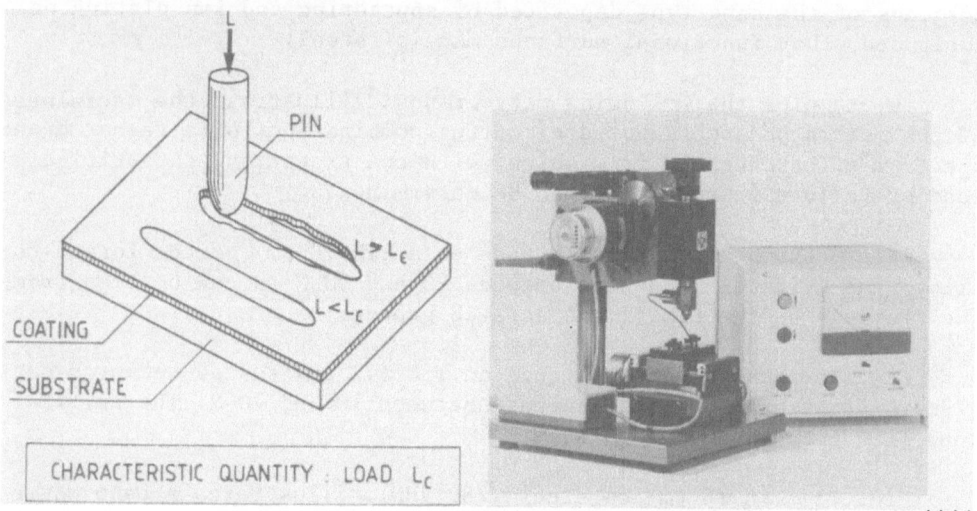

Fig. 7. Principle of the scratch test method.

Fig. 8: Scratch test instrument[11]

as a measure of the adherence of the coating on the substrate. The
principle of the test is presented in Fig. 7, the apparatus in Fig. 8.

According to the theory the critical load, L_c, of thin coatings depends on the coating/substrate adherence and on the substrate hardness. If the substrate deforms plastically, the following relation
describes the adhesive behavior:

$$\sigma_A = K \cdot \frac{1}{\sqrt{R^2 \pi}} \cdot \sqrt{L_c \cdot H}$$

σ_A : Adhesion (kp/mm^2)
R : Radius of the scratching pin (mm)
L_c : Observed critical load (kp)
H : Substrate hardness (kp/mm^2)
K : Coefficient, $0.2 < K < 1$

If the hardness of the substrate is not known one can use the
following approximation:

$$\sigma_A = K \cdot \frac{2 L_c}{\pi R \cdot b}$$

b : Scratch width (mm)

In the case of hard coatings on hard substrates, this critical
load is relatively well-defined and reproducible. In the past the method has proven to be particularly well suited for quality control,
and for studying the adhesion of TiC, TiN, Al_2O_3 coatings and combinations thereof on cemented carbide cutting inserts, but also for hard
coatings of the same type deposited by sputtering and ion plating on
tools and other functional surfaces made of steel.

As examples the following micrographs[9] illustrate the usefulness
of the method on some substrate/coating combinations of a rather broad
spectrum of hardnesses, from which, with one exception, the critical
load of failure in adherence can be determined.

Fig. 9 shows two scratches made on a TiN (2 μm) coated forged Co-
alloy. The critical load lies between 20 and 30 N as can be seen from
the flakings on the rim of the scratch groove.

Fig. 10 shows seven scratches on TiC (12 μm) coated cemented carbide in steps of 10 N load increase between 10 and 70 N. The critical
load lies between 40 and 50 N.

Fig. 11, TiC (6 μm) on steel (280 VHN), illustrates a case where
the fracture occurs almost entirely within the coating itself. The
coating cannot be detached from the substrate. In such cases no

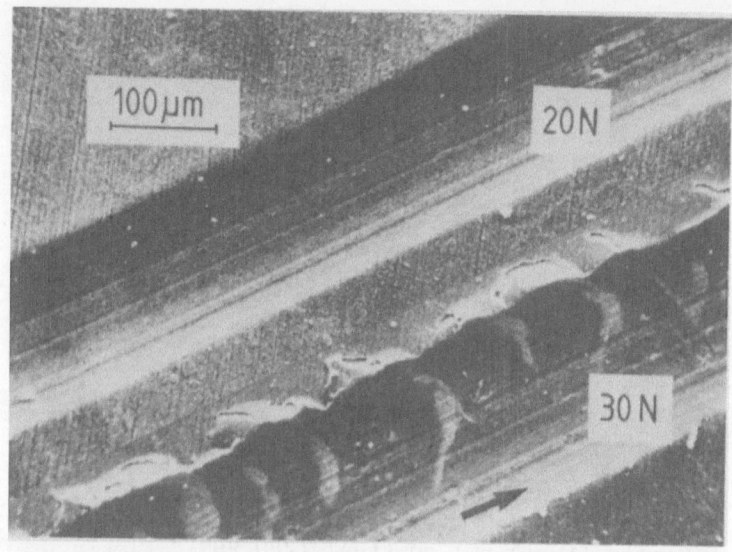

Fig. 9. TiN (2 μm) coating on a hardenable Co-alloy,
        20 N < L_c < 30 N.

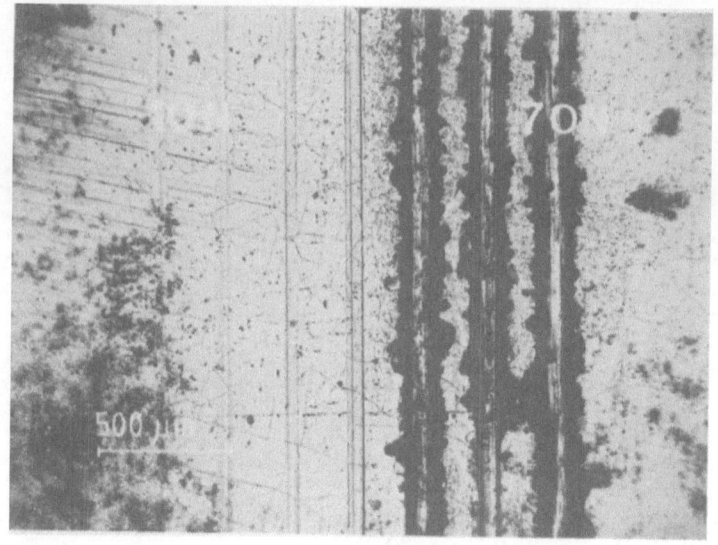

Fig. 10. TiC (12 μm) on cemented carbide, 40 N < L_c
         < 50 N.

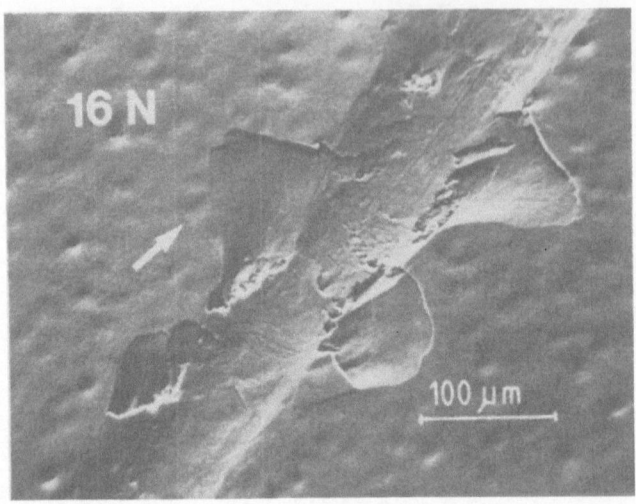

Fig. 11. TiC (6 μm) on steel (280 VHN). Flaking
within the layer

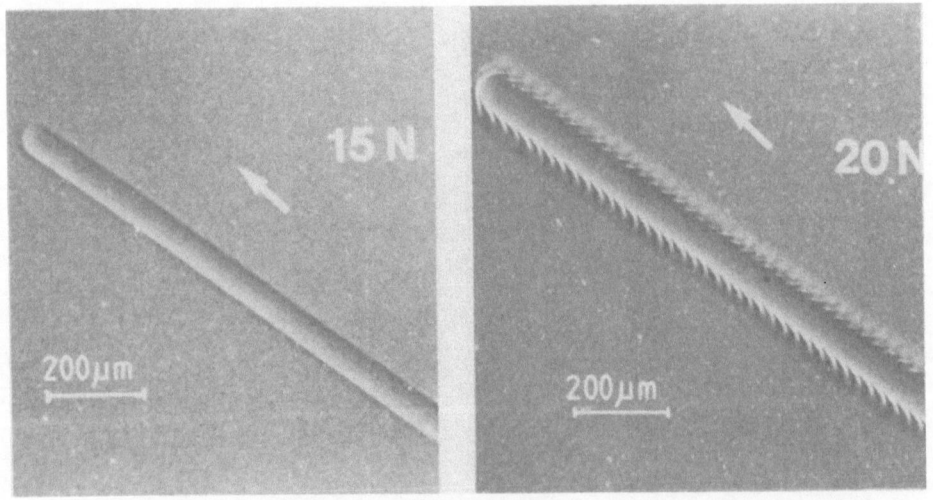

Fig. 12. Electrodeposited bright nickel coating (300 VHN) on steel
(250 VHN); $15\ N < L_c < 20\ N$.

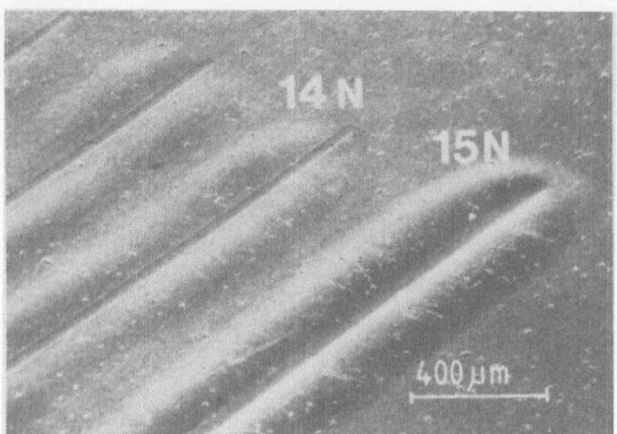

Fig. 13. Electroless Ni-P coating (10 µm) on
steel (250 VHN); calculated bond
strength: 450 MPa.

critical load can be determined since the adherence is better than
the coherence of the coating.

Figure 12 shows two scratches made under different loads on a
ductile coating, namely of bright Ni (300 VHN) on steel (250 VHN). The
critical load is attained, when the first undulations appear without
that the coating, however, chips off, in the present case 15 N < L_c
< 20 N. The calculated bond strength is 690 MPa.

In Fig. 13 the case of a low adherent coating is presented: electroless Ni-P (10 µm) on steel (VHN 250), load increase in steps of
1 N from 12 to 15 N. When adherence is low no undulations will be produced on moving the diamond pin across the surface. Rather, the coating is lifted up and forms a regular blister alongside the edge of
the scratch trace. The extent of the detached zone of the coating is
much larger than could be concluded from the width of the scratch
groove. The calculated bond strength is 450 MPa.

As a further and very sensitive criteria for the determination
of the critical load acoustic emission can be used[11-14]. An emission
of noise will be detected when the separation of the coating from the
substrate occurs by brittle fracture. This is the case for hard coatings on hardened steel or cemented carbides.

The acoustic emission is measured with an accelerometer mounted
directly sideways above the diamond pin. The amplified and rectified
signal is written on a recorder and transmitted by an earphone.

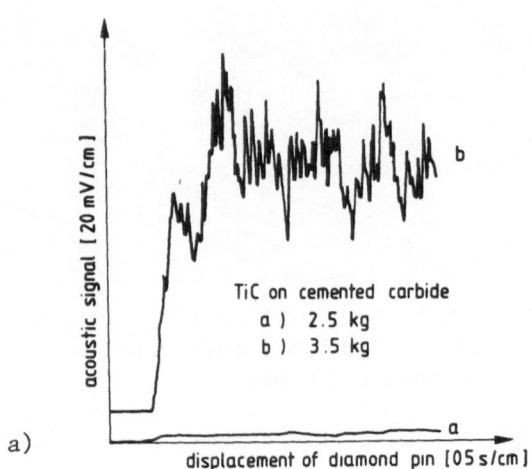

a)

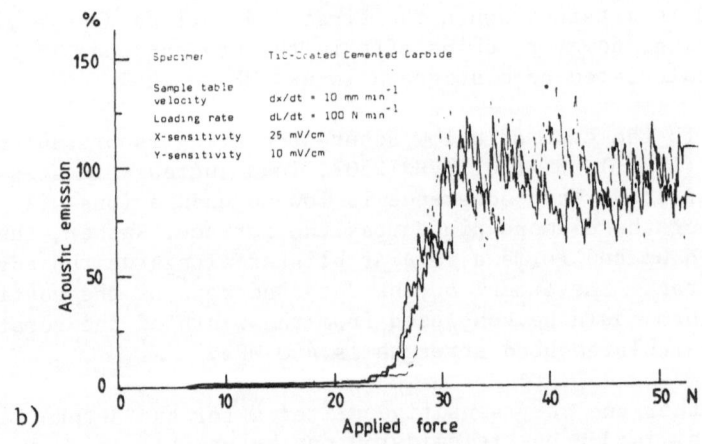

b)

Fig. 14. Acoustic emission of a TiC coated cemented carbide[11],
(a) stepwise load increase, (b) continuous load increase

At the onset of the critical load the amplitude of the signal will rise steeply and remain generally constant even when the charge is further increased.

Fig. 14 shows two recordings of the acoustic signal with increasing load on a TiC coated cemented carbide cutting tip[11].

Some adherence values determined according to the scratch test method are reported in Table 2.

Table 2. Adherence Values Determined According to the Scratch Test Method

Substrate	Coating	Substrate Hardness (kp/mm^2)	Adherence $K=1$ (kp/mm^2)
steel	galv. Ni	300	10
steel	PVD TiN	200	5-15
HSS	sputter TiC	1100	50
cem. carb.	CVD TiN, TiC	1500	→ 200
glass	evaporated metal	600	0.1-10

FATIGUE BEHAVIOR

On this subject results are available only on the fatigue behavior of TiC deposited according to CVD onto steel and cemented carbides(cc).Owing to their good tribological behavior and their role as a diffusion barrier TiC layers (and also TiN) are used in bearings[15] (slider, roller, tapered and ball bearings). The mechanical properties of TiC and other hard coatings such as hardness, Young's modulus, tensile strength, are very different from those of the substrate, steel or cemented carbide. To gain more understanding about the behavior of the substrate/coating composite, it was necessary to examine the mechanical properties which are important in obtaining reliable ball bearings. TiC coated AISI 440C steel (0.95-1.15 % C, 17 % Cr, 0.5-0.75 % Mo, balance Fe) and WC-6 % Co   cemented carbide samples were subjected to static, dynamic and oscillating loads through a ball. The main parameters were coating thickness, load, presence or absence of a Cr_7C_3 interlayer and time. For details concerning experimentation reference is made to[16].

Of these experiments the results of the oscillatory load tests are the most interesting. An oscillatory load at 80 Hz on a ball of 6 mm diameter pressed against a TiC coated sample, simulated at a

singular point the forces which in a ball bearing are spread out over
the whole race. Indentation profiles were used, the heat effects being
negligible owing to the low loads applied. An experiment consists of
eight tests with the same load but of different durations, i.e. between 1 and 70 h. After the tests, the profiles of the damaged zones
are recorded with a Talysurf apparatus. From these recordings the
following factors can be determined (Fig. 15 upper right hand corner):

- the maximum depth, D, of the indentation
- the surface, S, of the worn zone
- the height, h, of the surface wear debris deposits
- the surface, s, of the accumulated zone of debris.

Figure 15 shows furthermore two characteristic examples of damage.

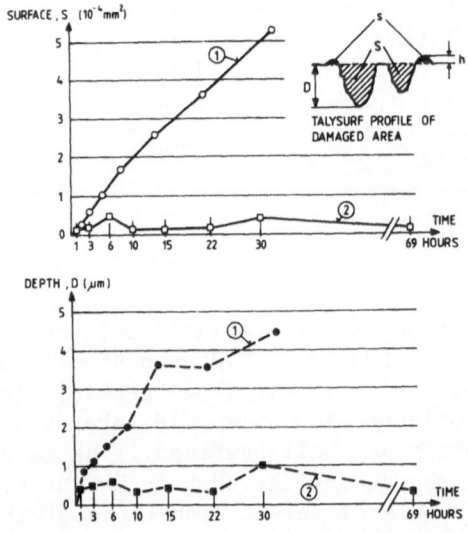

Fig. 15. Oscillating load tests, (a) max crater depth D and wear
         debris deposit height h, (b) worn surface of a damaged area
         S and surface s of the accumulated zone of debris, against
         duration for two different cases: (1) ruby ball against
         Cr_7C_3(1 μm)/TiC(5 μm) coated cemented carbide, load 75 %
         of elastic limit (87 N), (2) ruby ball against TiC (2 μm)
         coated cemented carbide, load 50 % of elastic limit (58 N).

In the first case the fatigue crater dimensions are almost proportional to the vibration time. In the second case, for identical ball and substrate materials, but with a thin TiC layer only, without a Cr_7C_3 interlayer, the fatigue damage is independant of time, i.e. of the number of oscillations. This type of wear, called the "zero wear regime" was encountered in most tests. The results are summarized in Fig. 16.

The presentation takes into account the depth (D) and the surface (S) of the excavation below the zero line and the wear debris thrown up above the zero line (h and s), as determined on the Talysurf profiles.

The two partner material combinations give different results:

- with the cemented carbide substrate the presence of an interlayer of Cr_7C_3 is unfavorable

- damage is more pronounced with the thicker TiC functional layers.

Figure 17 shows the damage produced by the oscillating movement against applied load. Under high load the steel substrate degrades slowlier than the cemented carbide (cc); under a low load the cemented carbide substrate presents a smaller degradation of fatigue.

From the static, dynamic and oscillating load tests on TiC coated cemented carbide and steel, the following conclusions in relation to applications of these coatings for ball bearing elements can be drawn:

(1) the best performance (resistance against fatigue and permanent deformation) is obtained with thinner TiC layers (about 2 μm) and without interlayers

(2) the presence of a Cr_7C_3 interlayer makes the cemented carbide less resistant to fatigue under stresses produced by oscillating and rolling contact loads, but is useful on steel when a thick TiC layer is necessary (abrasive wear)

(3) when cemented carbide is pressed against ruby, which is harder and almost rigid, the former can stand an oscillating load corresponding to 50 % of the elastic limit without deformation if it is not coated with an interlayer of Cr_7C_3; with a load of 75 %, however, extensive damage occurs in every case.

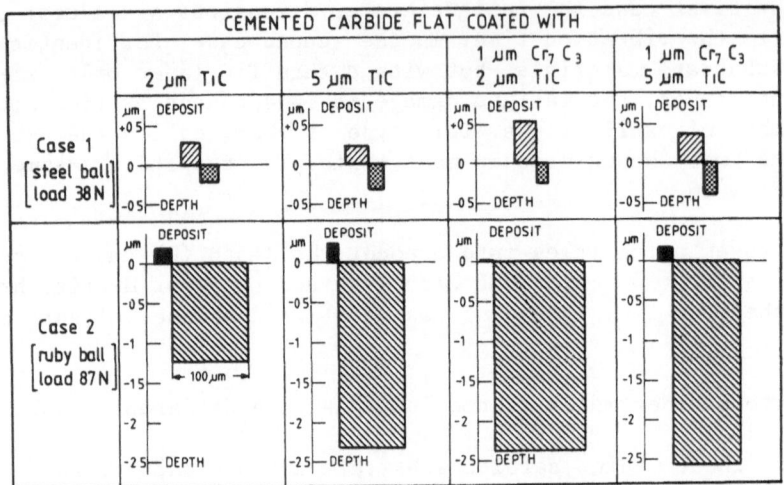

Fig. 16. Depth and worn surface averaged over eight different test
durations, (a) steel ball against coated cemented carbide,
load 95 % of elastic limit (38 N), (b) ruby ball against
coated cemented carbide, load 75 % of elastic limit (87 N).

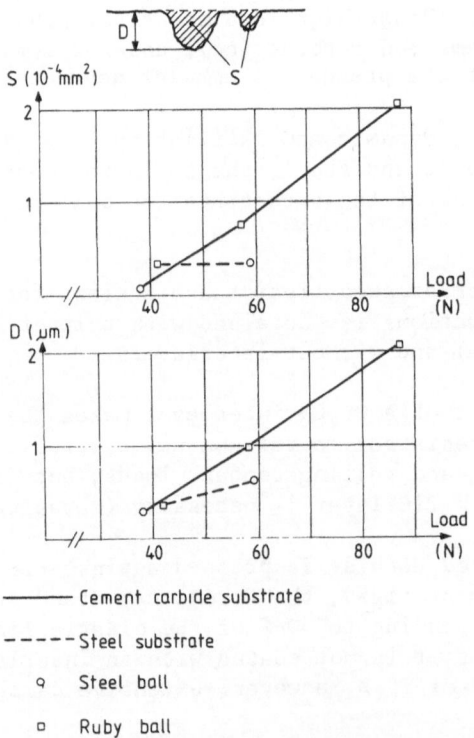

Fig. 17. Damage (fatigue crater D and S) as a function of applied load

Under a heavy load, the steel substrates show moderate degradation, under a light load, the most resistant substrate is the cemented carbide.

These results supported the experience already available from practice, that TiC coatings on bearing elements could improve friction, wear and lifetime of a bearing, especially those having to function under extreme conditions.

FRICTION/WEAR BEHAVIOR

Hard coatings in general have a low coefficient of friction (μ) against themselves and against ferrous materials and greatly reduce wear, as has been shown by different authors for CVD[17], TD[4], and PVD[18] coatings.

In the following figures the friction and wear behavior of CVD coatings as measured on a pin and disc equipment (Fig. 18) in dry and humid air are reported[19].

PVD coatings of the same kind behave tribologically similarly, provided they exert the same bond strength to the substrate.

The hard carbide coatings grown by the thermal diffusion process (TD) exert friction coefficients similar to the OVERLAY carbide coatings obtained by CVD. The wear behavior of these materials as measured on a flat pin/disc apparatus[4] is reported in Fig. 22.

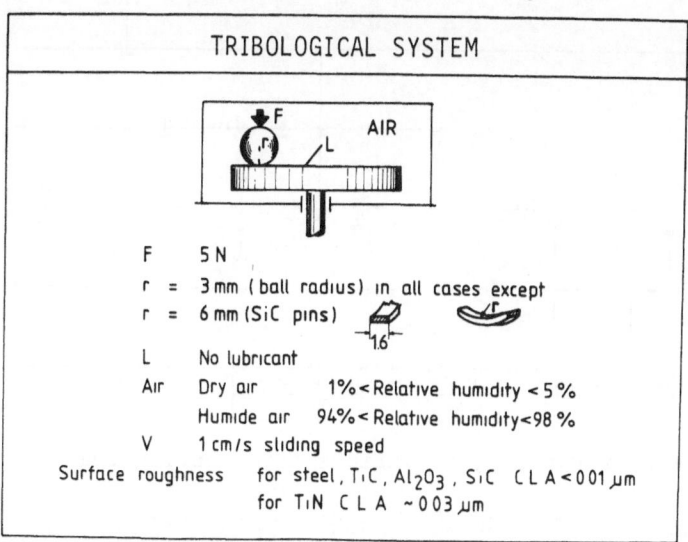

Fig. 18. Experimental conditions of pin/disc friction and wear tests

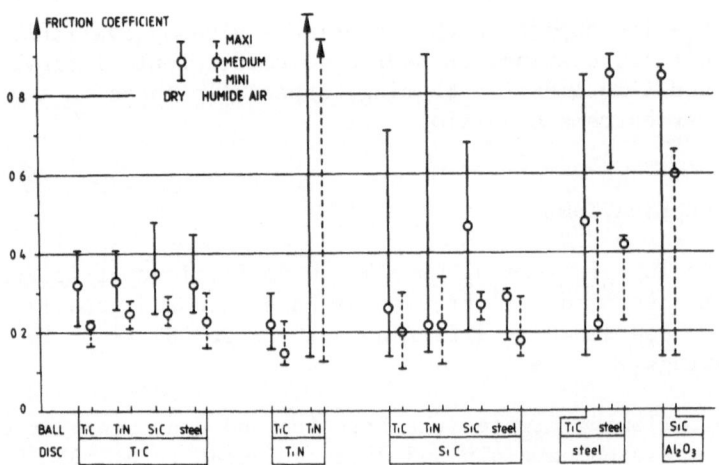

Fig. 19. Friction coefficient, μ, of various hard coating combinations

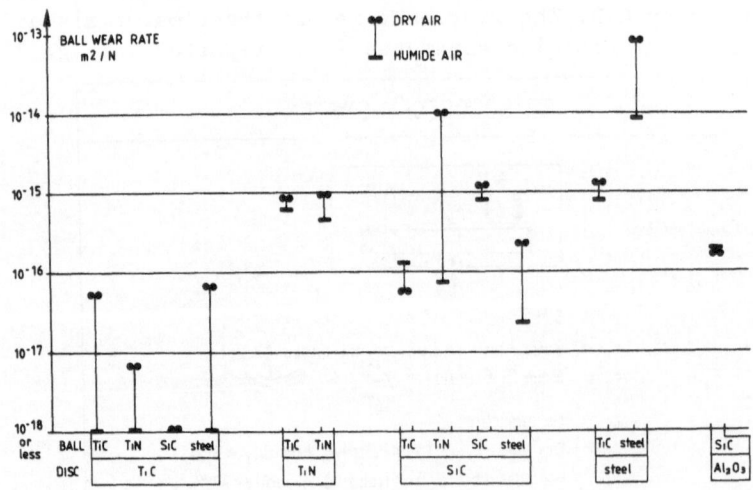

Fig. 20. Ball wear rate, W, of various hard coating combinations

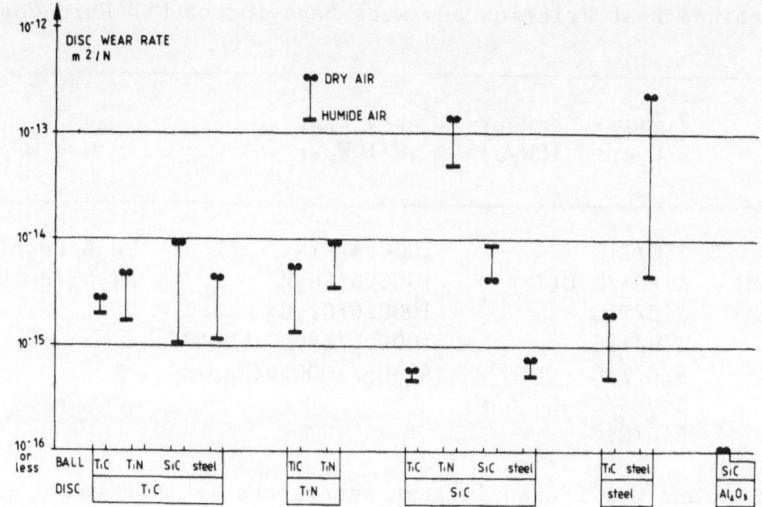

Fig. 21. Disc wear rate, W, of various hard coating combinations

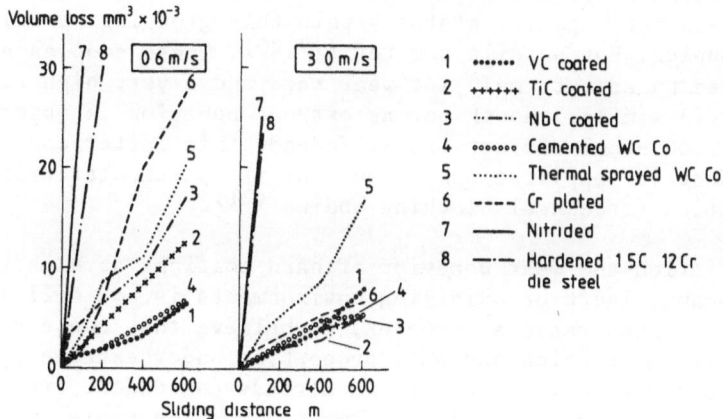

Fig. 22. Volume loss against sliding distance of TD carbide coated
steel, compared to other hard, wear resistant materials[4]

From the results reported in Fig. 19 to 21 one can deduce the
best friction couples, the best wear couples and from these groupings,
the best combined CVD-friction and wear couples. Those couples with
the best tribological behavior as measured on a pin disc machine
and within given limits of load, speed and environment are summarized
in Table 3.

Table 3.Combined Best Friction and Wear Behavior of CVD Hard Coatings

$\mu<2\cdot\mu_{ref}$ $W\approx W_{ref}$	$2\cdot\mu_{ref}<\mu<3\cdot\mu_{ref}$ $2\cdot W_{ref}<W<10W_{ref}$	$\mu>3\cdot\mu_{ref}$ $W>10W_{ref}$	$\mu>3\mu_{ref}$ $W\approx W_{ref}$
TiC/SiC	TiC/TiN	100Cr6/TiN	Fe_xB/Fe_xB(BAM)
TiC/TiC(BAM)	Al_2O_3/TiC(2kp)	100Cr6/Fe_xB	Al_2O_3/cc bor
TiN/TiN(BAM)	TiC/TiC	100Cr6/Cr_7C_3	
100Cr6/SiC	TiN/TiC	100Cr6/X205CrWMoV12 1	
100Cr6/TiC	SiC/TiC	Al_2O_3/100Cr6(2kp)	
	Cr_7C_3/Cr_7C_3(BAM)		
	SiC/SiC		

By weighing μ and W differently some inversions in this table are
quite possible.

Fe$_x$B against Fe$_x$B and Al_2O_3 against boronized cemented carbide
(cc bor) has a very special status within this grouping of friction
and wear couples. For details see ref.[17,20]. These couples are
characterized by an extremely low wear rate and a very high friction
coefficient. A similar though not as extreme behavior is observed
for AISI 52 100 steel against Cr_7C_3. Indeed, this latter couple is
used in industrial applications because of its particular tribological
characteristics (freewheel blocking bodies [21]).

The friction and wear behavior of hard coatings at high temperatures in vacuum, inert or oxidizing environments is not well investigated. We have good reasons, however, to believe that these coatings
have interesting friction and wear properties under extreme conditions
of temperature and environment, as we already can deduce from the
good results obtained with: coated cemented carbide tools, where local
temperatures of up to a 1000°C can occur; from deep drawing tools,
where the temperature might rise well above a hundred degrees C; and
also from bearings operating at moderately high temperatures in inert
gas atmospheres in nuclear reactors (250-350°C, He).

Within the prospecting programm for new materials with good tribological properties for use in the He-loop of nuclear reactors, the
variation of the coefficient of friction of TiC against TiC with temperature in a quasi-inert atmosphere (He) has been investigated.
Fig. 23 shows that the friction coefficient stays remarkably low with
increasing temperature up to 600°C; above this temperature the

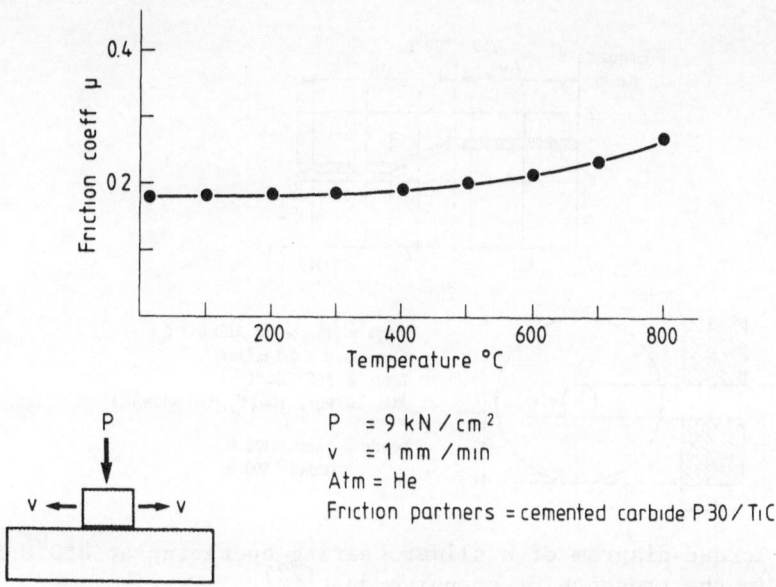

Fig. 23. Friction coefficient of TiC against TiC as a function of
         temperature in inert atmosphere

friction coefficient decrease is partly due to increased deformation
of the cemented carbide in the contact region.

The torque diagram of a slider bearing operating at 250°C in He
in the presence of graphite dust is given in Fig. 24. It shows very
low friction losses under these conditions and also that the influence
of the environment, air or He, is small. As can be derived from this
diagram, the bearing surfaces had only slight signs of wear after the
test. Knowing the abrasive character of graphite in the absence of air
and/or humidity, the TiC coating has shown in this application an outstanding resistance against both adhesive and abrasive wear.

CORROSION AND CORROSION PROTECTION

Very little systematic work has been published about the corrosion behavior of hard coatings. Qualitatively speaking, the carbide,
nitride and oxide hard coatings have very good corrosion resistance
towards acid, caustic and neutral salt aqeous solutions. An example of
the quantitative behavior of TiC in H_2SO_4 is given in Fig. 25.

From current density/potential curves obtained potentio-dynami-
cally at slow scan rates, it can be shown that dissolution occurs between about 850 and 1150 mV. This is followed by a passive region up
to about 1800 mV where severe corrosion is observed. Our studies

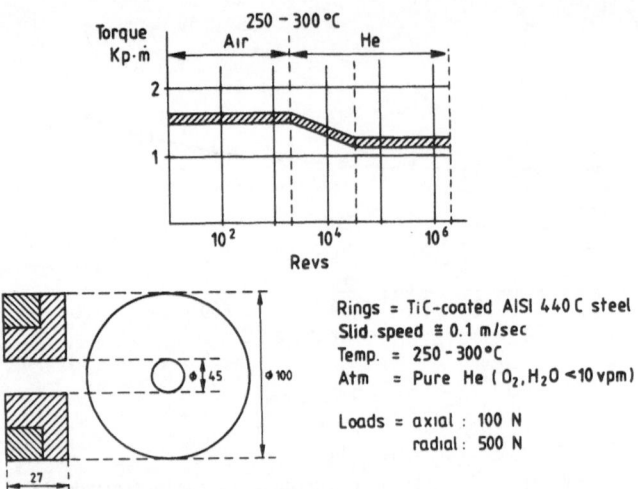

Fig. 24. Torque diagram of a slider bearing operating at 250°C in He in the presence of graphite dust[22]

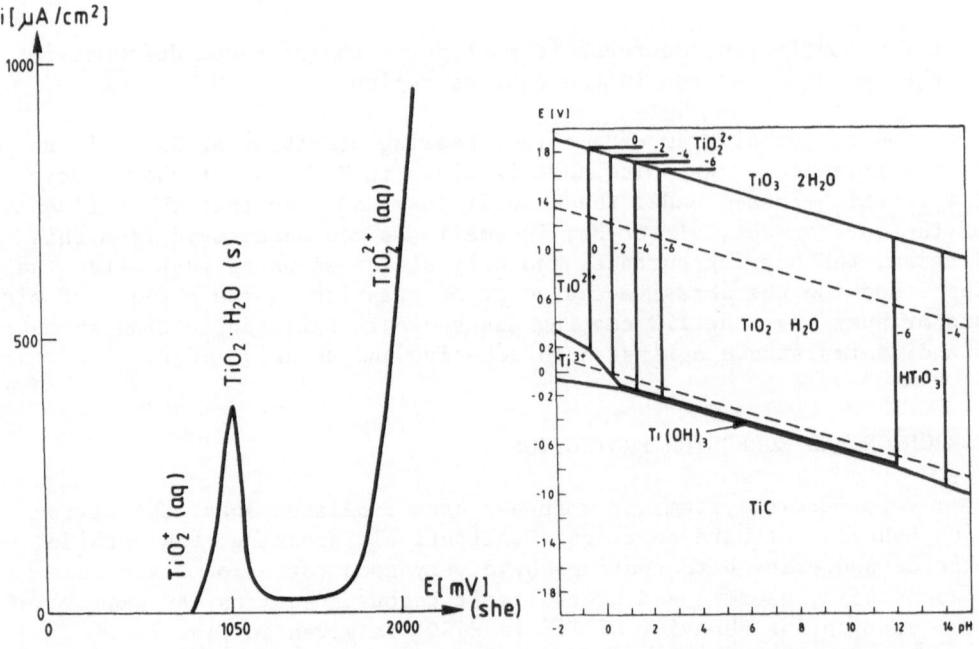

Fig. 25. Corrosive behavior of TiC in 2N sulfuric acid (10mV· min⁻¹, 23°C, 0.1 cm²)

Fig. 26. Equilibrium potential-pH diagram for the system TiC–H₂O at 25°C

suggest [23] that at about 800 mV, TiO^{2+}(aq) is formed together·with
CO and CO_2. At about 1000 mV, $TiO_2 \cdot H_2O$(s) is formed and at 1800 mV,
TiO^{2+}(aq) is formed. These results concur well with the Pourbaix diagram (Fig. 26) for Ti modified for the system $TiC \cdot H_2O$. Hence, it can
be deduced that TiC coatings should be highly corrosion resistant in
sea water or fairly neutral industrial waste waters, either by themselves or with a passivation layer of $TiO_2 \cdot H_2O$. This has been proven
by practical tests with TiC and TiN coated watch cases made of different steels. Table 4 summarizes the results:

Table 4. Corrosion Tests on TiC and TiN Coated Steels[24]

Substrate	Artificial seawater	Artificial perspiration
Stainless steel (AISI 316)	sufficient	good
Tool steel (AISI D2) 10 μm TiC	not sufficient	sufficient
Tool steel (AISI D2) 4 μm Cr_7C_3/4 μm TiC	good	good
Stainless steel (AISI 316) 4 μm TiN	good	good
Stainless steel (AISI 316) 1 μm TiN/ 4-5 μm TiC	good	good
Martensitic stainless steel (AISI 440C) 1 μm Cr_7C_3/4-5 μm TiN	good	good
Martensitic stainless steel (AISI 440C) 1 μm Cr_7C_3/4-5 μm TiC	good	good

From this table it can be seen that TiC and TiN have a remarkably
high corrosion resistance, in most corrosive mediums higher than
stainless steels and Ni-Co-superalloys. However, this only applies if
coatings are absolutely dense, i.e. without open pores and microcracks.
The first type of defect structure and morphology can form during
layer growth, e.g. by incorporating extremely fine dust particles
which are overcoated and the origin of protrusions which easily break
off; the second type can form due to stress release leading to micro-
fissuration as shown by electrographic printing (EGP) [25] for TiC,
TiN, HfC, TiC/Al_2O_3 coated cemented carbide cutting inserts,
(Fig. 27-31).

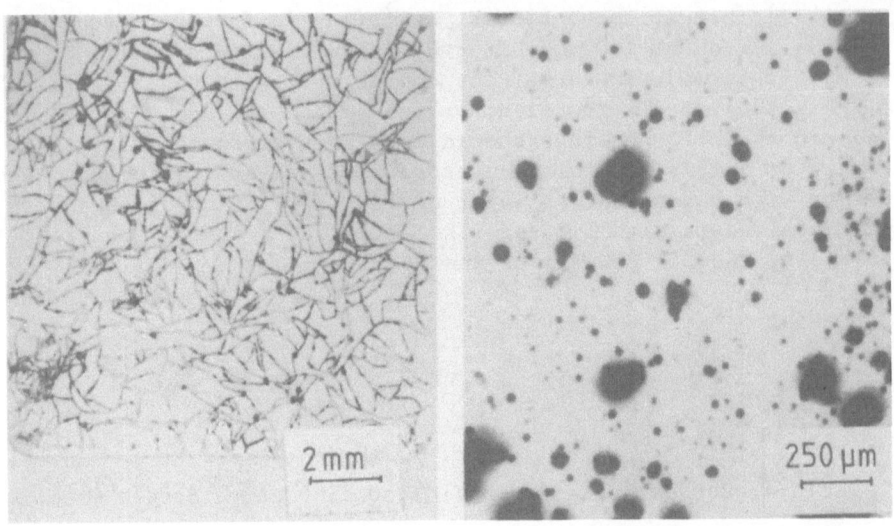

Fig. 27. EGP of TiC (4 μm) on ce-   Fig. 28. EGP of Al₂O₃ (1 μm) on cemented carbide substrate   mented carbide substrate
showing imaged cracks   showing pores

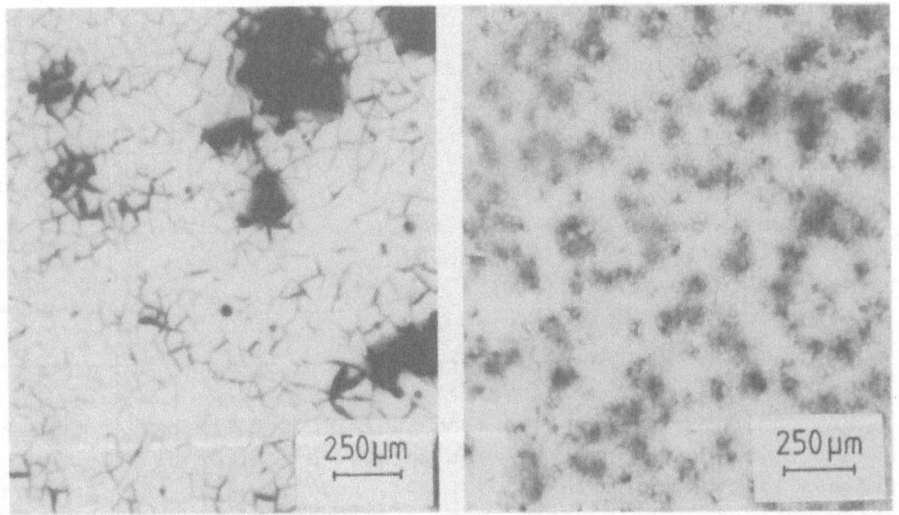

Fig. 29. EGP of TiC (4 μm)/Al₂O₃   Fig. 30. EGP of TiN (3.2 μm) on ce-
(1 μm) double coating on   mented carbide substrate
cemented carbide sub-   showing a crack pattern
strate showing cracks in   (observe the difference
the TiC coating and pores   with the TiC crack pattern,
in the oxide coating   Fig. 27)

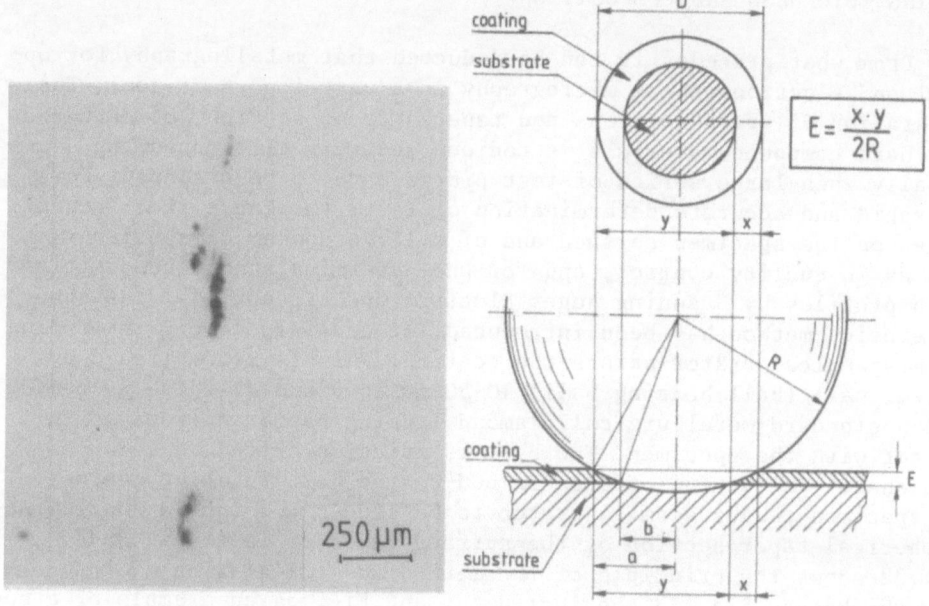

Fig. 31. EGP of CVD-Ta (25 μm) on steel substrate (inside wall of tube) revealing pores

Fig. 32. Principle of the spherical erosion method for coating thickness determination and composition depth profile analysis(by SAES,see text)

Fig. 33. Spherical erosion machine

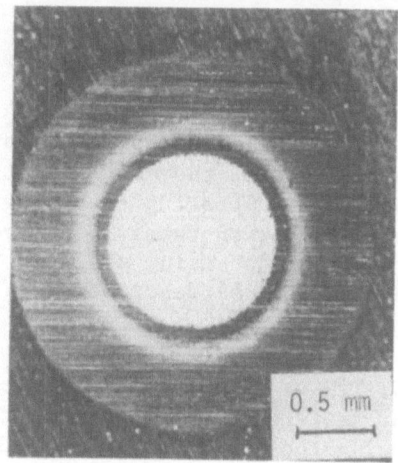

Fig. 34. Spherically eroded crater on a double coating of Cr_7C_3 (2 μm)/TiC (20 μm) on steel

COATING THICKNESS AND COMPOSITION

From what precedes it can be deducted that metallography for optical and electronoptical micrography is a very important tool. The
preparation of cross sections and tapered cross sections of hard and
very hard compound materials is tedious and very time consuming, especially when large series of test pieces have to be prepared. For
the rapid and accurate determination of layer thickness at different
places on the specimen surface and of a large number of specimens,
such as in quality control, and for the determination of composition
depth profiles by Scanning Auger Electron Spectroscopy[26] the spherical erosion method has been introduced. It consists simply in eroding
a hemispherical crater across the coating. This is accomplished by
a steel ball (ball bearing ball) 10-50 mm in diameter which is coated
with a standard metallurgical diamond lapping paste and rotated in
contact with the specimen. The ball is friction driven by a conical
axle shaft. The action of the diamond-coated ball rotating against
the specimen causes a small crater to be eroded and thus a shallow hemispherical taper section of the surface layer or layers is made.
Fig. 32 shows the principle of the method and calculation of the coating thickness, Fig. 33 the instrument and Fig. 34 an example of a double coating Cr_7C_3(2µm)/TiC(20µm) on steel (for details see ref 27).

The method is practically destruction-free for studies or controls where corrosion is not involved. For hard compound materials,
e.g. hard coated cemented carbides, the cratering by erosion per specimen is of the order a fraction of a minute, and thus many specimens
can be prepared and examined in a very short time.

APPLICATIONS

The coating of cemented carbide and high speed steel cutting
tools is one of the most important and successful industrial developments in hard coating technology. The lectures of this session will
deal mainly with theme. Therefore for illustration two industrial applications using steel as the substrate material for hard
coatings will be described.

Example 1: Forming tools. Ledeburitic chromium steels coated with
TiC, TiN or Ti(C,N) are used for deep drawing operations since high
wear resistance under high loading is required. The coated deep
drawing tools include punch and die assemblies for the manufacture of
brass cartridges for ammunition, for shock absorbers for vehicles,
for cups for electric storage batteries. For this latter case punches,
dies and blank holders were made of different materials and compared
against one another[21]: steel DIN 1.2601 (Fig. 4), ferrotitanite,

cemented carbide and TiC (4-6 µm) coated steel DIN 1.2601 ($\approx$ AISI D2).
The number of strokes and the tool costs are reported in Fig. 35. For
a twofold cost increase of the steel tool by coating, a lifetime increase by two orders of magnitude can be obtained. Thus the production
cost of the finished part is greatly decreased.

Example 2: Coated ball bearing elements. Ball bearings with
either TiC-coated races or TiC-coated balls (cemented carbide balls)
have been used successfully in hostile environments such as vacuum or
inert gases at room temperature or at more elevated temperatures
(200-300°C)[15,21,28,29]. Another domain of application with a potential of industrial manufacturing are navigation instrument bearings
used in gyroscope motors. These bearings are lubricated and one of
the prime objectives is that these bearings continue to operate when
the liquid lubricant circulation fails and that degradation of the lubricant under the action of the highly active debris formed is minimized.

The tests were carried out under the conditions as stated in
Table 5

Table 5. Test Conditions for Gyro Bearings

- Grooved bearings,   : 19/6 mm x 6 mm, 17-20° contact $\angle$;
  type RA-6190XHV-258 BB*   outer ring rotating
- Precision   : ABEC 5
- Environment   : air, room temperature
- Radial load   : 2 N
- Axial load   : 10 N
- Speed   : 24'000 rpm
- Total running time   : 25'000 and 20'000 h respectively
- Materials combinations   : a) - steel races AISI 440C, TiC (5 µm)
                                      coated
                                  - steel balls AISI 440C
                                  - phenolic cage, impregnated with oil
                                      SRG 160; 4-5 mg grease Andok C

                               b) - steel rings AISI 440C
                                  - steel balls AISI 440C (reference)
                                      and TiC coated cemented carbide balls
                                      (WC+6 % Co)
                                  - cage as under a) above

* Roulements Miniatures SA, CH-2500 Bienne, Switzerland

Excellent results have been obtained with both bearing material
combinations. The coatings had been applied by CVD and correspond to
the state of the art as mastered five and three years ago respectively.

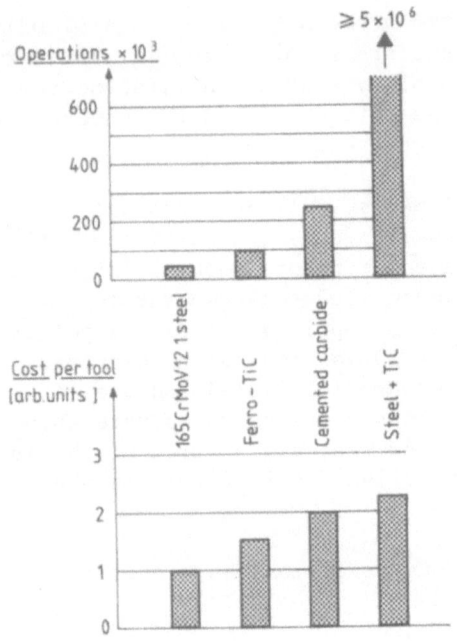

Fig. 35. Deep drawing tool lives against tool material

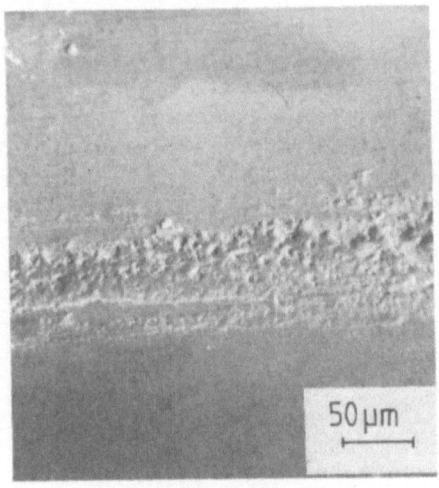

Fig. 36. Track on ball bearing raceway of uncoated bearing after 3000 h at 24'000 rpm

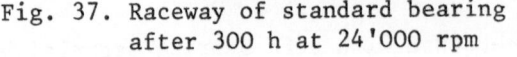

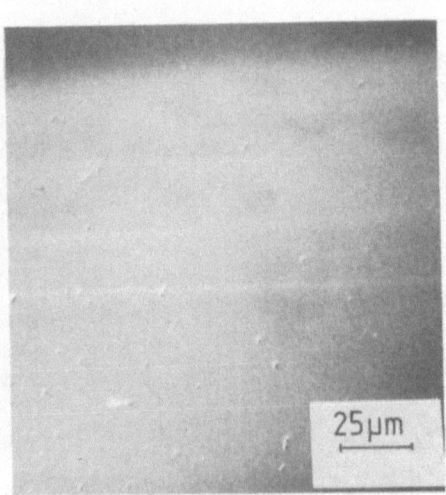

Fig. 37. Raceway of standard bearing after 300 h at 24'000 rpm

Fig. 38. TiC coated race of bearing after 25'000 h at 24'000 rpm

During the testing procedure the bearings were stopped at regular intervals for inspection of the mechanical components and the lubricant degradation. Deceleration measurements were made regularly from cruising speed to full stop. Before and after the test the vibration level of each bearing was controlled individually.

The results obtained from six bearings with TiC coated races according to Table 5 materials combinations a) can be summarized as follows[30]:

- The average level of vibration of the coated and uncoated bearings at the beginning of the test were about the same, i.e. 0.20 to 0.25 g. Already after several thousand revolutions at 24'000 rpm the coated bearings ran smoother than the uncoated ones.

- The deceleration of the coated bearings at different time intervals reflected the stable running of the TiC coated bearings. From 100 h until the end of the 25'000 h test, the deceleration times remained the same within 10 %.

- The lubricant of the TiC coated bearings remained fresh and unaltered compared to the uncoated bearings; no signs of adhesive wear between the balls and races were observed.

After 3'000 h of operation, tracks could be observed on the races of the standard (uncoated) bearings and on the balls (Fig. 36). It appears that the traces consist of wear debris transferred from the cage to the races, decreasing in this way the clearance between the races and the balls, and increasing the bearing vibration at higher speeds.

In order to examine what causes the deposit to take place on the uncoated races, tracks were examined after 300 and 1'250 h of operation. Fig. 37 shows a worn surface; there are no signs of wear deposit yet, but a deterioration of the surface due to adhesive wear between balls and races can be seen.

When TiC coated races were examined after 25'000 h, no traces of wear on the races could be observed; Fig. 38 shows the surface of the race after test. No signs of cage wear debris can be observed.

The blackening and deterioration of the lubricant as observed in the uncoated standard bearing is due to aging and decomposition resulting from chemical reactions between the lubricant, the highly active fresh bearing surfaces and the wear particles.

The results obtained from the bearings with TiC coated cemented carbide balls according to Table 5 materials combination b) can be summarized as follows[31].

- Due to the higher Youngs modulus of cemented carbide (cc) the bearings employing TiC coated cc balls showed a smaller deflection curve, i.e. a higher stiffness. Hence the load bearing capacity of such bearings is somewhat lower than for bearings with all components made of steel.

- Torque and vibration measurements under static and dynamic conditions on bearings with steel balls (reference) and TiC coated cemented carbide (cc) balls gave the same results.

- In the lifetime test over 20'000 h the bearings mounted with TiC coated cc balls ran without incident and would have run longer, whereas the regular bearings mounted with steel balls (reference bearings) failed already after 3'000 h due to too great of an increase of the torque and vibration level.

- The deceleration time measured at different intervals varied between 165 and 195 s for the bearings with the coated balls, for the steel ball mounted bearings the deceleration time decreased already after a few hundred hours and amounted to only some seconds after 3'000 h, thus indicating the destruction of the bearing.

- The loss of lubricant was 60 to 70 % after 3'000 h. Quantitatively, this corresponds to the loss encountered with standard uncoated bearings. Further loss between 3'000 and 20'000 h is negligible. The 30 to 40 % lubricant which remains in the bearing is sufficient to lubricate the friction couple "steel against TiC". This would not be the case for the friction couple "steel against steel".

- No signs of wear could be observed after 20'000 h running time, neither on the raceways nor on the cage. On the ball surfaces, however, spotted areas amounting to about 5 % of the total surface can be seen on the TiC coated cc balls (Fig. 39). These spots correspond to areas where the coating has either partially or totally spalled away. Nevertheless this deficiency did not lead to a bearing failure during the test. Microscopical studies on unused TiC coated cc balls treated in the same batch revealed similar surface defects (Fig. 40). The portion of the surface affected is however well below 5 %. For details see ref[31].

These deficiencies in the coating could be eliminated by a more stringent surface preparation and cleaning and avoiding dust deposition onto the surface during layer growth.

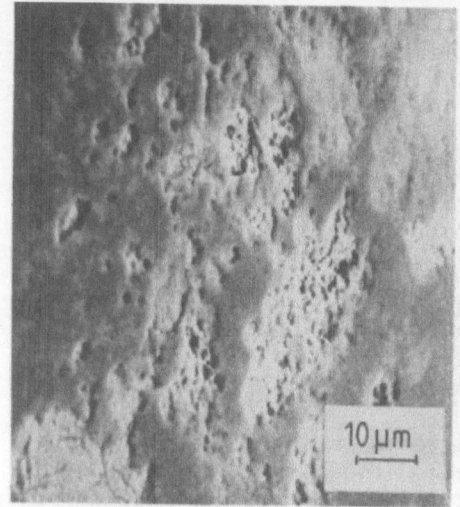 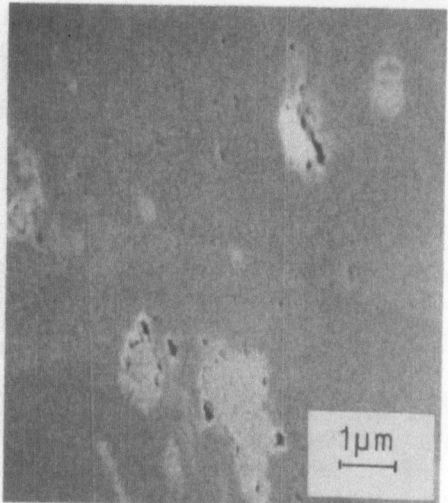

Fig. 39. SEM view of a spot on a   Fig. 40. SEM view of a spot on an
         TiC-coated cemented car-   unused TiC-coated cemented
         bide ball after 20'000 h   carbide ball
         running time

Generally it can be concluded for both types of material combination in gyro bearings that

- with either, TiC coated races or TiC coated balls good results were obtained,
- with either system the running behavior of ball bearings can be greatly improved and the lifetime extended,
- the liquid lubricant is no longer degraded,
- the bearings with the coated balls represent the technically more feasible and economically more interesting way for a future industrial production of bearings comprising hard coated bearing elements,
- in a further step of development the coated cemented carbide balls should be replaced by coated steel balls.

CONCLUSIONS

It has been shown that hard coatings deposited by CVD, PVD, TD on functional surfaces of steel, special alloys or cemented carbides can provide significant improvement in the tribological and corrosion behavior of tools and machine elements.

To achieve this, the coatings must be dense, free of open pores and microcracks, adhere well to the substrate and withstand high mechanical stress.

Acknowledgment: The provision of SiC samples for the tribological studies by the Centre d'Etudes Nucleaires de Grenoble (F) and the Societe Europeenne de Propulsion in Paris (F) is greatly appreciated.

REFERENCES

1. H. E. Hintermann, Tribology International 13:267 (1980)
2. Handbuch fur Werkzeugstahle, Rochling-Burbach GmbH, Volkingen, BRD No. 704.380/7.73, S. 118, 122, 128, 134.
3. A. Kulmburg, Oesterreichische Ing. Zeitschrift 19:185 (1976).
4. T. Arai, in: "Oberflachentechnik Proc. Internat. Congress on Surface Technology SURTEC 81," VDI-Verlag, Dusseldorf 331 (1981).
5. G. Dearnaley, these proceedings.
6. H. K. Pulker, A. Perry, and R. Berger, in: "Proc. Sem. on Surfaces and Interfaces," Fonds Natl Suisse, Bern (1980), Surface-Oberflache, 23, No. 2:(April 1982), to be published.
7. Bohrmeister Guhring, 13, Ausgabe 21 (1980).
8. L. Chollet and J. Beguin, Electron Microscopy 3:164 (1980).
9. P. Laeng, P. A. Steinmann and H. E. Hintermann, in: "Proc. Sem. On Surfaces and Interfaces," Fonds Natl. Suisse, Bern (1980), Surface Oberflache, 23, No. 2:(April 1982), to be published.
10. P. Benjamin and C. Weaver, Proc. Roy. Soc. 254A:163 (1960).
11. Manual to scratch test instruments LSRH (1981).
12. P. Laeng and P. A. Steinmann, in: "Proc. 8th Internat. Conference on CVD," The Electrochem. Soc., Pennington, NJ 723 (1981).
13. A. J. Perry, ibid, 737.
14. H. E. Hintermann, Schmiertechnik Tribologie 28, No. 5:159 (1981).
15. H. E. Hintermann, H. Boving, and W. Hanni, Wear 48:225 (1978).
16. M. Maillat, H. Boving and H. E. Hintermann, Thin Solid Films 64:243 (1979).
17. K.-H. Habig, W. Evers und H. E. Hintermann, Z. Werkstofftechnik 11:182 (1980).
18. T. Jamal, R. Nimmagadda, and R. Bunshah, Thin Solid Films 73:245 (1980).
19. H. E. Hintermann, M. Maillat, and C. Menoud, to be published.
20. R. Bonetti and H. E. Hintermann, in: "Proc. VI Intern. Conf. on CVD, Atlanta 9-14/10/77," The Electrochem. Soc., Princeton, NJ, 351 (1977).
21. H. E. Hintermann, VDI-Berichte Nr 333:53 (1979) and Tribology International 13:267 (1980).
22. H. E. Hintermann and H. Boving, in: "Proc. "NUCLEX 81", Basel, 6 -10/9/81," to be published.
23. R. D. Cowling and H. E. Hintermann, J. Electrochem. Soc. 117:1447 (1970); ibid 118:1912 (1971).
24. W. Hanni and H. E. Hintermann, in: "Proc. 10 International Congress of Chronometry, " Geneva 455 (1979).

25. A. Tvarusko and H. E. Hintermann, Surface Technology 9:209
    (1979).
26. V. Thompson, H. E. Hintermann, and L. Chollet, Surface Technology
    8:421 (1979).
27. H. Boving et R. Rocchi, in: "Actes 53 e Congres de la Soc.
    Suisse de Chronometrie," La Chaux-de-Fonds, 507 (1979).
28. H. Gass, H. E. Hintermann, G. Stehle, and H. M. Briscoe,
    in: "Proc. European Space Tribology Symposium, ESRO,"
    Frascati (1975).
29. J. Lammer, G. Jochimsen, and H. E. Hintermann, in: "Proc.
    ASME/ASLE Intern. Lubr. Conf.," San Francisco, A9-84 (18-
    21/8/80).
30. H. Boving, H. E. Hintermann and M. Stehle, Lubrication
    Engineering 37:534 (1981).
31. H. Boving, H. E. Hintermann, and M. Stehle, in: "Proc. ASME/ASLE
    Lubrication Conference, New Orleans (1981), to be published.

DISCUSSION

G. Dearnaley:

I'm afraid I do have to take issue with your remark that ion implantation can be regarded as a special instance of ion plating. I
elaborated the differences in a paper at the First Ion Plating and
Allied Techniques Conference in Edinburgh in '77. But just to give
you a brief example, the effects of implanting yttrium or cerium
into steel are totally different within the matrix from the effects
you would have by coating, or ion plating those metals onto the surface. And there are many other examples of fundamental and practical
differences between the two techniques. I know there has been a lot
lot of confusion and I thought I'd like to establish this point for
for our audience.

H. Hintermann:

Thank you very much. I take this as an instruction, a teaching
for myself. Thank you.

C. A. Brookes:

A comment about your scratch hardness measurements. In scratch
hardness with indentations, it has been found necessary that you
should measure the friction whilst carrying out the scratch. What
will determine the deformation process, is the resultant stress,
not the normal load. And I think it would be particularly important
when looking at thin surface layers on the substrate.

H. Hintermann:

I would welcome any other critique on that instrument because we use
it for a practical reason and the people who know more about indentation studies and scratch studies would be most welcome to make
their comments and maybe help improve this method which should become
a practical method for those who are studying flaking of such coatings.

C. A. Brookes:

The adhesion between the diamond slider and the various coatings will
vary from one specimen to another and will also vary from one load to
another. So I think to measure the friction would be very useful.

W. Williams (to C. A. Brookes):

Chris, I have always thought since the 60's that your measurements
showed the onset of the role of plastic deformation in controlling
the friction rather than diffusion because at 800°C, one does, in
fact, begin to get measurable gross plastic deformation in TiC crystals.

C. A. Brookes:

I think the important measurement to demonstrate the effect of the
diffusion mechanism is that if you measure the change in friction
force as a function of time, then below the critical temperature it's
virtually independent of time. But above the critical temperature,
the friction force goes up. So, the longer the time of contact, the
greater the friction force. And this surely reflects increase in
adhesion through the diffusion process.

W. Williams:

Could I just return to Dr. Hintermann's very educational review of
his pioneering work in the subject of coatings and ask a question
specifically about the process for thermal deposition through the
plasma, ion plating that is. In the case of a compound which is not
necessarily constant boiling, is there not a problem with the fact
that the particles arriving at the surface having been vaporized
thermally and passed through the plasma, that the rate of arrival of
these particles will not be identical for the two species in the
compound? Hence, one would get an excess of one or the other leading
to some precipitation or other problems in the coating?

H. Hintermann:

I think there are better specialists in the field here in this room
than I am, but I think you are right. There are two types of species.

The ones which are arriving on the surface with the thermal energy
and the others which have partially made collisions, caught the
charge for a certain time, and then they are accelerated towards that
surface and impinge on that surface penetrating it to some extent. I
think this is a very low percentage in comparison to the thermally
arriving species. So you would get two types of materials on that
surface. The good diffusion is partially due to the very clean surface which you get by sputtering away all kinds of impurities.

W. Williams:

Yes. What I'm thinking of, say a titanium carbide source, thermally
excited would produce fluxes in the vapor phase of titanium and carbon atoms which are not equal because of the heat of formation of the
compound and the rate of sublimation of the elements and so forth.
The two particles' fluxes are different and so the question is what
does that do to the coating that results? Maybe that's not a practical problem.

H. Hintermann:

I would guess that the one which arrives in a lower quantity is just
determining practically the deposition rate of the particular
compound you look at. If you want to put down titanium nitride or
titanium carbide and you have a deficiency in carbon, this deficiency
in carbon is determining actually the rate of growth of that species.
And you might co-deposit some material which is not titanium carbide
but say with some titanium in it.

W. Williams:

Well, that was my point.

H. Hintermann:

In fact, this is a very nice method to play around to make all kinds
of compositions while keeping the one more deficient than the other.

R. Sivan:

Early on in your paper you compared chromium-titanium carbide deposits with straight titanium carbide coatings. Could it not be that
the problem with the initial chromium carbide layer was a result of
the diffusion of carbon from the substrate and you get the usual
problem of the eta phase forming? Because chromium has an appetite
for carbon.

H. Hintermann:

That's exactly true. We use chromium carbide on steel in cases we
want to be sure that the corrosion resistance is good enough to put a
passivating layer in between. We thought it might also be helpful in
cemented carbides. But it is not. Chromium is a good carbide consumer and I think you get more eta phase.

A. T. Santhanam:

I have a general question to ask. It is often stated that in any new
technological development the best material and process are identified very early in the game and subsequent progress is obtained only
by process optimization. What is your opinion in the case of coatings on carbides?

H. Hintermann:

It might have come very much the same way. You had relatively fast
and decent coating. Then you move up with lots of effort to make it
industrially better and better. And that costs a lot of money and
time. At the beginning of a learning curve you don't bring up much
of a result, then it suddenly rises steeply up to say 80 or 90 per
cent of the confidence level and you have to work a long, long time
to make it a good industrial product. I think it is in many cases
that way. Otherwise you probably wouldn't pursue it if you did not
see, after a while, a relatively steep rise of progress.

A. T. Santhanam:

Do you see any rapid breakthrough in identification of entirely
unknown coatings for carbides in the next five years?

H. Hintermann:

If you will allow me this remark. I don't think that progress is
made by developing further and further new type of coatings. We can
develop coatings which might have 10 or 20 or even 50 per cent better
performance. But I think what the industry looks for is to take the
coatings they have, and to increase their confidence level in these
coatings. In other words, they would rather have among 10,000 coatings, almost 10,000 good ones, than to have a coating which is 50 per
cent better with a confidence level of only 98 or 97 per cent. So, I
think in these carbide coating developments, we look forward to
making a coating that is 2 to 3 times better than just 20 or 30 per
cent better, and in the meantime optimize what we have to the point
that industry can be really sure to get 100 good pieces. There is
not too much left to make it still better. I mean you have diamond
and boron nitride and maybe some other things that we don't know yet.
But there aren't too many other good candidates.

STRUCTURE/PROPERTY RELATIONSHIP OF CVD-TiC COATINGS ON WC-Co

V.K. Sarin

GTE Laboratories, Inc.
40 Sylvan Road
Waltham, MA   02254

INTRODUCTION

The advent of coated cemented carbides represents the greatest advance in cutting tool technology since the development of the tungsten carbide (WC-Co) tool itself. It can be said without exaggeration that coated tools have lead to expotential increases in metal cutting productivity. Since their introduction around ten years ago, several types of wear resistant coatings (i.e. TiC, TiN, Al_2O_3 etc.) have been developed and commercialized. Most of these developments have been achieved via monitoring the properties of these coatings using hardness, transverse rupture strength (TRS), and wear as characterized by machining.[1,2,3] Limited work on relating these properties to structure and morphology of the coating as determined by optical and scanning electron microscopy (SEM) has been reported.[4] Unfortunately these structural analyses have not provided more than a superficial insight into the several factors that can greatly affect the physical and wear resistant properties of these coatings, namely grain size and morphology, impurities, defects, stoichometry and concentration gradients. A better understanding of their roles in improving or degrading coating durability and performance may lead to further improvement of wear resistant coatings.

To a great extent wear resistant TiC coatings have been the basis for most of the investigations due to their unparalled use and wide range of application. CVD TiC coatings are generally deposited on WC-Co composites via heterogeneous gas reactions of volatile $TiCl_4$ and hydrocarbon in the temperature range of 800° to 1200°C using H_2 as a carrier gas. Under these conditions, TiC can form in a rather broad range of hydrocarbon to $TiCl_4$ ratios. The choice of hydrocarbon can greatly influence both the thermodynamics and kinetics of this reaction, yielding coatings varying in structure and morphology. It was the purpose of this investigation to study the influence of two hydrocarbon species, methane (CH_4) and propane (C_3H_8), on the structure, properties and performance of TiC coatings.

EXPERIMENTAL PROCEDURE

CVD TiC coatings were deposited on cemented carbide substrates using a conventional hot wall, resistance heated, vertical unit. The deposition process was performed at 1000°C and atmospheric pressure. The preheated reaction gas mixture, either CH_4, $TiCl_4$ and H_2 or C_3H_8, $TiCl_4$ and H_2, contacted specimens at several levels in the reactor. Scrap inserts were used on the top and bottom of each run to maintain a constant charge. Two types of substrates, designated as A and B, were coated. Each batch of substrates was identically processed. Table 1 summarizes critical characteristics of the coatings examined.

The morphology and structure of the coatings were examined using optical (OM) scanning (SEM) and transmission (TEM) microscopes. Coating and eta phase thickness measurements were made on polished cross-sections of all the samples. X-ray diffraction was used for phase identification and lattice parameter measurements. Microprobe and EDS analysis in the STEM mode were utilized for profiling concentration gradients. Abrasion wear resistance was evaluated with pin and wheel method. Both Vickers and Knoop hardness measurements were taken on the cross-section, as-coated, and polished surfaces of all the coatings examined.

Coating performance in machining was evaluated on SNG 432 inserts with equivalent edge hone in a 40 hp CNC lathe equipped with a direct tool wear monitoring system. The system registers

Table 1. Hydrocarbon source and characteristics of coatings.

Sample/ Substrate	Hydrocarbon	Coating Time hrs	Thickness Range TiC (μm)	Eta (μm)	Structure Designation
1A&B	C_3H_8	0.25	2-4	0.5-1.0	Fine (F)
2A&B	C_3H_8	0.5	8-9	0.5-1.0	Columnar (C)
3A&B	C_3H_8	1.0	12-14	0.5-1.0	Columnar (C)
4A&B	C_3H_8	3.0	25-30	1.0-1.5	Columnar (C)
5A&B	CH_4	0.5	2-3	0.6-1.2	Fine (F)
6A&B	CH_4	3.0	6-8	1.0-1.5	Equiaxed (E)
7A&B	CH_4	4.0	9-10	1.2-1.6	Equiaxed (E)

workpiece radius change (ΔR) due to the tool wear and deformation during machining. Conventional flank and crater wear measurements at several intervals into the cut were also made.

RESULTS AND DISCUSSION

Structure

Examination of the TiC coatings in cross-section on the TEM revealed several interesting features. Figures 1 and 2 show typical structures obtained using CH_4 or C_3H_8 respectively as the hydrocarbon source. It is clear that both types of coatings have two distinct grain morphologies, defined as Zone 1 and Zone 2, with a sharp transition between them. In both cases the grain size and morphology of Zone 1 and its width were found to be almost identical. The width of Zone 1 in all the samples investigated was observed to be in the range of 1.5 to 2μm. The remainder of the coating, (Zone 2) contained relatively uniform equiaxed grains for the coatings obtained using CH_4 (Figure 1) and large nonuniform columnar TiC grains (Figure 2) with C_3H_8. In both cases a zone of $M_{12}C(W_{12-x}Co_xC)$ type eta phase was observed at the interface between the coatings and substrates (Table 1).

From the observed structure of the coatings a hypothesis as to their growth kinetics can be drawn. The initial nucleation of TiC on WC-Co substrates is obtained on the binder phase (an alloy of Co with W and C in solution) and is therefore influenced to a great extent by the composition of the substrate [5]. Once nucleation has been achieved, the growth rate in Zone 1 seems to be controlled by the rate of carbon diffusion[6] from the substrate, independent of the hydrocarbon species in the reactor. Since the diffusion of carbon is increasingly hindered by the growing TiC layer, and its availability decreases at the interface of the substrate, a point is reached (at approximately 1.5 to 2μm) where the carbon activity of the gases in the reactor (CH_4 or C_3H_8) becomes rate controlling resulting in Zone 2 morphologies. At this point the growth rate of the TiC coating increases in proportion to the level of available carbon resulting in equiaxed grains with CH_4 and larger columnar grains with the higher levels of carbon from C_3H_8.

Another important factor that should be considered in the growth kinetics and therefore the morphological development of these coatings is the diffusion of tungsten [7] from the substrate to the growing TiC layer. This diffusion seems to occur along the TiC submicron grain boundaries and was only observed in Zone 1 for both the CH_4 and C_3H_8 coatings. It is difficult to predict whether this diffusion of tungsten occurs during the growth process and participates in the nucleation and growth of the coating in Zone 1, or easily diffuses along the submicron TiC grain boundaries as Zone 1 morphology is being developed. If this diffusion occurs simul-

Fig. 1. Cross-sectional TEM micrograph showing the grain
         morphology of a typical CVD TiC coating obtained
         with CH_4 as the hydrocarbon source.

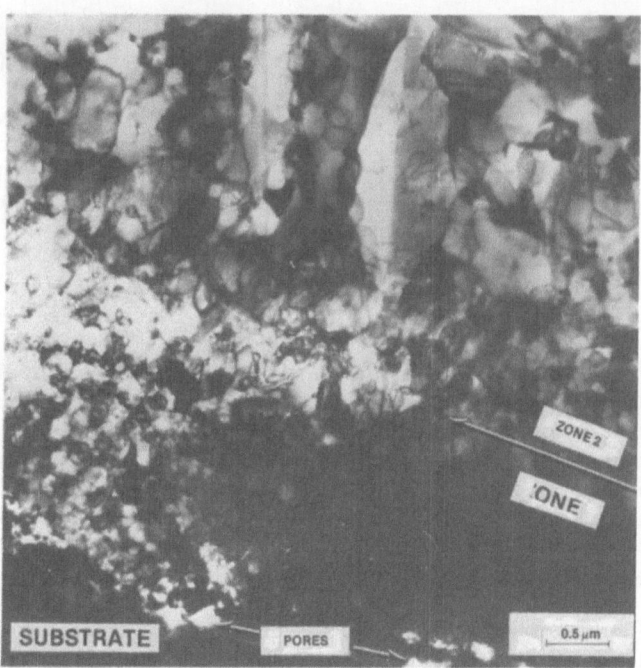

Fig. 2.   Cross-sectional TEM micrograph showing the grain morphology of a typical CVD TiC coating obtained with C₃H₈ as the
hydrocarbon source.

taneously as the coating grows, the presence of tungsten in (Zone
1) or its absence (Zone 2) could also be partly responsible for the
rate of nucleation of the TiC grains and therefore the difference
in grain size between these two zones.

Both microprobe and EDS analysis in the STEM mode on our specimens confirmed the diffusion of both carbon and tungsten from the
WC-Co substrate to the TiC layer. Trace amounts of titanium diffusion from the coating to the substrate were also observed. The
formation of porosity at the interface can be attributed to the
establishment of a "Kirkendall Effect" between the coating and the
substrate[8]. Furthermore, due to carbon deficiencies created in the
cemented carbide substrate by the diffusion of carbon to the coating
(Zone 1 growth) eta phase $(W_{12-x}Co_xC)$[2,9,10] zones are obtained between the coating and the substrate (Table 1). Since eta phase has
a higher density then the substrate it is formed from, pores can
result in this zone, due to volume shrinkage.

For the purpose of this investigation only the effect of grain
size and morphology of the TiC coatings on properties and performance are discussed. It should be noted that for such wear resistant coatings on cutting tools, the coating and substrate work as
an intergral system. Therefore, the properties and composition of
the substrate greatly influence the performance of coated tools.

Properties

Attempts to measure indentation fracture toughness (IFT) and
surface hardness of the TiC coatings yielded inaccurate results
due to the presence of cracks in the coatings. The linear thermal
expansion coefficient of TiC ($7.42 \times 10^{-6}/°C$) is considerably higher
than that of the substrates used, (3 to $5 \times 10^{-6}/°C$) and since
the coating was deposited at 1000°C, high tensile stresses are
developed during the cooling cycle. These tensile stresses results
in crack formation along the cross-section of the coatings. For
identical substrates thicker coatings contain more cracks and therefore the fracture toughness of the coating-substrate system should
decrease as the coating thickness increases.

Knoop indentations made at 900°C and room temperature on the
surface a C_3H_8 coating are shown in Figure 3. As has been previously reported[11], detection of cracks on the irregular surface of
the coatings is difficult, but cracks formed at the corner of these
indentations (Figure 3) are clearly visible. Also noteworthy is
that the coating seems to remain intact along the periphery and
depth of the indentation (Figure 4). This occurs due to the formation of a series of microcracks which are not visible. Polishing of the irregular coating surfaces enhanced the detection of
the inherent cracks. Hardness measurements on polished surfaces
(Figure 5) clearly indicate the dissipation of load along the cracks,

Fig. 3.   Crack formation at the edges of a Vickers hardness indentation made at 900°C (a) and room temperature (b).

Fig. 4.   SEM photomicrograph showing the presence of the TiC coat-
          ing along the periphery and depth of the Vickers inden-
          tation.

and shifting of cracked islands resulting in inaccurate values for
both hardness and IFT.   Therefore hardness values on the cross-
section of these coatings were the only ones considered (Table 2).

        Lattice parameter measurements (Table 2) indicated that gen-
erally very close to stoichiometric TiC coatings were obtained.
Abrasive wear measurements using a pin and wheel method could only
be performed on the relatively thick coatings with limited success.
Results indicate that both equiaxed (CH_4) and columnar (C_3H_8) coating have similar room temperature abrasive wear rates, in the range
of 8 to 9 x 10^{-5} cm^3/min.   Typical wear rates on substrate material
obtained under identical conditions were around 2 to 3 x 10^{-3} cm^3/
min.

Machinability

        Since the investigation was limited to studying the effect of
the hydrocarbon species on the structure, properties and performance of the coatings, the role of the substrate is not considered.
It is assumed that since substrates with identical processing histories and compositions were utilized, relative comparison between
these coatings could be safely made.   Coating and eta phase thickness ranges were matched, as closely as possible, for relative performance evaluations.   The effects of grain morphology (fine, equiaxed and columnar) on machinability characteristics of these coatings are discussed.

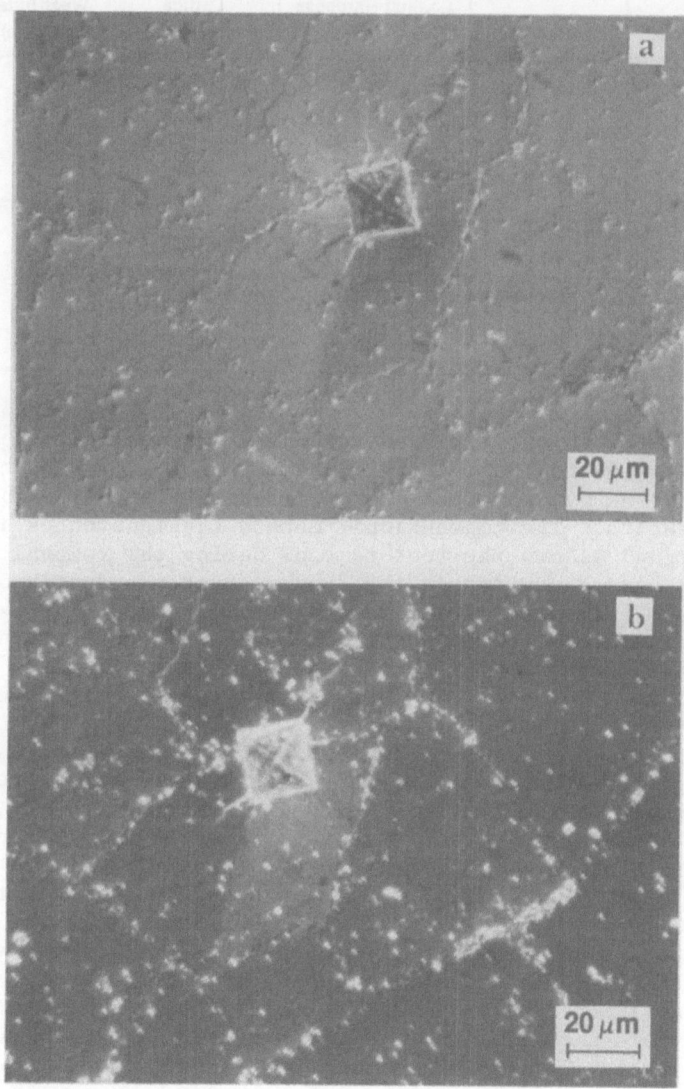

Fig. 5. Photomicrographs (OM) showing how cracked islands shift
to partially accommodate load obtained due to hardness
indentation (b) using polorized light.

Table 2. Cross-sectional hardness, lattice parameter and abrasive
          wear resistance of some of the TiC coatings investigated

Sample/Substrate	Hydrocarbon	Coating Hardness Knoop (kg/mm^2)	Lattice Parameters (Å)	Abrasive Wear Rate (cm^3/min)
1A	C_3H_8	2596	4.328	—
2A	C_3H_8	2657	4.327	—
3A	C_3H_8	2657	4.327	—
4A	C_3H_8	2700	4.329	$8-9 \times 10^{-5}$
5A	CH_4	2640	4.330	—
6A	CH_4	2657	4.328	—
7A	CH_4	2690	4.330	$8-9 \times 10^{-5}$

     The use of both conventional methods of measuring tool wear
(i.e. removing the tool at several intervals for microscopic measurements) and the dynamic tool wear monitoring system yield very
different results. The conventional method (designated as intermittent cutting) allows the tool to cool during the removal and re-
installation period whereas in the dynamic system a continuous cutting mode is obtained resulting in the cutting edge being exposed
to higher temperature for longer periods for the same cutting conditions. An increase in the cutting speed, since all tests were
performed without a coolant, is assumed to generate higher cutting
temperatures.

     In the intermittent cutting mode, flank wear (V$_{B}$max) resis-
tance of the TiC coating with equiaxed grain morphology (CH$_4$) was
found to be superior to that of counterpart columnar grained coatings (C$_3$H$_8$) on both substrate A (Figure 6) and B (Figure 7) at
400 and 500 sfpm. The opposite effect was observed at a cutting
speed of 600 sfpm (Figure 7). This reversal in performance can be
attributed to the relatively higher temperatures attained, even in
the intermittent mode, at this cutting speed. These results indicate a transition in the performance of TiC coatings between
equiaxed or columnar grains at a certain cutting temperature, the
equiaxed morphology having a superior resistance to flank wear at
the relatively lower cutting temperatures. TiC coatings with similar grain morphology (fine grained), independent of the hydrocarbon species used (CH$_4$ or C$_3$H$_8$) demonstrate almost identical
resistance to flank wear (Figure 6).

     In the continuous cutting mode (using the dynamic monitoring
system) phenomena other than attrition of the cutting edge masked
tool performance. Plastic deformation at the initiation of the

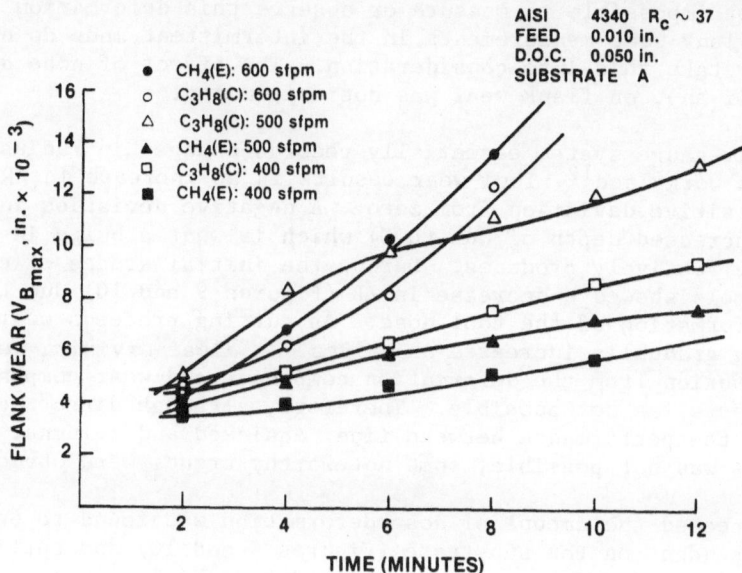

Fig. 6.   Machining performance of TiC coatings (substrate B) in
          the intermittent cutting mode.

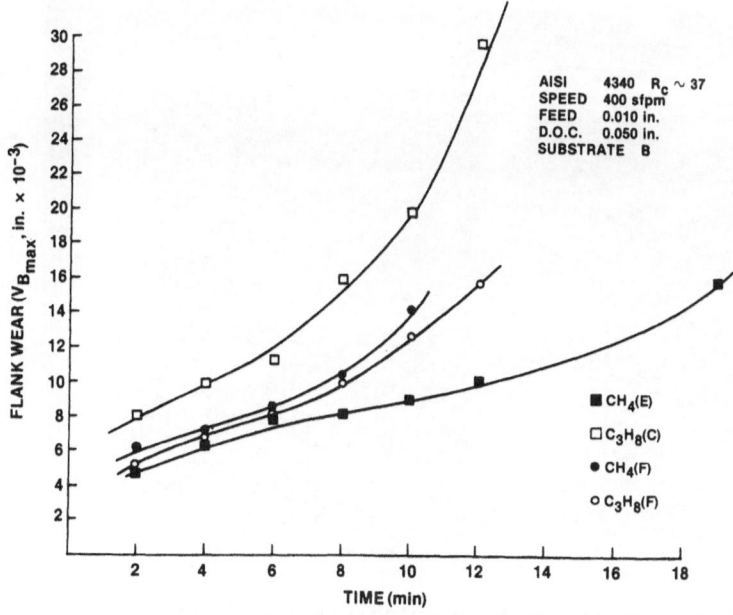

Fig. 7.   The effect of cutting speed on flank wear of TiC coatings
          (substrate B) in the intermittent cutting mode.

cut produces a bulge in the tool nose (Figure 8) which effectively
lowers the cutting point and moves it towards the workpiece. Since
it is almost impossible to measure or observe this deformation in
the (OM), flank wear measurements in the intermittent mode do not
effectively take this into consideration. The effect of nose deformation, if any, on flank wear was not determined.

The air-gauge system essentially records changes in radius
(ΔR) of the workplace[12]. Tool wear results in an increase in ΔR producing a positive deviation from zero. A negative deviation indicates an increased depth of cut (DOC) which is what a bulge in the
tool nose effectively produces. During the initial stages of cutting all tools showed a decrease in ΔR (Figures 9 and 10) due to
plastic deformation of the tool nose. As cutting proceeds wear of
the coating gradually increases ΔR values. A clear division between
the contribution from the deformation component and wear component
in these tests was not possible. Therefore, although direct comparison in the performance between fine, equiaxed and columnar grained coatings was not possible, some noteworthy trends were obvious.

As expected the amount of nose deformation was found to be
greatly dependent on the substrate (Figures 9 and 10) and cutting
conditions. The increase in temperature due to higher cutting speeds
results in higher negative ΔR values (corresponding to nose deformation). Substrate A with columnar grained (C_3H_8) coatings shows

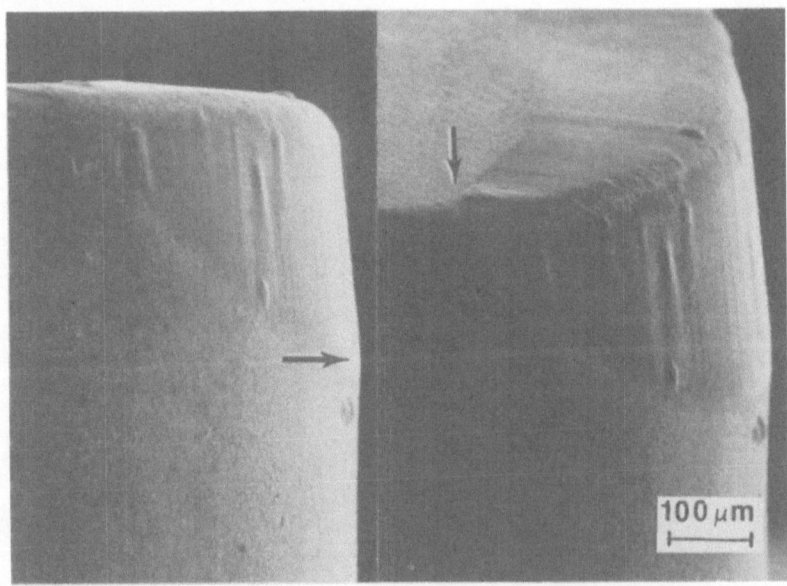

Fig. 8. SEM photomicrographs showing plastic deformation at the
         nose of a coated tool. Arrows indicate a bulge on the
         flank (a) and a dip in the nose on the rake surface (b).

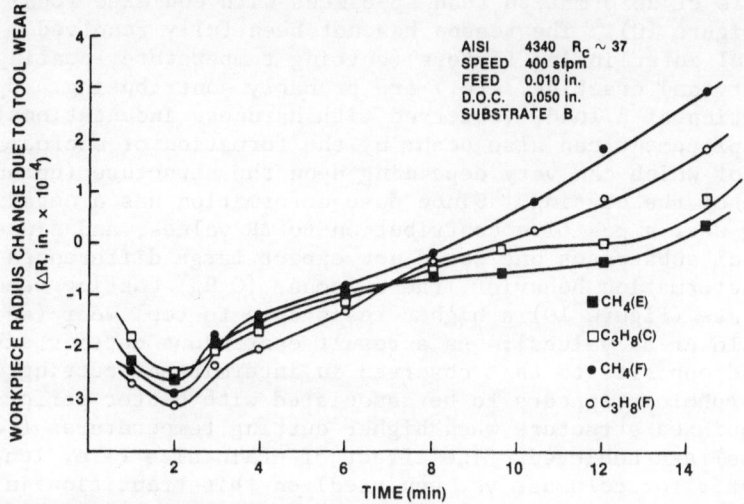

Fig. 9.   Machining performance of TiC coatings (substrate B) in
          the continuous cutting mode.

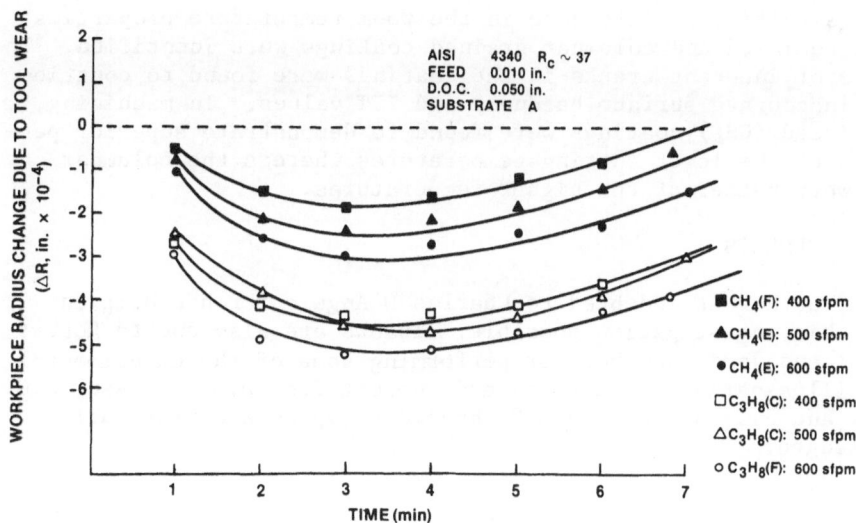

Fig. 10. The effect of cutting speed on ΔR (nose deformation plus
          wear) of TiC coatings (substrate A) in the continuous
          cutting mode.

higher levels of deformation than specimens with equiaxed (CH_4) coatings (Figure 10). The reason has not been fully resolved, since several interlinking factors (cutting temperature, coating deformability and cracking, etc.) are probably contributing. Upon the application of a load, (observed with hardness indentations) coating displacement can also occur by the formation of microcracks, the extent of which can vary depending upon the structure (columnar or equiaxed) of the coating. Since nose deformation has a negative and coating wear a positive contribution to ΔR values, and since for identical substrates one would not expect large differences in substrate deformation behavior, the columnar (C_3H_8) coating seems to demonstrate (Figure 10) a higher resistance to tool wear (as indicated by lower ΔR values). As a result continuous cutting exhibited a trend opposite to that observed in intermittent cutting. The columnar morphology appears to be associated with better performance than the equiaxed structure when higher cutting temperatures (longer cutting runs) are achieved. The effect of grain size (i.e. length to width ratio for columnar vs. equiaxed) on this transition in machining characteristics still requires clarification.

CONCLUSIONS

Cross-sectional grain morphology of CVD TiC coatings has shown that these coatings, independent of hydrocarbon source used, have two distinct growth zones. The use of either CH_4 or C_3H_4 results in an almost identical Zone 1 (submicron) structures, whereas CH_4 coatings have equiaxed and C_3H_4 columnar Zone 2 morphologies.

No significant difference in the room temperature properties between equiaxed and columnar grained coatings were identified. The presence of inherent cracks in the coatings were found to contribute to inaccurate surface hardness and IFT values. In machining, the equiaxed (CH_4) coatings were found to demonstrate superior performance at the lower cutting temperatures whereas the columnar (C_3H_8) were better at the higher temperatures.

ACKNOWLEDGEMENTS

The author is indebted to Charles D'Angelo for his help in making this investigation possible. Thanks are also due to Robert Wentzell and Jamie Sanchez for performing some of the experimental work. Illuminating discussions and support from D.M. Koffman, J.G. Baldoni and S.T. Buljan and K.E. Bucolo's typing are gratefully acknowledged.

REFERENCES

1. W. Schintleister, O. Pacher, and K. Pfaffinger, J. Electrochem. Soc. 123:924 (1976).

2. V. K. Sarin and J. N. Lindstrom, J. Electrochem. Soc. 126:1282
    (1979).
3. M. Lee and M. H. Richman, Metals Techol. 538 (1974).
4. M. E. Sjostrand, "VII CVD Int. Conf. Proceeding," T. D. Sedgwick
    and H. Lydtin, eds., The Electrochem. Soc. Inc. (Princeton,
    N.J) 452 (1979).
5. B. Lehtinen, Jernkont Ann. 155:446 (1971).
6. A. Hara, T. Yamamoto, and M. Tobioka, High Temp. - High Press.
    10:309 (1978).
7. W. Rix and G. Dix, U.S. Patent 3,558,445.
8. V. K. Sarin, "VII CVD Int. Conf. Proceeding," T. D. Sedgwick
    and H. Lydtin," The Electrochem. Soc. Inc. (Princeton, N.J.)
    476 (1979).
9. W. Ruppert, "III CVD Int. Conf. Proceeding," F. A. Glashe, ed,
    The Nuclear Society (Hinsdale, ILL.) 340 (1972).
10. W. D. Sproul and M. H. Richman, Met. Technol. 3:489 (1976).
11. A. Tvarusko and H. E. Hintermann, " VI CVD Int. Conf. Proceeding."
    L. F. Donaghey and P. Rai-Choudhury, ed, The Electrochem.
    Soc. Inc. (Princeton, N.J) 115 (1977).
12. J. G. Baldoni, S. T. Buljan, and V. K. Sarin, these Proceedings.

DISCUSSION

R. Warren:

Two comments on the problem of measuring mechanical properties of
these coatings. One might have better luck using Hertzian indentation. What one obtains in this case is a circular ring crack which
seems fairly independent of the flaws in the system. The other suggestion is that by varying the substrate composition you can vary
the substrate thermal expansion and therefore develop varying thermal stresses in the coating and possibly by observing the density
of the thermal cracks over a series of different thermal expansion
couples, you may be able to obtain some idea of the strength distribution in the coating.

V. K. Sarin:

Well, I think in the first case, yes. We have tried a certain
amount of that. But the coatings are typically 5 microns, and if
you apply a load where you get deformation of the substrate, you
can get cracking. Your second comment is one of the basic areas
of research. Work is in progress to match the substrate and the
coating in terms of expansion coefficient and other properties
to reduce these cracks and get better thermal expansion matching.

R. Warren:

In terms of measuring the properties of the coating you need to
produce a coating which is going to be the same, in each case, on
different substrates.

V. K. Sarin:

That's right. And we are starting to do that on other than cemented
carbide substances to study just the coating.

R. Sinclair:

How are you making your cross section TEM specimens? In our experience this isn't straightforward.

V. K. Sarin:

No, it's not. The idea here is to try and start the hole formation
in your foil away from the coating by masking the coating. And then
when the hole starts growing and gets closer to the coating, you remove the mask. That way you can get the coating to stay. Now, the
problem is to remove the foil and put it under the microscope.

The coating sticking out is extremely brittle and if you sneeze,
you can break it.

A. Doi:

I am pleased to see that you have succeeded in geting very thin
foil specimens of the coating for TEM observation. I have also
done some work in this area and found cobalt included in these
coatings. Could you comment on this?

V. K. Sarin:

Yes. In certain cases you can find cobalt on the surface of the
titanium carbide coating. The diffusion rates will vary depending
on your coating parameters and using certain coating parameters,
as I showed, you can get tungsten to go way up into the coating
along the grain boundaries.

A. Doi:

What factors most effect cobalt diffusion into the titanium carbide
layer?

V. K. Sarin:

Well, I think it's a difficult question to answer.

A. Horsewell:

I'd like to ask you about the columnar grains. You saw morphological changes; did you also see crystallographic changes between them?

V. K. Sarin:

The columnar grains were in the $\langle 110 \rangle$ direction most of the time, but they are non-uniform. I would say they are in the easy growth direction which would be $\langle 110 \rangle$.

A. Horsewell:

And for the other coating you found no preferred orientation.

V. K. Sarin:

In the methane coating it's randomly oriented equi-axed grains.

CHARACTERISTICS OF CVD HARD FACE COATINGS

R. Sivan and R. Israel

Iscar Ltd.
Nahariya
Israel

ABSTRACT

The Chemical Vapor Deposition process (CVD) is one of considerable complexity. This paper shows that in order to carry out meaningful quality control and development, sophisticated equipment is required. The problems and effects of CVD coating of carbides, nitrides and oxides on cemented carbide substrates are discussed principally with the aid of the Scanning Electron Microscope.

By an understanding of the physical properties of the coating and their microstructural requirements, an example of development work at ISCAR LTD on an improved alumina complex coating is described.

INTRODUCTION

CVD hard face coatings represent the major advance in cemented carbide technology over the last decade, although the first CVD coatings were produced by van Arkel as long ago as 1924![1]

The aim of CVD coatings is to marry the superior hardness of the coating with the bulk toughness of the substrate, and thus increase the cutting range and life of these inserts. Though we could consider any hard coating, in practical terms we could confine our remarks to a 4-15 micron thick single or complex coating containing the metals titanium, hafnium and aluminum together with their nitrides, carbides, oxides and/or combinations of these.

In figures 1, 2, and 3 we illustrate respectively the more
commonly accepted microstructures; a single phase TiC coating, a multiple TiC-TiCN-TiN coating, and a complex coating containing Al_2O_3 as
the surface layer.

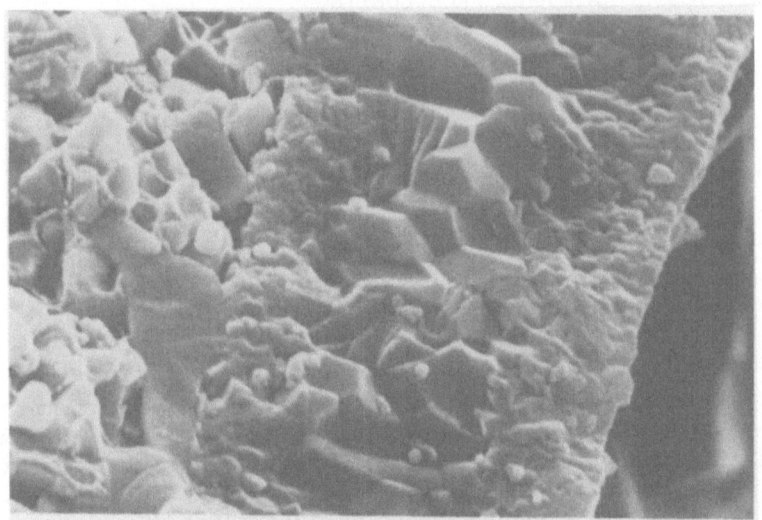

Fig. 1.   Iscar IC757, single layer TiC, x10K.

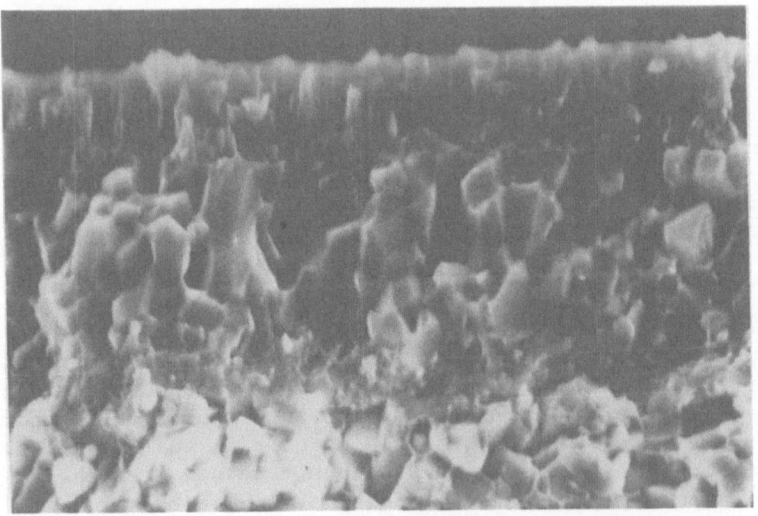

Fig. 2.   Iscar IC656, multiple TiN coating, x 10K.

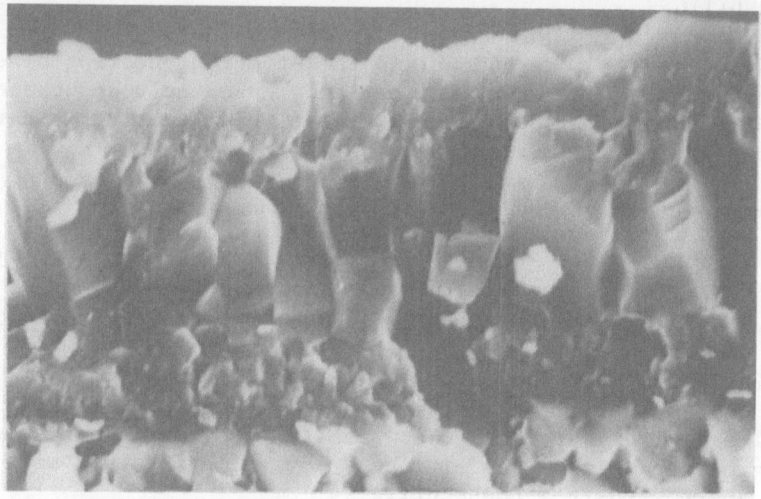

Fig. 3.   Iscar IC848, complex alumina coating, x10K.

Throughout the last decade, the production of hard coatings by CVD process has run into a series of metallurgical problems which one by one have been overcome in the inexorable progress of coating technology.   New tools have been introduced to arm the metallurgist: the β–scope for measuring coating thicknesses nondestructively, the Scanning Electron Microscope for easy observation of the coating structure at increased magnification, the Auger Scanning Microscope for surface studies of composition and the Energy Dispersive System for elemental analysis.   Today the quality control of coated carbides is a sophisticated exercise which involves most of these advanced techniques hand-in-hand with standard metallurgical testing such as optical metallography and machining tests.

Much has been written about the required properties of a good coating,[2,3] e.g., chemical stability with the workpiece and the machining environment; resistance to oxidation; high abrasive resistance; good adhesion to the substrate; and low thermal stress.   These all relate to its inherent properites; but a good coating must also be uniform to provide consistent tool life, have stable composition during machining, and of course be easily and economically produced. The chemical vapour deposition technique provides the best technique available; an excellent review of CVD coatings has been made by Yee.[4]

The quality of the coating is determined by:
    a.   the coating conditions
    b.   the substrate.

COATING CONDITIONS

Schintlmeister[2,3] has summarized the effect of coating conditions on the coating characterization, and these include: the composition of the gas media; the temperature at which coating is carried out; the range of gas flow; and the coating time.

The most common of the relevant reactions are

$$TiCl_4 + CH_4 \rightleftharpoons TiC + 4HCl \qquad\qquad (1)$$

$$2TiCl_4 + N_2 + 4H_2 \rightleftharpoons 2TiN + 8HCl \qquad\qquad (2)$$

$$2AlCl_3 + 3H_2 + 3CO_2 \rightleftharpoons Al_2O_3 + 6HCl + 3CO \qquad (3)$$

By altering the relevant quantities and combinations of gases containing oxygen, carbon and nitrogen, any number of hard metal coatings can be deposited consecutively.

In doing so, however, we come across a large number of problems associated with the following phenomena:

1. In TiC coatings: - very rough surfaces
                     - coarse dendritic grains
                     - porous structures.
2. In TiN coatings: - coarse columnar crystals.
3. In Alumina coatings: - rough surfaces
                         - uneven thicknesses
                         - powdery formations
                         - amorphous structures.

Since these defects tend to be greater in single layer than complex coatings, the tendency is for coatings to become quite complicated.

The trends towards more and more complex coatings are also a result of trying to satisfy the varied demands that are required of a hard coating:[5,6]

- almost universal use of aluminum oxide as at least one layer
  in a complex coating.
- chemical stability with respect to the workpiece, the atmosphere and the substrate. Here again alumina coatings are vital
  in providing a barrier to carbon diffusion, and so too are
  titanium nitride and carbonitride.
- resistance to thermal cracking, provided by titanium nitride
  which is said to behave as a "lubricant" and so prevent local
  ized overheating.
- edge strength, also provided by titanium nitride.

Hence the two principal coatings tend to be based on:

- titanium nitride, used at lower machining speeds where its thermal crack resistance and edge strength are more effective, and
- aluminum oxide, in complex coatings at higher machining speeds where resistance to abrasion, adherence, diffusion and oxidation are more relevant.

SUBSTRATE CONDITIONS

The condition of the substrate has been eloquently discussed.[7] The importance of the carbon balance in the substrate can be seen from the principal equation during the reaction between the first coating layer and the substrate:

$$5TiCl_3 + 6WC + 6Co + \frac{15}{2} H_2 \rightarrow 5TiC + W_6Co_6C + 15HCl \qquad (4)$$

As a result off this process, carbon diffuses across the coating/substrate interface and results in local carbon-deficient areas in the substrate leading to the formation of the interstitial compound, eta (W_6Co_6C). This compound is a source of weakness stemming both from its inherent brittle nature and from the internal stresses and porosity caused when this highly dense material is formed from its surroundings. The results can be cracks and shrinkage porosity which give the phase a "Swiss Cheese" appearance. However, this carbon diffusion is necessary for a strong interfacial bond[8] and is promoted by a higher proportion of tungsten carbide rather than by cobalt and/or titanium carbide.

Gass[9] has shown that the initial TiC layer is nucleated by diffusion of carbon from the substrate because the activity of carbon in the substrate is greater than that from the gas stream. As the growth of this initial TiC layer increases, the loss of carbon gradually decreases. Our work has shown that under normal conditions a maximum 1 micron interfacial TiC layer is formed in most cases.

TESTING PROCEDURES

The following test procedures are commonly carried out on coated carbide inserts:

(1) Physical tests to assess the substrate before coating for specific gravity, cobalt content, grain size.

(2) Beta-scope measurements for nondestructive measurement of total coating thickness.

(3) Optical microscopy for assessment of substrate microstruc-
     ture, porosity, eta phase, and individual coating thick-
     nesses. Here it is possible to detect by color the dif-
     ferent coating layers in that TiC is grey, TiCN purple and
     TiN gold. Al_2O_3 is difficult to identify.

(4) Scanning electron microscopy (Fig. 4) enables the identifi-
     cation of the individual layer microstructure and is par-
     ticularly invaluable for observing the alumina layers(s).
     At x10,000 magnification, photographs reveal abnormalities
     in structure, porosity, layer thicknesses, and surface
     condition.

(5) Energy dispersive analysis when connected to the SEM gives
     the added possibility of identifying the individual coating
     layers and eta phase.

(6) Machining tests are also common now and are carried out with
     standard turning and milling procedures.[10] Some of the
     conditions for these tests are:

Substrate type:	ISO M15
	Geometries TNMG 433, TNMA 432
Workpiece Type:	Turning/AISI 1045 steel
	Milling/AISI 1060 steel
Turning wear rate:	V_B mm/minute, wear measured after 2 and 16 minutes
Milling performance:	Breaking feed f_z mm/tooth.

Fig. 4. General view of SEM, EDS, and XRF equipment at Iscar Ltd.

     The results of machining tests correspond to microstructures
(Figs.1-21) are given in Table 1.

Table 1. Machining Tests of Coatings Whose Microstructures are Shown
          in Figs. 1-21 on Steel and Cast Iron

Fig. No.	Turning Steel Wear Rate V_B µm/min	Milling Steel Breaking Feed f_z m/tooth	Turning Cast Iron Wear Rate VB µm/min
1	18	0.7	--
2	16	0.9	5
3	6	0.9	4
5	23	0.7	--
6	19	0.7	--
7	25	0.7	--
8	29	0.6	--
9	20	0.8	--
10	28	0.7	20
16	18	0.8	10
17	8	0.8	4
18	22	0.8	--
19	7	0.5	4
21	23	0.5	15

COATING PHENOMENA

Associated with Coatings of TiC Only

     Figure 1 illustrates a good TiC coating with the sort of tech-
nical performance associated with it (Table 1). One can identify
three separate TiC layers:

(1) The fine equiaxed initial layer built up subsequently by
     carbon diffusion from the substrate. The rate of deposition
     of TiC depends on the source of carbon in the substrate.
     This layer usually develops a thickness of approximately
     1 micron, but when it becomes greater than 1.5-2 microns,
     cracks such as those shown in Fig. 5 can occur.
(2) The intermediate layer usually varies from 1-4 microns and
     is the coarsest of the three layers. Sometimes this coarse-
     ness is manifested as columnar crystals (Fig. 6), but we
     have no evidence to show that these are deterimental to
     cutting performance (Table 1).
(3) The final surface layer is a fine, medium-sized, equiaxed
     grain.
(4) Porosity both at the substrate/coating interface and the TiC
     intermediate layer can be a problem. Porosity at the
     interface (Fig. 7) is usually the result of dirty surfaces

Fig. 5.   Fracture in interfacial TiC layer, x10K

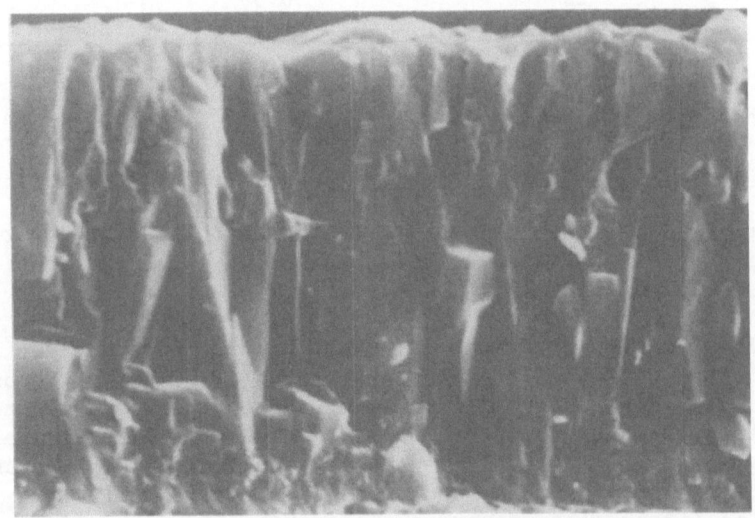

Fig. 6.   Coarse TiC coating, x10K

which can be prevented by correct cleaning techniques before
coating. Porosity in the TiC layer is a function of the conditions
for coating and can be eliminated with care (Figure 8).

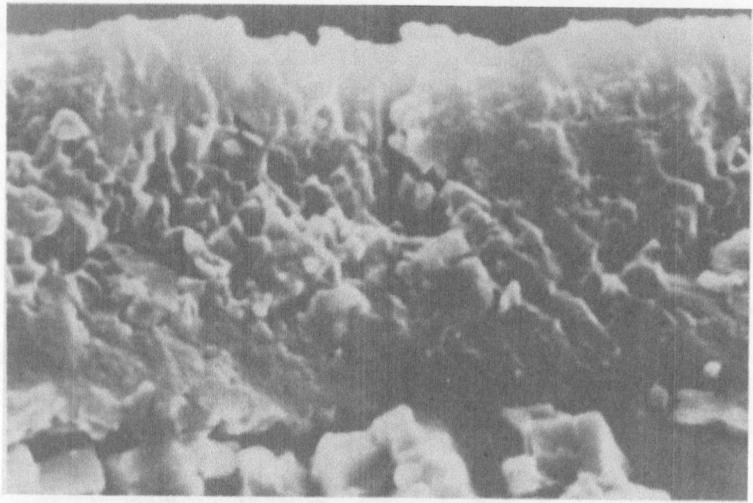

Fig. 7.   Porosity at interface, x10K

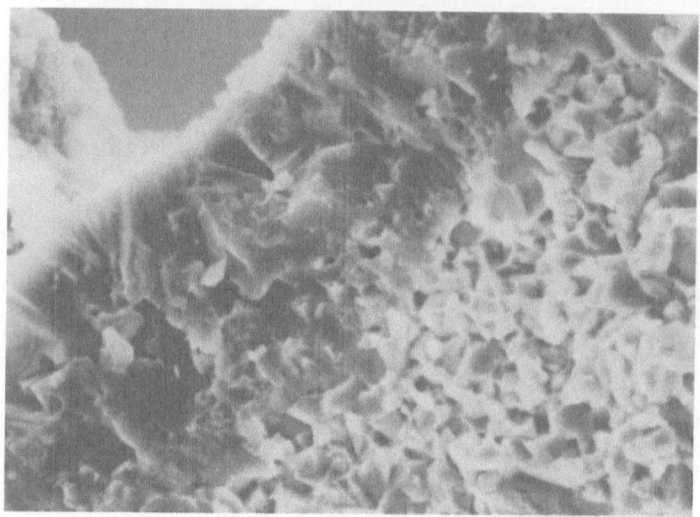

Fig. 8.   Porosity within TiC layer, x5K

Associated with TiN Coatings

The early TiN coatings consisted of a double layer of TiC, a surface layer of TiN, and between the TiC and TiN layers, a very fine layer of TiCN (Fig. 9). Today a variety of initial coating layers are being applied before the final surface TiN coating.

The problems associated with these coatings are:

(1) Coarse columnar structure of the TiN coating, which, when exaggerated, lead to poor performance (Fig. 9).

(2) The TiN coating should give a bright "gold" coloration, but variation may arise such as a matte finish, brown hue or even speckled coating. These are problems more of aesthetics than of performance.

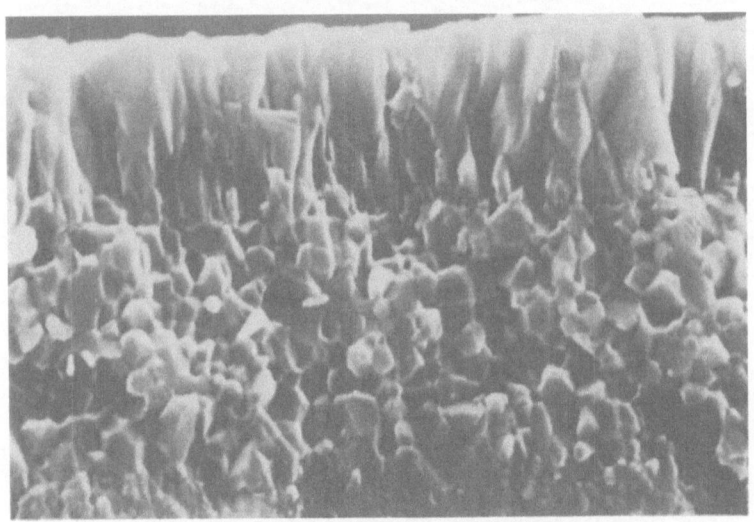

Fig. 9   Coarse surface TiN layer, x10K

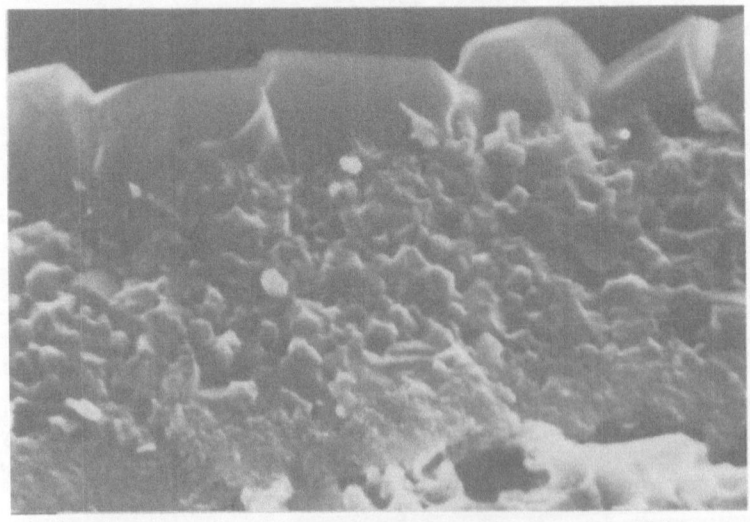

Fig. 10.   Coarse surface Al_2O_3 layer, x10K

Associated with Al_2O_3 Coatings

These are any coatings that contain an alumina layer, though generally we are referring to coatings where the Al_2O_3 layer is at the surface (Fig. 3). The problems met here are:

(1)   The grain structure of the alumina layer can vary from powdery, which usually gives rise to a very thin coating layer that does not perform well, to a thick, block-like, sometimes amorphous, structure, that leads to a flake-off in the early stages of work (Fig. 10).

(2)   Under very poor coating conditions, both of these phenomena can be seen to result in extremely uneven coating thickness. We believe that alumina coatings are much more susceptible to CVD furnace conditions than are either TiC or TiN Coatings.

(3)   Surface appearance is critical in alumina coatings. The coating should give a black or dark grey appearance, but "white" areas associated with gas conditions in the furnace can occur. "White" coatings are the result of unstable coating rates and give rise to a variety of interesting structures on the surface of the coated insert. These disasters are illustrated in Figs.11-15.

Fig. 11.   Alumina whiskers, x1500, "Grasshoppers"

Surface Appearance

   In addition to the regular quality control inspection of the
coating cross-section, we have instituted the inspection of surface
appearance. Those surfaces illustrated in Figs. 11-15 are the
exceptions. More common are equiaxed grains, varying from very fine
(Fig. 16) to very coarse (Fig. 18), with the best surface structure
being somewhere in between (Fig. 17).

Fig. 12.   Irregular alumina coating, x5000, "Ku Klux Klan"

Fig. 13.   Platelets on alumina coating, x1000, "Regatta"

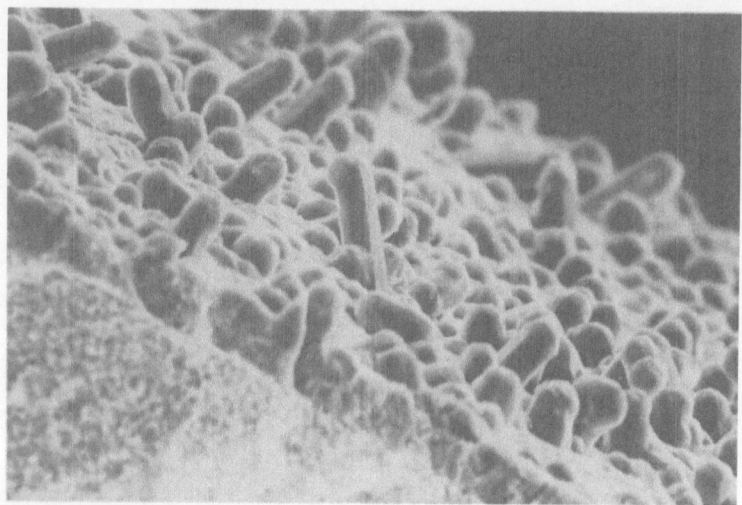

Fig. 14. Block-type alumina whiskers, x1000, "Seal Colony"

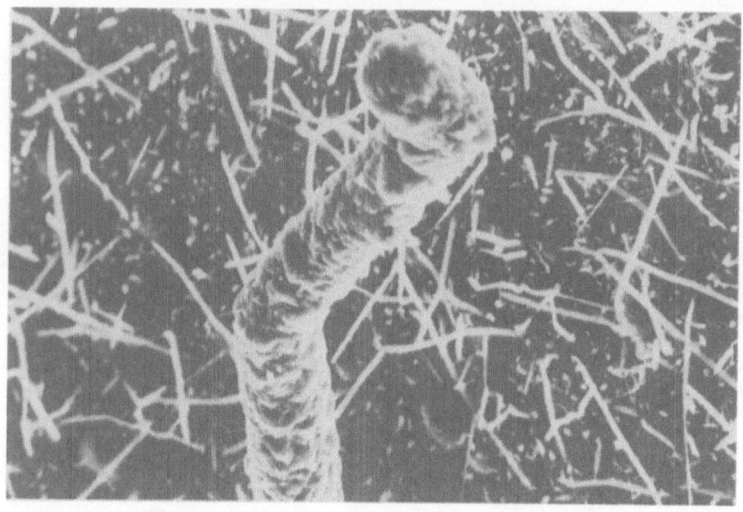

Fig. 15. Abnormal dendritic alumina growth, x1000, "Worm"

Associated with Substrate

Finally we come to problems that arise not from the coating layers at all, but from the substrate onto which they are deposited.

Eta Formation. As discussed previously, the result of carbon diffusion from the substrate to the coating is for the cemented carbide at the coating interface to become denuded; and, as a result,

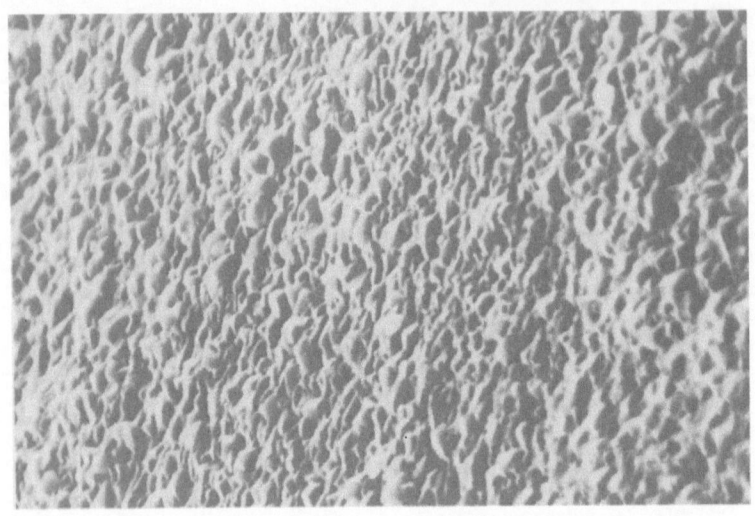

Fig. 16.   Iscar IC848, very fine Al_2O_3 coating surface, x5K

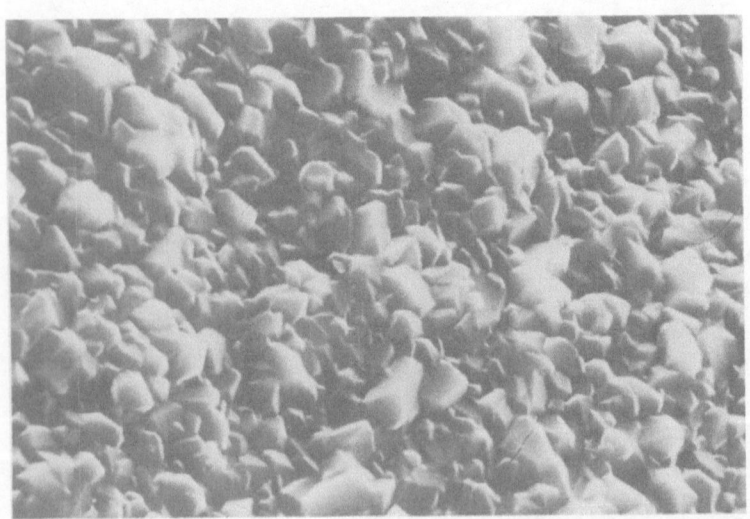

Fig. 17.   Iscar IC848, normal Al_2O_3 coating surface x5K

the brittle intermetallic eta phase (W_6Co_6C) can form. This can
easily be identified metallographically by preferential etching, but
not so easily using the SEM. However, if you know what you are
looking for, the problem becomes easier. Fig. 19 shows an agglom-
erate which could mistakenly be taken for WC, but illustrates a
colony of the eta phase as confirmed by EDS analysis, showing the

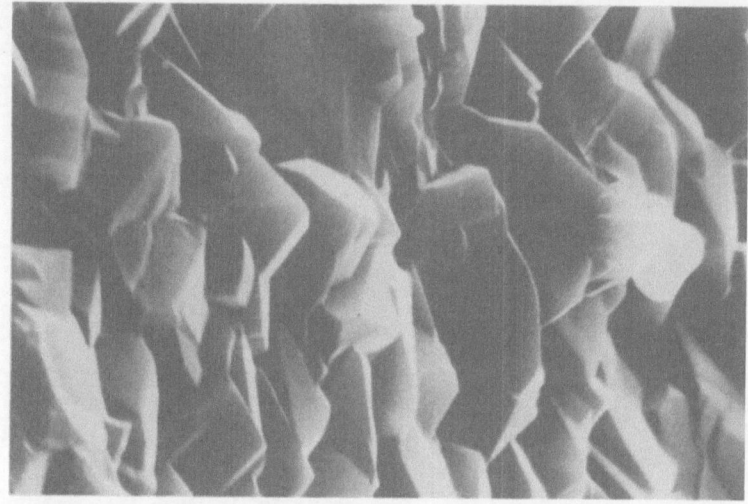

Fig. 18.   Iscar IC848, coarse Al_2O_3 coating surface x5K

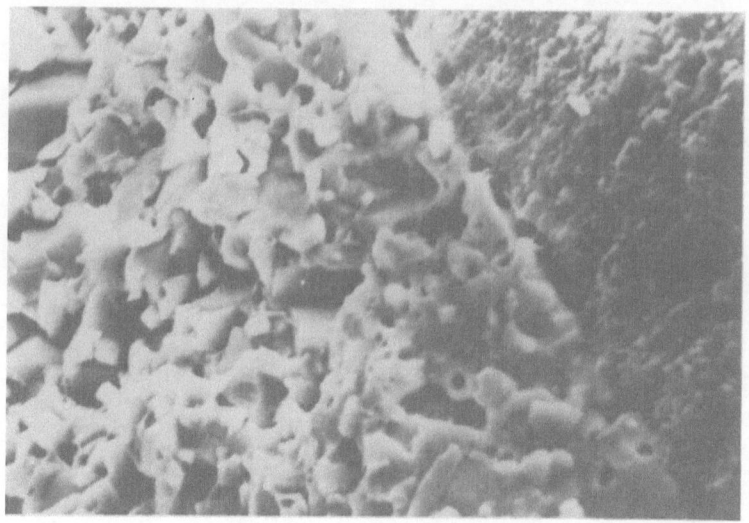

Fig. 19.   Eta phase, x10K

presence of cobalt together with WC (Fig. 20). The shape of the colony is also a "tell-tale." It resembles "Swiss cheese" in that it contains considerable porosity both within and at its perifery. This is explained as mentioned before as being caused by the shrinkage when eta is formed.

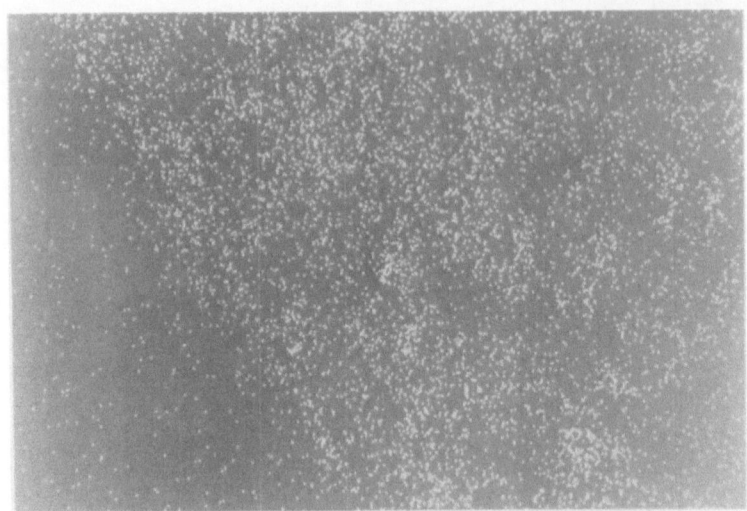

Fig. 20.   EDS of area shown in Fig. 19 showing cobalt distribution

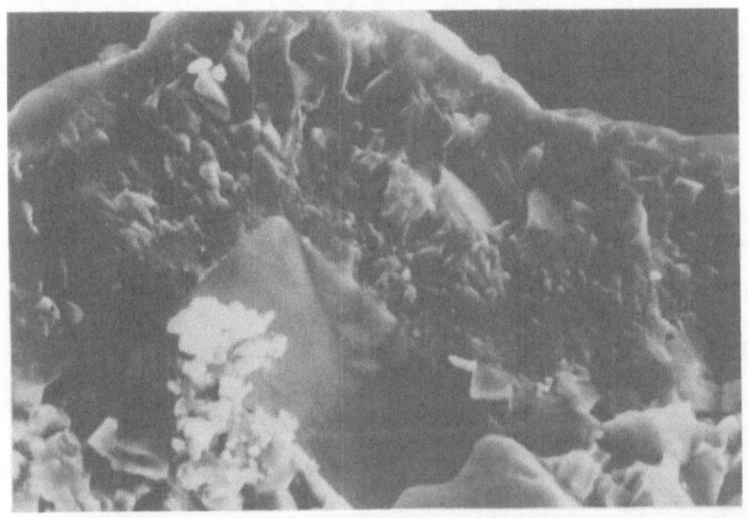

Fig.21   Irregular coating profile as a result of carbide grain
         growth, x10K

Grain Growth. Occasionally due to incorrect production techniques, the substrate shows abnormal grain growth of tungsten
carbide. When such growth does occur (Fig. 21), the cutting, and
more specifically the milling tests are disastrous (Table 1). Fig.
21 shows an extreme case where the WC grain growth has occurred at
the surface to be coated, causing a disruption in the insert

contour. The coating will then follow this contour and result in a protrusion in the coating surface leading to poor cutting test results.

DEVELOPMENT OF CVD COATINGS

There are of course many directions in which coatings are being developed at Iscar Ltd, we believe that by identifying the coating microstructure, understanding it and correlating it with the coating conditions and with the subsequent cutting performance, we shall be able to produce improved coatings more complex in appearance and with superior performance. One such development has been shown in Fig. 3 in which we have used

Al_2O_3 to limit carbon diffusion, to resist adhesion
             and abrasion wear, to provide oxidation resistance
TiN   to improve edge strength
TiC   To withstand adhesion wear.

Table 1 shows that this new combination has produced considerable improvements in both turning and milling performances on both steel and cast iron. The optimum sequence of coating has been developed by a consideration of both the inherent physical properties of the coating layers and the correlation of their microstructure with their cutting performance.[11,12]

CONCLUSION

We scarcely need to add that the CVD coating technique is extremely complex, and that it requires sophisticated testing equipment such as the SEM, EDS and machining tests in order to assess and characterize the hard metal coating.

In this paper we have tried to describe the use principally of scanning electron microphotographs to illustrate the structures arising from problematic coating and thus to characterize them.

Subsequently, from an understanding of the problems that arise, it has been possible to overcome them and eventually develop an improved complex CVD hard face coating.

REFERENCES

1. A. E. van Arkel, Physica 4:286 (1924).
2. W. Schintlmeister, Pwd. Met. Int. 13, 2:71 (1981).
3. W. Schintlmeister, Metall. 28:690 (1974).
4. R. K. Yee, Int. Met. Rev. 1:19 (1978).
5 W. E. Hintermann and H. Gass, Schweizer Archiv. 35:157 (1967).

6. J. R. Peterson, J. Vac. Sci. Technol. 11:715 (1974).
7. V. V. Sarin, J. N. Lindstrom, Jnl. Electrochem. Soc. 126:281
    (1979).
8. J. N. Lindstrom and R. T. Johannesson, Jnl. Electrochem. Soc.
    123:555 (1976).
9. H. Gass, et al. "5th Int. Conf. on CVD," 99 (1975).
10. E. Lenz, O. Pnveli, and L. Rozeanu, Wear 53 (1979)
11. R. Porat, Israeli Patent Appl. No. 58548 (79).
12. R. Porat, "Thermal Properties of Coating Materials and their
    Effects on the Efficiency of Coated Cutting Tools," 8th Int.
    Conf. on CVD, France 1981.

DISCUSSION

A. Horsewell

Your swiss cheese eta phase, those pores, could they possibly be
pullouts from the fracture surface?

R. Sivan:

No. In my opinion, definitely not. You can detect pullout in
scanning electron microscopy. These were continuous cracks between
a certain region which I claim to be eta and the substrate. With
pullout, I wouldn't expect them to be continuous like they are.

V. K. Sarin:

I have a couple of comments. It is fairly well established that
certain amounts of eta are beneficial to coating adherence. The
second comment is about the interfacial porosity which you mention
as bad surface preparation, could possibly be caused by a Kirkendall
effect. Since carbon and tungsten are diffusing from the substrate
to the coating, and small amounts of titanium diffuse from the coating to the substrate, you can cause this effect, creating pores. The
eta phase, of course, because of shrinkage as you have mentioned, can
give you pores within itself.

R. Sivan:

I don't think it is the Kirkendall effect because the porosity is
very massive and continuous. But it's open for discussion.

F. Rymas:

Can you tell me what causes a dark brown, mottled surface on a nitride coating versus a nice gold coloration?

R. Sivan:

I'll just say it's incorrect coating procedure. I don't really want
to go into details of the coating technique.

C. Dodd:

I wonder if you or one of the other speakers could make a few comments on the comparison of reactively sputtered titanium nitride
coatings as compared with the CVD coatings, both titanium nitride
and titanium carbide.

R. Sivan:

Well, I, at this particular stage in time, would not like to comment.

L. Toth:

One of the things I noticed about these last three talks is that
there is a lot of detail being given about the substrates and the
coating but no comparable detail about the process itself. I
don't know if this is because it is an industrial process. But to
give a plug to some of our grantees, we are interested in things
like mathematical modeling of the CVD process. And this, I think,
would really help people like yourselves to understand what's happening in the process.

R. Sivan:

I agree. If you'd like to give me the information I would be very
happy.

W. Williams:

So far the only remarks about eta phase near the interface have
indicated that it is beneficial, because of improved adherence.
On the other hand, there is other evidence, as I recall from Hara's
work in Japan and some work at Sandvik, that the eta phase is detrimental in terms of the transverse rupture strength of the coated
tool. We showed some time ago the carbon deficiency at the surface
and indicated how that could be treated to avoid the production of
eta phase, I'm wondering what your experience is along those lines.

R. Sivan:

I believe you need eta phase to prove that in fact you had good
carbon diffusion. The problem arises when you get continuous eta
phase formed. And we, as a general rule, and I don't think it's

very much of a secret, would say that if you have continuous eta
phase at the interface, it should not be more than about 2 microns.
Or, if it is intermittent, it should not be more than about 5 microns. But these are rule of thumb findings. I still confirm,
and I claim, that you need eta there. It gives you the best coating
growth characteristics. However, you can go too far. You can have
too much eta. And I agree with you, with too much eta, you get problems of tensile rupture.

TEM STUDY OF MICROSTRUCTURE AND CRYSTALLOGRAPHY

AT THE TiC/CEMENTED CARBIDE INTERFACE

S. Vuorinen[*] and A. Horsewell[*+]

Laboratory of Applied Physics I[*]
Technical University of Denmark
DK-2800 Lyngby, Denmark
and
Metallurgy Department[+]
Risø National Laboratory
DK-4000 Roskilde, Denmark

ABSTRACT

Titanium carbide coating layers on cemented carbide substrates have been investigated by transmission electron microscopy. Microstructural variations within the typically 5 μm thick chemical vapour deposited (CVD) TiC coating were found to vary with deposit thickness such that a layer structure could be delineated. Close to the interface further microstructural inhomogeneities were observed, there being a clear dependence of TiC deposition mechanism on the chemical and crystallographic nature of the upper layers of the multiphase substrate.

Coherent layers of TiC grains are found on the dominant $\{10\bar{1}0\}$ and $\{0001\}$ surfaces of WC grains. Three different orientation relationships are found on $\{10\bar{1}0\}$ and one on $\{0001\}$. Misfit accommodation is discussed.

INTRODUCTION

The life of cemented carbide cutting tools is substantially improved by titanium carbide coatings produced by chemical vapour deposition (CVD). In the CVD process, a gaseous mixture of $TiCl_4$, H_2 and CH_4 is passed over the cemented carbide substrate at temperatures of the order of $1000^{\circ}C$. The structure and morphology of the, typically 5 μm thick, TiC coating has been shown by a number of investigators using optical, scanning and transmission electron microscopy techniques to be complex [1,2,3,4,5]. Since the mechanism of nucleation and growth of the coating is both substrate and vapour phase controlled, microstructural variations occur with deposit thickness. Additionally, changes within the upper regions of the substrate occur both as a result of decarburization and diffusion between cemented carbide component phases.

Cemented carbides are basically two phase materials resulting from the liquid phase sintering of cobalt and tungsten carbide. The structure is thus that of a relatively ductile binder phase of a cobalt-rich solid solution of Co, W and C and a relatively brittle hard phase of angular grains of pure hexagonal WC. Cubic transition metal carbides, referred to as γ-carbides, are often added in small quantities to improve wear resistance. The bulk substrate microstructure so described is subject to surface decarburization during the CVD deposition of TiC coatings. This results primarily in the formation of a brittle carbide of tungsten and cobalt referred to as η-carbide. Sub coating layers of η-carbide, which may be up to 5 μm thick, are greatly reduced by the pre-carburization of cemented carbide substrates[6].

In the present paper we report on transmission electron microscopy observations of the TiC coating and in particular on the microstructure of the "interfacial carbide" region close to the cemented carbide substrate. Further, the TiC/cemented carbide interface studied here is shown to be a complex three-dimensional combination of seven sub-interfaces; these involve TiC and the substrate constituent phases as well transformation products resulting from the deposition process.

The TEM observations of the TiC/WC interface are of particular interest. Direct observation of the microstructure together with selected area diffraction evidence has revealed that TiC is initially CVD-deposited as a coherent layer adopting one of a limited number of orientation relationships with the WC grain onto which it is deposited.

EXPERIMENTAL PROCEDURE

All foils for transmission electron microscopy were prepared from commercially available CVD TiC-coated cemented carbides (Sandvik Coromant GC 1025) containing approximately 6 wt % Co, 5 wt % γ-carbides (Ti, Ta, Nb)C), balance WC.

Specimens were sectioned by low speed diamond sawing and then carefully thinned by diamond polishing to obtain wafers approximately 30 μm thick. Electron transparent specimens were obtained by ion-beam milling which was predominantly carried out from the polished wafer surface opposite to that covered by the TiC coating. The final stages of thinning were controlled by intermittent observation in the scanning electron microscope. For cross sectional specimens, the coating could be preserved during ion-beam milling by mounting two mechanically polished wafers in the ion-beam thinner holder so that the TiC coated surfaces lay edge-to-edge. Transmission electron microscopy was carried out using a JEOL 200A operating at 200 kV.

MICROSTRUCTURAL OBSERVATIONS

The observations reported here are concerned with the microstructural variations within the CVD deposited TiC coating, the microstructures developed in the upper region of the hard metal substrate and the details of the TiC/substrate interface.

The results were obtained from electron transparent regions of thin foils containing the TiC/substrate interface or from thin regions entirely within the 6 μm thick TiC coating. Individual TEM micrographs from positions within the coating and interfacial layers are located by reference to SEM observations of perpendicular sections through the coating (fig. 1).

The structure of the TiC coating

The TiC coating is composed of two distinct regions. There is thus a layer close to the substrate and extending to a thickness of 1.5-2 μm which is composed of fine, equiaxed TiC grains as shown in fig. 2. In this "interfacial carbide" layer the average grain size is of the order of 0.1 μm but is by no means uniform. Regions of extremely fine grain size (~ 0.01 μm) could be found, predominantly in close connection with the cobalt areas of the substrate surface (fig. 3). In the areas of interfacial carbide immediately above γ-carbide grains, the TiC grain size is increased to 0.2 μm-0.5 μm.

The uppermost layer of the TiC coating is distinguished from the interfacial region alone by virtue of grain size, there being an order of magnitude increase. This upper layer is typically 2-4 μm thick with grain size reaching 1μm at the top of the coating (fig. 4)

Voids were observed in all regions of the TiC coating. The voids occur mainly at the boundaries between TiC grains and appear not to be arranged in any network. Void size (<0.1 μm) and distribution remained largely constant through-out the coating thickness.

Many examples of voids pinning grain boundaries are seen typically in the uppermost part of the coating (arrowed, fig. 4).

The structure of the TiC/substrate interface

The region between the coating and the original hard metal surface is here referred to as the TiC/substrate interface. The multiphase nature of the substrate results in there being TiC/WC, TiC/

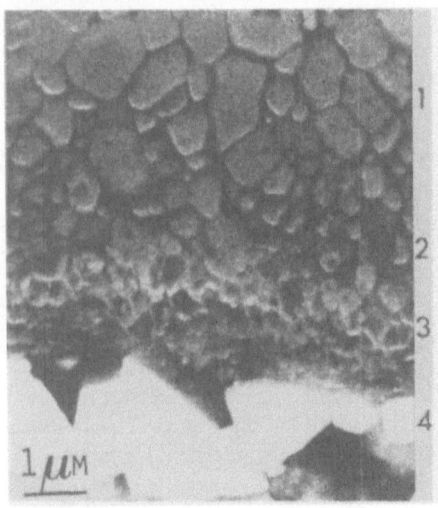

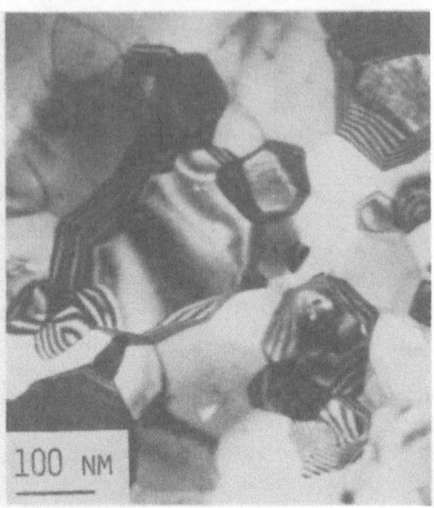

Fig. 1. TiC coating on cemented
carbide substrate. SEM. A multi-
layer structure may be delineated

Fig. 2. Bright field transmission electron micrograph of CVD
TiC in the "interfacial carbide"
layer.

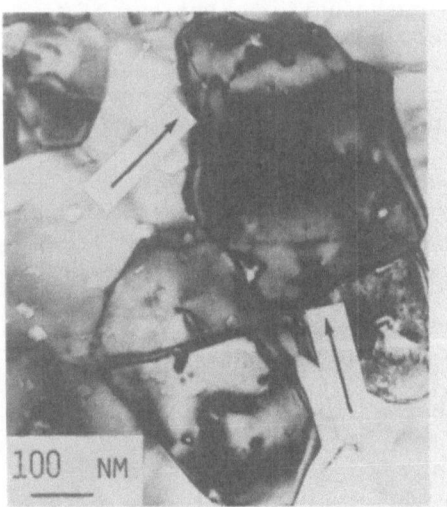

Fig. 3. Extremely fine-grained
TiC in vicinity of Co regions at
substrate surface.

Fig. 4. Uppermost region of TiC
coating showing large grain size.
Many voids may be seen, often
pinning grain boundaries (arrowed)

binder and TiC/γ-carbide contacts. Since formation of $M_{12}C$, η-carbide also occurs as a result of binder phase decarburization during TiC deposition, the interface also contains TiC/$M_{12}C$ contacts. Observations of the grain size variations within the initially deposited TiC coating ("interfacial carbide") are reported above. Additional TiC grain-size changes are found in contact with γ-carbide grains, fig. 5. Here, the TiC grains in direct contact with γ-grains have an extremely fine grain size (<0.05 μm). There occurs an abrupt increase however in the TiC grain size in the adjacent regions (fig. 5) such that the grain size becomes 0.2 μm-0.5 μm which is considerably larger than the average "interfacial carbide" grain size of 0.05 μm-0.1 μm.

Fig. 6 shows TiC in connection with both $M_{12}C$, η-carbide and tungsten carbide. Since there is direct TiC/WC contact, concurrent with η-carbide formation, a continuous layer of η-carbide on the original substrate is precluded.

Voids could be found at all interfaces, being clearly facetted at TiC/WC and TiC/γ interfaces. Void size and density were significantly reduced in the vicinity of cobalt binder and $M_{12}C$ phases.

Centred dark field micrographs of "surface" WC grains, fig. 7 and 8, gave clear indication that an orientation relationship exists between the first layer of deposited TiC and WC grains. The images result from common reflecting planes across the interface. These TiC layers were found to possess a limited number of orientation relationships with WC grains. Since the characteristically angular WC grains were observed to be bounded by $\{10\bar{1}0\}$ or $\{0001\}$ planes, orientation relationships were found to be determined by the plane type on to which deposition occurred. The crystallography of the TiC/WC interface is considered below.

DISCUSSION

The microstructure of CVD coated hard metal substrates may be seen as a complicated multi-layer structure which may be directly related to the diffusion and transformation processes occurring during high temperature deposition. The broadly parallel layers may be delineated from the scanning electron micrograph, fig. 1.

The layers are:
  1) upper part of coating
  2) interfacial carbide layer,
  3) TiC/substrate interface,
  4) upper part of substrate,
  5) bulk substrate.

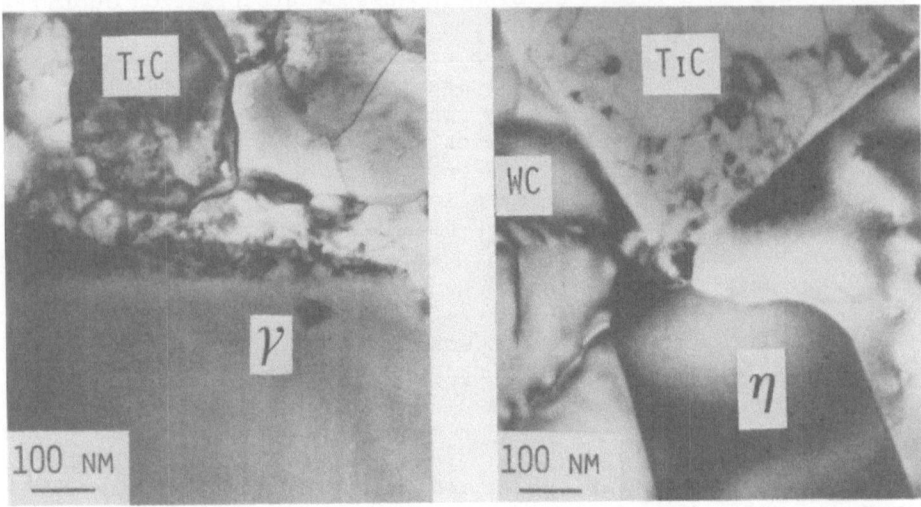

Fig. 5. Fine-grained TiC at the γ-carbide/TiC interface.

Fig. 6. Direct TiC/WC in the presence of η-carbide formation shows directly that η-carbide is not in the form of a continuous layer.

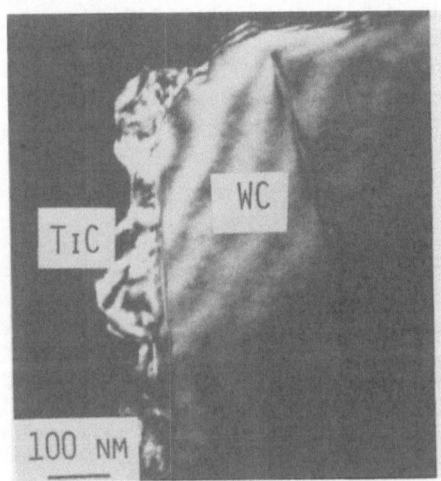

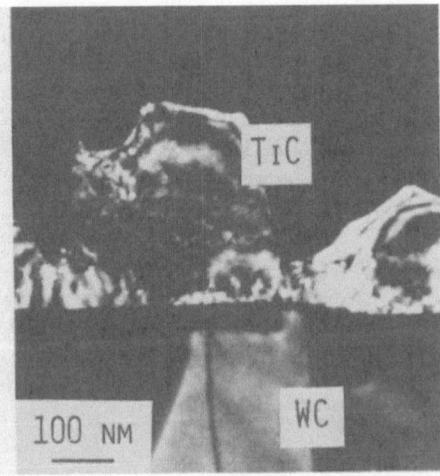

Fig. 7. Coherency of the first layer of TiC grains on a WC grain Centred dark field. Common reflection.

Fig. 8. Coherent TiC grains on {10$\bar{1}$0} surface of WC.

The TiC Coating

The upper part of the TiC coating is distinguished from the interfacial carbide layer principally in terms of grain size. It has been
previously shown that both nucleation[1] and growth rate[7] occur rapidly
close to the substrate as a result of good carbon supply from the
substrate. It is thus clear that the fine grained interfacial carbide layer is associated with TiC deposition with carbon diffusion
from the substrate. As the coating layer increases in thickness it
hinders the supply of carbon from the substrate and carbon supply
from vapour becomes dominant with a consequent drop in growth rate
and an increase in TiC grain size. In addition to this variation between upper TiC coating and interfacial carbide, a clear result of
the present study is that further microstructural variations occur
within the interfacial carbide layer. Interfacial carbide is thus
strongly influenced by the inhomogeneous nature of the substrate.

The interfacial carbide deposition may be directly related to
the various components within the substrate. In carbon rich regions,
i.e. binder and η-carbide, very fine TiC grains are observed being
a result of the simple diffusion of carbon from these phases. Adjacent
to WC and γ-carbides, however, decomposition or stoichiometry changes
are required before carbon can be supplied to the growing TiC layer.
Since WC is highly stoichiometric it is less able to act as a carbon
donor, as has been proposed for γ-carbides[3]. The increased growth
rate results in large (0.2-0.5μm) TiC grains in the vicinity of
γ-carbide grains.

It is apparent that the upper region of the substrate undergoes
microstructural changes which are largely due to decarburization
during TiC formation. The dominant transformation here is the formation of the $M_{12}C$ type η-carbide due to carbon loss from the cobalt
binder phase (Co(W,C)). The present observation of large, substrate
related, microstructural changes within the interfacial carbide layer
indicates that η-carbide does not occur as a continuous layer on the
substrate surface during the early stages of deposition as has been
previously suggested. Rather, as is seen directly in the micrograph
fig. 6, η-carbide is non-continuous on the substrate surface.

The interface

The true nature of the interface between hard metal substrate
and TiC coating is here identified as being a three dimensional combination of altogether seven sub-interfaces reaching a thickness of
0.5 μm. The subinterfaces can be divided into three groups; the interfaces between TiC and the substrate directly (1-3), the interface
between TiC and the $M_{12}C$ type η-carbide (4) and interfaces between
$M_{12}C$ and the substrate (5-7).

(1) WC/TiC
(2) Co/TiC
(3) γ /TiC
(4) $M_{12}C$/TiC
(5) $M_{12}C$/WC
(6) $M_{12}C$/Co
(7) $M_{12}C$/γ

The structure and distribution of these sub-interfaces is naturally important in determining adherence and wear properties of the TiC coating as well as being of particular importance during the nucleation and growth of the TiC layer.

The results of the present investigation clearly demonstrate that nucleation differences on the cemented carbide substrate are due to both the chemical and crystallographic differences of the phases comprising the substrate. Special orientation relationships of the TiC/WC interfaces have thereby been observed (following section). Local differences in growth rate at the interface, within the deposited TiC layer, are seen to be determined by chemical differences, principally the supply of carbon from the substrate. Away from the interface, growth rate appears more uniformed and not substrate dependent.

Voids were frequently observed at all sub-interfaces as well as TiC/TiC grain boundaries. Rather fewer voids were observed in the fine grain regions near the TiC/Co and TiC/$M_{12}C$ carbide interface which suggests a close dependence of void formation on the general processes of vapour deposition nucleation and growth. Facetted voids were frequently observed at the TiC/WC interfaces, being an indication both of the high annealing temperatures undergone during deposition and the orientation relationship at this interface.

CRYSTALLOGRAPHY OF THE TiC/WC INTERFACE

As was seen above, centred dark field micrographs of "surface" WC grains (fig. 7 and 8) may also include a first layer of deposited TiC grains because orientation relationships between the WC and TiC grains can result in common reflecting planes.

Determination of the orientation relationship is carried out using selected area diffraction across the TiC/WC interface. A schematic diffraction pattern is shown in fig. 9 where it may be seen that the direction, B, in TiC and WC is 111 and 0001 respectively. Constructing superimposed TiC and WC stereographic projections in these orientations (fig. 10) allows the orientation relationship to be found by inspection

$$[111]_{TiC} \; || \; [0001]_{WC}$$

$$(11\bar{2})_{TiC} \; || \; (10\bar{1}0)_{WC}$$

which occurs on the $(10\bar{1}0)$ plane of WC. In addition to this, two other orientation relationships may be found on the $\{10\bar{1}0\}$ type WC surfaces as follows

$$[011]_{TiC} \; || \; [0001]_{WC}$$

$$(001)_{TiC} \; || \; (1010)_{WC}$$

and

$$[011]_{TiC} \; || \; [0001]_{WC}$$

$$(01\bar{1})_{TiC} \; || \; (10\bar{1}0)_{WC}$$

Finally, for deposition onto the (0001) surface of WC grains, co-herent TiC is also found. In this case with the following orienta-tion relationship

$$[011]_{TiC} \; || \; [12\bar{1}0]_{WC}$$

$$(01\bar{1})_{TiC} \; || \; (0001)_{WC}$$

Interfacial Structure

Having obtained the orientation relationships between the ob-served coherently deposited first layer of TiC and the WC grain sur-faces it is found instructive to postulate a model of the possible atomic positions on either side of the interface. These models are presented in figs. 11a, b, c and 12 for the three $(10\bar{1}0)_{WC}$ and $(0001)_{WC}$ interfacial planes respectively. In these projections we have taken the maximum titanium carbide lattice parameter which is that for sub-stoichiometric $TiC_{0.86}$ as reported by Storms[8]. This allows maximum mis-fit to be considered. Although the orientation relationships shown are given by selected area diffraction experiments, no information can naturally be obtained for the exact positioning of the projected layers. Those drawn in figs. 11 and 12 are those which give the most reasonable metal/carbon coordination across the interface.

Misfit and Misfit Accommodation

The projections of figs. 11 and 12 are drawn in the unrelaxed configuration and it can easily be seen that misfit between the abut-ting crystals increases markedly away from the centre of the projec-tions. In the real interface this misfit naturally cannot accumulate and must lead to misfit accommodation. Misfit may be accommodated in this case either by misfit dislocations or changes in lattice parameter due to nonstoichiometry in the titanium carbide (as al-

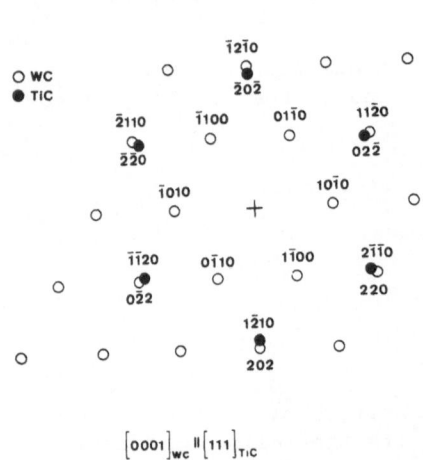

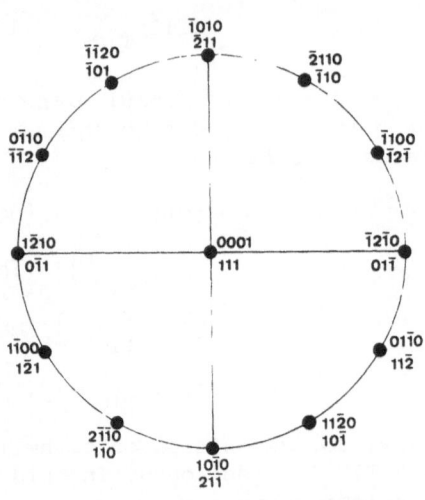

Fig. 9.   Schematic diffraction
pattern taken across the TiC/WC
interface giving B = 111_{TiC}:0001_{WC}

Fig. 10.   Superimposed stereographic projections from Fig.9.

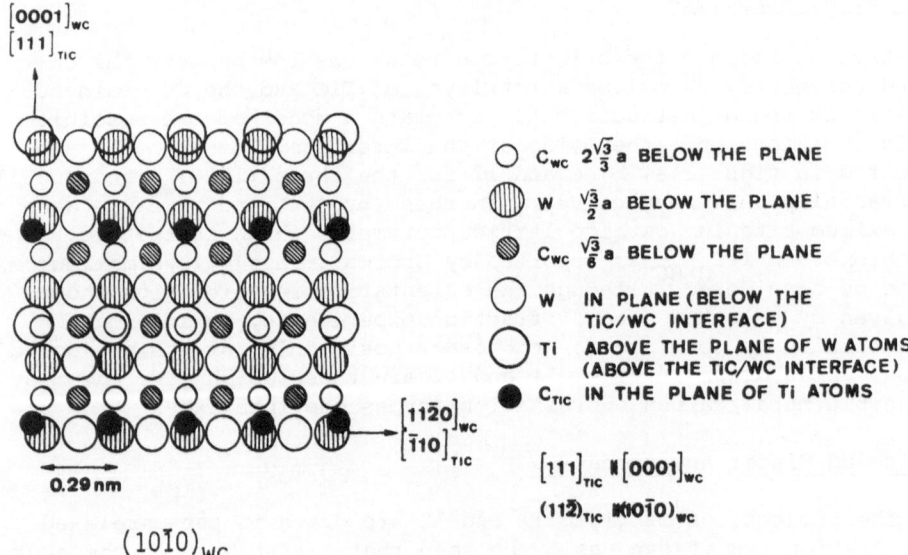

Fig. 11 a,b,c.   Projections of atomic positions on either side of
the $(10\bar{1}0)_{WC}$ interface for the indicated orientation relationships.

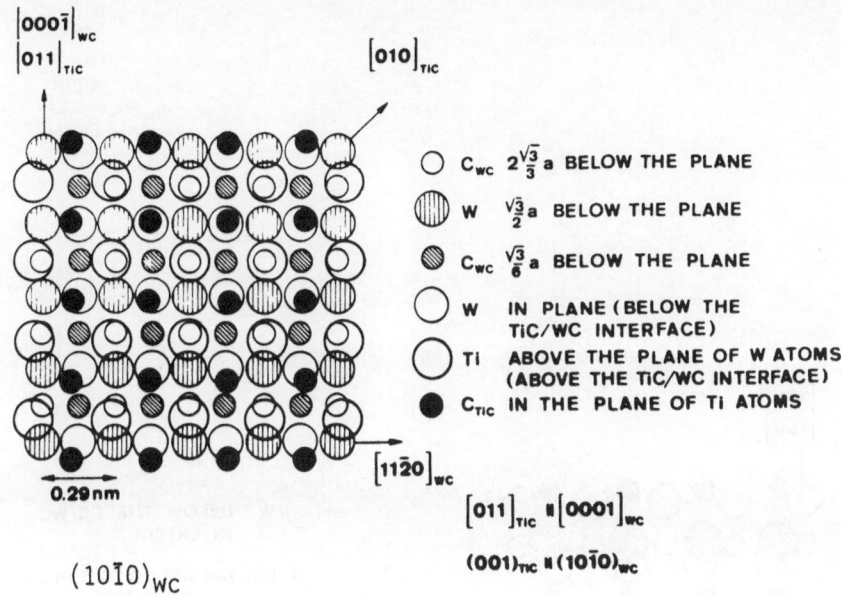

Fig. 11 b.

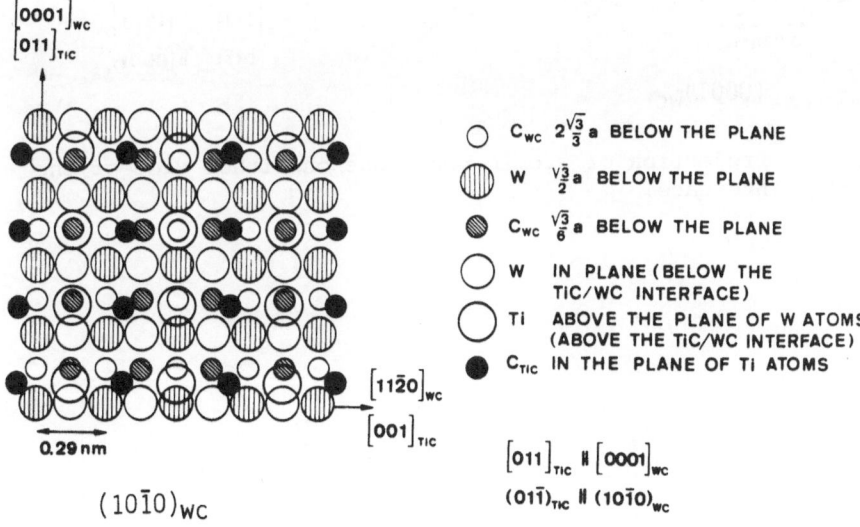

Fig. 11 c.

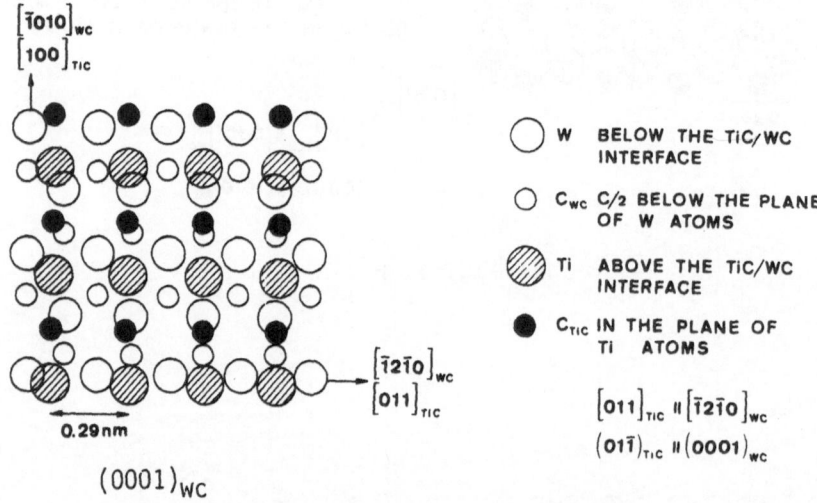

Fig. 12.   Projection of atomic positions on either side of the
(0001)$_{WC}$ interface.

TABLE 1.

SUBSTRATE	OVERGROWTH	ORIENTATION	f_o	m/n	F_o
$(10\bar{1}0)_{WC}$	$(001)_{TiC}$	$[011]_{TiC}$ // $[0001]_{WC}$	− 0.073	14/13	− 0.0022
		$[01\bar{1}]_{TiC}$ // $[\bar{1}2\bar{1}0]_{WC}$	− 0.051	15/14	0.0160
$(10\bar{1}0)_{WC}$	$(011)_{TiC}$	$[011]_{TiC}$ // $[0001]_{WC}$	− 0.073	14/13	− 0.0022
		$[100]_{TiC}$ // $[\bar{1}2\bar{1}0]_{WC}$	− 0.329	3/2	0.0066
$(10\bar{1}0)_{WC}$	$(112)_{TiC}$	$[111]_{TiC}$ // $[0001]_{WC}$	− 0.243	8/3	0.0086
		$[\bar{1}10]_{TiC}$ // $[\bar{1}2\bar{1}0]_{WC}$	− 0.051	15/14	0.0160
$(0001)_{WC}$	$(011)_{TiC}$	$[011]_{TiC}$ // $[\bar{1}2\bar{1}0]_{WC}$	− 0.051	15/14	0.0160
		$[100]_{TiC}$ // $[\bar{1}010]_{WC}$	0.162	6/7	− 0.0037

ready indicated) and/or through diffusion of tungsten into the TiC
lattice. The misfit to be accommodated for each of the determined
orientation relationships is summarized in Table 1. In those cases
where the misfit, f_o, exceeds ~ 10 % then this misfit parameter
(Frank and Van der Merwe[9]) is no longer realistic. The possibility
then exists for the coincidence lattice approach (Matthews[10]) to
be adopted. Then the misfit parameter, F_o, may be greatly reduced.
These parameters are calculated according to the following equations.

CONCLUSIONS

    a) The CVD TiC coating is composed of a fine-grained region
close to the interface and an upper, large grained, region. This
layer structure is associated with carbon supply, being initially
controlled by diffusion from the substrate and thereafter by the
vapour phase reaction. In addition to the layer structure (which
may be readily seen in the optical and scanning electron micrographs)
finer scale variations may be seen in TEM within the interfacial carbide layer. Such grain size and morphology changes appear directly
related to the chemical and structural differences of the substrate
component phases. It is suggested that these inhomogeneities may be
explained with reference to the ease of carbon supply from the component substrate phases. TiC microstructure near the interface therefore depends on the phase onto which, or near to which, it nucleates
and grows. Changes in substrate composition or in the distribution
of substrate phases are therefore expected to result in microstructural changes in this first layer of deposited TiC.

    b) The true nature of the TiC/cemented carbide interface has
been emphasized, it being composed of a total of seven sub-interfaces. Transmission electron microscopy observations have shown
that the first deposited layers of TiC on WC normally occur coherently,
there being a restricted number of observed orientation relationships.
The orientation relationships predominate at the TiC/cemented carbide
interface because the majority of WC surfaces are {10$\bar{1}$0} or {0001}
type, both of which show coherency with TiC. The orientation relationship is restricted to the first layer of TiC on individual WC
grains over a thickness of approximately 0.1 - 0.2 μm. Thereafter
randomly oriented, equiaxed TiC grains are observed. Misfit between
WC and TiC may be accommodated by misfit dislocations and/or non-
stoichiometry in the TiC as well as diffusion of tungsten into the
TiC lattice.

    c) Defects in the deposited TiC coating consisted generally of
few isolated dislocations. Voids were observed at all sub-interfaces
as well as TiC/TiC grain boundaries.

ACKNOWLEDGMENTS

Financial support was received from the Nordic Industrial Fund and the Academy of Finland. Specimens were provided by Sandvik AB.

REFERENCES

1. H. E. Hinterman, H. Gass, and J. N. Lindstrom, in: "Proceeding of the 3rd International Conference on Chemical Vapor Deposition," F. A. Glaski, ed., American Nuclear Soc., Hinsdale, Illinois, 352 (1972).
2. W. Rupert, ibid, 340.
3. N. K. Sharma, W. S. Williams, and R. Gottshall, Thin Solid Films 45:265 (1977).
4. E. Breval and S. Vuorinen, Mat. Sci. and Eng. 42:361 (1980).
5. V. K. Sarin, in: "Proceedings of the 7th International Conference on Chemical Vapor Deposition," The Electrochemical Soc., Princeton, NJ (1979).
6. V. K. Sarin and J. N. Lindstrom, J. Electrochem. Sci. 126:1281 (1979).
7. K. G. Stjernberg, H. Gass, and H. E. Hintermann. Thin Solid Films, 40:81 (1977).
8. E. K. Storms, "The Refractory Carbides," Academic Press, New York (1967).
9. F. C. Frank and J. H. van der Merwe, Proc. Roy. Soc., A198: 205
10. J. W. Matthews, IBM Res. Rep. RC 4266 (no. 19084) (1973).

DISCUSSION

M. Brun:

Did you claim that these interfaces were coherent or not?

S. Vuorinen:

Yes. The titanium carbide is coherent on tungsten carbide.

M. Brun:

Isn't it surprising that you have three different orientations and they are all coherent? Wouldn't you instead expect to have one particular orientation that would be the least energy type?

S. Vuorinen:

I wouldn't say so. I have found these three from (110) planes and
titanium carbide has been found on other substrates to grow in this
direction. I think that's why it would be a logical result to find
these three orientation relationships. The orientation relationship
observed on (0001) tungsten carbide grains is seldom found.

A. T. Santhanam:

Did you mention any orientation relationship between solid solution
grains and titanium carbide?

S. Vuorinen:

I was speaking about orientation relationships between titanium
carbide and tungsten carbide grains. I have not found any orientation relationship at the other interfaces.

A. T. Santhanam:

What was the substrate?

S. Vuorinen:

It was commercial cemented carbide made by Coromant, CC-1025.

D. Quinto:

In the last two papers there was mention of microporosity in the
upper part of the TiC coating. What is the significance of that?

Is it perhaps related to the inherent cracking that was observed in
the first paper by Sarin?

S. Vuorinen:

I think it is because of the different specimen preparation we have
used. I have done parallel sections to the substrate and have not
been able to look at the porosity. I'm not able to answer the
question.

A. T. Santhanam:

The porosity found in your micrographs, could it be unreacted
chlorides?

S. Vuorinen:

You are speaking about voids in the coating. I have tried to analyze
it but no results from that.

A. Horsewell:

We speculated that perhaps voids in the upper regions of the titanium carbide coating could possibly be beneficial by reducing grain growth. You could see they are pinning grain boundaries. I don't know what your feelings would be on that, but they may even be beneficial in some circumstances.

ALTERATION OF SURFACE PROPERTIES BY ION IMPLANTATION*

C.J. McHargue and M.B. Lewis
Metals and Ceramics Division
B.R. Appleton, H. Naramoto, C.W. White and J.M. Williams
Solid State Division
Oak Ridge National Laboratory
Oak Ridge, TN 37830

INTRODUCTION

Some exploratory experiments involving the bombardment of semiconductor materials by high-energy ions to alter the electrical properties of these materials were conducted in the 1950s. Since then, ion implantation has become a standard processing technique in the semiconductor industry to introduce dopants into a wide range of materials. During the 1970s interest in this technique was extended to modification of the chemical or mechanical properties of metals and the optical and electrical properties of insulators. Reference 1 contains a set of reviews covering studies outside of semiconductor technology. This paper describes studies to extend the use of ion implantation techniques to modify the mechanical properties of structural ceramics.

Because of the nature of the ion implantation process, elements can be introduced into the near-surface regions of solids in a manner that is often independent of normal equilibrium constraints (e.g., diffusivities and solubilities). The process is microscopically nonequilibrium in nature, hence, compositions and structures unattainable by conventional methods may be obtained.

As an energetic ion impinges onto a target material, atoms are displaced from their normal lattice sites by atomic collisions. Sputtering of the outermost layers may occur. Point defects are produced over depths of a few hundred to a few thousand atomic

---
*Research sponsored by the Division of Materials Sciences, U.S. Department of Energy, under contract W-7405-eng-26 with the Union Carbide Corporation.

distances from the surface, after which the injected atom(ion) comes to rest.

Figure 1 illustrates the deposited energy and deposited ion profiles for 1 MeV Fe^+ in Al_2O_3 as calculated with the E-DEP-1 computer code of Manning and Mueller.[2] This code used theoretical stopping powers calculated by the Lindhard, Scharff, and Schiott (LSS) theory. Since point defects are produced by the deposited energy, the figure illustrates the rate of defect production and the distribution of implanted ions. Most of the ions come to rest in the region where defect production is high. In analyzing property or structural changes, a number of effects must be considered: (1) alloying effects of the injected ions; (2) point defect configurations; (3) alloy-defect interactions; (4) nonequilibrium lattice sites of the injected ions; and (5) charge imbalance.

In the present study, three hard ceramic materials of widely varying chemical and bonding nature were chosen. They were single crystalline Al_2O_3, single and polycrystalline α-SiC, and polycrystalline TiB_2. Injected ions included those which form both extensive solid solutions and limited solid solutions under equilibrium conditions.

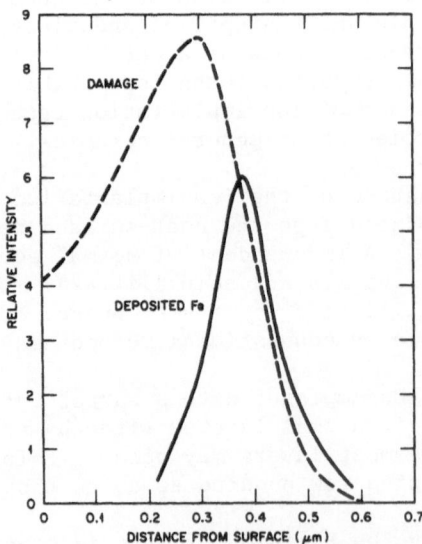

Fig. 1.   Calculated deposited energy and deposited ion profiles for 1 MeV Fe in Al_2O_3. Point defects production corresponds to the deposited energy profile.

The use of ion implantation to inject metal ions into these particular materials has been limited. Drigo and co-workers[3,4] determined the lattice disorder created by implanting lead in Al_2O_3. There have been a number of studies of the defect structure produced by implantation of gaseous species such as H, He, Ar, Kr as well as self-ion implantation of Al or O.[5-10] Although no data on mechanical properties have been reported, a group at Sandia National Laboratory has calculated the surface stress[5-6] produced in Al_2O_3 from a measurement of the deflection of a cantilever beam specimen during implantation.[11]

Most of the studies on implanted SiC have dealt with optical and electrical properties or the structure of the implanted region. The bombarded volume has been reported to become amorphous after a relatively low irradiation fluence.[12-16]

EXPERIMENTAL PROCEDURE

Single crystals of Al_2O_3 were obtained from two sources, Union Carbide Corporation* and Crystal Systems, Inc.** Both materials had high-purity, low-dislocation density and were oriented within $\pm 2°$ of (0001). Before implantation, each specimen was annealed for five days at 1200°C in air to remove any mechanical damage introduced during sample preparation.

Polycrystalline sintered α-SiC was prepared by Carborundum using boron and carbon as densification agents. The microstructure (Fig. 2a) was fine grained (2 to 10 μm diameter) and equiaxed. Single crystals of SiC were grown during synthesis of SiC in an Acheson furnace at Carborundum Corporation.

The polycrystalline TiB_2 was prepared at ORNL from Starck powder which contained $\sim$1% oxygen as an impurity. Specimens having 98.4% theoretical density were prepared by vacuum hot pressing at 2050°C under 25 MPa for 4 h. The average grain size was 23 μm, but the microstructure (Fig. 2b) contained some grains as large as 75 to 100 μm, indicating that considerable grain growth had occurred during processing.

A Varian-Extrion 200 kV implanter was used for the chromium implantations. Fluences of 10^{16} and 10^{17} ions-cm^{-2} at a particle energy of 300 keV were used for some Al_2O_3 specimens. A graded energy implantation of 95, 190, and 280 keV to a total fluence of 2×10^{16} ions-cm^{-2} was used for the remaining Al_2O_3 specimens and the SiC specimens. Use of the graded energy implantations widens the implantation zone. No attempt was made to control the

*Union Carbide Corporation, Crystal Products Division, 888 Balboa Avenue, San Diego, CA 92123.
**Crystal Systems, Inc., P.O. Box 1957, Salem, MA 01970.

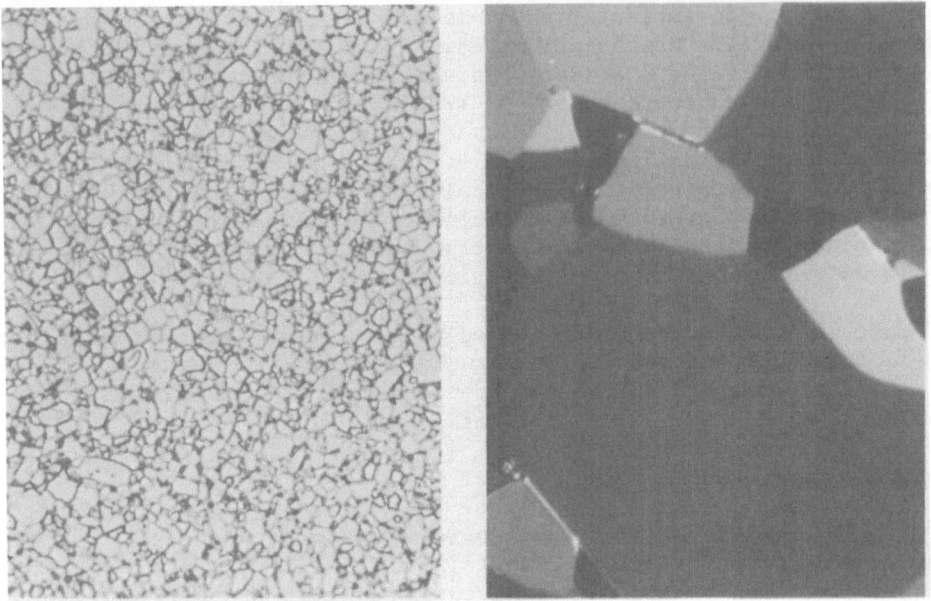

Fig. 2.   Optical photomicrographs of starting polycrystalline
          materials (a) α-SiC, (b) TiB$_2$.

implantation temperature, and beam heating was estimated to give a
specimen temperature < 300°C.

    The Ni$^+$ and Fe$^+$ implantations were carried out using the ORNL
5 MV Van de Graaff facility. The particle energy was 1 MeV.
Nickel fluences of 0.5×10^{17} and 1×10^{17} ion-cm^{-2} were used for
Al$_2$O$_3$ and 1×10^{17} ions-cm^{-2} for TiB$_2$. The doses of Fe$^+$ in Al$_2$O$_3$
were 2×10^{17} and 4×10^{17} ions-cm^{-2}.

    A summary of the implantation conditions is given in Table 1.
The average composition of the implanted zone is given as the
ratio of number of implanted ions to the number of aluminum atoms
contained in an undamaged volume of the same size. Since the profile of the implanted species is approximately Gaussian (Fig. 1),
the width of the implanted zone was taken as the width at half
maximum and all the ions were considered to be located in a zone
having this width.

    Mechanical properties were evaluated by microhardness,
fracture toughness, and scratch-wear measurements. In all cases,
only half of a specimen surface was implanted so that property
measurements of unimplanted and implanted regions could be made
under identical conditions. The mechanical properties will be

Table 1. Implantation Conditions

Material	Implant Species	Energy (keV)	Fluence (ions-cm^{-2})	Composition* (Implanted Ion/Al)
Al$_2$O$_3$	Cr^{++}	300	0.1 x 10^{17}	0.055
		95,190,280	0.2 x 10^{17}	0.0425
		300	1.0 x 10^{17}	0.55
	Ni$^+$	1000	0.5 x 10^{17}	0.0875
			1.0 x 10^{17}	0.175
	Fe$^+$	1000	2.0 x 10^{17}	0.350
			4.0 x 10^{17}	0.70
SiC	Cr^{++}	95,190,280	0.2 x 10^{17}	0.05
TiB$_2$	Ni$^+$	1000	1.0 x 10^{17}	0.20

*Composition was calculated by considering the implanted ions to be uniformly distributed in a region centered on the peak and contained within the calculated half width of distributions such as shown in Fig. 1.

presented as relative values, i.e., ratio of implanted value to unimplanted value. The Knoop microhardness technique was used with a force of 0.147 N in order to confine the impression to the near surface region. Typical impression depths of 3000 Å exceed the depth of the implanted layer for the chromium implantations. As a result, the relative hardness values indicate the direction of property changes, but they may underestimate the magnitude of the effects in the implanted region.

Fracture toughness was determined by calculating the apparent K_{IC} from indentation cracking. Vickers microhardness indentations were made with 0.49, 0.98, 1.47, and 1.96 N forces and the ratio of crack length to diagonal determined. Two models were used for the calculations.[17,18] The treatment by Marion[17] consistently gave higher values for both implanted and unimplanted regions than

that of Evans,[18] however, the ratio of implanted to unimplanted
values were similar for the two methods. Results reported in this
paper are based on Evans' model.

Wear resistance was determined from a test in which the tangential force was measured as a diamond indenter moved across a
surface at a fixed rate and normal force. The apparatus was built
from drawings provided by J. J. Wert of Vanderbilt University. It
has been shown that the specific energy of material removal is
directly proportional to the tangential force and the ratio of the
tangential force to the normal force given an abrasive friction
coefficient.[19,20]

Rutherford backscattering-channeling (RBS) techniques were
used to determine the lattice location of chromium implanted in
Al_2O_3 as well as the amount of damage created in each sublattice.
Used with annealing experiments they enabled us to separate the
effects of implantation damage from alloying. Specimens of Al_2O_3
and SiC were examined by transmission electron microscopy (TEM).

RESULTS

$\underline{Al_2O_3}$

Chromium was chosen for the major portion of this work
because it exhibits total solid solubility under equilibrium conditions and there are data on the hardness of Al_2O_3-Cr_2O_3 solid
solutions.[22,23]

Rutherford backscattering analysis showed large amounts of
damage to both the Al and O sublattices in the as-implanted state.
Figure 3 contains the results for 10^{17} Cr^{++} ions-cm^{-2} in Al_2O_3
(Cr to Al = 0.55). Notice that, although the implanted-aligned
curve is displaced upward from the virgin-aligned crystal, it has
not reached the implanted-random values. If an amorphous layer
were formed, as has been reported for Kr implantation,[10] the
aligned and random curves would overlap. Analysis of the backscattering from the implanted Cr reveals that near the surface
(i.e., at low concentration and low damage), Cr occupies substitutional sites, but near the peak concentration those ions are
randomly distributed among interstitial and substitutional sites.

Specimens from the Cr-graded energy implantation sample were
prepared for examination by transmission electron microscopy by
ion milling from the unimplanted face. Thinned specimens were
examined at 1 MeV accelerating potential in the ORNL Hitachi 1000
microscope. The electron diffraction pattern showed the implanted
zone to be crystalline despite the large amount of damage. The
TEM images contained a high density of "black spots" similar to
those produced by low-temperature neutron radiation damage (Fig. 4).

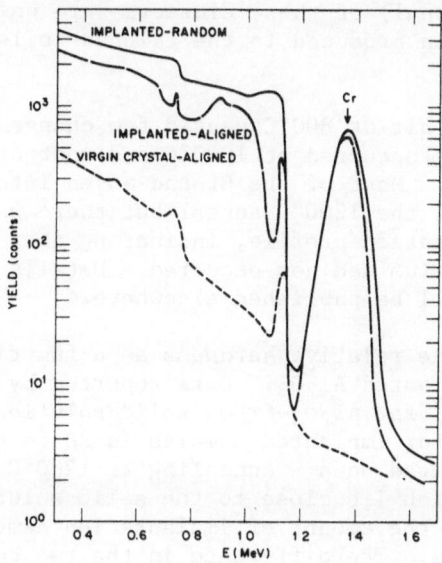

Fig. 3. Rutherford backscattering-channeling curves for Cr
implanted to a fluence of 10^{17} ions-cm^{-2} in Al$_2$O$_3$.

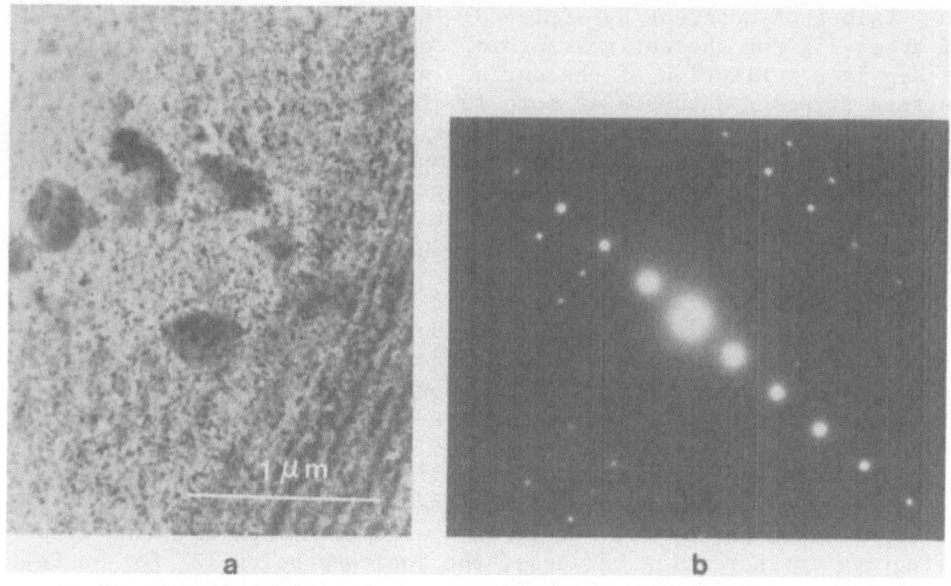

Fig. 4. a) Transmission electron micrograph and b) electron diffraction pattern of Cr-implanted Al$_2$O$_3$. The implanted region is
crystalline and contains point defect clusters.

Such spots generally represent images of the strain fields of point defect clusters. Attempts to identify the character (i.e., interstitial or substitutional) of these clusters were unsuccessful due to the large distortion produced in the thinned foils by the residual stresses.

Annealing 1 h in air at 800°C caused few changes in the RBS curves. Some recovery occurred at 1000°C and most of the damage was removed at 1200°C. Most of the Cr had moved into substitutional lattice sites after the 1200°C anneal but there was little change in the concentration profile, indicating that long-range diffusion of the chromium had not occurred. Details of the annealing studies by RBS will be published elsewhere.[21]

Figure 5 shows the relative hardness as a function of composition for the Cr-implanted Al_2O_3. Data reported by Bradt[22] and Ghate and co-workers[23] for Al_2O_3-Cr_2O_3 solid solutions are also shown. The hardness for implanted samples is 28 to 45% greater than that for unimplanted ones. Annealing at 1200°C reduces the hardness to values which lie close to the solid solution curve. As might be expected, the amount of implantation damage increased with increased fluence. The difference in the two curves, then, is caused by one or more of the factors: (a) chromium in interstitial lattice sites, (b) point defect clusters, or (c) chromium-defect interactions.

Values of apparent K_{IC} (Fig. 6) indicate a constant increase of about 15% for the entire range of composition. If this value of K_{IC} is a reflection of the surface stress, it appears that the surface stresses saturate at some low fluence and remain constant. Arnold, Krefft, and Norris[5] found that the integrated stress resulting from volume dilation in non-implanted cantilever beams of Al_2O_3 became approximately constant for gas ion fluences between 10^{15} and 10^{16} ions-cm^{-2}. Annealing at 1200°C returned the value of the apparent K_{IC} to that of the virgin crystal.

The wear resistance of the graded-energy chromium implant (Cr:Al = 0.0425) was measured in the scratch-wear test. For normal forces of 0.196 and 0.294 N, the specific energy of material removal increased by 25% as the moving indenter passed from the unimplanted surface to the implanted region.

The relative hardnesses of the Ni^+ and Fe^+ implanted Al_2O_3 is shown in Fig. 7. The hardness increased by 22% for a Ni to Al ratio of 0.088 and by 50% for a ratio of 0.17. Iron was much less effective in increasing hardness, the increase being 12% for an Fe to Al ratio of 0.35. The high fluence Fe region contained 1.4 Fe ions for each molecule of Al_2O_3 and exhibited a hardness increase of 20%. Annealing experiments will be conducted in an attempt to separate "alloy" effects from "damage" effects. At these

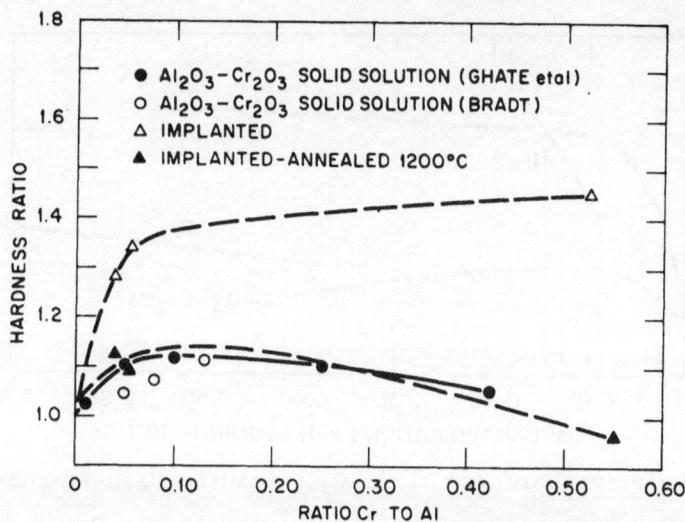

Fig. 5. Relative hardness of Cr-implanted Al_2O_3 alloys. The composition was calculated from the ratio of Cr to Al in the implanted volume. Data for conventional solid solutions from references 22 and 23 are included for comparison.

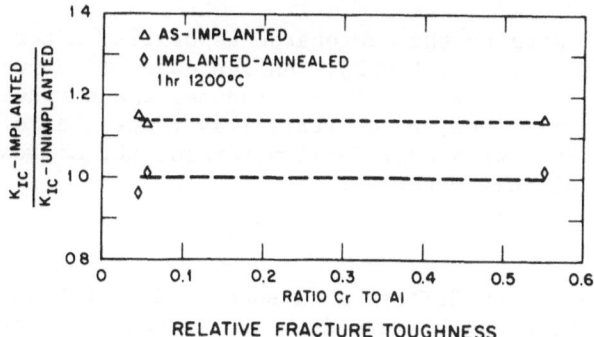

Fig. 6. Relative indentation fracture toughness for Cr-implanted Al_2O_3.

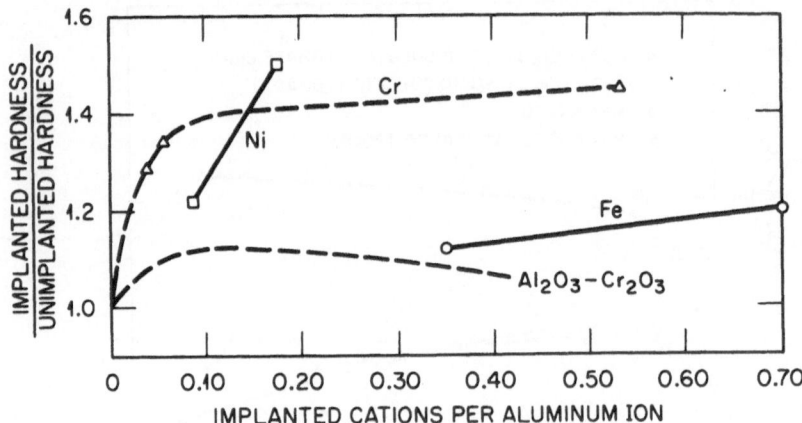

Fig. 7.   Relative hardness of Al₂O₃ implanted with Fe and Ni.

concentrations, however, precipitates of second phases may form
since these specimens greatly exceed the solid solubility limit.

The apparent fracture toughness increased more for the Ni^+
and Fe^+ implants than for the Cr^{++}. Figure 8 shows increases of
35 and 43% for the Ni^+ specimens and 35 and 30% for the Fe^+.

TiB_2

Figure 9 illustrates the microhardness of TiB_2 after implantation to a Ni to Al ratio of 0.20. An increase of 69% was
observed. Annealing for 2 h at 1425°C reduced the hardness from
the as-implanted value but it was still 1.44 times that of unimplanted sample. Further studies will determine the microstructure
and damage state of this material.

SiC

Implanting SiC with Cr^{++} to a fluence of 2×10^{16} ions-cm^{-2}
(Cr:Si = 0.05) caused the bombarded region to become amorphous, as
was determined by electron microscopy. Back-thinned specimens
were prepared by ion milling from the unimplanted face. Specimens
which allowed the damage along the ion range to be studied in profile were prepared by ion milling in a direction perpendicular to
the implantation direction.

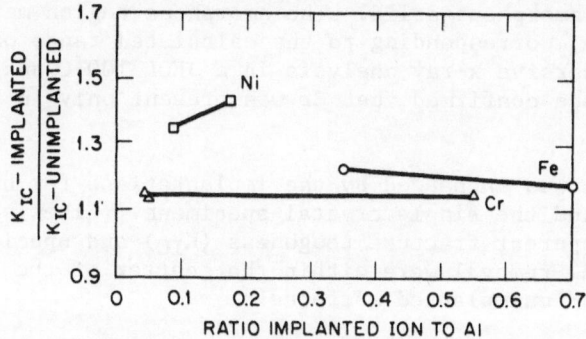

Fig. 8.   Relative indentation fracture toughness of Al_2O_3 implanted with Fe and Ni.

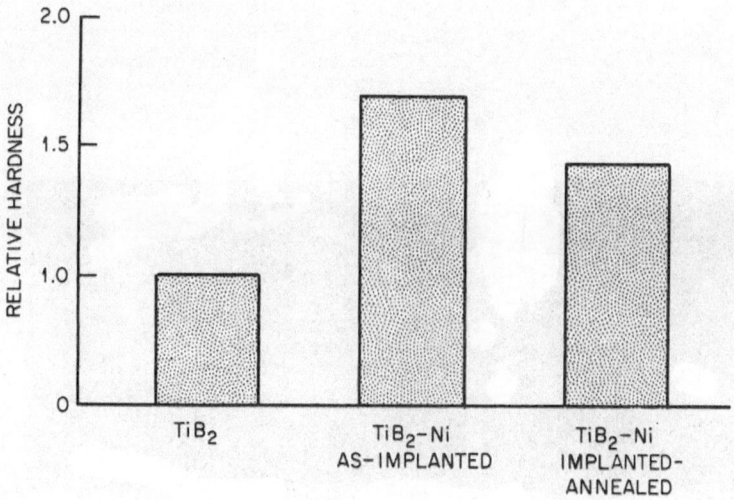

Fig. 9.   Relative hardness of TiB_2 implanted with 1 MeV Ni.

Figure 10 contains a TEM photograph which shows the amorphous surface layer on the crystalline substrate. The insert in the photograph shows the halo pattern characteristic of electron diffraction from an amorphous solid. The amorphous region measures about 2500 Å wide, corresponding to the calculated range of the Cr.[2] Energy dispersive x-ray analysis in a JEOL 100-C analytical electron microscope confirmed that Cr was present only in the amorphous layer.

The hardness was unchanged by the implantations for both the polycrystalline and the single crystal specimens. Likewise, values for the apparent fracture toughness (K_{IC}) and specific energy of material removal were within the scatter of the data for both implanted and unimplanted surfaces.

Surface profilometry detected a large (700 Å) step height in moving from the unimplanted region to the implanted area. Since this increase in volume is much greater than that occupied by the injected ions, this amorphous layer had a much lower density than crystalline SiC. Scanning electron microscopy gave evidence that there were regions of cracking at the free surface.

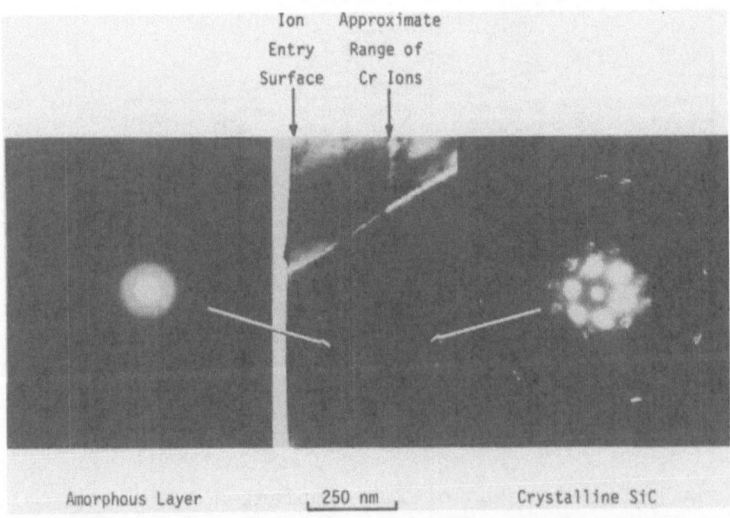

Fig. 10.   Transmission electron micrograph and electron diffraction pattern of Cr-implanted SiC. The implanted region is amorphous.

SUMMARY

It has been demonstrated that ion implantation can be used to form both stable and non-equilibrium surface alloys on ceramics. The amount of lattice damage appeared to be related to the degree of covalent bonding, the higher the covalency, the greater the amount of disorder introduced.

Mechanical properties of Al_2O_3 and TiB_2 were increased by 20 to 60% by implantation. The annealing experiments provided a basis for separating "alloy" effects from "damage" effects. The alloy effects in Al_2O_3 were stable upon annealing to at least 1200°C whereas much of the implantation damage was removed at this temperature. Although α-SiC was made amorphous by the implantation, its mechanical properties (microhardness and indentation fracture toughness) were unaltered.

ACKNOWLEDGMENTS

The authors gratefully acknowledge the aid and support of a number of colleagues including V. J. Tennery and P. F. Becher for discussions on mechanical property determination, B. C. Leslie for metallographic support, L. L. Hall, S. B. Waters, and C. E. Zachary for specimen preparation, and C. S. Yust, P. S. Sklad, and J. T. Houston for transmission electron microscopy.

REFERENCES

1. J. K. Hirvonen, ed., "Ion Implantation," Vol. 18, *Treatise on Materials Science and Technology*, Academic Press, New York (1980).
2. I. Manning and G. P. Mueller, "Depth Distribution of Energy Deposition by Ion Bombardment," *Comp. Phys. Comm.* 7:85 (1974).
3. A. V. Drigo, S. L. Russo, P. Mazzoldi, P. D. Goode, and N.E.W. Hartley, "Lattice Disorder in Implanted Insulators: Pb Implantation in α-Al_2O_3," *Rad. Eff.* 33:161 (1977).
4. A. Carnera, A. V. Drigo, and P. Mazzoldi, "Atom Location in Complex Lattices: Pb in α-Al_2O_3," *Rad. Eff.* 49:29 (1980).
5. G. W. Arnold, G. B. Krefft, and C. B. Norris, "Atomic Displacement and Ionization Effects on the Optical Absorption and Structural Properties of Ion-Implanted Al_2O_3," *Appl. Phys. Lett.* 25:540 (1974).
6. G. B. Krefft and E. P. EerNisse, "Volume Expansion and Annealing Compaction of Ion Bombarded Single Crystal Al_2O_3 and Polycrystalline Al_2O_3," *J. Appl. Phys.* 49:2725 (1978).

7. T. F. Luera, J. A. Borders, and G. W. Arnold, "Studies of Radiation Damage Produced by Ion Implantation in Sapphire," in: *Ion Implantation in Semiconductors—1976*, F. Chernov, J. A. Borders, and D. K. Brice, eds., Plenum Press, New York (1976).

8. G. B. Krefft, W. Beezhold, and E. P. EerNisse, "Effect of Ionizing Radiation on Displacement Damage in Ion-Bombarded Single Crystal α-Al_2O_3 and α-SiO_2," *IEEE Transactions on Nuclear Science*, NS-22:2247 (1975).

9. B. D. Evans, H. D. Hendricks, F. D. Bazzarre, and J. M. Bunch, "Association of the 6-eV Optical Band in Sapphire with Oxygen Vacancies," *Ion Implantation in Semiconductors—1976*, F. Chernov, J. A. Borders, and D. K. Brice, eds., Plenum Press, New York (1976).

10. H. M. Naguib, J. F. Singleton, W. A. Grant, and G. Carter, "Lattice Disorder in Alumina Single Crystals Produced by Ion Bombardment," *J. Mat. Sci.* 8:1633 (1973).

11. E. P. EerNisse, "Sensitive Technique for Studying Ion-Implantation Damage," *Appl. Phys. Lett.* 18-581 (1971).

12. R. R. Hart, H. L. Dunlap, and O. J. Marsh, "Disorder Produced in SiC by Ion Bombardment," *Rad. Eff.* 9:261 (1971).

13. V. V. Makarov, T. Tuomi, K. Naukkarinen, M. Luomajarri, and M. Riihonen, "Laser Induced Recrystallization and Defects in Ion Implanted Hexagonal SiC," *Appl. Phys. Lett.* 35:922 (1979).

14. R. B. Wright and D. M. Gruen, "Raman Scattering Study of Ion Bombardment Induced Amorphization of SiC," *Rad. Eff.* 33:133 (1977).

15. R. R. Hart, H. L. Dunlap, and O. J. Marsh, "Backscattering Analysis and Electrical Behavior of SiC Implanted with 40 keV In," *Ion Implantation in Semiconductors*, I. Ruge and J. Graul, eds., Springer-Verlag, Berlin (1971).

16. A. B. Campbell, J. Schewchun, D. A. Thompson, J. A. Davies, and J. B. Mitchell, "N Implantation in SiC: Lattice Disorder and Foreign Atom Location Studies," *Ion Implantation in Semiconductors*, Plenum Press, New York (1975).

17. R. H. Marion, "Use of Indentation Fracture to Determine Fracture Toughness," *Fracture Mechanics Applied to Brittle Materials*, S. W. Freiman, ed., American Society for Testing and Materials, Philadelphia, PA (1979).

18. A. G. Evans, "Fracture Toughness: The Role of Indentation Techniques," *Fracture Mechanics Applied to Brittle Materials*, S. W. Freiman, ed., American Society for Testing and Materials, Philadelphia, PA (1979).

19. A. B. vanGroenou, N. Maan, and J.D.B. Veldkamp, "Scratching Experiments in Various Ceramic Materials," *Philips Res. Rep.* 30:320 (1975).

20. A. B. vanGroenou and J.D.B. Veldkamp, "Grinding Brittle Materials," *Philips Tech. Rev.* 38:105 (1978/79).

21. C. J. McHargue, B. R. Appleton, H. Naramoto, C. W. White, and J. M. Williams, "The Structure of Chromium-Implanted Al_2O_3," Materials Research Society Annual Meeting, November 1981.

22. R. C. Bradt, "Cr_2O_3 Solid Solution Hardening of Al_2O_3," *J. Am. Ceram. Soc.* 50:54 (1967).

23. B. B. Ghate, W. C. Smith, C. H. Kim, D.P.H. Hasselman, and G. E. Kane, "Effect of Chromia Alloying on Machining Performance of Alumina Ceramic Cutting Tool," *Ceram. Bull.* 54:210 (1975).

IMPROVEMENT OF WEAR RESISTANCE IN CEMENTED TUNGSTEN CARBIDE BY ION
IMPLANTATION

G. Dearnaley

N.P.Division
AERE Harwell
Didcot, England, OX11 ORA

INTRODUCTION

Cobalt and tungsten are among the few most important strategic materials in the West. For this reason, any process which produces a large increase in the life of cobalt-cemented tungsten carbide tools is to be welcomed, particularly when the economic justification is good.

The process of ion implantation [1] consists of the injection of energetic atoms into the surface layers of a material. It is carried out in vacuum by accelerating ions to energies generally around 100 keV. During the past decade good progress has been made in the use of this technique for the improvement of mechanical properties of steels, titanium and other alloys[1]. This paper deals with the present status of our tests and understanding of the behavior of ion-implanted Co-cemented tungsten carbide.

Equipment for carrying out the process under industrial conditions has now been developed, on the basis of this work at Harwell, and the first machines are due to be installed this year (1981). Up to a thousand wire-drawing dies per day can be implanted, and larger facilities are now coming into operation. The paper will include a brief description of these machines, which differ radically from those developed for semiconductor device implantation.

RESULTS

In contrast with the results of tests made under industrial conditions, several previous laboratory tests on nitrogen-implanted

cemented tungsten carbide have failed to show any improvement in
mechanical properties, and the reasons for this help us to understand
why this is not always the case.

For example, the standardised test which consists of bringing
into contact a rapidly rotating wheel of brass or steel[2] has failed
to show any reduction in size of the wear scar as a result of ion
implantation. However, since the duration of the test is only a few
minutes it is easy to understand that migration and segregation of
implanted nitrogen may not be rapid enough to take effect.

Other accelerated wear tests under abrasive conditions against
silicon carbide grit were equally unsuccessful[3], and this test was
also relatively short in duration.

Much more gentle abrasive wear tests have been carried out by
Charter and Minter[4] using an ultra-fine suspension of 500 Å alumina
particles in water, in a vibratory polisher. The quantity of material
removed over several days of abrasion was determined by microscopic
examination of Knoop diamond indentation marks punched into the
original polished surface of carbide. The length of each indentation
was measured as a function of time, and no difference was observed as
a result of the standard nitrogen implantation, i.e. 4.10^{17} N^+ ions/
cm^2 at about 100 keV energy.

Dearnaley and Charter[5] reported the first laboratory test
procedure to show a very significant improvement in wear resistance
after nitrogen implantation. This was a pin-on-disc test in which
a loaded 0.5 mm diameter Co-cemented WC pin rested on a rapidly
rotated disc of the same material. A flow of light hydrocarbon
spirit ("white spirit", a paint thinner) was used to carry away wear
debris and to provide a mild coolant. The wear depth, i.e. the down-
ward displacement of the pin and the horizontal frictional force were
measured by transducers and recorded continuously. The material for
these tests was 6% Co, 94% WC with a grain size about 1 μm supplied
by Wimet Ltd., Coventry, England (Wimet 'N' grade) polished smooth
with 0.25 μm diamond dust.

In the unimplanted carbide the initial friction coefficient is
high. At sliding speeds of 1 m/sec and loads of 50 - 100 N on the
pin a transition occurs after typically 20 hours of testing: the
frictional force drops to a value approximately half its original
magnitude, but this condition is unstable and there are many upward
excursions in friction. Often these are preceded by a brief
increase in friction, which may be due to localized adhesion.
Figure 1 shows an example of the friction trace at the start, and
Figure 2 that after many hours of testing.

When the disc has been implanted with nitrogen (4.10^{17} ions/cm^2
at 90 keV) the behavior is very different. The transition to a low-

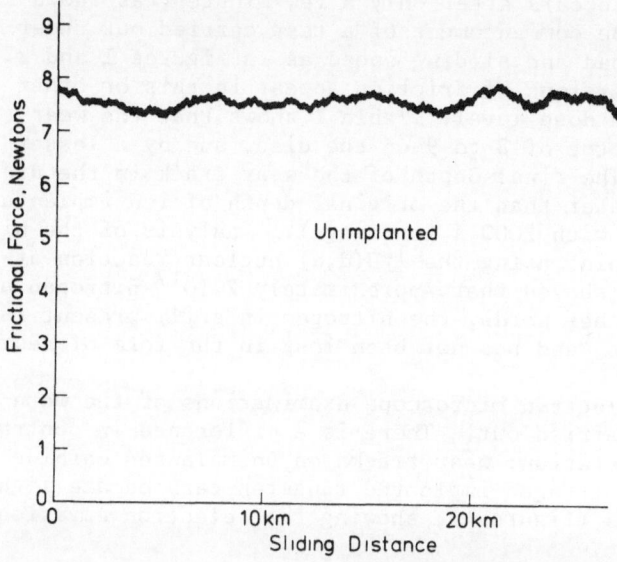

Fig. 1.   Commencement of a pin-on-disk friction test of cemented
          tungsten carbide with a similar material for both pin and
          disk.   The load was 75 N, pin diameter 0.5 mm, and sliding
          speed 1 m/sec.

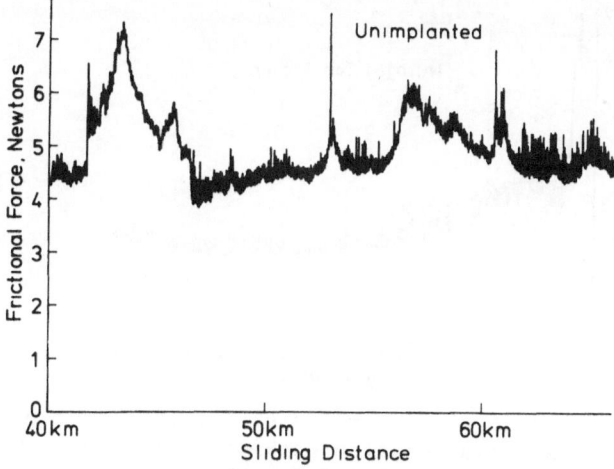

Fig. 2.   A later stage in the pin-on-disk testing of Co-cemented
          tungsten carbide, under the same conditions as in Fig. 1.

friction regime occurs after only a few minutes, as shown in figure
3 which shows the commencement of a test carried out under the same
conditions of load and sliding speed as in figures 1 and 2.   No
significant reversions of friction appear in this or other such tests
at this nitrogen dose level.   Table I shows that the wear rate is
reduced by a factor of 8 to 9 on the disc, and by a lesser factor on
the pin also.   The final depth of the wear track in the disc was
appreciably greater than the original depth of ion implantation (about
6000 Å compared with 2000 Å ion range).   Analysis of the nitrogen at
the tip of the pin, using the $^{14}N(d,\alpha)$ nuclear reaction at 2.4 MeV
deuteron energy showed that approximately 2.10^{17} nitrogen atoms/cm^2
remained.   In other words, the nitrogen is still present in the two
carbide surfaces, and has not been lost in the form of debris.

Scanning electron microscope examinations of the worn carbide
surfaces were carried out.   There is a difference in contrast as a
result of implantation: wear tracks on unimplanted carbide are dark,
while those on nitrogen-implanted tungsten carbide are lighter than
the unworn areas (figure 4), showing that electron emission levels
have increased.

A closer examination shows how little wear has taken place on
the nitrogen implanted cemented carbide disc (figure 5), the original

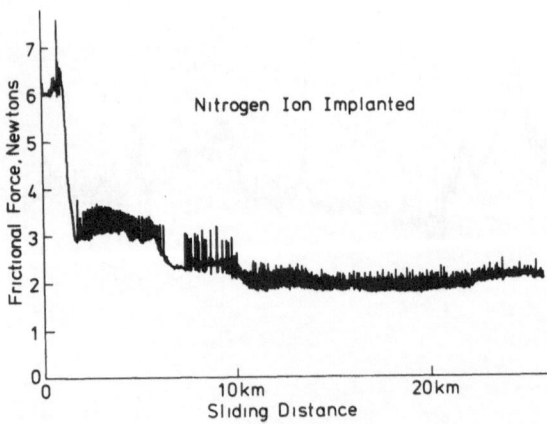

Fig.3.   The commencement of a pin-on-disk friction test of ion
         implanted cemented tungsten carbide, under the same
         conditions as in Figs. 1 and 2.   The disk received
         4.10^{17} N$^+$ ions/cm^2 at 100 keV.

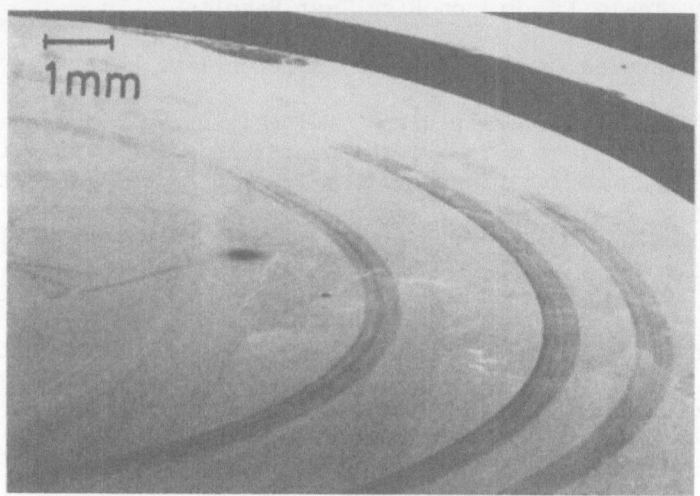

Fig. 4(a). Pin-on-disk wear tracks on unimplanted Co-cemented tungsten carbide. SEM photograph taken at 70° tilt angle.

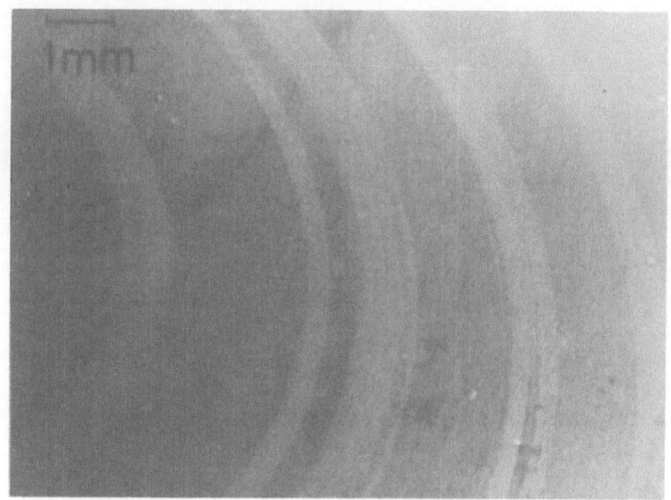

Fig. 4(b). Pin-on-disk wear tracks on Co-cemented tungsten carbide, ion implanted with 4.10^{17} nitrogen ions/cm^2. Tilt angle 50°. Note that the contrast in this case is opposite to that in Fig. 4(a), indicating differences in electron emission.

Table 1. Pin on Disc Test Results

Cemented Tungsten Carbide (6% Co), like on like Materials

	Volumetric disc wear rate x 10^9 cm^3/cm pin travel	Volumetric pin wear rate x 10^9 cm^3/cm pin travel	Average friction force, Newtons
Untreated Discs (1)	5.9 ± 0.5	5.5 ± 0.7	21
(11)	6.4 ± 0.5	–	22
Implanted Disc	0.7 ± 0.3	1.5 ± 0.5	10

Test Conditions:-   0.5 mm diameter cylindrical untreated pin
19.3 mm diameter wear track
2000 rpm rotational speed
100 Newton pin load
White spirit coolant

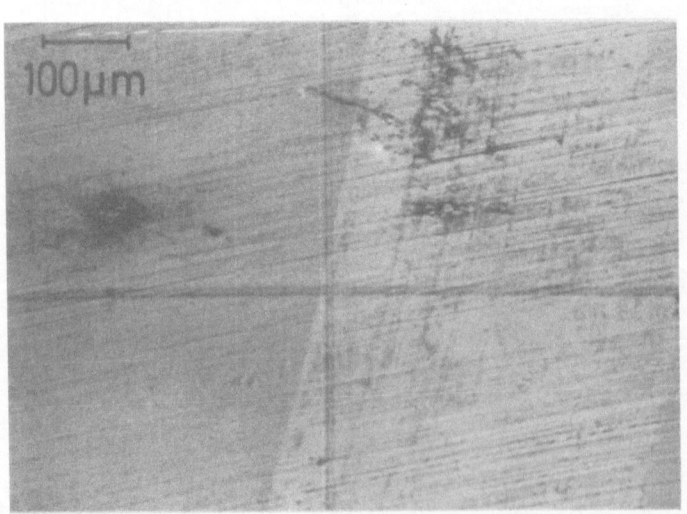

Fig. 5. Wear track in nitrogen-implanted cemented tungsten carbide.
Note that the original surface grinding marks are still
visible, since so little wear has taken place.

polishing marks being visible right across the wear track, with a noticeable absence of grain pull-out. By contrast, the unimplanted disc tested under similar conditions shows none of the original surface finish and grooving and grain removal are both evident (figure 6).

Longitudinal sections were cut from the wear tracks and examined (figure 7) but apart from the surface grooving in the unimplanted case there was little difference in the sub-surface zones. There has been no change in the carbide grain structure.

In one tungsten carbide disc which received a lower nitrogen dose (about $2.10^{17}/cm^2$) there was an interesting friction behavior (figure 8). After a very rapid initial drop in friction the frictional force continued to fall until a series of periodic increases in friction occurred, at about 5 minute intervals, over a 6 hour span. For a brief period, intense stick-slip adhesion occurred, only to revert again to the relatively smoother frictional behavior. Both this and other traces (figures 2,3) show minor stick-slip effects that are absent in the high-friction state that is characteristic of unimplanted carbide.

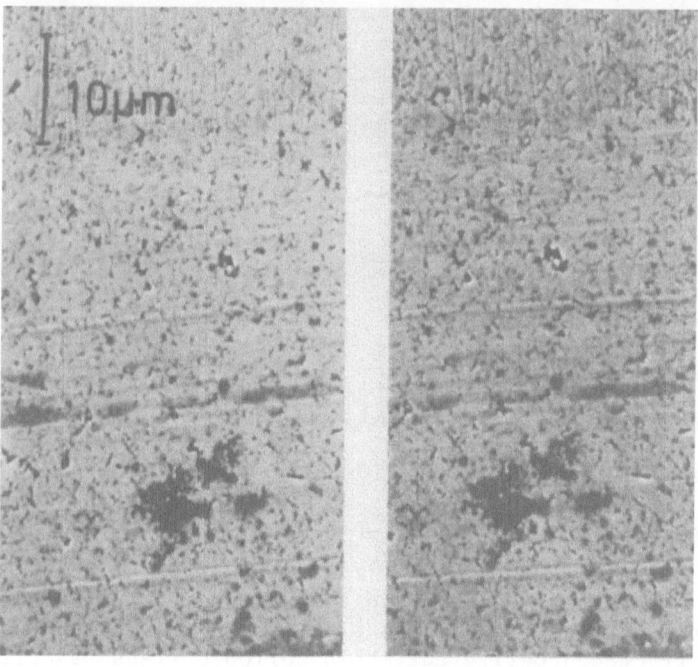

Fig. 6. Stereo pair SEM photograph of the wear track in unimplanted Co-cemented tungsten carbide. The original surface finish has disappeared, and grooving and grain detachment are visible in the track.

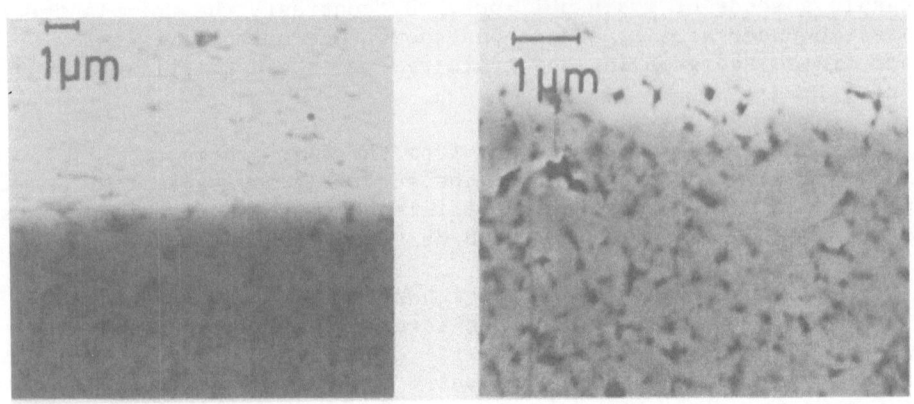

Fig. 7.   SEM photographs of longitudinal sections cut from wear
          tracks in (left) unimplanted and (right) nitrogen implanted
          WC/Co. Each had received a similar exposure to wear.

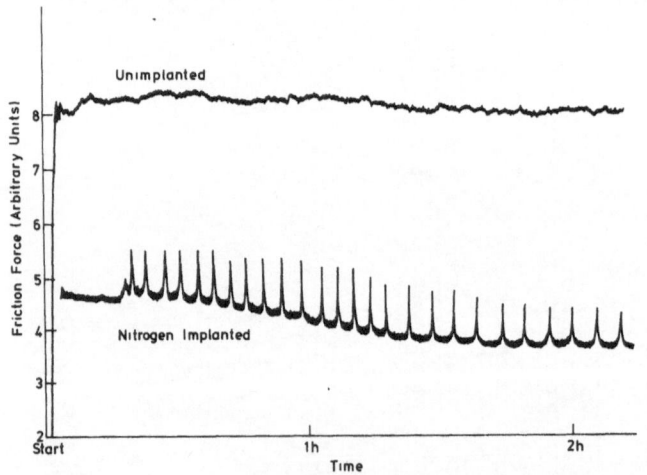

Fig. 8.   Initial part of the wear test on a tungsten carbide disk
          that received a lower dose of 2.10^{17} nitrogen ions/cm^2.
          The friction coefficient shows periodic rises associated
          with brief periods of stick-slip adhesion.

MECHANISMS

In the case of steels there is now a reasonably good understanding of how the implantation of light interstitial atoms such as
nitrogen, carbon or boron brings about improved resistance to wear
and fatigue[6]. These atoms have the property of segregating, at
relatively low temperatures, to dislocations forming so-called
Cottrell atmospheres[7]. This hampers the movement of the dislocations
which consequently renders the surface harder and resistant to fatigue
and the fatigue-like mechanisms which precede the formation of a
particle of wear debris. This property of interstitial impurities
has long been recognized as the cause of strain ageing in steel[7].

Under conditions of sliding wear, two other mechanisms occur,
which may well account for the persistent effects of ion implantation,
observable well beyond the stage at which the original and shallow
implanted layer has been eroded away[6]. The first is the production
and forward propagation of fresh dislocations by the load exerted at
contacting asperities, which must deform plastically until they
support the load. A dense dislocation network is generated ahead of
the worn surface, and this may drag forward associated impurity atoms,
according to the ideas of solute drag described by Haasen[8]. The
second important effect is frictional heating, which occurs at the
same time, and at the same small regions on the surface. Rowson and
Quinn[9] have demonstrated that under typical conditions local hot-spot
temperatures may reach 600° - 700°C. Pipe diffusion of impurities,
driven by the enormous concentration gradient arising from the non-
equilibrium ion implantation process, is then very likely to occur.
Our present experiments in steels are aimed at distinguishing the
relative importance of solute drag and thermally-activated pipe
diffusion.

The coefficient of friction is also reduced by ion implantation
into steel and this may be due to two effects:- (i) junctions formed
between the two surfaces will be more brittle, due to lessened
dislocation movement, and there is evidence that they break off to
leave a smoother, burnished surface; and (ii) an oxide film is more
likely to be retained under these circumstances and the presence of
the oxide may reduce metal-to-metal adhesion at the interface.

However, when we consider the composite material, cobalt-
cemented tungsten carbide, the situation is very different and even
more complex. Quite different factors seem to be important, and it
may be argued that ion implantation has a very important role to
play in furthering our understanding of these mechanisms (just as it
has proved to be of value in the field of thermal oxidation[10]).

The first difference is that cobalt has little chemical affinity
for nitrogen or carbon, unlike iron, and no stable compounds exist.
The nitrogen or carbon introduced by ion implantation is therefore

likely to be in an atomically-dispersed and non-equilibrium form, and it is to be expected that its properties will be unusual.

Since the material is a composite, we must first determine whether the implantation process affects mainly the carbide grains, or the cobalt binder, or possibly the interface between the two. This is by no means certain as yet, but the indication is that the dominant mechanisms occur within the binder phase. The influence of ion implantation again persists to depths which are large compared with both the ion range (<< 1 μm) and the carbide grain size (∿ 1 μm) and it is easier to conceive of a forward transport taking place within the binder matrix.

The periodic effects in friction behavior, and in particular the relaxation oscillator observed in figure 8 are evidence for a thermally-driven process in which a brief high-friction state induces a thermal effect which leads to a lower friction condition again, which is then subject to wear which eventually exposes a high-friction surface. Possible thermal effects are (i) local melting and redistribution of cobalt (ii) oxidation of cobalt or (iii) some effect in the carbide grains. This last mechanism is thought to be less likely because of the refractory nature of the material.

Homer and Dearnaley[11] found that implantation of nitrogen has no effect on the thermal oxidation of cobalt at about 800°C. However, the brighter appearance of the wear tracks in figure 4 would be consistent with the presence of oxide, since generally oxides have a greater secondary electron emission coefficient than the parent metal.

Cobalt has a low stacking fault energy, and the binder phase in WC/Co contains significant quantities of tungsten which further lowers the SFE[12]. The behavior of nitrogen in cobalt is not established, but if it were to lower the SFE still more it would, according to the ideas of Hirth and Rigney[13], stimulate the formation of a thin polishing layer, or Beilby layer, of cobalt over the surface of the composite. Dislocation interactions leading to a more lamellar microstructure of strains would favour this mechanism (figure 9). Page and Roberts[14] have reported that nitrogen ion implantation into cobalt induces a significant softening, as measured by the Meyer index in diamond indentation tests. This suggests a lowering of the SFE, allowing the metal to deform by extensive microtwinning.

Another consequence of the tungsten present within the binder phase is to raise the martensitic transformation temperature of cobalt, for the hcp → fcc transition. If the high concentration of nitrogen serves to raise still further this transition temperature the effect would be beneficial in terms of wear, because hcp cobalt has a near-perfect c/a ratio and a tendency to orient for basal slip parellel to the wear surface. Further work needs to be done before this can be established.

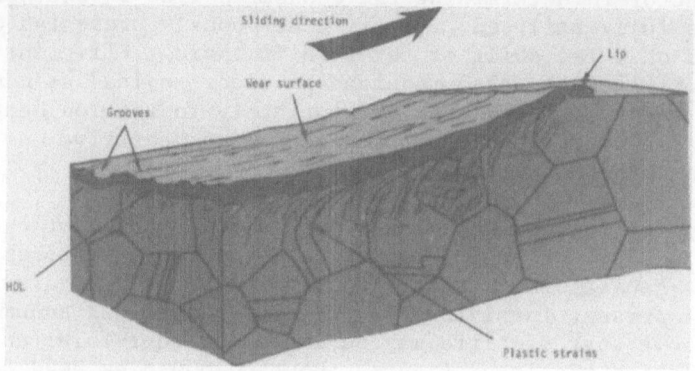

Fig. 9.   A schematic illustration of the highly deformed layer (HDL)
or Beilby layer formed by polishing a metal, particularly
those with low stacking fault energies, such as cobalt.

If the result of nitrogen is to favour the production of a very
thin polishing layer of cobalt across the surface of the composite,
then the lowered friction condition is explained. As this layer
wears away, local increases in friction may occur, which in turn
would lead to a softening of the binder and a replenishment of the
film. Tests to establish this by means of Auger electron spectroscopy
are now under way.

In recent work on the wear of martensitic stainless steel
against cemented tungsten carbide at temperatures up to $500^{\circ}C$,
Sullivan and Petraitis[15] have observed significant reductions in
wear rate above about $400^{\circ}C$. Analysis of wear debris showed the
presence of cobalt oxides (Co_2O_3 and Co_3O_4) and WO_3 under these
conditions, and the authors conclude that these oxides may form
favorable bearing surfaces. Frictional heating may lead to surface
oxidation, but it is not clear what influence nitrogen or carbon
implantation may have on this mechanism, unless a mixed oxynitride
is produced with superior bearing properties. No information
regarding such phases has been found, and X-ray diffraction
experiments similar to those of Sullivan and Petraitis are necessary.

THE ECONOMICS OF THE PROCESS

Ion implantation is a capital-intensive process and so an

important factor in determining the economics lies in whether the
equipment is fully utilised, i.e. is continuously presented with work
and operated on a two-shift or three-shift basis. Efficient work-
handling facilities are also required in order to load as many items
as possible and to present them appropriately to the ion beam. Under
these favorable conditions, the process is a competitive one, as a
few examples can demonstrate.

     The capital cost of the plant, with a 60 cm cube workchamber,
is around $200,000. A useful criterion is how long it takes to
recover this capital investment. Consider first a wire-drawing
operation at present involving 240,000 die changes per annum. We
have seen above that die life may be increased four-fold, and often
six-fold. Taking the lower figure, about 60,000 dies need to be
implanted each year and this is within the capability of the machine
if operated on a 3-shift basis. Assuming the cost of a new 28 mm
overall diameter die to be $10, and that it is repolished seven
times at a cost of $1 each time the annual expenditure on dies will
be $127,500 compared with $510,000 at present, i.e. a saving of
$382,500 p.a. The operating cost of the implantation facility would
be around $150,000 p.a. and the capital outlay would be recoverable
within the first year.

Fig. 10. A set of Co-cemented tungsten carbide wire dies
          successfully treated by nitrogen ion implantation. The
          life of such dies can be extended by factors of 4 to 6.

Fig. 11.   Punch and die set of a type that shows a considerable
           increase in life as a result of nitrogen ion implantation.
           It is reported that this improvement persists throughout
           successive regrinds of the end face of the punch.

Solid tungsten carbide drills retail for about $2 each and can
be implanted in batches of 1000 or more at a time for a unit cost of
less than 40¢. Since the life of the drills is doubled, this is
economic both for the user and for the supplier of premium tools.

Precision tungsten carbide punches and dies (figure 11) are much
more expensive and since a life increase by a factor of six has been
reported the economic benefit is even more favorable. The cost of
nitrogen ion implantation is probably not more than 30 per cent of
the cost of such a tool, even on a one-at-a-time treatment basis.

The reduction in consumption of cemented tungsten carbide for
the above tools could be significant, since major increases in life
are reported for relatively large tools.

The potential benefits for users of cemented tungsten carbide
tools, within the range of acceptable tool surface temperatures, are
likely to be considerable. This is now a significant driving force
towards a better understanding of how it is that ion implantation
can improve the life of this material.

EQUIPMENT

Two implantation machines have been developed at Harwell for
the treatment of metal and carbide tools, and these are the proto-

types of equipment now being manufactured commercially[*] under licence.

Unlike the expensive and complex machines that have been developed for the implantation of semiconductor devices, these machines have no magnetic analysis of the ion beam and the entire output of the ion source is beamed downwards towards the work. This helps to make the equipment easy to operate by unskilled staff. Other features are:-

   (i)   relatively intense beam currents combined with a long ion source life;
  (ii)   large beam areas at the work-piece surface;
 (iii)   versatility in regard to work handling, enabling the treatment of heavy single items or large arrays of small ones;
  (iv)   reliability, combined with easy maintenance of ion source and high voltage supplies;
   (v)   compact design, economising on floor space;
  (vi)   safety at all times eg as regards X-ray emissions and electrical hazards.

Figure 12 shows the smaller machine, to which the name PIMENTO has been given (prototype implantation machine for engineering tools). This delivers up to 5 mA of gaseous ions, such as nitrogen, into an area of about 100 cm^2 within a work-chamber with a floor size of 45 cm x 45 cm. The operating pressure is about 5.10^{-5} torr, and the vacuum system is automatically operated from a single 'start' button. The high voltage unit, also developed at Harwell, is of the Cockcroft-Walton type and is very resistant to surges or overload. Over the past four years the equipment has performed with a high degree of reliability, being available for use approximately 90 per cent of the time. The useful life of the plant is also very long, especially by comparison with that used for high temperature thermochemical treatments, eg boron diffusion. The present total operating cost is about $80 per hour, and the equipment has been used for the commercial treatment of many tools and components, besides providing a test facility for further ion source developments.

A much larger facility is now in operation (figure 13) in order to treat larger components, such as moulding tools nearly 1 m across, and to explore the problems and benefits of a full scale implantation process. The work chamber is 2.5 m in diameter and 2.5 m long, evacuated by an Edwards 50 cm oil diffusion pump and Roots pump, aided by a liquid nitrogen cooled cryogenic baffle. This combination achieves a base pressure of 10^{-7} torr, but with gas supplied to the large ion source the operating pressure is about 3.10^{-5} torr. At present the maximum energy available is 60 keV, shortly to be uprated

------

[*] by Hawker Siddeley Dynamics Engineering Ltd., Manor Road, Hatfield, UK.

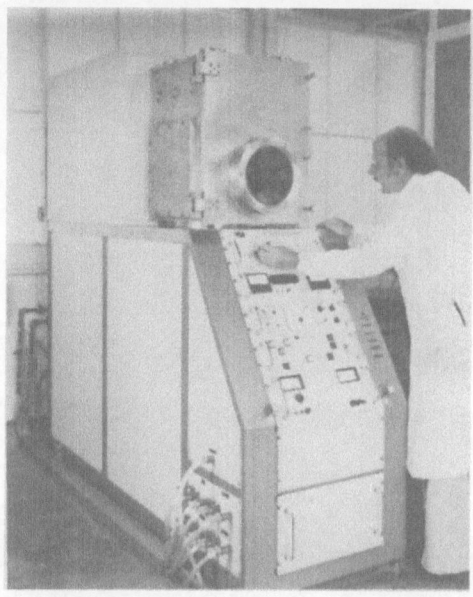

Fig. 12. 100 keV industrial implanter developed at Harwell for the implantation of gaseous ions.

to 100 keV with an appropriate isolation transformer. The design beam current is 25 mA, and about 10 mA is achievable at this stage over a total beam area of approximately 1000 cm^2. The figure shows one piece of heavy duty work-handling equipment, and other items e.g. for the rotation of cylindrical components are being installed. The machine runs with the minimum of supervision, i.e. it is possible to switch it on in the morning and leave it unattended for the rest of the day.

In the area of manufactured equipment, designed for the ion implantation of large numbers of identical components, it is possible to envisage dedicated machines designed for the ion implantation of many identical items at the rate of about 8 cm^2 per minute. Examples of tools and components for which this represents an economically attractive process have been given by Charter et al. in a recent paper[16]. They include wire drawing dies, punch and die sets, back extrusion punches, solid carbide drills and slitting knives. In some cases we are already treating the entire output of UK companies with a single prototype machine, and jobbing facilities available on a commercial basis are to commence early in 1982. Perhaps the only widely-used tungsten carbide tool for which the process would be economically unattractive is also the one for which ion implantation is technically unsuccessful due to the high temperatures involved: this is the throw-away cutting tip.

Fig. 13.   Large ion implantation facility at Harwell, 2.5 m long by
           2.5 m in diameter.   The beam area is 1500 cm^2.

ACKNOWLEDGEMENTS

     The author wishes to thank his colleagues, S.J.B.Charter,
P.D.Goode, F.J.Minter and R.E.J.Watkins for discussions and
permission to reproduce results.   He is most grateful to Steve
Roberts for the SEM photographs of worn tungsten carbide specimens.
Some of this work was funded by the Engineering Materials
Requirements Board of the UK Dept. of Industry.

REFERENCE

1.   G.Dearnaley, J.H.Freeman, R.S.Nelson and J.Stephen "Ion
         Implantation" North Holland Publishing Co., Amsterdam (1973).
2.   S.Harper, British Non-Ferous Metals Technology Centre, priv.
         comm. (1977).
3.   W.Bettles, Wimet Ltd., priv.comm. (1978).
4.   S.J.B.Charter and F.J.Minter, AERE, unpublished work (1980).

5.   G.Dearnaley and S.J.B. Charter, Proc.Int.Conf.on New Frontiers
        in Tool Materials for Metal Cutting & Forming, London,
        Engineers' Digest, 120 Wigmore St. London (1981).
6.   G.Dearnaley, Radiation Effects, to be published (1982).
7.   A.H.Cottrell, "Introduction to Metallurgy" E.Arnold, London
        (1973), p.383.
8.   P.Haasen, "Physical Metallurgy" Cambridge University Press,
        Cambridge (1978).
9.   D.M.Rowson and T.F.J.Quinn, J.Phys.D. 13:209 (1980).
10.   G.Dearnaley, Nucl.Instr. & Methods, 183:899 (1981).
11.   R.Homer and G.Dearnaley, AERE, unpublished work (1977).
12.   C.T.Sims, J.Met. Dec. (1969) p.27.
13.   J.P.Hirth and D.A.Rigney, Wear, 39:133 (1976).
14.   T.F.Page and S.Roberts, Proc.Conf. on Surface Modification of
        Materials by Ion Implantation, Manchester 1981, to be
        published by Pergamon Press, Oxford (1982).
15.   J.Sullivan and S.Petraitis, Proc.Int.Conf. on Wear of Materials,
        San Francisco 1981, to be published (1981).
16.   S.J.B. Charter, L.R.Thompson and G.Dearnaley, Proc.Int.Conf.
        on Met.Coatings, San Francisco 1981, to be published in
        Thin Solid Films (1981).

DISCUSSION

R. Rice:

I gather that you attribute the primary effect of the implantation
to its effect on the cobalt. The reason I make this surmise is
that we have done some initial work on ion implantation of ceramics,
for example, silcon nitride for bearing applications, and the initial results showed little or no improvement. We felt it to be
not too surprising since the increase in hardness that you can
achieve by implanting material that is already quite hard is limited.
So, am I correct in saying that it is primarily due to the effects
on the cobalt?

G. Dearnaley:

I'm very glad you asked that question. The effect of bombardment
is to introduce into the tungsten carbide a considerable excess,
10 to 20 atomic percent, of carbon and nitrogen, plus many displacements creating many vacancies in the tungsten sublattice. Now, we
know that cobalt has a tendency to migrate up vacancy concentration
gradients. In simple terms, it exchanges with other cations more
quickly than the neighboring host atom So we picture that the effects are taking place on the surface of the tungsten carbide pro-

viding very favorable sites at which the cobalt can be located and
held. We believe this initiates and stabilizes the cobalt thin
film across the surface. So the implantation has no effect on the
cobalt binder. It prepares the surface of the carbide grain by
this bombardment creating vacancies in the tungsten sublattice
which the cobalt then populates.

H. Hintermann:

This is indeed a very nice technology. I would like to ask could
the nitrogen be exchanged by boron? Could you also boride the
surface?

G. Dearnaley:

I believe you could. We haven't done this. But we have tested with
carbon and nitrogen in tungsten carbide. The effects are similar
and in the case of steel we have shown that carbon, nitrogen and
boron are equally effective. The reason why we have perhaps ignored
boron is that of the three species it's the most difficult to produce as an intense ion beam. So, for practical purposes, one could
forget it. But I would be very surprised if it behaved differently
from the other two light interstitial anions.

H. Hintermann:

Do you think that the method would be applicable to bearings? I
don't talk only about ball bearings or slide bearings, but about
instrument bearings and instrument points and watch bearings. I
think that would be a very nice application for that.

G. Dearnaley:

I would expect that to be an application which would work. I
should also mention slitting knives, dies, extrusion punches, punch
and die sets - a whole variety of tools, mainly large lumps of tungsten carbide. And it's appropriate, I'm sure, in this country to
mention that cobalt and tungsten are two of the most strategic materials in the western world, particularly in the United States.
And so any process which extends the life of large tungsten carbide
tools must be a good thing.

A STEM MICROANALYTICAL INVESTIGATION OF NITROGEN IMPLANTED CEMENTED WC-Co

J. Greggi and R. Kossowsky

Westinghouse R&D Center
1310 Beulah Road
Pittsburgh, Pennsylvania 15235

INTRODUCTION

The surface modification of materials by high fluence ion implantation is rapidly becoming accepted as a potentially important technological process. In fact, recent conferences[1,2] and publications[3] have emphasized solely this particular application of ion beam technology. Of the numerous surface properties which may be modified by ion implantation, one of the more promising applications occurs in the area of surface hardening for improved wear or fatigue resistance. Hartley[4] has recently reviewed the literature in this area and points out that the most dramatic increases in wear resistance are obtained in the cemented WC-Co composites. Although suggestions have been made by Dearnaley and Hartley[5] and Hartley[6] to account for this increased wear resistance, no microstructural evidence exists to support these conclusions. Therefore, we have instituted a microstructural and microchemical study of ion implanted cemented WC-Co composites with the ultimate goal of understanding the improved wear resistance in these materials. This paper presents the first results of this study.

EXPERIMENTAL PROCEDURE

Specimens suitable for STEM evaluation were prepared from cemented WC-Co cylinders with a convenient diameter of ≈3mm. Slices ≈0.5 mm thick were ground to ≈0.08 mm on silicon carbide papers ending with 600 grit. Final thinning was accomplished by ion milling in a Commonwealth Scientific IMMI miller using argon ions at 12 kv. The specimens were subsequently implanted with 75 keV N ions (actually 150 keV N_2^+) in a linear accelerator equipped with magnetic separation to a total fluence of 10^{17} N/cm^2 at a flux of 2 x 10^{14} $N/cm^2/$ sec. Implants were made simultaneously into both pre-thinned specimens and specimens which had not received the final thinning operation.

Range-energy and energy deposition calculations for 75 keV N into both WC and Co were performed with a modified EDEP-1 code of Manning and Mueller[7] and the results of these calculations are summarized in Table 1. The results are comparable in both cases. For WC and Co, respectively, the calculated projected ranges (Rp) are 51 nm and 73 nm, the deviations in the projected ranges (ΔRp) are 21 nm and 31 nm, and the maximum in the energy deposition profile (X_m) occurs at 24 nm and 41 nm. For 10^{17} N/cm^2 the average concentration of N in both WC and Co is ≈13at%. A convenient unit for the damage created by the nuclear energy losses is the displacement per atom or dpa. For both the WC and Co the maximum dpa is ≈30, although this calculation assumes an "average" atom for the WC phase.

TABLE 1
Relevant Parameters for 75 KeV N ($10^{17}/cm^2$) in WC-Co

	Projected Range (Rp) (nm)	Deviation (ΔRp) (nm)	Damage Peak (nm)	Maximum Damage (dpa)	Average N Conc. (at%)
WC	51	21	24	27	13.0
Co	73	31	41	35	12.5

STEM investigations were performed on a Philips 400T electron microscope equipped with a KEVEX system X-ray energy dispersive spectrometer (EDS) and GATAN electron energy loss spectrometer (EELS). Presently, the most extensive STEM investigation has been made on the reference specimens and the pre-thinned implanted specimens. To our knowledge, this is the first comprehensive study of the microstructure of implanted WC-Co composites.

EXPERIMENTAL RESULTS

A survey was made of the prominent microstructural and microchemical features of the non-implanted material to establish a reference for the implanted condition. Figure 1a shows a representative low magnification micrograph of the fine grain structure of the cemented WC-Co composite. The major microstructural features

are the darker grains of the carbide phase and the smaller, lighter
colored grains of the Co based binder phase. This binder phase· is
indicated at the position marked (A) at the junctions of three or
four WC grains. Importantly, for this study, the dislocation
density in the carbide phase is low. At the very most, a few
random dislocation segments or small loops per grain are observed
as shown in Fig. 1b. Under strong diffracting conditions or close
to a major zone axis, the binder phase exhibits the linear fault-
like structure shown in Fig. 2a. The faults run continuously
across the binder phase up to the boundary. Convergent beam
microdiffraction patterns from a number of these grains readily
revealed the cubic pattern shown in Fig. 2b indicating that this
phase is the metastable cubic form of Co. This pattern also shows
a number of weaker reflections possibly associated with the faulted
structure. Although these faults have not been characterized in
this study, they may well represent a localized cubic to hexagonal
Co transformation.

     Electron energy loss spectra from this phase readily revealed
both the Co M edge and $L_{2,3}$ edge but did not indicate the presence
of any C. However, EDS analysis showed this phase to contain
≈3at% W. The migration of point defects created by the irradiation
process can often result in chemical segregation at various sinks.
For this reason, the non-implanted reference material was checked
for any naturally occurring chemical segregation at various
boundaries. In the Co binder phase the W concentration was
generally constant within a distance of ⪰25 nm from a carbide-Co
interface. At closer positions to the WC-Co boundary, the measured
concentration of W in the Co phase increased, but the results were
so inconsistent and scattered that it was concluded that the
adjacent WC phase was influencing the results. A similar situation
occurred when attempting to determine the Co concentration in the
WC phase immediately adjacent to the boundary. Therefore, segregation of either Co or W to this boundary could not be determined.

     EDS analysis of the bulk carbide phase showed the Co concen-
tration in this phase to be ≈2at% Co (relative to W, i.e., C not
included) and the minimum level of detectability of Co in this
phase for the counting rates and times employed· was determined to
be ≈0.8-1.0at% Co. In two instances where WC-WC boundaries were
checked for Co segregation, a slight enhancement of the Co level
was observed as shown by the Co concentration profile in Fig. 3a;
however, a larger number of WC-WC boundaries would have to be
analyzed to determine the consistency and reliability of these
results. Co concentration profiles extending into the WC phase
from the Co phase are shown in Fig. 3b for three separate traces.
Although there is a slight indication that the Co concentration
increases somewhat in the close proximity of the boundary, this
result cannot be considered significant in view of the Co phase.

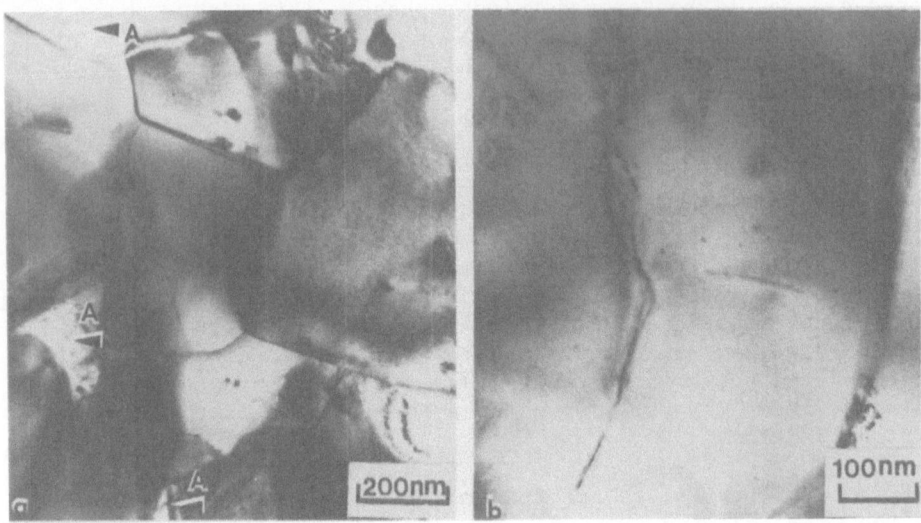

Fig. 1.   TEM micrographs of non-implanted WC-Co showing
          (a) WC grains and Co binder at A and (b) dislocation
          density in WC.

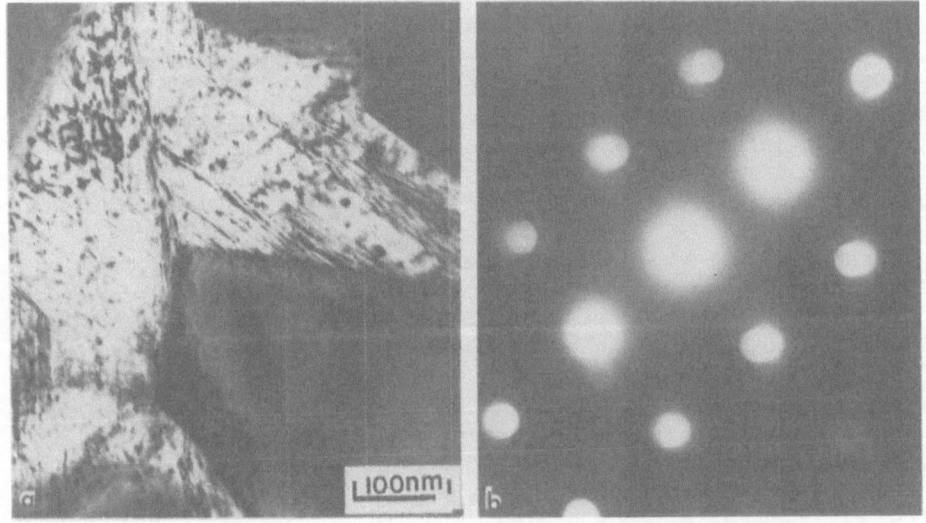

Fig. 2.   (a) TEM micrograph of faulted structure in Co phase,
          and (b) typical microdiffraction pattern from this
          phase.

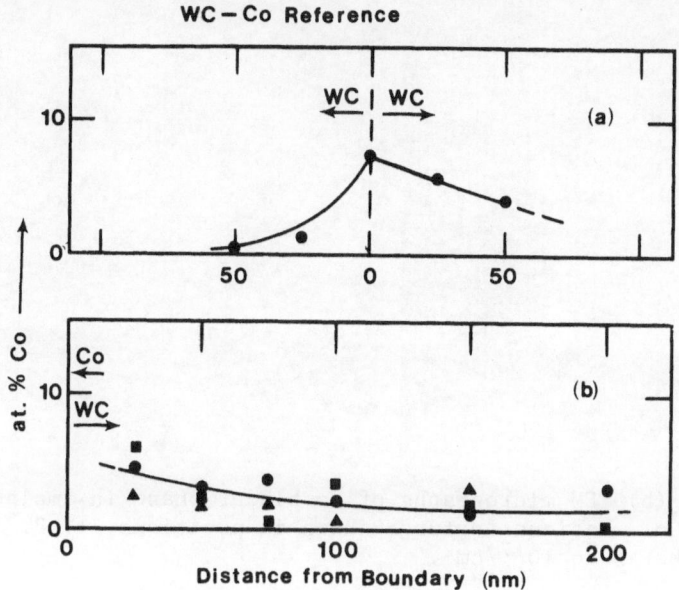

Fig. 3   Co concentration profile relative to W.   (a) across
WC-WC boundary and (b) into WC grain at WC-Co boundary.

In summary, for the non-implanted reference material, our
results show that the binder phase is essentially cubic Co,
although highly faulted, and contains ≈3at% W, and the carbide
phase is WC containing ≈2at% Co. A slight indication exists for
Co segregation to the boundaries of adjacent WC grains, but
segregation of either W or Co to the boundaries of adjacent WC-Co
phases, if any segregation exists at all, is limited to the
immediate boundary region.

For specimens implanted to $10^{17}N/cm^2$, the most complex microstructural and chemical changes occur in the Co binder phase.
Figures 4a and 4b are high magnification micrographs of this
phase taken under strong diffracting conditions. A microstructural
feature always observed in this phase is indicated at position (1)
in both figures. This feature is essentially a rim of material
≈25 nm in width completely outlining the Co phase. Tilting through
large angles fails to reveal any microstructural features or contrast changes in this rim of material. Such in insensitivity to
orientation usually implies an amorphous structure. Within the
bulk of the binder phase two microstructural features are observed.
The feature indicated at positions marked (2) in Figs. 4a and 4b
exhibits a fine fringe contrast. Although this fringe contrast

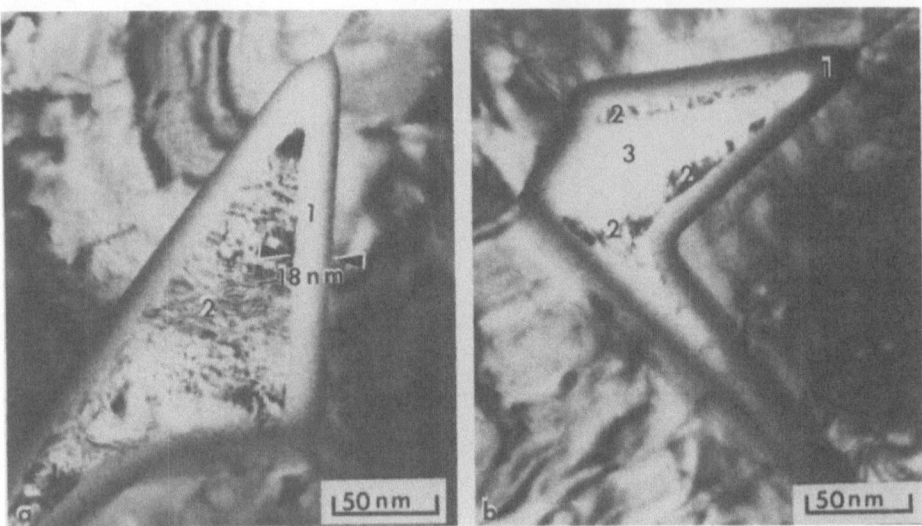

Fig. 4.   (a) (b) TEM micrographs of Co binder phase in implanted
           WC-Co. Various features shown in positions, 1, 2, 3.
           75 KeV N to $10^{17}/cm^2$.

may result from a damage induced substructure, the complexity of the
faulted microstructure in the non-implanted Co phase does not make
this conclusion obvious. This fringe structure is not always
uniform throughout the Co phase but often occurs in patches. The
remaining areas generally show considerably less contrast such as
the region marked by (3) in Fig. 4b. Figure 5 shows another of
the Co based binder grains exhibiting all of the previously
mentioned microstructural features and associated convergent beam
microdiffraction patterns from each feature. The diffraction
pattern from location (1) shows the total absence of any diffraction
spots confirming the amorphous nature of this layer as suggested
by the tilting experiments. The diffraction pattern from location
(2) in the "fringe" region of the Co phase shows the cubic form of
this phase as observed in the non-implanted condition; however,
the pattern from location (3) in the region showing less contrast
shows a different orientation from that in location (2). These
regions, therefore, are undergoing an irradiation induced phase
transformation or an orientation change. Unfortunately, the
pattern from location (3) can represent either the cubic or hcp
form of Co and we have not collected enough patterns as yet to
remove this ambiguity. The amorphous nature of the rim, however,
was confirmed on a number of different grains. In an adjacent WC
grain a crystalline pattern is again observed, indicating that
the amorphous region is localized to the rim.

     The most significant microstructural feature which is found
in the implanted carbide phase is a considerably large dislocation

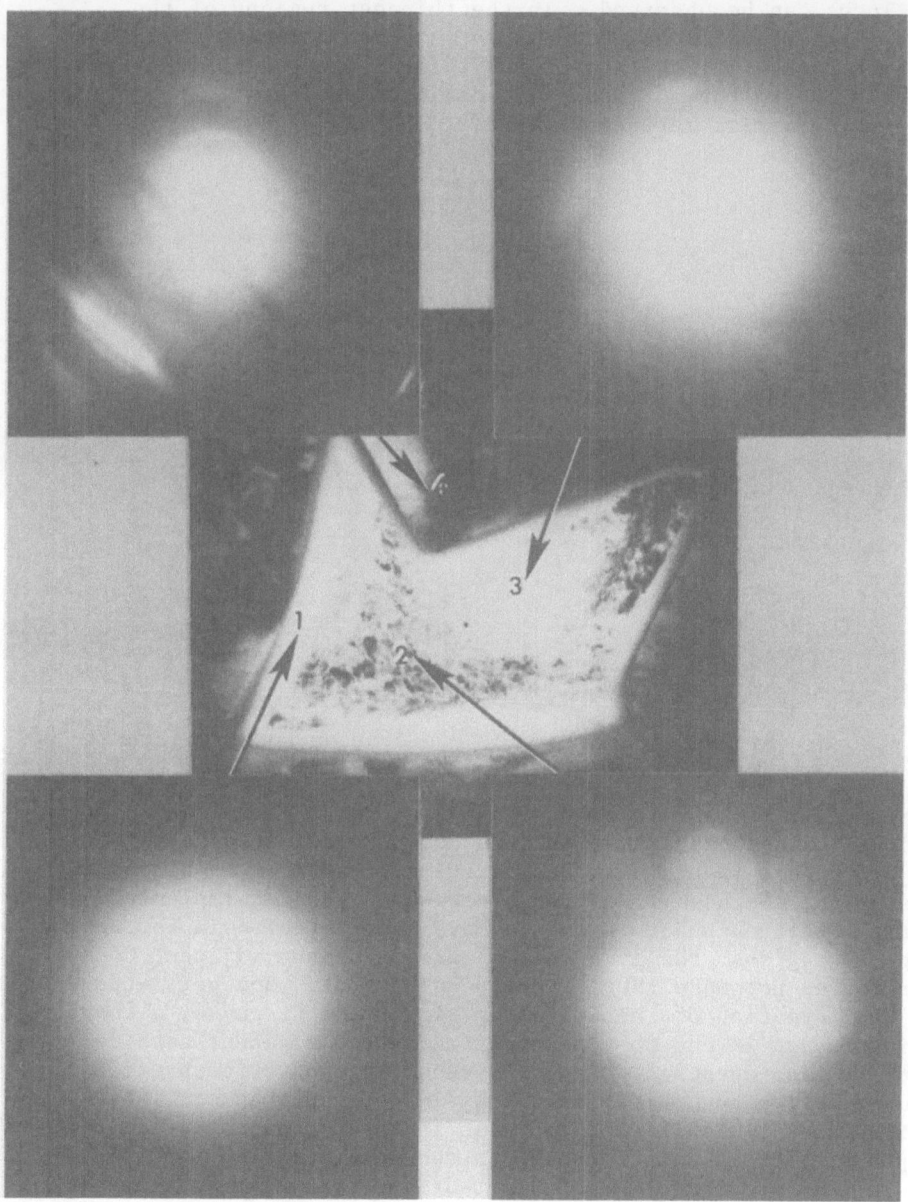

Fig. 5.   TEM micrograph of Co binder phase in implanted WC-Co
          and microdiffraction patterns from various features.

density as seen in the two representative micrographs in Figs. 6a
and 6b. This dislocation substructure consists of a uniformly
dispersed but aligned dislocation segments although occasional
small loops can be observed. In the thinnest regions of the
grains at the foil edges the density is often reduced, indicating
that either the dislocations or the point defects that generate
them have run to the surface. The dislocation densities were
estimated by a line intercept method to be $\gtrsim 10^{11}/cm^2$.

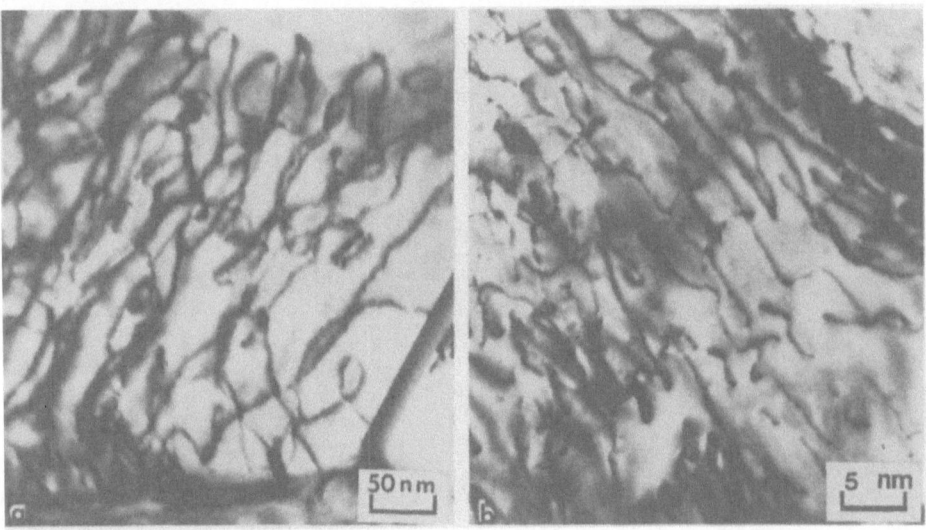

Fig. 6.   (a) (b) TEM micrographs of dislocation substructure in
          WC in N implanted WC-Co. (75 keV N to $10^{17}/cm^2$).

     Chemical analyses using EDS techniques indicated no signifi-
cant difference in the bulk chemistry of either the Co binder
phase or the WC phase from that observed in the non-implanted
material. Although Both W and Co have been detected in the narrow
amorphous regions, the immediate proximity of the adjacent Co and
WC grains do not make these results meaningful. However, attendant
with the formation of the amorphous rim in the Co phase, a corresponding change in the Co concentration in the adjacent carbide
grains is also detected. Figure 7 shows concentration profiles
of Co (relative to Co + W) extending into the WC phase. These
profiles show a significant increase in the Co concentration over
a depth of ≈ 100 nm when compared to the non-implanted condition.
It is highly probable, therefore, that the rim which develops in
the Co phase is, in part, related to the movement of Co from this
region into the WC grains. At the present we have been unable
to detect any microstructural features associated with the implanted
nitrogen.

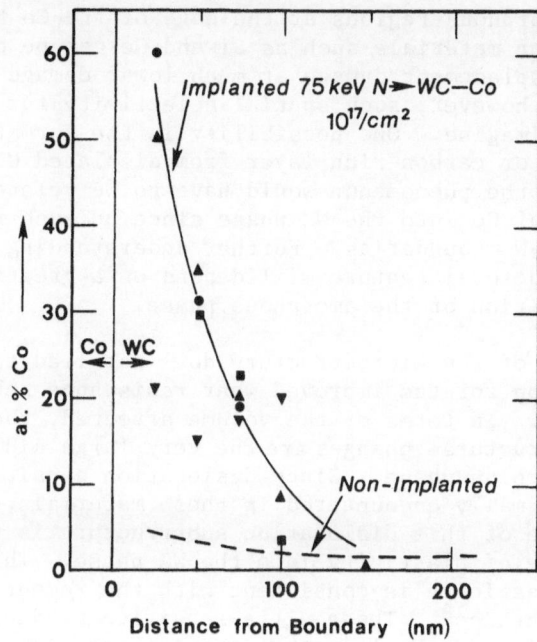

Fig. 7.   Co concentration profiles relative to W in WC phase in
          N implanted WC-Co (75 keV N to $10^{17}/cm^2$).

DISCUSSION

This study shows that a complex range of microstructural and
chemical changes occur in cemented WC-Co composites after implantation with 10^{17} N/cm^2. These physical changes encompass a range of
phenomena previously observed for the implantation process including
damage induced substructures, radiation induced or assisted phase
transformations, and solute redistribution. In the WC phase, the
damage clearly results in a dense dislocation substructure probably
through the generation, growth and intersection of dislocation
loops. In the Co binder phase, the form that the displacement
damage takes is not quite as obvious. Both microstructural observations and microdiffraction patterns show that a phase transformation is occurring in this phase. Although further characterization
is needed for this transformation, a transformation from the metastable cubic form of Co to the stable hcp form is consistent with
our results and, in fact, to be expected. One possible mechanism
for this transformation is migration of the point defects to the
faults with structural arrangements there resulting in growth of
the hcp phase.

One of the more interesting microstructural changes is the
formation of the amorphous regions at the edge of the Co binder
phase. Semiconductor materials such as Si and Ge can be driven
amorphous by the displacement damage at much lower damage levels
than employed here; however, such spatial selectivity for this
process is hard to imagine. One possibility is the formation of
an amorphous carbon or carbon-rich layer from displaced C in the
WC phase. As such, the phenomenon would have to be related to the
observed migration of Co into the WC phase since no such regions
were observed at WC-WC boundaries. Further understanding of this
particular microstructural feature will depend on a greater knowledge of the composition of the amorphous phase.

The complexity of the microstructure does not lead to an
immediate explanation for the improved wear resistance achieved
by ion implantation. In terms of the volume affected, the most
significant microstructural changes are the very large dislocation
densities in the carbide phase. Since dislocation densities of
$10^{11}/cm^2$ are not normally encountered in these materials, one
probable consequence of this dislocation substructure is the induction of some measure of plasticity into the WC phase. This dislocation induced plasticity is consistent with the recently reported
hardness tests of Pathica[8]. These ultra-sensitive hardness measurements with penetration depths less than the implantation affected
zone actually show a softening in WC-Co after implantation.
Although this implies that the tools themselves would be softer,
improved wear properties might be realized by an increased resistance to crack propagation. However, based on observations by
Blombery et al.[9] that wear of cemented Co-WC tools is initiated by
loss of the binder phase, Hartley[6] has suggested that the implantation improves the flow properties of the binder, thus improving
retention of this phase. Formation of Co-N complexes was suggested
as the potential strengthening mechanism of the Co phase. We
cannot comment here on the formation of such complexes, but a
possible cubic to hexagonal phase transformation is observed and,
of course, attendant changes in the flow properties of the binder
are expected with this transformation.

On a finer scale, there is still the chemical and structural
changes at the WC-Co interfaces to consider. Although these regions
only compromise a small fraction of the total implantation affected
zone, they may influence both the retention of the binder phase or,
possibly, the early stages of crack initiation and propagation once
this phase is lost.

SUMMARY AND CONCLUSIONS

Implantation of 10^{17} N/cm^2 into the surface of cemented WC-Co
composites induces a variety of complex microstructural and

microchemical changes. These include formation of a large dislocation substructure in the WC phase, a phase change in the Co binder
phase probably cubic to hcp Co, and formation of an amorphous layer
in the Co binder at WC-Co interfaces accompanied by a considerable
enrichment of Co in the WC phase at these locations. These microstructural and chemical changes are predominantly associated with
the damage component of the implantation process. The complexity
of the implanted microstructure does not lead to an immediate or
simple explanation for the enhanced wear resistance in the implanted
cemented WC-Co. However, the damage induced dislocation substructure in the WC phase is the major volume affect and provides an
explanation for the recently measured softening after implantation.

ACKNOWLEDGMENTS

     The authors are grateful for the assistance of W. J. Choyke
and N. J. Doyle of Westinghouse Research and Development Center in
performing the N implants and to J. A. Sutila also of Westinghouse
R and D for the range energy calculations. The assistance of
C. W. Hughes for both specimen preparation and STEM evaluation is
also acknowledged.

REFERENCES

1. C. M. Preece and J. K. Hirvonen, eds. "Ion Implantation
       Metallurgy," The Metallurgical Society of AIME, New York
       (1980).
2. International Conference on "Modification of the Surface
       Properties of Metals by Ion Implantation," UMIST, Manchester,
       U.K., 23-26 June (1981), to be published.
3. J. K. Hirvonen, ed. "Treatise on Materials Science and Tech-
       nology," Vol. 18, Ion Implantation, Academic Press, New
       York (1980).
4. N.E.W. Hartley, Tribological and Mechanical Properties, in:
       "Treatise on Materials Science and Technology," Vol. 18,
       Ion Implantation, J. K. Hirvonen, ed., Academic Press,
       New York (1980).
5. G. Dearnaley and N.E.W. Hartley, in: Proc. Conf. Appl.
       Small Accelerators, Denton, U.S.A., Oct. (1976), IEEE,
       20-29.
6. N.E.W. Hartley, in: "Surface Treatments for Protection,"
       Vol. 3 (10), Inst. Metall., London, 197-209 (1978).
7. I. Manning and G. P. Mueller, Computer Physics Communications,
       7:85 (1974).
8. J. B. Pethica, Microhardness Tests with Penetration Depths
       Less than Ion Implanted Layer Thickness, in: "Int. Conf.
       on Modification of the Surface Properties of Metals by Ion
       Implantation," UMIST, 23-26 June (1981), to be published.

9. R. I. Blombery, C. M. Perrot, and P. M. Robinson, Similarities
       in the Mechanisms of Wear of WC-Co Tools in Rock and Metal
       Cutting, Wear, 27:382 (1974).

DISCUSSION

A. T. Santhanam:

When you pre-thin your samples and then follow by ion implantation,
I wonder how relevant the structural observations are from those
that were implanted and then thinned for microscopy.

J. Greggi:

It depends on what the effect of the pre-thinning is. I can probably
name three major factors. One, it's impossible to heat-sink a pre-
thinned sample. That means there is probably a little bit of annealing going on at the same time. You now have two free surfaces
instead of one. One of them could be very close to the nitrogen
distribution and that's probably why we are losing nitrogen in conjunction with a little bit of thermal annealing. Three, and this
is probably the most important, though it's really not been discussed
that much, is that at these fluence levels you can introduce very
severe compressive strains into the surface and this can affect
the way the microstructure evolves and can affect the properties.
If you don't have the constraints of the substrate, then you probably
don't have those surface stresses. So, it depends on what you're
looking at.

A. Santhanam:

What is the role of image forces on the production of defects?

J. Greggi:

They can play a significant role in the production of defects whether
it's a pre-thinned or a non pre-thinned sample. For instance, you
can't help various components of the dislocation substructure moving
to the surface under the influence of the image forces. And I think
that's been shown quite significantly in a lot of studies on radiation damage.

G. Dearnaley:

I don't disagree with anything that was said. It's very nice to be
in this position at this early stage in research. But I would make
a point about the damage related properties. In the microstructure
he has observed that damage is produced equally by bombardment with
nitrogen, carbon, or inert species like neon. But when it comes to
the model I presented in my talk, which occurs during sliding wear,
and the necessity for favorable sites for the cobalt layer, then
there's a very big difference. We observed no significant improvement in the wear resistance of cemented tungsten carbide bombarded
with neon. So, it's not the damage process. I'm arguing, therefore,
that it's the big nonstoichiometry, the introduction of these anion
species, nitrogen and carbon, which create a great number of vacancies and defects in the tungsten sublattice which are more suitable
for the cobalt than would be otherwise. So, I think that when it
comes to wear it is much more critical what you bombard with.

J. Greggi:

I won't disagree with any of that.

R. Sinclair:

Do you know the character of the dislocations that were introduced?
You need $\langle 11\bar{2}3 \rangle$ type dislocations for slip to occur in tungsten
carbide.

J. Greggi:

No. I didn't go through with that.

P. Sklad:

You mentioned that you were not able to detect carbon with the your
EELS measurements, in the unimplanted case. What about in these
cobalt grains after implantation?

J. Greggi:

There is a slight problem there and I think Geoff Dearnaley will back
me up. It's almost impossible in any implantation system at the
moment to get rid of carbon. Even if you work with a nice laboratory
system at 10^{-7}, you can't get rid of carbon on the surface from just
smashing it onto the surface. I did see carbon after implantation
but I don't know where it comes from.

EROSION BEHAVIOR OF HARD SURFACE COATINGS/INSERTS

Alan V. Levy and Thomas W. Bakker

Lawrence Berkeley Laboratory
Materials and Molecular Research Division
Berkeley, CA 94720

INTRODUCTION

The wear resistance requirements of some of the components in the emerging energy systems necessitates the use of hard materials of the refractory hard metal family, i.e., carbides, nitrides, borides, silicides, to serve at the wear surface. They are used either as deposited coatings on structural metal surfaces or as separately fabricated inserts that are assembled into a structural metal retaining area. There has been considerable study of the wear behavior of carbides, nitrides and borides in rubbing and sliding wear and in abrasive wear. However, there has been very little research conducted to determine their resistance to wear by erosive particles directed at the surface by a gas stream. In several of the newer energy conversion and utilization systems, particularly those that use coal, the mechanism of erosive wear is an active one that must be addressed.

The purpose of this work was to determine the basic erosion behavior of several of the most promising refractory hard metal coatings and bodies that are currently either in development or commercial use. A representative group of materials was obtained from a few of the suppliers of hard surface materials and tested at room temperature in an air blast tester. The materials selected were meant to be a sample and not a definitive representation of all of this type of material available. The tests were done at room temperature only to establish an initial basis for understanding the nature of the erosion process and not to attempt to simulate any regime of service conditions. With this screening work completed, the continuing effort will incorporate additional materials and test conditions more nearly simulating service conditions.

499

Table 1

Material Designation	Composition	Substrate	Fabrication Method	Surface Condition	Source
CNTD SiC (Hard)	Silicon Carbide	Graphite	Chemical vapor deposited	as deposited	San Fernando Laboratories
CNTD SiC (Soft)	Silicon Carbide	Graphite	Chemical vapor deposited	as deposited	San Fernando Laboratories
LW-5	Tungsten Carbide	Stainless Steel	Detonation Gun sprayed	ground	Union Carbide Linde
LW-15	Tungsten Carbide	Stainless Steel	Detonation Gun sprayed	ground	Union Carbide Linde
ROKIDE C	Chromium Oxide	Black iron	Oxy-acetylene sprayed	as sprayed	Norton Co.
NC-132	Silicon Nitride	None	Hot pressed	as pressed	Norton Co.
NC-203	Silicon Carbide	None	Hot pressed	as pressed	Norton Co.
NC-403	High purity silicon carbide + silicon	None	Reaction sintering + densification of slip cast material	as sintered	Norton Co.

EXPERIMENTAL CONDITIONS

Flat, rectangular specimens of the order of 3cm x 2cm x 1/2cm were used. Table 1 lists the materials tested. Since several of the materials tested are still in development or initial production applications, their proprietary nature precludes a detailed description of their composition, structure or method of processing.

The specimens were placed in an air blast tester[1] and eroded incrementally with up to 280gm of 200μm, angular SiC particles, carried in an air stream at 100fps at room temperature. The velocity was determined using a rotating disc method.[2] The angle of impingement between the direction of the particles out of the nozzle and the flat target surface was 30°, 60° or 90°. Total test time ranged from 8 min. to 15 min. (approximately 5 sec/gram) depending upon when a steady state erosion rate was reached. A steady state erosion rate is defined as that condition of the target surface where each succeeding batch of particles causes the same amount of weight loss of the specimen as the previous batch.

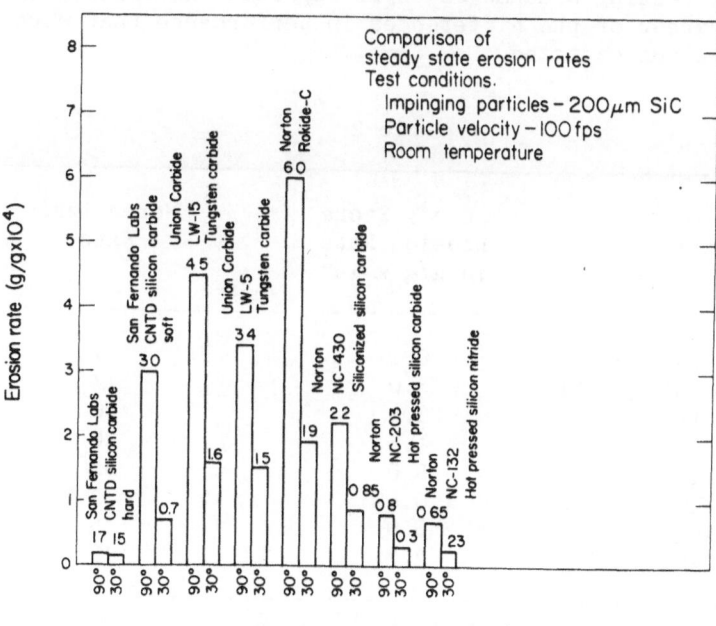

Fig. 1. Bar chart of steady state erosion rate of each material.

The specimens were blasted with small amounts of particles in each erosion increment that were increased as the steady state condition was approached, as can be seen by the weight loss curves. Weighing was done on a balance accurate to ±0.1 miligram.

RESULTS

The steady state erosion rate of each material tested at 30° and 90° impingement angles is shown in Fig. 1. A wide range of performance occurred for the group of test materials. In all instances the materials eroded more at 90° angle than at 30°, which is typical of brittle type materials. Some materials showed relatively little difference in erosion between the two impingement angles, especially the hard CNTD silicon carbide material. However, others showed three to four times greater erosion at 90° than at 30°. There was a wide variation in the erosion behavior of silicon carbide depending on its type, fabrication method and source. The two tungsten carbide materials behaved in a similar manner, especially at the 30° impingement angle.

Table 2 compares the performance of the materials at an impingement angle of 90° by normalizing them with respect to the performance of the CNTD hard silicon carbide. The rather wide spread in their performance can be easily seen. Since all of the materials were procured for testing because of their reported excellent wear resistance, the extent of the differences in performance that were measured were not expected.

Table 2

Material	Steady State Erosion Rate in g/g x 10^4	Normalized Rate
CNTD SiC (Hard)	0.17	1
CNTD SiC (Soft)	3.0	17
LW-5 WC	3.4	20
LW-15 WC	4.5	26
ROKIDE C	6.0	35
NC-132	0.65	3.8
NC-203	0.9	5
NC-430	2.2	13

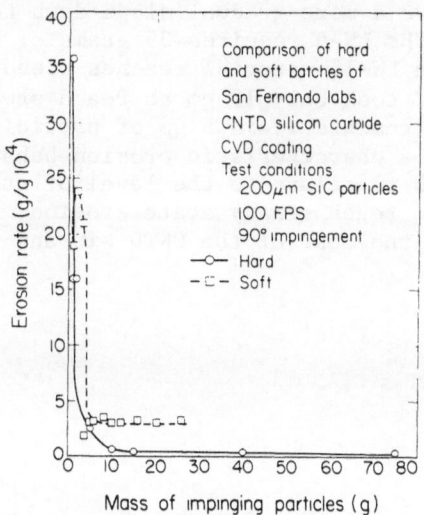

Fig. 2.   Incremental erosion rate of CNTD SiC, hard and soft.

Fig. 2 shows the pattern of the incremental erosion of the hard
and soft CNTD silicon carbide that was chemically vapor deposited on
graphite. It can be seen that both materials rapidly reached an
erosion rate peak after the initiation of erosion and then rapidly
decreased to a low steady state erosion rate. The soft SiC peak is
lower than that of the hard SiC and its rise could be measured while
the rate of the hard SiC was at a peak at the first increment of
one gram of particles. The hard SiC reached a considerably lower
steady state rate which accounted for its lower overall steady state

of erosion.   The hard SiC took somewhat longer to reach a steady state condition.

      Figures 3 and 4 show the incremental erosion curves of the LW-5 and LW-15 detonation gun sprayed tungsten carbide coatings on a stainless steel substrate.   The curves are similar in shape to that of the CNTD hard SiC, but have a more gradual slope down to their steady state erosion rate.   The LW-5 requires 35 grams of particles to reach steady state while the LW-15 material reaches steady state in only 15 gm.   The CNTD hard SiC took only 10 gm to reach steady state erosion and the CNTD soft SiC reached it in 5 gm of particles.   The time to reach steady state is a characteristic erosion behavior property of materials.   It appears to relate to the level of steady state erosion, the longer it takes to reach steady state erosion, the lower is the final erosion rate in the case of the CNTD SiC and the sprayed WC.

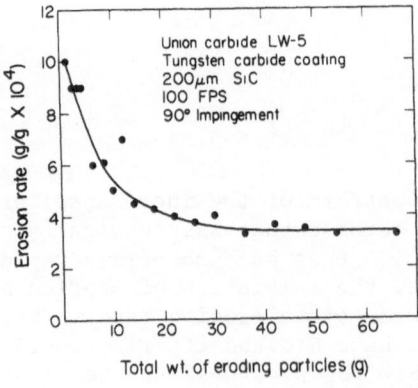

Fig. 3.   Incremental erosion rate of LW-5 sprayed WC.

The hot pressed NC-132 silicon nitride and Ni-203 hot pressed silicon carbide had low rates of erosion at steady state and incremental erosion rate curves that were different from those of the deposited materials. Figures 5 and 6 show that the nature of the erosion was one of an increasing erosion rate up to a steady state rate, similar to that which occurs in ductile metals.

A comparison of the incremental erosion curves for the several types of materials is shown in Fig. 7. The initial behavior varies somewhat, but each material reaches a steady state conditon in a

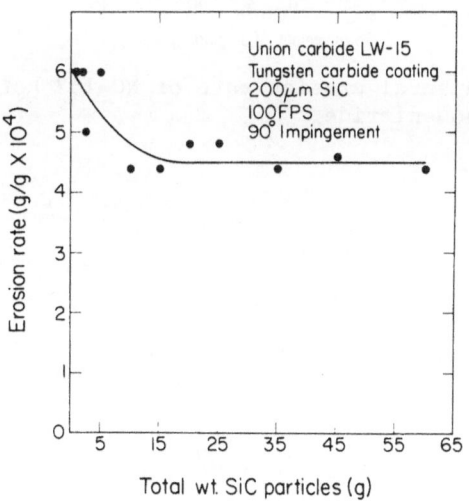

Fig. 4.   Incremental erosion rate of LW-15 sprayed WC.

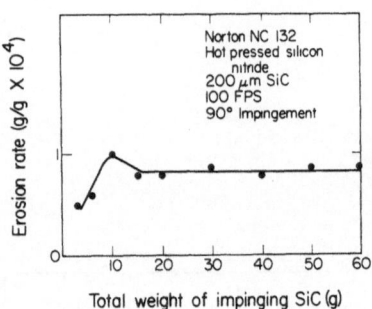

Fig. 5.   Incremental erosion rate of NC-132 hot pressed
          silicon nitride.

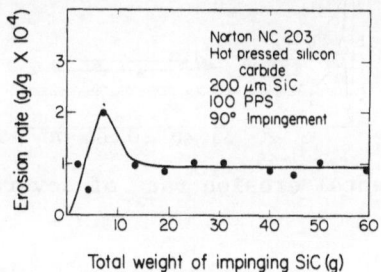

Fig. 6.   Incremental erosion rate of NC-203 hot pressed
silicon carbide.

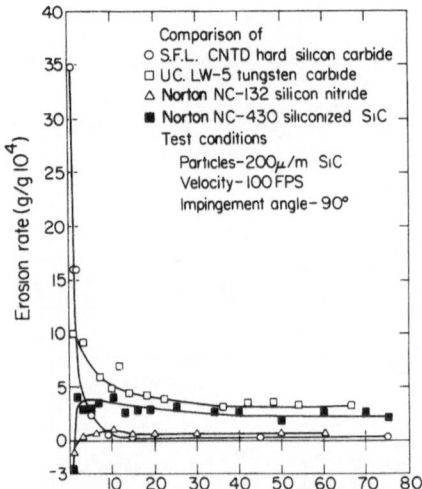

Fig. 7.   Incremental erosion rate of several materials.

relatively few grams of impacting particles. The negative initial
readings for the hot pressed silicon carbide and nitride materials
probably are due to embedded particles of erodent in the surface.
The very high initial erosion rate readings relate more to the mechanism of initial erosion than to steady state erosion behavior as the
highest peak material, the CNTD hard SiC, has the lowest steady state
erosion rate.

CNTD SILICON CARBIDE

Steady State Erosion

HARD   SOFT

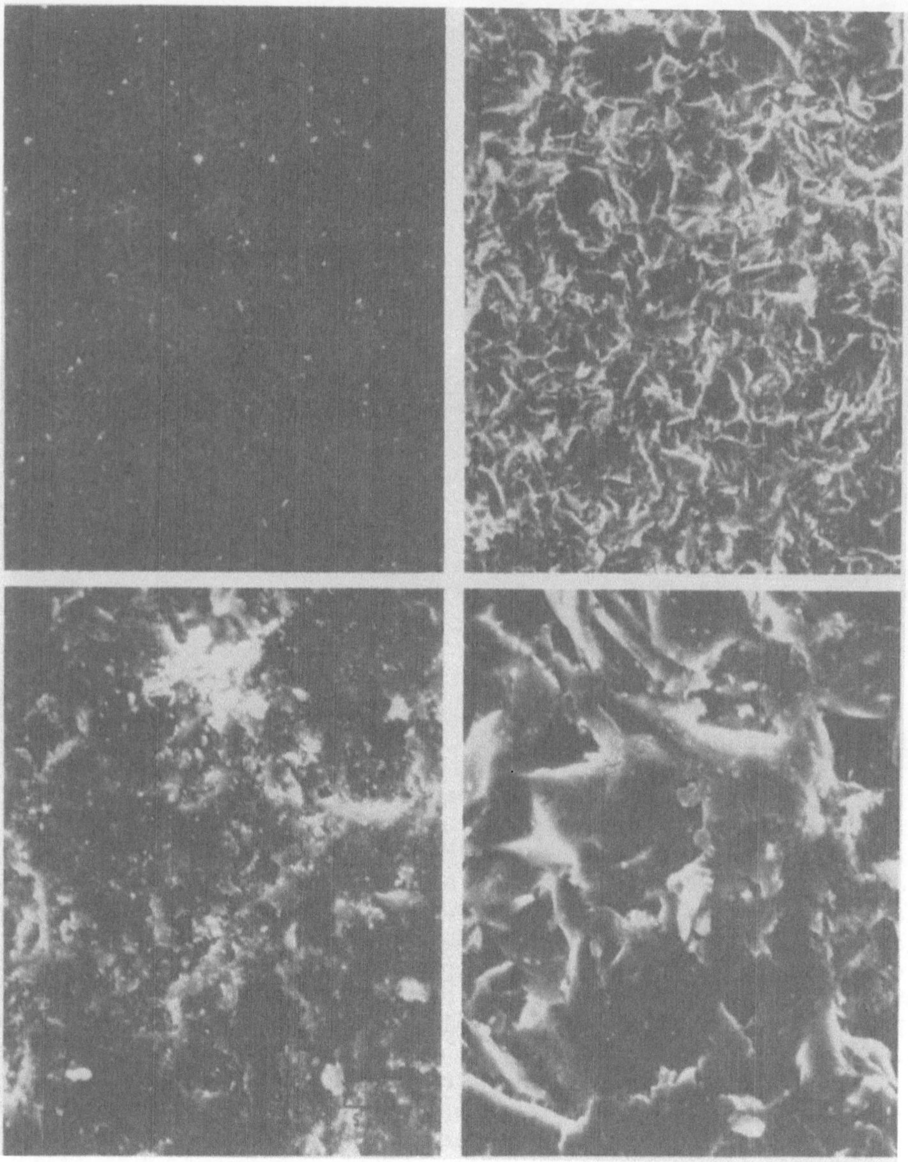

Fig. 8.   CNTD silicon carbide eroded surfaces at steady state.

CNTD SILICON CARBIDE

Peak Erosion

Hard   Soft

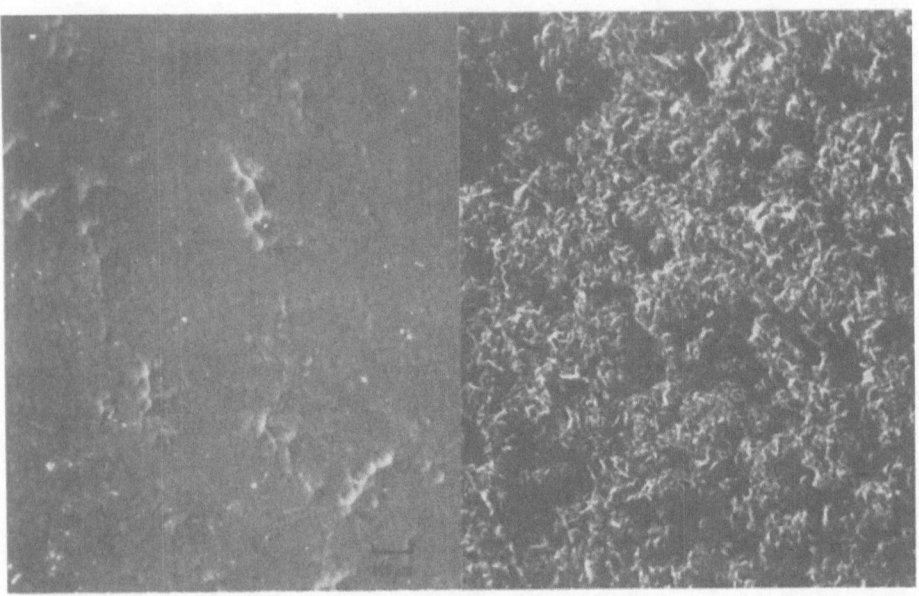

Fig. 9.   CNTD silicon carbide eroded surfaces at peak erosion rate.

METALLOGRAPHIC ANALYSIS

The scanning electron microscope (SEM) was used to study the nature of the physical deformation that occurred on each material as the result of the erosion process. Fig. 8 shows scanning electron microscope (SEM) photos of the CNTD silicon carbide coatings eroded surfaces after steady state conditions were reached. The uneroded surfaces of the two coatings were essentially alike. After erosion there is a great difference in the appearance of the surfaces at both lower and high magnifications. The hard SiC appears to be eroding by the loss of fine chips of materials, representative of a very fine grain size. The soft SiC on the right hand side of Fig. 8 is eroding by a mechanism of combined cleavage of crystallites of a considerably larger grain size than that of the hard SiC and some plastic deformation of material that appears to have some small degree of ductility.

Fig. 9 shows the appearance of the eroded surfaces of the hard and soft CNTD silicon carbide at the time of the peak erosion rate as shown in Fig. 2. It can be seen that considerably more surface has

Fig. 10. Large grained areas of hard CNTD silicon carbide near coating–substrate interface.

been affected in the soft SiC than in the hard SiC even though the
peak erosion rate of the hard SiC is higher at this early point in
the erosion of the two surfaces.

Fig. 10 shows a phenomenon which occurred only in the hard CNTD
silicon carbide near the coating-substrate interface. The coating
preferentially eroded in areas which appear as grooves in the left
side photo. At high magnification on the right side, the grain size
at the root of the cracks can be seen to be considerably larger than
that of the major part of the coating. This resulted in a preferential erosion pattern along the paths of the larger grains. There
may even have been some porosity present in the regions of the
apparent cracks to further reduce the erosion resistance of the area.
The chipping away of small grains that were typical of the hard SiC
material can be seen in the regions on either side of the large
grained area.

Fig. 11 shows the steady state erosion surface of the LW-5 tungsten-carbon coating. The appearance of the material at the surface
indicates that considerable plastic deformation had occurred along
with some lesser amount of brittle fracture or chipping. The degree
of plastic deformation is considerably more than was seen on most of
the other coatings but is representative of those with some amount
of metal binder content. The nature of the platelets formed is
similar to those formed when ductile metals are eroded.

DISCUSSION

The erosion behavior of the hard materials tested varied over a
relatively wide range as is shown in the bar graph, Fig. 1. The
variation in hardness of the various refractory hard metals tested
was too small to relate to the differences in measured erosion rate.
Therefore, the erosion rates must be attributed to a combination of
characteristics such as composition, amount and type of binder
material, grain size, and other factors which combine to absorb and
distribute the kinetic energy of the impacting particles. All of
the materials tested had the characteristic erosion behavior of
brittle materials, i.e., the erosion rate was greater at the 90°
impingement angle than at the more shallow 30° impingement angle.
The role of such binder materials as silicon metal in the material
systems does not modify the basic mode of erosion although it does
modify the sensitivity of the material to erosion with the more
intimately mixed silicon-silicon carbide or nitride materials having
the best erosion resistance. Since several of the materials tested
are highly proprietary and their grain structures very fine, the
distribution of silicon in them is not known without further analysis
or information from the supplier.

The very low erosion rate of the CNTD SiC (hard) from San

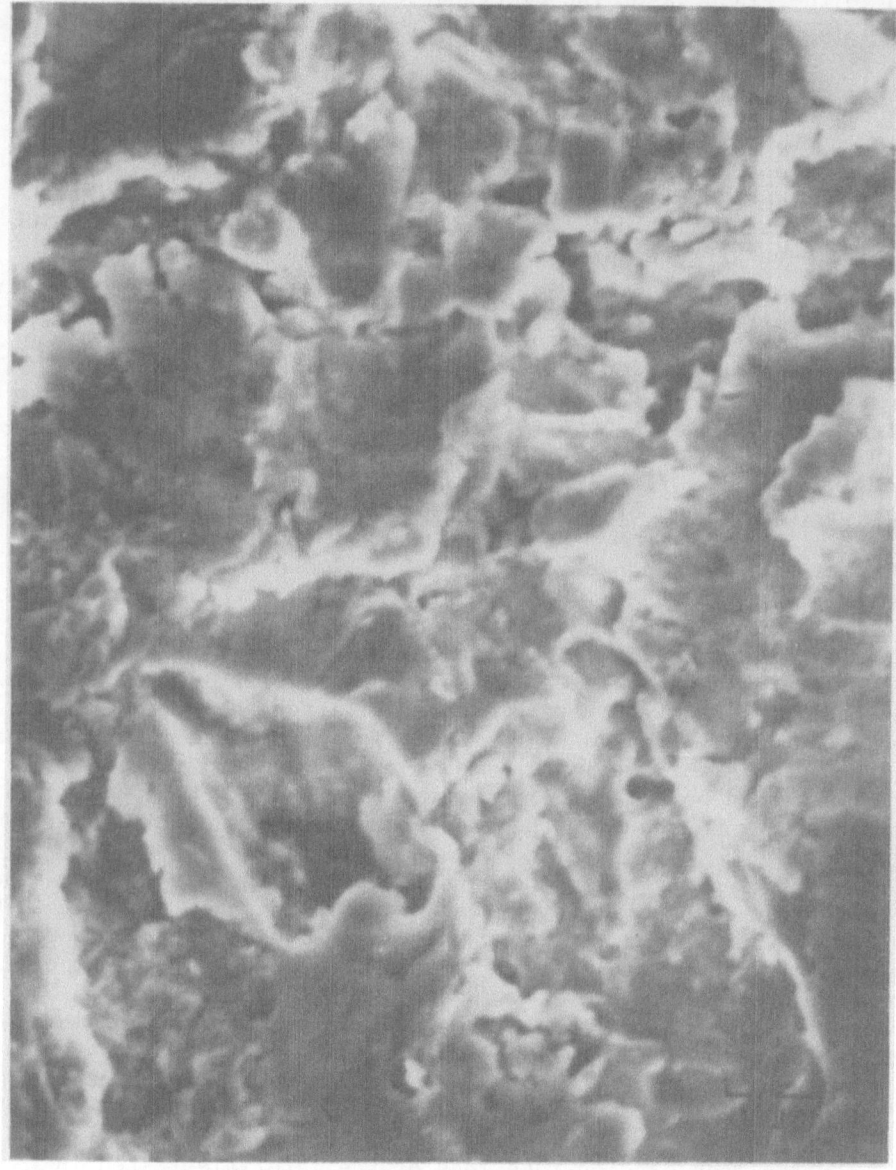

Fig. 11.   Steady state erosion surface of LW-5 tungsten-carbon
coating.

Fernando Laboratories and its insensitivity to the impingement angle is due to the fineness of the distribution of the binder phase and the small grain size of the material. This resulted in material loss by chipping away of very small pieces. A modification of this material, the CNTD SiC (soft) has considerably different grain structure and a marked difference in the erosion mechanism as can be seen in Figures 8 and 9. This resulted in a considerably higher erosion rate as can be seen in Fig. 1. Scanning Auger Microscopy (SAM) analysis of the two CNTD materials indicated that ·the soft, higher erosion rate material had a considerably higher oxygen content which could also have affected its erosion rate.

The relatively low erosion rates of the hot pressed silicon carbide and silicon nitride from the Norton Co. also relates to the fine grain size and binder distribution that can be achieved by this type of processing. Hot pressing is generally limited to producing wear resistant bodies or inserts. The ability of the chemical vapor deposition process to deposit the coating of CNTD SiC on large surfaces with such a fine structure and low erosion rate shows the promise of this method of developing wear resistant material systems.

The erosion rate peaks that some of the materials experienced, as shown in Fig. 2 for the CNTD SiC, is typical of the erosion of some brittle materials. In the work of Zambelli and Levy[3] to determine the erosion behavior of NiO formed on CP nickel, the same type of peaks were observed. They are due to the initial loss of material in the outer layers of the brittle material after the surface area has been thoroughly cracked by the impacting particles without pieces being removed. After the initial loss, cracks penetrating into the material separate out pieces of material for removal at a considerably lower rate, sharply reducing the erosion rate to a much lower value.

This cracking mechanism accounts for the difference in the shape and peak height of the erosion rate curves for the hard and soft CNTD SiC. The smaller grained, more strongly bonded hard material could undergo considerably more initial surface cracking without loss of material than the soft material. Hence, when the crack pattern has been completed in the surface layers, an initial high rate of loss occurs in the hard SiC and a lower initial rate in the soft material. The much lower steady state loss of the hard material compared to the soft can be seen in the curves.

The normalized erosion rate loss of the materials tested in Table 2 indicates the superiority of the CNTD SiC (hard) in the type of erosion test carried out. At other test conditions of velocity and particle size, shape, and composition other relative behaviors could occur among the materials tested.

The erosion rate curves of the detonation gun applied tungsten-carbon coatings LW-5 and LW-15 is similar to that for the CNTD SiC (hard), but the curves fall off much more gradually to a higher steady state condition because of their different structure and composition. Within the same composition, the more gradual the slope of the curve to steady state erosion, the lower is the steady state erosion rate. However, the comparison does not appear to apply between different materials. The CNTD SiC materials' erosion rates fall off to steady state considerably faster than do the tungsten-carbon coatings; yet are considerably lower. The difference in the erosion mechanism between the CNTD SiC and the LW-5 materials appears to undergo considerably more plastic deformation at the eroding surface than does the CNTD SiC material.

The hot pressed bodies of silicon carbide, NC-203, and silicon nitride, NC-132, have erosion rate curves that are considerably different from the previously discussed materials. They do reach a peak erosion rate after the initiation of erosion, but undergo a lower but measureable erosion rate prior to reaching the peak rate. In the case of the silicon nitride, the peak rate is very near the steady state erosion rate. A comparison of several of the curves of materials tested is shown in Fig. 7. The reasons for these differences in the shapes of the curves have still to be determined.

CONCLUSIONS

1. All of the materials tested eroded in a brittle manner, undergoing more erosion at a 90° impingement angle than a 30° angle.

2. The CNTD SiC (hard) had the best erosion resistance, apparently due to the fine grained microstructure of the carbide phase and the fine distribution of the silicon rich phase.

3. The coating materials had a peak erosion rate at the beginning of the erosion process, which has been observed for other brittle coatings on substrates. The hot pressed bodies had an increasing erosion rate up to a steady state value, which is typical of metals.

4. The amount of apparent plastic deformation that occurred in some of the materials can be related to the amount and condition of the metallic phases in the materials, but more work is required to establish this relationship.

5. The large grain size near the coating-substrate interface of the CNTD SiC (hard) that eroded perferentially could be related to an instability in the deposition process that occurred near the initiation of deposition.

REFERENCES

1. A. Levy, The Solid Particle Erosion Behavior of Steel as a
   Function of Microstructure, Wear 68:269 (1981).
2. A. Ruff and L. Ives, Measurement of Solid Particle Velocity in
   Erosive Wear, Wear 35:195 (1975).
3. G. Zambelli and A. Levy, Particulate Erosion of NiO Scales,
   Wear 68:305 (1981)

ACKNOWLEDGMENTS

This work was supported by the Director, Office of Energy Research,
Office of Basic Energy Sciences, Materials Sciences Division of the
U.S. Department of Energy under Contract Number W-7405-ENG-48.
Selection and procurement of the materials that were tested were made
by Donald Boone. Pauline Chik prepared the metallographic specimens.

DEFORMATION CHARACTERISTICS AND MECHANICAL PROPERTIES OF HARDMETALS

E. A. Almond

National Physical Laboratory
Teddington
Middlesex,U.K.

ABSTRACT

Fundamental deformation characteristics of transition metal interstitial-compounds are briefly reviewed, and reference is made to experimental and proven metallic cermet systems using WC/Co hardmetals as a basis for comparison. Emphasis is given to the need for a critical approach to mechanical testing of hardmetals and for caution in interpreting results for empirical and intrinsic property measurements. Deformation mechanisms and microstructural-mechanical property correlations are examined for elastic and plastic properties, strength, fracture, fracture toughness and fatigue.

Some theoretical models for the microstructural dependence of strength and toughness of hardmetals are discussed in relation to experimental observations and their generality is examined. An attempt is made to define the role played by the fundamental properties of the constituents in determining the mechanical behaviour of metallic cermets. A summary is given of the present state of understanding of mechanical properties, and directions are specified for future research to improve the effectiveness of design and application of hardmetals.

INTRODUCTION

The development of current compositions for hardmetals, probably owes more to trial and error selection, than to a scientific understanding of the microstructural and mechanical requirements for tool materials to be used in metal cutting, rock mining

Table 1. Physical and mechanical properties of hardmetal constituents and hardmetals[1-5]

Compound or hardmetal	Crystal structure	Hardness HV*	Young's modulus E kN mm^{-1}	Poisson's ratio ν	Thermal exp. coef. α 10^{-6} K^{-1}	Thermal cond. k Wm^{-1} K^{-1}	Therm. shock parameter** R K^{-1} m^2 s^{-1}
a) TiC	fcc	3000	450	0.19	7.4	34	0.011
TaC	fcc	1800	285	0.24	5.5	23	0.003
Cr$_3$C$_2$	orthorhombic	1400	373		9.9	19	0.004†
Mo$_2$C	hex	2500	533		6.7	22	0.005†
WC	hex	2200	696	0.18	5.2	35	0.008
Al$_2$O$_3$	hex	3000	400	0.23	5.5	34	0.012
TiB$_2$	hex	3300	480		8.0	25	0.005†
b) 89WC–Co		1250	575	0.22	5.6	120	0.03
56WC–19TiC–16TaC–Co		1700	480	0.22	6.8	25	0.01
18WC–60TiC–7Mo$_2$C–5Cr$_3$C$_2$–Co–Ni		1800	440	0.22	7.7	20	0.005

*Measured with 50g load on compounds   **Defined as $\dfrac{(1-\nu)k}{E\alpha}$ † For $\nu = 0.2$

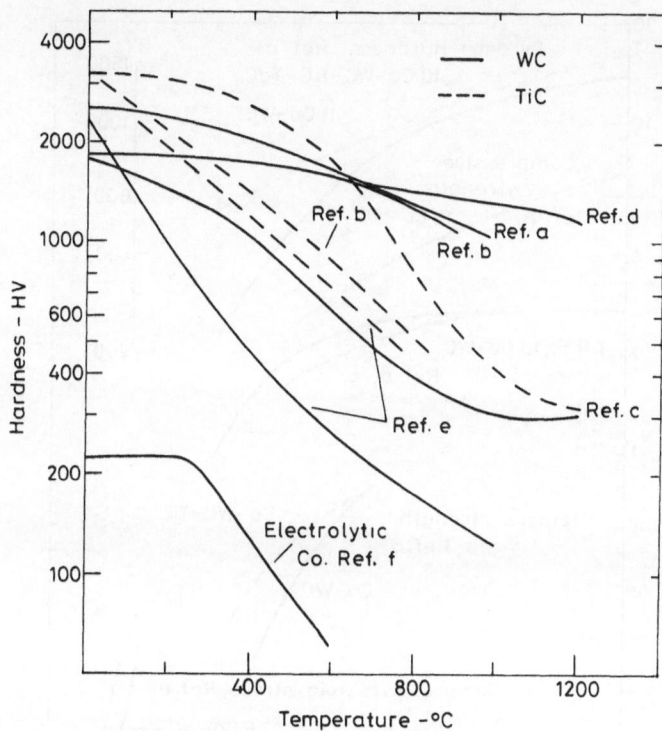

Fig. 1. Temperature dependence of hardness of WC, TiC and Co.
Ref a) Dawihl and Mal[8], b) Westbrook and Stover[9], c) Myoshi and Hara[10], d) Schwab and Krebs[11], e) Atkins and Tabor[12], f) Cobalt Monograph.[13]

and wear applications. If regarded as a composite material, it would be expected that the property requirements for the hard constituent would be fulfilled by transition metal carbides, nitrides and borides; while for the binder phase the Group VIII metals, Fe, Co and Ni, would have the required ductility, toughness, and wettability, and have the benefit of being weak carbide formers. To some extent, simple composite theory works well in the initial selection of particle and binder constituents for room temperature mechanical properties and thermal shock resistance[1-4] (Table 1). For example, the parameter for the latter property[5,6] (the product of the parameter K' in Table 1 and the tensile strength) is higher for WC than for other potential hard constituents in both the sintered compact and cemented form, and this is translated into superior shock resistance in service[7]. However, some of the problems in adopting this simple approach become evident when judging the suitability of different constituents for high

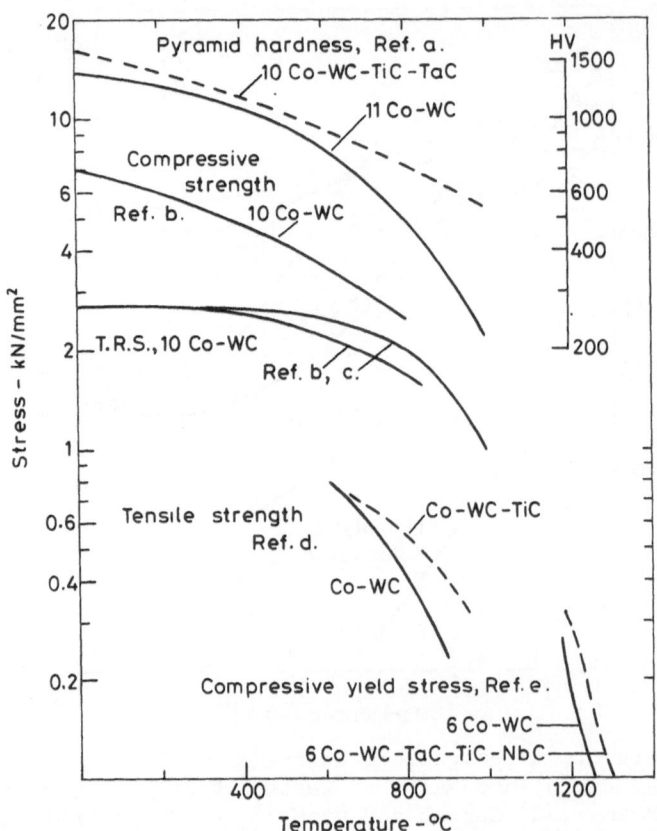

Fig. 2. Temperature dependence of hardness, compressive strength,
transverse rupture strength, tensile strength, and
compressive yield stress for Co-WC hardmetals, and Co-WC-
cubic-carbide hardmetals. Ref a) Dawihl and Mal,[8]
b) Aschan et al[14], c) Ueda[15], d) Trent and Carter[16],
e) Schenck et al.[17]

temperature applications. For here it is found that there are
anomalies and contradictions in reported data on the hot hardness
of carbides[8-12] (Fig.1) which are also at variance with results on
the temperature dependence of the mechanical properties of the
respective hardmetals[8],[14-17] (Fig.2). Thus, for example, the hardmetals are harder than they would be expected from some of the data
for their constituents; also it is found that TiC and TaC, individually or together, confer extra hardness and strength to WC hardmetals at temperatures above 600 °C, rather than to soften them as
is suggested by some data in Fig.1. The discrepancy between

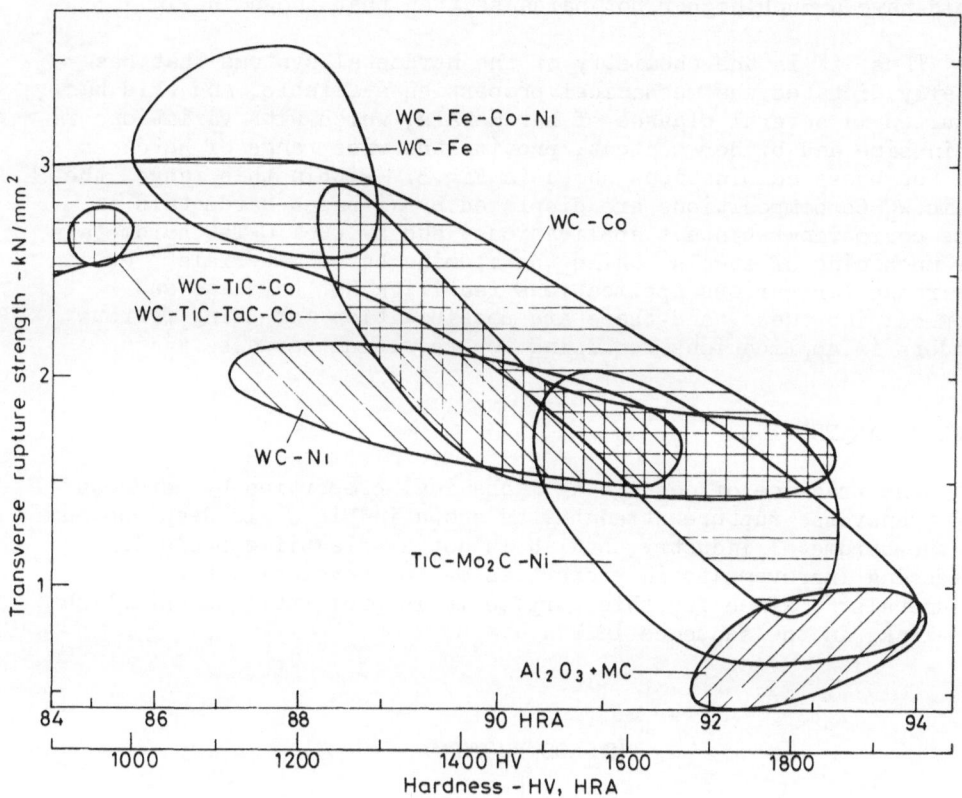

Fig. 3. Typical transverse-stength and hardness ranges for hardmetals of different compositions.

predicted and real properties is partly explained by the necessity
to take into account chemical effects that cause the cubic carbides,
TiC and TaC, to form mixed carbides with WC, and for these to be
disposed in regions of high contiguity that probably confer extra
rigidity on the microstructure at high temperature.   Similarly,
dissolution of WC in the binder-phase in Co- and Ni- base hardmetals
produces an alloy that is 50% harder than the pure metal[19], and
would have a much higher hot hardness than that shown in Fig.1.[13]

    Thus, it is the chemistry of the hardmetal systems that has
chiefly dictated the mechanical properties available, and this has
resulted in several classes of hardmetals, which with variations in
grain-size and binder content, provide the wide range of hardness
and toughness combinations shown in Fig.3.   Within this range, the
basic WC-Co compositions are displaced by Ni-based hardmetals in
some corrosion-resistant applications, and by WC-TiC-TaC hardmetals
for machining of steels; while the alumina-based materials[20] extend
the range for various applications requiring hot hardness and
chemical inertness, and there are possibilities for using ferrous
binders in applications requiring low to medium hardness.[21,22]

MECHANICAL TESTING

    The practice of describing mechanical properties by hardness
and transverse rupture strength, as shown in Fig.3, is deep rooted
in the hardmetal industry, but it is not a scientific basis for
assessing improvements in properties or for developing better
hardmetals[23], since for this purpose it is preferable to establish
the links in the sequence in Fig.4.

MICROSTRUCTURE

|

DEFORMATION CHARACTERISTICS

|

MECHANICAL PROPERTIES

|

PERFORMANCE

Fig. 4.   The basic sequence for relating microstructure to
          performance

Though apparently straightforward, the forging of a sound chain of
understanding in this sequence, has been hindered in hardmetals by

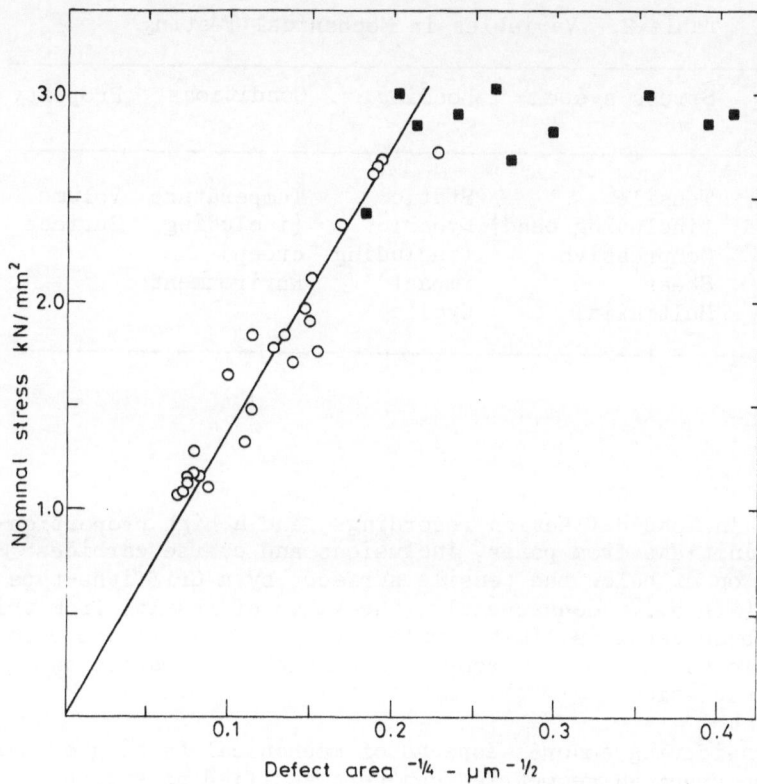

Fig. 5. Nominal stress at fracture initiation site versus defect-
size in transverse rupture tests on fine-grained 11% Co-WC
hardmetals.[25]
Defect symbols: O - pores and inclusions, ■ - large WC
grains.

two notable weaknesses in interpreting data. Firstly, properties
are frequently measured in relation to their variation with carbide
grain-size and binder-phase content, while ignoring equally important effects from variations in specimens' porosity, binder-phase
composition, and grain-size distribution. Secondly, transverse
rupture strength has been treated as though it were intrinsic in
origin (see below), whereas it is dependent on specimen size and
shape, and is falsely based in the formula for its derivation,
which assumes that failure initiates at the tensile surface by
fully elastic deformation. In practice, a wedging correction should
be applied to the standard formula[24], the tensile surface often
undergoes significant plastic strain that would not always be

Table 2. Variables in Mechanical Testing

Specimen type	Stress system	Loading	Conditions	Property
Smooth Precracked	Tensile (including bend) Compressive Shear Multiaxial	Static Dynamic (including impact) Cyclic	Temperature (including creep) Environment	Volume Surface

detectable in load-deflection recordings, and a high proportion of
fractures initiate from pores, inclusions and coarse carbides[25] at
positions, on or below the tensile surface, by a Griffiths-type
mechanism (Fig.5.). Consequently, the value of results from this
and other bend tests is limited unless they pertain to the onset of
plastic deformation, or are complemented with information on
fracture initiation.

In considering general aspects of mechanical testing of hardmetals, the specimen geometries can be classified as smooth or pre-
cracked depending on whether they are used for measuring strength
or fracture mechanics parameters respectively, while the properties
that are measured are either intrinsic or empirical in origin.
Intrinsic properties are based on atomic structure and microstructure, and include such properties as elastic moduli, plane-
strain fracture toughness and yield strength. Theoretically they
can be calculated from a knowledge of atomic structure and deformation mechanisms; they are temperature and strain-rate dependent,
and their magnitude is independent of specimen-size and geometry,
within a range of defined limits. It is this latter characteristic
that distinguishes the important role played by intrinsic properties
in the sequence in Fig.4, since not only are they essential for
relating microstructure to mechanical properties but they are also
the basis for deriving a scientific approach to the design and
prediction of performance of tools.

In examining the relevance of individual tests to the property
requirements for hardmetals, it is important to remember that in
the majority of applications the stresses are mainly compressive,
and that a need for strength in tension only arises if significant
tensile stresses are generated as a result of tool geometry or
mishandling. The most important variables in mechanical testing
are listed in Table 2. Of the smooth-specimen tests used for

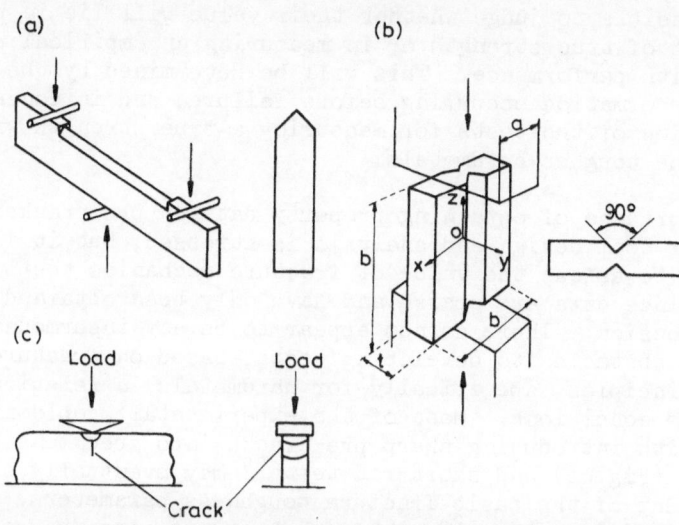

Fig. 6.   (a) and (b) Specimens and loading geometries for measuring:
(a) limiting strength in tension of hardmetals[27];
(b) strength under different ratios of compressive to
tensile stress.[28] (c) Arrangement for precracking
fracture-toughness specimens by wedge indentation.[29]

measuring intrinsic properties, compressive and compressive-fatigue
tests reproduce reasonably closely the stress conditions met with
in many metal forming tools, and the former test is valuable for
providing basic information about yield stress, work hardening
behaviour, and compressive strength. The value of torsion or shear
tests for assessing hardmetals is difficult to judge since although
shear stresses are undoubtedly important in drills and cutting tools,
it has not yet proved possible to devise an inexpensive practicable
test system that would enable shear strengths to be measured
accurately for tool materials.[26]

The greatest gap in the testing armoury for hardmetals is in
strength tests. Progress has been made in developing a test for
measuring the tensile strength as governed by microstructure, for
materials that would otherwise exhibit significant defect-initiated
fracture in conventional bend tests[27]. (Fig.6a). Also, a test[28]
has been devised that enables the strength of materials to be
measured under various ratios of compressive/tensile stress that
can be altered to simulate the stress conditions in the tool. (Fig.6b)
It is probable that with minor modifications in specimen geometry
these two tests could answer some of the problems of measuring the
strength of tool materials; however, until more-detailed stress

analyses are performed and more experience is gained in their use, it is not possible to judge whether their value will lie in providing a measurement of true strength or in measuring an empirical property correlated with performance. This will be determined by the amount of plastic deformation occurring before failure, and in consequence the application of the tests for measuring a true strength will be limited in the tougher hardmetals.

The importance of obtaining property data on pre-cracked specimens for tool design and analysis is stressed, but it is not yet possible to assess the value of fracture mechanics tests for hardmetals since data are sparse and have only been obtained for testing in tension. There do not appear to be any insurmountable experimental obstacles to developing tests, based on fracture mechanics principles, specifically for hardmetals in relation to their service conditions. Most of the experimental problems[29] associated with introducing sharp pre-cracks into specimens have been solved, (Fig.6c) and short-rod tests[30] may eventually simplify the measurement of the basic fracture toughness parameters. Thus, specimens containing a single pre-crack that are already used to determine static K_{IC} values, could also be used in studying mechanical, environmental and thermal fatigue, while tests on specimens containing shock-induced microcracks[6] would provide information on the validity of extending fracture mechanics principles to small crack-sizes in hardmetals.

Although hardness is not recognised as an intrinsic property, it deserves special mention, since, despite the multiplicity of scales, the majority of tests are based on measuring the load to produce unit area of surface deformation, and as such, the results are interrelated by the compressive flow stress of the test material. The ratings of hardmetals by the various tests can be different because of effects from differences in indenter geometry, work-hardening rates of test materials, and the corrections for expressing the results, but provided these limitations are realised, the test gives useful information on the mechanical properties and condition of material close to the specimen surface.

Another family of empirical tests for hardmetals is based on measurements of the surface cracking that occurs during indentation in a hardness test. The cracks are mainly circumferential and radial for sperical and cone indenters, and mainly radial for pyramid indenters (Fig.7) and efforts have been made to relate the degree of surface cracking to an intrinsic mechanical property, such as the plane-strain fracture toughness.[31] Some success has been claimed with ceramic materials[31], but the mathematical analysis of the stress field around a pyramid indenter is difficult. Also, it is unlikely that such a relationship would be expected for hard-metals, since the radial cracks propagate through the deformed material around the indentation[32], and this material would not

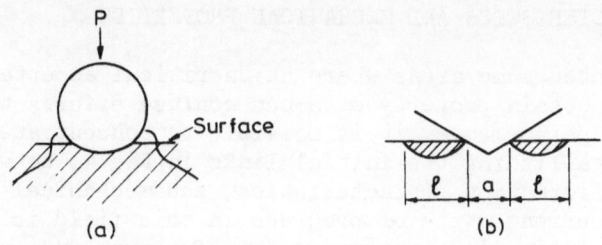

Fig. 7. (a) Hertzian cone crack beneath surface indented with ball
indenter.
(b) Palmqvist cracks beneath surface indented with
pyramid indenter.

present the same resistance to fracture as virgin material. Nevertheless, studies of indentation cracking produce information
relevant to impact erosion and abrasion wear mechanisms,[33] and if
the tests are used judiciously they may have a role to play in
quality control or product development, either in the original form
of the test or with modifications.[34]

Of the remaining aspects of mechanical testing of hardmetals,
the following considerations merit special attention in performing
existing tests and devising new test procedures.

The importance of using reliable testing techniques, for
example, for pre-cracking K_{IC} specimens, and of using careful
specimen preparation cannot be over-emphasised. Ironically, many
of the problems in this area occur in research performed outside
of industry, since most tool manufacturers and users have considerable experience and facilities for precision machining of specimens
and equipment components.

Despite their deficiencies, most existing tests for hardmetals
use specimens with simple shapes that are relatively easy and
inexpensive to manufacture, and that present minimal alignment and
gripping problems. These criteria also need to be met by new
mechanical tests for low ductility materials in general.

The wide range of temperatures and strain-rates encountered in
tool applications, make it essential that both variables are taken
into account when deciding on the appropriate conditions for
testing. Hot hardness tests provide an economical method for
qualitatively assessing probable compressive strength in high
temperature service, but the absence of a reliable test for
measuring resistance to impact or high strain rate deformation is
a major weakness of hardmetal testing.

DEFORMATION CHARACTERISTICS AND MECHANICAL PROPERTIES

Having indicated some areas where an uncritical acceptance of
the tests used to obtain property data can confuse efforts to relate
microstructure to performance, it is possible to concentrate more
effectively on establishing the initial links in Fig.4, between
microstructure, deformation characteristics, and mechanical
properties. The current state of progress in this field is that:
- plastic deformation in compression, and fracture in compression
  and tension, have been characterized in WC-Co hardmetals,
  partly in cubic-carbide hardmetals, and in various constituents
  of these materials;
- indirect observations have been made of high-temperature
  deformation-mechanisms by interpreting data for parameters
  for creep-rate and the temperature dependence of various
  properties;
- sufficient measurements have been made of compressive and
  tensile properties, and fracture toughness to provide a good
  indication of their microstructural dependence in WC-Co
  hardmetals at room temperature, but data are not comprehensive
  for other hardmetals, for temperature- and strain-rate
  dependence, and hardly exist for fatigue properties.

A consequence of the sparsity of information is that microstructural models for deformation behaviour are constantly being
revised or discarded as new observations are made, and the present
section will describe these newer observations.

Deformation Characteristics

Of the transition-metal carbides, usually associated with
established hardmetal compositions, single crystal and hot-pressed
TiC and TaC have been subjected to the greatest amount of research
into deformation characteristics.[35-37] Plastic deformation occurs
on the $\{111\}$ <1$\bar{1}$0> slip system, but is not significant below 800 °C
and TaC is softer and more ductile than TiC, which fractures by
cleavage on $\{100\}$ planes. As in other materials, much of the
information about deformation mechanisms has been inferred from
results of high temperature tests by obtaining the best fit for
parameters in expression of the form:[37]

$$\sigma \ = \ A \exp - (U_1/kT) \qquad\qquad\qquad \dots 1$$

$$\text{and } \ \dot{\varepsilon} \ = \ B \, \sigma^n \exp - (U_2/kT) \qquad\qquad\qquad \dots 2$$

where σ is the flow-stress variable relevant to the type of test (tension, fatigue, etc), $\dot{\varepsilon}$ is the creep-rate, k and T are Boltzmann's constant and the absolute temperature respectively, A and B are constants, while U_1, U_2 and n are the material parameters that are derived from the data.

Other parameters, such as V, the activation volume, can be derived from these expressions, but differ in their definition depending on the form of equation 2. Application of this type of analysis to the properties of TiC below 1700 °C has given values for U_2 of 5 eV/atom, which is in good agreement with the activation energy for diffusion of C in TiC.[36] This has prompted the suggestion that the proposed glide mechanism involving the dissociation of <110> dislocations, is controlled by thermally-activated kink formation at C vacancies.[36]

Few data exist on the deformation characteristics of WC single crystals, and this is probably related to the difficulty of growing crystals free of Co inclusions. In fact, variability arising from Co contamination and porosity probably makes a major contribution to the disparity in the results for WC in Fig.1. Nevertheless, studies of slip traces in single crystals and of hardness anisotropy provide strong evidence of a $\{01\bar{1}0\}$ $<2\bar{1}\bar{1}3>$ slip system in WC.[38]

Thin foil electron microscopy studies of the dislocation structures in WC grains in WC-Co hardmetals, have been mainly confined to network analysis,[40] and the accumulated evidence and proposed dislocation reactions can be summarized as follows:[39]

$$\frac{1}{6}\,[11\bar{2}3] + \frac{1}{6}\,[11\bar{2}3] \qquad \ldots \text{ 3a}$$

$$\frac{1}{3}\,[11\bar{2}0] + [0001] \rightarrow \frac{1}{3}\,[11\bar{2}3]$$

$$\frac{1}{6}\,[02\bar{2}3] + \frac{1}{6}\,[20\bar{2}3] \qquad \ldots \text{ 3b}$$

where reaction 3b (Fig.8) is the subject of controversy.

With the development of techniques for examining large areas of microstructure by transmission electron microscopy,[41-44] it was soon established that dislocations were not uniformly distributed throughout the WC structure.[45,46] High dislocation densities observed in a proportion of the grains were probably inherited from deformation introduced during milling of the powders (Fig.9a) and had survived the sintering treatment.[47] Thus it was found that raising the sintering temperature decreased the number of such grains.[46] In general it is observed in as-sintered structures, that WC grains contain curved dislocations, stacking faults, or are dislocation-free, while the Co-based binder phase is a heavily

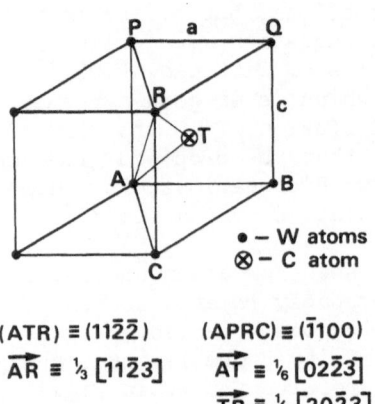

WC hexagonal unit cell $\frac{c}{a} = 0.975$

• — W atoms
⊗ — C atom

$(ATR) \equiv (11\bar{2}\bar{2})$ $(APRC) \equiv (\bar{1}100)$

$\overrightarrow{AR} \equiv \frac{1}{3}[11\bar{2}3]$ $\overrightarrow{AT} \equiv \frac{1}{6}[02\bar{2}3]$

$\overrightarrow{TR} \equiv \frac{1}{6}[20\bar{2}3]$

Fig. 8.   The WC unit cell and directions of the Burgers vectors in the proposed reaction:

$$\frac{1}{3}[11\bar{2}3] \rightarrow \frac{1}{6}[02\bar{2}3] + \frac{1}{6}[20\bar{2}3]$$

faulted fcc structure (Figs. 9b – d). Progress in understanding the nature of contiguity of WC grains at WC/WC interfaces in WC-Co hardmetals has recently been made with the observation by Sharma et al[48] by X-ray analysis in scanning transmission electron microscopy, that in a 2 nm region at the interface the Co/W ratio is three times higher than in WC remote from the interface.

There are few reported data on the dislocation structure in cubic-carbide hardmetals. Dislocations are observed in TaC-WC-Co hardmetals, but the density is much lower than in WC-Co hardmetals (Fig.9a), while the addition of TiC leads to the formation of regions of (Ta, Ti, W)C (Fig.9f) which are usually surrounded by WC grains, and featureless except for dislocation arrays at the boundaries of a TiC-rich core region.[49,50] Observations on (V,Ti) C-(Ni,Mo) hardmetals[51] and Ni-based WC hardmetals,[52] provide further evidence of the low dislocation density in cubic-carbide phases compared with that in WC.

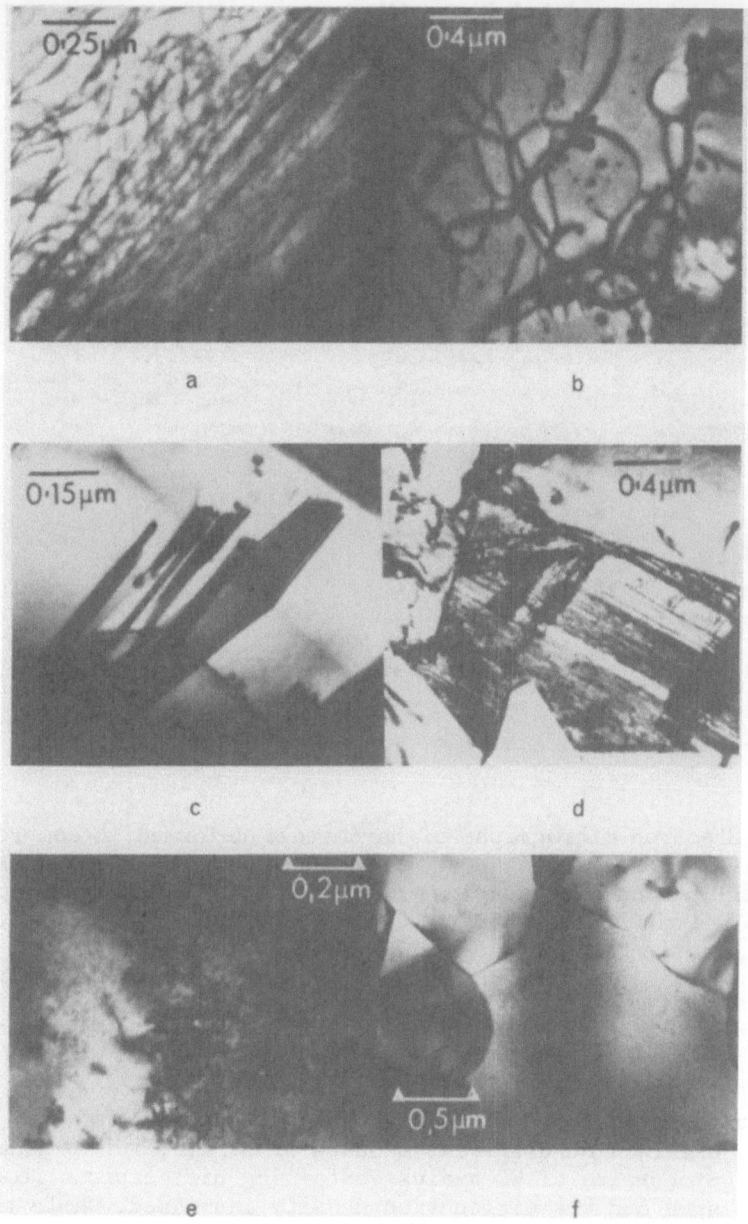

Fig. 9. Electron micrographs showing: (a) dislocations in WC
grain in spirit-milled powder, and (b) dislocations in
WC grain in sintered structure of coarse-grained 11% Co-
WC hardmetal: (c) stacking faults in WC grains, and (d)
in binder phase, in fine-grained 11%Co-WC hardmetal: (e)
dislocations in TaC grain in 2%TaC-10%Co-WC hardmetal:
(f) (Ta,Ti,W)C grain in 10%TaC-2%TiC-10%Co-WC hardmetal.

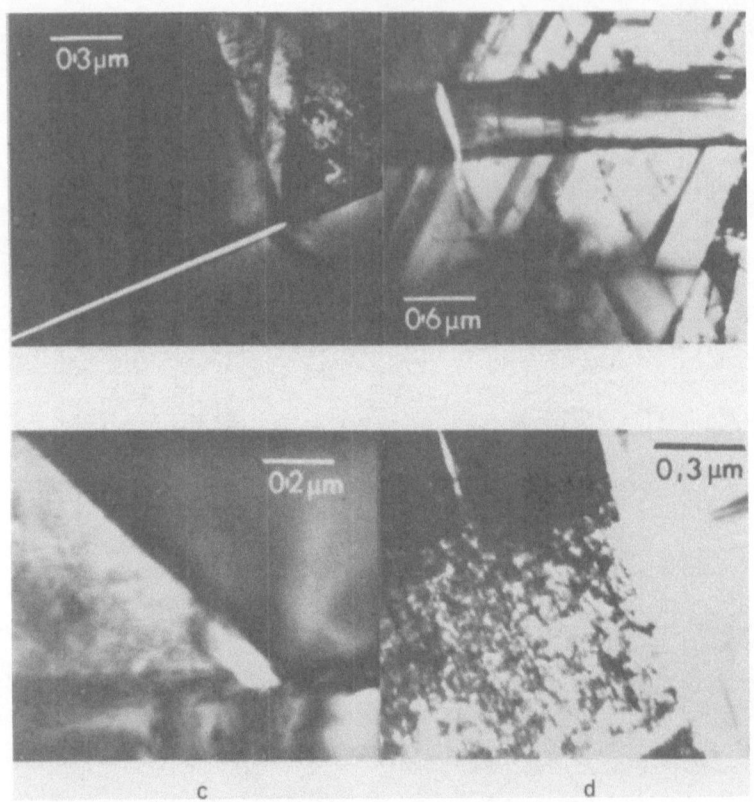

Fig. 10.   Electron micrographs of hardmetals deformed in compression:
           (a) microcrack at WC/WC interface, and (b) microcrack at
           slip-band intersection in WC grain, in coarse-grained
           11%Co-WC hardmetal, (c) cavity in binder phase at WC/WC
           junction in coarse-grained 9%Co-WC hardmetal:
           (d) dislocations and interface-microcrack in TaC grain
           in 2%TaC-10%Co-WC hardmetal.

        In compression tests on WC-Co hardmetal specimens with 6, 9
and 11%Co, examined at different amounts of strain, it was found
that: the proportion of WC grains containing dislocations remained
almost the same and the dislocation density increased, while the
proportion of grains containing stacking faults increased from 6%
to 25% of the total, presumably at the expense of initially-
unfaulted grains;[46,53] the ratio of hcp to fcc structure in the
binder-phase increased with strain, and microcracks initiated at
WC/WC interfaces (Fig.10a) until more than 50% were cracked at a
strain of 4% (Fig.11). Transgranular-cracks initiated at slip band
intersections were sometimes observed but there was no characteristic

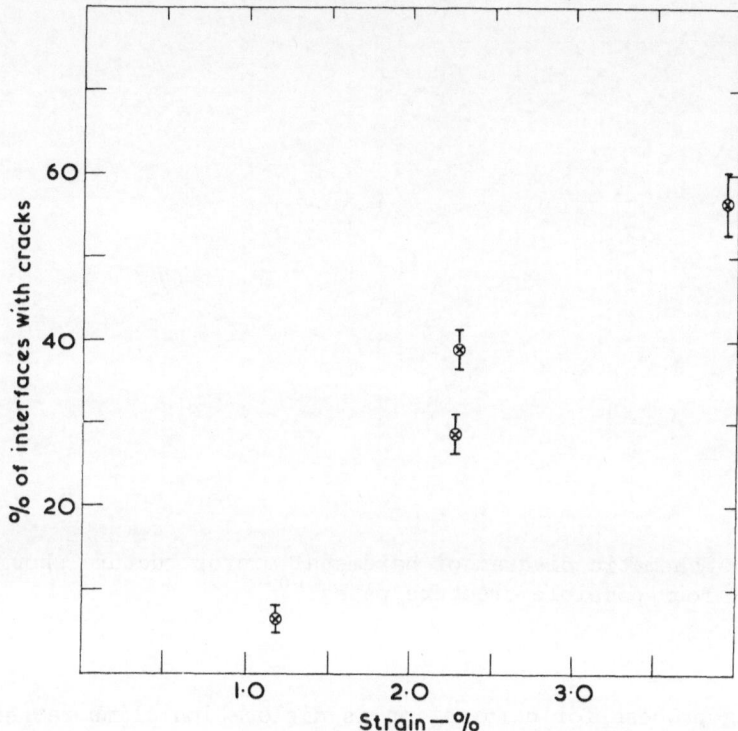

Fig. 11.   Variation in percentage of cracked WC/WC interfaces
versus strain, in compression specimens of coarse-
grained 11%Co-WC hardmetal.[47]

crystallographic cleavage-plane. (Fig.10b)   Cavitation of the binder
phase occurred in regions of high strain intensity (Fig.10c).

The deformation processes in compression of TaC-WC hardmetals
are probably similar to those above, with an increase in dislocation density and the initiation of microcracks at carbide-carbide
interfaces (Fig.10d). Similarly, Gottschall et al[45] have observed
deformed mixed-carbide grains (Ta, Ti, Nb, W)C in compression tests
on 6%Co hardmetals at 1100 and 1400 °C, while rounding of WC grains
in compression at the higher temperature was attributed to
dissolution and reprecipitation occurring in the solvus region.

There are few direct observations of deformation processes in
hardmetals in the temperature range 700 to 1000 °C. Activation
energies derived for these temperatures from expressions 1 and 2
for WC- and mixed-carbide hardmetals[8,54,55] tend to be higher than
the 2.9eV atom for Co diffusion, and it is likely that the rate

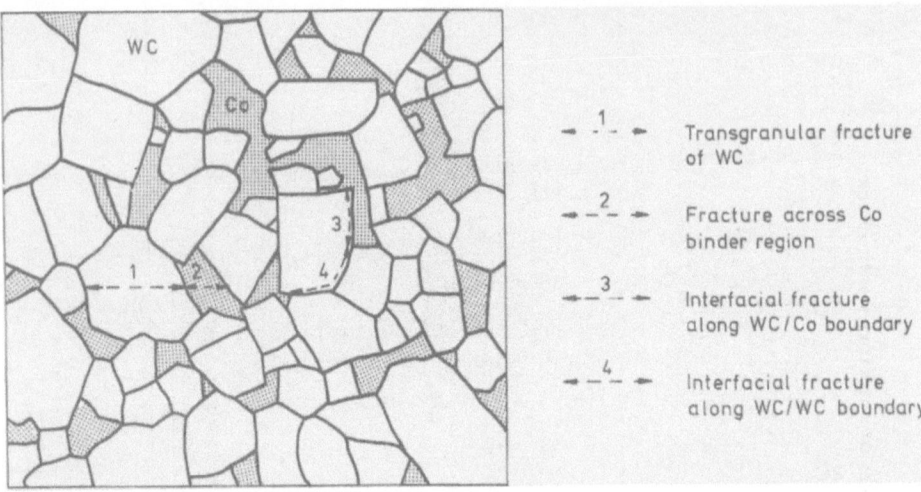

Fig. 12. Schematic diagram of hardmetal microstructure showing
          four possible fracture paths.[60]

controlling process for deformation is dislocation climb rather
than diffusional creep.[55]

    A prominent feature in room-temperature tensile-deformation
of hardmetals[25] is the initiation of fracture from pores,
inclusions or coarse-carbide grains (Fig. 13a). In WC-Co hardmetals,
the ratio of transgranular/intergranular fracture of WC increases
with increase in grain-size,[56] but there has been contention over
the relative contribution of the WC and Co phases. Thus, there
are four possible fracture paths (Fig. 12), and a suggestion that
the predominant path was through the binder phase, was corroborated
by results from fracture replicas, scanning electron fractography
and Auger spectroscopy.[58,59] However, by using Auger spectroscopy
on uncontaminated fracture surfaces in 6% and 12% Co-WC hardmetals,
Lea and Roebuck[60] demonstrated that about 50% of the fracture
propagates through WC/WC interfaces, and the latter contain
approximately a monolayer of Co atoms, which is in good agreement
with the results of Sharma et al on the nature of WC/WC contiguity.
Furthermore, the fracture observations demonstrated that less than
25% of the fracture would pass through the binder phase.

    In compression fatigue on a 15.6%Ni-WC hardmetal it was
observed that damage by dislocation mechanisms was associated with
precipitation reactions and accrued mainly in the binder phase.[52]
Dislocation structures have not been studied in fatigue-crack-growth

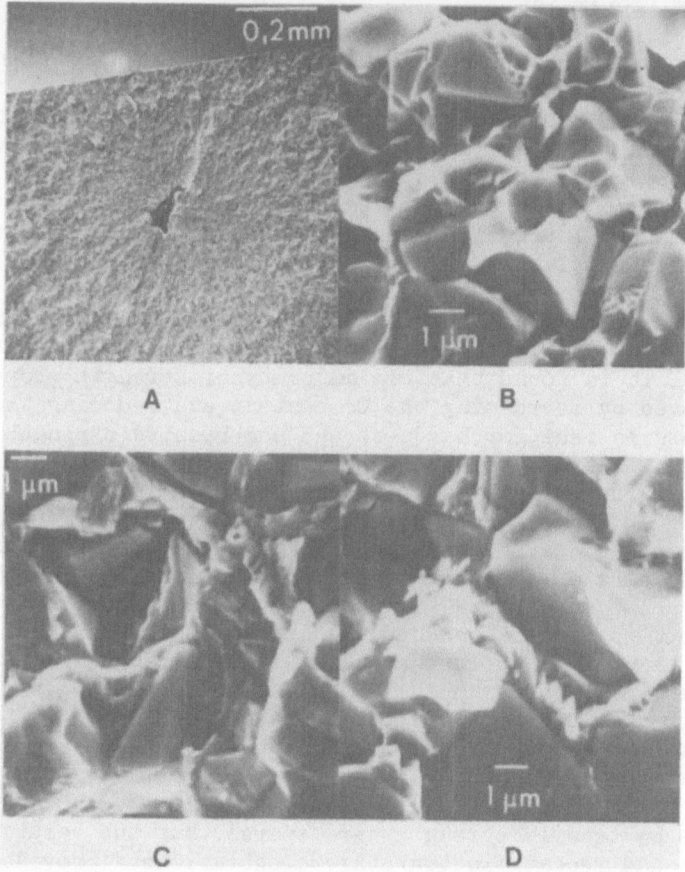

Fig. 13. Electron micrographs of: (a) fracture initiation source in transverse-rupture-test specimen of fine-grained 11%Co-WC hardmetal: (b)-(d) fracture path in fatigue specimen of a coarse-grained 11%Co-WC hardmetal, showing fracture surfaces corresponding to (b) unstable crack propagation, and (c) and (d) fatigue crack growth.

tests. but in the fatigue-fracture path in a coarse-grained 11%Co-WC hardmetal, it was found that fracture of the binder phase regions was brittle and crystallographic in appearance,[61] in contrast to unstable fracture where ductile rupture is observed (Figs. 12b - d). Similar observations are reported for 6% and 10% Co/WC hardmetals.

Mechanical Properties

In examining how the deformation characteristics of hardmetals are manifested in the their mechanical properties, it is found that the majority of reported data concern the effect of variations in Co content and WC grain-size, mainly in relation to transverse rupture strength. These results are described by various authors elsewhere,[63,65] while the purpose of the present analysis is to examine how the intrinsic mechanical properties are affected by metallurgical variables.

A summary of room temperature mechanical behaviour in smooth-specimen tests on WC-Co hardmetals is presented in Fig.14A-E, and in general it is found that the majority of strength properties are improved by decreasing the Co content and reducing the WC grain-size. This is true for hardness and compressive strength in static and fatigue loading, where the outstanding properties of hardmetals ensure their superiority over other materials in many applications (Fig. 14B-D). Strength, however, is not always the most important requirement, and although average compressive strengths have improved in recent years as a result of better materials and improved testing techniques (Fig.14C), ductility in compression is often needed, but is usually reduced when strength is increased. Another property that often suffers as a result of strength increases, is reproducibility; for example in the compressive-fatigue results for hardmetals shown in Fig.14D, the number of cycles to failure can vary by more than two orders of magnitude at an average life of 10^5 cycles.[68,69]

Data on tensile strengths are sparse, but the results of Gurland[70] and recent fractographic observations[25] provide a reasonable explanation for observations that tensile strength and bend strength reach a maximum at low Co contents and fine WC grain-sizes. Thus, the peaks in properties shown in Fig.14E probably represent the intersection of results for two different fracture mechanisms: on the left hand side of the peak, the majority of fractures occur under fully elastic conditions and are probably initiated at pores, inclusions and coarse WC grains by a Griffiths-type mechanism; on the right hand side of the peak, fracture is preceded by a significant amount of plastic straining, and the strength levels are a truer reflection of effects from micro-structural variables. Consequently, higher tensile strengths can be obtained by reducing porosity by hot-isostatic-pressing, and by reducing impurities and narrowing the grain-size distribution. A bend test has been devised to determine the improvements that can be obtained by taking these measures.[27]

Microstructural requirements for obtaining good room-temperature properties are not always compatible with those for producing good high-temperature performance. As is shown in Fig.2, strength

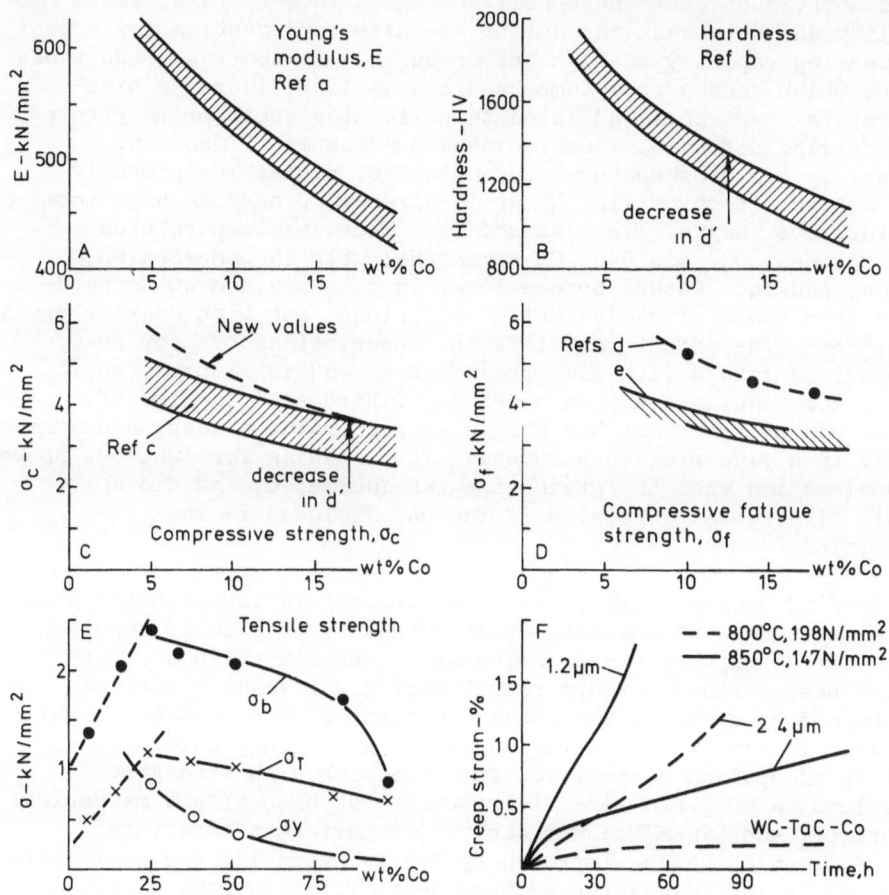

Fig. 14A-E.   Dependence of various mechanical properties of WC-Co
hardmetals on Co content and WC grain size, 'd'
(research results supplemented by commercial data):
A - Young's modulus; B - Vickers hardness; C - Ultimate
compressive strength; D - Ultimate compressive
strength in fatigue; E - Bend strength, σ_b; tensile
yield stress, σ_y; tensile strength, σ_t (Nishimatsu[70]
and Gurland); F - Creep strain in tension in 10%Co-
WC hardmetal at different WC grain-sizes, temperatures,
stresses, and with 10%TaC addition (Ueda[55] et al).
Refs: a) Lardner and McGregor[66]; b) Lee and Gurland[67];
c) Kreimer[63]; d) Rogan and Parry[68]; e) Johansson et
al[69].

levels start to decrease rapidly at temperatures between 600° and
900 °C in WC-Co hardmetals and this effect is most marked at high
Co contents, such that the benefits of good room-temperature
ductility and toughness may not be adequate compensation for a poor
load-bearing capacity at high temperatures. In contrast, additions
of TiC, which reduce room-temperature ductility, increase high-
temperature strength in addition to performing their primary role
of conferring diffusion-wear resistance. The most important
reversal in a microstructural dependence of properties probably
concerns the effect of fine WC grain-sizes in producing high room-
temperature strength (Fig. 14B and C). Thus, as temperatures
reach between 800° and 900 °C, strength levels in compression,
tension, and bend tests, become lower in fine-grained WC-Co hardmetals than those of equivalent compositions, but with coarser grain-
sizes.[55,71] The effect parallels the observations of poor creep
behaviour of metals with fine-grain sizes, and is illustrated in
results for tensile creep of a 10% Co-WC hardmetal in Fig.14F.
Here it can be seen that the strain at a constant stress, accelerated
rapidly in a fine-grained hardmetal after loading for 10h, but showed
no acceleration when the grain-size was increased, and did not
exhibit significant straining if TaC was included in its
composition.[55]

     To find data on effects of metallurgical variables other than
WC grain-size and Co content, it is necessary to examine results
on transverse rupture strengths, despite the limitations of the
test. Thus, various workers report that significant variations can
be obtained by altering the total C content of the hardmetal, while
remaining in the two-phase region, and that it is possible to
identify an optimum composition for obtaining high strength.[65]
Variations in WC grain-size distribution can also affect mechanical
properties, and Exner and Gurland[65,67] report that transverse
rupture strength can be improved by 15% by narrowing the grain-size
range. In using transverse rupture tests for measuring other
properties the work of Suzuki is distinguished by the attention
paid to reporting details of variability as an important property[73],
and expressing results as a cumulative distribution. Measurements
of effects of changing the rate of loading have failed to reveal
a strain rate dependence in 6, 7 and 10% Co-WC hardmetals[72,74]
but this subject merits further investigation by other test procedures
since materials that exhibit a temperature dependence of mechanical
properties are usually strain-rate sensitive.

Fracture Mechanics Parameters

     The plane-strain fracture toughness, K_{IC} value, of WC-Co
hardmetals is increased by raising the Co content and increasing
the WC grain-size. This is shown in Fig.15 for data obtained on
specimens with natural precracks,[29,75-77] but similar trends have
been observed for specimens with spark-eroded notches.[78]   A

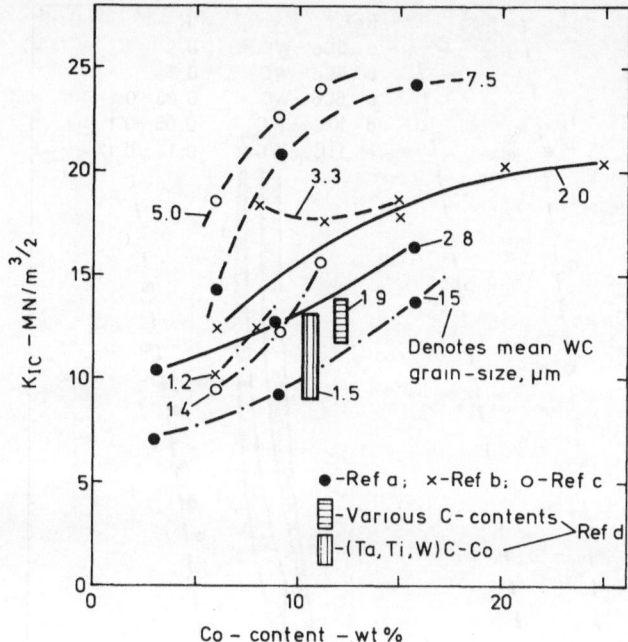

Fig. 15. Variation in plane strain fracture toughness with Co
content for WC-Co hardmetals with different WC grain-
sizes. Ref. a) Lueth[75], b) Ingelstrom and Nordberg[76],
c) and d) Almond and Roebuck.[29,77]

dependence of K_{IC} on C- content has been found in two-phase WC-Co
hardmetals, while additions[77] of TiC and TaC tended to lower K_{IC}.

There are few data for fatigue-crack growth in hardmetals.
The few available results[61,62,80] are plotted on a logarithmic scale
in Fig. 16 for da/dN, the crack growth rate, versus ΔK, the stress-
intensity range, according to convention for a Paris-law
relation:[79]

$$\frac{da}{dN} = C\Delta K^m \qquad\qquad \dots 4$$

where C is a constant, and the value obtained for m is often
interpreted as an indicator of the operative crack growth rate
mechanism. It can be seen that crack growth rate appears to
increase considerably with increase in R-value (ratio of minimum
load to maximum load) and is lower at higher Co contents for a
given ΔK value. The results give values for m (Table 3) that

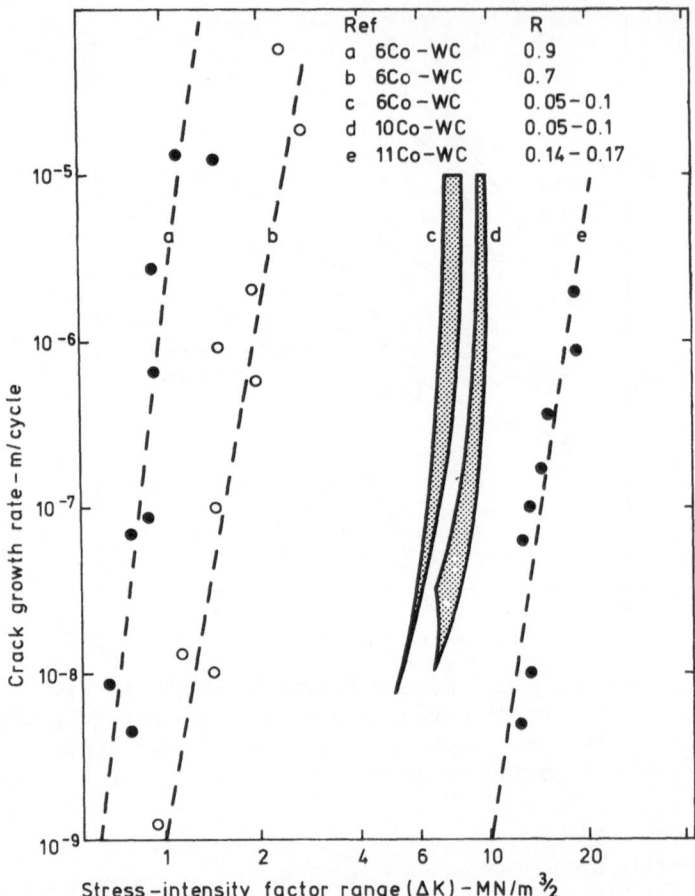

Fig. 16. Crack growth rate versus stress-intensity-factor range
          for fatigue tests on various WC-Co hardmetals.
          Refs: a) and b) Evans and Linzer[80]; c) and d) Fry[62];
          e) Almond and Roebuck.[61]

are much higher than those observed in metals, and such high values
are often associated with microstructurally-dependent crack growth.[79]
Evidence to support this theory has been described earlier (Fig. 12)

MECHANISMS, MODELS AND OBSERVATIONS

       Progress in understanding how and why hardmetal microstructure
affects mechanical aspects of tool performance will be made more

Table 3.   Materials, Conditions and m-Values for Fatigue Crack
Growth Measurements on WC-Co Hardmetals

Co %	Grain-size	Frequency Hz	R	m	Ref
6	fine	5	0.05-0.10	12.7-31.3	62
10	fine	5	0.05-0.10	11.4-19.5	62
11	coarse	20	0.14-0.17	10	61
6	-	10-350	0.9	15	80
6	-	10-350	0.7	11.6	80

All tests on double torsion specimens except
Ref 61 which used single-edge notched beam
specimens.

rapidly if a basic understanding is established first in relating
microstructure to intrinsic mechanical properties. Attempts to do
this, have been confined almost entirely to WC-Co system and have
become more realistic in their modelling recently, now that it is
accepted that plastic deformation of WC grains can be considerable
and is exhibited as a bulk phenomenon rather than being confined
to surfaces, and that the parameter known as the mean binder-
thickness, d_b , used in early theories, is an inadequate
representation of the true binder spacing in real microstructures
(d_b is defined as $V_b . d_w / V_w$, where subscripts b and w refer
to binder and WC phases respectively, V is the volume fraction,
and d_w is WC grain-size).

Models exist for explaining the yield stress, tensile strength,
and fracture toughness of WC-Co hardmetals but all are semi-
empirical in the relationships they propose, partly relying on
experimentally-observed dependences of microstructure on properties.
Consequently, they cannot be tested for other hardmetal systems
where the equivalent amount of data does not exist. No theories
are able to explain all the deformation characteristics of the
WC-Co system, and some of the deficiencies will be discussed in the
following sections on room-temperature properties. High temperature properties are omitted since there have been insufficient
experimental observations on the deformation characteristics of the
constituents in the microstructure to test the assumptions of the
theories.

Strength Models

In considering the onset of plastic deformation, attention has been paid to the load transfer problem of describing the distribution of stresses and strains between the WC and binder-phase constituents by continuum mechanics theory. Early analyses of elastic behaviour of Hashin and Shtrikman[81] indicated that at low Co contents, the measured elastic moduli were closer to those for a soft dispersion in a continuous stiff matrix than for other possible dispersions. This observation was confirmed in a finite element analysis of a simulated hardmetal structure by Jaensson and Sundström.[82]

In examining the distribution of stresses between the constituents, Gurland[83] made a distinction between the contiguous and non-contiguous components of the WC structure, having volume fractions CV_w and $(1-C)V_w$ respectively where C is the contiguity factor. An expression for σ_y the yield stress of the hardmetal was derived in the form:

$$\sigma_y = \sigma_{be} (1 - CV_w) + \sigma_w CV_w \qquad \ldots 5$$

where σ_{be} and σ_w are respectively the 'in situ' yield stress of the binder and stress in the contiguous WC structure. The expression is similar to that for parallel fibres of volume fraction CV_w oriented parallel to the stress axis. Elastic-plastic behaviour was also examined by Sundström[84] who used a finite element analysis of a simulated hardmetal structure to demonstrate that during the early stages of yielding relatively high plastic strains in the binder can be accommodated by elastic deflection of the WC structure, and this can be used to explain the hysteresis in the stress-strain curves of these materials.

For large plastic deformations, Lee and Gurland[67] demonstrated that the hardness, H could be expressed as:

$$H = H_b(1 - CV_w) + H_w CV_w \qquad \ldots 6$$

where H_b and H_w are respectively the 'in situ' hardnesses of the binder and WC structures.

A weakness in continuum mechanics theories is that WC/WC interfaces are treated simply as bridges between carbide grains, and they are not assigned specific properties. Sliding at WC/WC interfaces was proposed as a possible deformation mechanism by Chermant and Osterstock in reviewing dislocation models for WC/Co hardmetals. They suggested that this might be one of several explanations for the experimentally observed dependence of yield stress on λ_b, the mean free path in the binder phase. They also commented on the constancy of σ_y/C values for hardmetals for low

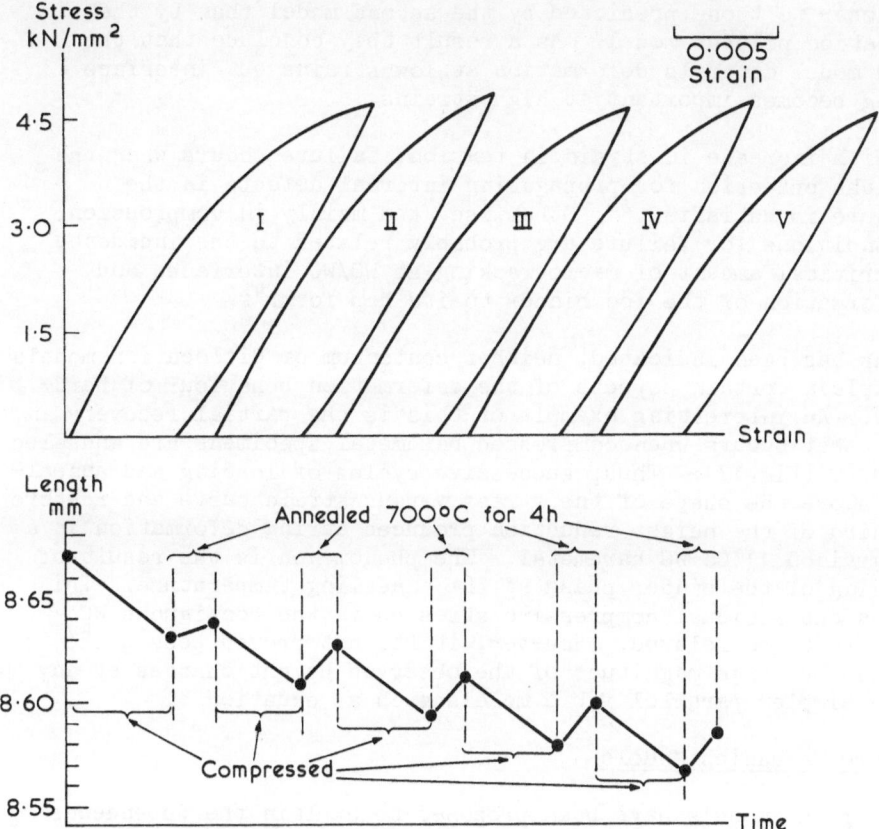

Fig. 17.   The effect of repeated cycles of compression and
annealing at 700 °C for 4h on the length and stress
versus strain curves of a fine-grained 11%Co-WC
hardmetal.[87]

Co contents, and suggested that this indicates that the applied
load is supported by the contiguous carbide structure. Two other
mechanisms, examined to explain the dependence of σ_y on λ_b were
those involving shear of WC grains under the stress from a
dislocation pile-up in the binder, and strengthening by the
geometrically necessary dislocations that are generated to accommodate the deformation. The need for the existence of the latter
mechanism to explain strengthening has been challenged by
Fischmeister and Karlsson[86] who postulate that continuum mechanics
theories are adequate for this purpose. However, Chermant and
Osterstock find that the observed slopes of σ_y versus $\lambda_b^{-\frac{1}{2}}$ graphs

are closer to those predicted by the second model than by the dislocation pile-up model. As a result they conclude that the second model controls deformation at low strains but interface sliding becomes important at high strains.

With increase in strain in tension, failure occurs when the Griffiths criterion for propagating internal defects in the structure is satisfied.[25] Otherwise, and mainly in compression, the conditions for failure are probably related to the incidence of a critical amount of microcracking at WC/WC interfaces and transformation of the fcc binder to its hcp form.[42]

As has been indicated, neither continuum or dislocation models can explain certain aspects of the deformation behaviour of hard-metals. An interesting example of this is the partial recovery in height that occurs when compressed hardmetal specimens are annealed at 700 °C (Fig.17). Thus, successive cycles of loading and anneal-ing restore the shape of the stress versus strain curve and restore one third of the height reduction produced during deformation in a fine-grained 11%Co-WC hardmetal. The phenomenon is the result of softening of the binder phase at the annealing temperature. This enables the residual compressive stresses in the contiguous WC structure to be relaxed. However, it has not proved possible[87] to explain the large magnitude of the observed height changes by any of the simpler parallel fibre models such as equation 6.

Fracture Mechanics Models

Various models have been proposed to explain the dependence of fracture toughness on microstructural parameters in WC-Co hard-metals; but the situation is confused, because the various authors have used different theoretical approaches and have tested their models against sets of experimental observations that have often displayed different trends to those used by other authors. The following four typical analyses illustrate the different approaches.

Lindau[88] based his model on a criterion that a critical strain, had to be reached at a distance λ_b ahead of the crack tip before a crack can propagate. This leads to the result:

$$K_{IC}^2 = 6\pi\lambda_b \, \sigma_y^2 \left(\frac{EV_b\varepsilon_c}{\sigma_y}\right)^{n+1} \qquad \qquad \ldots \, 7$$

where n is the work hardening exponent of material ahead of the crack. Using values of 0.25 and 6% for n and ε_c respectively, he found his experimental results for K_{IC} were linearly related to the parameter on the right hand side of equation 7.

Pickens and Gurland[78] suggested that a criterion based on ductile fracture of the binder would be compatible with crack extension occurring when the crack opening displacement reached a value of the order, λ_b , and concluded that:

$$G_{IC} = \alpha \sigma_{be} \lambda_b \qquad \qquad \dots 8$$

where α is a constant, and σ_{be} is the 'in situ' yield stress of the binder phase (equation 5). They observed a linear dependence of G_{IC} on λ_b in their results but they could not explain the low values for α of 0.24 that they obtained.

In Chermant and Osterstock's[89] model, the cleavage of carbide grains at the crack tip is regarded as a critical event in crack extension, and occurs by the conversion of the stored energy in a dislocation pile up of length λ_b in the binder phase being converted into the surface energy of the cleaved carbide grains. Accordingly this gives the relation,

$$2\,\gamma_w = A\,\frac{\lambda_b^2}{d_w} \qquad \qquad \dots 9$$

where γ_w is the surface energy of WC, and A is a constant. Their experimental observations show that G_{IC} is linearly dependent on λ_b^2/d_w.

In common with the previous analysis, Nakamura and Gurland[90] used an energy balance, but they considered the contributions from the energy required to produce the three different fracture paths observed on the fracture surface (Fig. 12).

They obtained the expression:

$$G_{IC} = 2\gamma_T\,V_w^{\frac{2}{3}} + \alpha''\,\lambda_b\,V_w^{\frac{2}{3}}\,\frac{(1-V_w^{\frac{2}{3}})}{V_b}\int_o^{\varepsilon_c}\sigma(\varepsilon)d\varepsilon \qquad \dots 10$$

where γ_T is the total effective surface energy from the three energy contributions, α'' is a constant, and the integral represents the energy required for ductile rupture of the binder phase. Their experimental values for G_{IC}, together with those from earlier work, were found to vary linearly with λ_b for

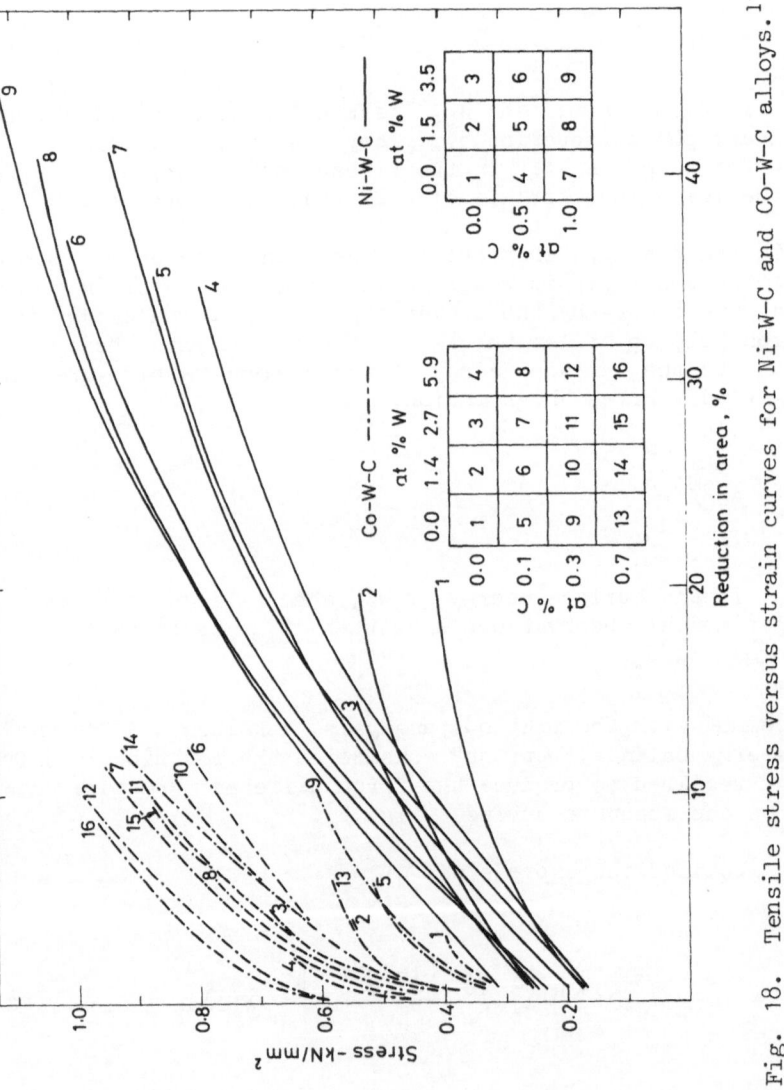

Fig. 18. Tensile stress versus strain curves for Ni-W-C and Co-W-C alloys.[19]

constant values of V_w.

There are many weaknesses in the theoretical derivations of equations 7 - 10. However, they are also based on assumptions that do not agree with experimental observations, and these mainly concern the nature of the fracture path and the value for the integral in equation 10 and similar expressions. Thus, as has been emphasized by Viswanadham et al,[59] the fracture propagates through a different path in the microstructure from that observed on a planar section through the microstructure, and the energy requirements will be different from those assumed in the various models described above. Thus if the major source of energy dissipation is that for ductile rupture of the binder then the cracked region at the tip of a sharp precrack will have a different areal composition from that observed on a metallographic section, and crack extension will involve the energy balance:

$$G_{IC} \simeq A_b \, \lambda_b \int_0^{\varepsilon_c} \sigma(\varepsilon) d\varepsilon \qquad \qquad \ldots 11$$

where A_b is the areal fraction of ductile rupture on the fracture surface (path 2 in Fig. 12) Lea and Roebuck[60] have demonstrated that A_b is probably the same as the areal fraction of binder on a metallographic section, but that WC/WC interface-fractures accounted for about 50% of the fracture surface. A consequence of this observation is that to obtain reasonable agreement with experimental values for G_{IC}, the integral in equation 11 would need to be much higher than is commonly assumed. This however is quite likely, since existing models have failed to take into account, the very high work hardening rate of Co-based binder phases[19] (Fig. 18).

Thus some of the weaknesses of the existing models can be reduced by incorporating the above observations, but the expressions relating fracture toughness to microstructure, are still too empirical for them to be used in quantitatively predicting how fracture toughness could be improved without affecting other microstructurally dependent properties.

PRACTICAL CONSIDERATIONS

Although fundamental explanations have not yet been obtained for the microstructural dependence of mechanical properties of hardmetals, significant advances have been made in elucidating the microstructural dependence of tool failure mechanisms. This has been achieved mainly by the application of scanning electron microscopy which, for example, has enabled abrasive wear mechanisms in mining tools to be rationalized. Concurrently, in the drive for binder-phase substitution in Co-based hardmetals, trials on new compositions have provided useful information about the metallurgical requirements for hardmetals for various applications. As a result, tool technology is reaching a stage where it will be possible to design simple tools on the basis of laboratory data on mechanical properties, and WC-Co hardmetals will probably be replaced in many applications where they have been the only hardmetal used since their inception.

Applications

The property requirements for tool applications are usually expressed in terms of the tool materials' resistance to various wear mechanisms, but apart from hardness, no mechanical property is stipulated as being important for an individual application. Nevertheless, it is possible to define some of the simpler requirements without discussing wear mechanisms in detail.

Thus, the introduction of coatings for cutting-tools has eliminated many of the chemical and attrition wear problems, but the substrate still needs to be creep-resistant at high cutting-temperatures,[91] and sufficiently tough to resist the propagation of microcracks in the coating.[92] Also in coated and uncoated tools thermal-fatigue resistance needs to be combined with creep resistance in interrupted-cutting applications.[7]

In examining abrasion resistance, particularly in rotary drilling of rocks, recent observations[93-97] indicate that for relatively soft rocks it may be necessary to have separate property requirements for the binder-phase and the contiguous WC structure in WC-Co hardmetals. Thus, when the binder has been eroded-away due to its softness relative to the abrasive and the WC structure, the toughness of the latter determines abrasion resistance. Consequently, resistance to abrasion decreases with increase in fracture toughness of the hardmetal (increased Co-content) but would increase with toughness of the WC contiguous structure.

At high drill-speeds in general, and especially with non-abrasive rocks, temperatures in the range 500 - 1000 °C can be generated at the tool surface,[97] and the tool material needs to be resistant to thermal fatigue and creep distortion. As a result, the finer grained hardmetals, chosen for low-temperature abrasion-resistance, suffer high wear-rates, and this is an area where the development of creep resistant hardmetals would be beneficial.

In percussive drilling and in metal-forming, it appears that impact toughness is an important property, and in the former

application it has been observed that wear resistance and trans-
verse-impact strength show a similar dependence on Co content
(Fig.14E). This indicates that at low Co-contents, better wear
resistance could be obtained by reducing porosity and other defects,
and this has been proved.[98] Furthermore, if residual defects are
regarded as a potential source for tool failure, it is possible to
use their existence as the basis for tool-life prediction and
tool-design.[99] It has been demonstrated how this can be achieved
by using probabilistic fracture mechanics on a microstructural
scale to analyse simple shapes, such as countersink dies, and to
predict their lives (Fig.19). This illustrates that although the
task of designing tools for metal-shaping and for mining appears
formidable, it should be possible to use a scientific approach to
design to improve their resistance to some of the simpler failure
mechanisms.

Improved Properties

It is encouraging that recent improvements in mechanical
properties of hardmetals have been scientifically based, as for
example in the application of hot isostatic pressing, which not
only removes gross-defects but also removes smaller defects that
can behave as incipient cracks. Similarly, in searching for a
substitute for Co in the binder phase of hardmetals, various workers
have demonstrated the inadequate strength of Ni and have shown that
gamma-prime strengthening, used in superalloys, can be employed in
Ni-based binder phases in WC and TiC hardmetals to give equivalent
and possibly superior mechanical properties to those of Co-based
hardmetals.[49,100-102]

Such work is effective because it recognizes that even the
simplest binder-phase is a complicated alloy with a wide-range of
possible properties, all completely different to those of the pure
metal base (Fig.18). This is especially true in hardmetals with
Fe-Ni based binders, where the hardness of the partly-martensitic
binder confers good abrasion resistance,[21] and high temperature
strength can be obtained by additions of Mo and Cr.[103] High
temperature properties suitable for metal cutting, is obtainable
in a number of commercial TiC - Al_2O_3 grades,[2-20] while TiB_2
hardmetals with Fe and Ni binders have good high temperature
properties and will be a commercial proposition if their room-
temperature toughness is improved.[104]

The possibility of producing improved properties in WC-Co
hardmetals by precipitation hardening of the binder phase through
alloying additions, does not appear to have been investigated in
detail. The potential for this mechanism has been demonstrated
in various experiments on WC-Co hardmetals with W-rich binder
phases, where it has been shown that precipitation of Co_3W can be
used to increase hardness, where this is a service requirement.[105,106]

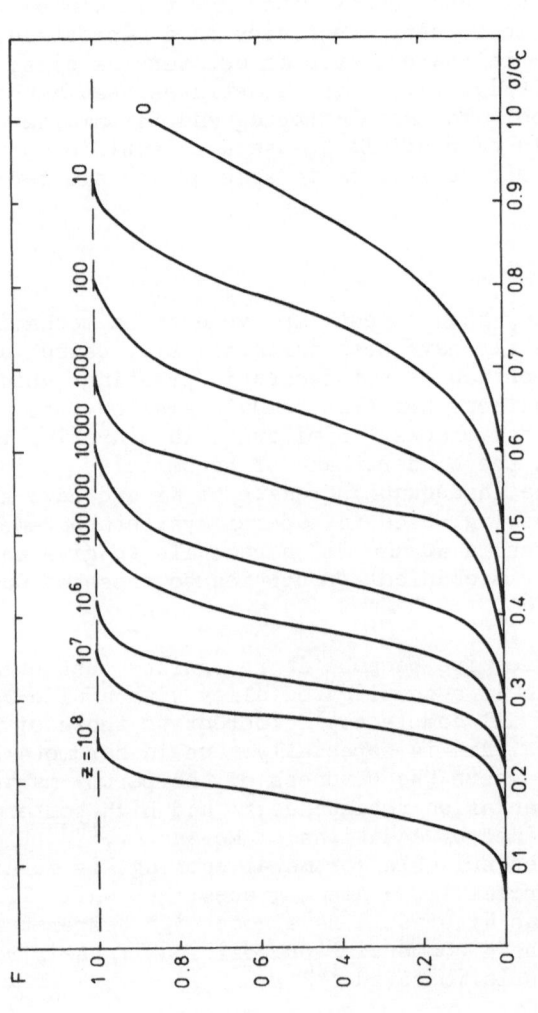

Fig. 19. Predicted[99] cumulative-failure-probability, F, of hardmetal countersink-dies at different ratios of applied stress to ultimate strength, for various numbers of loading cycles, Z.

However, reduced toughness and the loss of the hardness increment at high temperatures are characteristic of this and many other precipitation processes, and consequently the range of applications may be limited.

FUTURE REQUIREMENTS

It is evident that a considerable amount of experimental work needs to be done before deformation mechanisms in hardmetals are sufficiently understood to be able to determine the limits to improving the mechanical properties and to apply the knowledge in improving tool performance. Most current knowledge is related to the dependence of mechanical properties on mean carbide grain-size and binder-phase content in WC-Co hardmetals, and does not explore the possibilities offered by altering grain-size distribution and binder-phase composition.

The techniques are available for studying deformation characteristics and microstructural variables, and it is suggested they should be employed in investigating mechanisms of plastic deformation and fracture in both WC based cubic-carbide hardmetals tested at low and high temperatures. Micro-analytical techniques for surfaces should be employed to determine the fracture path in static and fatigue loading. Absence of such information is a major handicap in attempts to model the mechanisms of deformation and fracture.

Data on crack-growth in mechanical, thermal and corrosion fatigue are needed for simple and complex loading to see whether the relevant properties are related to service performance, and to provide information for attempting design of simple tool shapes.

In the field of substitution, a detailed investigation is required of all the potential strengthening mechanisms that can be employed to improve the mechanical properties of the binder phase in hardmetals for high and room temperature applications. A detailed study of the contiguous WC structure in WC-Co hardmetals could provide information on the nature of WC/WC bonds, and indicate whether such bonding is advantageous, and could be obtained in other hardmetal systems.

ACKNOWLEDGMENTS

Much of the unpublished work in this paper was performed in collaboration with Dr B Roebuck. Helpful discussions with Dr B Roebuck, Dr M G Gee and Dr R Morrell of NPL, are gratefully acknowledged.

REFERENCES

1. H. E. Exner, Physical and chemical nature of cemented
    carbides <u>Int. Met. Rev.</u> 24:149 (1979).
2. K. J. A. Brookes, "World Directory and Handbook of Hardmetals",
    Engineers' Digest, London, 2nd Ed. (1979).
3. C. Bonjour, New developments in cemented carbide cutting
    tools, <u>Wear</u>, 62:83 (1980).
4. R. Morrell. National Physical Laboratory.
5. W. R. Buessem, Thermal shock testing, <u>J. Am. Ceram. Soc.</u>,
    38:15 (1955).
6. Y. W. Mai and A. G. Atkins, Fracture strength behaviour of
    tool carbides subjected to thermal shock, <u>J. Am. Ceram.
    Soc.</u>, 54:593 (1975).
7. C. S. G. Ekemar, S. A. O. Iggström , and G. K. A. Heden,
    Influence of some metallurgical parameters of cemented
    carbide on the sensitivity to thermal fatigue cracking
    at cutting edges, <u>in</u> "Materials for Metal Cutting",
    ISI Publication 126, Iron and Steel Inst., London (1970).
8. W. Dawihl and M. K. Mal, Contribution to the study of the
    deformation behaviour and structure of WC-TiC-TaC-Co
    alloys, <u>Cobalt</u>, 26:25, (1965).
9. J. H. Westbrook and E. R. Stover, Carbides for high-temper-
    ature applications, <u>in</u> "High Temperature Materials and
    Technology", Wiley, New York (1967).
10. A. Miyoshi and A. Hara, High temperature hardness of WC,
    TiC, TaC, NbC and their mixed carbides <u>J. Jap. Soc.
    Powder and Powder Met.</u>, 12:24 (1965).
11. G. M. Schwab and A. Krebs, Measurement and theory of the
    variation in hardness with temperature of transition
    metal carbides, <u>Plansee fur Pulvermet</u> 19:91 (1971).
12. A. G. Atkins and D. Tabor, Hardness and deformation of
    solids at very high temperatures, <u>Proc. Roy. Soc.</u>,
    292A:441 (1966).
13. Cobalt Monograph, Centre d'Information du Cobalt, Belgium
    (1960).
14. L. J. Aschan, I. Johansson and L. E. Gustafsson, High
    Temperature properties of WC-Co cemented carbides, <u>in</u>
    "Proc. 4th Nordic High Temp. Symp. - NORTEMPS 75".
15. F. Ueda, H. Doi, F. Fujiwara and H. Masatomi, Bend
    deformation and fracture of WC-Co alloys at elevated
    temperatures, <u>Trans. J.I.M.</u>, 18:247, (1977).
16. E.M. Trent and A. Carter, Sintered titanium carbide alloys,
    <u>in</u> "Symposium on Powder Metallurgy", ISI Special Report
    58, Iron and Steel Inst., London (1954).
17. S. R. Schenck, R. J. Gottschall and W. S. Williams,
    Deformation of cemented carbides: high temperatures,
    stress relaxation, and strain-rate dependence,
    <u>J. Mat. Sci.</u>, 32:229 (1978).
18. A. G. Atkins, High temperature hardness and creep, <u>in</u>
    "The Science of Hardness Testing and its Research
    Applications", Am. Soc. Met., Metals Park, Ohio (1973).

19. B. Roebuck and E. A. Almond, A comparison of Co and Ni alloys
     containing small amounts of W and C, in "Proc. 10th Plansee
     Seminar" Vol. 1. Metallwerk Plansee, Austria (1981).
20. R. P. Wahi and B. Ilschner, Fracture behaviour of composites
     based on Al_2O_3-TiC, J. Mater. Sci. 15:875 (1980).
21. D. Moskowitz, M. J. Ford and M. Humenik, Jr., High strength
     tungsten carbides. Int. J. Powder Met., 6:55 (1970).
22. L. Prakash, H. Holleck, F. Thummler and G. Spriggs, WC-
     cemented carbides with improved binder alloys, in
     "Towards Improved Performance of Tool Materials" to be
     published, Metals Society, London.
23. E. A. Almond, Towards improved tests based on fundamental
     properties, in: "Towards Improved Performance of Tool
     Materials", to be published, Metals Society, London.
24. S. Timoshenko and J. N. Goodier, "Theory of Elasticity",
     McGraw Hill, London (1951).
25. E. A. Almond and B. Roebuck, The transverse rupture test for
     hardmetals, Met. Sci., 11:458 (1977).
26. M. J. Kerper, L. E. Mong, M. B. Stiefel and S. F. Holley,
     Evaluation of tests and correlation of results for brittle
     cermets, J. Res. NBS, 61:149 (1958).
27. B. Roebuck, The tensile strength of hardmetals, J. Mat. Sci.,
     14:2837 (1979).
28. E. A. Almond, R. S. Irani and B. Roebuck, A square
     indentation test for tool materials Mater. Sci. Eng.
     44:173 (1980).
29. E. A. Almond and B. Roebuck, Precracking of fracture tough-
     ness specimens of hardmetals by wedge indentation,
     Met. Technol., 5:92 (1978).
30. L. M. Barker, A simplified method for measuring plane strain
     fracture toughness, Eng. Fract. Mech., 9:361 (1977).
31. B. R. Lawn and E. R. Fuller, Equilibrium penny-like cracks
     in indentation fracture, J. Mat. Sci., 10:2016 (1975).
32. E. A. Almond and B. Roebuck, Some observations on indentation
     tests for hardmetals, in: "Conf. on Recent Advances in
     Hardmetal Production", Loughborough U. of Tech. and Met.
     Powder Rep. (1979).
33. R. K. Viswanadham and J. D. Venables, A simple method for
     evaluating cemented carbides, Met. Trans., 8A:187 (1977).
34. E. A. Almond and B. Roebuck, Extending the use of indentation
     tests, in: "International Conf. on Science of Hard
     Materials", to be published, Plenum, New York.
35. D. J. Rowcliffe and W. J. Warren, Structure and properties of
     TaC crystals. J. Mat. Sci. 5:345 (1970).
36. J. L. Chermant, G. Leclerc and B. L. Mordike, Deformation
     of TiC at high temperatures. Z. Metallkde, 71:465 (1980).
37. G. E. Hollox, Microstructure and mechanical behaviour of
     carbides, Mater. Sci. Eng. 3:121 (1968).
38. S. B. Luyckx, Slip system of tungsten carbide crystals at
     room temperature, Acta Met 18:233 (1970).

39. S. Hagege, J. Vicens, G. Nouet and P. Delavignette, Analysis of structure defects in tungsten carbide. Phys. Stat. Sol., (a) 61:675 (1980).

40. T. Johannesson and B. Lehtinen. The analysis of dislocation structures in WC by electron microscopy Phys. Stat. Sol. 16A:615 (1973).

41. E. A. Almond and B. Roebuck. Ion beam thinning applied to electron microscopy of hardmetals Metall. Mater. Technol. 5:184 (1973).

42. V. K. Sarin and T. Johannesson, On the deformation of WC-Co cemented carbides, Met. Sci., 9:472 (1975).

43. A. Hara, T. Nishikawa and T. Nishimoto, Transmission electron microscopy of WC-Co alloys. J. Jap. Soc. Powder and Powder Met., 16:310 (1970).

44. R. Arndt, The plasticity of WC-Co hardmetals, Z. Metallkd., 63:274 (1972).

45. R. J. Gottschall, W. S. Williams and I. D. Ward, Microstructural study of hot-deformed cemented carbides, Phil. Mag. 41A:1 (1980).

46. E. A. Almond and B. Roebuck. The origin of WC substructure and the effect of processing on the microstructure of WC-Co hardmetals in: "Proc. 10th Plansee Seminar" Vol 1, Metallwerk Plansee Austria 1981.

47. B. Roebuck and E. A. Almond. Microstructural events preceding fracture in compression in hardmetals in: "Conf. on Recent Advances in Hardmetal Production", Loughborough U. of Tech. and Met. Powder Rep. (1979).

48. N. K. Sharma, I. D. Ward, H. L. Fraser and W. S. Williams, STEM analysis of grain boundaries in cemented carbides, J. Am. Ceram. Soc., 63:194 (1980).

49. E. A. Almond and B. Roebuck. Unpublished work.

50. O. Rudiger and H. E. Exner, Application of basic research to the development of hardmetals, Powder Met. Int., 8:7 (1976).

51. R. K. Viswanadham. Carbide-binder interface ledges in (ViTi)C + (Ni,Mo) cermets Met. Trans., 10A:1631 (1979).

52. E. F. Drake and A. D. Krawitz, Fatigue damage in a WC-Ni cemented carbide composite, Met. Trans., 12A:505 (1981).

53. R. Greenwood, University of Birmingham, private communication.

54. M. J. Murray and D. C. Smith. Stress induced cavitation in Co bonded WC. J. Mater. Sci. 8:1706 (1973).

55. F. Ueda, H. Doi, F. Fujiwara, H. Masatomi, and Y. Oosawa, Tensile creep of WC-10%Co and WC-10%TaC-10%Co alloys at elevated temperatures, Trans. Jap. Inst. Met. 16:591 (1975).

56. M. J. Murray, Fracture of WC-Co alloys: an example of spatially constrained crack tip opening displacement, Proc. Roy. Soc. 356A:483 (1977).

57. J. L. Chermant, A. Deschanvres and A. Iost, Fracture mechanics, statistical analysis and fractography of carbides and metal carbides composites, in: "Fracture Mechanics of Ceramics, Vol. 1", Plenum, New York (1974).

58. R. K. Viswanadham and T. S. Sun, Determination of fracture modes in cemented carbides by Auger electron spectroscopy, Scripta Met., 13:767 (1979).

59. R. K. Viswanadham, T. S. Sun, E. F. Drake and J. A. Peck, Quantitative fractography of WC-Co cermets by Auger Spectroscopy, J. Mater. Sci., 16:1029 (1981).

60. C. Lea and B. Roebuck, Fracture topography of WC-Co hardmetals, Met. Sci., 15:263 (1981).

61. E. A. Almond and B. Roebuck, Fatigue-crack growth in WC-Co hardmetals, Met. Technol., 7:83 (1980).

62. R. Fry, University of the Witwatersrand, private communication.

63. G. S. Kreimer, "Strength of Hard Alloys", Consultants Bureau, Plenum, New York (1968).

64. P. Schwarzkopf and R. Kieffer, "Cemented Carbides", Macmillan, New York, (1960).

65. H. E. Exner and J. Gurland, A review of parameters influencing some mechanical properties of tungsten carbide-cobalt alloys, Powder Metal. 13:13 (1970).

66. E. Lardner and N. B. McGregor, Determination of elastic constants and stress strain relationship to fracture of sintered WC-Co alloys, J. Inst. Met., 80:369 (1951).

67. H. C. Lee and J. Gurland, Hardness and deformation of cemented tungsten carbide, Mater. Sci. Eng., 33:125 (1978).

68. J. Rogan and J. S. C. Parry, Fatigue strength of cemented tungsten carbides and tool steels subjected to cyclic compressive stresses, in: "6th AIRHPT. High Pressure Conf."Plenum, New York (1978).

69. I. Johansson, G. Persson and R. Hiltscher, Determination of static and fatigue strength of hardmetals, Powder Metall. Int., 2:119 (1970).

70. C. Nishimatsu and J. Gurland, Experimental survey of the deformation of the hard-ductile two-phase alloy system WC-Co, Trans. Am. Soc. Met., 52:469 (1960).

71. J. T. Smith and J. D. Wood, Elevated temperature compressive creep behaviour of WC-Co alloys, Acta Met., 16:1219 (1968).

72. J. Gurland, The influence of basic design and processing variables on sintered WC-Co alloys, in: "Powder Metallurgy", Interscience, New York (1961).

73. H. Suzuki, K. Hayashi, and H. Sakanoue, Mechanical properties of high strength WC-10%Co cemented carbides, Powder and Powder Metall, 16:83 (1969).

74. H. Doi, F. Ueda, Y. Fujiwara and H. Masatomi, The effect of boundaries on strength and toughness of cemented carbides, in: "Grain Boundaries in Engineering Materials" Claitors, Baton Rouge, La. (1974).

75. R. C. Lueth, Determination of fracture toughness parameters for WC-Co alloys, in: "Fracture Mechanics of Ceramics, Vol 2" Plenum, New York (1974).

76. N. Ingelstrom and H. Nordberg, The fracture toughness of cemented tungsten carbides, Eng. Fract. Mech., 6:597 (1974).

77. E. A. Almond and B. Roebuck, Precracking of fracture toughness specimens by wedge indentation, in: "Fracture Mechanics Methods for Ceramics", ASTM STP 745, to be published.

78. J. R. Pickens and J. Gurland, The fracture toughness of WC-Co alloys in SENB specimens precracked by electro-discharge machining, Mater. Sci. Eng., 33:135 (1978).

79. J. F. Knott, "Fundamentals of Fracture Mechanics" Butterworths, London (1973).

80. A. G. Evans and M. Linzer, High frequency cyclic crack propagation in ceramic materials, Int. J. Fract., 12:217 (1976).

81. Z. Hashin and S. Shtrikman, A variational approach to the theory of the elastic behaviour of multiphase materials, J. Mech. Phys. Solids, 11:127 (1963).

82. B. O. Jaensson and B. O. Sundström, Determination of Young's modulus and Poisson's ratio for WC-Co alloys by the finite element method, Mater Sci. Eng., 9:217 (1972)

83. J. Gurland, A structural approach to the yield strength of two-phase alloys with coarse microstructures, Mater. Sci. and Eng., 40:59 (1979).

84. B. O. Sundström, Elastic-plastic behaviour of WC-Co analysed by continnum mechanics, Mater. Sci. Eng., 12:265 (1973).

85. J. L. Chermant and F. Osterstock, Elastic and plastic characteristics of WC-Co composite materials, Powder Metal. Int., 11:106 (1979).

86. H. Fischmeister and B. Karlsson, Plasticity of two-phase materials with a coarse microstructure, Z. Metallkd., 65:311 (1977).

87. E. A. Almond, M. G. Gee and B. Roebuck. Unpublished work.

88. L. Lindau, On the fracture toughness of WC-Co cemented carbides, in: "Proceedings of 4th International Conference on Fracture", Univ. Waterloo Press (1977).

89. J. L. Chermant and F. Osterstock, Fracture toughness and fracture of WC-Co composites, J. Mater. Sci., 11:1939 (1976).

90. M. Nakamura and J. Gurland, The fracture toughness of WC-Co two phase alloys - A preliminary model, Metal. Trans. 11A:141 (1980).

91. E. M. Trent, Evaluation of tool materials for metal cutting, Eng. Digest. 38:15 (1977)

92. K. J. Stjernberg, Fracture toughness of TiC-coated cemented carbide Met. Sci. 14:89 (1980).

93. J. Larssen-Basse, Wear of hardmetals in rock drilling: a survey of the literature, Powder Met., 16:1 (1973).

94. R. I. Blomberry, C. M. Perrot and P. M. Robinson, Abrasive
     wear of WC-Co composites. Wear mechanisms, Mater. Sci
     and Eng., 13:93 (1974).

95. K. H. Zum Gahr and A. Fischer, Influence of microstructure
     of WC-Co hardmetals on strength and abrasive wear, Metall.,
     35:38 (1981).

96. J. Larssen-Basse, Some mechanisms of abrasive wear of
     cemented carbide composites, Mataux Corros. Ind., 54:8
     (1980).

97. H. Jonsson, Wear of cemented carbide bits during percussive
     drilling in magnetite-rich ore, Planseeber. Pulvermet.,
     24:108 (1976).

98. E. Lardner, Cemented carbides for coal mining, Colliery
     Guardian, 919 (1977).

99. E. A. Almond, M. G. Gee and L. N. McCartney, The application
     of reliability technology to tool design, in: "Towards
     Improved Performance of Tool Materials", to be published,
     Metals Society, London.

100. H. Doi and K. Nishigaki. Binder phase strengthening through
     precipitation of intermetallic compound in TiC base
     cement with high binder concentration, Mod. Dev. Powder
     Metal., 10:525 (1977).

101. R. K. Viswanadham, P. G. Lindquist and J. A. Peck,
     Preparation and properties of WC-(Ni, Al) cemented
     carbides, in: "International Conf. on Science of Hard
     Materials" to be published, Plenum, New York.

102. D. Moskowitz, Cemented TiC tool for intermittent cutting
     tool applications. U.S. Patent 4019874 (1977).

103. D. Moskowitz, Abrasion resistant Fe-Ni bonded WC, Mod. Dev.
     Powder Metal 10:543 (1977).

104. C. F. Yen, C. S. Yust and G. W. Clark, Enhancement of
     mechanical strength in hot-pressed TiB_2 composites by

     addition of Fe and Ni, in: "New Developments and
     Applications in Composites", Met. Soc. AIME (1979).

105. H. Suzuki, M. Sugiyama and T. Umeda, Aging mechanisms of
     sintered WC-Co hardmetals, J. Jap. Inst. Met., 29:467
     (1965).

DISCUSSION

R. K. Viswanadham:

I have two comments and two questions. We also tried to do quantitative Auger spectroscopy on WC-Co samples that were fractured
in situ in vacuum. The fracture surfaces were not exactly flat.
In spite of that, the cobalt: tungsten ratios on the fracture surface were high and they did not decay to the polished values in

anything like a monolayer. The distance estimated is more in the
order of 300 to 400 Å. This is true in the case of fracture surfaces
that were produced in air as well. And the second comment is, in
our fracture toughness model we did not treat the binder as a bulk
material. A constraint factor was included. I think by now it is
fairly obvious that there are tremendous differences in the mechanical behavior of a binder between bulk and under constraint.

E. Almond:

You are completely correct about that. I just wanted to illustrate
that there was this integral term which we think is extremely important as well. And about the Lea and Roebuck paper, I think as
you pointed out and as I discussed with you, we did get very flat
fracture surfaces and I think the kind of things that enter into
this kind of analysis are too detailed to go into here.

R. K. Viswanadham:

You said that you would characterize hardness as an empirical property because it is shape and volume dependent. Could you elaborate
on that?

E. Almond:

I put hardness at the borderline. Theoretically, if you knew enough
and you could do the finite element analysis with strain hardening
rates and everything put into it, you could say that hardness is
an intrinsic property because it's calculable from the basic atomic
structure and deformation characteristics of the material. But the
mathematics is beyond anybody at present; to do it properly and com-
prehensively. It's certainly beyond anyone to do it for the Vickers
indenter, where the sharp corners of the indenter produce stress discontinuities in any attempted analysis. I don't think anybody who
is going to attempt that will succeed for an awful long time until
different mathematical approaches are developed. So, I'd say that
hardness is on the borderline of being between an intrinsic and ex-
trinsic property, in that it is mathematically impossible at present
to explain it from basic fundamental atomic structure or microstructure.

R. K. Viswanadham:

Am I correct in understanding then, that when you say it is shape
and volume dependent, you are talking from a modeling point of
view and not from a measurement point of view?

E. Almond:

The shape and volume dependence wasn't in all those properties. I

was talking about bend tests where if you increase the volume and
change the span ratio you get different answers to your bend strength.
Of course, this is a popular thing to try when you're manufacturing
steel, to decrease the span and decrease the volume and get tremendously high transverse rupture strengths.

R. K. Viswanadham:

My second question concerns the fracture morphology change between
catastrophic and fatigue fracture. Did you observe this morphological
change at all ΔK levels, or only at some?

E. Almond:

We saw it mainly when the crack growth increment was comparable to
that of the carbide and binder phase separation. I didn't elaborate
on this because I didn't have time, but that makes me think that
this is possibly the only example seen where cobalt is behaving in
a constrained manner. But it's behaving in a brittle manner because
it's constrained to deform by the very small plastic zone at the low
ΔK values. At high ΔK you have a larger plastic zone and you're
looking more to a volume process and the plastic zone extends into
several binder phase regions.

R. Rice:

I have two comments which are based primarily on much more extensive
studies in ceramics than in cermets, but I think are appropriate
here and I welcome the author's comments on them. I think that in
all of these areas we need much closer coordination between three
sets of observations for understanding mechanical strength. One
is the usual strength test. I think that has to be closely coordinated with fracture mechanics tests and I would emphasize that they
have to be appropriate to the particular scale of flaw size we're
dealing with which is not always recognized. And third, I think
the area which I'm pleased to see has occurred recently in the cemented carbides but I think a great deal more needs to be done
and that is to do fractography, particularly to see if, in fact,
the fracture mechanics measurements and the strength measurements
are self consistent. I think the interrelationship of these three,
the strength, the K_{IC}, and the fracture origin discrimination, are
critical. The second comment I would make is that I think one has
to be very careful, certainly in pure ceramic materials, I know
this definitely to be the case, in taking general observations of
fracture mode and relating those too closely with fracture behavior.
The reason I say this is that we can demonstrate in many cases that
the fracture mode may change significantly over the fracture path.
And the fracture mode that one wants to concentrate on is that in
the immediate area of the flaw out from the failure-initiating defect. For example, in partially stabilized zirconia, we've shown

recently that many of the polycrystalline bodies that are available
today fail from a single grain boundary and yet all other fracture
in the body, including that immediately surrounding, is all transgranular. If one characterized the failure mode of that body, you
would say that it is all transgranular. Yet that's totally imper-
tinent to what is initiating failure. I think that may be very
important in your questions of the amount of cobalt and tungsten
carbide contiguity.

E. Klimek:

I've got one comment and ten thousand questions. You started your
talk by saying, "What do I tell my mechanical or design engineer
as to how to make this shape or form?" I'm still waiting for that.

E. Almond:

Microstructural engineering, not tool engineering.

E. Klimek:

When you talked about finite element analysis, was this microstructural or macrostructural?

E. Almond:

Talking about the materials having average properties and treating
it as a continuum with those properties. But not breaking up the
carbide grains into little finite elements.

A. Horsewell:

I would just like to make a comment about the martensitic transformation in cobalt. As we know, it can be formed in two ways: one
on cooling down through the M_s temperature and secondly on deformation. But something which is not normally said in the context of
cemented carbides is that these two types of transformation product
can be different. For cooling down through the M_s temperature we
have the three possible a/6 [112] partial dislocations passing at
120° to each other on any chosen set of {111} planes giving zero
shape change, and many different {111} planes are favorable. This
gives the interlocked structure that you showed in your micrograph.
On deformation, of course, only one of those a/6 [112]s would be
favorable and there would be a very large total shape deformation
in one direction on one set of {111} planes, and it could be this
which leads to tearing of the binder.

E. Almond:

I agree with what you say. The problem of presenting anything on

the binder phase is that you can only do it by interpreting diffraction patterns. I think this is a very longwinded way of doing
it, even though it is the only way available to us at the moment.

A. Horsewell:

But I think it shows up very well in the microstructure and it
showed up very well in your pictures.

E. Almond:

Yes, to some extent. But I think to put anything quantitative
into it you've got to use diffraction patterns because we have
tried measuring stacking fault energies of binder phase and you
just can't get anywhere.

H. Fischmeister:

I'd like to congratulate Dr. Almond on this review. I was also
very glad to see the word "load transfer" at the top of your modeling table and this gives me cause to say one word about dislocation
microscopy of tungsten carbide. If the strain partitions completely
to the cobalt phase, and we have models which give a good description
of the properties assuming that (it's an oversimplication but apparently an admissible one) then, of course, the deformation of the
tungsten carbide plays no role at all. And all the dislocation microscopy is for the birds. Now really the situation is not that bad.
But what I would like to say as a warning to dislocation microsco-
pists, and perhaps also as a goal, is they have the tools to determine the strain, the local strain of the carbide gains which is an
extremely important property for modeling. But when they do that,
they should try to make quantitative estimates and they must also
look for strain concentrations at points where the skeleton of the
carbide grains receives the maximum strain. This is a very difficult
task, but I think this is where one has to go if one wants to create
a valid model.

E. Almond:

I agree with what you say about the fiber model for hard metals
and equal strains. In trying to explain those compression recovery
results, the fiber model did not give us the right answer and you
have to invoke reverse dislocation motion in the binder phase to
explain how you could get such a big recovery effect. You can't
do it on continuum mechanics alone.

INDENTATION TESTING OF A BROAD RANGE OF CEMENTED CARBIDES

Richard Warren* and Hansjoachim Matzke**

*Chalmers University of Technology
S-412 96 Gothenburg, Sweden

**European Institute for Transuranium Elements, CEC
D-7500 Karlsruhe, W. Germany

ABSTRACT

Vickers diamond indentation and Hertzian indentation were used as a means of investigating the fracture toughness of a wide range of unbonded and metal-bonded, cemented, carbides. Diamond indentation provides a sensitive and reproducible method of toughness testing; the results can be related empirically to conventional toughness parameters. Hertzian indentation is a complement to diamond indentation, permitting measurement on brittle materials not always amenable to diamond indentation.

INTRODUCTION

The Palmqvist test, based on the measurement of cracks forming at the corners of a Vickers hardness indentation, is becoming increasingly accepted as a method for assessing the toughness of hardmetals. Although it has been demonstrated that the relationship between indentation cracking and more well-defined toughness parameters is complex[1,2,3], the test is attractive because of its simplicity and because of its proven sensitivity to the state of the sample. It can be related empirically to other properties but should perhaps also be considered as a surface toughness parameter in its own right. The usefulness of the test will no doubt increase with experience.

563

The phenomenon of cracking at indentations is not restricted
to diamond pyramid indentations in cemented carbides; recently
the subject has been dealt with more generally by e.g. Lawn and
Wilshaw[4] and Evans and Wilshaw[5]. Three types of indentation
cracking will be treated in the present work. Illustrated schematically in Fig. 1, these are: (i) Hertzian cone cracks (HCC)
formed in brittle solids around spherical or cylindrical indentors,
(ii) Palmqvist cracks (PC), (iii) semicircular median cracks (SMC)
formed under a sharp indentor. When formed under a pyramid diamond
in opaque solids the latter are not immediately distinguishable
from Palmqvist cracks. However, as will be seen below the distinction is significant when indentation is applied in testing. SM
cracks form in brittle solids while Palmqvist cracks are observed
in materials exhibiting slightly more plasticity such as cemented
carbides. Hertzian cone cracks occur during elastic indentation,
i.e. in the absence of plastic deformation. The crack is very
stable even in very brittle solids.

In the present work, the application of the above types of
indentation cracking in the testing of a large range of cemented
carbides is investigated. Since the different types of cracking
are applicable at different degrees of plasticity they should
compliment each other. This possibility is explored by investigating both bonded and unbonded carbides. Before presentation of
experimental findings, indentation cracking will be examined in
more detail with respect to its relationship to fracture behaviour
in general.

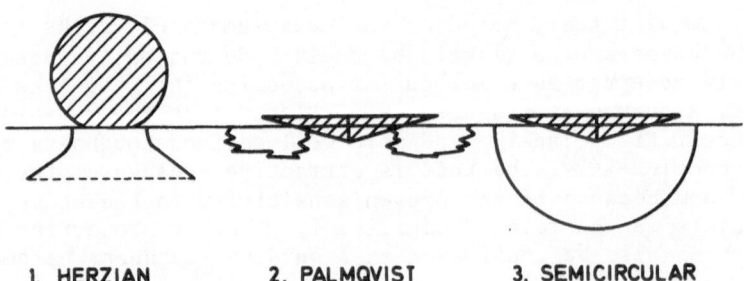

1. HERZIAN   2. PALMQVIST   3. SEMICIRCULAR
   CONE CRACK   CRACK   MEDIAN CRACK

Fig. 1. Three types of indentation cracking; sectioned verti-
        cally in median plane.

RELATIONSHIPS BETWEEN INDENTATION CRACKING AND FRACTURE TOUGHNESS
PARAMETERS

The Palmqvist Test

Typical Palmqvist crack profiles are shown in Fig. 2. They
are normally relatively shallow and begin a little way behind the
tip of the indentation. This has been confirmed for several grades
of hardmetal[6,7] and similar behaviour has been reported in tool
steels[8]. On the basis of these observations, Fig. 3 represents the
crack geometry and includes definitions of the terms to be used
here.

Experimentally, it is observed that the total crack length
($L_T = 4L$) increases linearly with the indentation load. It has
been suggested that for carefully prepared, unstressed surfaces,
the L_T vs. P plot passes through the origin[9,10], i.e. that

$$P = W L_T \tag{1}$$

where W is a constant for the material. This is very convenient
since the material can be characterised at a single indentation
load, W being taken as a measure of the crack resistance. In a
number of cases the L_T vs. P plot has been found not to pass
through the origin[3,11] in spite of careful sample preparation i.e.

$$L_T = (P - P_c)/W^* \tag{2}$$

where P_c is an initial load required before cracking is observed.
The analysis by Perrott[1,2] indicates that this can be expected.
Often P_c is relatively low so that measurements made at sufficiently high loads are consistent with eqn. (1) to a good approximation.

Although W can be considered as a measure of surface toughness in its own right it is desirable to know how it is related
to more well-known fracture parameters. For example, if the crack
is approximated to a through-crack in an infinite plate[12], if the
tensile stress level in the surface is assumed to be directly
proportional to the indentation load and to fall off as $1/x^2$ with
the distance x from the centre of the indentation, then the critical stress intensity factor is given by:

$$K_{IC} = 2(L/\pi)^{1/2} k P \int_{L}^{C_R} [1/x^2 (C_R^2 - x^2)^{1/2}] dx \tag{3}$$

where k is the constant relating the stress level to the load P.
Integrating gives:

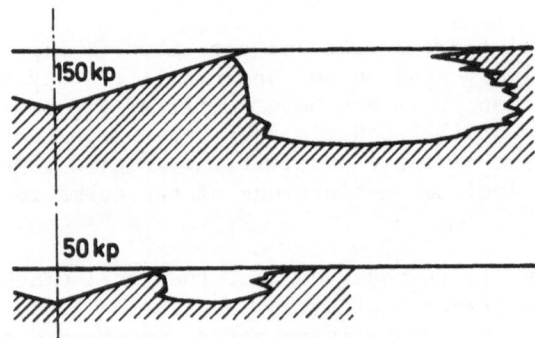

Fig. 2. Vertical section through Palmqvist cracks in cemented carbides. Results of Palmqvist reported in ref. 6.

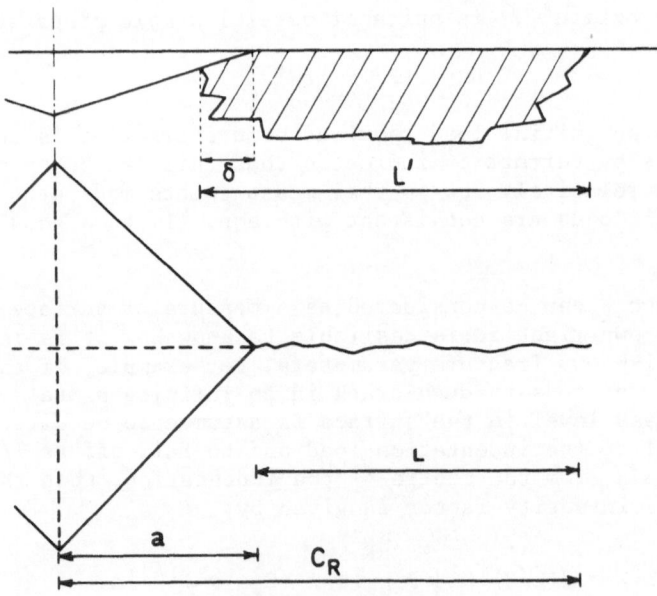

Fig. 3. The geometry of cracking around a Vickers indentation.

$$K_{IC} = 2(L/\pi)^{1/2} k P \left[\sqrt{C_R^2 - a^2} / C_R^2 \, a \right] \qquad (4)$$

Putting $C_R = L+a$, this becomes for $L \gg a$:

$$K_{IC}^2 = 4 k^2 P^2 / \pi L a^2 = 8.64 k^2 H P/L \qquad (5)$$

where H is the indentation hardness. Thus the surface can be assigned a toughness related to indentation parameters:

$$K_{IC} = const. \sqrt{H W} \qquad (6)$$

A similar equation has been given by Lawn and Wilshaw[4]. It is to be noted that the equation predicts that W is constant.

In Fig. 4, literature-derived values of $\sqrt{HW}$ are plotted against conventionally-measured K_{IC} for WC-Co alloys*. In some cases the two sets of measurements were made on the same material[3,13,14]. In other cases, corresponding values of $\sqrt{HW}$ and K_{IC} from different sources could be linked where sufficient microstructural information was given. A relationship between the two can also be obtained via the empirical expressions given by Perrott for a range of alloys[2]:

$$1/W = 1.85 \ (H - H_c) \times 10^{-16} \qquad (7)$$

$$H = (31 - 3 \ln G_{IC}) \times 10^9 \qquad (8)$$

where G_{IC} is the fracture energy and $H_C = 10.6$ GN/m². The linearity predicted by eqn. (6) applies to a good approximation for alloys with K_{IC} up to about 18 MN/m$^{3/2}$ and can be used in this range as a means of estimating K_{IC} by indentation. The inclusion of the results of Zum Gahr[8] for tool steels indicates that the equation can be extended to materials covering a wide range of hardness.

The justification of equation (6) is largely empirical. It is a considerable oversimplification neglecting as it does the complex nature of indentation; its failure at high K_{IC} values (small L) is a reflection of this. However, since it requires only measurement of H and W it can be applied directly to a wide range of materials. The application of rigorous analyses such as that of Perrott requires knowledge of further parameters.

A modification of eqn. (6) by use of a crack length $L' = L + \delta$ (see Fig. 3) with $\delta \propto a$ was investigated empirically. However, because of the scatter of the available data, this refinement could not be justified.

---

* Footnote: Where reported data deviated from eqn. (1), W values were derived from the results obtained at relatively high loads (P > 500 N). Results exhibiting large deviation were omitted.

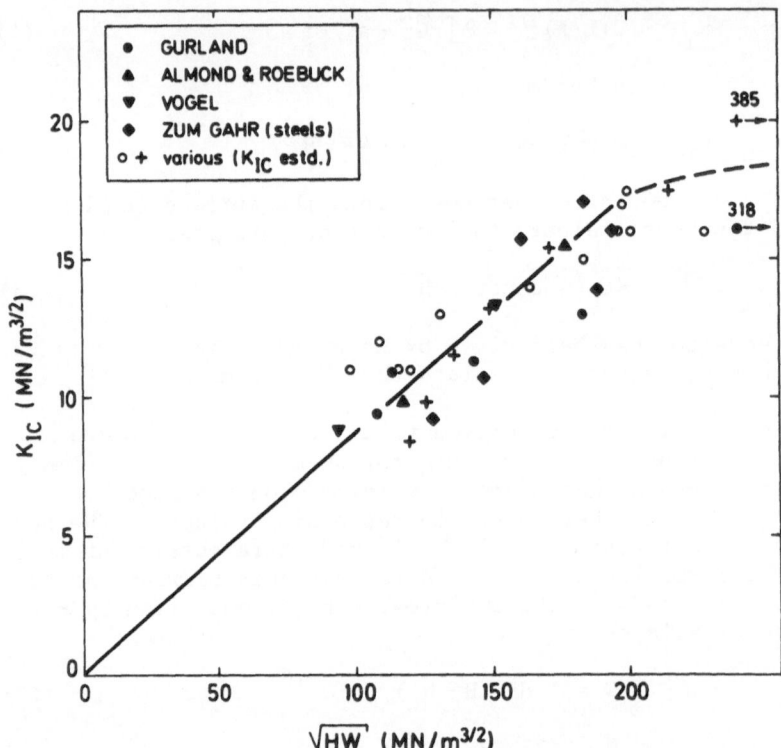

Fig. 4. Relationship between K_{IC} and $\sqrt{HW}$ for WC-Co alloys and tool steels. Filled symbols indicate K_{IC} and $\sqrt{HW}$ measured on same samples[3,8,13,14]. Open symbols are results from different sources[9,10,15-24] but on material with similar microstructure. Crosses are points derived from eqns. (7) and (8).

Semicircular Median Cracks

In materials exhibiting very little plasticity, sharp indentation is expected to lead to the development of a crack under the indentor and subsequently to a roughly semicircular median crack. In their investigation of such cracks, Lawn and Fuller[25] arrived at the following relationship:

$$P^2/C_R^3 = \text{const. } E \; G_{IC}$$

giving:

$$K_{IC} = \text{const. } P/C_R^{3/2} \qquad (9)$$

The predicted linear dependence of $C_R^{3/2}$ on load has been confirmed for glass[25], silicon nitride[14], zinc sulphide[5] and $Al_2O_3/$TiC ceramics[14]. The few examples of parallel K_{IC} and indentation data are plotted in Fig. 5 and found to be in satisfactory agreement with eqn. (9).

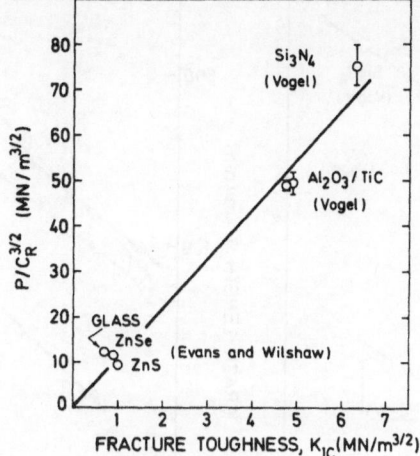

Fig. 5. Relationship between K_{IC} and $P/C_R^{3/2}$ for brittle solids. The results are limited to examples in which K_{IC} measurement and indentation were performed on identical or very similar samples[5,14].

In opaque materials SM cracks are not directly distinguishable from Palmqvist cracks. Usually the two cases can be recognised in log-log plots of P versus C_R and L respectively (e.g. see Fig. 6).

Hertzian Indentation

Although Hertzian cone cracks also grow stably with increasing load, the growth is into the bulk of the sample and so cannot be easily observed in opaque materials. The crack *initiation*, however, is a recognisable event, occurring as a ring-crack around the indentor. A number of analyses[4,26,27] indicate that the load required to initiate the crack passes through a minimum value as a function of the crack depth (length). The crack depth that is possible in practice and therefore the observed critical load for crack initiation, P_c, can be assumed to depend on the size of available flaws in the material. This led Warren to propose[27] that, provided the material contained an adequate flaw population, the observed value of P_c would always be the minimum value, given by:

$$P_c = B \, G_{IC} \, R \qquad (10)$$

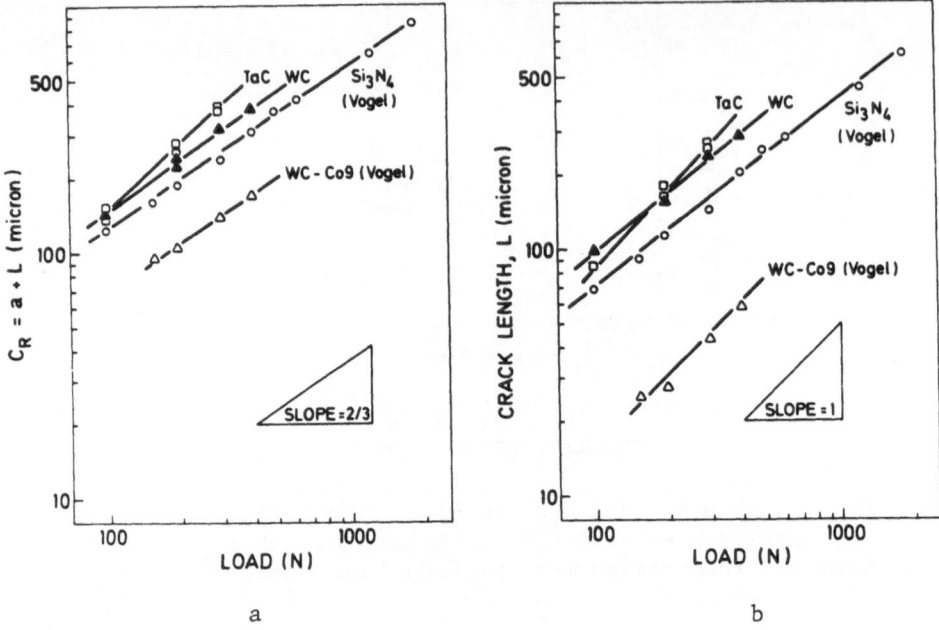

Fig. 6. Crack length versus diamond indentation load for selected
materials. In 6a) the crack length is C_R (SMC behaviour)
in 6b) it is L (PC behaviour). The WC-Co results of Vogel[14]
are consistent with both.

where R is the radius of the spherical indentor and B is dependent
on the elastic properties of indentor and sample as well as the
location of the ring crack in relation to the indentor contact
area[27]. Equation (10) has been applied in the measurement of the
fracture toughness of a number of carbides, and oxides[27,28,29].
The results obtained are self-consistent and show reasonable agreement with values of K_{IC} measured by other methods, where available.

An important assumption of the above analysis is that no
permanent indentation of the sample occurs. Thus the method is
limited to relatively brittle solids.

Crack Length versus Hardness Plots

Viswanadham and Venables noted for a large series of WC-Co
alloys that the Palmqvist crack length at a given load was directly
proportional to the hardness[10]. Thus the indentation results could
be expressed by a relationship such as that given by eqn. (7)
regardless of the composition and microstructure of the alloys.
(Ti,V)C-Ni alloys behaved similarly but with a different value of

H_c (the hardness below which cracks are no longer observed). In a later work on VC-Ni alloys Viswanadham and Precht[37] found that the slope of the W^{-1} vs. P plots could in fact vary in the same alloy system and that this slope was related to the fracture mode.

Such plots provide a very convenient way of presenting indentation data allowing a quick assessment of specific alloy systems. Thus in developing successful alloys, systems are sought with high values of H_c and low slopes. Results can be presented and used without any prior assumption about the nature of the cracking. Perrott has explored the nature of these plots in terms of indentation theory[1,2].

PROCEDURE

Though many of the results presented here have been obtained in the authors' laboratories, we alse borrow freely from published work in order to better investigate the versatility of indentation testing. The materials and techniques used in our own work have mostly been described in earlier publications and will only be given briefly here.

Materials

Details of the investigated carbides and cemented carbides are presented in Table 1. The pure carbide samples were prepared by hot pressing[27,31]. In the NbC_{1-x} series the carbon content was adjusted by small additions of free Nb or C. These materials were given an homogenization anneal (50 h, 2300 K) after hot pressing.

The cemented carbide samples were prepared by mixing carbide and binder metal powders, cold pressing and sintering in vacuum (ref. 32-37). Except for the alloys with a pure Ru binder, liquid phase sintering was employed. In all cases the microstructures of the samples were characterised by measurement of carbide grain size, carbide contiguity and volume fractions.

Indentation Testing

The samples were sectioned and polished with diamond down to 1 micron paste. Diamond indentation was made with a standard Vickers hardness testing machine at loads between 5 and 100 kg. Total crack lengths, L_T, were measured on at least three indentations for a given sample and load. The general validity of eqn. (1) was confirmed for selected samples from each carbide/binder system. Thereafter the value of W for individual specimens was based on a single load only (> 20 kg).

Table 1. Properties of the Carbide Phase in Pure Carbide and
Cemented Carbide Samples.

Carbide	C: Metal atom ratio	Grain size* $\bar{d}$ (microns)	Volume fraction porosity	Micro-hardness (kg/mm^2)	Assumed Young's modulus** $(GN\ m^{-2})$
Pure carbides					
TiC	0.96	25	0.06	~3000	450
ZrC	0.45	15	0.08		400
VC	0.91	10	0.11	~2300	415
NbC	0.97	25	0.08	~2000	480
"	0.90	12	0.07		410
"	0.87	15	0.047		410
TaC	0.99	17	0.08		540
WC	1.01	2	0.005	2100	690
Cemented carbides					
TiC–Co	0.95	5–8	~0	2550	
TiC–Ni	0.95	5–8	"		
TiC–Ru	0.95	1–3	<2%		
VC–Co	0.91	8–35	~0	2250	
NbC–Co	0.98	8–26	"	1750	
TaC–Co	1	3–12	"	1500	
TiC–22 m/o WC		5–8	"	2200	
TiC–42 m/o WC		3–5	"		
VC–73 m/o NbC		3–7	"		
VC–20 m/o NbC		5–12	"	2100	
NbC–5 m/o TaC		6–15	"	1900	
NbC–62 m/o TaC		5–14	"	1660	

*   Grain size in the cemented carbides varies depending on
     sintering conditions. ($\bar{d}$: mean intercept length)

** Values derived from literature data for fully dense material.

     Hertzian indentation was carried out on testing machines
modified for the purpose[27-29]. At least three indentator radii
were used to confirm the linearity predicted by eqn. (10) (Auer-
bach's law). The value of P_c was found as the average of the value
of at least four "fresh" areas of the sample. The porosity in the
carbides gave an adequate defect population for the minimum P_c
condition assumed in Warren's model[27].

Evaluation of Results

     All the cemented carbide alloys have been assumed to exhibit
Palmqvist behaviour. The indentation results are presented in terms
of $\sqrt{HW}$ in accordance with eqn. (6) with an empirical constant of

proportionality obtained from Fig. 4, thus

$$K_{SC} = 0.087 \sqrt{HW} \tag{11}$$

Thus we equate the surface toughness K_{SC} with the bulk K_{IC} of the material, this interpretation is convenient but arbitrary since the toughness of the surface is not necessarily expected to be identical with that of the bulk.

As well as presenting the indentation data in the above form the results for selected cemented carbide systems are presented and evaluated in terms of W^{-1} versus H plots.

In the diamond indentation tests on the pure carbides, load vs. crack length data were first tested against eqns. (1) and (9). Where SMC behaviour was indicated, K_{SC} was estimated from eqn. (9) using an empirical constant of 0.092 obtained from Fig. 5. As far as possible, results from other sources have been evaluated in a similar manner. In certain cases, cracking behaviour can seemingly be consistent with both PC and SMC as is demonstrated in Fig. 6. Cemented carbides showing this duality were assumed to exhibit Palmqvist behaviour.

RESULTS

The surface toughness of various WC alloys is presented in Figs. 7a-c as a function of the volume fraction binder, V_B. It should be noted here that fracture toughness is not necessarily expected to be linearly related to volume fraction. K_{SC}, as defined in Fig. 4, can be expected to rise sharply with V_B above $K_{SC} \approx$ 18 $MN/m^{3/2}$. Nevertheless, the toughness of WC-Co alloys measured by indentation (Fig. 7a) shows the same trends as conventionally measured K_{IC}. Though it is difficult to compare microstructural data from different sources with accuracy, a tendency for toughness to increase with grain size is noticeable, at least up to $\bar{d} = 3$ microns. Fig. 7b indicates that Co can be replaced by substantial amounts of Fe and Ni without significant loss of toughness. The results of Brabyn et al.[19] (Fig. 8) suggest that Ni can even produce significant improvements in toughness. This is supported by the conventional K_{IC} measurements of Prakash[38].

Unbonded WC could be successfully tested with both Hertzian and diamond indentation; in the latter, SMC behaviour was observed (Fig. 6). The two independent methods are in good agreement with each other and both are consistent with an extrapolation of the cemented carbide results (see also Table 2).

Fig. 7c illustrates the poorer toughness of alloys based on cubic WC-TiC solid solution carbides (see also Fig. 13).

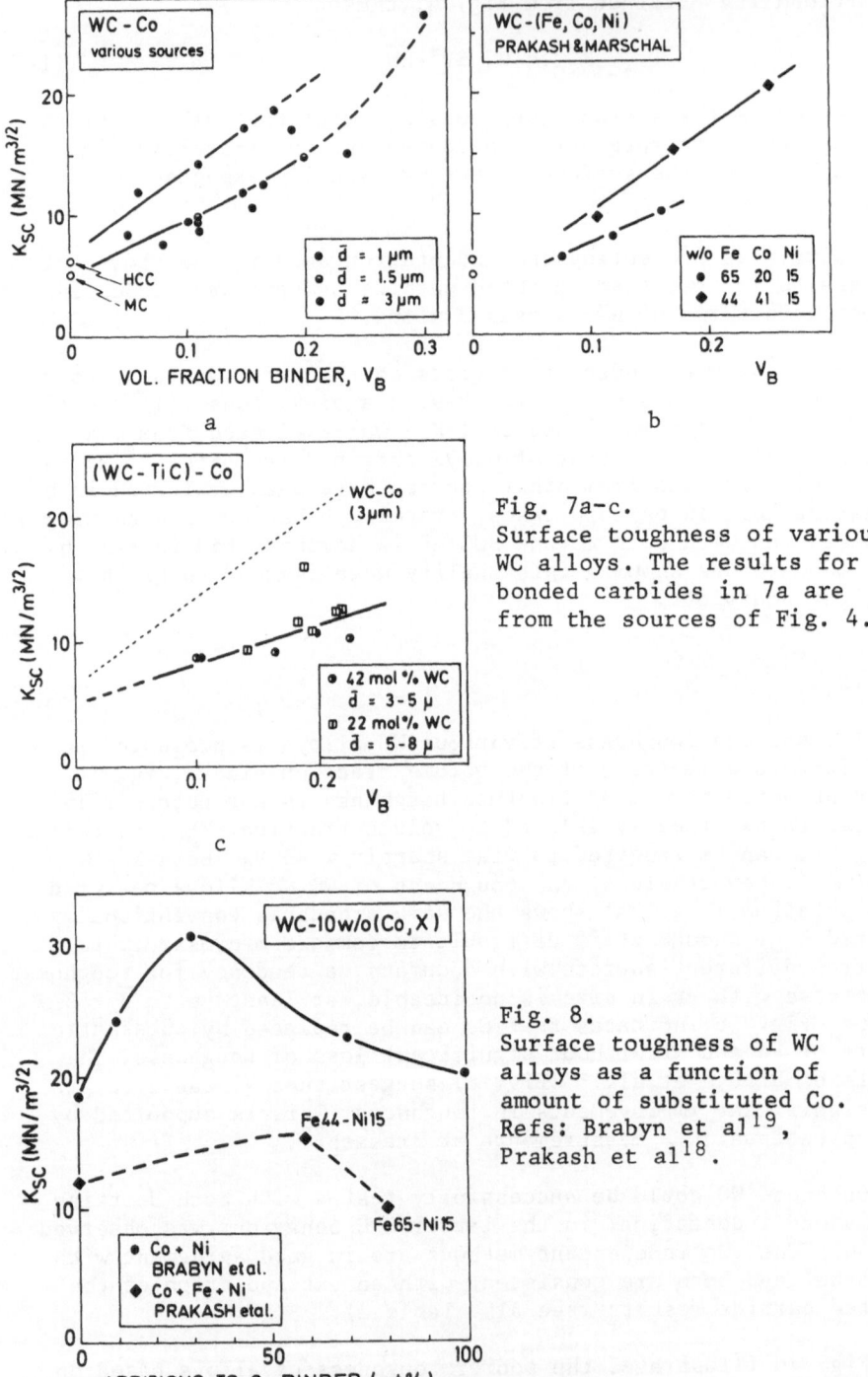

Fig. 7a-c.
Surface toughness of various
WC alloys. The results for
bonded carbides in 7a are
from the sources of Fig. 4.

Fig. 8.
Surface toughness of WC
alloys as a function of
amount of substituted Co.
Refs: Brabyn et al[19],
Prakash et al[18].

Fig. 9 shows the surface toughness of several TiC-based alloys as a function of volume fraction. The toughness of TiC-Co and TiC-Ni alloys is comparable to that of WC based alloys. Indentation reveals the poorer toughness of TiC-Ru alloys and TiC/Al$_2$O$_3$ ceramics. All the alloy systems give similar toughness for uncombined TiC by extrapolation and this agrees well with the value obtained by HCC. Diamond indentation on the unbonded TiC did not give satisfactory results. The TiC/Al$_2$O$_3$ ceramics exhibited SMC behaviour[14]; thus the results of Fig. 9 indicate that the three indentation approaches are mutually consistent.

Fig. 10 shows how the toughness of TiC-Ru alloys is improved by additions of Ni. Similarly, an improvement in TiC-Ni alloys by means of Ru additions is indicated.

The surface toughness of TaC-Co alloys is shown in Fig. 11. Hertzian indentation could not be performed on the unbonded TaC because of plastic deformation. With diamond indentation this carbide exhibited Palmqvist behaviour (see Fig. 6). The carbide grain size of these alloys ranged from 3 to 10 microns. There was a marked *fall* in toughness above $\bar{d}$ = 4 microns. Above $\bar{d} \approx 5$ microns any effect of grain size was not detectable.

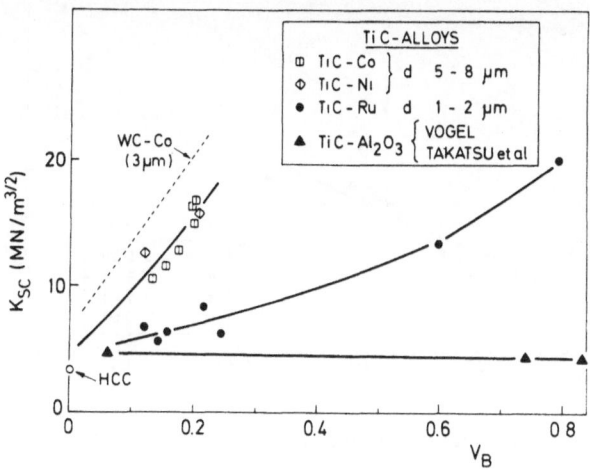

Fig. 9. Surface toughness of various TiC based alloys.

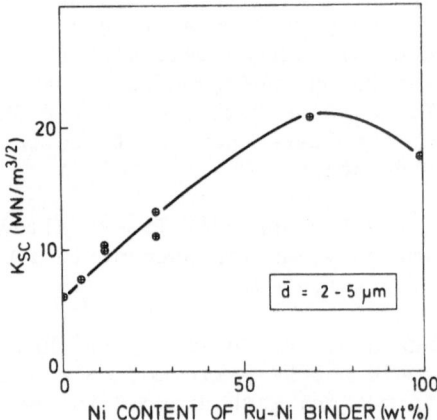

Fig. 10. Effect of Ni-content in the Ru-Ni binder on the surface
toughness of TiC-20 v/o (Ru,Ni) alloys.

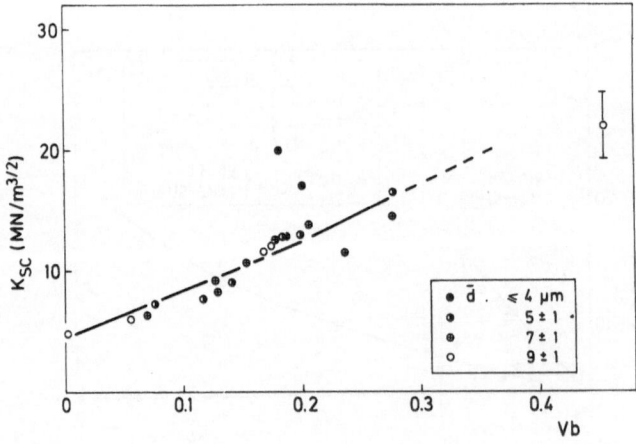

Fig. 11. Surface toughness of TaC-Co alloys.

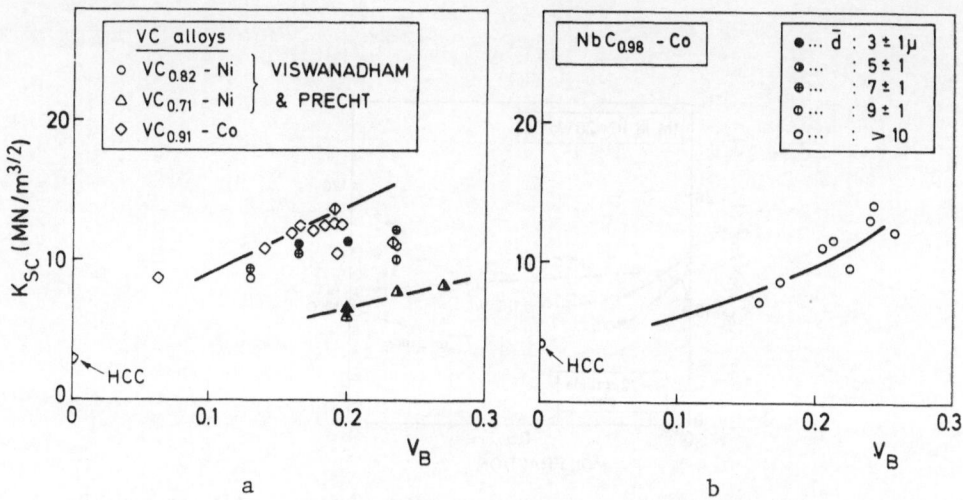

Fig. 12. Surface toughness of VC and NbC cemented carbides.
The grain size symbolism in b) apply also to VC alloys in a).

The results for VC and NbC alloys are presented in Figs. 12a
and b. Again the results of the Palmqvist test on the cemented
carbides are consistent with the Hertzian measurement on the pure
carbide. The results of Viswanadham and Precht[37] for VC-Ni alloys
reveal a marked influence of the carbon content. There is no rec-
recognisable dependence on grain size.

The results of tests on mixed carbide alloys are summarised
in Fig. 13 as the effect of carbide composition in carbide/cobalt
alloys with fixed Co content (20 v/o). Both the TiC-WC and VC-NbC
exhibit minima in the toughness. It should be emphasised that this
does not necessarily reflect a corresponding minimum in the unbonded mixed carbides; carbide/binder interactions must also be
considered.

Table 2 summarises the results of indentation for unbonded
carbides. The carbides, TiC, ZrC, VC and NbC were too brittle to
give satisfactory indentation with the diamond indentor (e.g.
because of surface chipping, complex crack patterns etc), but exhibited ideal Hertzian behaviour. TaC indented plastically under
the spherical indentor but instead exhibited Palmqvist behaviour
in diamond indentation. WC could be indented successfully with
both indentors, showing SMC behaviour with the diamond indentor
(see Fig. 6). Thus a value for the surface toughness could be obtained for each carbide with at least one of the indentation
methods. The quoted values have been adjusted to full density by
assuming that the fracture energy, G_{IC}, is directly proportional
to the relative density. As has already been noted, the values are
broadly in agreement with estimates obtained by extrapolating the
toughness of cemented carbide alloys to zero binder content.

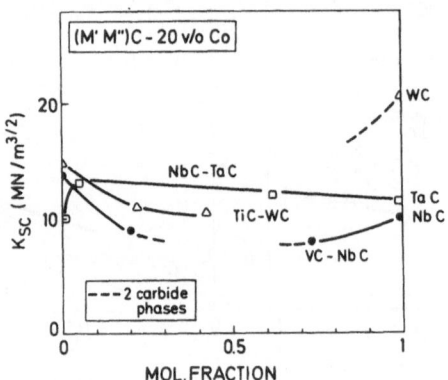

Fig. 13. Surface toughness of mixed carbide alloys with 20 v/o Co.

Table 2. The fracture toughness of unbonded carbides

Carbide	K_{IC} MN/m$^{3/2}$ (adjusted to full density)					
	HCC	SMC	PC	Extrapolated values		
				K_{SC} vs V_B	K_{IC} vs X*	W^{-1} vs H
TiC$_{0.96}$	3.5	x	x			
TiC$_{0.95}$	-	-	-	4 ±0.5		⩽6.4
ZrC$_{0.95}$	2.6	x	x,			
VC$_{0.91}$	3.3	x	x	~6		
VC$_{0.8}$	-	-	-	4-5		⩽4.6
NbC$_{0.97}$	3.7	x	x			⩽4.2
NbC$_{0.90}$	3.7					
NbC$_{0.80}$	3.5					
TaC	x	xx	5.2	~4		⩽4.4
WC	6.2	5.1	xx	6.5	5-7	⩽6.7

x = unsatisfactory indentation
xx = not appropriate
* Conventional K_{IC} reported as a function of an appropriate
   variable such as V_B or l_B, the mean free path of the binder.

Crack Length/Hardness Plots

Palmqvist data for various cemented carbides are presented in the form of W^{-1} vs. H plots in Fig. 14. A K_{SC} scale derived from Fig. 4 has been superimposed to relate the plots to fracture toughness. Results are included for VC-Ni alloys exhibiting two fracture modes (I and III) as found by Viswanadham and Precht[37] and two distinct sets of WC-Co data. The results of this study can be interpreted in terms of a series of linear plots as were the VC-Ni results[37] and as is illustrated schematically in Fig. 15.

A linear relationship between W^{-1} and H must indicate that the crack resistance, W, has a similar but inverse dependence on the same microstructural parameters as hardness; e.g. decreasing grain size, increasing contiguity, decreasing volume fraction binder and increasing hardness of the binder. Changes in the slope of the relationship must indicate a change in this correlation. For VC-Ni alloys the change in slope from I to III was associated with a change in fracture mode from intergranular to transgranular[37]. As the alloy series approaches the unbonded carbide composition, a transition to fracture dominated by transgranular carbide failure is bound to occur. Thus the line representing mode III fracture can be expected to end at a point defined by the values of W^{-1} and H of the unbonded carbide (Fig. 15). This interpretation is supported by extrapolation of the steepest lines for each alloy in Fig. 14 to the hardnesses of the unbonded carbides; the corresponding values of K_{SC}, included in Table 2, are found to be consistent with measured values.

Favourable properties are represented by low W^{-1} and high hardness; thus it is desirable to keep the transition point X as far to the right as possible by suppression of such undesirable fracture modes. This can be accomplished by optimisation of the microstructure and binder properties.

The importance of the microstructure and binder phase is illustrated in Fig. 16, which includes the toughness of cemented carbides and their parent carbides. For example, comparing the WC-Co and TaC-Co alloys indicates that there exists an optimum carbide grain size. In comparing mode I and mode III VC-Ni alloys, both excessive grain size and a high alloy content in the binder were identified as the cause of the fracture transition[37]. The importance of the intrinsic properties of the binder is also shown by comparison of TiC-Co and TiC-Ru alloys. Unbonded VC and NbC have similar toughness; the superiority of VC-Co alloys over NbC-Co must originate in microstructural differences, e.g. the lower contiguity[32] of the former.

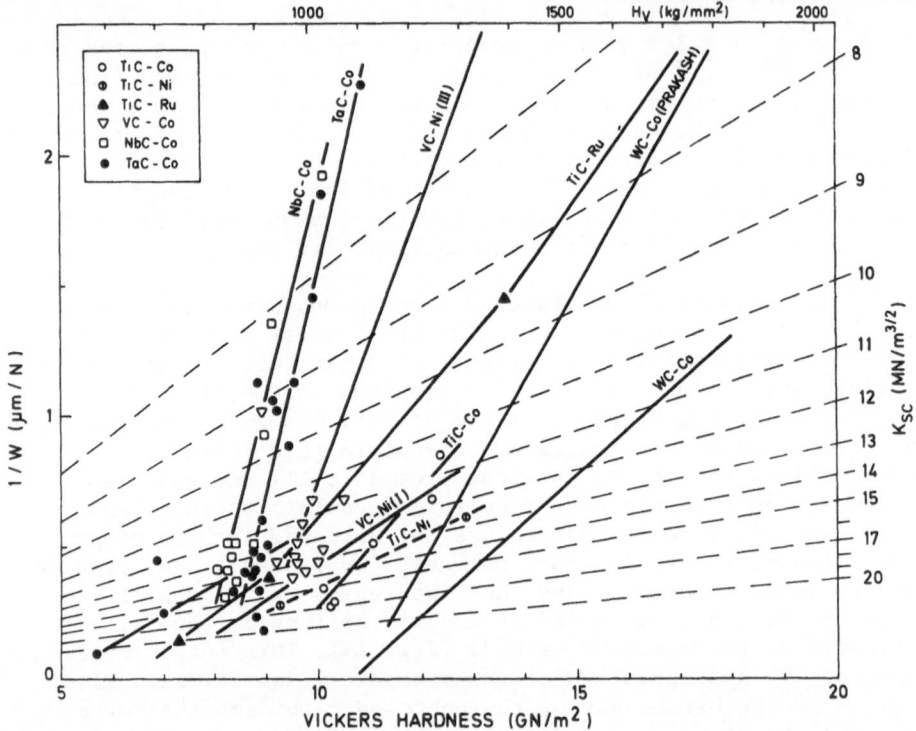

Fig. 14. Hardness/crack length plots for various cemented
        carbides. The line labeled Prakash is from ref. 18.

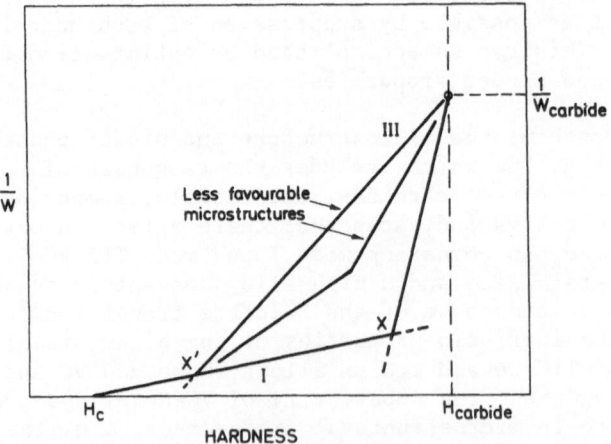

Fig. 15. Schematic representation of hardness/crack length
        plots for a given carbide binder system.

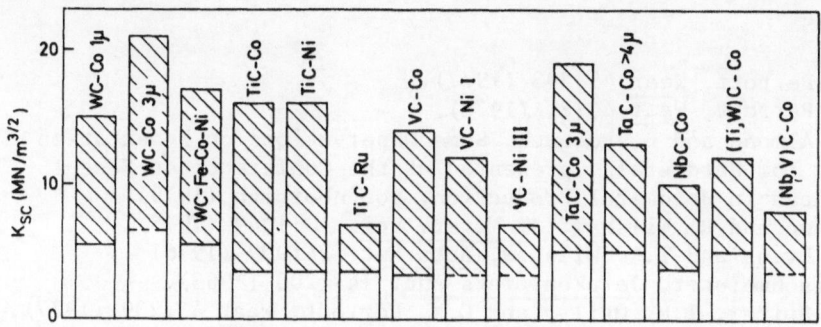

Fig. 16. Surface toughness of various cemented carbides with
         20 v/o binder. Unshaded areas indicate toughness of
         unbonded carbide.

CONCLUSIONS

     Indentation testing provides a simple and sensitive method
for assessing the toughness and fracture behaviour of a broad
range of cemented carbide alloys. Crack resistance in indentation
can be used as a measure of toughness in its own right or related
empirically to conventional fracture toughness.

     Diamond indentation can sometimes also be applied to unbonded
carbides and other ceramics to give results consistent with measurements on cemented carbides. However, for materials that are too
"brittle", diamond indentation may be unsuccessful; in these cases
Hertzian indentation provides consistent estimates of toughness.
Here, the latter method has been used to obtain the fracture toughness of the transition-metal carbides TiC, ZrC, VC, NbC and WC.

     Crack length/hardness plots for cemented carbide alloys of a
given carbide/binder combination can give added insight into the
fracture behaviour of the system. For example they indicate changes
in fracture mode.

     For a given carbide, modification of the microstructure and
the binder phase was shown to have considerable effect on the fracture toughness as measured by indentation.

ACKNOWLEDGEMENTS

     The authors would like to thank Dr. R.K. Viswanadham for
valuable discussions. This work was financed partly by the Carl
Tryggers Research Foundation, Stockholm.

REFERENCES

1. C.M. Perrott, Wear 45:293 (1977).
2. C.M. Perrott, Wear 47:81 (1978).
3. E.A. Almond and B. Roebuck, Some Observations on Indentation
    Tests for Hardmetals, presented at the Conference on Recent
    Advances in Hardmetal Production, Loughborough (1979).
4. B. Lawn and R. Wilshaw, J. Mater. Sci. 10:1049 (1975).
5. A.G. Evans and T.R. Wilshaw, Acta Met. 24:939 (1976).
6. H. Fischmeister, Jernkontorets Ann. 147:200 (1963).
7. J.W. Suiter, I.M. Ogilvy and C.M. Perrott, Wear 43:239 (1977).
8. K.H. Zum Gahr, Z. Metallk. 69:534 (1978).
9. H.E. Exner, Trans. TMS-AIME, 245:677 (1969).
10. R.K. Viswanadham and J.D. Venables, Met. Trans. 8A:187 (1977).
11. P.O. Snell and E. Pärnama, Planseeber. Pulvermet. 21:27 (1973).
12. B.R. Lawn and T.R. Wilshaw, "Fracture of Brittle Solids"
    Cambridge University Press, Cambridge (1975).
13. E.L. Exner, J.R. Pickens and J. Gurland, Met. Trans. 9A:736
    (1978).
14. W. Vogel, Project Report, Inst. Werkstoffwissensch., University
    of Erlangen-Nürnberg (1980).
15. W. Böhlke and K. Voigt, Neue Hütte, 18:4 (1973).
16. W. Dawihl and G. Altmeyer, Z. Metallk. 55:231 (1964).
17. J.R. Pickens and J. Gurland, Matls. Sci. Eng. 33:135 (1978).
18. L. Prakash and A. Marschall, KfK 2993B, Kernforschungszentrum,
    Karlsruhe 63 (1980).
19. S.M. Brabyn, R. Cooper and C.T. Peters, in "Proc. 10th Plansee
    Seminar", H.M. Ortner, ed., Metallw. Plansee, Reutte, 675 (1981).
20. N. Ingelström and H. Nordborg, Eng. Fracture Mech. 6:597 (1974).
21. R.C. Lueth, in "Fracture Mechanics of Ceramics – vol. II",
    R.C. Brandt et al., ed., Plenum, New York (1974).
22. J.L. Chermant and F. Osterstock, J. Matls. Sci. 11:1932 (1976).
23. M.J. Murray, Proc. R. Soc. Lond. A, 356:483 (1977).
24. L. Lindau, Doctoral Thesis, Chalmers Univ. of Technology (1976).
25. B.R. Lawn and E.R. Fuller, J. Matls. Sci. 10:2116 (1975).
26. A.G. Evans, J. Amer. Ceram. Soc., 56:405 (1973).
27. R. Warren, Acta Met. 26:1759 (1978).
28. H. Matzke, T. Inoue and R. Warren, J. Nucl. Matls. 91:205 (1980).
29. H. Matzke, J. Matls. Sci. 15:739 (1980).
30. T. Inoue and H. Matzke, J. Amer. Ceram. Soc., to be published.
31. H. Matzke and C. Politis, unpublished report.
32. R. Warren and M.B. Waldron, Powder Metallurgy, 15:166 (1972).
33. R. Warren, Planseeber. Pulvermet. 20:299 (1972).
34. G. Grathwohl and R. Warren, Matls. Sci. and Eng. 14:55 (1974).
35. R. Warren and M.B. Waldron, Powder Met. Internat. 7:18 (1975).
36. J.S. Jackson, R. Warren and M.B. Waldron, Powder Met. 17:255
    (1974).
37. R.K. Viswanadham and W. Precht, Met. Trans. 11A:1475 (1980).
38. L.K. Prakash, " WC Hartmetallen mit Fe-Basis-Bindelegierungen",
    Doctoral Thesis, Kernforschungszentrum, Karlsruhe (1980).

EFFECTS OF PRECOMPRESSION ON THE

SURFACE TOUGHNESS OF WC-5% Co

Silvana Bartolucci Luyckx

Boart Basic Research Group
Physics Department
University of the Witwatersrand
Johannesburg, South Africa

INTRODUCTION

During the 1960's and early 1970's several papers were published describing changes in the properties of WC-Co alloys resulting from prior compressive loading. The studies most relevant to the present work are those which describe the effects of: (a) plastic strains of up to 0.5% in WC-6% Co on resistivity, coercivity, bending strength and yield stress, by Ivensen and co-workers[1]; (b) plastic strains, in uniaxial compression, of up to 1% in WC-5.9% Co on hardness, coercivity and h.c.p./f.c.c. cobalt ratio, by Hara and Ikeda[2]; (c) plastic strains of more than 10% under non uniform triaxial compressive stress on resistivity, hardness, coercivity and compressive strength, by Pelepelin[3]; (d) uniaxial compression applied to WC-6% Co on the residual tensile stress in the cobalt, by Tumanov and co-workers[4].

A technique which has proved to be very sensitive to changes caused by pre-compression is the Palmqvist cracks technique, in which the lengths of microcracks nucleated at the corners of Vickers indentations are measured. This technique has been used by Exner and Gurland[5] and by Coster[6]. Exner and Gurland applied a compressive stress of 27 kbars to WC-10% Co samples. They reported that the Palmqvist crack lengths changed after compression up to more than 30% and that the lengths varied with the direction of the cracks. Coster applied a non-uniform triaxial compressive stress to WC-3% Co samples. He found that the Palmqvist crack lengths increased with increasing compressive prestress up to about 60 kbars, then decreased.

In the present work the Palmqvist crack technique has been

used to measure changes in crack propagation resistance (toughness) occurring on surfaces cut from WC-5% Co samples, after cyclic pre-compression. The samples were components used for high pressure applications but, for commercial reasons, no details can be given on sample geometry and on type of loading.

EXPERIMENTAL

Specimens of WC-5% Co having an average carbide grain size of 1-1.5 μm were cyclically loaded in compression, from zero to a maximum compressive stress ranging from 10 to 50 kbars. Two batches of samples were tested: those from Batch A underwent 30, 630 or 730 compressive cycles, while those from Batch B underwent 8, 80, 320 or 680 cycles. One sample from each batch was kept in the as-received condition. All samples were sectioned through the axis parallel to the direction of the applied compressive stress, were polished and then indented by applying a 15 kg load to a standard Vickers diamond indentor. The surface preparation prior to the indentation tests followed Exner's recommendations[7], which are essential to obtain Palmqvist crack lengths reproducible and representative of the true surface toughness of the material.

The indentor was oriented in such a way that one of the diagonals of the resulting indentations would be parallel to the direction of the compressive prestress. The Palmqvist cracks nucleated at the corners of the indentations were roughly in the direction of the diagonals. The cracks in the direction of the compressive prestress will be called "vertical cracks", the ones normal to that direction will be called "horizontal cracks". The length of the cracks was measured optically at a magnification of 800X. The highest magnification available (1500X) would have been preferable, except that the use of an oil-immersion objective was very time-consuming, since it involved recleaning the surface after the measurement of each set of four cracks. Time had to be one of the deciding factors in a project involving hundreds of crack length measurements, therefore the lower magnification was selected after realizing that the loss of resolution introduced an error which was small compared to the scatter in the crack length results. The large scatter in crack length results was caused by the non-homogeneous microstructure of WC-Co. The scatter increased with increasing indenting load, therefore the indenting load selected (15 kg) represents a compromise between a relatively low scatter in Palmqvist crack lengths and cracks long enough to keep the percentage error low.

It was found that the total error in the crack length measurements was about 10%, i.e. ± 4 μm, which agrees with the error quoted by Exner and Gurland[5] and by Coster[6]. This large error allowed only qualitative correlations between Palmqvist crack length, compressive prestress and number of compressive cycles.

The set of samples which had been subjected to a total of 30 cycles was annealed at 800°C for 2 hours. Palmqvist crack lengths were measured before and after annealing.

One as-received sample and one sample which underwent 320 cycles at a maximum stress of 20 kbars were indented at loads ranging between 5 and 50 kg. This determined the scatter at each indenting load and the ratio between indenting load and sum of Palmqvist crack lengths, a parameter which is often used to measure crack propagation resistance on a surface, i.e. surface toughness (e.g. Reference[8]).

RESULTS

1. The Vickers hardness of the material did not change significantly after compression.
2. In the as-received specimens the average Palmqvist crack length was about 50 μm in Batch A and 55 μm in Batch B, although the two batches were nominally the same. Since the two batches had been manufactured months apart, small differences in manufacturing conditions may have caused small differences in the microstructure, e.g. in the grain size distribution.
3. Figure 1 summarizes the results obtained from samples belonging to Batch A. The average Palmqvist crack lengths (average of "vertical" and "horizontal" cracks) have been plotted against the stress at which the samples were precompressed. Each line joins points representing results from samples which were precompressed for an equal number of cycles. For the sake of simplicity the 10% error has not been indicated in any of the graphs.

    In the samples which underwent 730 cycles the average Palmqvist crack lengths do not vary appreciably with increasing prestress and, up to 20-30 kbars, do not show any appreciable difference from the results obtained after 630 cycles. Above 20-30 kbars the average Palmqvist crack lengths after 730 cycles are consistently longer than after 630 cycles. This was probably due to microcracks forming in the material somewhere between 630 and 730 cycles.

    Figure 2 summarizes the results obtained from samples belonging to Batch B. In this batch the average Palmquist crack lengths decreased appreciably with increasing the number of cycles but remained statistically constant with increasing the compressive stress. The line joining the points representing results from samples which underwent 680 cycles has a similar behaviour to the line corresponding to 730 cycles in Figure 1.

    Figure 3 summarizes the results from both batches. The approximate percentage reduction in crack length was calculated for each number of cycles at the prestress of 30 kbars. Figure 3 indicates that most of the crack length reduction occurs during the first 200 to 300 cycles.

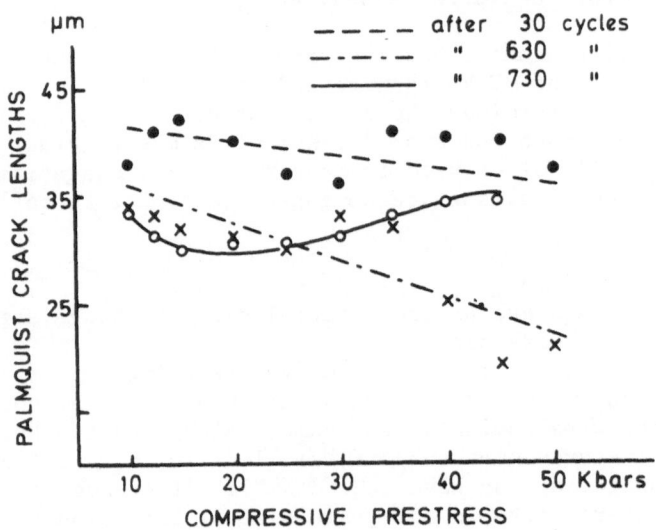

Fig. 1. Effect of compressive prestress on the average Palmqvist
         crack lengths in WC-5% Co samples of Batch A. The lines
         join points representing results obtained after ·30, 630
         and 730 compressive cycles.

4.   Figures 1 and 2 show the results obtained by averaging "verti-
     cal" and "horizontal" cracks, i.e. cracks in the direction of
     the compressive prestress and cracks normal to that direction.
     In the as-received samples the difference between vertical and
     horizontal cracks nucleated at the corners of indentations
     were consistently within the 10% error, therefore the as-
     received material was isotropic. In the precompressed samples
     the differences between average vertical and average horizontal
     cracks were substantial, i.e. generally larger than 10%,
     therefore the material was anisotropic, the vertical cracks
     being consistently longer than the horizontal cracks.

     As a general trend, the anisotropy of precompressed samples
     appeared to increase with increasing the number of cycles.
     For example, after 8 cycles the average vertical crack lengths
     were higher than the average horizontal crack lengths but the
     difference between the two averages was seldom larger than
     8 μm.

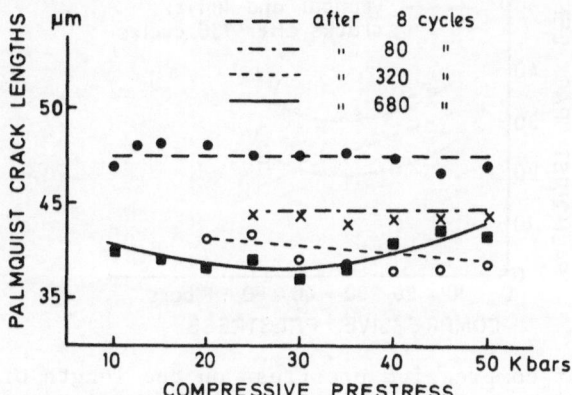

Fig. 2. Effect of compressive prestress on the average Palmqvist
         crack lengths in WC-5% Co samples of Batch B. The lines
         join points representing results obtained after 8, 80,
         320 and 680 compressive cycles.

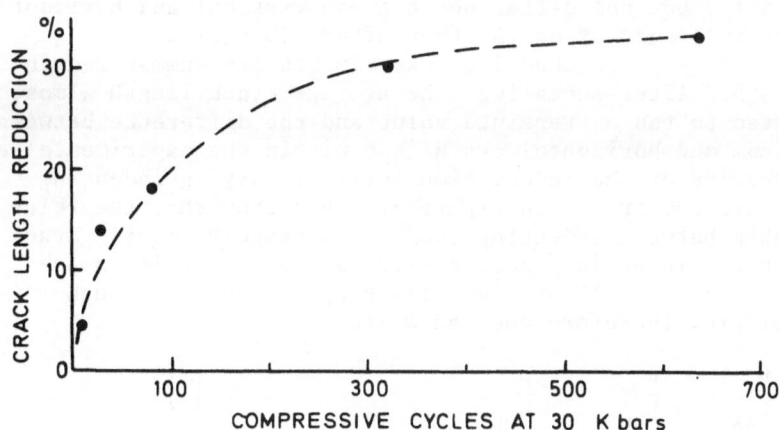

Fig. 3. Effect of the number of compressive cycles on the percen-
         tage reduction in Palmqvist crack length in samples from
         Batches A and B, at 30 kbars.

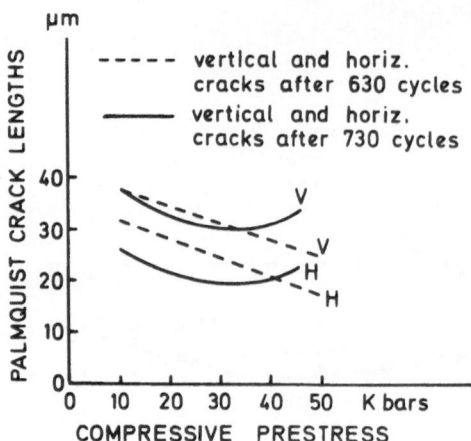

Fig. 4.   Effect of compressive prestress on the length of "vertical"
          and "horizontal" Palmqvist cracks in samples compressed
          for 630 and 730 cycles.

On the other hand, after 730 cycles the difference between
average vertical and average horizontal Palmqvist crack
lengths was never smaller than 12 μm. Figure 4 shows
typical results from sets of measurements on samples which
underwent 630 and 730 cycles. The total average crack lengths
of the two sets of samples are approximately equal (see also
Figure 1), but the difference between vertical and horizontal
cracks is larger after 730 than after 630 cycles.

5.   The results of the annealing experiments are summarized in
      Figure 5. After annealing, the average crack length almost
      reverted to the as-received value and the difference between
      vertical and horizontal cracks was within the experimental error.

6.   The results of the indentation tests at varying indenting
      loads are summarized in Figure 6. They show that the rela-
      tionship between indenting load and average Palmqvist crack
      length is linear in precompressed samples as it is in as-
      received ones. All straight lines appear to go through
      the origin, therefore one can write:

$$\frac{P}{L}_{AR} < \frac{P}{L}_{VC} < \frac{P}{L}_{HC} \qquad\qquad [1]$$

where: P = load applied to the Vickers indentor, i.e. indent-
           ing load;
       L_{AR} = average length of the four Palmqvist cracks
           nucleated at the corners of each Vickers indentation
           in the as-received sample;

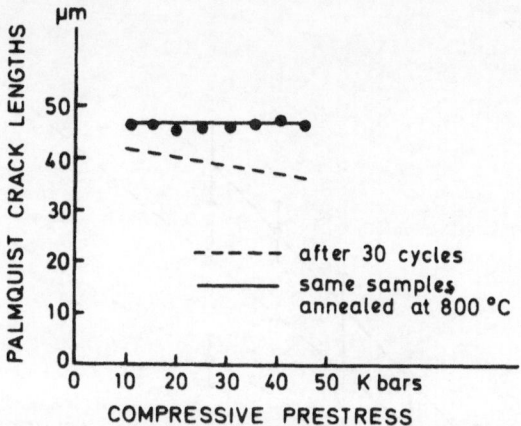

Fig. 5. Effect of compressive prestress on the average Palmqvist
crack length in samples which underwent 30 compressive
cycles. The dotted line represents the results before
annealing, the continuous line after annealing.

L_{VC} = average length of the two "vertical" Palmquist
cracks nucleated at the corners of each Vickers
indentation in the precompressed sample;
L_{HC} = average length of the two "horizontal" Palmqvist
cracks in the precompressed sample.

Relationship [1] indicates that in order to produce a crack
of fixed length a lower indenting load is required for as-
received specimens than for precompressed specimens. Precompressed specimens appear to have a higher resistance to
surface cracks than as-received specimens and in precompressed
specimens cracks appear to meet a higher resistance when propagating normally to the direction of the compressive prestress.

DISCUSSION

     Before attempting to interpret the present results, it is use-
ful to compare them with the results found in the literature.

a)   As mentioned in the Introduction, Exner and Gurland[5] observed
     Palmqvist crack length reductions of up to 30-35% after com-
     pressing WC-10% Co samples at a stress of 27 kbars. In the
     present material (WC-5% Co) a reduction of 30-35% at a stress
     of about 30 kbars has only been observed after about 300 cycles
     (Figure 3).

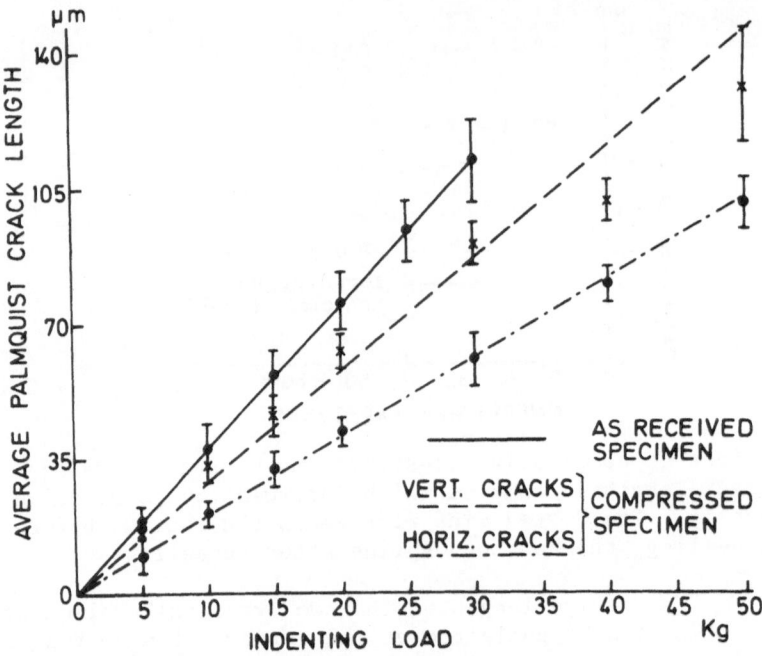

Fig. 6. Effect of the indenting load on the average Palmqvist
crack lengths in an as-received sample and in a sample
which was compressed at 20 kbars for 320 cycles.

Exner and Gurland also observed anisotropy after compression:
in the compressed samples the cracks propagating in the direction of the compressive prestress were generally longer than
the cracks in the normal direction. In their material they
observed differences between the two types of cracks of up to
35% after only one cycle, while in the present material differences of that magnitude were only observed after about
300 cycles.

Exner and Gurland attempted to explain the effects of pre-
compression by assuming that precompression decreases the
residual tensile stresses in the cobalt phase. This, however,
does not agree with the experimental results of Tumanov and
co-workers[4] who found that the residual tensile stresses in the
cobalt increase after compression.

b)   Coster[6] tested WC-3% Co samples and found that Palmqvist
crack lengths increased with increasing compressive prestress
up to about 60 kbars. These observations are in disagreement
with the work reported here. Moreover, Coster did not observe

a difference in length between "vertical" and "horizontal" cracks. Perhaps, more than one compressive cycle is required to induce anisotropy in a 3% Co grade.

c) Ivensen and co-workers[1], Hara and Ikeda[2] and Pelepelin[3] reported increases in resistivity, coercivity and hardness after inducing relatively high plastic strains by compression. Generally, they explained these effects by assuming the formation of microcracks. This appears to agree with the results obtained here (Figures 1 and 2), since the increases in crack lengths after 730 and 680 cycles at stresses higher than 20-30 kbars are also thought to be due to microcracks.

In order to attempt to interpret the observed results, it is necessary to note that in the present material cracks are mostly intergranular and that, according to Gurland[9], in a fine-grained WC-5% Co grade the contiguity of the carbide grains is expected to be at least 0.6. This means that at least 60% of the surface of an average carbide grain is in contact with other carbide grains and that the carbide skeleton can be considered as continuous. Therefore, the energy required to propagate an intergranular crack is spent partly in the cobalt and partly in fracturing WC-WC "bridges". As a result, an increase in crack resistance may be accounted for by changes occurring in the cobalt phase and/or in the WC-WC bridges.

As far as the cobalt phase is concerned, it has been observed by Hara and Ikeda[2] that the h.c.p./f.c.c. cobalt ratio increases with increasing plastic prestrain, therefore the transformation f.c.c. Co $\rightarrow$ h.c.p. Co is expected to occur in the present tests. However, in a material which has a high carbide contiguity the applied compressive load is transmitted mainly through the contacts between the grains, therefore plastic strain and allotropic transformation are expected to be induced in the cobalt only after the carbide has yielded.

As far as the WC-WC bridges are concerned, it is common experience that the probability that a grain boundary is on the path of a propagating crack increases with decreasing the angle between the boundary and the direction of the crack (Figure 7). Therefore, the WC-WC boundaries found on the path of "horizontal" cracks are mostly "horizontal" boundaries, i.e. boundaries normal to the direction of the prestress, while the path of "vertical" cracks is mostly through "vertical" WC-WC boundaries.

As a result, the crack reduction and the anisotropy observed in this work may be explained by assuming that the compressive prestress increases the crack resistance of the WC-WC boundaries of amounts which increase with increasing the angle between the boundary and the stress direction. For boundaries of equal size one

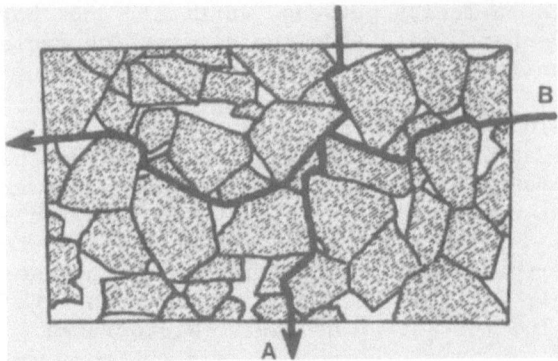

Fig. 7. "Vertical" intergranular cracks, such as A, propagate
         mainly along almost "vertical" WC-WC boundaries, while
         "horizontal" intergranular cracks, such as B, propagate
         mainly along almost "horizontal" WC-WC boundaries.

may assume a relationship of the type:

$$r = C \sin^n\psi + r_o \qquad\qquad [2]$$

where: r = crack resistance of a WC-WC boundary (see Figure 8)
              after compression,
       C,n = constants,
         ψ = angle between the WC-WC boundary and the direction of
              the compressive stress,
         r_o = average crack resistance of WC-WC boundaries before
              compression.

The increase in crack resistance can be explained in terms of
an increased residual compressive stress in the carbide bridges,
in agreement with the experimental results of Tumanov and co-workers[4].

It is reasonable to assume that the residual compressive stress
is largest in WC-WC boundaries which are normal to the compressive
stress, since those are sites of largest stress concentrations.

The effects of residual compressive stresses would be expected
to tend to a maximum value, after a certain number of cycles, as a
result of strain hardening, in agreement with the results in
Figure 3. Strain hardening would be expected to be followed by
the formation of microcracks, in agreement with the results in
Figures 1 and 2, for 730 and 680 cycles at compressive stresses

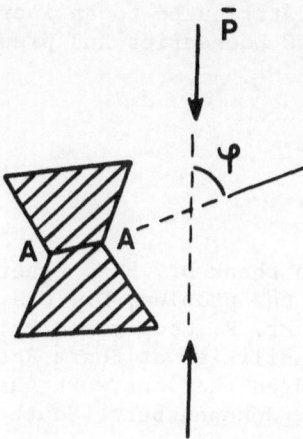

Fig. 8.   Schematic representation of two WC grains having a common
           boundary, AA. The angle between the boundary and the com-
           pressive stress $\bar{P}$ is ψ, which appears in [2].

higher than 20-30 kbars. Furthermore, residual compressive stresses
due to precompression would be expected to be relieved by annealing,
which is also in agreement with the results reported in Figure 5.

CONCLUSIONS

      The following conclusions can be drawn:

(a)   The crack propagation resistance, measured by the Palmqvist
      crack technique on surfaces cut from WC-5% Co samples, is in-
      creased by cyclic precompression.

(b)   The degree of crack propagation resistance increases rapidly
      with an increase in the number of cycles up to about 300, then
      increases slowly up to a maximum value. At stress levels higher
      than 20-30 kbars the crack propagation resistance initially in-
      creases then, after 600-700 cycles, starts to decrease, probably
      due to the formation of microcracks.

(c)   Cyclic precompression converts WC-5% Co into an anisotropic
      material. The anisotropy appears to increase with increasing
      the number of compressive cycles.

(d)   All the observed results seem to agree with the interpretation
      that the toughening effect is mostly due to an increased crack
      resistance of the WC-WC boundaries. The increase is attributed
      to residual compressive stresses introduced by precompression.

The anisotropy effect is attributed to an increasing crack resistance as the angle between WC-WC boundaries and prestress direction increases.

ACKNOWLEDGMENTS

The author wishes to thank Dr. H.E. Exner and Professor C. Dimitriou for discussing the problem, Dr. G.J. Rees and Mr. R. Fry for helpful suggestions, Mr. R. Cooper and Dr. C. Peters for making available experimental facilities at Boart Research Centre. It is also gratefully acknowledged that the work has been sponsored by Boart International Ltd., Johannesburg, South Africa.

REFERENCES

1. V.A. Ivensen, V.A. Chistyakova and O.N. Eiduk, Changes in the properties of WC-Co hard alloys during deformation, Poroshk. Met. 129: 39 (1973).
2. A. Hara and T. Ikeda, Behaviour of compressive deformation of WC-Co cemented carbide, Trans.Jap.Inst. Met.13:128 (1972).
3. V.M. Pelepelin, Effect of plastic deformation on the physico-mechanical properties of WC-Co alloys, Poroshk.Met. 35: 933 (1965).
4. V.I. Tumanov, A.A. Cheredinov and K.F. Kuznetsova, Magnetic investigation of microstress in hard alloys, Poroshk.Met. 165: 710 (1976).
5. H.E. Exner and J. Gurland, The effect of small plastic deformations on the strength and hardness of a WC-Co alloy, J. Mater. 5: 75 (1970).
6. M. Coster, Etude du grossissement des grains et du comportement sous fortes pressions de composites WC-Co, Thèses, Université de Caen (1969).
7. H.E. Exner, The influence of sample preparation on Palmqvist's method for toughness testing of cemented carbides, Trans. AIME 245: 677 (1969).
8. R.K. Viswanadham and J.D. Venables, A simple method for evaluating cemented carbides, Metall. Trans. 8A:187 (1977).
9. J. Gurland, A structural approach to the yield strength of two-phase alloys with coarse microstructures, Mat. Sci. Eng. 40: 59 (1979).

DISCUSSION

A. T. Santhanam:

At what temperature did you do the pre-stressing?

S. B. Luyckx:

Room temperature.

A. T. Santhanam:

What was the yield strength of your material?

S. B. Luyckx:

Unfortunately I can't give details on the mechanical properties
of this material.

A. T. Santhanam:

Could you even say whether the maximum pre-stress was higher or
lower than the yield stress?

S. B. Luyckx:

It comes very close to the yield stress.

T. Hall, Jr.:

I think we're interested in your results for the same reason that
your sponsor probably is. Is it possible that the upturn in the
crack length can give you a prediction of failure?

S. B. Luyckx:

Unfortunately in this material most of the fractures that we observed
were initiated at flaws.

T. Hall, Jr.:

Were these materials cycled to failure in the experiment, or is
that part of an ongoing effort?

S. B. Luyckx:

Yes. Part of the ongoing.

R. Rice:

I'd like to ask if it is a possibility that micro-cracking might
be another explanation for your results. In pure ceramic materials
frequently the fracture toughness will go through a maximum as the
amount of micro-cracking increases. Therefore, the upturn in the
crack length that you suggested in connection with microcracking,
could that possibly just be a greater degree of micro-cracking
which would correspond to a reduction in toughness. One would expect micro-cracks to form preferentially in the vertical direction
in compressive loading and hence be more limiting on the horizontal
cracks.

S. B. Luyckx:

I see. So, you would expect that these micro-cracks would relieve
tensile stresses and lead to crack reductions.

H. Fischmeister:

It strikes me that the effect that you have observed might be explained by work hardening of the binder phase. Most theories of
fracture toughness contain the work hardening of the binder phase
and the tougher the binder becomes, the higher the fracture toughness.
I think one could also argue that horizontal layers of binder would
work harden in a different way than vertical ones, because they are
exposed to different stress oscillations. I'm not suggesting this
is better than your explanation, I am only saying that it might be
worth while to consider this as an alternative.

S. B. Luyckx:

Thank you.

EXTENDING THE USE OF INDENTATION TESTS

E. A. Almond and B. Roebuck

National Physical Laboratory
Teddington, Middlesex, UK

ABSTRACT

Results are described of experiments performed using indenters of various geometries on a range of WC/Co hardmetals, with hardnesses ranging from 1000 to 1800 HV30.

The deformation characteristics of hardmetals show a similar dependence on indenter angle to that shown by other materials for lubricated and unlubricated pyramid indenters. As indentation temperature decreases, relative increases in hardnesses are similar, but Palmqvist cracking shows a greater temperature dependence in the coarse-grained cemented carbides. Stress-corrosion cracking, as measured by increases in the length of Palmqvist cracks and surface flaking, occurs in most acidic environments, and is dependent on the carbon content of the hardmetal.

When ball indentations are made at decreasing distances from the square edge of specimens, flakes break away from the edge. The mechanisms of crack growth change when hardness increases above 1375 HV30. Damage produced by flaking is small on the indented face but large on the side face: flake size increases with increase in hardness.

INTRODUCTION

For materials with low ductility such as hardmetals, hardness is frequently the only mechanical property measured by industry, while some manufacturers use the degree of radial cracking

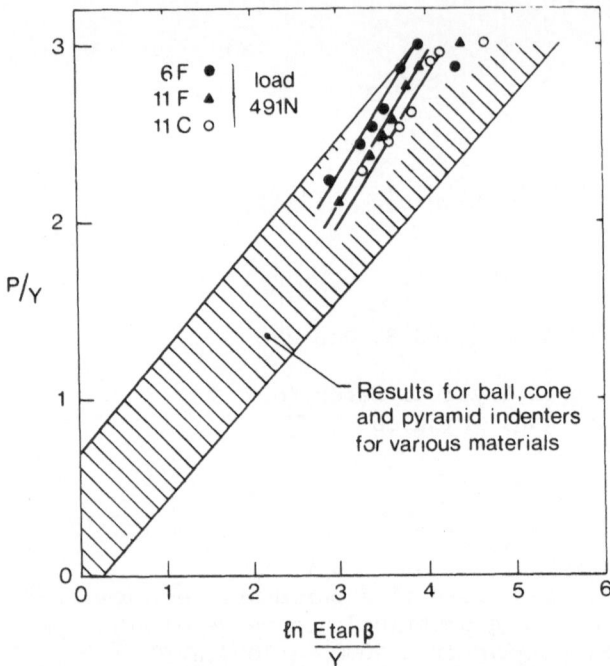

Fig. 1. Variation of ratio of indentation pressure to flow
         stress, with change in indenter angle for 6F, 11F and
         11C compared with results on other materials using
         various indenter geometries.

produced during indentation as a guide to the toughness of the
material. The response of hardmetals to indentation by diamond
pyramid indenters is in good agreement (Fig. 1) with expanding-
cavity models[1], where plastic displacements are accommodated
elastically outside the plastic zone according to:

$$\frac{P}{Y} = C + D \ln \frac{E \tan \beta}{Y}$$

where P is the indentation pressure, Y is the flow stress, C
and D are constants, E is Young's modulus, and β is the angle
of inclination of the faces of the indenter with the specimen
surface.

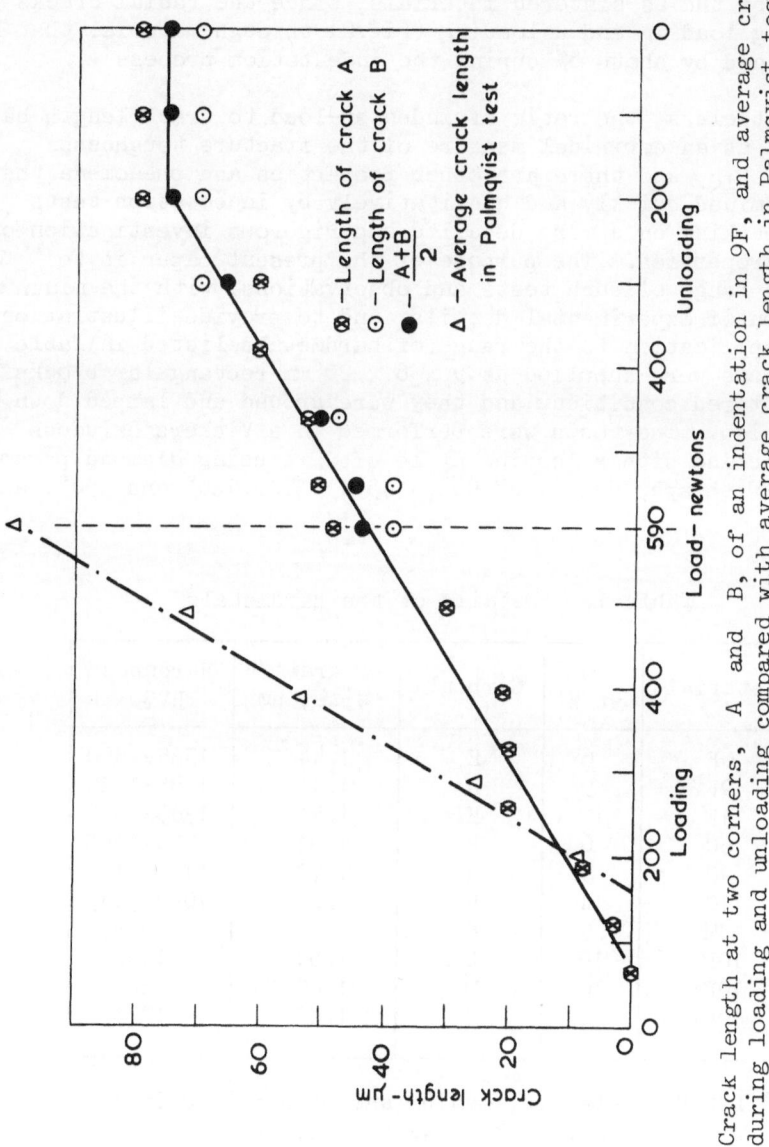

Fig. 2.   Crack length at two corners, A and B, of an indentation in 9F, and average crack length during loading and unloading compared with average crack length in Palmqvist test.

In contrast, radial or Palmqvist[2] cracking in hardmetals
cannot be explained satisfactorily either in terms of crack
geometry or crack length, by theories for growth of median vents
in brittle materials[3] or by other models. Nor can Palmqvist parameters be related intrinsically to the plane strain fracture
toughness of the as-sintered materials, since the radial cracks
grow during loading and unloading (Fig.2) through material that has
been strained by about 8% during the indentation process[4].

Nevertheless, the ratio of indenter-load to crack-length has
been used[5] as an empirical measure of the fracture toughness·
parameter G_{IC}, and there are other properties and phenomena that
can be examined quickly and qualitatively by indentation tests
before embarking on a more detailed and rigorous investigation of
specific properties. The purpose of the present paper is to
describe a range of such tests and observations, with the minimum
description of experimental details, and to provide illustrations
of their application to the range of hardmetals listed in Table 1.
The specimens were supplied as 5 x 6 x 20 mm rectangular blocks in
the as-sintered condition, and they were ground and lapped down to
a 1 μm finish. The tests were performed in a Vickers hardness
testing machine with a loading cycle of 12 s using diamond pyramid
indenters with apex angles of 90°, 115°, 122°, 136° and 150°, and

Table 1.   Details of the Hardmetals

Material	Co wt %	Carbon*	WC grain** -size, μm	Hardness HV30
6F	6	H	1.44	1535-1560
9F	9	H	1.40	1385-1420
11F	11	H	1.41	1305-1355
6C	6	H	4.81	1225-1245
9C	9	H	4.70	1120-1140
11C	11	H	5.20	1090-1115
3M	3	H		1925
10F2	10	L	1.80	1390
10F5	10	M	1.80	1365
10F7	10	H	1.80	1275

* L,M,H denote low, medium and high respectively. No
graphite or eta-phase were present.

** Mean-linear-intercept value multiplied by 1.5.

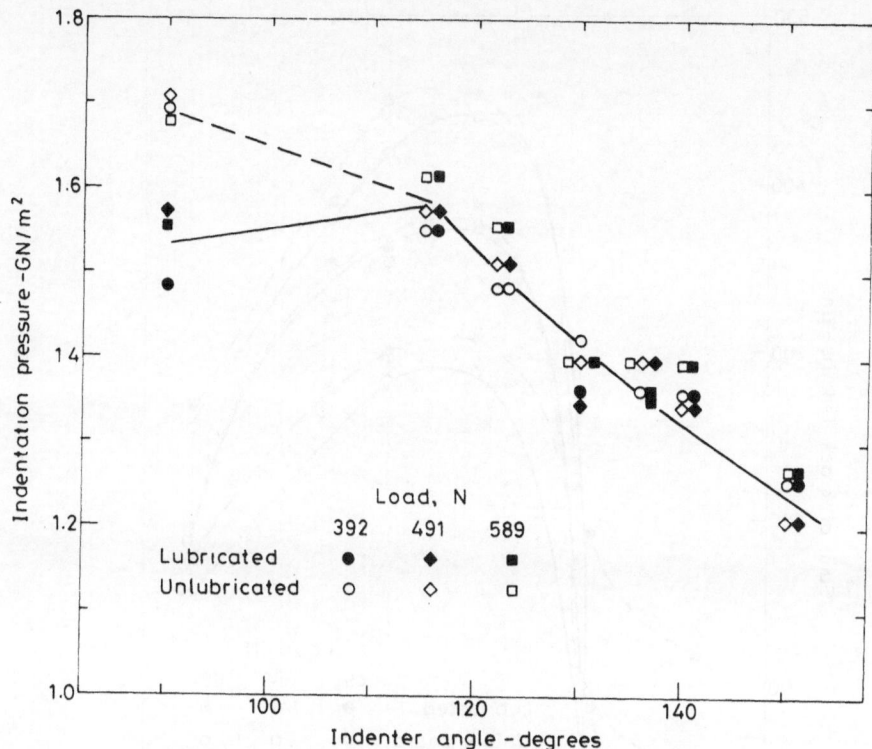

Fig. 3. Variation of indentation pressure with angle of pyramid
         indenter for 11F with and without lubrication.

ball indenters of 2, 3, 4 and 5 mm dia, manufactured from a
6 wt%Co/WC hardmetal.

THE EFFECT OF LUBRICATION

     For studying the effect of lubrication on indentation, 11F
specimens were indented initially on unlubricated surfaces, and
then after the surfaces had been brushed with an extreme pressure
lubricant of graphite in alcohol. The indentation pressure was
not altered by lubrication for indenter angles in the range
115°-150°, but it was reduced by about 10% at an indenter angle of
90° (Fig.3). This observation, together with the results in Fig.1,
indicates that for indenter angles less than 115°, material flows
upwards along the faces of the indenter during indentation[6], and
the resistance to flow can be reduced by lubrication.

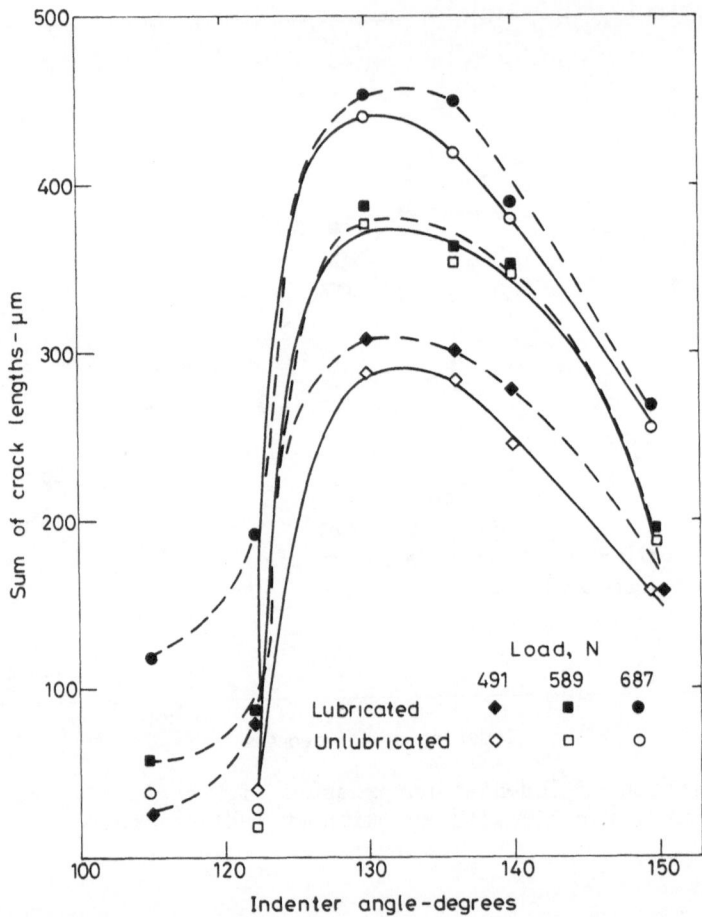

Fig. 4. Variation of total crack length with angle of pyramid
indenter for 11F with and without lubrication.

Lubrication produced increases in crack length of about
5-10% when using indenters with angles of 130°-150°; while with an
indenter angle of 122°, the increment was much larger, such that
significant cracking was observed in specimens that were almost
crack-free after indentation in the unlubricated condition (Figs 4
and 5). The mechanistic explanation for the enhanced crack growth
is probably mechanical rather than chemical, and it could be due
to: (a) the effect of oil entrapped in the crack, producing a
wedging action beneath the indenter corners at the mouths of the
radial cracks, or (b) lubrication of the crack faces, enabling
them to move past each other more easily when the crack path
deviates from a radial direction. However, neither mechanism

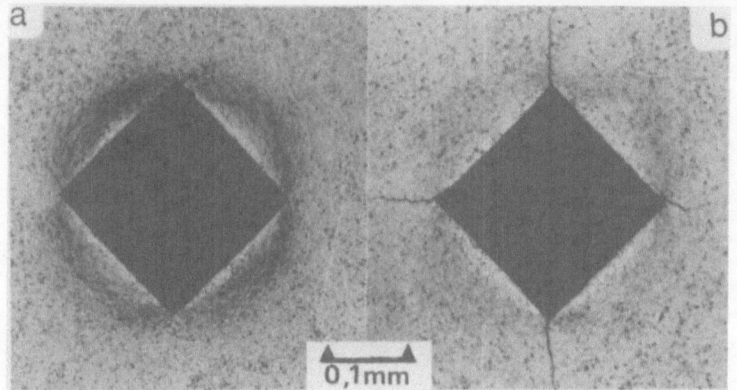

Fig. 5.   Indentation produced in 11F with 122° indenter and 687N
load, a) without and b) with surface lubrication.

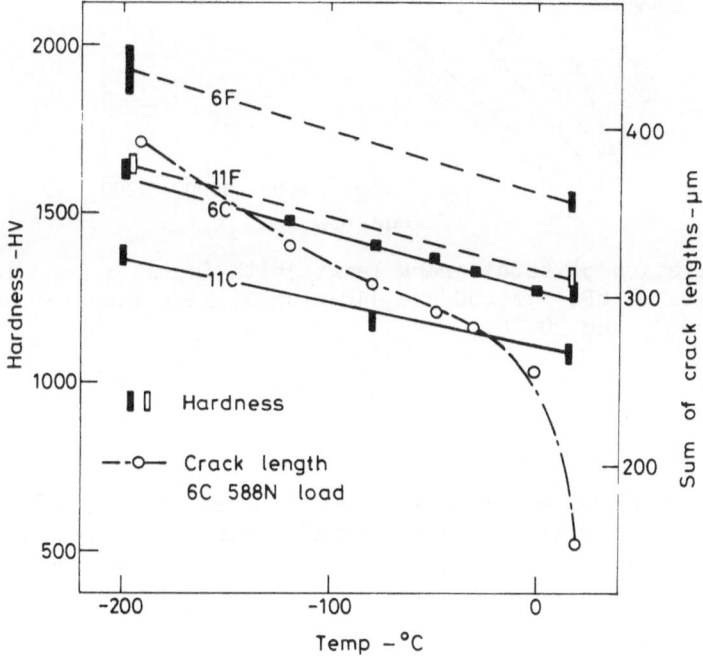

Fig. 6.   Variation of hardness and total crack length with
temperature for tests on 6F, 11F, 6C and 11C with Vickers
indenter.

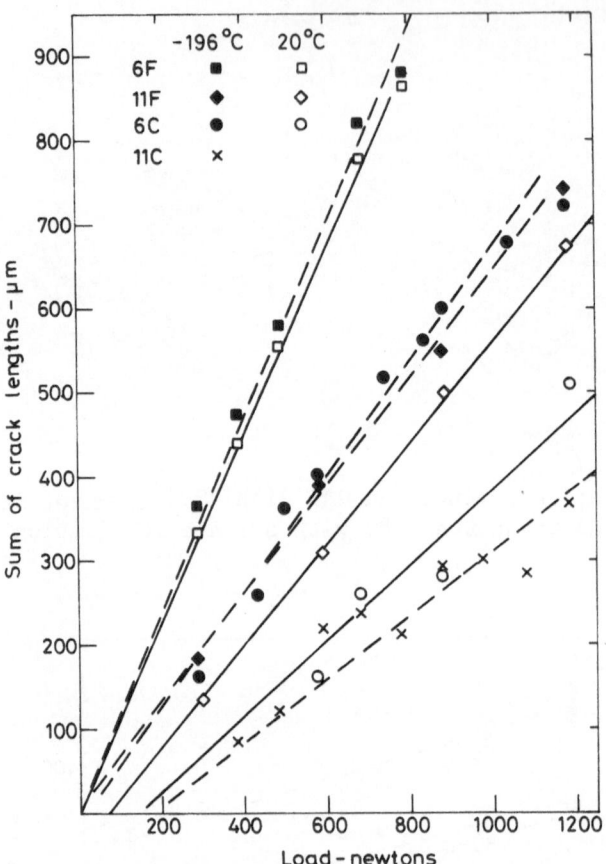

Fig. 7.   Variation of total crack length with indenter load for
          tests on 6F, 11F, 6C and 11C with Vickers indenter at
          – 196° and 20 °C.

explains the enhanced crack initiation shown in Fig. 5, and this
observation deserves further investigation since it suggests that
lubrication might promote cracking in tools of certain geometries.

LOW TEMPERATURES

     Although hot hardness measurements are being used increasingly
to assess high temperature properties of tool materials, there is
little information on the properties of hardmetals at temperatures

below 0 °C. In most metallic materials the resistance to plastic deformation increases and toughness decreases as the testing temperature is lowered. This was also found to be true for hardmetals, as can be seen in Figs. 6 and 7 for tests on the 6% and 11% Co/WC materials with a Vickers pyramid indenter. The temperature dependence of hardness for the four hardmetals shown in Fig. 6 is in good agreement with that observed on a 10% Co/WC material[7], and it appears that the variation with temperature was not dependent on grain size or Co content. However, there was a marked grain size dependence in the effect of temperature on Palmqvist cracking (Fig. 7). Thus, in the fine-grained materials the cracks produced at - 196 °C were only marginally longer than those produced at room temperature, whereas the cracks in the coarse-grained 6% Co material were 50% longer at - 196 °C compared with the room temperature values, and significant cracking was induced in the 11% Co material even though cracks could not be produced at room temperature. Various annealing and low temperature cycling experiments were also performed prior to indentation to ensure that the results had not been affected by cooling stresses or by temperature and stress induced transformations of the binder phase from its fcc to hcp form.

The observations are of limited practical value since hardmetals are rarely used at operating temperatures below 0 °C; however, for the purpose of laboratory testing, low temperature indentation may provide the solution to the problem of introducing precracks into some of the tougher hardmetals[8]. The fundamental relevance of the results is that they show that at low temperatures, grain size is more important in determining the temperature dependence of cracking than it is in its effect on the temperature dependence of yielding, and this needs to be taken into account in any mechanistic models for the fracture and deformation of hardmetals.

STRESS CORROSION CRACKING

The use of indented specimens to study stress corrosion cracking was proposed in earlier work[9] where it was found that the length of Palmqvist cracks in the 6F hardmetal had increased significantly after exposure to HF vapour for 16h. In further tests on the hardmetals listed in Table 1, it has been found that:

i) On exposure to the vapour of concentrated HCl, H_2SO_4 and HNO_3 acids, crack length increased and the slope of the crack-length versus indentation-load curve remained linear.

ii) In an environment of HF vapour, crack growth was significantly greater in cracks produced by high indentation loads (Fig. 8).

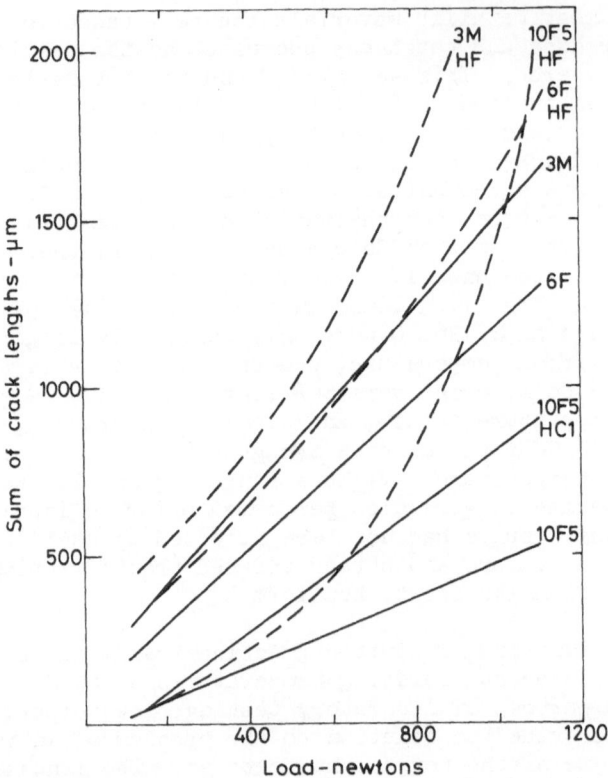

Fig. 8.   Schematic representation of variation of total crack
          length with indenter load for 3M, 6F and 10F5 before and
          after exposure to HF for 16h, and for 10F5 after
          exposure to HCl for 24h.

iii) In tests on the 3M hardmetal, crack length increased with
time of exposure in moist H_2S gas, but did not increase in the
presence of steam or dry hydrogen gas at a pressure of 150 bar.

iv) Corrosion assisted crack growth was accompanied by a surface
displacement around the indentation, similar to that produced by
annealing to remove residual stresses (Fig. 9).

These and previous observations indicate that crack growth was
the result of the combined effect of the residual stress field
around the indentation, and the specific action of individual
corrosive media. Thus the phenomenon does not appear to be a
simple hydrogen assisted crack growth process, but it is dependent
on the general corrosion properties of the materials. To obtain

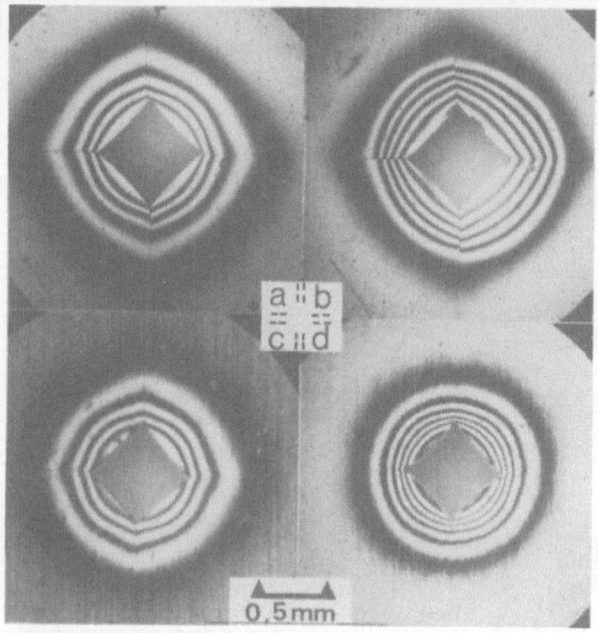

Fig. 9. Interference patterns showing displacements around Vickers
indentations; a) and b) 6F indented with 298N, a) before,
and b) after annealing at 800 °C for 1h: c) and d) 3M
indented with 392N, c) before and b) after exposure to
moist H_2S for 70h.

further information for this interpretation, three hardmetals 10F2, 10F5 and 10F7, with identical grain-sizes and Co
contents, but with different carbon contents, were indented and
exposed to HCl and HNO_3 vapour for various times. Measurements
were made of the gradient of the crack-length L versus indentation-
load P after various exposure times, and it was found that the
P/L value decreased with time as shown in Figs. 10 and 11. The
P/L versus time curves indicate that in HCl vapour, the extent of
stress corrosion cracking did not vary significantly with carbon
content, but in HNO_3 vapour the degree of cracking increased with
increase in carbon content. These observations are consistent
with results obtained on a wide range of hardmetals, where it has
been found that the corrosive attack by a 50% HCl/50% HNO_3 mixture
increases as the carbon content of the hardmetal is increased[10].

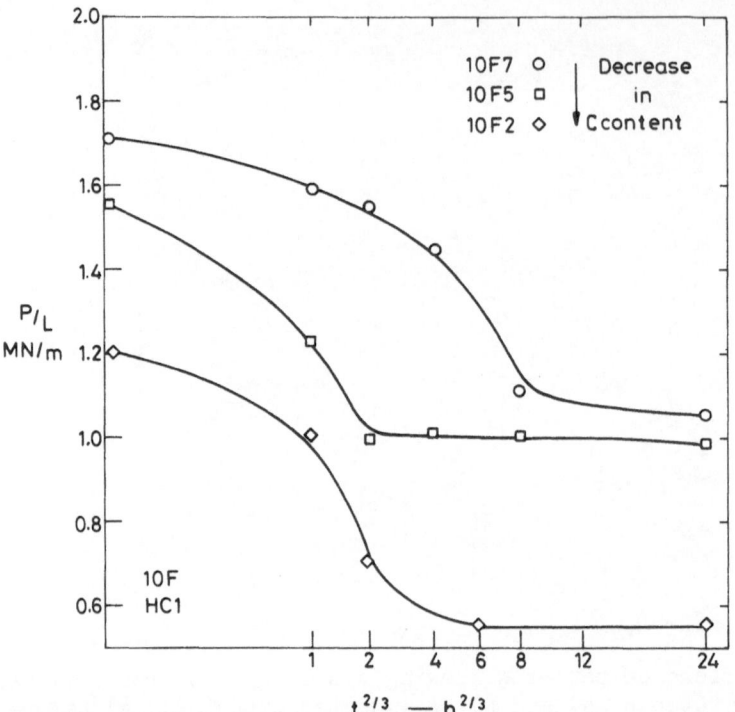

Fig. 10.   Variation of the load to crack-length ratio with time of
          exposure to HCl for 10F2, 10F5 and 10F7.

     In general, in the majority of experiments performed so far,
crack growth was probably caused by stress corrosion of the
binder phase of the hardmetal, except in experiments in HF vapour
where the existence of extensive surface corrosion indicated that
the WC constituent had also been significantly attacked.   The
present test may assist in the selection of hardmetals for various
corrosion resistant applications.   Initial comparisons of crack
growth in Ni-bonded and Co-bonded hardmetals in simple acidic
environments have not revealed any significant differences in
susceptibility to stress corrosion cracking; however, tests need
to be performed in environments that are more representative of
service conditions before the value of the test can be assessed.

EDGE FLAKING

     Indentation of hardmetals with spherical indenters produces
circumferential cracks commonly referred to as Hertzian, and

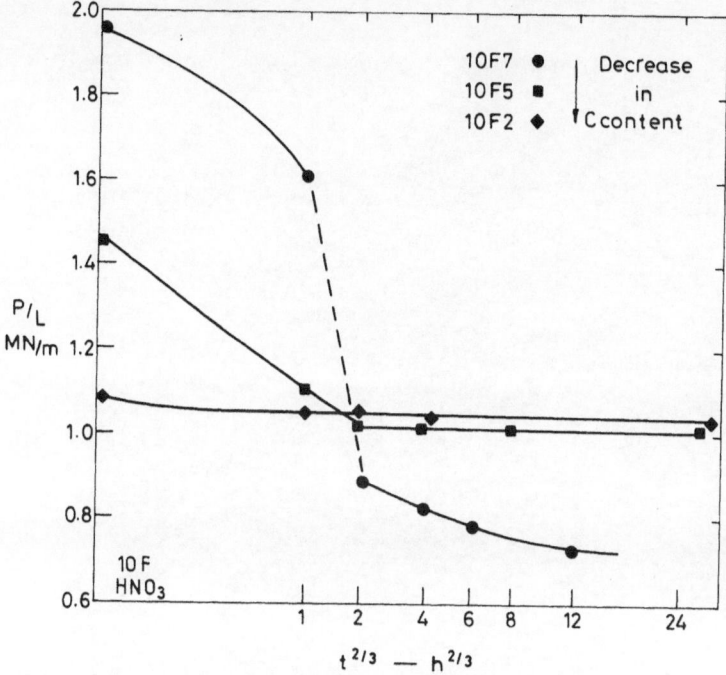

Fig. 11. Variation of the load to crack-length ratio with exposure to HNO$_3$ for 10F2, 10F5 and 10F7.

although the extent of cracking is probably related to toughness, it is not practicable to use Hertzian-crack parameters for routine testing of hardmetals. However, it was suggested by Trent[11,12] that the fracture properties of edges of hardmetal tools could probably be assessed by indenting a specimen with a ball indenter at decreasing distances from an edge, and by measuring the distance at which a flake starts to break away from the edge as a result of the indentation. The following tests were performed to examine this idea.

Initially various combinations of ball diameters and indentation loads were investigated to determine conditions that would produce edge-flaking within the load limits imposed in a Vickers hardness machine. For this purpose, specimens were made with rectangular edges with an edge-radius of less than 0.1 µm, and were mounted in a rig which was attached to the table of the hardness machine (Fig. 12). The indenters were made of a fine-grained 6% Co/WC hardmetal, and they were over-loaded before they were used in tests so that they would not deform plastically during testing.

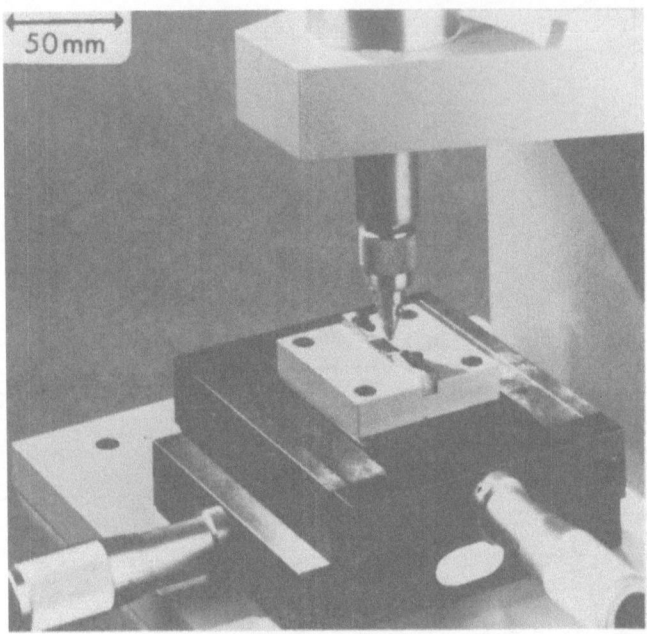

Fig. 12. The test fixture for edge flaking experiments.

     Experiments with 3 and 4 mm dia ball indenters on hardmetals
with a range of hardnesses, showed that for an indenter load of
784 N, the sequence of events was as follows. As the distance of
the indentation from the edge was decreased, the size of the
indentation increased and bulging was observed on the side face of
the specimen (Fig.13c). Cracks were produced around or within the
indentation, and they were radial or Hertzian for materials with
hardnesses below or above 1375 HV30 respectively (Fig.13a and b).
The distance of the indentation from the edge when cracks were
initiated did not vary significantly in the different hardmetals   (Fig.14). Flaking occurred in most but not all the hardmetals, when the indenter was brought sufficiently close to the
edge of the specimen.

     To increase the possibility of producing edge-flaking
throughout the hardness range, the diameter of the indenter was
decreased to 2 mm and the load was increased to 1079 N. As a
result, edge flaking was obtained in all the materials, and crack
initiation was predominantly radial as shown in Fig.13b. The
distance t of the indentation from the edge, when flakes formed,
did not vary significantly with hardness of the hardmetal; however,
the width w and the depth d of the flake, as measured on the

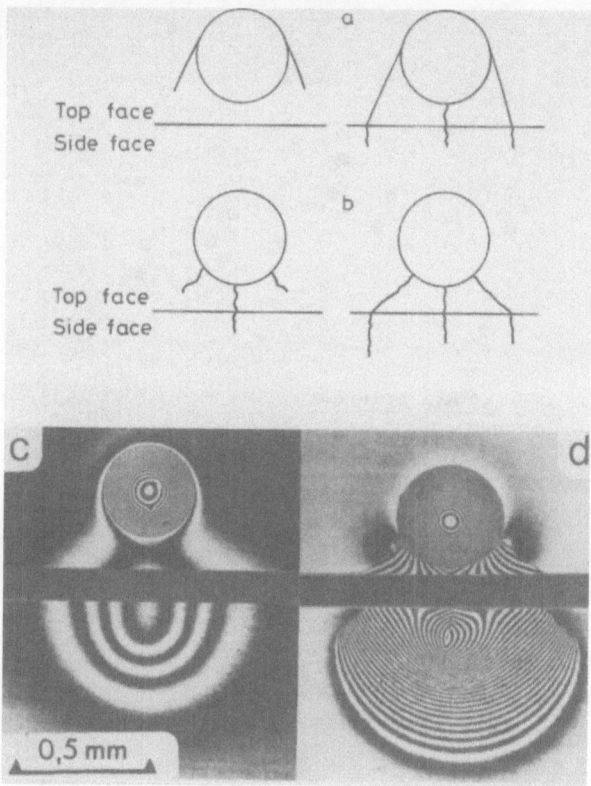

Fig. 13.   Indentation with 3 mm ball and 785 N load. Showing
           crack initiation in materials with hardnesses a) greater,
           and b) less than 1375HV; c) bulging on side face and
           d) edge-flaking in 6C specimens.

side face of the specimen, increased with increase in hardness.
Measurements of the mass of the flakes, and calculations of the
product of d, t, w and the material's density (defined as the
flake weight product) showed that both quantities were approximately
linearly dependent on hardness. The latter is shown in Fig.15.

     The results indicate that edge flaking is dependent on the
hardness of the hardmetal but this may be due to an apparent
relationship between hardness and fracture toughness, and the test
is not yet sufficiently discriminative to determine which is the
factor governing edge-flaking. Amongst many applications for a
fully developed test would be the determination of the optimum
edge geometries and hardmetals for providing resistance to
edge-flaking and improving edge-retention in cutting tools. There

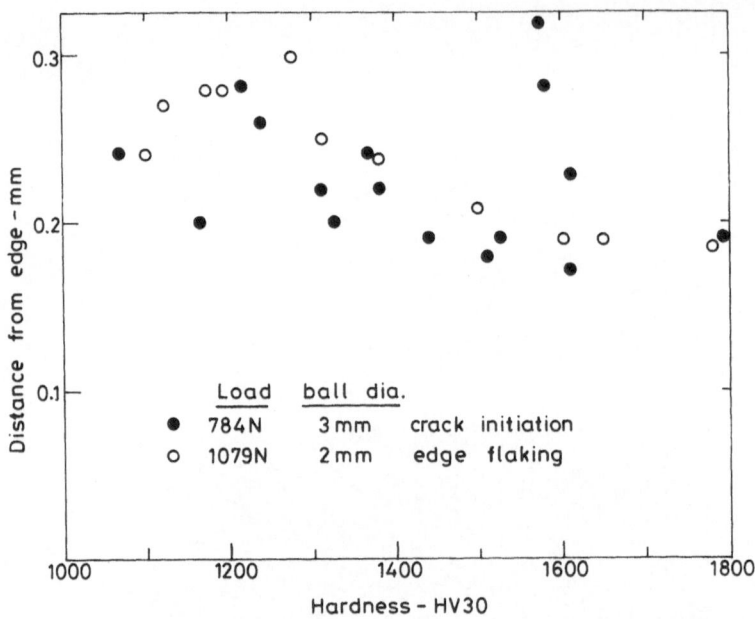

Fig. 14.   Distance of ball indenter from edge, for crack initiation
and for edge flaking in tests on hardmetals with
different hardnesses.

may also be applications in the use of the test to measure the
edge properties of coated tool tips and the ultra-hard tool
materials.

CONCLUSIONS

     Both the initiation and the growth of Palmqvist cracks were
enhanced in tests involving surface lubrication and the lowering
of specimen temperature, while crack growth alone was assisted in
cracked specimens exposed to acidic environments. The increases
in crack length produced by the various test conditions were
probably due to mechanical effects in lubricated specimens, and to
stress corrosion of the binder phase in the corrosion experiments,
but there is no simple explanation for the high temperature-
dependence of cracking in coarse-grained materials compared with
the low temperature-dependence of hardness in tests below 0 °C.

     Observations on the edge-flaking phenomenon indicate that
hardness may be an important factor in determining both the crack

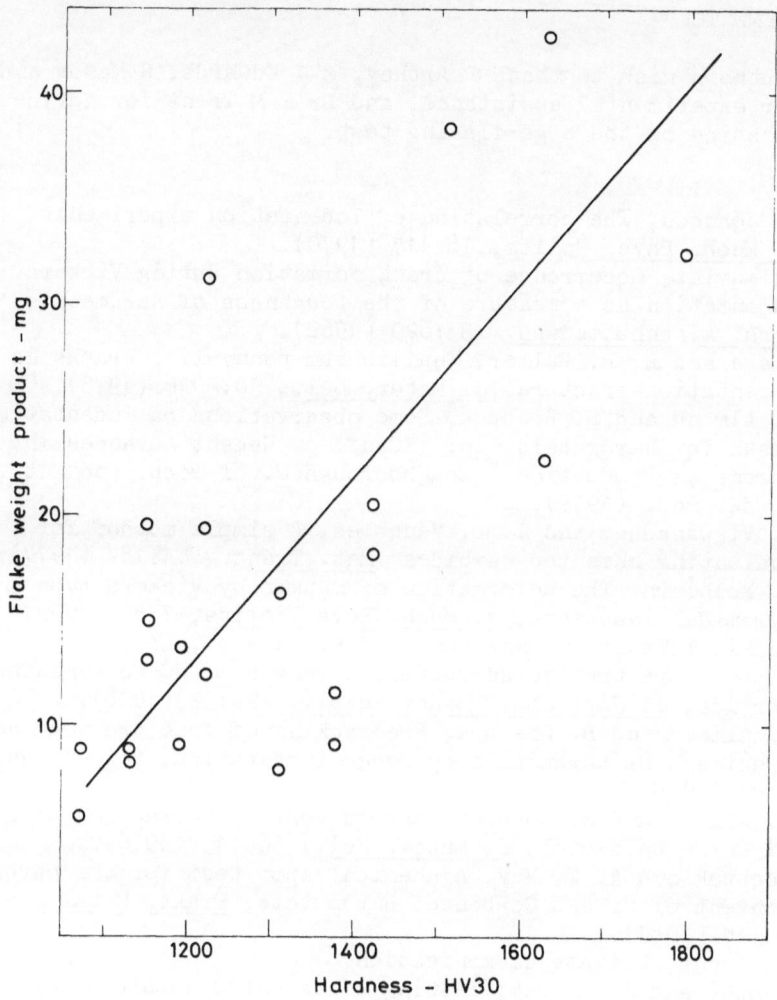

Fig. 15. Variation of flake weight product with hardness for
edge-flaking tests with 1079 N load and 2 mm dia ball.

initiation mechanism and the size of the flake that forms on the
side face. A stress analysis for ball-indentation near an edge
and a more discriminative testing system, possibly incorporating
harder indenters, would help to separate the roles of fracture
toughness and hardness in determining edge-flaking susceptibility.

   All the tests may have some practical application, either in
material selection or in investigating tool geometry, and may be
useful for testing coated tool tips and ultra hard tool materials.

ACKNOWLEDGEMENTS

The authors wish to thank R Arthey, M J Edwards, S Mason and R Patel for experimental assistance, and Dr E M Trent for advice and consultation on the edge-flaking test.

1. K. L. Johnson, The correlation of indentation experiments, J. Mech. Phys. Solids, 18:115 (1970).
2. S. Palmqvist, Occurrence of crack formation during Vickers indentation as a measure of the toughness of hardmetals, Arch. Eisenhuttenwes. 33:629 (1962).
3. B R Lawn and E. R. Fuller, Equilibrium penny-like cracks in indentation fracture, J. Mater. Sci., 10:2016 (1975).
4. E. A. Almond and B. Roebuck, Some observations on indentation tests for hardmetals, in: "Conf. on Recent Advances in Hardmetal Production," Loughborough U. of Tech. and Met. Powder Rep. (1979).
5. R. K. Viswanadham and J. D. Venables, A simple method for evaluating cemented carbides, Met. Trans. 8A:187 (1977).
6. T. O. Mulhearn, The deformation of metals by Vickers-type pyramidal indenters, J. Mech. Phys. Solids, 7:85 (1959).
7. H.Suzuki, T.Tanase, F.Nakayama and K. Hayashi, Low temperature transverse-rupture strength of WC-Co cemented carbide, J. Jap. Soc. Powder Metall. 25:32 (1978).
8. E. A. Almond and B. Roebuck, Precracking of fracture toughness specimens of hardmetals by wedge indentation, Met. Technol. 5:92 (1978).
9. E. A. Almond and B. Roebuck, Stress corrosion cracking of a 6% Co/WC hardmetal, J. Mater. Sci., 565:11 (1976).
10. B. Roebuck and A. T. May, A chemical spot test for the carbon content of Ni and Co-bonded hardmetals, Prakt. Metallogr. 31:18 (1981).
11. E. M. Trent, Private communication.
12. M. Dlouhy and J. Houdek, Testing of cemented carbides for dynamic toughness, Pokroky Práškové Met. 9 (1971).

SOME ASPECTS OF THE FRACTURE OF WC-Co COMPOSITES

Frédéric Osterstock and Jean-Louis Chermant

Equipe Matériaux Microstructure
Laboratoire de Cristallographie et Chimie du Solide
L.A. 251, ISMRA-Université, 14032 Caen Cedex, France

INTRODUCTION

During the last decade, fracture mechanics measurements on cemented carbides became of current use. Critical stress intensity factor values obtained in different laboratories with different specimen geometries show good agreement for similar batches (1)(2)(3)(4)(5). By this way it appeared that tungsten-carbide is rather tough if compared to other carbides (6)(7). Furthermore, this toughness can be increased by cobalt additions. Combination of critical stress intensity factor and rupture stress, mostly obtained in bending, allows one to evaluate the size of the critical defect, a_c, i.e. the size of the initiating defect at the moment where crack propagation becomes catastrophic. The calculated values of a_c are low compared to other carbides or single phased ceramics (7). This arises from liquid-phase sintering, however it shows clearly that mechanical strength of cemented carbide may be affected by very small defects like porosity or carbide crystal bundles... Influence of structural heterogeneities was already evidenced by S.B. Luyckx (8) when she noted that fracture origin is generally associated with the presence of impurities or inclusions. The statistical analysis of rupture made by P. Anderson (9) gave the same conclusion. H. Suzuki and K. Hayashi (10) made fractographic analyses of broken specimens, located the initiating defect and measured the size. By this way, they showed that there exists an accurate relation between the value of the stress acting on the defect and the defect size. Following, E.A. Almond (11) proposes a modification of the Weibull theory to account for the various nature of the defect whereas B. Roebuck (12) proposes a new bending specimen configuration in order to eliminate the influence of the sintering defects and so, to study the "microstructural" origins of rupture, i.e. those arising from the interaction between the two phases or from one of the two phases only. Hot isostatic pressing eliminates some types of defects and may result in a rupture stress increase of 100 % without variation of the toughness (13), but the "microstructural" values of 4000 MPa (10)(12) are not reached.

Thus, defect size may be controlled by a good monitoring of the sintering process. However defects always exists or may arise by fracture of the carbide crystals.

The aim of this paper is to analyse the initiation and the behavior of such defects. To do this we studied mostly materials located on the "ductile" side of the rupture stress-cobalt content variation, i.e. materials containing more than 20 % in volume. This arises from the observation that the value of a_c increases with cobalt content although sintering defects should decrease in size as the amount of binder is increased (14). Thus, to look for evidence of sub-critical crack growth became the main purpose.

MATERIALS AND EXPERIMENTAL PROCEDURES

The materials used were WC-Co composites manufactured by the Eurotungsten Company by presintering in vacuum at 900°C and final sintering at ∿1450°C. These materials were present with cobalt volumic ratios of 5 %, 10 %, 16 %, 22 %, 30 % and 37 %. For each volumic ratio the mean diameters of the carbide crystals were 2.2 µm, 1.1 µm and 0.7 µm. The microstructural parameters were measured using quantitative metallography and image analysis on polished specimens which were etched in a Murakami solution (5). This etching technique was also used with specimens to be observed in optical or scanning electron microscope.

Rupture stresses at room temperature in air were measured in three point bending on ground or polished specimens with dimensions of $2 \times 5 \times 20$ mm³. Knife supports were 16 mm apart (15).

Critical stress intensity factor measurements with SENB specimen at room temperature in air were also made in three point bending. Notches were introduced by spark erosion. Specimen dimensions were $4 \times 8 \times 40$ mm³ and the knife supports were 32 mm apart (14).

To measure the critical stress intensity factor and rupture stress in liquid nitrogen the bending support was positionned in a vessel with a double wall. The vessel was made with stainless steel and the vacuum between the double wall was maintained with a primary rotary pump. An isolating cover limited evaporation of the nitrogen. A hole in the cover permits the passage of the upper loading knife. The bulk of the knife was 22 mm in diameter and was in three parts. The two ends were in stainless steel and the middle part

was in teflon to avoid thermal loss. Only the rupture load was measured with this set.

Sub-critical crack growth was measured using DCB specimen. The
specimens were 3 x 30 x 80 mm^3. A slot, 1 mm deep, was made on one
side of the specimen in order to keep the notch in the center. Figure 1 shows the experimental arrangement used to make these measurements. A spark eroded notch was first introduced into the specimen and then sub-critical crack-growth was induced to produce a natural sharp crack. A clip gauge was used to measure the displacement of the loading points. The specimen was loaded to just below
the critical fracture load, the crosshead was then stopped and the
relaxation of the load was recorded. The load-displacement were recorded independently with a Hewlett-Packard 7005B flat-bed recorder.

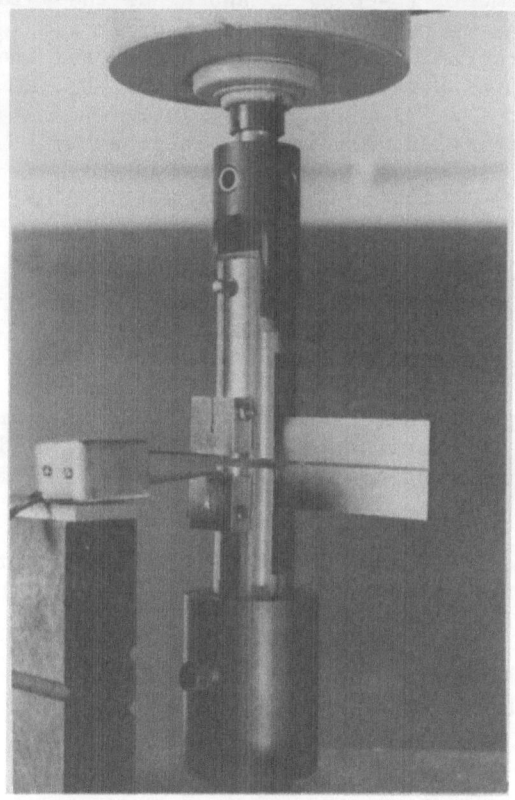

Fig. 1 : $K_I V$ measurement with DCB specimens set-up.

K_I-crack velocity ($K_I V$) diagrams were then evaluated according
to the method proposed by A.K. Virkar and R.S. Gordon (16). The value of the stress intensity factor is given by :

$$K_I = \frac{3.46 \, Fa \, [1 + 1.32 \, (h/a) + 0.532 \, (h/a)^2]^{1/2}}{(eb)^{1/2} \, h^{3/2}}$$

with   F : applied load.
        a : crack length.
        h : half width of the specimen.
        b : thickness of the specimen.
        e : thickness of the specimen at the slot.

The compliance of the specimen, C(a), is given by :

$$C(a) = \frac{24}{Ebh^3} \ (0.33 \ a^3 + 0.66 \ ha^2 + 0.542 \ h^2 a)$$

where E is the Young modulus of the alloy tested.

The compliance of the specimen can be calculated from the loading point displacement, $Y = C(a) \ F$, and hence the value of the crack length, a, can be read off a diagram or it can be calculated using a computer program. The applied load, F, and the crack length yield the value of the stress intensity factor, K_I, at each instant :

The crack propagation velocity, V, is given by :

$$V = (C_M + \frac{24 \ P(a)}{Ebh^3}) \ \frac{dF}{dt} / \frac{24 P'(a)}{Ebh^3} \ F$$

with C_M : compliance of the machine.
        $P(a)$ = $(0.333 \ a^3 + 0.66 \ ha^2 + 0.542 \ h^2 a)$.
        $P'(a)$ = $(a^3 + 1.32 \ ha + 0.542 \ h)$.
        dF/dt : load relaxation rate.

Most room temperature measurements like K_{IC} measurement with SENB specimen or $K_I V$ determinations were run using a Tinius Olsen, type Locap, universal testing machine with a loading capacity of 10 000 daN.

Loading rate effects and liquid nitrogen temperature measurements were made using an Instron universal testing machine, with a loading capacity of 10 000 daN.

Fractographic investigations were made using a Jeol 20 TS scanning electron microscope.

RESULTS AND DISCUSSION

To begin with,we shall define more accurately the purpose of this paper. First we assess the values of the critical stress intensity factor, K_{IC}, and of the rupture stress in bending to obtain the best estimation of the critical defect size, a_c. The variation of a_c with cobalt volumic ratio will thus be discussed and the reasons to look for the existence of sub-critical crack growth at room temperature be assessed.

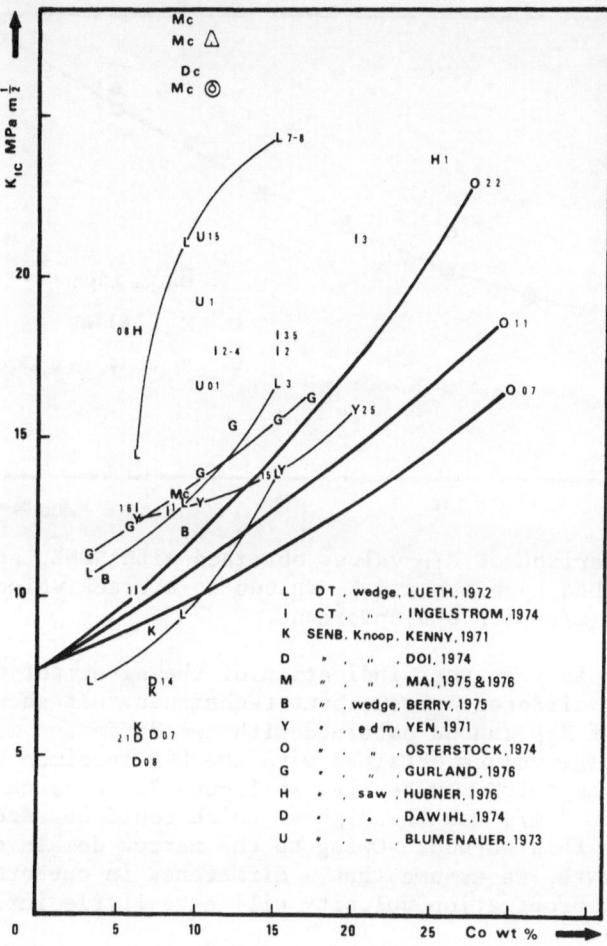

Fig. 2 : Variation of K_{IC} as a function of cobalt content and measurement and notching techniques (wedging, Knoop indentation, electron discharge machining or sawing with diamond saws).

Figure 2 is a comparison of the values of K_{IC}, measured up to the year 1976, as a function of cobalt content and of different measurement and notching techniques. The mean diameter of the carbide crystals is noted beneath the experimental line or point. One can see that a rather good agreement exists between most of the values (1)(2)(3)(4)(5)(17)(18)(19)(20)(21)(22)(23). A general trend is that toughness increases with increasing cobalt content and this increase is the sharper the larger the carbide crystal size. This is an indication of the work which is done during crack propagation to deform and to tear the cobalt.

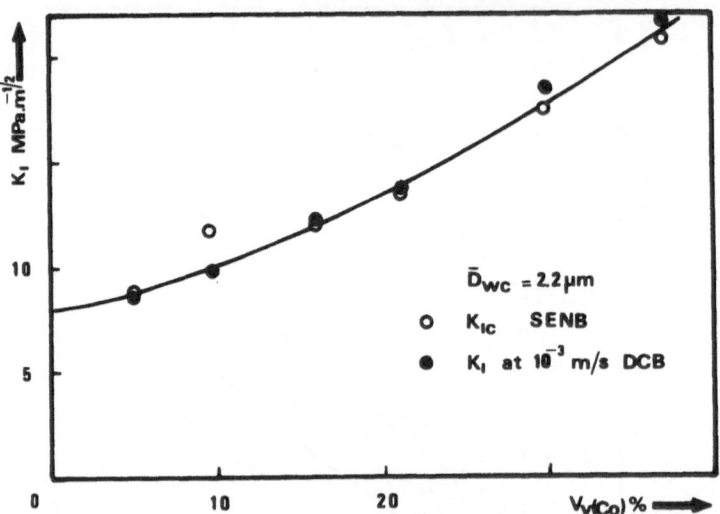

Fig. 3 : Comparison of K_{IC}-values obtained with SENB, spark erosion
notched specimen, and K_I-values at a crack velocity of
10^{-3} m/s with DCB specimens.

Figure 3 is a further indication of the agreement which may be
obtained with different measurement techniques. It shows that the
same values of K_{IC} can be obtained with two different measurement
techniques. The values obtained with the DCB specimen were read off
the $K_I V$ diagram which may be seen in figure 7. A crack propagation
velocity of 10^{-3} m/s was the highest which could be recorded with
accuracy with this method. Owing to the narrow domain of sub-critical crack growth, we assume that a difference in one order of magnitude in crack propagation velocity will have little influence on the
K_I-value.

Variation of rupture stress in bending with cobalt content have
already been largely published (24)(25). As the cobalt content increases, the rupture stress passes through a maximum. On the low
binder content side the behavior may be called brittle, whereas on
the high binder content side it may be called ductile. In Table 1
are listed the numerical results for the critical stress intensity
factor, K_{IC}, the critical strain energy release rate, G_{IC}, and the
rupture stress in bending σ_r. The critical size of the defect, a_c,
can thus be calculated owing to (26) :

$$a_c = (1/1.21\pi) \ (K_{IC}/\sigma_r)^2$$

The values of a_c are also plotted in table 1 and the variation
of a_c with cobalt volumic fraction are shown in figure 4.

Table 1 : Values of rupture parameters of the various batches of
          WC-Co alloys used.

$\bar{D}_{WC}$ μm	$V_V(Co)$ %	K_{IC} MPa.$m^{1/2}$	G_{IC} J.m^{-2}	σ_r MPa	a_c μm
	5	8.8	115	1200	14.0
	10	11.8	215	2300	6.9
2.2	15	11.9	236	2700	5.0
	22	13.5	347	2800	6.1
	30	17.3	644	2650	11.2
	37	20.7	991	2500	19.6
	5	9.1	124	1060	19.6
	10	11.9	219	1690	13.0
1.1	15	13.4	304	2280	9.1
	22	12.0	276	2530	5.9
	30	16.0	549	2750	8.9
	37	16.0	592	2670	10.2
	5	8.8	114	1940	5.2
	10	8.8	120	2250	4.0
0.7	15	10.9	200	2500	5.0
	22	11.4	252	2700	4.7
	30	13.3	379	2910	5.5
	37	14.5	486	3090	5.8

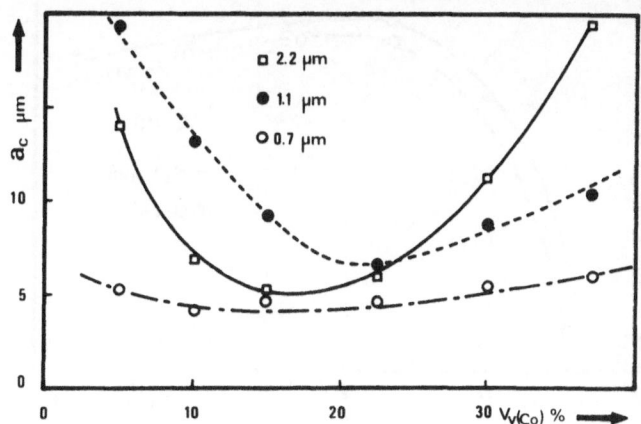

Fig. 4 : Variation of the critical defect size, a_c, as a function
         of cobalt volumic ratio, $V_V(Co)$.

The two coarse grained materials exhibit a very marked minimum. The
value of a_c first decreases as the cobalt content increases. This
is a consequence of pore and carbide crystal accumulation disparition
as the binder phase is increased. One would thus expect that the
value of a_c either remains at the value of the minimum or so,or con-
tinues to decrease as the binder content is further increased.
Instead, we observe an increase of a_c. Two possible explanations
may be proposed for this phenomenon: either there exists another
type of defect which increases with binder content, or the same defects, i.e. sintering defects, are present and they experience sub-
critical crack extension during loading of the specimen.

        If the second hypothesis is correct one should verify it by
rendering the material more brittle, that is more susceptible to
smaller defects. This was done by running a series of measurements
in liquid nitrogen. A second advantage of working at much lower
temperature is that sub-critical crack growth, if it exists at room
temperature should be suppressed at the temperature of liquid nitrogen, as it is mostly a thermal activated phenomenon. The results
obtained are shown in figures 5 and 6 as a function of cobalt volumic ratio and are compared to room temperature results. Surprisingly, it appears that both rupture stress in bending, σ_r, and critical stress intensity factor, K_{IC}, are not affected by the temperature
change. Such an effect has already been observed by M. Nakamura and
J. Gurland (27). The consequence in the present work is that the
variation and the values of a_c remainsapproximatively the same as
at room temperature and that is was not possible to show whether
sub-critical crack extension exists or not.

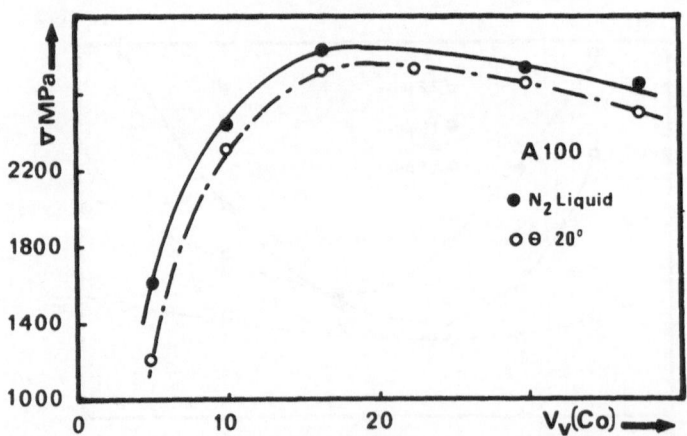

Fig. 5 : Variation of the rupture stress in bending, σ_r, measured
        in liquid nitrogen, as a function of cobalt volumic ra-
        tio,V_V(Co), and comparison with room temperature values.

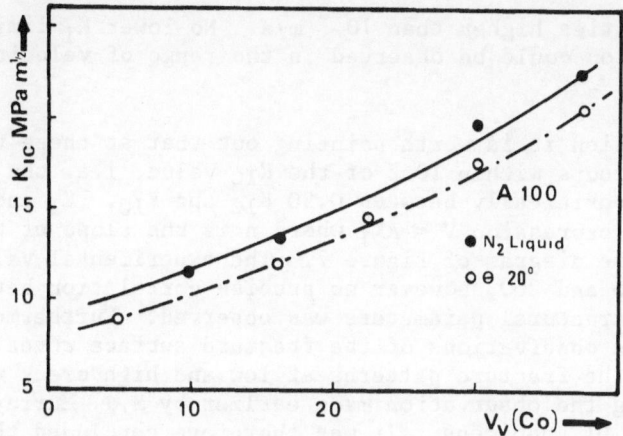

Fig. 6 : Variation of the critical stress intensity factor, K_{IC},
measured in liquid nitrogen, as a function of cobalt
volumic ratio, $V_V(Co)$, and comparison with room temperature values.

Following, stress intensity factor-crack velocity, or K_IV
diagrams were thus drawn directly from measurements at room temperature on DCB specimens. Figure 7 shows the K_IV diagrams obtained
using this method. The crack velocity is between 10^{-7} m/s and
10^{-2} m/s but difficulties were encountered when attempts were made

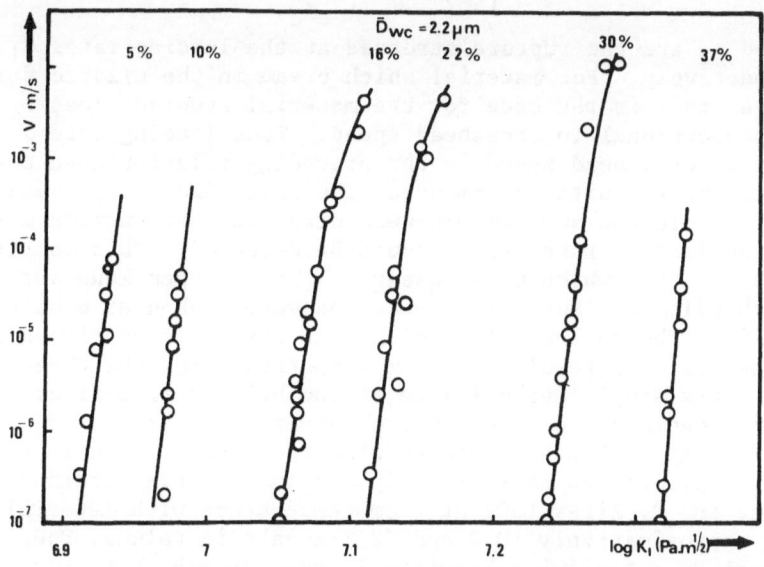

Fig. 7 : K_IV diagrams as obtained from DCB specimen measurements.

to reach velocities higher than 10^{-3} m/s. No lower K_I-limit for
crack propagation could be observed in the range of velocities studied.

   In addition it is worth pointing out that at these velocities
crack growth occurs within 10 % of the K_{IC} value, i.e. the crack
propagates sub-critically between 0.90 K_{IC} and K_{IC}. K_I and V are
related by the expression $V = AK_I^n$ where n is the slope of the
straights in the diagram of figure 7. The experimental values of n
lie between 140 and 200, however no precise correlation between n
and the microstructural parameters was observed. Furthermore, electron microscope observations of the fracture surface revealed no
difference in the fracture patterns at low and high crack velocities
thus confirming the observation made earlier by M.J. Murray and C.M.
Perrot (28) on DT specimens. It was therefore concluded that the
same fracture mechanism is operating at low and high crack velocities.

   That the values of n are high means that sub-critical crack
growth of the initiating defect is limited. However an attempt was
made to make further evidence of this possible event. This was done
by measuring the influence of sub-critical crack growth on the rupture stress when the specimen is loaded at different rates. The
higher the loading rate the more rapidly K_{IC} is attained at the tip
of the defect,and less time the defect has to growth. This means
that the increase in loading rate should provide an increase in
measured rupture stress. This effect may be quantified by the relation (29) :

$$(\sigma_1/\sigma_2) = (\dot{\delta}_1/\dot{\delta}_2)^{1/n+1}$$

where σ_1 and σ_2 are the rupture stresses at the loading rates $\dot{\delta}_1$
and $\dot{\delta}_2$ respectively. For material which breaks in the elastic deformation range, this is the case for the material studied, loading
rate $\dot{\delta}$ is proportional to crosshead speed. Thus loading rate $\dot{\delta}$ can
be replaced by crosshead speed in the preceding relation when the
measurements are run with specimens of the same size. Crosshead
speeds of 5 μm/min and 5000 μm/min were used with the Instron machine, and no loading rate effect could be detected. This confirms
the high values of n which were measured. Twenty specimens were
used at each velocity, further 5 specimens were broken at a velocity
of 500 μm/min. The absence of loading rate effect is confirmed by
the fact that all the results at room temperature for the three loading rates fit a single Weibull line,as can be seen from figure 8.
This confirms the absence of extensive sub-critical growth of the
rupture initiating defects at these binder contents. These results
are also a complement to those already obtained by P.M. Braiden,
R.W. Davidge and R. Airey (30) at room temperature with materials
containing approximatively 10 % and 22 % cobalt in volume. They also
made evidence of extensive sub-critical crack growth at temperatures
of 850°C and 950°C.

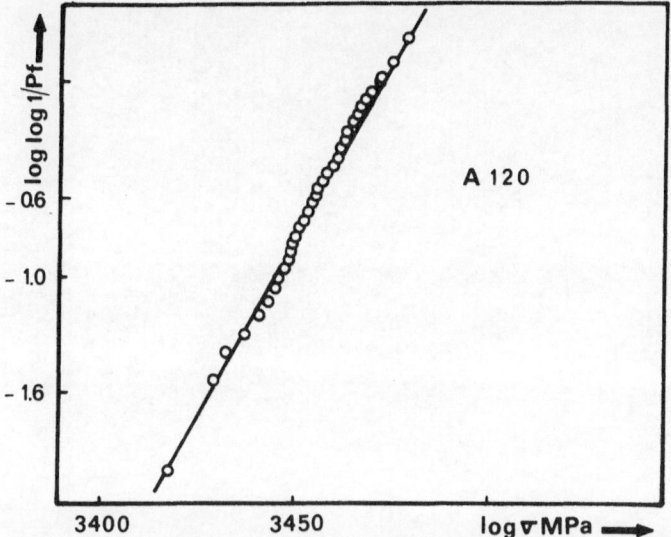

Fig. 8 : Weibull plot of the rupture stresses measured at various
loading rate with a material containing 37 % cobalt in volume and with $\bar{D}_{WC}$ = 2.2 µm.

Concerning the nature of the fracture initiating defects,frac-
tographic investigations give some results. Figure 9 shows the surface submitted to tension of a specimen broken at room temperature,
near the fracture path. It can be seen that there exists a series of
microcracks which seem to initiate and grow independently. The smallest of these defects have the size of some carbide crystals, thus
approximately the value of the calculated critical defect size, a_c.
The micrographs in figure 10 show other aspects of these microcracks.
Figure 10a confirms that they grow independently and do not result
from branching of the main crack,since the apparent branching exists
in two opposite directions for the same microcrack.Figure 10b is a
magnification of the zone of figure 10a and shows the transgranular
rupture through the carbide crystals.

It appears thus that the fracture initiating defect in the
high binder content part of the critical defect size comes from
the rupture of carbide chains as they can no longer accommodate the
deformation which is imposed to the specimen. The variation of a_c
at the higher cobalt content describes thus the size of the chains
which break and propagate in a catastrophic manner at once. This
explains why rupture stresses are the same at room and liquid nitrogen temperatures,and are not affected by changes in loading rate.
These ruptures of carbide crystals may be called the "microstructural" defects,and the fact they are carbide chains rather than individual crystals may explain why the expected values of the rupture
stress of 4000 MPa (10)(12) may not be reached. A further argument
to this comes from analyses of fracture toughness parameters (32)

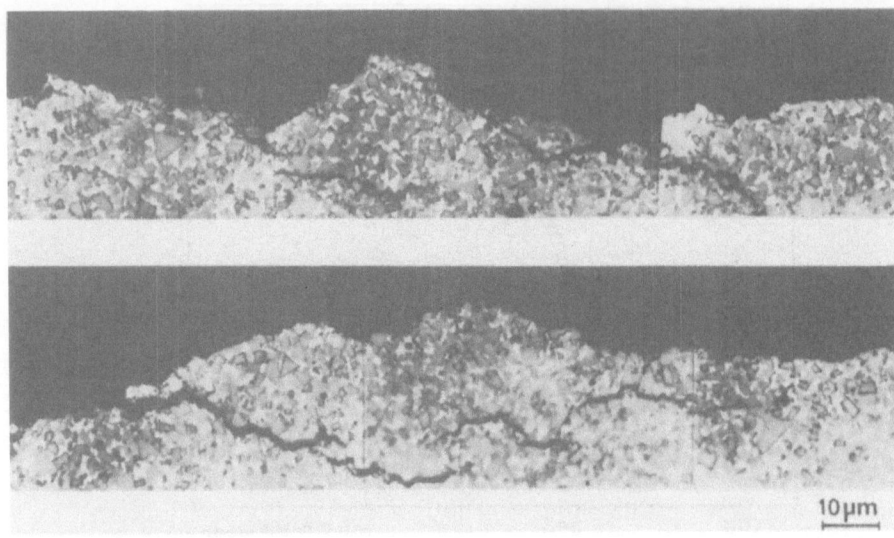

Fig. 9 : Surface submitted to tension near the fracture path of a
         specimen broken in three point bending at room temperature
         ($V_V(Co) = 30$ % and $\bar{D}_{WC} = 2.2$ μm).

(33). Measurements of the notch root stress of WC-Co materials
shows that the stress acting in the process zone, which is assumed
to contain no sintering defects, are between 2500 MPa and 3000 MPa
over the whole domain of cobalt content. Rupture of carbide crystals chains is possible since it has been shown that carbide-
carbide boundaries exist and are often in coincidence (31).

CONCLUSION

        In this paper we have measured fracture mechanics parameters
of tungsten carbide cobalt alloys at room and liquid nitrogen temperature. The results indicate that rupture for the low cobalt volumic ratios is the propagation of preexisting defects,whereas at
higher binder contents the propagating defect must be initiated by
the rupture of carbide chains. These ruptured carbide chains may be
called the microstructural defects and they propagate in a catastrophic way directly after initiation.

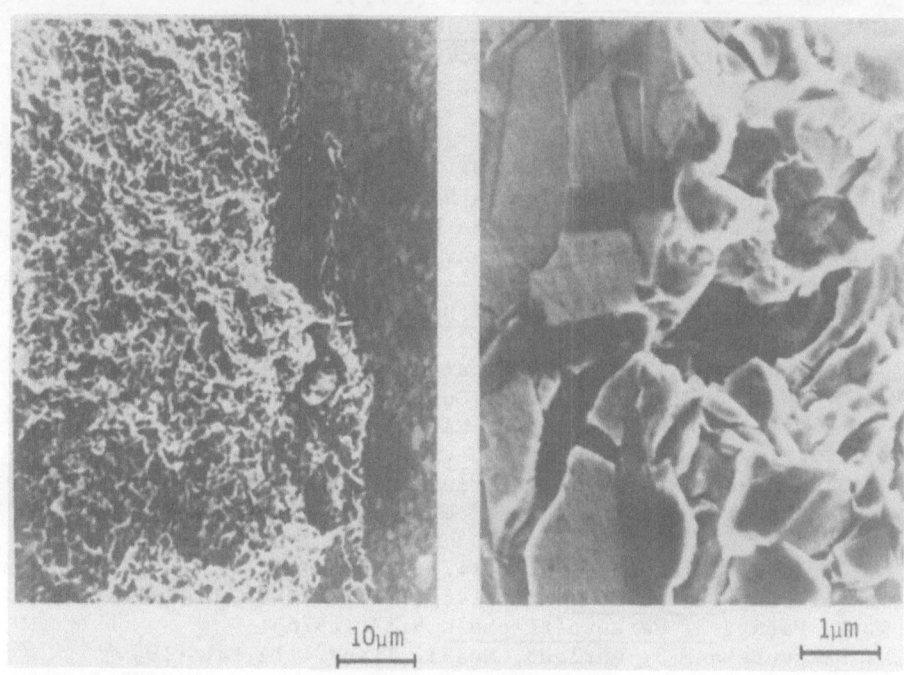

Fig. 10a   Fig. 10b

Fig. 10 : Scanning electron micrographs showing initiating microstructural defects (V_V(Co) = 30 % and $\overline{D}_{WC}$ = 2.2 µm).

REFERENCES

1. S. S. Yen, Thesis of Master of Science, Lehigh University, USA,
    (1971).
2. R. C. Lueth, Ph.D. Thesis, Michigan State University, USA,
    (1972).
3. N. Ingelstrom and H. Nordberg, Eng. Fract. Mech. 6:597 (1974).
4. H. Hubner, Z. Metallkde. 67:507 (1976).
5. F. Osterstock, These de Docteur Ingenieur, University of Caen,
    France (1975).
6. J. L. Chermant, A. Deschanvres, and A. Iost, "Fracture Mechanics
    of Ceramics," R. C. Bradt, D.P.H. Hasselman, F. F. Lange,
    eds., Plenum Press 1:347 (1974).
7. R. Moussa, These de 3e Cycle, University of Caen, France,
    (1981).
8. S. B. Luyckx, Acta Met. 23:109 (1975).
9. P. P. Anderson, Planseeb. Pulverment. 15:180 (1967).
10. H. Suzuki and K. Hayashi, Planseeb. Pulvermet, 23:24 (1975).
11. E. A. Almond, Metal. Sci. 12:587 (1978).

12. B. Roebuck, J. Mat. Sci. 14:2837 (1979).
13. U. Engel and H. Hubner, J. Mat. Sci. 13:2003 (1978).
14. J. L. Chermant and F. Osterstock, J. Mat. Sci. 11:1939 (1976).
15. J. L. Chermant, A. Deschanvres, and F. Osterstock, Powd. Met.
      2:63 (1977).
16. A. V. Virkar and R. S. Gordon J. Amer. Ceram. Soc.59:68 (1976).
17. P. Kenny, Powd. Met. 14:22 (1971).
18. M. Doi, and F. Ueda, in "Grain Boundaries in Engineering
      Materials," Proc. 4th Bolton Landing Conf., June 1974, J. L.
      Water, J. H. Westbrook, D. A. Woodford, eds.,Claitov's Pub.
      (1975).
19. Y. W. Mai and A. G. Atkins, J. Mat. Sci. 10:1904 (1975).
20. G. Berry, on "New Tool Materials and Cutting Techniques,"
      London, April 25. 1975, Session I, paper 4.
21. H. C. Lee, J. Pickens, and J. Gurland, VIII Workshop on Hard
      Materials, NSF, Baltimore, MA., June 1976.
22. W. Dawihl, Arch. Eisenhuttenwes. 45:729 (1974).
23. H. Blumenauer, W. Bohlke, W. Flurschutz, R. Kohlermann, "5th
Int. Powder Metallurgy Conference," Dresden, October 23-25,
      1975,
24. J. Gurland, Jernkont. Ann. 1:147 (1963).
25. J. Gurland, Trans. AIME 227:1146 (1963).
26. R. F. Pabst, Z. Werkstofftechnik 6:17 (1976).
27. M. Nakamura and J. Gurland, Meall. Trans. 11A:141 (1980).
28. M. J. Murray and C. M. Perrot, "Proceedings of the 1976 Int.
      Conf. on Hard Materials Tool Technology," Carnegie Mellon
      Inst., June 22-24, 1976.
29. R. W. Davidge, in "Mechanical Behaviour of Ceramics," University
      Press, Cambridge (1979).
30. P. M. Braiden, R. W. Davidge, and R. Airey, J. Mech Phys.
      Solids 25:257 (1977).
31. S. Hagege, Thesis, University of Caen, France (July 1979).
32. F. Osterstock, Thesis, University of Caen, France, (December
      1980).
33. F. Osterstock, in "Fracture Mechanics of Ceramics," IIIrd
      Symposium held at Penn State University, July 15-17, 1981.

DISCUSSION

H. E. Exner:

One very important question in cemented carbides is what tensile
strengths will you get when you don't have any flaws or any pores.
I think you have answered this question quite nicely by saying
that if you don't have flaws in your material, cracks will start
at carbide grain boundaries. This will be the limiting stress.
Is this so?

F. Osterstock:

Yes. That's right. This means that in your structure you have
chains of carbide crystals and some lower size of these chains
will determine the fracture stress rather than only one crystal.
If you made the calculation using K_{IC}, critical defect size and
rupture stress, you will have much higher values of the rupture
stress if you consider the size of only one crystal. So you have
to accept the fact that there are some full chains which break at
the same time.

T. Hall, Jr.:

Would your model suggest that if you had a bi-modal distribution
in your grain size that you might have increased fracture toughness?

F. Osterstock:

I am not very sure. If you have a bimodal distribution it is the
larger crystals which will determine the rupture stress. For its
effect on fracture toughness perhaps you should choose other models.
But rupture stress should be discussed in terms of defect size and
toughness and not just toughness.

MISMATCH STRESS EFFECTS ON MICROSTRUCTURE-FLAW SIZE DEPENDENCE OF K_{IC} AND STRENGTH OF METAL BONDED CARBIDES

Roy W. Rice

Naval Research Laboratory
Washington, D.C. 20375

ABSTRACT

This paper addresses two related effects of the mismatch stresses due to differences in thermal expansion between metal and carbide phases. First, a mathematical model, developed for the microstructural dependence of fracture energy, due to mismatch strains that exist between noncubic grains in a single phase body, is extended to such mismatch strains in two phase bodies, such as Co bonded WC. The combination of mismatch strains and the applied stress in the vicinity of a crack tip are treated as the source of microcracking in, or between, WC grains and local plastic deformation in the Co phase. Resultant calculations show good comparison with literature data on fracture energy, as well as strength data for flaws sufficiently larger than the microstructure.

The second effect the paper addresses are deviations from the strength behavior predicted by K_{IC} measurements or calculations at smaller crack sizes when cracks no longer encompass a complete range of grain misorientations and hence mismatch stresses. Some cracks will then be associated with grain combinations having net tensile components, which will add to the applied stress to aid failure. Recent literature data is examined, showing support for the occurrence of this mechanism.

I. INTRODUCTION

Substantial effort has been appropriately devoted to the study of microstructural effects on the mechanical behavior of metal bonded carbides, especially WC-Co. While such study has documented effects of important variables such as grain size (G), volume fraction metal binder (β), and binder mean free path (λ), a general model for the interrelation of these variables has been lacking. This has been due in part to uncertainty and controversy on the roles of plastic deformation in the metal phase, and especially of stresses due to mismatches in elastic properties and especially thermal expansion between the ceramic and metal phases.[1]

Recent studies of ceramics give insight into the role of mismatch stresses on fracture energy (γ) and flexure strength (σ), and how they can be related to other microstructural parameters.[2-4] The purpose of this paper is to apply these concepts to WC-Co. This gives a comprehensive model for the microstructural dependence of γ, and hence fracture toughness, K_{IC}, and σ, which includes effects of plastic deformation and mismatch stresses. The latter are in fact central to the development, and represent a departure from previous treatment of their effects.

The paper is divided into two parts. The first part deals with the development of a model for γ, and hence K_{IC} and σ for the case where cracks of concern are sufficiently large in comparison with the microstructure such that the statistical average of the microstructural and the mismatch stresses along the fronts of typical strength controlling cracks can be neglected. The second part of the paper deals with cases where such cracks are no longer sufficiently large in comparison to microstructural factors so that statistical variations of these factors, especially mismatch stresses for different cracks become significant. In both parts, the analysis is shown to compare well with experimental data.

II. MICROSTRUCTURAL MODEL FOR γ AND σ

A. Model Background

Recently, Rice and Freiman[2] developed a model to quantitatively predict the grain size dependence of γ for single phase ceramics of noncubic structure based on stresses and cracking between grains due to effects of thermal expansion anisotropy. This model can be adapted to strain mismatches between different phases in multiphase bodies such as cermets. This section adapts their model to cermets, primarily WC-Co.

The essence of this model is that local tensile stresses such as those due to thermal mismatches can add to applied stresses to create local microcracks (Fig. 1). The formation of these microcracks commonly absorbs mechanical energy from the applied stress field and hence can increase fracture energy. The ease of microcracking increases with increasing dimensions of grains and intergranular phases between which mismatch stresses develop due to greater contribution of the local stresses between them. This leads to two opposite trends of γ with increasing microstructural dimensions e.g., grain size (G). First, as G increases, more microcracks are formed tending to increase γ. However, second, each microcrack absorbs less applied strain energy, hence decreasing γ, since the local mismatch stresses increasingly contribute to the microcrack formation, i.e., a grain size is eventually reached where the body spontaneously microcracks with no applied stress.[3] These two opposing trends normally give rise to a maximum of γ at some G. However, in a two or multiphase body, a simple maximum may not occur due to differing effects of the changes in the sizes of each phase. Further, in cermets, plastic deformation of the metal phase can be important.

B. Model Development

In the following model development, a number of simplifying approximations will be made both to keep the mathematics tractable and to more clearly illustrate the essence of the model and its applicability. In general, it is believed that the effects of the assumptions are within the uncertainty of some of the model parameters.

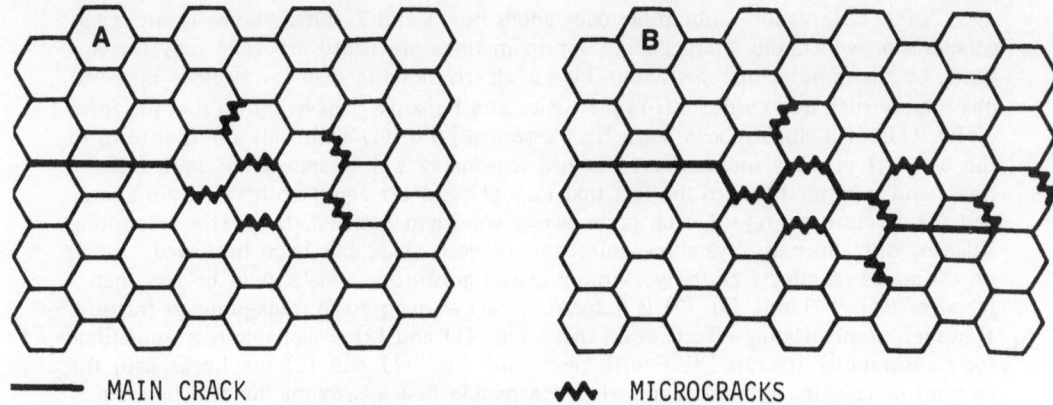

MAIN CRACK   MICROCRACKS

Fig. 1 — Schematic sketch of microcracking concept. A) initial microcracks generated around crack. B) propagation of crack (in part by connecting microcracks) with generation of more microcracks. Note some transgranular propagation is shown.

The model starts with a concept introduced by Davidge and Green[5] for spontaneous fracture around isolated second phase spherical ceramic particles in silicate glasses. They assumed that such fracture would occur when the strain energy stored in that particle and the surrounding matrix due to the particle-matrix mismatch stresses just equaled the surface energy for fracture around the particle. The volume dependence of the strain energy equated to the surface dependence of fracture energy gave a particle size dependence of spontaneous fracture in reasonable agreement with observations. Applied to a two phase body, such as a cermet where the metal phase occurs mostly as a thin layer of average thickness λ between the grains, the strain energy, D, is approximated as:

$$D = 2\left[4\pi\left(\frac{G}{2}\right)^2 \frac{\lambda}{2} \frac{E_\lambda(\Delta\epsilon)^2}{2} + \frac{4\pi}{3}\left(\frac{G}{2}\right)^3 \frac{E_G(\Delta\epsilon)^2}{2}\right]. \tag{1}$$

The energy need for fracture around a grain, Γ, is:

$$\Gamma = (2)4\pi\left(\frac{G}{2}\right)^2 \gamma_B. \tag{2}$$

Here, uniform spherical grains of diameter G have been assumed, and the volume of the metallic phase is approximated as a spherical shell. E is Young's modulus (subscripts λ and G are respectively for the metallic intergranular phase and the grain phase) $\Delta\epsilon$ the rms or absolute average mismatch strain between the two phases, and γ_B the fracture energy for the formation of a crack between two grains in the metal layer, or between one grain and the metal layer.* The factor 2 outside the bracket of Eq. (1) is based on the observation that the strain energy due to a particle in a matrix is about equally divided between the matrix and the particle. The factor 2 on the right side of Eq. 2 represents the formation of two fracture surfaces.

---

*Note the use of these average γ_B and especially $\Delta\epsilon$ values implicitly assumes a representative statistical sampling by the crack of the microstructure as noted earlier.

A few observations should be made about Eqs. 1 and 2. First, the basic stress calculations on which the strain energy approximations are based are valid only for isolated, i.e., noninteracting, particles. This is clearly not the case for stresses in a two
phase body such as a cermet. However, Rice and Pohanka[3] have shown that the form
of Eq. (1) for a single phase body (i.e., essentially Eq. (1) with only the first term in
the bracket) predicts spontaneous fracture reasonably well despite grain-grain interactions, and attributed this to the fact that Eqs. (1) and (2) count both the strain energy
and the fracture energy of each grain twice, so, when equated, this extra accounting
balances out. Second, crystalline anisotropy of each phase has been neglected. Based
on estimates of effects of fairly extreme elastic anisotropy, this should be less than a
factor of two.[4] Third, Eq. (2) is based on inter- as opposed to trans-granular fracture.
However, compensating effects again make Eqs. (1) and (2) reasonable representations
for transgranular fracture.[3] Fourth, assuming Eqs. (1) and (2) are linear with the
fraction of cracking around a grain is a reasonable first approximation and hence pertinent to partial fracture around a grain.[3]

Equating Eqs. (1) and (2) gives the condition for fracture around part or all of a
grain due only to mismatch stresses. This gives a critical grain size (or mean free path)
for spontaneous fracture, G_s (or λ_s). However, of greater interest here is the average
local stress (σ_i) causing spontaneous fracture. Since σ_i is simply estimated as:

$$\sigma_i \sim E_c \Delta \epsilon \tag{3}$$

where E_c = the Young's modulus of the body, i.e., the composite modulus of the two
phases, this can be used to eliminate $\Delta \epsilon$. In solving for σ_i, it is useful to multiply the
numerators and denominators of Eq. (1) by E_c^2, giving:

$$\sigma_i = \sqrt{\frac{4 E_c \gamma_B}{\lambda_s \dfrac{E_\lambda}{E_c} + \dfrac{G_s}{3} \dfrac{E_G}{E_c}}}. \tag{4}$$

Now, following Rice and Freiman, Eq. (4) is generalized by observing that at $G <
G_s$ microcracking can still occur if a stress applied to the body, σ_a, can add to the local
mismatch stresses, σ_i, to satisfy the fracture condition; i.e.,

$$\sigma_m = \sqrt{\frac{4 E_c \gamma_B}{\lambda \dfrac{E_\lambda}{Ec} + \dfrac{G}{3} \dfrac{E_G}{E_c}}} = \sigma_a + \sigma_i \tag{5}$$

where σ_m is the stress for microcrack formation. Note that a factor k reflecting possible stress concentration, e.g., of the applied stress, σ_a, near the tip of a crack whose
propagation is of concern can be included in Eq. (5), but is generally not significant.[2]
Next, they[2] assumed that the fracture energy absorbed due to microcracking (γ_μ)
around a grain is simply the product of the number (N) of microcracks formed per unit
area of main crack propagation and the average energy (W) the formation of each
microcrack absorbs from the applied stress field and showed that reasonable approximations are simply:

$$N \alpha \frac{\sigma_i}{G^2 \sigma_a} \tag{6}$$

and

$$W \alpha \; \frac{\sigma_a^2 G^3}{E_c}. \tag{7}$$

Then, solving Eq. (5) for σ_a, substituting this in γ_μ and substituting for σ_t from Eq. (3), one obtains:

$$\gamma_\mu = M(\Delta\epsilon) G \left[\sqrt{\frac{4 E_c \gamma_B}{\dfrac{E_\lambda}{E_c} \lambda + \dfrac{G}{3} \dfrac{E_G}{E_c}}} - (\Delta\epsilon) E_c \right]. \tag{8}$$

M, which incorporates the proportionality constants and any stress concentration (as $1/k$) has been shown to be ~ 2.5.[2]

Again, following Rice and Freiman, the total fracture energy, γ is assumed to be:

$$\gamma = \gamma_{pc}(1 - G/G_s) + \gamma_\mu \tag{9}$$

where γ_{pc} = the fracture energy in the absence of microcracking. Equation (9) simply assumes that microcracking reduces the amount of material the main crack has to fracture (e.g., Fig. 1), and hence the contribution of γ_{pc} (which goes to zero at G_s where in principal the body is totally microcracked) while absorbing energy as reflected in the γ_μ term. (Note negative values of γ_μ are not used since they are not physically meaningful.)

Finally, two modifications are made to account for the nature of a cermet. First the thickness of the metallic layer is assumed to be related to its volume fraction (β) and G by the common stereological formula for grains separated by films:

$$G = \frac{3\lambda \, (1 - \beta)}{\beta}. \tag{10}$$

Second, to account for plastic deformation of the metal phase, the common procedure for bulk metals of addition a term, γ_p to γ_B is followed. For convenience, this added term will be given as $Q = \gamma_p/\gamma_B$ so Eq. (8) becomes:

$$\gamma_\mu = M \Delta\epsilon \, G \left[\sqrt{\frac{4 E_c \gamma_B (1 + Q)}{\lambda \dfrac{E_\lambda}{E_c} + \dfrac{G}{3} \dfrac{E_G}{E_c}}} - \Delta\epsilon E_c \right]. \tag{11}$$

While Q might be considered a constant, it should actually be a function of β and λ; Q should increase with the amount of metal present, e.g., $Q \alpha \beta$. However, for a fixed β, Q should also increase with increasing λ. Since the yield stress is a function of $1/\sqrt{\lambda}$ and the plastic strain should be proportional to λ, Q should vary approximately as $\sqrt{\lambda}$. Thus, letting:

$$Q = B\beta \sqrt{\lambda} \tag{12}$$

where B is a constant is reasonable.

C. Testing of the Model

Turning now to the evaluation of the above equation, we first focus on Eq. (11) since γ_μ will often be the dominant factor in γ and hence also in tensile strength (σ) which will be considered later. In order to evaluate γ_μ for WC-Co, typical values of E_λ (2.1 × 10^5 MPa, 30 × 10^6 psi) amd E_G (7 × 10^5 MPa, 100 × 10^6 psi) are used, and values of E_c are taken from Paul's compilation.[6] Other parameters are estimated where reasonable guidance is available. However, since there must be some uncertainty about some of the parameters, further guidance will be obtained by comparison of equation predictions to data compiled by Pickens[7] from his own studies and those of others, e.g., Kenny,[8] Lueth[9] and Chermant and Osterstock.[10]

Utilizing the above values in Eqs. (11) and (12), a value of B was sought to give reasonable values of γ_μ (and hence γ) as a function of λ and G. A value of $B = 10^5$, i.e., ten times as much energy absorbed in plastic flow as by fracture itself for $\lambda = 1 \mu$ and $\beta = 0.1$ was found (Fig. 3). Since $\gamma_B = 3$ J/m^2 is probably conservative, B may in fact be lower because B decreases nearly inversely with γ_B for constant values of γ_μ. Note that besides generally fitting the range of data using reasonable values of various parameters, Eqs. (11) and (12) are also consistent with higher values of γ for higher β values.

Next some variations in $\Delta\epsilon$ and B were considered. It is reasonable to expect that $\Delta\epsilon$ would decrease some as either or both β and λ increase due to increased stress relaxation. French[13] has reported decreasing WC-Co expansion mismatch stresses as β increased based on x-ray measurements; i.e., averaged over an area of 1 mm dia. Therefore, one variation considered was to decrease $\Delta\epsilon$ from 8 × 10^{-3} at β ~0 as β increases parallel with French's measured stresses decrease, i.e., to $\Delta\epsilon$ ~4 × 10^{-3} at β = 0.3. As shown in Figs. 2 and 3, this results in similar trends. Also, since $\Delta\epsilon$ may vary inversely with the yield of the metal phase, $\Delta\epsilon$ decreasing from 8 × 10^{-3} as $1/\sqrt{\lambda}$ was considered. This basically gave trends between those for fixed $\Delta\epsilon$ and $\Delta\epsilon$ decreasing with β after French's measurements.

While Eq. (11) was considered the most reasonable physically, two variations were considered to check this. First, $Q = B'\beta$ was considered. B' values giving γ_μ curves as a function of λ or G that agreed well with data were not found. Generally $Q = B'\beta$ gave (1) to rapid a rise of γ at low values of λ and G, and; (2) very broad maxima or plateaus at larger λ or G values. However, B' values of the order of 100, i.e., again giving γ_p/γ_B ~10 at β ~0.1, gave closest approach to the data. The second variation, $Q = B''$, a constant, was found to give γ_μ increasing essentially linearly (for $\Delta\epsilon$ constant) or with some concavity toward the λ or G abscissa (for $\Delta\epsilon$ decreasing parallel with French's residual stress measurements) but with γ_μ first increasing with β then later decreasing with β at higher β values, at higher values of λ (but not G) for $\Delta\epsilon$ constant. B'' values of ~25 × 10^6, i.e., giving γ_p/γ_B ~2.5 at β = 0.1 indicated reduced plastic flow for such Q dependence, thus suggesting the possibility that in further development a Q dependence between $\lambda^{1/2}$ and λ be considered.

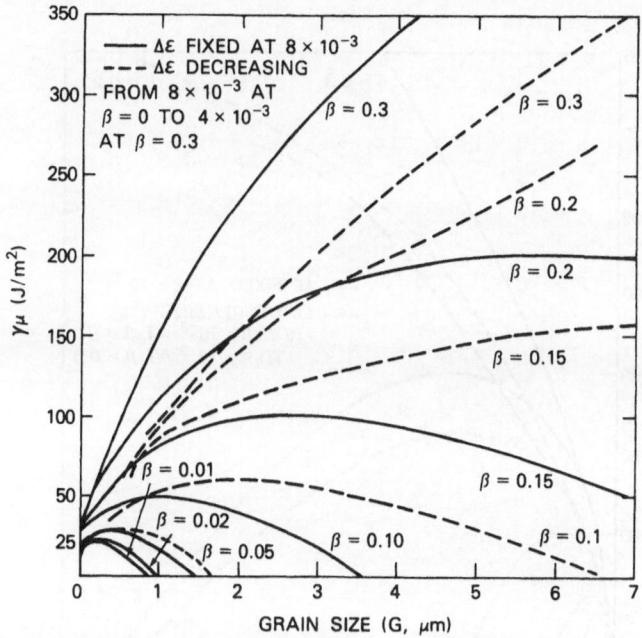

Fig. 2 — Microcracking contribution to fracture energy, γ_μ vs G for WC-Co using a constant or changing mismatch strain (as shown), and plastic flow contribution factor B of 10^5. Note $\Delta\epsilon$ of 8×10^{-3} is based on an average expansion difference $(\Delta\alpha)$ between WC and Co of $\sim 10 \times 10^{-6}$ $°C^{-1}$ and no significant stress relief below ~ 800 °C; i.e. $\Delta\epsilon = \Delta\alpha\Delta T = (10 \times 10^{-6})$ (800).

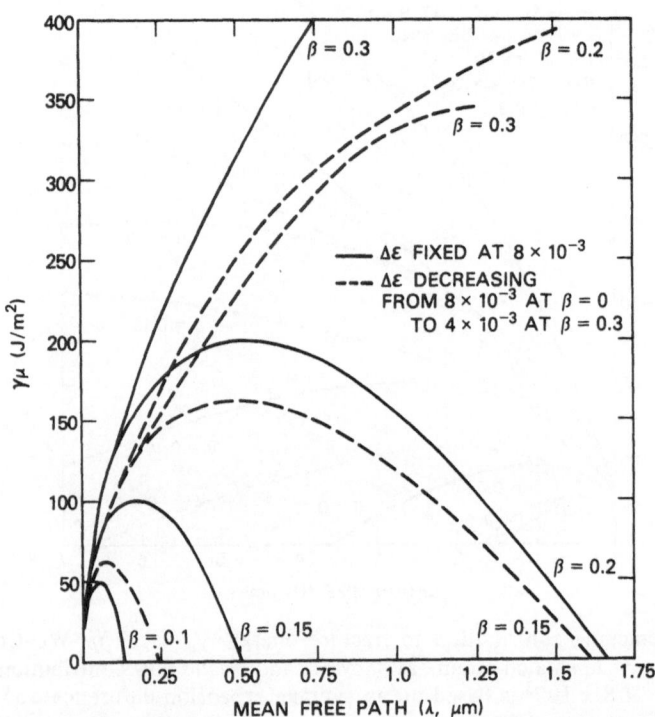

Fig. 3 — Microcracking contribution to fracture energy, γ_μ vs λ for WC-Co using the same condition as for Fig. 2

Next, calculations of flexural strengths (σ) were made using:

$$\sigma = A \sqrt{\frac{E_c \gamma}{a}} \tag{13}$$

where A and a are respectively the flaw geometry and size parameters, and γ as calculated from Eq. (9). Results are shown in Figs. 4 and 5 for $\Delta\epsilon$ = constant for $A \sim 1$ and $a \sim 40$ μm, i.e., similar to many ceramics, and in the range reported by Suzuki and Hayashi.[14] Results for $\Delta\epsilon$ decreasing with increasing β were similar; the main differences being somewhat lower values at higher G and β values and β = 0.3 curve instead of progressively getting further from the β = 0.2, crosses it after passing through its maxima. These curves show many of the features of reported strength trends. Thus, besides giving reasonable values of σ, they generally show maxima σ as G or λ increase, and the curves are more closely spaced to the left of their maxima than to the right. Further note that the maxima shown in Figs. 4 and 5 are much broader than shown in earlier data (e.g., Ref. 2), but is much more consistent with recent tests of hot isostatically pressed WC-Co supporting the observation that the earlier, sharper maxima reflect in part composition dependent defects.[15–17]

D. Extensions of the Model

Three modifications of the model can be foreseen with further development. The first major modification is accounting for the fraction contiguity (C) of the grains with one another. This can readily be done by combining Rice and Freiman's equation for the contiguous grains with that developed in this paper for grains separated by metal. Thus

$$\gamma_\mu = CM(\Delta\epsilon_G)\left[\sqrt{9E_G\gamma_{BG}G} - (\Delta\epsilon_G)E_G G\right]$$

$$+ (1 - C)M(\Delta\epsilon_c)G\left[\frac{4E_c\gamma_{BC}\left(1 + B\beta\,\dfrac{\sqrt{\lambda}}{1 - c}\right)}{\left(\dfrac{\lambda}{1 - c}\right)\dfrac{E_\lambda}{E_c} + \dfrac{G}{3}\dfrac{E_G}{E_c}} - (\Delta\epsilon_c)E_c\right] \tag{14}$$

$$\gamma = C\left[\sigma_{PCG}\left(1 - \frac{G}{G_{SG}}\right) + \gamma_{\mu G}\right]$$

$$+ (1 - C)\left[\gamma_{pcc}\left(1 - \frac{G}{G_{Sc}}\right) + \gamma_{\mu c}\right] \tag{15}$$

where a final subscript G or C has been added to designate respectively appropriate properties of the grain themselves or the composite.

The second, and more complicated major modification is to account for possible anisotropy of each phase. For all phases elastic anisotropy exists, but can commonly be neglected without serious error. Thermal expansion anisotropy (i.e. of noncubic grains)

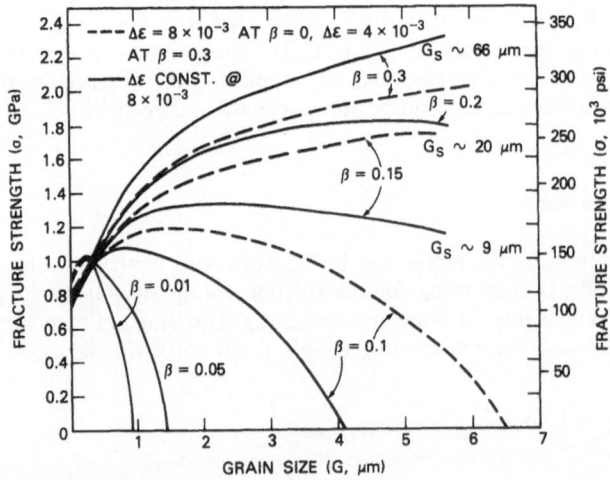

Fig. 4 — Fracture strength of WC-Co vs WC grain size, G. Calculations were based on an ~half penny shaped surface flaw size (i.e., depth a) ~40 μm, and fracture energy data and parameters from Fig. 2. Note G_s values shown for curves of fixed $\Delta\epsilon$ that do not interest the G axis, i.e., that do not directly show G_s. Note a value of 15 J/m² for γ_{pc} of eqn. 9 was based on fracture energies of pure polycrystallatine carbides.

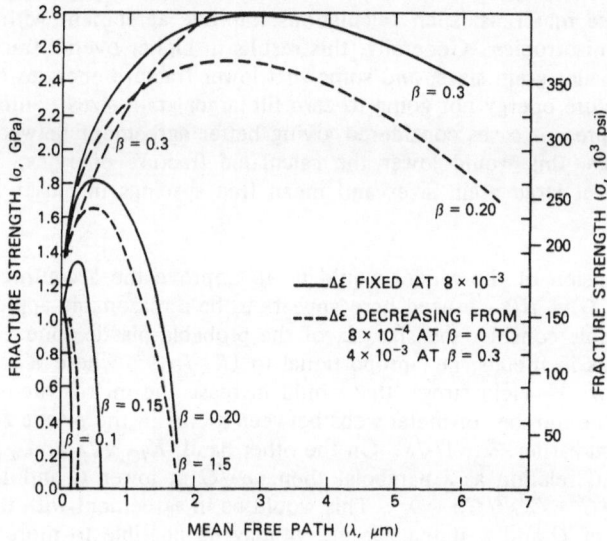

Fig. 5 — Fracture strength of WC-Co versus Co mean free path (i.e. thickness), λ. Calculations were based on an ~half penny shaped surface flaw of depth ~40 μm, and fracture energy data and parameters from Fig. 3.

can be more serious, especially where the anisotropy is substantial. Thus, for WC, the two single crystal expansion coefficients are $\gamma_a \sim 20 \times 10^{-6}$ °C^{-1} and $\gamma_c \sim 2 \times 10^{-6}$ °C^{-1}. Ideally, equations of the form developed here should be integrated over the spectrum of misorientations and resulting $\Delta\epsilon$ values. This, however, would at best be very difficult. As shown by Rice and Freiman, a useful step in this direction is to determine the contribution of each of the principal strain mismatches (i.e., $\Delta\epsilon_i = \epsilon_i - \epsilon_p$ where the ϵ_i are the three principle crystal strains and ϵ_p is the average or polycrystalline strain), i.e., ·

$$\gamma_\mu = \frac{\gamma_{\mu a}}{3} + \frac{\gamma_{\mu b}}{3} + \frac{\gamma_{\mu c}}{3} \tag{16}$$

where the second subscript after each γ_μ refers to use of $\Delta\epsilon_a$, $\Delta\epsilon_b$ and $\Delta\epsilon_c$ where a, b, and c refer to the three principal crystal axes. (Note that none of the $\gamma_{\mu i}$ terms are allowed to go negative and substract from γ_μ itself since this is not physically meaningful.) For single phase materials, such calculations improve agreement with the data, especially for large anisotropies. Generally, this results in higher overall fracture energies, especially at smaller grain sizes, and somewhat lower fracture energies at larger G values, but with fracture energy not going to zero till larger grain sizes. Similar results would occur in the present cases considered giving better agreement between calculations and data. Thus, this would lower the calculated fracture energies, and hence strengths of WC-Co at large grain sizes and mean free spacings in closer agreement with the data.

The third extension of the model would be to improve the accouting of plastic deformation. While $Q = B\beta\sqrt{\lambda}$ used here appears to be a reasonable approximation, it does not for example consider the size, L, of the probable plastic zone. If L has a similar dependence as in metals, i.e., proportional to $(K_{IC}/\sigma_y)^2$, where $K_{IC} = $ critical stress intensity and $\sigma_y = $ yield stress, this would increase the microstructural dependence of Q. Thus, the number of metal webs between grains in the plastic zone would be $\alpha L^2/G^2$. As noted earlier, $\sigma_y \alpha 1/\sqrt{\lambda}$. On the other hand, $K_{IC} \alpha\sqrt{\lambda}$ and γ, and if we approximate the $\gamma - G$ relation as a parabola, then $\gamma\alpha\sqrt{G}$ at lower G and $1 - \sqrt{G}$ at higher G. Thus, $L^2/G^2 \alpha G^2\lambda^2/G^2 = \lambda^2$. This would be in agreement with the possible greater dependence of Q and λ noted earlier. It may be possible to more accurately include the K_{IC} dependence of the zone size by iteratively solving for γ with progressively more accurate Q terms based on the previous $\gamma - G$ or $\gamma - \lambda$ solution. Potentially even more important is to more accurately account for the possible dependence of $\Delta\epsilon$ on β and λ.

III. DEVIATIONS DUE TO CRACKS NOT STATISTICALLY SAMPLING THE MICROSTRUCTURE

When cracks are sufficiently large to represent a good statistical sampling of the microstructure two factors are important. First, the average of all the σ_i values, $<\sigma_i>$ for all grain-metal film interaction close to the crack tip is zero, and hence can be neglected in considering the failure stress, as in the previous section. Second, the local σ_i's can result in microcracking as previously discussed. As crack sizes decrease so they are no longer large enough relative to the microstructure, the average of the σ_i values affecting the crack tip will no longer average to zero, and less microcracking may occur, or it may not be randomly oriented relative to the crack. While both factors are important, the former is probably dominant and is the most straightforward to handle. Thus all further development and analysis will treat this as the single deviation.

Clearly as crack sizes progressively decrease relative to the microstructural scale σ_i values affecting individual crack tips will progressively increase in value. Some cracks will experience net tensile values of $<\sigma_i>$ and others net compressive values (for a net average of zero over all cracks). In general, cracks with the highest net tensile $<\sigma_i>$ will dominate failure. Thus one can easily modify the Griffith equation to account for the effect of $<\sigma_i>$ adding to the applied stress σ to cause failure, i.e.,

$$\sigma + <\sigma_i> = A \sqrt{\frac{E\gamma}{a}} \qquad (17)$$

where $<\sigma_i>$ will be zero above some value of a for a given microstructure, and begin to progressively increase as a decreases relative to that microstructure.

A test of this is to use Eq. (13), the normal Griffith equation, i.e., using only σ_a, to calculate γ. A plot of the resultant γ as a function of a should progressively decrease below the normal polycrystalline value as a progressively decreases below the value where $<\sigma_i>$ is no longer zero. As seen in Fig. 6 data of Suzuki and Hayashi[14] and Almond and Roebuck[18] clearly shows such a trend. Next assuming γ really remains at the normal polycrystalline value, one can calculate $<\sigma_i>$ as a function of a. Extrapolation of $<\sigma_i>$ values to $a \sim G$ gives values of the order of those estimated from $\sigma_i \sim E\Delta\epsilon$, i.e., 3-6 GPa, providing good support for this concept.

When the scale of the flaws is that of individual grains, e.g., larger grains in the body, then the fracture energy will decrease. This occurs since energies for fracture along individual grain boundaries or along preferred surfaces in individual grains are intrinsically lower, as has now been demonstrated in several ceramics.[19] However, this only appears to become a possible factor at quite small flaw size due to the typically small WC grain sizes. Thus for example many of Suzuhi and Hayaski's samples do not appear to meet the criteria for their intrinsic transition to lower fracture energies. This, combined with the good agreement indicated by the above analysis and with similar more extensive analysis of failure in noncubic ceramics,[20] strongly supports the concept of increasing contribution of mismatch stress to failure as the flaw size decreases.

V. DISCUSSION AND SUMMARY

A model for the dependence of fracture energy, and hence K_{IC} or strength on microstructure of metal bonded carbide bodies has been proposed. The central theme of this model is the superposition of applied stresses and the local stresses due to mismatch strains between adjacent grains or phases to cause microcracking, i.e., rupture, between the grains. This rupture is seen to form extra microcracks in addition to those linked together to form the ultimate failure surface. In the case of cermet type materials considered here plastic flow of the metallic phase can be an important factor in absorbing energy during the formation of microcracks.

It should be noted that the model does not explicitly depend on total intergranular failure, which is important since WC-Co can exhibit substantial transangular failure.

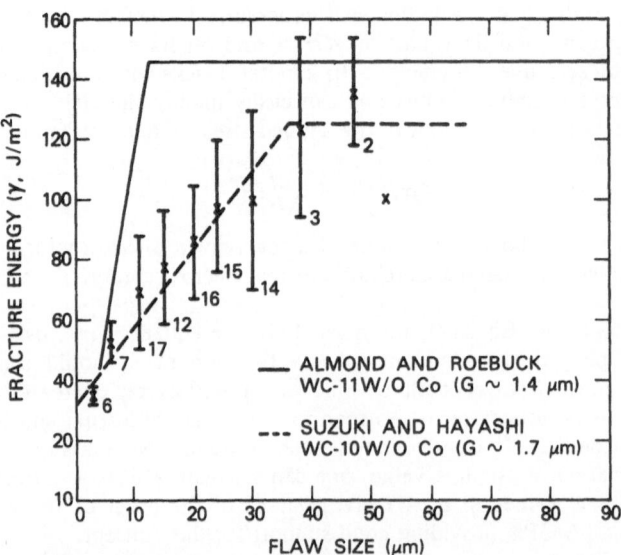

Fig. 6 — Plot of γ calculated from the Griffith equation as a function of flaw size, *a*. The decrease at smaller flaw sizes is attributed to increasing contributions of mismatch stresses as discussed in the text.

Two factors should be noted. First, the model is concerned with the fracture energy for fracture initiation, where the fracture mode can often be more intergranular. Thus, in ceramics, the author has often observed complete integranular failure at the fracture origin with subsequent total transgranular crack propagation. The latter is not necessarily unexpected since deflection of a moving crack around a grain may be more difficult with increasing crack velocity. Second, as noted earlier, the form and parameters of the equations do not appear to be changed drastically by transgranular failure. It is clear that the local stress fields due to mismatch strains can be quite complex and could also aid transgranular failure, especially in materials with preferred cleavage such as WC.

Despite varying only two parameters, $\Delta\epsilon$ and B within expected bounds, and in relatively coarse steps, rather reasonable property trends were calculated. Clearly, iteration of all parameters, and in finer steps should give closer fits to the data. Modification of the equations, e.g., as discussed in a previous section should also be quite helpful. While both of these should be persued, it is also important to further test the model, and especially to conduct experiments to better estimate or measure important parameters, e.g., γ_B and γ_{pc}.

How exact the model can be made by either obtaining more accurate parameters or just "better curve fitting" is uncertain. However, it is suggested, that even in its present simple state, it provides first a new view of the effect of the local internal stresses, and second guidance on important parameters for good cermets. Thus, for example, it provides some quantitative guidance for the importance of good bonding between phases, i.e, γ_B, and ductility of the metal, i.e., B, which have probably been qualitatively appreciated before. However, the value of a large $\Delta\epsilon$, i.e., a large strain mismatch may not have been appreciated before, but is consistent with the general higher strength and toughness of WC based bodies amongst metal bonded carbides and other "hard metals." Further, growing capability of finite element and other computer analysis may be able to answer many questions such as the dependence of B and $\Delta\epsilon$ on β and G. Thus it is suggested that the concepts of the present model also provide a basis for more sophisticated analysis.

ACKNOWLEDGMENT

The extensive aid provided by Dr. J. Pickens (then of Brown Univ.) in providing data and aid in computation by Messrs. A. Gonzalez, T. Donahue, and C. Parnham of the Naval Research Lab. has been most helpful. Partial financial support from the Naval Air Systems Command (P. Weinberg contract monitor) is gratefully acknowledged, as in review of the manuscript by D. Lewis.

REFERENCES

1. H.E. Exner and J. Gurland, "A Review of Parameters Influencing Some Mechanical Properties of Tungsten Carbide-Cobalt Alloys," Powder Metallurgy, Vol. 13, No. 25, 13-31 (1970).

2. R.W. Rice and S.W. Freiman, "Grain Size Dependence of Fracture Energy in Ceramics: II, A Model for Non-Cubic Materials" J. Am. Ceram. Soc. 64(6) 350-54 (1981).

3. R.W. Rice and R.C. Pohanka, "The Grain Size Dependence of Spontaneous Cracking in Ceramics," J. Am. Ceram. Soc. 62(11-12) 559-63 (1979).

4. R.W. Rice, "Grain Size Dependence of Fracture Energy of Ceramics, III Effects of Elastic Anisotropy," submitted for publication.

5. R.W. Davidge and T.J. Green, "The Strength of Two-Phase Ceramic/Glass Materials," J. of Mat. Sci. 3, 639-634 (1968).

6. B. Paul, "Prediction of Elastic Constants of Multiphase Materials," Trans. Met. Soc. AIME 218, 36-41 (1960).

7. R. Pickens and J. Gurland, "The Fracture Toughness of WC-Co Alloys Measured on Single-Edge Notched Beam Specimens Precracked by Electron Discharge Machineing," Mat. Sci. & Eng. 33, 135-42 (1978).

8. P. Kenny, "The Application of Fracture Mechanics to Cemented Tungsten Carbides," Powder Metallurgy, Vol. 14, No. 27, 22-38 (1971).

9. R.C. Lueth, "Determination of Fracture Toughness Parameters for Tungsten Carbide-Cobalt Alloys," Fracture Mechanics of Ceramics 2, 791-806, ed. by R.C. Bradt, D.P. Hasselman, and F.F. Lenge, Plenum Press, New York (1974).

10. J.L. Chermant and F. Osterstock, "Fracture toughness and fracture of WC-Co composites," J. Mat. Sci. 11, 1939-1951 (1976).

11. Hu-Chul Lee, "Hardness and Deformation of Cemented Tungsten Carbide," Mat. Sci. and Eng. (33) 125-33 (1978).

12. H.W. Newkirk, Jr., and H. H. Sisler, "Determination of Residual Stresses in Titanium Carbide-Base Cermets by High-Temperature X-ray Diffraction," J. Am. Ceramic Soc., Vol. 41, No. 3, 93-103 (1958).

13. D. N. French, "X-Ray Stress Analysis of WC-Co Cermets: II, Temperature Stresses," J. Am. Ceramic Soc., Vol. 52, No. 5, 271-275 (1969).

14. Hisashi Suzuki and Kozi Hayashi, "Strength of WC-Co Cemented Carbides in Relation to their Fracture Sources," Planseeberichte für Pulvermetallurgie, Bd. 23, 24-36 (1975).

15. U. Engel, H. Hübner, "Strength improvement of cemented carbides by hot isostatic pressing (HIP)," J. Mat. Sci. 13, 2003-2012 (1978).

16. E. Lardner and D.J. Bettle, "Isostatic hot pressing of cemented carbides," Metals and Materials, 540-545 (1973).

17. E. Lardner, "Recent progress in cemented carbide," Powder Metallurgy, No. 2, 65-70 (1978).

18. E.A. Almond and B. Roebuck, "Defect-Initiated Fracture and Bend Strength of Tungsten Carbide-Cobalt Hardmetals," Mat. Sci. 11(10) 458-61 (1977).

19. R.W. Rice, S.W. Freiman, and J.J. Mecholsky, Jr., "The Dependence of Strength-Controlling Fracture Energy on the Flaw-Size to Grain-Size Ratio," J. Am. Ceram. Soc. 63(3-4) 129-36 (1980).

20. R.W. Rice, R.C. Pohanka, and W.J. McDonough, "Effect of Stresses from Thermal Expansion Anisotropy, Phase Transformations and Second Phases on the Strength of Ceramics," J. Am. Ceram. Soc. 63(11-12) 703-10 (1980).

Note added in Proof

The model of this paper is based on additions of energies. Thus, it does not depend on the specific sequence of events; i.e. whether fracture and plastic deformation occur simultaneously or sequentially. Therefore, the model is consistent with the ligament failure concept discussed at this conference, i.e. where a grain boundary fractures, then the Co ligaments associated with it subsequently fail.

DISCUSSION

H. E. Exner:

I think it's always interesting to see ideas from one material transferred to another one. What I was very interested in is that you put put a lot of emphasis on thermal stresses. I would like to go back to the paper Mrs. Luyckx presented yesterday without even mentioning the thermal stresses. If you put a plastic deformation on, you expect that something will happen in the material because it changes the stress state of the binder phase and the carbide phase. Several years ago we put a plastic deformation of about 0.2% by compression on simple transverse rupture strength bars and found that the crack lengths became anisotopic. I also expected that the transverse rupture strength in the different testing directions would be different. It was not. It was within the experimental scatter. I think you gave a good explanation. Transverse rupture strength is just not sensitive to those stresses you were pointing out.

R. Rice:

Yes, if the crack size is sufficiently large, then they should average out. The average along the crack front should then be zero.

S. B. Luyckx:

I would like to comment that this model envisions fracture in this material to be discontinuous taking place by sort of cracks ahead of the main crack which has been observed.

R. Rice:

Yes, I was very encouraged. In attempting to apply this, I, of
course, tried to read more in the literature than I normally do
in the cemented carbide area and I also talked with several people
and I was particularly encouraged by Dr. Fischmeister's comments
on noting both the effects of plastic deformation and micro-cracking
prior to failure, which would certainly be very consistent with this
type of approach.

A STUDY OF THE FRACTURE PROCESS OF WC-Co ALLOYS

Joonpyo Hong and Joseph Gurland

Division of Engineering
Brown University
Providence, Rhode Island 02912

INTRODUCTION

The important role of the deformation of the cobalt-rich
binder phase in the fracture process of sintered WC-Co alloys has
been noted in many recently published papers, either by observation
of the deformation and rupture of cobalt on the fracture faces of
the alloy (1-10) and/or by incorporating the work of plastic deformation of the binder phase into an appropriate fracture theory
(1-4, 6, 8, 11). In general, the dominant contribution of the
cobalt to the fracture resistance of these alloys is made obvious
a priori by considering that 1) the values reported for the
fracture energy of sintered WC-Co alloys (10^2-10^3 Jm^{-2}) are evidence of considerable plastic work during fracture when compared to
the typical cleavage energies of tungsten carbide and other brittle
materials (10^{-1}-10^1 Jm^{-2}), and 2) the primary variable controlling
the fracture toughness is the thickness of the binder phase layers,
as shown by many investigators (1, 4, 6-9, 11, 12).

However, the deformation process of the cobalt in the alloy
is not well understood, and it is not certain yet that the thickness of the cobalt layers is the only controlling variable, since
additional microstructural effects attributed to the carbide
particle size have also been reported (3, 6, 12). Further development of the fracture theory of cemented carbide requires an accurate
description of the fracture mechanism. A detailed examination of
the fracture process has therefore been carried out. The results
of the fracture observations are expected to further the correlation of fracture toughness values with microstructural parameters.

OBSERVATION OF FRACTURE PROCESS

The fracture processes of a number of WC-Co alloys were studied
by examining: 1) the crack growth on the surface of double-
cantilever-beam (DCB) specimens, 2) internal crack profiles in DCB
specimens, and 3) matching fracture surfaces from single-edge
notched beam (SENB) specimens. The specimens were produced from
commercial tungsten carbide and cobalt powders. Standard polishing
and etching procedures were used for the preparation of metallographic sections.

1)   Surface crack growth

The step-by-step surface-crack growth was observed in-situ by
scanning electron microscopy (SEM) on small wedge-loaded DCB specimens, using a specially constructed loading stage in the electron
microscope. The loading stage was described previously (13).

The observed fracture sequence of a specimen of high cobalt
content and coarse WC grain size is shown in Figure 1. Propagation of the fracture begins with the discontinuous cracking of WC
grains by cleavage and cracking of WC/WC grain boundaries. Ductile
tearing of the binder phase follows, sometimes along shear bands.
Cracking or debonding along the WC/Co interfaces occurs both in the
early stages of fracture (Figure 2a) or after considerable deformation (Figure 2b).

Examination of crack propagation in alloys of different compositions and different particle sizes showed similar fracture processes, except for a change in the relative frequency of occurrence
of various components. In Figure 3, for instance, one notices more
WC/WC boundary cracking in the specimen of medium cobalt content
and small grain size than was observed in the specimen of Figure 1.
As will be discussed later, this is attributed to the larger
number or density of WC/WC interfaces. In general, one observes
a tendency for the WC/WC grain boundaries to break first, provided
that these are favorably oriented with respect to the macroscopic
crack direction and that they lie within a short distance from the
crack plane.

2)   Interior fracture profiles

The observation of the fracture process at the surface was
supplemented by the examination of fractures interior to the specimens, revealed by sectioning specimens which were epoxy-infiltrated
under load. As seen in Figure 4, for the same specimen shown in
Figure 1, the process of fracture inside the material is qualitatively the same as that on the surface, i.e., WC particle cracking
or WC/WC boundary cracking, followed by rupture of the ligaments
of cobalt.

a) Discontinuous cracking of WC grains and WC/WC boundaries

b) Ductile tearing of cobalt ligaments

Figure 1.   Fracture propagation in specimen with high cobalt and
            coarse WC grain size (60 vol. % WC - 40 vol. % Co,
            d_{WC} = 3.6 μm).

a) Early stage of crack propagation
    (69.5 vol. % WC - 30.5 vol. % Co, d_{WC} = 2.96 μm)

b) After appreciable deformation - optical micrograph
    (60 vol. % WC - 40 vol. % Co, d_{WC} = 3.6 μm)

Figure 2.   Interfacial WC/Co debonding

Figure 3.   Fracture propagation in specimen of medium cobalt and
            small WC grain size (80 vol. % WC - 20 vol. % Co,
            d_{WC} = 0.89 μm).

Figure 4.   Fracture appearance in sectioned epoxy-infiltrated
            specimen. Optical micrograph. Same specimen as
            shown in Figure 1.

Measurements of crack-segment lengths were performed on the sections of partially broken, wedge-loaded DCB specimens. Interesting results are shown in Figure 5, namely that the average length of the fracture segments in WC particles is somewhat smaller than the mean intercept length of the particles, and that the mean fracture length in the cobalt binder phase is somewhat larger than the mean free path length of cobalt. These results support the preceding conclusion that the WC particles break before the cobalt ligaments rupture. The cobalt rupture path is not necessarily the path of minimum energy since its local orientation and length is imposed by the precursor events, i.e., the cracks in the carbide phase.

3)   Matching fracture surfaces

SEM fractographs of matching fracture surfaces of broken SENB specimens, pre-cracked by electro-discharge machining, were obtained on alloys representing four combinations of cobalt content (high, low) and WC particle size (large, small). The four basic modes of local fracture are identified by comparing the appearance of equivalent sites on the matching surfaces of each specimen, as follows:

Cleavage fracture of WC particles: matching particle shapes and, sometimes, river patterns.

Brittle fracture of WC/WC boundaries: smooth non-matching WC particle facets.

Ductile tearing of cobalt binder phase: matching fibrous-type fractures.

Interfacial fracture at WC/Co boundaries: WC particle facet on one face, smooth or fibrous cobalt on other face.

The fracture appearance is classified in more detail into several types, with particular attention to the fibrous fracture of cobalt:

Type A (Figure 6), multiple interfacial decoherence of WC particles leaves a pattern of ruptured cobalt ligaments on the larger WC particle face on one fracture surface, and the opposite fracture surface shows several smooth WC particles surrounded by cobalt ridges.

Type B (Figure 7), in the case of intergranular or transgranular cracking of WC particles, the broken or debonded particles are surrounded by rough ridges of cobalt which match well on the two fracture surfaces. If the WC particle is broken by cleavage, it has an identical shape,

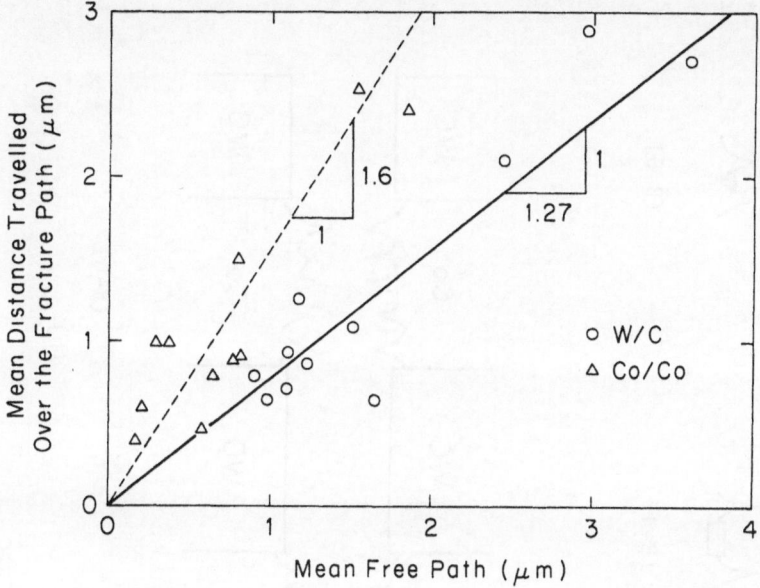

Figure 5. Mean fracture segment length in WC vs. mean free path
           in WC particles (circles) and mean fracture segment
           length in cobalt vs. mean free path in cobalt
           (triangles).

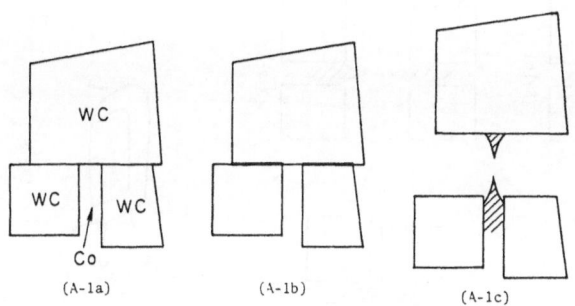

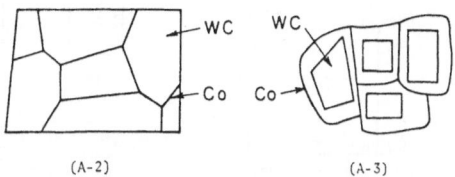

Figure 6. A-type fracture

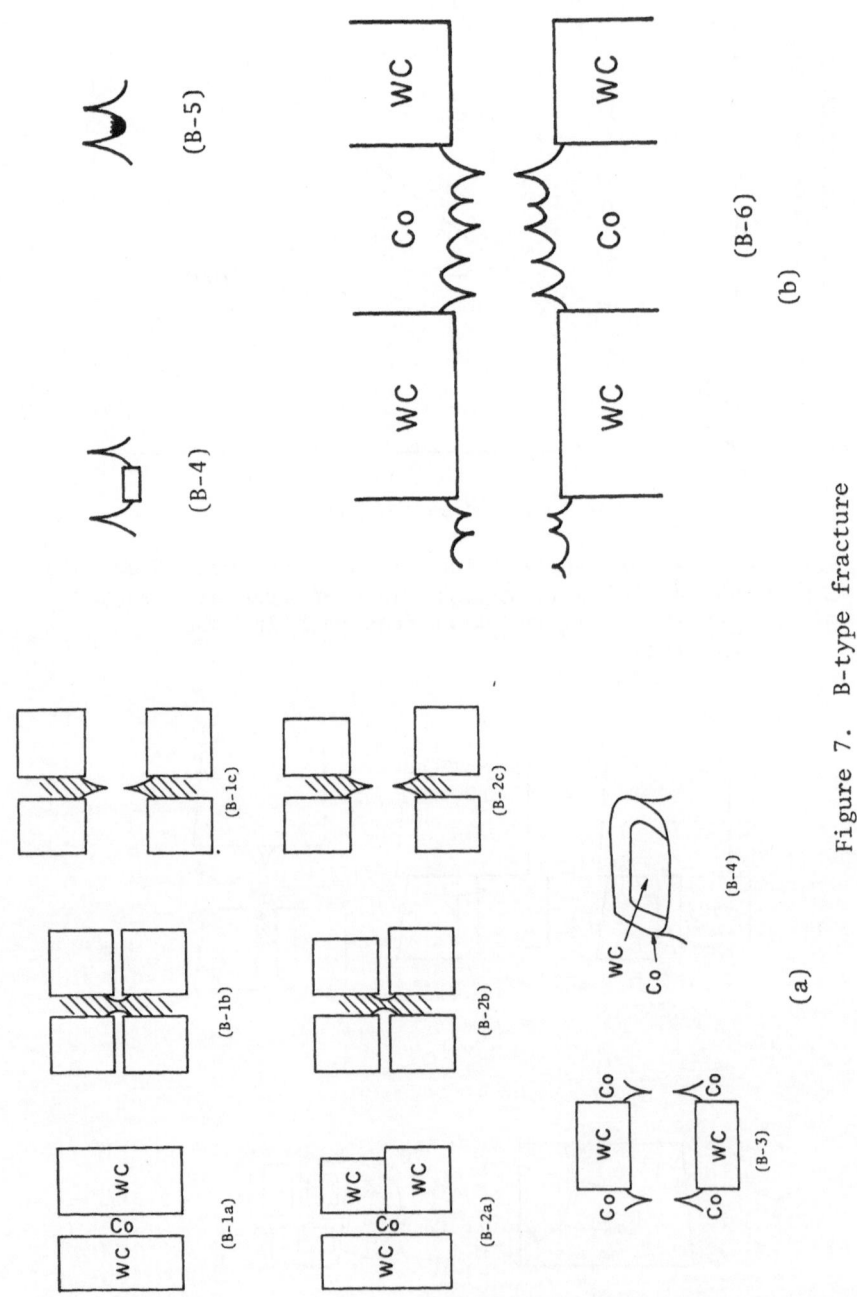

Figure 7.   B-type fracture

and, sometimes, matching river patterns, on the two fracture surfaces. If the separation occurred by WC-WC interfacial decohesion, the particle shapes will be different on opposite sides.

Type C (Figure 8), rupture of a cobalt layer of non-uniform thickness leaves a pattern of dimples or tear ridges on both surfaces. In general, the thinner the cobalt layer, the smaller is the size of the dimples. This type usually includes a portion of the surface of cracked or decohered WC particles.

Type D (Figure 9), WC-Co interface failure after appreciable deformation leaves the WC and cobalt faces with a pattern of tearing ridges, as sketched in Figure 9D-1, 2. WC-Co interface failure in the initial stage of deformation leaves relatively clean, i.e., featureless WC and cobalt faces (Figure 9D-3).

Type E (Figure 10), rupture of continuous cobalt layers between WC particles will show matching dimples or ridges of size varying in proportion to the thickness of the cobalt layer.

The fracture characteristics of the four representative specimens are now described in terms of the observed fracture types, as a function of cobalt content and tungsten carbide grain size:

i) Low cobalt and small carbide grains (refer to Figure 11): The typical fracture is an A-type fracture (around points b and d), with, in addition, considerable WC cleavage (point e). There are also some C-type fractures, as seems to be the case at point c in Figure 13, and there is also an E-type fracture (point a).

ii) Low cobalt and large carbide grains (refer to Figure 12): The typical fracture mode is a combination of type A and E (around points b and c). There are many WC cleavage cracks (as in particle a).

iii) High cobalt and small carbide grains (refer to Figure 13): Although this sample contains small carbide grains, there are quite a few broken carbide particles, as seen at points a, b and e. These particles have fractures of type B-4. At point d, WC particles are debonded, causing a type A fracture. The parts just to the left of c and f are E-5 type failures. Overall, the cobalt failure shape is more complicated than for a low cobalt case (Figures 11 and 12) because more deformable cobalt is available.

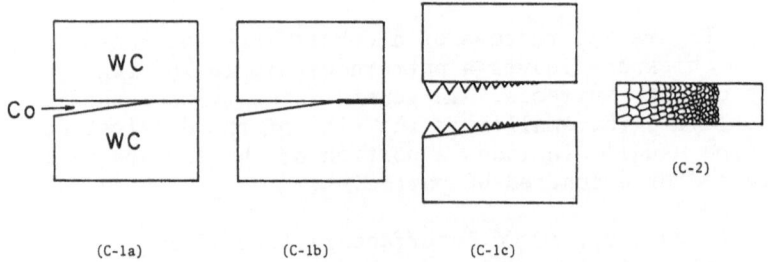

Figure 8.   C-type fracture

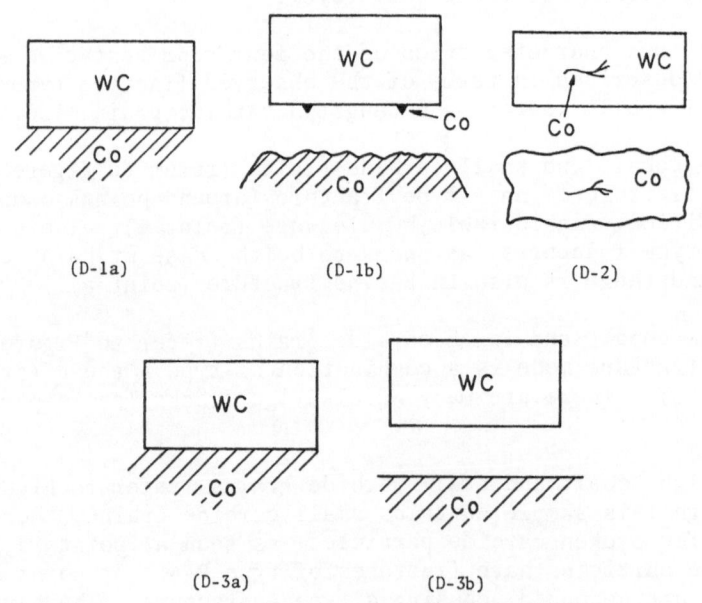

Figure 9.   D-type fracture

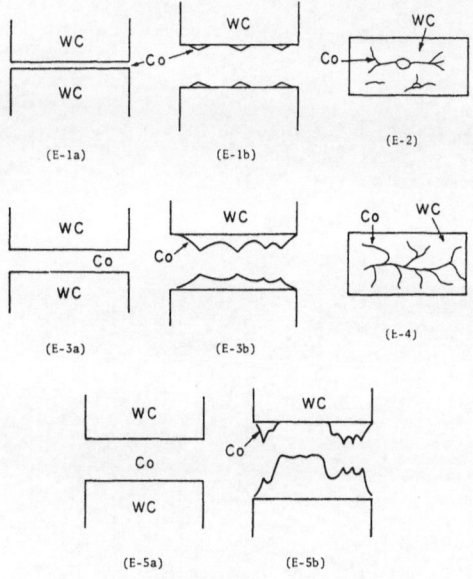

Figure 10. E-type fracture

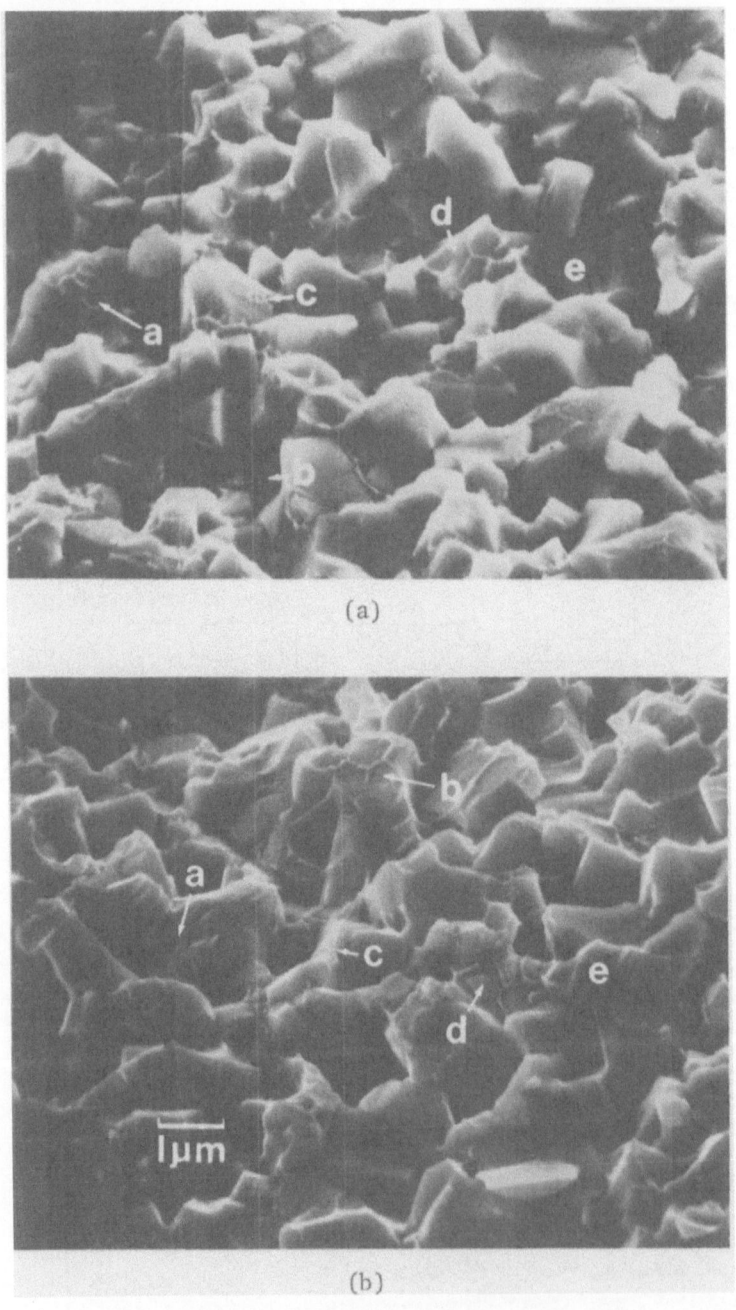

(a)

(b)

Figure 11.   Matching fractographs of broken specimen with low
             cobalt and small WC grains (90 vol. % WC - 10 vol.
             % Co, d_{WC} = 0.7 μm).

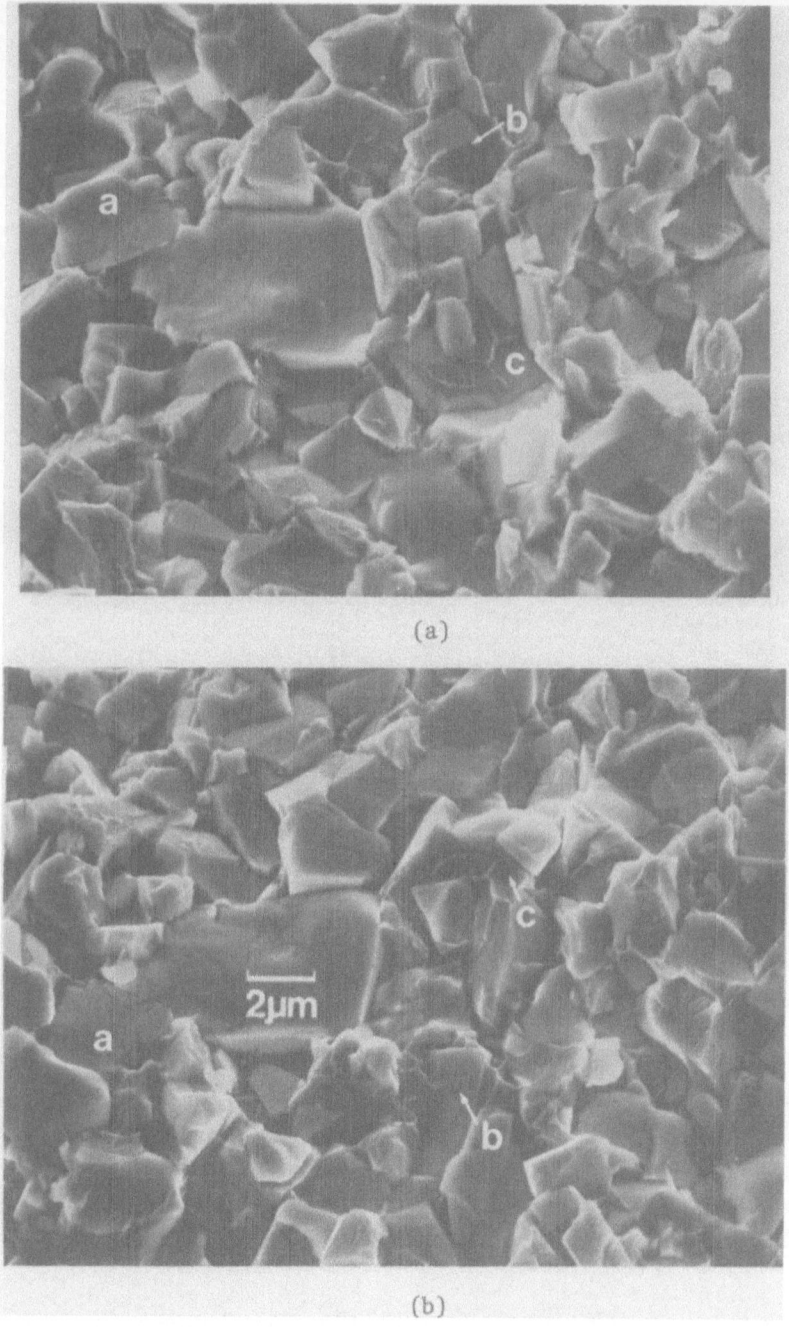

Figure 12.   Matching fractographs of broken specimen with low
             cobalt and large WC grains (90 vol. % WC - 10 vol.
             % Co, d_{WC} = 2.09 μm).

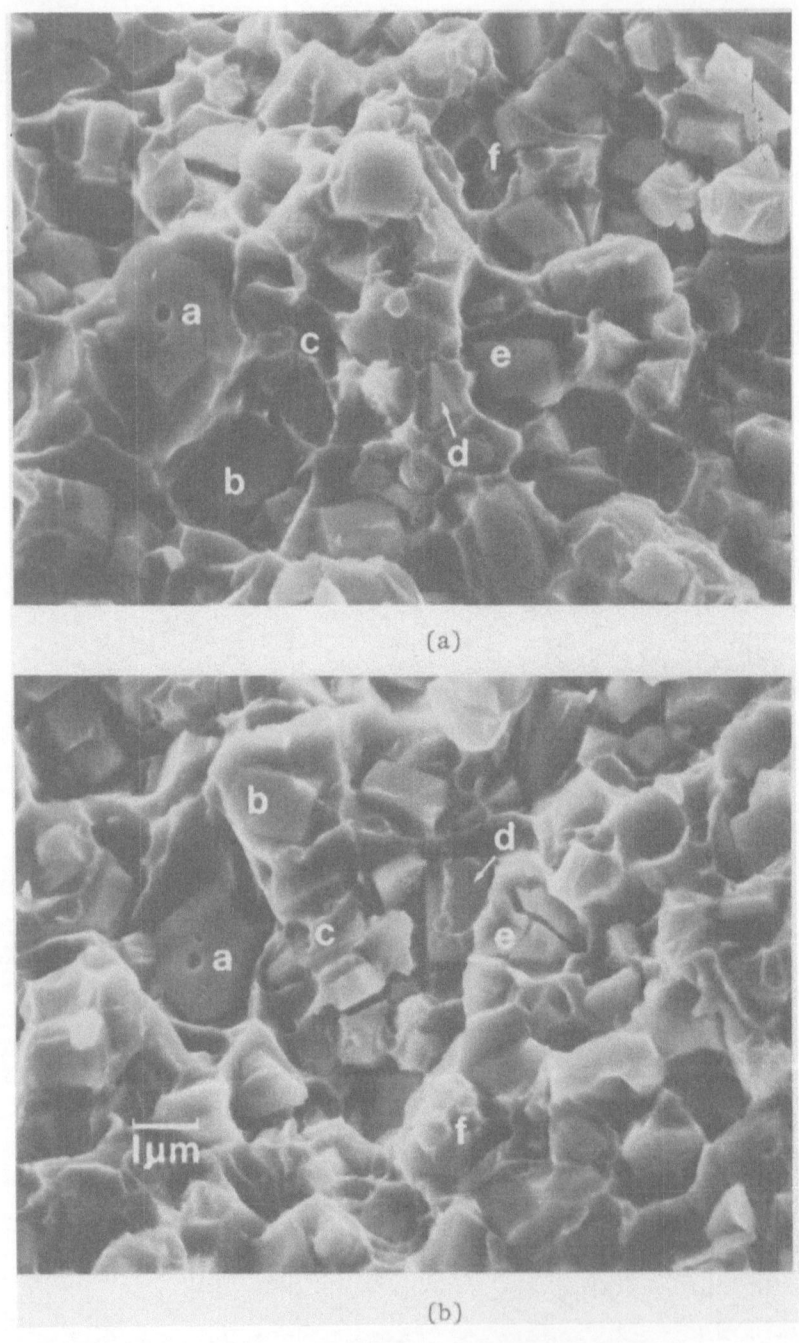

(a)

(b)

Figure 13.   Matching fractographs of broken specimen with high
             cobalt and small WC grains (60 vol. % WC - 40 vol.
             % Co, d_{WC} = 1.22 μm).

iv) High cobalt and large grain size (refer to Figure 14):
There are a considerable number of broken WC particles (for example
at a) and some WC/WC debonding (b, d and f). The fibrous fracture
around point c is type E-1. Near point d is an A-type fracture.
Point e is type D-1. Also noted, but not shown, are fractures of
types B-3 or B-6, and debonding between WC and cobalt of type D-3.

The predominant fracture types in each of the four representative specimens are listed in Table 1. Although certain types are
more prevalent than others in a given alloy, depending on its composition and particle size, examples of each type of fracture can
be found in each of the specimens. For instance, it is reported
in the literature (5, 7 and 10) that large WC particles ($d_{WC} > 5\mu m$)
generally fracture transgranularly and small particles ($d_{WC} < 2\mu m$)
are fractured intergranularly, but, as shown in Figures 12 and 13,
many cleaved WC particles are noted in specimens of small average
particle size. However, in general, the tendency for WC particle
cleavage increases with larger particle sizes and higher cobalt
contents. The frequency of WC/WC intergranular fracture is highest
at low cobalt contents and small WC particle sizes. These tenden-
cies may be the result of a geometric effect, namely that in order
for an interface between contiguous carbide grains to be debonded
it is necessary that this interface be favorably oriented and
located near the macroscopic fracture path. The probability of
occurrence of these conditions decreases with increasing carbide
particle size and increasing cobalt content, which in turn favor
carbide particle cleavage.

As far as the cobalt rupture mode is concerned, type A is
predominant at low cobalt content, due to thin binder layers and
high WC contiguity. With increasing cobalt content, the fracture
mode of the binder phase changes first to B-1, B-2 and E-1, E-3
and eventually to B-6 and E-5 at high cobalt content.

COMMENTS AND CONCLUSIONS

The results of the preceding section show that the overall
fracture process is basically similar in all of the specimens,
namely WC/WC boundary debonding and/or WC cleavage followed by
cobalt rupture. The former is a pre-condition for the latter.
The energy required for the propagation of the intergranular or
transgranular cleavage cracks in WC is believed to be much smaller
than that required by the plastic deformation and ductile rupture
of the binder. This argument, based on the observation of the
fracture process, indicates that the plastic deformation of the
cobalt indeed contributes the major portion of the dissipative
work during the fracture of WC-Co alloys.

In general, our observations support and justify the applicability of linear elastic fracture mechanics to these alloys,

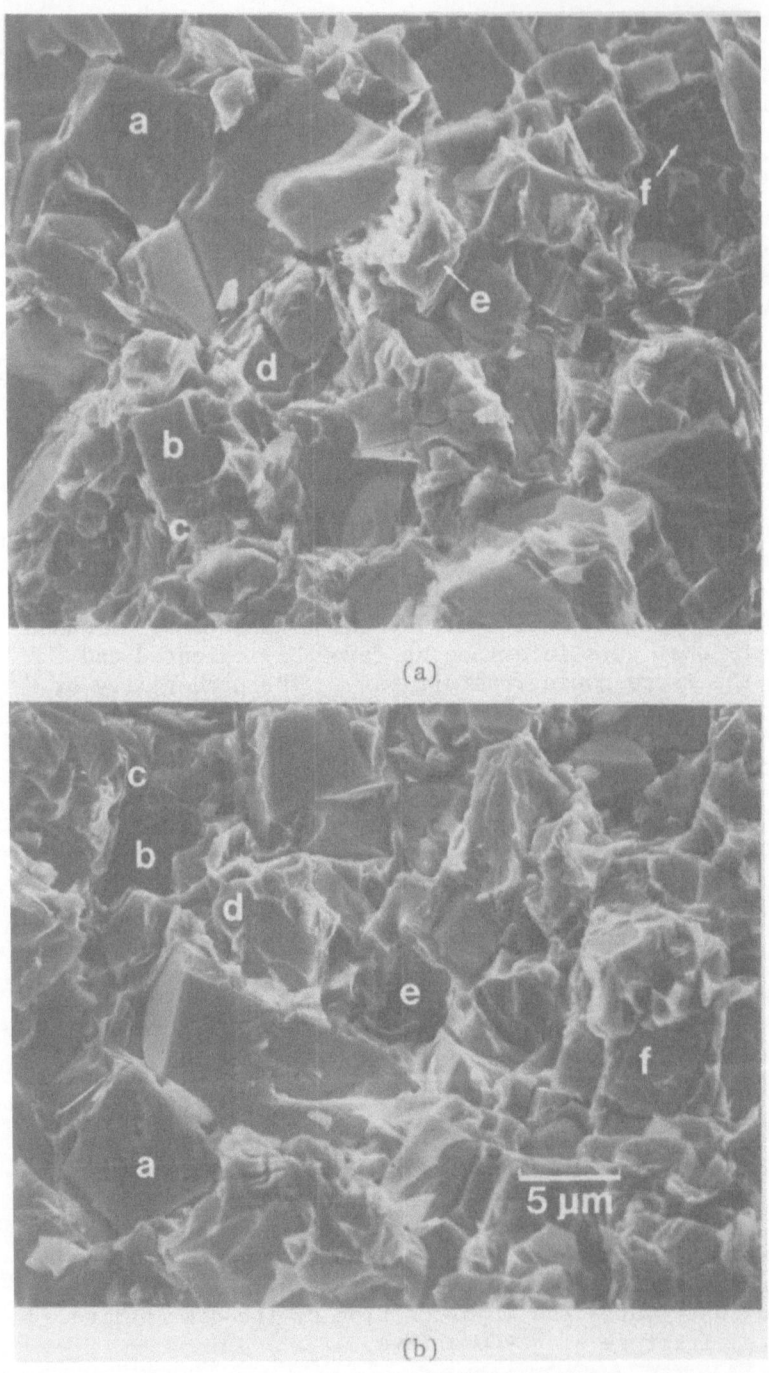

Figure 14. Matching fractographs of broken specimen with high
            cobalt and large WC grains (60 vol. % WC - 40 vol.
            % Co, d_{WC} = 3.6 μm).

Table 1: Typical fracture modes for four representative alloys.
          (Mode in parentheses is second most frequent.)

	Typical Cracking Mode		
Specimen	Carbide	Cobalt	WC/Co
Low Cobalt Small Carbide Particles	WC/WC(W/C)	A type (C-type)	---
Low Cobalt Large Carbide Particles	W/C(WC/WC)	A type, E-1 type	---
High Cobalt Small Carbide Particles	W/C, WC/WC	B-3 type	E-5 type
High Cobalt Large Carbide Particles	W/C(WC/WC)	B-6 type; B-3 type (All types)	D-1 type

under conditions of small scale yielding, since plastic yielding
is highly restricted to the ductile binder phase near the crack
tip, without greatly affecting the deformation of the bulk of the
material which deforms elastically.

ACKNOWLEDGEMENT

    This work was supported by the National Science Foundation
under Contract No. DMR77-08638-01 and by the Materials Research
Laboratory of Brown University, funded by the National Science
Foundation.

REFERENCES

1. B. Nidikom and T. J. Davies, Planseeb. Pulvermet. 28:29 (1980).
2. P. K. Viswanadham, T. S. Sun, E. F. Drake, and J. A. Peck,
    J. Mater. Sci. 16:1029 (1981).
3. M. Nakamura and J. Gurland, Met. Trans. 11A:141 (1980).
4. J. R. Pickens and J. Gurland, Mater. Sci. Eng. 33:135 (1978).
5. M. J. Murray and C. M. Perrott, "Proc. 1976 Int. Cong. on Hard
    Materials Tool Technology," p. 266 (1976).
6. J. L. Chermant and F. Osterstock, J. Mater. Sci. 11:1939 (1976).
7. J. L. Chermant, A. Deschanvres, and A. Iost, "Fracture Mechanics
    of Ceramics," R. C. Bradt, D.P.H. Hasselman, and F. F. Lange,
    eds., Plenum Press, New York , 1:346 (1974).
8. R. C. Lueth, "Fracture Mechanics of Ceramics," R. C. Bradt,
    D.P.H. Hasselman, and F. F. Lange, eds., Plenum Press,
    New York, 2:791 (1974).
9. S. S. Yen, M.S. Thesis, Lehigh University (1971).
10. A. Hara, T. Nishikawa, and S. Yazu, Planseeb. Pulvermet.
    18:28 (1970).
11. L. Lindau, "Fracture 1977," Waterloo, Canada, 2:19 (June 1977).
12. N. Inglestrom and H. Nordberg, Eng. Fract. Mech. 6:597 (1974).
13. J. Hong and J. Gurland, Metallography (in press).

DISCUSSION

H. E. Exner:

I wonder if it would be useful to match the two surfaces and find
out how large the deformation is in the crack zone by quantitative
topography and to look if the two surfaces fit or not. Would this
be a reasonable thing to do in cemented carbides? Just to take the
coordinates of all the surface in the crack region, to measure the
surfaces and to look if they fit or not.

J. Gurland:

Yes. I would say it would be interesting to do that. By the way, I should mention in this connection that this measurement would be a very useful contribution to all the theories, which use the plastic deformation of cobalt as the main energy absorbing process. To measure the actual topography of the fracture surface and thereby estabilish the deformation of the surface layers would be quite useful.

E. Almond:

You spoke about crack branching when you said that some of the cracks seemed to go nowhere. But they will still absorb some energy in deforming cobalt. How great was that contribution?

J. Gurland:

Well. It doesn't really enter into our theory whether the crack branches or not. I really don't know. I haven't followed that up. We just observed it. And I don't think I said it doesn't go anywhere. It may very well continue to extend. I don't know what happens to the branch. But crack branching does occur, that's all I can say.

C. A. Brookes:

My question really is a follow-up to Dr. Almond's question in the sense that the crack branching could contribute to energy dissipation and could perhaps explain the reason why in fracture toughness measurements using one method we get different predictions or answers to that in another. For example, if we do fracture toughness measurements in the bending case, we may get different answers to that in other processes. Is this reasonable?

J. Gurland:

Yes. I think that crack branching is very important. I know the mechanics people talk of their concern about crack branching. I just don't know anything about it. But, this theory, as well as all the other theories used in carbides only talk about the energy absorbed per unit area and don't say anything about crack branching. Perhaps they should, as an elaboration of these theories. The energy absorbed in crack branching could be very important and may make the difference between various compositions and various tests. I think it should be studied.

R. Rice:

I'd just like to follow up on the last comment. Crack branching
is being increasingly recognized in testing of ceramic materials
which often times may result from micro-cracking. Different types
of fracture mechanics tests, as well as the specific parameters of
a given test, may well alter the results you get if crack branching
is occurring. For example, in a notched beam test, if you have too
deep a crack, then the fact that a neutral axis which would suppress
micro-cracking or crack branching may give you a lower fracture
toughness result. When you start getting into these more complex
crack phenomena, not only do your fracture mechanics results differ
from test to test, but can depend on specific parameters such as the
crack depth or specimen size that you're dealing with. A good reference for this is, for example, to look at the extensive work on rock
mechanics where you can see huge dependences of the fracture mechanic
results, for example, on specimen size and crack size where you have
very extensive micro-cracking and crack branching results.

J. Gurland:

I appreciate your comment. I think the first thing I will do is
remove that one slide from my set to avoid further comment on crack
branching.

R. K. Viswanadham:

This is more of a comment than a question. I think that you observed
crack branching and secondary cracking, more frequently in microstructures that were not typical of commercial materials. The average roughness or the amplitude of roughness of the fracture surface
was much less than the grain size of the carbide in those microstructures. In typical commercial materials, the amplitude of roughness is of the order of the grain size. So, you can still keep that
slide because crack branching and all the related effects become more
dominant in what I like to call as the gray area. If $\lambda/d \ll 1$, the
binder is highly constrained and I think these effects become less
and less important. I think to assume that the dominant contribution
to fracture toughness comes from plastic deformation of the cobalt
is a valid assumption, because without that, there would not be any
need for cobalt.

J. Gurland:

Yes. I might make one comment on this. At least, to me it seems
obvious that it must be so. If we look at the values of fracture

energies for these materials we get G_{IC} values of the order of 10^2-10^3 J/m^2 which are indicative of considerable plastic work. If we look at the fracture energy of the carbide alone, or typical ceramic materials, you would get values one or two orders of magnitude less. So, that in itself indicates that some dissipative mechanism such as plastic deformation must be dominant.

HIGH TEMPERATURE CREEP OF SOME WC-Co ALLOYS

Frédéric Osterstock

Equipe Matériaux-Microstructure
Laboratoire de Cristallographie et Chimie du Solide
L.A. 251, ISMRA-Université, 14032 Caen Cedex, France

INTRODUCTION

The high temperature deformation behavior of tungsten carbide-cobalt alloys has been little investigated since the works of G. Altmeyer (1), W. Dawhil (2) and J.T. Smith and J.D. Wood (3). Moreover they are mostly concerned either with low cobalt volumic ratio or with temperatures less than 1000°C : the controlling processes which were proposed were diffusional vacancies flow in the binder and dislocation activity in the carbide phase. Only recently, some work were performed at temperatures as high as 1400°C by W.S. Williams and al. (4)(5)(6) with samples containing up to 10 % of cobalt in volume. Deformation of the carbide phase was proposed as a possible controlling mechanism, but transmission electron microscopy did not allow any hypothesis about the mechanism acting in the WC crystals. In most cases the results are limited, due to lack of deformation-sometimes less than 2 % - and the still high stresses which must be applied. Furthermore it is difficult to compare the results since they concern materials which have different cobalt volumic ratios, different mean diameters of the carbide crystals and sometimes additions of cubic carbides.

The aim of the present work is to study the creep behavior of WC-Co composites in a temperature range where the cobalt and the carbide phases have very different physical properties,and at cobalt volumic ratios which allow plastic deformation up to 15-20 %. These high deformations are needed to assess the deformation rate of steady state creep. In the temperature domain, which was chosen (1000-1300°C) cobalt is likely to deform at low stresses by vacancies flow and dislocations climb only begins to set up in the carbide crystals.

671

MATERIALS AND EXPERIMENTAL PROCEDURES

The materials used were WC-Co composites manufactured by the Eurotungsten Company by pre-sintering in vacuum at 900°C and final sintering at 1400°C. These materials were present with various cobalt volumic ratios of 22 %, 30 % and 37 % and, for each of these, the mean diameters of the carbide crystals, $\bar{D}_{WC}$, were either 2.2 μm, 1.1 μm or 0.7 μm. A material containing 16 % cobalt in volume and having a mean diameter of the carbide crystals of 2.2 μm was also used. The microstructural parameters were measured using quantitative metallography and image analysis.

The creep deformation was performed on parallelopipedic specimens approximately 8 mm high and with a section between 3 x 3 mm^2 and 4 x 4 mm^2. All the specimens were ground with a diamond grinding wheel. They were cut, by spark erosion, from blocks 4 x 8 x 45 mm^3 in size. These blocks were diamond ground previously to cutting : this method allows one to obtain best parallelism and perpendicularity between the faces.

The specimen is positioned between two loading pins, 20 mm in diameter. The pin were in tungsten and, in order to avoid diffusion of cobalt into the tungsten, the specimen was isolated via two dense alumina platelets.

The experiments were run in vacuum ($\approx 10^{-5}$ torr) in a Sesame-type furnace. Heating elements were two semi-cylindrical shells in tantalum. Shields are in tantalum, molybdenum and steel respectively, as one moves off from the center of the furnace. Temperature is measured with a tungsten-rhenium 6-26 % thermocouple fixed directly on the lower loading pin near the specimen. A second thermocouple, fixed on the first shield, is connected with a regulator which allows the temperature to be maintained constant at $\pm$ 2°C during the measurements.

The furnace is placed between the screws of an universal Tinius-Olsen, type Locap, testing machine with a capacity of 10 000 daN. Deformation is done under a load which was maintained constant using a TD A3 programmator. Displacement is measured outside of the furnace on the upper loading pin, using a Penny & Gilles inductive transducer.

Load and displacement signals were simultaneously recorded as a function of time on a Linseis 0490L recorder. Heating to the fixed temperature was done automatically and load was applied only after one hour the temperature has been reached. Deformation is measured as a function of time in a random order of loads. This could have been done since it proved that the same results were obtained. After steady state creep has been reached a deformation of $\sim$ 1 % was recorded.

True stress, σ, and strain, ε, were obtained using :

$$\varepsilon = \ln(1/l + \varepsilon_i)$$
$$\sigma = \sigma_i(1 - \varepsilon_i)$$

where ε_i and σ_i are respectively the values calculated from load and initial dimensions of the specimens.

RESULTS

High temperature creep rate, $\overset{\circ}{\varepsilon}$, is usually related to the applied stress by the relation :

$$\overset{\circ}{\varepsilon} = K \, \sigma^m$$

where m is the stress exponent.
K is a constant which constains effects of temperature and microstructure.

To determine the values of m, the results have been plotted in a Log $\overset{\circ}{\varepsilon}$ vs Log σ diagram, as can be seen in figure 1. This figure shows the results obtained with a material containing 37 % cobalt in volume and a mean carbide grain size of 2.2 μm at temperatures between 1000°C and 1300°C. A sigmoïdal behavior is observed, the stress exponent values being 6.7 , 2.7 and 4.5 as the stress increases. The region with the low m-value will be called stage II in the following. Whereas the low-stress region, with m ≈ 6.7 will be called stage I and the high-stress region, with m ≈ 4.5 stage III.

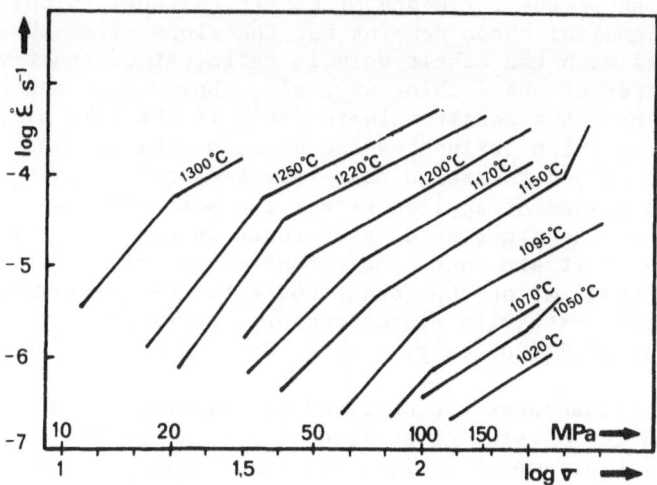

Fig. 1 : Stress-strain rate dependance as observed between 1000°C and 1300°C for a material containing 37 % cobalt in volume and with $\bar{D}_{WC}$ = 2.2 μm.

     In the following, only stage II will be considered because it
exhibits high microstructural dependance. This should allow one to
decide whether the cobalt or the carbide phase is controlling the
deformation.

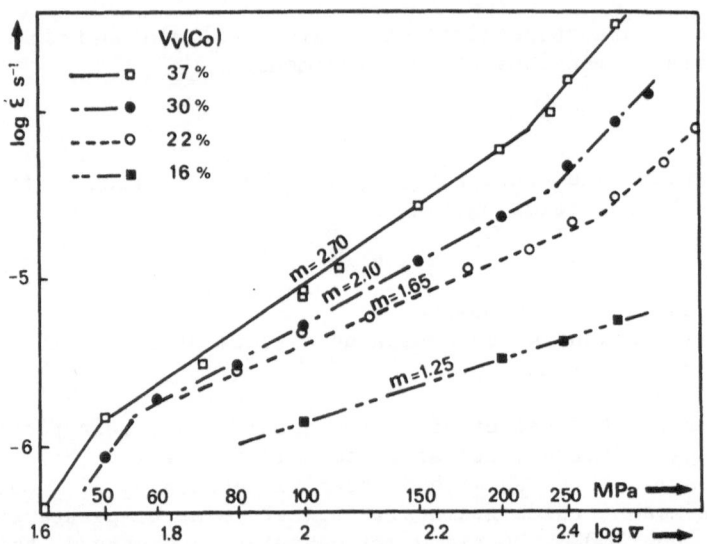

Fig. 2 : Cobalt volumic ratio dependance of the stress exponant, m,
         at given values of temperature (1150°C) and mean diameter
         of the carbide crystals ($\bar{D}_{WC}$ = 2.2 μm).

     Figure 2 shows the influence of cobalt content, V_V(Co) on sta-
ge II. The sigmoïdal shape remains but the slope, i.e. the value
of m, increases with the cobalt volumic ratio, at constant value of
the mean diameter of the carbide crystals. However it can be seen
from figure 3 that the cobalt volumic ratio is the only microstructural parameters which influences the value of the stress exponent.
As a general trend, deformation rate increases with cobalt volumic
ratio at fixed values of applied stress and mean diameter of carbide crystals. On the other hand, a decrease in grain size results
in an increase in strain rate. As a consequence of the dependance
of stress exponent, m, on the cobalt content, the quantitative variation of strain rate with microstructural parameters was determined
at fixed values of V_V(Co) only, i.e. V_V(Co) = 37 %, 30 % and 22 %.

     Figure 4 illustrates the observed dependence. This dependence
was determined for a value of applied stress of 150 MPa. It was the
only value which allowed the microstructural influence to be determined for all the materials tested. As a result, the creep behavior
in stage II can be described by the following expressions relating
strain rate, $\mathring{\varepsilon}$, microstructural parameters and applied stress :

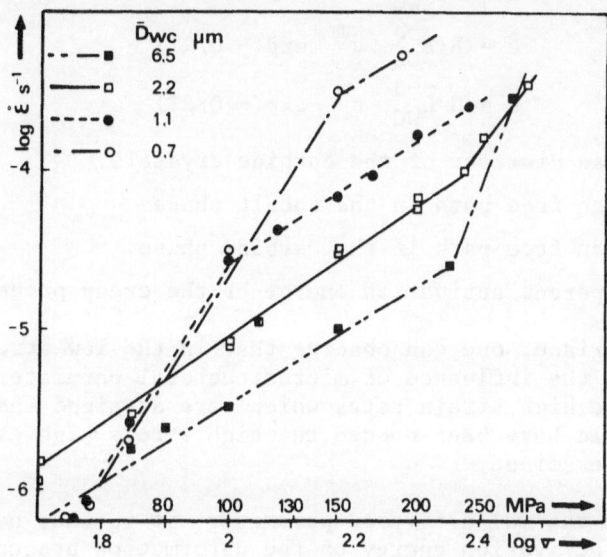

Fig. 3 : Influence of the mean diameter of the carbide crystals, $\bar{D}_{WC}$ on strain rate and stress exponent for a given value of the cobalt volumic ratio, $V_V(Co)$. Here $V_V(Co) = 37$ % and temperature of 1150°C.

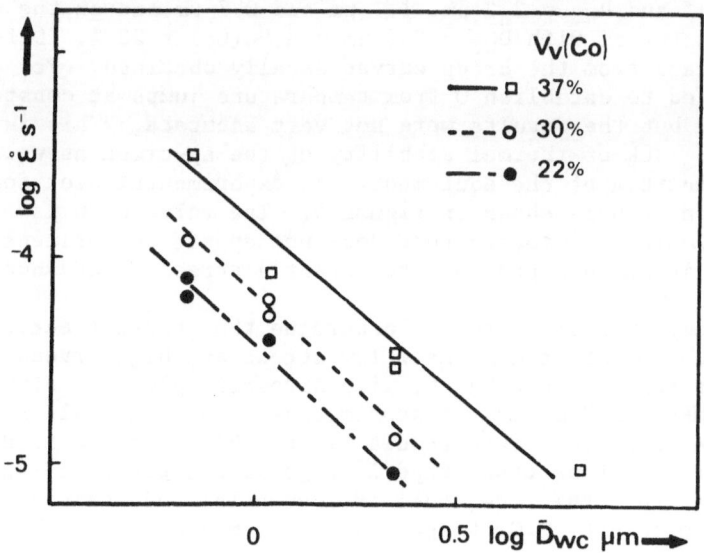

Fig. 4 : Influence of mean diameter of the carbide crystals, $\bar{D}_{WC}$, on strain rate for various values of cobalt volumic ratio at a stress of 150 MPa and a temperature of 1150°C.

$$\overset{\circ}{\varepsilon} = A \ \bar{D}_{WC}^{-2} \ \sigma^m \ exp(- \ Q/RT)$$

$$\overset{\circ}{\varepsilon} = B \ \bar{L}_{Co}^{-3} \ \sigma^m \ exp(- \ Q/RT)$$

$$\overset{\circ}{\varepsilon} = C \ \bar{L}_{WC}^{-3} \ \sigma^m \ exp(- \ Q/RT)$$

with $\bar{D}_{WC}$: mean diameter of the carbide crystals.

$\bar{L}_{Co}$: mean free path in the cobalt phase.

$\bar{L}_{WC}$: mean free path in the carbide phase.

Q   : apparent activation energy of the creep process.

As a comparison, one can observe that in the low stress–high exponent domain the influence of microstructural parameters is narrow. Due to the high strain rates which were attained and the high loads which could have been needed, the high stress–high exponent domain was not examined.

Creep is a thermal activated parameter, it is thus necessary to measure the activation energy of the deformation process. The apparent activation energy, Q, was measured following :

$$Q = R(\partial \ ln \ \overset{\circ}{\varepsilon}/\partial \ 1/T)_\sigma = - \ RT(\partial \ ln \ \overset{\circ}{\varepsilon}/\partial \ ln \ T)_\sigma$$

The values of Q were determined for four batches ; on the one hand $V_V(Co) = 37$ % and $\bar{D}_{WC} = 2.2$ µm, 1.1 µm and 0.7 µm, and on the other hand for a material with $\bar{D}_{WC} = 2.2$ µm and $V_V(Co) = 22$ %. Determination was made from the creep curves usually obtained. For one batch we tried to establish Q from temperature jumps at constant applied load but the results were not very accurate. This is probably due to lack of thermal stability of the specimen as well as to thermal inertia of the equipment. An experimental plot for the determination of Q is shown in figure 5. The value of Q is 540 ± 40 kJ/mole. Microstructure does not appear to influence the scatter and it was not possible to detect a stress dependance of Q.

An attempt was also made to determine the apparent activation energy of the two other domains : low stress and high stress. This was possible for only one batch, with Arrhenius plots at stress levels of 50 MPa and 300 MPa. Approximatively the same values as for stage II are obtained. This result is possible if one considers the shift of the transition stage I–stage II and stage II–stage III with temperature within the model of sequential and individual processes as proposed by T.G. Langdon and F.A. Mohammed (7).

Observation of deformed material was made using either a scanning electron microscope (Jeol TS 20) with polished and deformed specimens or a transmission electron microscope (Jeol 120 CX) with foils thinned by ion milling. Figure 6 shows the appearence of a

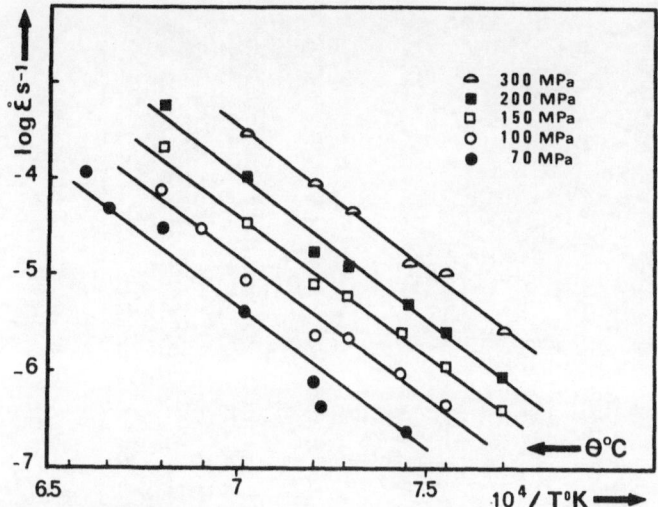

Fig. 5 : Experimental plot of $\overset{\circ}{\varepsilon}$ vs 1/T for the determination of the apparent activation energy, Q. Data are obtained from creep measurements at constant temperature. Material tested $V_V(Co) = 37 \%$; $\bar{D}_{WC} = 2.2$ µm.

specimen polished and deformed at two stages of deformation. Some cobalt evaporated during the deformation but the following remarks can be made :

- there is displacement of the carbide as well as individual crystals or as groups of crystals.

- what seems to be carbide-carbide boundary behaves differently : one may observe as well decohesion as keeping up their coherence.

- grain boundary sliding is not well developped as can be observed from polishing scratches.

Transmission electron microscopic observations have been made on foils cut from specimens deformed up to 30 %. The foils were cut following an angle of 45° with compression axis. Comparison was made with "as sintered" specimen. Figure 7 shows some features. One notes the absence of dislocation cells in the cobalt ; however there is a high density of stacking fault due to the Co hcp → fcc transformation which takes place at approximatively 400°C. Concerning the carbide crystals no regular dislocation pattern is visible. There exist simultaneously crystals without or with few dislocations and crystals containing several forms of dislocations structures ; pile-ups, bundles or individual dislocations. Furthermore some

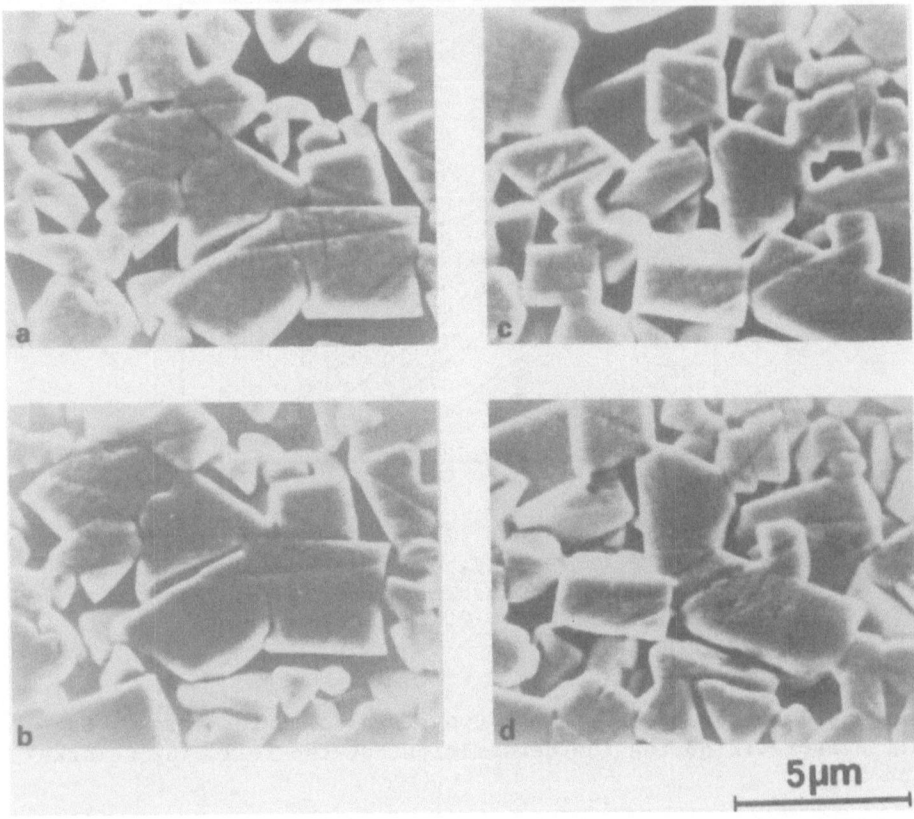

Fig. 6 : Two groups of carbide crystals at different deformation
state of the specimen, ε_p, (a and c, ε_p = 8 % ; b and d
ε_p = 30 %) $\bar{D}_{WC}$ = 2.2 μm and V_V(Co) = 37 %.

crystals may be separated by a low angle boundary and exhibit very
different structures (16).

DISCUSSION

From the results which were obtained one can make the following remarks :

- increase in testing temperature and in cobalt content does not
result in a great increase of strain rate. This illustrated in
figure 8 where we have plotted some of our results and those
from the litterature for WC-Co alloys and pure cobalt. Our results compare very well with those of J.S. Smith and J.D. Wood
(3) and the experimental points obtained by M.J. Murray and D.C.
Smith (8) in tension. The experimental points for the pure co-

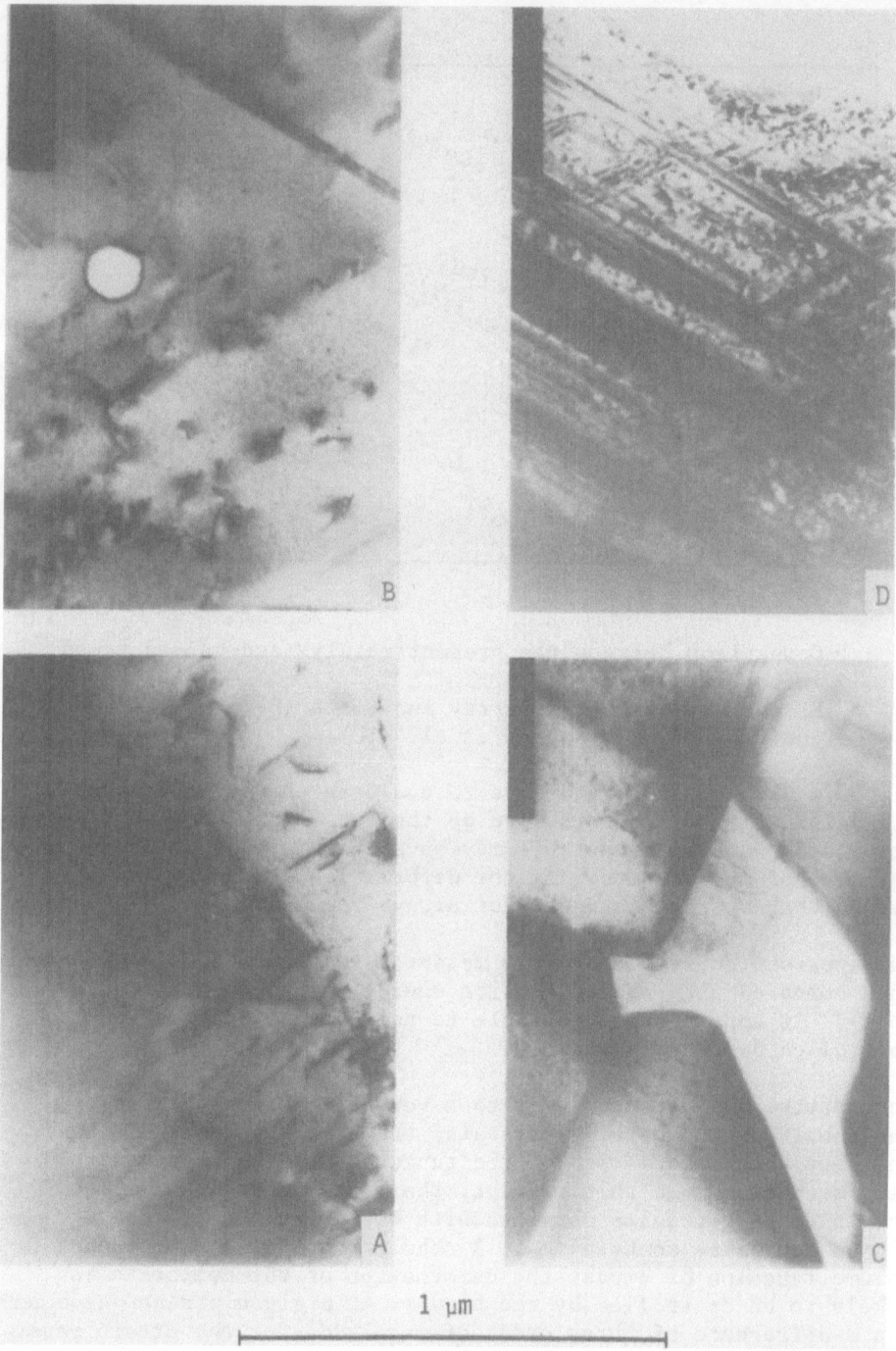

Fig. 7 : Dislocations in the tungsten carbide crystals and stac-
         king faults in a specimen deformed up to 30 % at 1150°C.

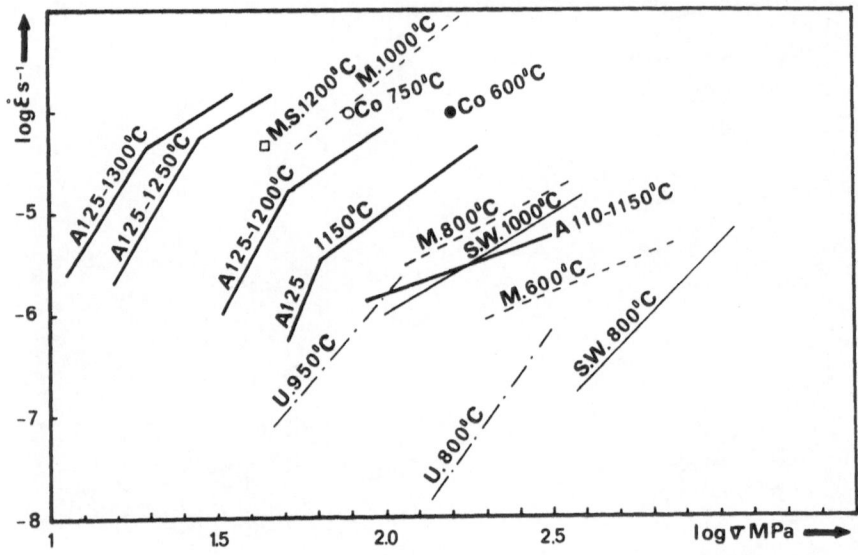

Fig. 8 : Comparison between the present results and values found
         in the litterature.
         M = Mal (9) ; M.S. = Murray and Smith (8) ; S.W. = Smith
         and Wood (3) ; U = Ueda et al. (10).

balt are plotted as a reference to evaluate the strengthening if
the deformation were controlled by the cobalt phase. The dis-
crepancies with other authors may have their origin in an overestimate of strain rate. As the deformation is mostly low it
may be that steady state was not always reached.

- the apparent activation energy measured ($Q \simeq 540$ kJ/mole) is high
  when compared with self-diffusion energy in the cobalt (260 kJ/
  mole). It appears thus possible to assume that deformation is
  controlled by tungsten carbide.

- furthermore and alternatively to a very high strengthening of
  the cobalt by the carbide crystals, the high stress levels also
  indicate that deformation of the tungsten carbide is a controlling mechanism. In this respect, the plots in figure 9 allows
  one to compare results obtained with WC-Co with those from a
  TiC-Co composite containing 50 % cobalt in volume. For such a
  volume fraction of binder the deformation of the composite is
  likely to be controlled by the binder. At a given stress one observes a difference of three order of magnitude for the strain rate.

- finally one may consider the evolution of the value of the stress

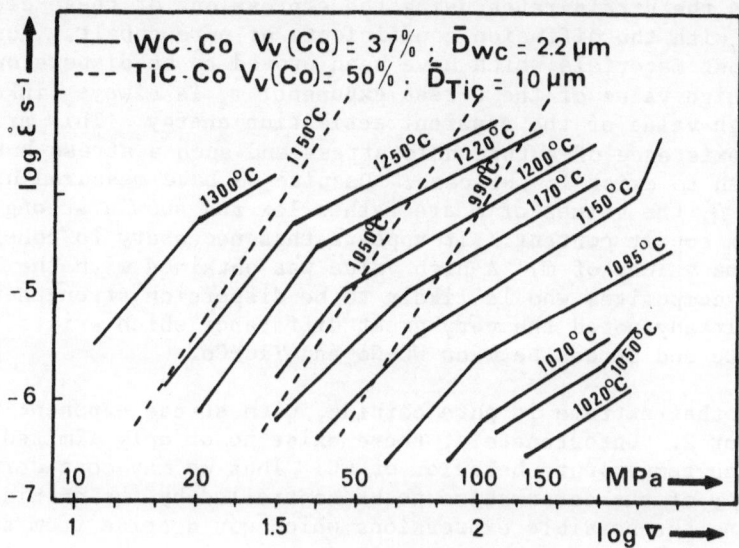

Fig. 9 : Comparison of creep curves of WC-Co ($\bar{D}_{WC}$ = 2.2 μm ;
V_V(Co) = 37 % and TiC-Co ($\bar{D}_{TiC}$ = 10 μm ; V_V(Co) = 50 %).

exponent. The value of this exponent increases with cobalt volumic fraction. The value of m tends to 1 for pure carbide and
seems to be described by the expression : $m = 1/(1 - 1.75\ V_V(Co))$.
This relation describes also the value m = 6 obtained with the
TiC-Co composite. It appears thus that m can be influenced by
stress or strain repartition in the composite.

It is now possible to discuss whether the cobalt or the carbide controls the deformation of the WC-Co alloys.

Cobalt is a possible parameter since the mean free path in
the cobalt appears in strain rate expression :

$$\dot{\varepsilon} = B\ \bar{L}_{Co}^{-3}\ \sigma^m\ \exp(-\ Q/RT)$$

The alloys should thus be considered as dispersion-strengthened.
Testing temperature, 1150°C, is approximatively 0.8 $T_{M(Co)}$, where
T_{MCo} is the melting temperature of the cobalt. Diffusion flow is
thus a possible mechanism and, owing to the $\bar{L}_{Co}^{-3}$ dependence, this
is a Coble-type diffusive creep (11). However at these temperature a Nabarro-Herring type diffusion creep, i.e. bulk diffusion,

is more likely to take place (12)(13). This can be verified if one
calculates the strain rates using the expressions of these creep
phenomena with the diffusion coefficients of pure cobalt. Furthermore in most materials which have been proved to be dispersion-strengthened, a high value of the stress exponent, m, is always linked
with a high value of the apparent activation energy. This arises
from the existence of a threshold stress and such a stress has not
been proved to exist in our case. Despite we have measured high
values of Q, the values of m are rather low and show a strong dependence with cobalt content. It appears thus necessary to consider
the extreme values of m. A high value was obtained with the TiC-Co
50 % vol. composites who is likely to be dispersion strengthened.
We have already noted the very great difference which exists in
strain rate and stress between WC-Co and TiC-Co.

The other extreme is pure carbide, with stress exponent values
of m = 1 or 2. Unfortunately, there exist no or only limited data
on the high temperature behavior of WC. Thus we may consider the
possibility of the deformation to be controlled by the carbide phase
and examine the possible expressions which would arise from such an
assumption.

The testing temperature is approximatively $0.45 - 0.50\ T_{MWC}$,
where T_{MWC} is an extrapolated melting temperature for the tungsten
carbide. These temperatures correspond to a domain in which dislocation climb becomes possible. This mechanism will thus be examined.

Climb velocity , V_c, is given by (14)(15) :

$$V_c \propto \frac{D\ \sigma_e\ b^3}{h\ k\ T}$$

with   D : diffusion coefficient in the carbide.
       σ_e : effective stress acting on the climbing dislocation.
       b : Burgers vector of the climbing dislocation.
       h : distance of climb.
       k and T are respectively the Boltzmann constant and the
             temperature.

Strain rate is given by :

$$\dot{\varepsilon} = \frac{b}{L}\ \frac{V_c}{h}$$

where $\varepsilon = b/L$ is a deformation unit.

Climb may take place either under the action of a dislocation
pile-up or not.

If there is dislocation pile-up of length, l, the effective

stress, σ_e, is $\sigma_e = \sigma^2$, where σ is the applied stress. Deformation rate becomes :

$$\overset{o}{\varepsilon} = \frac{\sigma^2}{h^2}\frac{1}{L}\quad\frac{b^4}{kT}D$$

If there is no dislocation pile-up, we have $\sigma_e = \sigma$ and $\overset{o}{\varepsilon}$ becomes :

$$\overset{o}{\varepsilon} = \frac{\sigma}{h^2 L}\quad\frac{b^4}{kT}D$$

Depending on the assumption which can be made on the existance of pile-ups and the signification which can be given to parameters like, h, 1 and L, various expressions relating strain rate, applied stress and microstructural parameters may be obtained.

As an example, climb under the action of a pile-up gives, with L, h and 1 proportional to $\bar{D}_{WC}$:

$$\overset{o}{\varepsilon} \propto \sigma^2 \bar{D}_{WC}^{-2}$$

Another possibility is climb without pile-up which can give :

$$\overset{o}{\varepsilon} \propto \sigma \bar{D}_{WC}^{-3} \text{ or } \overset{o}{\varepsilon} \propto \sigma \bar{L}_{WC}^{-3}$$

Further assumptions can be made concerning the nature of the carbide-carbide boundaries which have been shown to be often in coincidence (16). So various dimensional dependences may be obtained.

The important point, however, is the fact that the stress exponents which are obtained, are of the same order as those which were measured or extrapolated for pure carbide. Furthermore, the apparent activation energy, Q = 540 kJ/mole, is of the same order of magnitude as those measured for other carbides. Some examples may be cited : diffusion of carbon in SiC (550 kJ/mole) (17), diffusion of carbon in TiC (480 kJ/mole) or diffusion of titanium vacancies in TiC (710 kJ/mole) (18)(19) and diffusion of carbon in TaC (540 kJ/mole) (20). From these results it follows that the deformation of the composite is governed by the deformation of the carbide phase even at high cobalt volumic ratios.

One may thus propose the following behavior for stage I and stage II. The relative disposition of the log $\overset{o}{\varepsilon}$ - log σ plots may result from sequential processes (7). For stage I the stress exponent has a high value (m = 6) and microstructure has little influence, a deformation of the WC crystal involving bulk diffusion is thus a possible one. As the measured apparent activation energies are equal in the two stage, deformation may take place by annihilation of dislocations of opposite sign during climb. Various mecha-

nisms have been proposed for such a process (21)(22)(23)(24). For
stage II grain sliding accomodated by dislocation climb may explain
the grain movements which were observed (Fig. 6). The sliding and
the rotation of the crystals may be rendered possible by local deformation of the crystals. Models proposed to describe accomodation
during superplastic behavior may be applied here (25)(26)(27)(28)
(29)(30)(31)(32).

However the absence of well defined dislocation structures
does not allow further development of this examination.

CONCLUSION

Tungsten carbide-cobalt composites have been deformed at temperature between 1000 and 1300°C. Influence of the microstructure
has been studied at 1150°C and it appears that the deformation of the
composite is controlled by the carbide phase. Various mechanisms
have been proposed but lack of physical data on WC and the impossi-
bility to determine the deformation pattern of the dislocations with
respect to the undeformed material does not allow one to choose.
Further work is thus needed, especially on pure tungsten carbide.

ACKNOWLEDGMENTS

The author gratefully thanks Prof. A. Deschanvres for help-
full discussions. The CNRS is also thanked for financial support
(ATP-CNRS n° 3728).

REFERENCES

1. G. Altmeyer, O. Jung, Z. Metallkde. 52:576. (1961)
2. W. Dawhil, G. Altmeyer, Z. Metallkde. 54:645 (1963).
3. J. T. Smith, J. D. Wood, Acta Met. 16:1219 (1968).
4. J. G. Baldoni, W. S. Williams, Ceram. Bull. 57:1100 (1978).
5. S. R. Schenk, R. J. Gottschall, W. S. Williams, Mat. Sci. Eng.
       32:229 (1978)
6. R. J. Gottschall, W. S. Williams, I. D. Ward, Phil. Mag. A,41:1
       (1980).
7. T. G. Langdon, F. A. Mohammed, J. Aust. Inst. Met. 22:188
       (1977).
8. M. J. Murray, D. C. Smith, J. Mat. Sci. 8:1706 (1978).
9. M. K. Mall, Thesis, University of Sarreland, F. R. G. (1964).
10. F. Ueda, H. Doi, F. Fujiwara, H. Masatomi, Y. Oosawa, Trans.
       Jap. Inst. Met. 16:591 (1975).
11. R. L. Coble, J. Appl. Phys. 34:1679 (1963).
12. F.R.N. Nabarro, "Rep. Conf. Strength Solids" 1947/1948, p. 75.
13. C. Herring, J. Appl. Phys. 21:437 (1950).
14. J. Friedel, "Dislocations," Pergamon Press (1964).

15. M. Suery, Thesis, University of Metz, France (June 1979).
16. S. Hagege, Thesis, University of Caen, France (July 1979).
17. M. H. Hon and R. F. Davis, J. Mat. Sci. 14:2411 (1979).
18. S. Sarian, J. Appl . Phys., 40:3515 (1969).
19. D. L. Kohlstedt, W. S. Williams, J. B. Woodhouse, J. Appl. Phys.
    41:4476 (1970).
20. J. L. Martin, Thesis, University of Paris-Nord, France
    (February 1972).
21. J. Weertman, Trans. ASM 61:681 (1968).
22. A. H. Cottrell, M. A. Jaswon, Proc. Roy. Soc. A199:104 (1949).
23. J. Bardeen and C. Herring, "Atom Movements," (1951).
24. F.R.N. Nabarro, Phil. Mag. 16:231 (1967).
25. K. A. Padmanabhan, Mat. Sci. Eng. 29:1 (1977).
26. A. Arieli and A. K. Mukherjee, Mat. Sci. Eng. 9:191 (1979).
27. R. C. Gifkins, Met. Trans. 7A:1225 (1976).
28. R. G. Gifkins and T. G. Langdon, Mat. Sci. Eng. 40:293 (1979).
29. K. A. Padmanabhan, Mat. Sci. Eng. 40:285 (1979).
30. R. G. Gifkins and T. G. Langdon, Mat. Sci. Eng. 40:293 (1979).
31. A. Ball and M. M. Hutchinson, J. Mat. Sci. 3:1 (1979).
32. A. K. Mukherjee, Mat. Sci. Eng. 8:83 (1921).

DISCUSSION

A. Krawitz:

I'd like to make a comment. Possibly a useful method for documenting
and studying the rotation or possible alignment of WC crystals with
the stress axis, would be a preferred orientation measurement of a
bulk sample. You can do this very effectively with neutron diffraction.

F. Osterstock:

I thank you for the comment.

R. Warren:

Looking at the absolute values of your strain rates, I think that
they are several orders of magnitude higher than what one would
expect from creep in pure unbonded tungsten carbide. I am inclined
to draw a parallel with a similar system that we have been studying,
the tungsten-nickel system. If we poison pure tungsten wire with
small amounts of nickel, we get several orders of magnitude increase
in the creep rate. The absolute values are very similar to what you
see in your system. This leads me to think that possibly you are

getting a creep of the carbide skeleton which is activated in some way by small amounts of cobalt. I would say that even if you had very small volume fractions of cobalt you'd get similar creep behavior.

F. Osterstock:

I did not do experiments at low cobalt fractions due to experimental difficulties. The only role the cobalt could play in this material, in my opinion, is that it avoids cavity formation and premature fracture of the sample. When you compare the creep of TiC-Co and WC-Co at the same temperature, for the same stress, there is a three orders of magnitude difference for a difference in cobalt volume of only 13 per cent. So, that's why I do believe that it's the carbide.

R. Warren:

Yes. I'm in agreement with you, in that I believe that tungsten carbide is the controlling component. But it is being affected by the cobalt.

H. Fischmeister:

Let me first say that I think this type of work is extremely important because creep properties will probably determine the suitability of substrates for coated cutting tools. And we know far too little about creep, so far. I would suggest, though, thinking about your results not only in terms of tungsten carbide control. It seems suggestive to me that you get a power exponent of 1 extrapolating to zero binder content. This means that in the limit of very, very, thin cobalt layers, you have an exponent which is characteristic of Coble creep. That is to say, diffusional deformation of the cobalt phase between rigid tungsten carbide grains could be a mechanism which might be reconcilable with your results. As you increase the cobalt content and the cobalt mean free path, you are bound to get contributions of parallel creep in the cobalt and the stress exponent will be a mixture between the Coble exponent of 1, and the parallel exponent of 4. The degree of mixture, depending on the length of the mean free path.

F. Osterstock:

I have calculated the creep rates of pure cobalt using the mean free path of cobalt at these temperatures assuming diffusion values which I have found in the literature. The creep rates that I have obtained from this calculation are orders magnitudes higher than the creep rates I have measured here. I think at 1100°C you have creep rates of 1 per second for pure cobalt irrespective of whether you assume Coble or Nabarro-Herring.

H. Fischmeister:

I am not suggesting that at real cobalt contents, that is to say,
of the order of 10 to 20%, Coble creep is still the active mechanism.
This is only in the extrapolation for very very thin cobalt layers.
At the cobalt contents you have and the mean free path you have, if
cobalt should be the rate controlling phase, then it is bound to be
a dislocation mechanism which would give you a high power exponent.

F. Osterstock:

At these temperatures you have no dislocation mechanism. It is
near the cobalt melting point, so dislocation activity is quite
unlikely.

FRACTURE TOUGHNESS AS AN AID TO ALLOY DEVELOPMENT

Tom McLaren   John B. Lambert

Fansteel Inc.   Fansteel Inc.
VR/Wesson Division   North Chicago, Ill.   60064
Waukegan, Ill.   60085

ABSTRACT

With the portent of raw material shortages for the manufacture of conventional WC-Co cutting tool materials, a study
to develop new materials for the hard metal industry was
undertaken. Fracture toughness testing, employing short-rod
fracture toughness samples on a Fractometer*, was a valuable
tool for alloy screening. In this study four material matrices
were examined:

(1) WC-Co alloys
(2) Steel-cutting cemented carbides
(3) (Ti, Mo)C-Ni compositions
(4) Composite Al_2O_3-TiC ceramics

Compositional and microstructural variations were introduced
to find, first, the effect on mechanical properties, including
fracture toughness, and, second, the effect on application
and tool performance. Although fracture toughness helped rank
material candidates within a system as to relative impact
resistance, tool life is dependent also on other factors,
such as hardness and oxidation resistance.

---

*Trade name for short-rod fracture toughness tester,
Terra-Tek, Inc., Salt Lake City.

INTRODUCTION

The greater power and higher speed capabilities available in modern machine tools demand continual improvement in the performance of cutting tools and the materials from which they are made. This improvement can be achieved through design changes in the tool itself or, more commonly, through compositional changes in existing tool materials. Increasing raw material costs have further intensified the search for alternative, high performance cutting tool systems. The introduction of new or less expensive raw materials into hard metal or ceramic systems must be accomplished without compromising certain critical material properties. These are:

(1)   hot hardness and resistance to degradation through thermal cycling,
(2)   chemical inertness, and
(3)   impact strength and resistance to chipping.

The objective of this work was to determine the usefulness of fracture toughness measurements, in combination with other easily measured physical properties, in assessing the relative service performance of candidate compositions. A demonstration of the utility of fracture toughness in explaining the effects of compositional and processing changes on the phase constitution of hard metal alloys was also of interest.

Transverse rupture strength (TRS) measurements and cutting tool shock tests are frequently used to evaluate the toughness and impact strength of tool materials. Although the TRS measurement (ASTM B405-70) provides a rapid measure of the shear strength of laboratory samples, it is sensitive to both sample preparation and material inhomogeneities. Surface defects, porosity, inclusions, or other microstructural irregularities may reduce the measured value significantly. Though useful as an indicator of lot to lot variability, TRS values often exhibit poor reproducibility. Machining shock tests are costly and time-consuming. Such tests are generally unsuitable for screening experiments. In addition, shock tests are sensitive to cutting parameters and tool edge preparation, and generally require multiple replication to establish statistical confidence.

In materials research and development it is the intrinsic toughness of a candidate material which is of primary concern. Fracture toughness is generally recognized as a material property which is independent of the method of preparation and the presence of defects. As an example, Engel and Hübner[1] have reported that the elimination of B-type porosity in a C-2 cemented carbide increased the strength by a factor of two but did not affect the hardness or fracture toughness. Fracture toughness tests on alloys with similar processing histories typically yield standard deviations of less than 3%[2]. Recent studies by Farb[3] and Barker[4] suggest that fracture toughness data are useful also in obtaining a more fundamental understanding of WC-Co hard metal systems. Farb showed that compositional changes which occur in the binder film during quenching from elevated temperatures can alter the mechanism of crack propagation. Barker suggested a technique for measuring macroscopic residual stresses in WC-Co systems.

In this paper the application of fracture toughness in the development of four experimental alloys systems will be discussed. These systems are:

(1) WC-Co and WC-Co/Ni compositions for use in coal mining roof bits,
(2) WC-NbC-TiC-Co steel cutting compositions wherein NbC was used as a replacement for TaC,
(3) TiC-Ni-Mo compositions containing additions of titanium nitride, chromium, and molybdenum carbide, and
(4) TiC-Al$_2$O$_3$ ceramic cutting tool compositions.

EXPERIMENTAL PROCEDURE

Material Preparation

All powders were prepared by blending, milling, and drying. Hard metal compositions, both WC- and TiC-base, were pressed and vacuum-sintered at temperatures ranging from 1380°C to 1480°C, depending on the binder content. Ceramics were hot pressed in graphite dies at temperatures in the range, 1600-1700°C.

Test Procedures

Hardness (R$_a$), transverse rupture strength (TRS), magnetic saturation, and abrasion resistance were determined in accordance with ASTM standards. Grain size was measured using the linear intercept method with a minimum of three determinations per sample. For the very fine WC-Co grades (< 1 micron grain size), SEM photographs at 2400X were used.

The tool shock test was performed in two ways. In one method, inserts were used in a single-tooth flycutter on a 10-HP horizontal mill. In the second method, a slotted log was mounted in a 20 HP lathe. Tool life was measured as inches of material removed under the test conditions until failure. A minimum of five samples were used for each test. Standard deviations in tool life were lower for the slotted log tests than for the flycutter. Wear life was also determined by turning with this lathe. The failure criterion was machining time to 0.020 inch wear land.

    For fracture toughness, the short rod fracture toughness
tester, developed and trade-named the Fractometer by Terra-Tek, was
used. Ground specimens, 0.750 inches long by 0.500 inches in
diameter, were slotted with a diamond saw to form the required
V-notch.[2] Slots were 0.250 inches deep and formed a 58° angle at
the notch root. The crack propagation was accomplished by placing
a stainless steel bladder in the notch which was then filled with
mercury until the specimen split. A typical Fractogram, in which
force is plotted versus crack displacement is shown in Fig. 1. The
value of the short-rod, plane-strain fracture toughness, K_{ICSR}, was
then determined from the formula:

$$K_{ICSR} = \frac{AF_c}{B^{3/2}(1-\nu^2)^{1/2}}$$

    where: F_c = the maximum force executed by the bladder

            B = the rod diameter

            ν = Poisson's ratio

    and A = a scaling factor for the apparatus

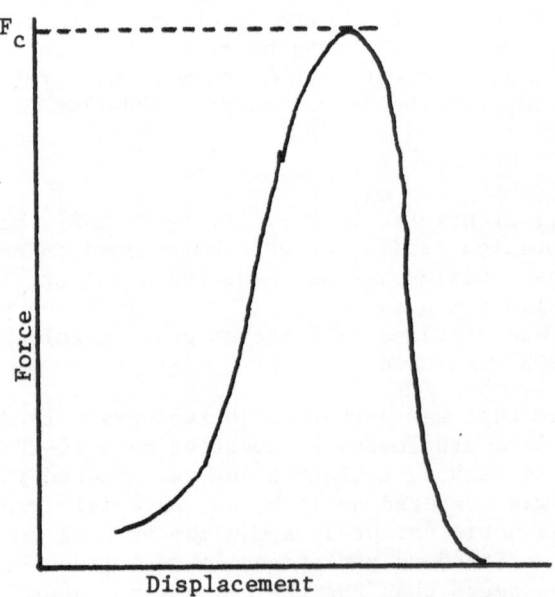

Fig. 1. Typical Fractogram for WC-Co

RESULTS

WC-Co and WC-Co/Ni Systems

The effect of grain size and cobalt content on fracture toughness was measured. The properties of experimental compositions with
cobalt ranging from 6% to 18% (by weight) and WC grain sizes from
less than 1 to 7.5 microns are listed in Table 1. Figures 2 and 3
show the short rod fracture toughness, plotted against cobalt mean
free path and cobalt content. The fracture toughness values for the
6% and 10% cobalt compositions are shown in Fig. 4 as a function of
WC grain size. The increase in K_{ICSR} with increasing grain size is
more rapid at cobalt content of 6% than 10%. Photomicrographs of
the 10% cobalt compositions are given in Fig. 5. An empirical
regression equation was developed to fit K_{ICSR} as a function of
grain size and cobalt binder content,

$$K_{ICSR}, \text{ MPa } \sqrt{m} = 4.65 + 0.35(\text{wt. } \% \text{ Co}) + 2.29\,\bar{d} - 0.15\,\bar{d}^2.$$

The relationship illustrates that fracture toughness is more strongly
influenced by grain size than by cobalt content.

Experiments in which nickel was substituted for cobalt in 10%
binder compositions are summarized in Table 2. In this series
grain size was held constant. Fracture toughness reached a maximum
at 50% substitution of nickel for cobalt. Hardness decreased
slightly with increasing nickel content, but abrasion resistance was
virtually unaffected. A second series of 6% mixed nickel-cobalt
compositions was field tested in sandstone roof bit drills. The
binder compositions and test data are given in Table 3. Tool performance was determined by measuring wear at the corner of the bit
after drilling 30 inches of sandstone. It was concluded that in
this application, a partial substitution of nickel for cobalt does
not affect the impact resistance or abrasion resistance of the tool
bit.

Steel Cutting Hard Metal Compositions

Secondary carbides, typically mixtures of TiC and TaC, are
customarily added to WC-base hard metals to improve hot hardness
and crater resistance. A dramatic increase in the market price of
TaC from $45/lb. in early 1979 to $190/lb. in mid-1980, however,
resulted in an industry-wide effort to replace TaC in steel cutting
formulations. Since NbC and TaC possess similar crystal structures,
solubilities, and physical properties, compositions containing NbC
as a replacement for TaC are of considerable current interest. A
series of such compositions in which various TiC/NbC ratios were
substituted for TaC in a typical steel cutting composition (Lot 1)
are given in Table 4. Physical properties and comparative machining

Table 1. Physical Properties of WC–Co Compositions

% Co	$\bar{d}^{(1)}$ (μm)	Co Mean Free Path, $\lambda^{(2)}$	R_a	TRS (MPa)	K_{ICSR} (MPa $\sqrt{m}$)
6.0	7.5	0.83	87.8	2230	15.6
6.0	3.6	0.40	90.9	1960	12.8
6.0	1.5	0.17	92.0	1930	10.6
7.0	0.5	0.07	92.1	1965	8.3
10.0	5.0	0.98	88.0	2500	15.1
10.0	3.2	0.61	88.5	2320	12.7
10.0	0.5	0.10	91.6	2660	8.6
13.0	4.0	1.06	88.0	2680	17.2
14.0	2.0	0.58	88.5	2780	15.1
16.0	2.2	0.25	87.0	2680	13.5
18.0	1.0	0.39	87.7	3400	12.8

(1) $\bar{d}$, avg, WC grain size = L/n

> where L = length of the intercept
>   countline, microns
>   n = the number of WC grains
>   intercepted

(2) λ, mean free path = $f\bar{d}/(1-f)$

> where f is the volume fraction cobalt

Table 2. Effect of Ni Substitution (10% Binder)

Binder Composition (Wt. %) Co	Ni	TRS (MPa)	R_a	K_{ICSR} (MPa $\sqrt{m}$)	Abrasion Resistance, cc/rev x 10^5
10	–	2800	88.6	11.59	20.0
7.5	2.5	2620	88.1	12.12	19.4
5.0	5.0	2580	87.8	12.47	21.4
2.5	7.5	2580	87.7	12.41	21.7
–	10	2410	88.0	10.71	21.0

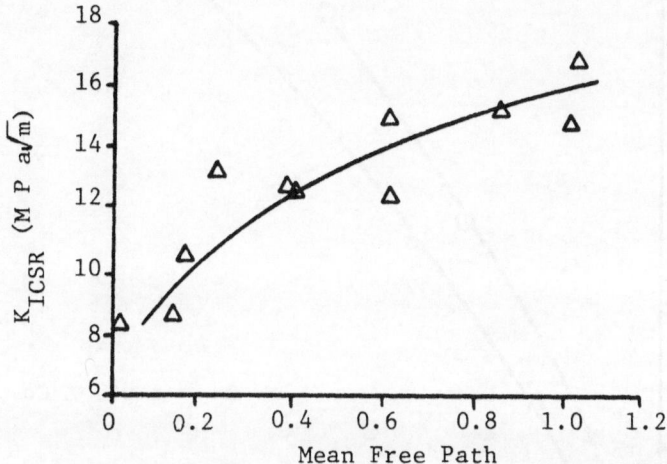

Fig. 2. K_{ICSR} vs. Mean Free Path (WC-Co)

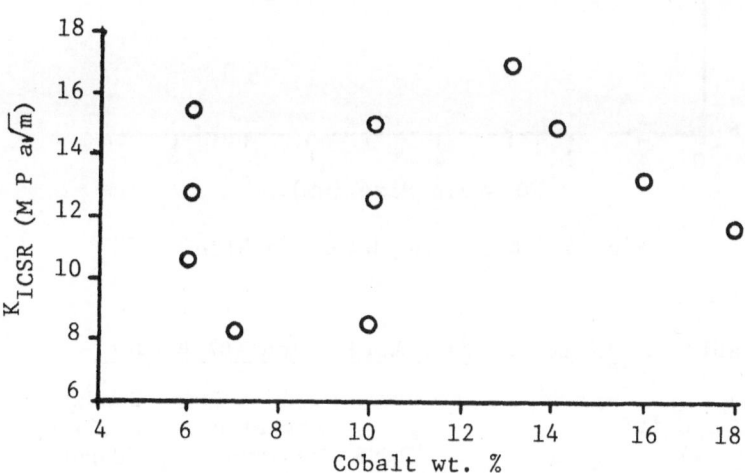

Fig. 3. K_{ICSR} vs. % Cobalt (WC-Co)
       - Grain Size Variable

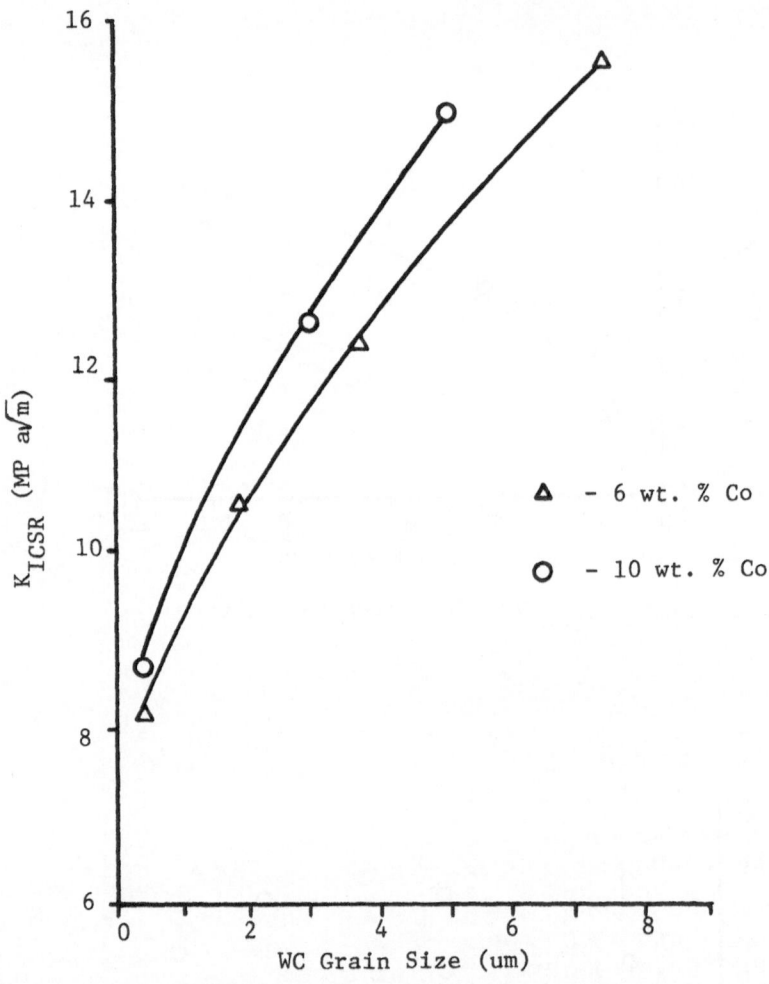

Fig. 4. K_{ICSR} vs. WC Grain Size

Table 3. <u>Effect of Ni Substitution (6% Binder)</u>

Binder Composition (Wt. %)		$\bar{d}$ (μm)	R_a	K_{ICSR} (MPa$\sqrt{m}$)	Abrasion Resistance, cc/rev x 10^5	Sandstone Drilling, Corner Wear (mm)
Co	Ni					
6.0	–	7.5	87.8	15.6	25.9	2.82
4.5	1.5	8.4	87.8	16.8	21.6	2.77
4.5	1.5	5.9	88.0	15.7	22.3	2.64
6.0	–	3.5	90.4	12.8	9.8	–
4.5	1.5	3.6	90.2	12.6	13.3	–

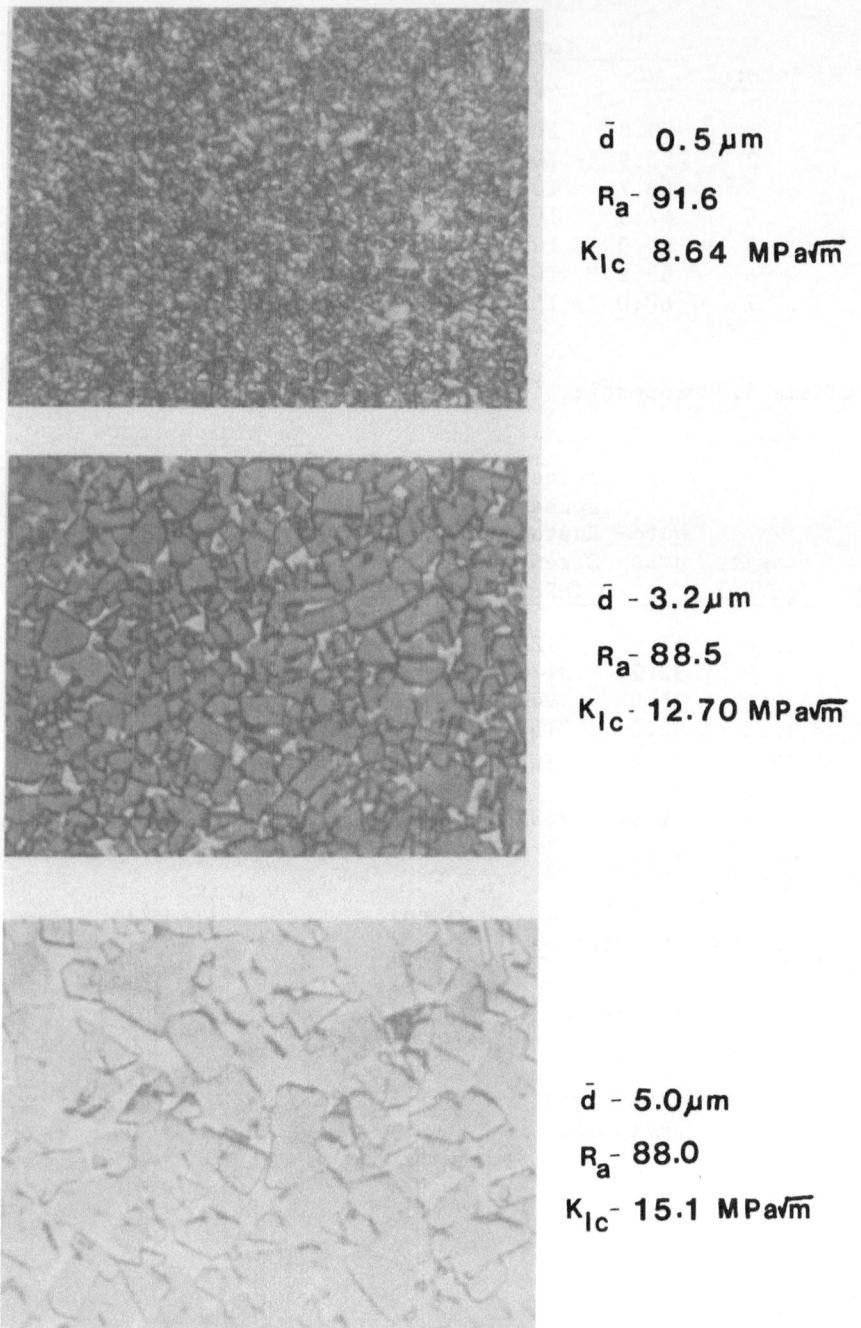

$\bar{d}$ 0.5 μm

R_a- 91.6

K_{Ic} 8.64 MPa√m̄

$\bar{d}$ - 3.2 μm

R_a- 88.5

K_{Ic}- 12.70 MPa√m̄

$\bar{d}$ - 5.0 μm

R_a- 88.0

K_{Ic}- 15.1 MPa√m̄

Fig. 5.   Microstructure and Properties (90% WC-10% Co)

Table 4.   Steel-Cutting Compositions

Lot	Composition (Vol. %)				
	WC	Co	TaC	NbC	TiC
1	59.8	10.8	8.8	–	20.5
2	55.9	10.1	–	15.0	19.0
3	63.7	10.8	–	16.3	9.2
4	67.0	10.8	–	4.2	18.0
5	65.9	10.8	–	8.5	14.8
6	59.2	10.4	–	15.5	13.9
7	60.0	10.8	–	8.6	20.5

Table 5.   Properties for Steel-Cutting Compositions

Lot	Density (gm/cc)	Hardness Ra	Transverse Rupture Strength (MPa)	K_{ICSR} (MPa$\sqrt{m}$)	Machining Tests	
					Relative Wear Resistance (Turning)[1]	Flycutter (In. to failure)[2]
1	12.6	91.9	3122	10.0	1.0	7.4
2	11.8	92.2	3178	9.2	0.8	5.3
3	12.6	91.9	2940	10.3	0.5	15.8
4	12.6	92.0	3150	9.6	0.8	5.3
5	12.7	91.7	3108	9.8	0.7	10.5
6	12.1	91.9	2436	9.7	0.7	10.5
7	12.1	91.9	2632	9.7	0.9	5.3

Machining Test Conditions:	[1] Turning	[2] Flycutter (6 in. dia.)
Work material	4140 steel	4140 steel
Speed in sfpm (m/min.)	500 (152)	420 (128)
Depth-of-cut, in. (mm)	.075 (1.91)	.150 (3.81)
Feed, in./min. (mm/min.) or in./tooth (mm/tooth)	4.42 (112.3)	.015 (.38)
Failure criteria	0.020" (.5 mm) wear land or	>0.020" (.5 mm) dia. chip

data for these compositions are listed in Table 5. A regression analysis of the K_{ICSR} values as a function of the volume percentages of the compositional ingredients yielded the following relationship:

$$K_{ICSR}, \text{ MPa } \sqrt{m} = -354.4 + 3.60(WC) + 3.96(Co) + 3.57 (TiC) + 3.66(TaC) + 3.58(NbC)$$

The standard error of the estimate was 0.082. The grain sizes of all carbide phases were essentially the same. The effect of grain size was not considered in the analysis. The relative effect of each constituent on fracture toughness is given by the magnitude of the regression coefficient. Within the scope of the experiment, cobalt content has the greatest effect on K_{ICSR}, while, among the carbides, TaC content was most important.

Although there were inconsistencies in the machining data, several conclusions can be drawn. A simple substitution of NbC for TaC on a volume basis (Lot 1 vs. Lot 7) resulted in a decrease in both K_{ICSR} and shock resistance. The wear resistance of all compositions containing NbC was less than that of the TaC-containing control even though hardness values (ambient temperature) were relatively unchanged. Because the TaC-containing composition had the best wear resistance, it may, therefore, be inferred that TaC is more effective than NbC in retaining hot hardness.

(Ti, Mo)C-Ni Hard Metals

Because (Ti, Mo)C-Ni alloys are more susceptible to chipping and fracture than WC-Co hard metals, their use has been limited to turning alloy steels at light depths-of-cut and moderate speeds. There are several reasons:

(a) The hot hardness of TiC is lower than WC even though at room temperature the TiC is harder.

(b) The thermal conductivity of WC-Co alloys is higher than for TiC-Ni. Temperature gradients in TiC-base inserts are therefore, higher, and cracking tendency is greater because of higher thermal expansion stresses.

(c) For equivalent volume percent binder and hardness, the WC-Co materials have higher fracture toughness than unmodified (Ti, Mo)C-Ni.

Compositional modification is necessary to obtain improvements in the TiC-base hard metal system. The composition changes tried in this study were:

(1) Partial replacement of TiC with TiN,

(2) Systematic changes of the Ti/Mo ratio,

(3) Changes of the C/Mo ratio by additions of C, Mo, Mo_2C, and

Table 6. Experimental Alloys in the (Ti, Mo)C-Ni System

		Composition (Wt. %)					Mole Ratios		Properties				Abrasion Resistance,
		TiC	TiN	Mo (as Mo_2C)	Ni	Cr	Ti/Mo	C/Mo (excl. TiC)	R_a	TRS (MPa)	Magnetic Saturation (e.s.u.)	K_{ICSR} (MPa $\sqrt{m}$)	cc/rev x 10^5
A.	TiN Replacement												
	A1	70	0	18	12	–	6.64	0.5	92.2	1360	1.05	6.68	24.98
	A2	50	20	18	12	–	6.54	0.5	93.1	1450	<0.1	7.68	19.98
	A3	66	0	16	18	–	7.02	0.5	90.7	1686	3.15	8.37	27.95
	A4	46	20	16	18	–	6.96	0.5	91.6	1967	<0.1	11.19	20.99
B.	Ti/Mo Ratio												
	B1	64.0	20	0	16	–	–	0.5	87.5	1264	3.82	–	
	B2	61.3	20	2.7	16	–	50.8	0.5	88.0	1336	2.72	12.20	
	B3	58.7	20	5.3	16	–	23.0	0.5	88.9	1400	1.81	10.31	
	B4	56.1	20	7.9	16	–	16.3	0.5	88.9	1457	0.77	10.80	
	B5	53.4	20	10.6	16	–	11.7	0.5	89.9	1514	0.11	10.51	
	B6	50.7	20	13.3	16	–	8.9	0.5	90.6	1478	0.06	10.53	
	B7	48.1	20	15.9	16	–	7.2	0.5	91.3	1529	0.03	9.70	
C.	C/Mo Ratio												
	C1	59.8	16	11.7	12.5	–	10.9	0.25	91.6	1022	<0.1	–	
	C2	59.8	16	11.7	12.5	–	10.9	0.50	91.6	1134	<0.1	7.39	
	C3	59.8	16	11.7	12.5	–	10.9	0.75	91.9	1225	0.6	7.63	
	C4	59.8	16	11.7	12.5	–	10.9	1.00	92.0	1449	6.8	7.69	
	C5	59.8	16	11.7	12.5	–	10.9	1.25	91.9	1239	15.0	8.25	
D.	Cr Addition												
	D1	55	15	18	12	–	6.58	0.5	92.1	1386	3.17	7.40	
	D2	55	15	18	11	1	6.58	0.5	92.6	1293	0.05	6.80	

mixtures thereof,
(4)   Substitution of Cr metal for Ni in a portion of the binder.

Physical properties for the experimental compositions are compared in Table 6. Substitution of a portion of the TiC by TiN
increases hardness, transverse rupture strength, and fracture toughness (Fig. 6). X-ray diffraction shows that TiN forms a solid
solution with TiC. Photomicrographs also indicate the TiN acts as a
grain refiner since the typical TiC grain size without TiN is about
5 microns and, with 20% TiN in the formula, is 1 micron. In work
reported by Kieffer[5], Rudy[6], and Asai[7], the increase in
fracture toughness found with TiN additions correlated with improvements in tool shock life.

Moskowitz[8] and Snell[9] showed, when either carbon or the
Ti/Mo ratio is low, titanium is found in the binder phase. Solid
solution strengthening of the nickel binder occurred until the
titanium content was about 6%, at which level γ'(Ni_3Ti) formation
began. In this study, K_{ICSR} increased with C/Mo ratio (Fig. 7) and
Ti/Mo ratio (Fig. 8). Magnetic saturation likewise increased with
both ratios. Since the magnetic property is a measure of unalloyed
nickel, K_{ICSR} is also clearly dependent on the amount of unalloyed
binder. In this context, it is likely that Fig. 8 depicts two stages:

(1)   the elimination of Ni_3Ti (5 < Ti/Mo ratio < 10) and
(2)   removal of titanium from solid solution (Ti/Mo ratio > 25).

The chromium addition was made with the aim of improving thermal
shock and oxidation resistance. Although K_{ICSR} was somewhat less
with chromium, wear life in machining (Fig. 9) improved markedly.

Composite Ceramics - Al_2O_3/TiC(N)

Two compositional variants were evaluated. Physical properties
and machining data are compared in Table 7. The substitution of TiN
for a portion of the TiC was found to increase both hardness and
K_{ICSR} while other properties were substantially unaffected. Tool
performance improved in both wear and shock. These effects parallel
results previously described for the (Ti, Mo)C-Ni system. Possible
explanations for the improvements with TiN are:

(1)   The hot hardness of the TiC-TiN solid solution is higher
       than for TiC,
(2)   Grain boundary energy for the composite phases is reduced.

In this same system, Van der Voort and Hinton[10] showed that
tool performance in turning steel rools correlated with hot fracture
toughness determined by the Palmquist method. Room temperature
tests did not discriminate between materials. It may be surmised
that better understanding of TiN and other additives would also result if K_{ICSR} measurements were made at temperatures of tool use.

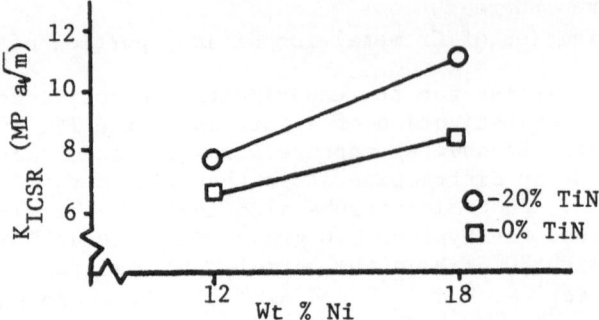

Fig. 6. K_{ICSR} vs. % Nickel (Ti, Mo) C-Ni

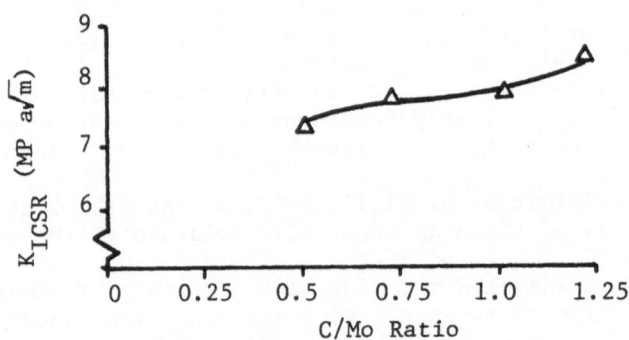

Fig. 7. K_{ICSR} vs. C/Mo Ratio (Ti, Mo) C-Ni

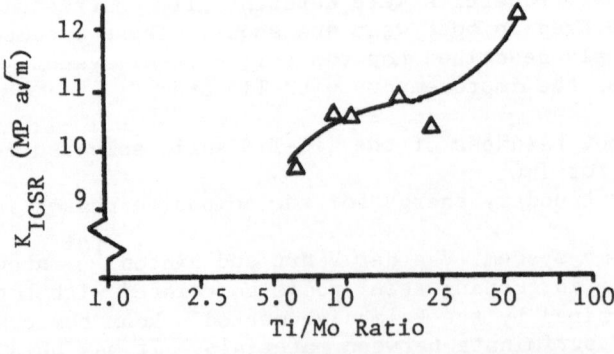

Fig. 8. K_{ICSR} vs. Ti/Mo Ratio (Ti, Mo) C-Ni

Table 7. Properties and Performance of Composite Ceramics

	Composition (Wt. %)			Density (gm/cc)	Hard ness, Ra	TRS MPa	K_{ICSR} MPa$\sqrt{m}$	Machining Tests	
								Wear [1] Life. (min.)	Fly- [2] cutter, in. cut
Lot	Al_2O_3	TiC	TiN						
1	70	30	0	4.28	95.0	79	3.77	4.2	51.8
2	70	20	10	4.34	95.4	81	4.53	4.8	70.5

Machining Conditions:	[1] Insert	[2] Flycutter
Insert Material	SNG 433 with 30°x 0.006 in. K-land Class 60 cast iron	Class 60 cast iron
Speed, sfpm (m/min.)	1200 (368)	1386 (422)
Feed, in./min. (mm/min.)	12 (305)	5.25 (133)
Depth-of-cut, in. (mm)	0.075 (1.91)	0.150 (3.81)
Failure criteria	0.015 in. flank wear	0.015 in. flank wear or chip size

SUMMARY

The short rod fracture toughness test is a useful and repro- ducible tool for experimental alloy screening in hard metal or cermet development. Since other properties, such as hardness and chemical stability, are equally important to tool performance, field evaluation of promising alloy candidates is always necessary.

However, fracture toughness is an important property from which to gauge tool performance, and, in this study, we have shown:

(1) The short rod fracture toughness, K_{ICSR}, in tungsten carbide-cobalt compositions increased with cobalt content and grain size, the latter being the dominant variable. Partial substitution of nickel for cobalt did not affect fracture toughness or tool performance in a sandstone drilling application.

(2) For steel cutting materials containing secondary carbides, a regression analysis showed the relative effects of the components on K_{ICSR} to decrease in the order: Co > TaC > WC > NbC > TiC.

(3) For the (Ti, Mo)C-Ni alloy system, partial substitution of TiN for TiC was verified to favorably affect tool wear and impact resistance. Fracture toughness also increased with TiN content. The binder film composition is affected by

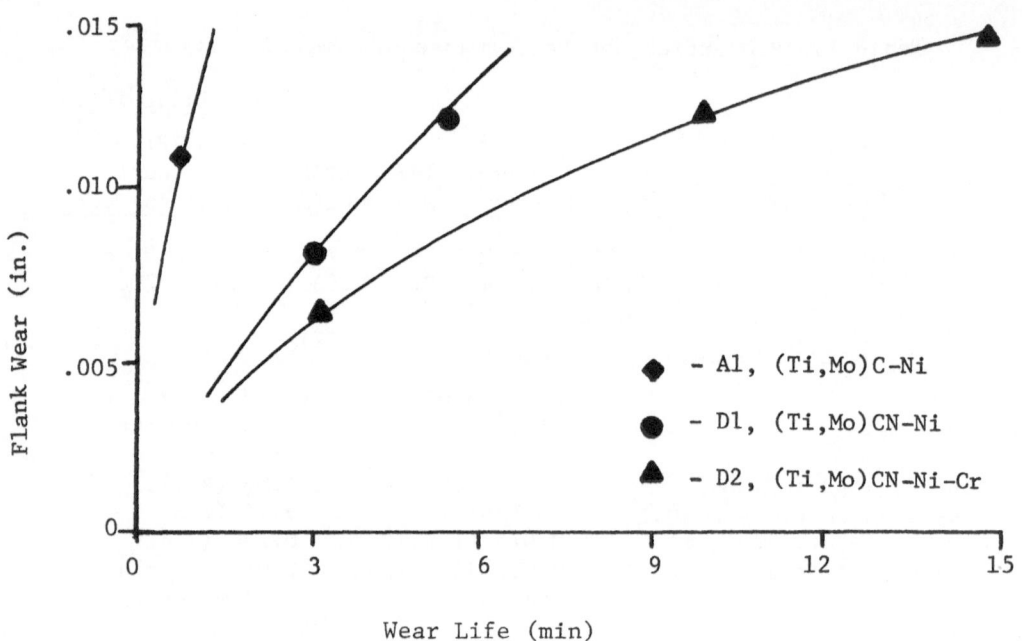

Fig. 9   Wear Life Curves for (Ti, Mo)C-Ni Compositions

the compositional ratios, C/Mo and Ti/Mo. Higher K_{ICSR} values were found as these ratios increased because the titanium was stabilized as TiC, preventing alloy formation with the binder. The addition of 10% chromium to the binder, however, improved tool performance but caused K_{ICSR} to decrease.

(4)   Modification of the 70% Al_2O_3/30% TiC composite ceramic with 10% TiN in place of TiC increased hardness and K_{ICSR}. Tool wear and impact resistance increased correspondingly.

One of the main difficulties in correlating laboratory test data with field performance is that unreasonable temperature extrapolations are often implied. Future studies should devote attention to defining the temperature dependence of mechanical properties, such as hardness and fracture toughness.

ACKNOWLEDGEMENT

The assistance of co-workers, Messrs. R. Cutler and J. Tingle, in obtaining some of the test data is acknowledged. Dr. R. Peters was also helpful in reviewing the text.

REFERENCES

1. U. Engel, and H. Hubner, Strength Improvement of Cemented Carbides by Hot Isostatic Pressing, J. Mat. Sci. 13:2003 (1978).
2. L. M. Barker, A Simplified Method for Measuring Plane Strain Fracture Toughness, Eng. Frac. Mech. 9:361 (1977).
3. J. Farb, Fracture Toughness of Cobalt Cemented Tungsten Carbides, M.S. Thesis, Univ. of Utah (1978).
4. L. M. Barker, Residual Stress Effects on Fracture Toughness Measurements, International Conf. on Fracture, ICF5, Cannes, France (1981).
5. R. Kieffer, P. Ettmayer, and M. Frendbofmier, in: "Modern Developments in Powder Metallurgy," 5:201 (1971).
6. E. Rudy, Journal of the Less-Common Metals 70:33 (1973).
7. T. Asai, T. Nomura, T. Yamamoto, and A. Hara, The Performance of a New Cermet Cutting Tool, Bull. Jap. Soc. of Proc. Eng. 12, No. 1:47 (1978).
8. P. P. Moskowitz, M. Humenik, Jr., Effect of Binder Phase on the Properties of $TiC-22.5Ni-MoC_x$ Tool Materials, Int. Jour. of Pow. Met. and Pow. Tech. 14, No. 1:39 (1978).
9. P. Snell, Planseeber. Pulvermetallurgie 22:91 (1974).
10. G. F. Van der Voort and R. Hinton, Metallographic Study of Factors Influencing the Toughness of 70% Al_2O_3/30% TiC Cermet Inserts, Proceedings of 1979 Conference of Productivity Improvement Through New Tools and Applications, Chicago, Ill., 153-158 (1979).

DISCUSSION

R. Cooper:

I am particularly interested in the results that you presented on the rock drilling grade of hard metal where you substituted nickel for cobalt because this seems to tie in closely with what we found. What I'd like to ask is, did you chose that particular combination of nickel and cobalt for the binder empirically or systematically? Could you give us a few more details on the rock drilling conditions you were using?

T. McLaren:

We chose the 25% substitution because there was only a slight decrease in hardness and a slight increase in the fracture toughness
results. We measured the corner wear on the tool bit after drilling
30 inches of sandstone.

D. Moskowitz:

Can you tell me if you did any comparison with respect to similar
volume per cent of binder for two basic systems say the WC-Co and
the TiC-(Ni, Mo) with or without a TiN replacement in terms of
fracture toughness. Was there any relative comparison of the two?

T. McLaren:

For the 12 wt.% nickel grade with the addition of TiN, we had fracture toughness values of approximately 8, which corresponds to just
about 8 volume per cent. For 10 volume per cent cobalt, we had
fracture toughness values of around 10 I believe. It's becoming
comparable, but I don't have exactly the same volume per cent.

R. Sivan:

With respect to the fracture toughness of steel cutting grades, as
I understand it, you are saying that you cannot replace tantalum
carbide.

T. McLaren:

We're not saying you cannot.

R. Sivan:

No. I'm saying you get a reduced fracture toughness. In your
opinion, what is the sort of maximum percentage substitution of
niobium carbide in these alloys?

T. McLaren:

Oh, that's a cost-performance trade-off, I believe, and it should
be made according to the confidence level of whoever is making in
the substitution.

G. Baldoni:

According to your linear regression analysis, fracture toughness
is related only to cobalt content and average grain size. Is it
justified to disregard the distribution in grain size?

T. McLaren:

I'm sure there's a limiting factor. But within the scope of the work we did here, we could.

H. Fischmeister:

To me the main merit of this paper is in filling in gaps in our knowledge of fracture toughness for all these complicated alloy systems. I'm not so sure I wholeheartedly agree with the application of fracture toughness arguments to wear problems. For instance, in the case of substituting titanium carbide by titanium nitride, there was a paper from one of the Japanese groups last year at Cincinnati that demonstrated a very good correlation between the thermal conductivity of the tools and the wear. And maybe part of the correlation that you seem to find is really due to some conductivity effects. In rock drilling, I think fracture toughness is much more germane to the issue, because there you have cracks extending over appreciable lengths in the abrasion process. In metal cutting wear, where you pull out individual grains, I don't think fracture toughness, which is a property averaging over carbide/carbide and carbide/cobalt interfaces is the property that should be invoked.

THE EFFECT OF LONG TERM STORAGE ON COBALT USED FOR CEMENTED CARBIDE

PRODUCTION

James J. Oakes

Teledyne Firth Sterling
LaVergne, Tennessee, U.S.A.

INTRODUCTION

Five samples of African Metals extra-fine cobalt powder produced between 1964 and 1981 were tested to determine if an effective shelf life existed for their direct use in cemented carbide production. The powders were tested for morphology, phases present, oxygen content and as binders in a WC-Co piece by Teledyne Firth Sterling. The tests showed that prior to 1974 the powders were probably produced differently than the newer powders. However, this did not effect shelf life which could be at least 15 years.

TESTING PROGRAM

The program carried out at Teledyne Firth Sterling consisted of the following tests on the powder samples listed in Table 1.

1) Morphological examination of the Cobalt powders by scanning electron microscopy (SEM).

2) Determination of the relative amounts of α (hcp) and β (fcc) phases present in each Cobalt powder by x-ray diffraction (XRD).

3) Oxygen content determination for the Cobalt powders by weight loss measurement after reduction in H_2 at 1050°C for 60 minutes.

4) Evaluation of 93.75% Tungsten Carbide, 6.25% Cobalt cemented carbide pieces made from a single Tungsten Carbide lot and each of the Cobalt lots. The Tungsten Carbide lot used had an average particle size (FSSS) of 1.6 microns, a total Carbon content of 6.14% and a free Carbon content of 0.04%. All grade powders

Table 1. Cobalt Sample Listing

Sample	Lot	Date Received	Quantity in Can, Jan. '81	Comments
1a	4879	10-22-64	5 lbs.	From top of container
1b	4879	10-22-64		From bottom of container
2a	816	10-04-67	9 lbs.	From top of container
2b	816	10-04-67		From bottom of container
3	394	6-05-74	1 lb.	Mixed
4	623	12-14-77	1/2 lb.	Mixed
5	371	1-01-81	275 lbs.	From top of container

Table 2. Data for Cobalt Powders

Sample	Hcp(%) by XRD	H_2 Loss (%) at 1050°C
1a	2.8	1.68
1b	2.2	1.71
2a	3.1	1.92
2b	5.0	2.13
3	21.0	1.30
4	16.0	1.19
5	22.0	0.69

were fabricated by vibratory milling where the Carbon/Tungsten
Carbide ratio was adjusted to 6.11% via a Tungsten powder addition. Test pieces pressed at 10 Tsi were vacuum dewaxed and
sintered at 1415°C for 60 minutes.

RESULTS AND DISCUSSION

The results of the morphological study on the SEM are presented
in Figure 1. Here the structure of powder samples 1a, 2a, 3, 4 and
5 is shown. Samples 1b and 2b were identical in appearance to 1a
and 2a and hence omitted. This data clearly shows powders 1 and 2
have a finer structure than 3, 4 and 5. This fact suggests a lower
temperature was used for the decomposition of the Cobalt compound
used to make the powders 1 and 2.[1]

The relative proportions of α(hcp) and β (fcc) phases between
powders 1 and 2 and the other three are listed in Table 2. This data
also indicates different processing for the two groups. Significant
reductions in the amount of α phase were shown to occur following a
stabilization treatment of 10 minutes at 530°C on Cobalt powders
made from various Cobalt compounds.[1] Hence it seems probable that
powders 1 and 2 were made by decomposing a Cobalt compound at a
lower temperature than powders 3, 4 and 5 followed by a stabilization
treatment above 500°C.

The apparent differences in powder manufacture mentioned above
did not appear to effect the oxidation rate of the powders. The
Oxygen content for each powder, listed in Table 2, is primarily
related to the powder's age although there is the exception of powder
sample 2. This discrepancy could be the result of different storage
conditions for this powder. It should also be noted that there is
little difference in the Oxygen content of powder samples taken from
the top and bottom of the containers where powders 1 and 2 were
stored.

In Table 3 the results of the testing on the cemented carbide
parts made from the Cobalt powders are presented. Unlike the Oxygen
content values, these data do not show much variation which correlates
strictly with the age of the Cobalt. For example, there seems to be
a slight decrease in the hardness and coercive force with powder age
at first observation. However, a closer examination reveals a slight
rise in the density of the harder pieces as well. This fact suggests
that there is probably slightly less Cobalt in the harder pieces.
Similarly the slight decrease in average strength with the age of
Cobalt seems less significant when the 95% confidence levels are considered. Finally, the percentage of magnetic Cobalt (computed from
the ratio of magnetic saturation of one test piece to another from
the same powder lot which is saturated with Carbon) which should be
most affected by oxidation shows little variation dependent on Cobalt
powder age.

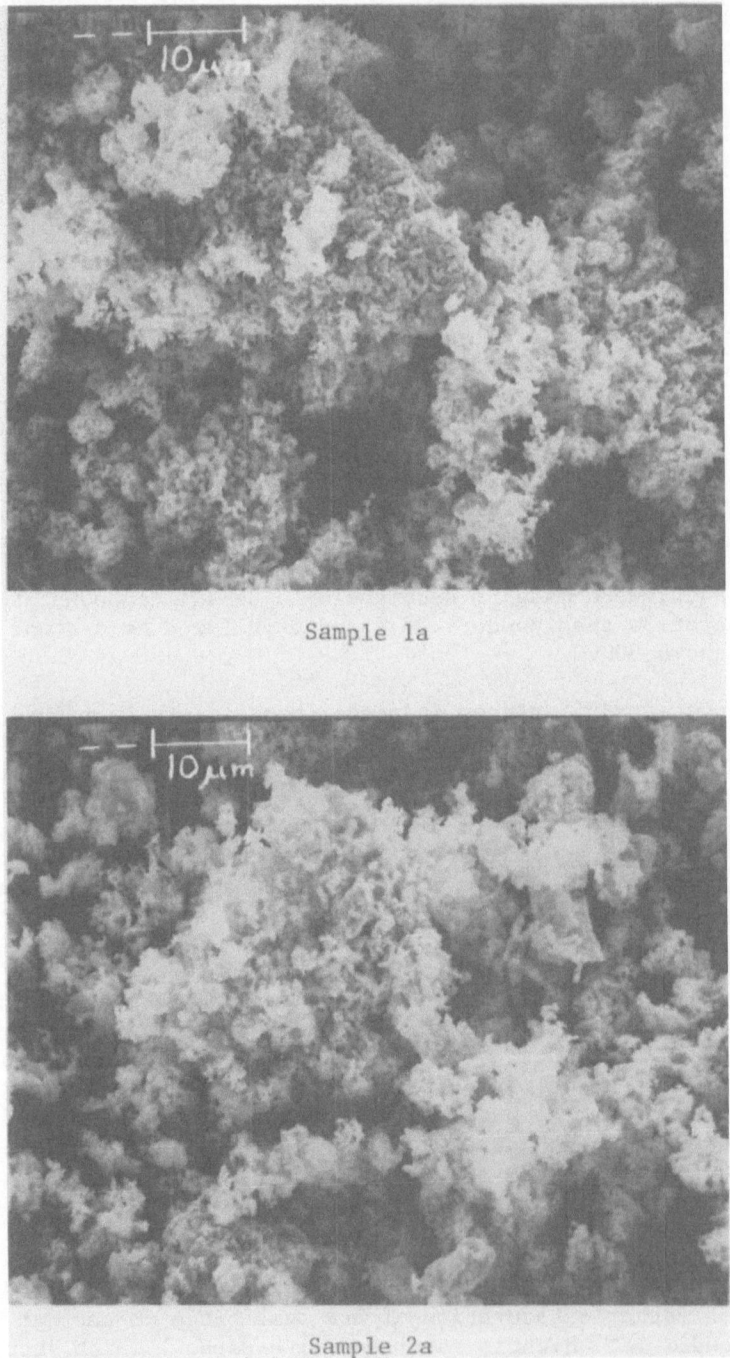

Sample 1a

Sample 2a

Figure 1.   SEM Micrograph of Cobalt Powder (1500X)

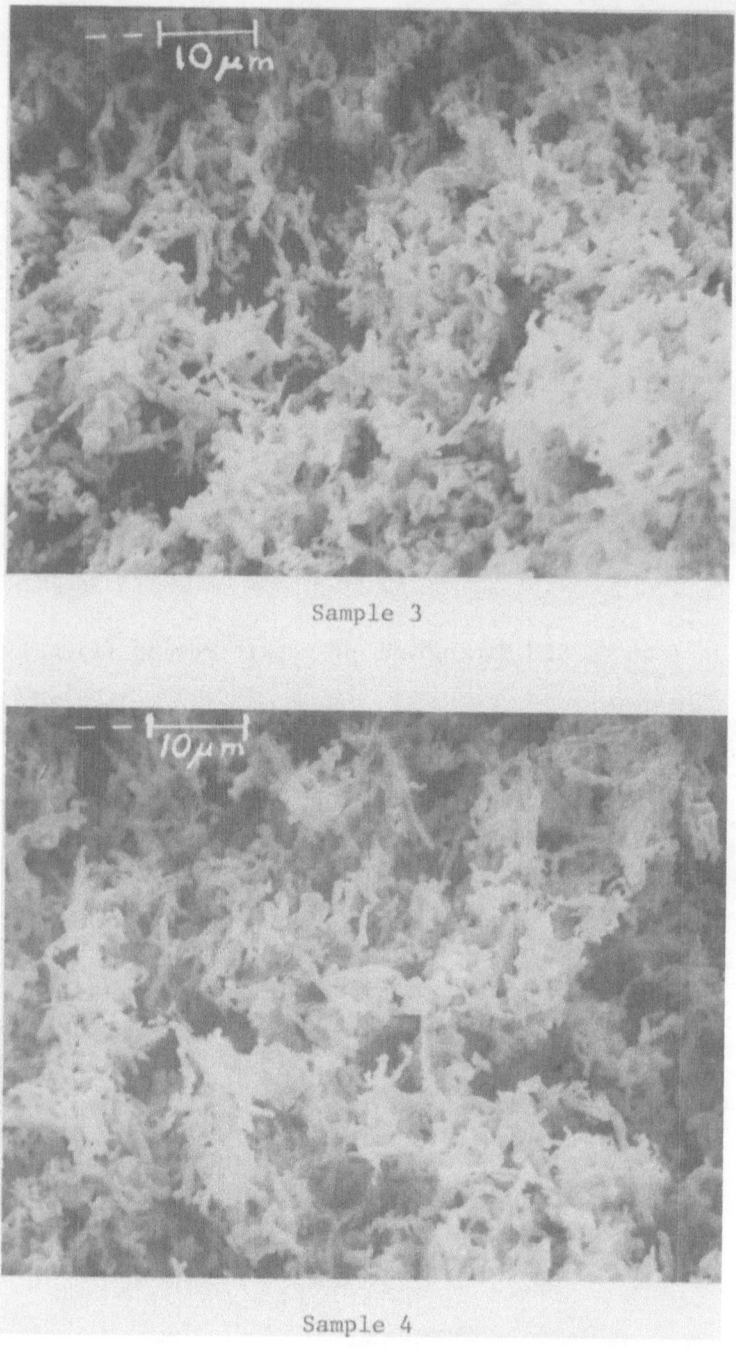

Figure 1. (con't) SEM Micrograph of Cobalt Powder (1500X)

Sample 5

Figure 1. (con't) SEM Micrograph of Cobalt Powder (1500X)

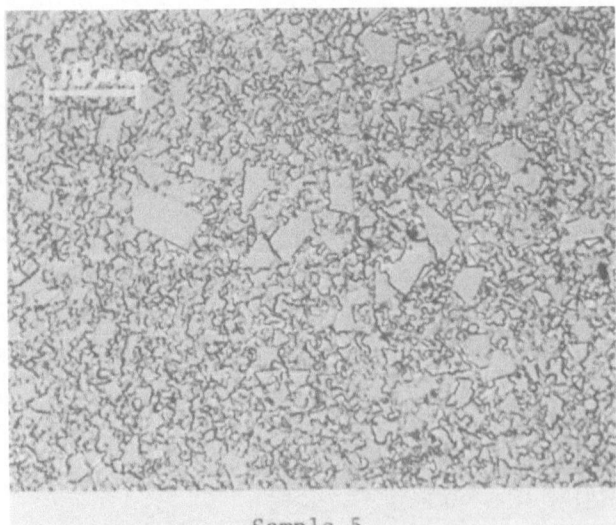

Sample 5

Figure 2. Microstructure of Cemented Carbide Parts Made from
           Various Cobalt Samples (1500X)

Sample 1a

Sample 1b

Figure 2. (con't) Microstructure of Cemented Carbide Parts
            Made from Various Cobalt Samples (1500X)

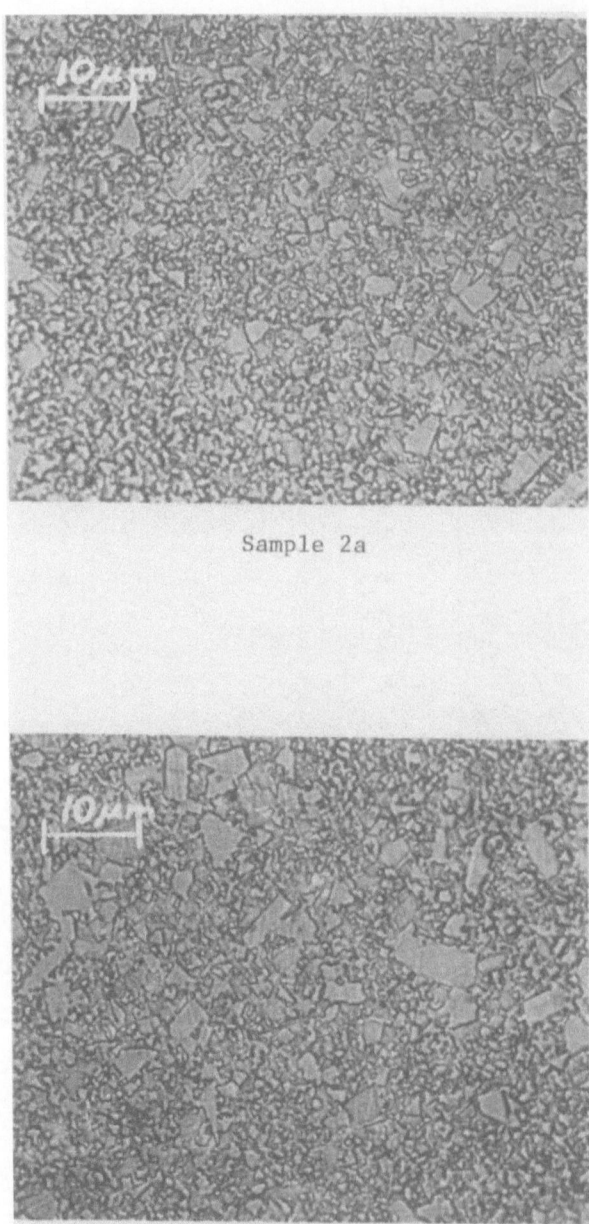

Sample 2a

Sample 2b

Figure 2. (con't) Microstructure of Cemented Carbide Parts
Made from Various Cobalt Samples (1500X)

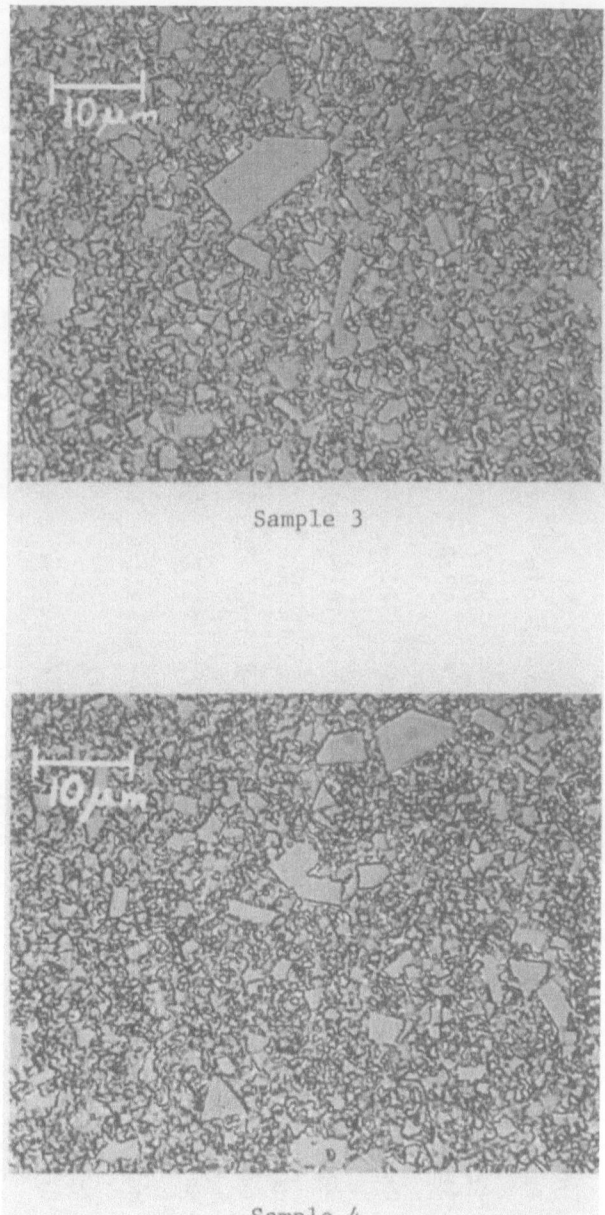

Sample 3

Sample 4

Figure 2. (con't) Microstructure of Cemented Carbide Parts
Made from Various Cobalt Samples (1500X)

Table 3. Properties of Cemented Carbide Pieces
Made From Various Cobalt Powders

		1a	1b	2a	2b	3	4	5
Density (g/cc)		14.91	14.94	14.91	14.95	14.94	14.95	14.95
Linear Shrinkage Factor at 10 Tsi		.816	.815	.816	.813	.818	.817	.820
Transverse Rupture Stress (Ksi)	$\bar{x}$	304.0	277.2	276.6	303.2	303.0	294.8	331.8
	s	18.5	29.8	18.0	17.2	43.7	26.0	16.8
Hardness (R_A)		91.1	91.3	91.3	91.4	91.5	91.5	91.5
Coercive Force (Oe)		187	189	191	192	196	198	194
Magnetic Saturation (cgs/g)		127.5	121.9	128.8	120.8	124.2	121.9	124.1
Magnetic Cobalt (%)		91.3	91.7	93.7	90.9	95.1	93.3	93.4
Macrodefects (40X)		None	None	lt. slits	None	med. por.	lt. slits	None
Apparent Porosity (200X) – ASTM B–276		1–0–0	1–0–0	2–1–0	1–0–0	2–0–0	1–0–0	1–0–0
% Cobalt from Physical Data		6.94	6.61	6.83	6.61	6.49	6.50	6.61

The microstructures corresponding to the samples in Table 3 are shown in Figure 2. All samples show some secondary recrystallization of the Tungsten Carbide. Again, however, there is no clear cut relationship to the age of the Cobalt powder.

SUMMARY AND CONCLUSIONS

The results of this study support the following conclusions.

1) Extra-fine Cobalt powder which has been stored for up to 15 years can be used directly from the container to produce good quality parts.

2) Powder samples 1 and 2 appear to have been produced differently from powders 3, 4 and 5.

3) The oxidation resistance of the powders was not significantly changed by any process changes.

4) Oxidation occurs to approximately the same extent for powder at the top and bottom of a container.

REFERENCES

1. M. H. Tikkanen, A. Taskinen, and P. Taskinen, Characteristic Properties of Cobalt Powder Suitable for Hard Metal Production, Powder Metallurgy 18, No. 36:259 (1975).

DISCUSSION FROM THE FLOOR

J. D. Knox, Vermont American Corporation, U.S.A.

The genesis of this paper should have some explanation as it is a replacement for papers whose authors unfortunately could not get here. The United States Government maintains a stockpile of strategic materials for use in an emergency. It has been suggested that extra-fine Cobalt powder be stockpiled. The responsible people have hesitated to consider the proposal seriously because of fears that there would be serious degradation over long storage times. Fortunately, Professor Gurland of Brown University saved Cobalt powder remaining from theses work over the years and has generously provided samples. Volunteer companies from among the members of the Cemented Carbide Producers Association, U.S.A., are testing these powders for oxidation, agglomeration, ratio of hcp to fcc and suitability for making cemented carbide. Jim Oakes' work was completed just two weeks ago and, fortunately, he is here.

E. R. Kimmel, GTE Products, U.S.A.

The hcp structure is the stable room temperature form of Cobalt.
Normally the extra-fine cobalt powder as produced will be predomi
nantly fcc. It has been suggested that the fcc converts to the hcp
on long term storage and that this change may be of significance in
the production of cemented carbides. Our experience in producing
fine Cobalt powders is that the ratio of hcp to fcc is almost entirely
dependent on the reduction temperatures. Our x-ray diffraction measurements on the Brown samples are in good agreement with those reported by Oakes and do not support the idea that there is a conversion
to hcp with long time storage. Even if there was the suggested conversion, I am not sure of its significance. The milling of the WC-
Co powders is reported in the older literature to convert the fcc
powder to hcp given sufficient milling time. A substantial part of
the conversion takes place within time spans common to production.
In fact, one of the theories advanced to explain the unique virtue
of Cobalt for making cemented carbide is precisely this: As it converts to the hcp it becomes more friable, breaks down to smaller
particles and is more uniformly distributed.

T. W. Penrice, Teledyne Firth Sterling, U.S.A.

The oxygen levels found in the samples are minimal considering
the fineness of the Cobalt powder, the time of storage, and the
conditions of storage. It suggests that Metallurgie Hoboken-Overpelt
(the producer) has in some way passivated the powder surfaces. A
possible procedure might be to cool the Cobalt after reduction in an
atmosphere of Carbon Dioxide. An absorbed CO_2 layer could inhibit
the rate of oxidation. Unfortunately, the MHO process is proprietary,
so we can only speculate.

E. Lardner, Wimet Limited, U.K.

All producers of cemented carbides have as part of their final
densification furnacing a dewaxing cycle which normally is, or can
easily be modified to take place in the presence of hydrogen. This
cycle is completed by the time 450° C is reached. Coincidently, it
is the time and temperature required to hydrogen reduce any Cobalt
oxide without adversely affecting the rest of the system. Therefore,
reasonable degrees of oxidation of the Cobalt powder can be accommodated with little or no adverse economic or metallurgical effects.

Haskell Sheinberg, Los Alamos Scientific Laboratory, U.S.A.

With long term storage there was a slight shift in the particle
size distribution (determined by sedimentation) of our two samples,

indicating formation of loose agglomerates which would doubtless
break up with conventional blending and milling procedures. There
was a discernable shift in the proportion of fcc/hcp, but effects on
processing product properties would likely be imperceptible.   I
would agree with Lardner that a slight increase in oxidation on long
term storage would not necessarily adversely affect product properties.

...

THE INFLUENCE OF THERMO ELECTRIC CURRENT ON THE WEAR OF TUNGSTEN

CARBIDE TOOLS

L.J. Bredell

Physics Department
University of Pretoria
Pretoria, South Africa

INTRODUCTION

Apart from the obvious benefits from machining studies, it is also used as a technique to study material properties, reactions between clean solid state surfaces and to gain information concerning the relative bonding strengths between surface atoms[1,2,3]. One of the interesting phenomena mentioned in the literature of the last two decades is that of the effect of the thermo electric current resulting from the EMF generated between tungsten carbide tool and workpiece, on the wear-rate of the tool. The literature survey can briefly be summarized as follows: Reduction of wear can be obtained by insulation and/or current compensation with an external current source[4-8]. No satisfactory explanation has yet appeared although Opitz's[4] electron microscopic analyses suggested that the thermo electric current aided the transfer of carbon atoms in the direction "hard metal to steel". A few workers claimed that thermo electric effects did not play a role in tool wear[9].

Experimental

Machining experiments were done in a standard lathe using disposable tungsten carbide tips on mild steel. Fig. 1 shows the tool and billet schematically. The flank wear on the tool tip is shown in Fig. 2 and it has been shown that the quantity a b (i.e. the length times the width of the worn area) plotted as a function of the total area A machined at constant cutting speed, yields a straight line. The slope of this line is taken as a measure of the wear-rate[3].

723

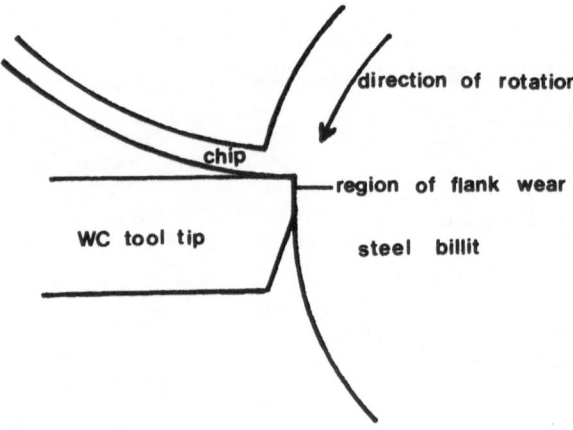

Fig. 1. Schematic representation of tool and billet.

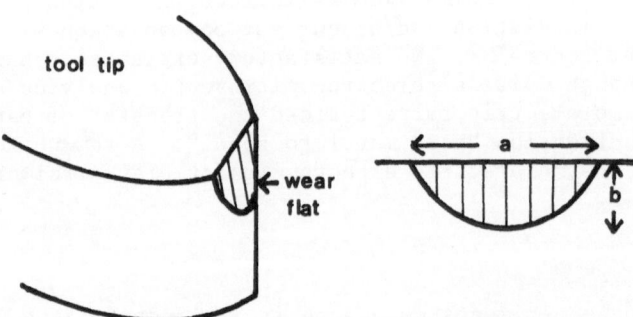

Fig. 2. Flank wear on tool tip and measurement of worn area.

The complete experimental set-up is shown in Fig. 3. Both tool and
workpiece were insulated from the lathe and the external electric
circuit was connected via a special slip ring system with gold
plated contacts, thus eliminating the uncertainty of lathe resistance.
By means of the external current source the effect of electric current on tool wear could be investigated in both positive and negative directions; the positive direction being taken as the direction of current flow caused by the thermo electric EMF generated
during machining. The polarities are marked in Fig. 3.

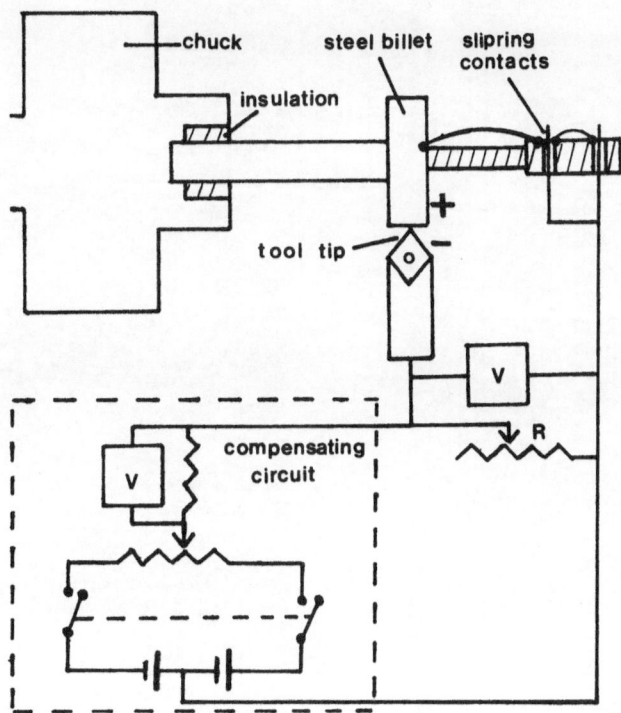

Fig. 3.   Apparatus to investigate thermo electric effects during
          a machining process.

Results

    In plates I and II the flank wear platterns after the indica-
ted number of passes are compared in the insulated and short circuited conditions respectively.

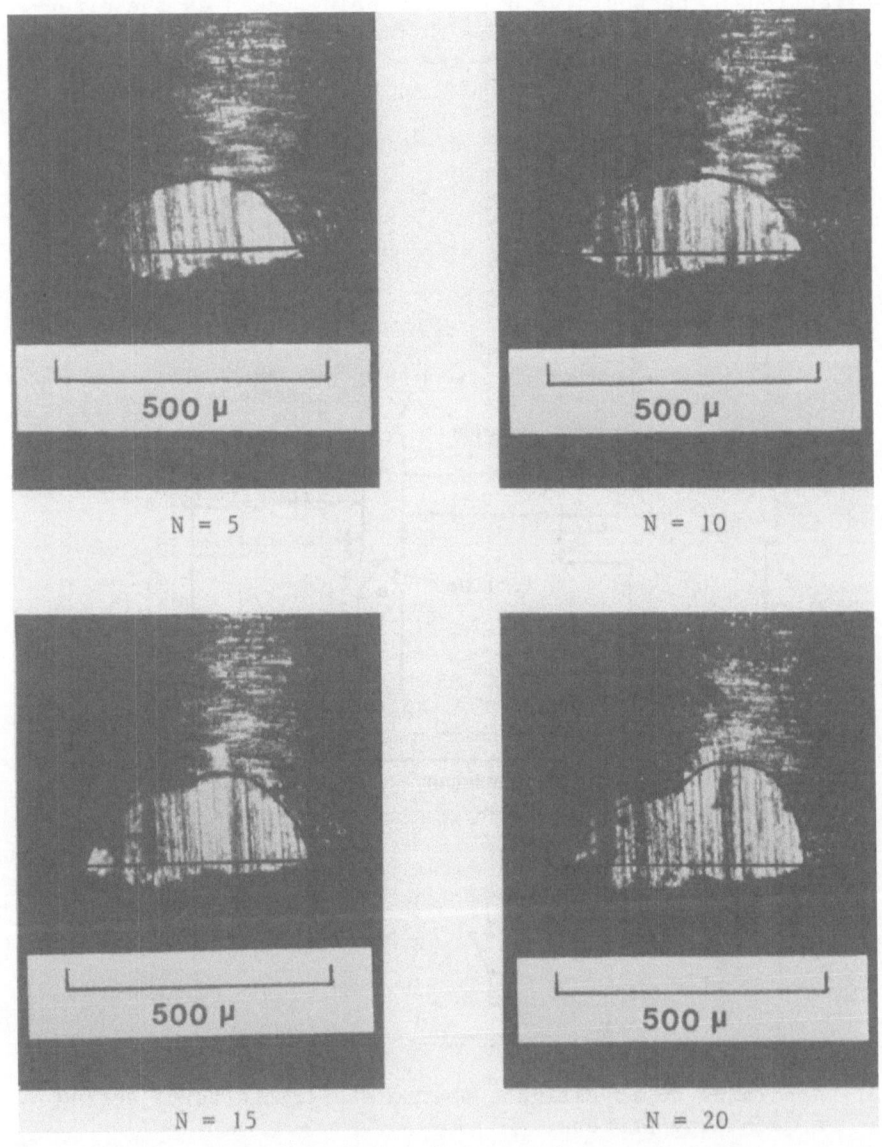

Plate I. Flank wear patterns after the indicated number of passes
(N) over the billet. (Insulated)

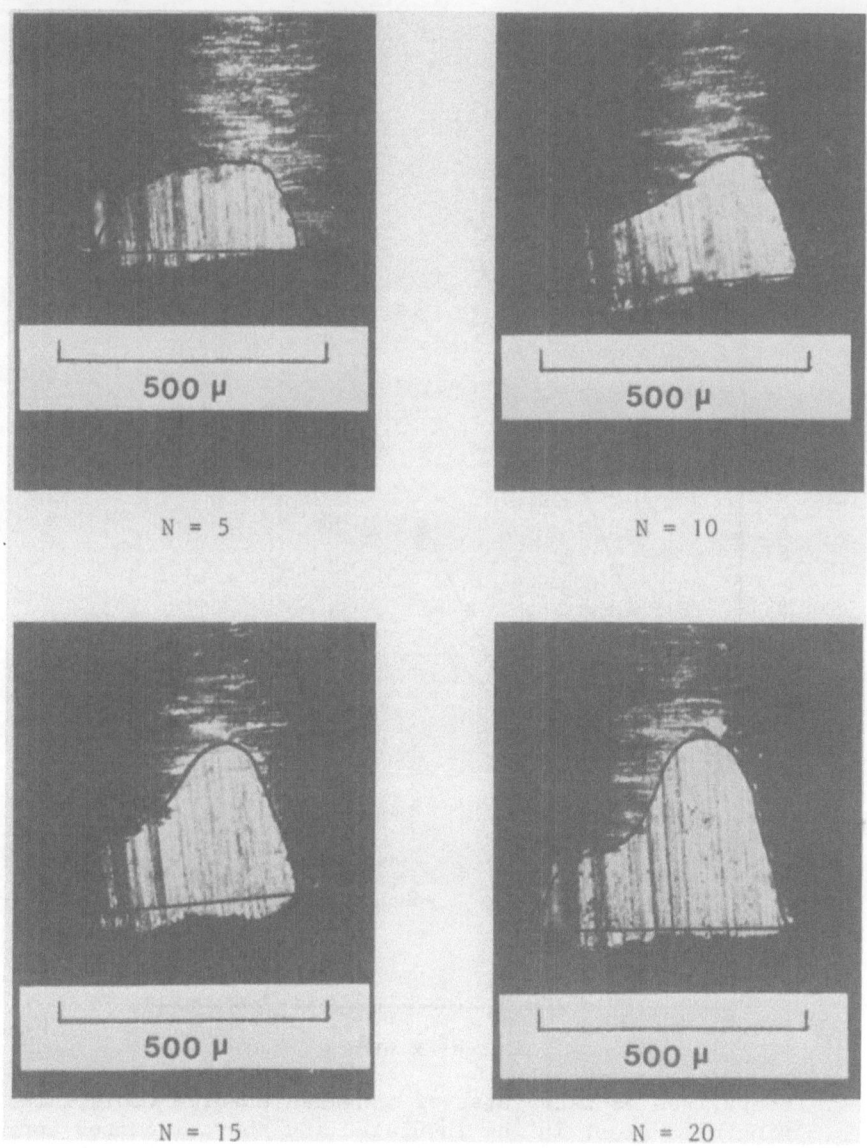

Plate II. Flank wear patterns after the indicated number of passes
(N) over the billet (Shortcircuit).

Continuation of the experiment in the short circuited state with
a tool tip which was initially insulated resulted in an abrupt
change in slope. (Fig. 4 represents the results graphically).

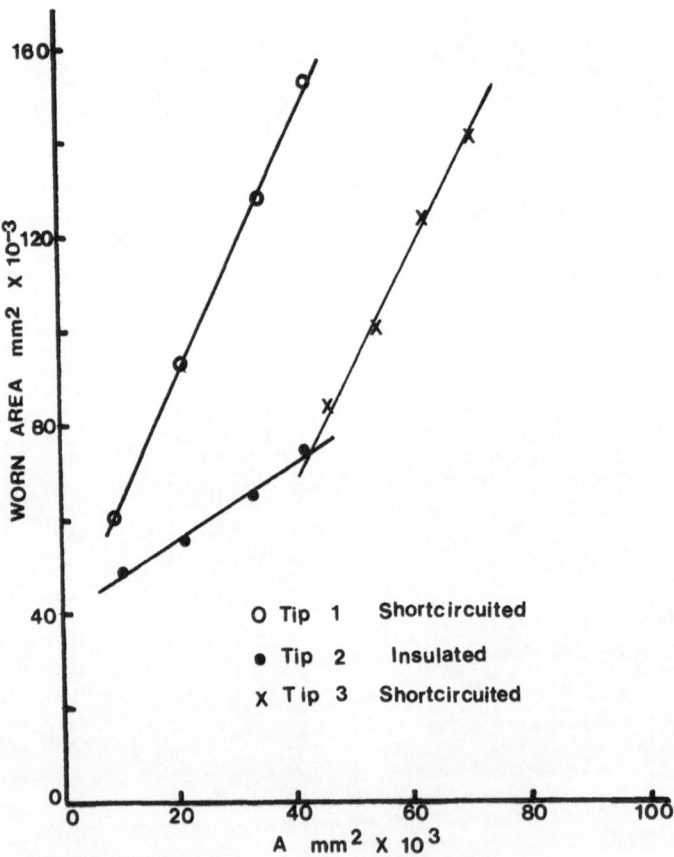

Fig. 4.   Comparison of wear rates of tungsten carbide tools turning mild steel in the insulated and shortcircuited conditions. Area machined (A) = N × average diameter × π ×
width of billet.

The wear rate was also measured as a function of thermo electric current by varying the external resistance. By using the external current source, wear rates were also obtained in the current
range −30 mA to +30 mA. Fig. 5 shows the wear rate as a function of
current. The electrical behaviour of the thermo electric circuit
is shown in Figs. 6(a) and 6(b) which indicate that it reacts
similarly to a conventional E M F source being short circuited.

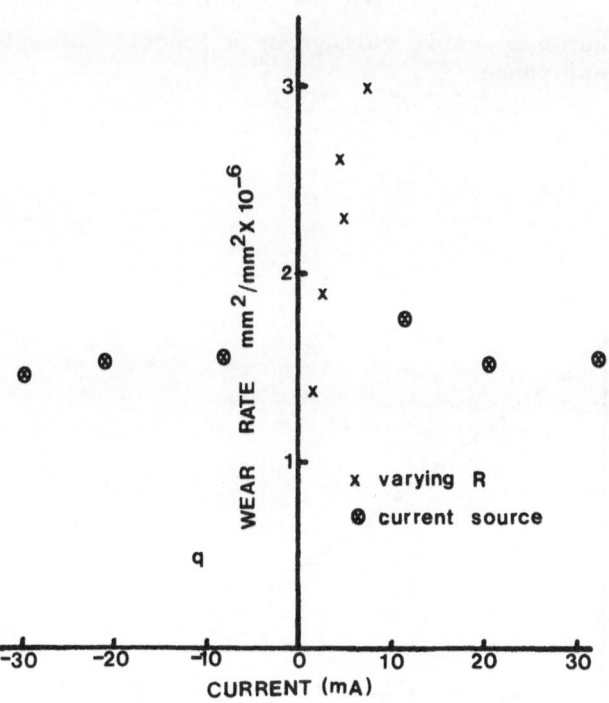

Fig. 5.   Wear rate versus current relationships obtained by
          varying the external resistance (R) or by using the
          external current source.

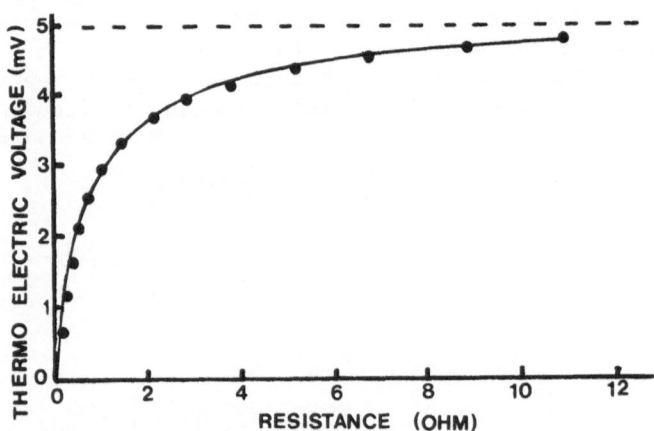

Fig. 6(a). Thermo electric voltage as a function of external
resistance (R).

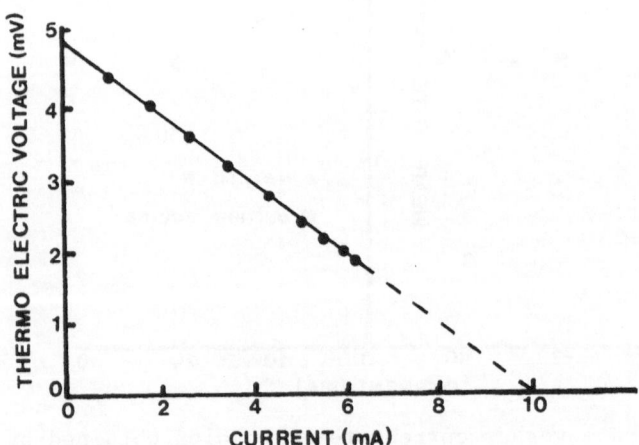

Fig. 6(b). Thermo electric voltage versus current obtained by
decreasing R.

Discussion of results

Wear mechanisms

    Any interpretation of results must neccessarily be done in terms of the known mechanisms of tool wear, namely:

(a) Adhesive and diffusion wear
(b) Abrasive wear
(c) Corrosive wear
(d) Fatigue wear.

    Of these four processes mainly a combination of the first two is likely to be responsible for flank wear on the tool, while a voltage and/or current effect could only possibly influence adhesive and diffusion wear.

Insulated and short circuited conditions

    In Table 1 the wear rates under these conditions are compared and analyzed statistically. It is clear that the variation as given by the standard deviation can be large enough to yield a factor between 1 and 3 for the ratio wear rate (shorted)/wear rate (insulated).

Table 1.   Comparison of wear rates in the insulated and short-circuited conditions.

Wear rate (insulated), $mm^2/mm^2 \times 10^{-6}$	Wear rate (shortcircuited) $mm^2/mm^2 \times 10^{-6}$
1,01	2,24
1,69	2,90
1,16	2,44
1,22	2.53
1,52	2,83
1,54	2,40
0,53	1,39
1,34	1,25
—	2,29
—	1,81

Average ± standard deviation	Average ± standard deviation
1,25 ± 0,37	2,21 ±0,56

Wear rate shortcircuited/Wear rate insulated

$$1,77 \begin{array}{c} + 1,38 \\ - 0,75 \end{array}$$

The reason for such a large variation is not quite clear but probably correlates with the different wear behaviours of individual tool tips.

The wear rate versus current behaviour

Current compensation in both positive and negative directions yielded intermediate wear rate values. This excluded a mechanism where the current contributed to mutual diffusion of ions between tool and workpiece because if this was the case, an increase in wear rate would have resulted from increasing positive current and a decrease from large negative current values.

If there was any heating effect due to the current, a symmetric function of wear rate versus current around the origin would have been expected. It can also be shown that the Peltier effect is negligible compared to the heat generated during machining.

Speculative explanation

It is difficult to conceive other mechanisms by which the observed data can be explained but one possibility may be the strong electric field that is bound to exist in the contact zone due to the difference in contact potentials between tool and workpiece materials. This could possibly govern the diffusion wear process as a consequence of the charge distribution at the interface being modified by the thermo electric current. This idea is still in the speculative stage and clearly more experimental and theoretical work is needed to support it.

Acknowledgements

The work discussed in this paper was done during the author's sabatical leave at the Clarendon Laboratory in Oxford.

References

1. J.F. Prins, Microcutting: A method to study surface plasticity at very high strain rates," Surface effects in crystal plasticity", Noordhof, Leyden (1977).
2. L.J. Bredell and J.F. Prins, Microcutting of steel using pyramidal diamonds with different apex angles, (To be published).
3. A.G. Thornton and J. Wilks, Tool wear and solid state reactions during machining, Wear 53 : 165 (1979).
4. H. Opitz, About wear on cutting tools, Proc. Conf. Lubrication and Wear, Inst. Mech. Eng. 664, Oct. (1957).
5. V.A. Bobrovskii, Extending tool life by breaking the thermoelectromotive force circuit, Russian Eng. J. 46 : 70 (1966).
6. J. Ellis and G. Barrow, Tool wear in metal cutting and its relationship with the thermoelectric circuit, Ann C.I.R.P. XVII : 39 (1969).

7. H.S. Shan and P.C. Pandey, Wear of cutting tools: Thermo-electric effects, Wear 32 : 167 (1975).

8. T. Hehenkamp, Untersuchungen über den elektrisch kompensierbaren Verschleiss an Drehmeisseln aus Hartmetallen, Archiv Eisenhueten. 29 : 249 (1958).

9. G. Engstrand, "Influence of thermoelectric currents on cutting tool wear", Kungl. Tekniska Hogskolans Handlinger, Sweden (1959).

DISCUSSION

W. Williams:

I'm relieved to hear you present evidence that the effect is probably not electro-migration of carbon in and out of the tool into the work piece. That was an idea some years ago. But from experiments we did on electro-migration of carbon in titanium carbide in 1969 I would guess that it would be impossible to make such an explanation work here. Because that effect is very subtle, very small, requires hundreds of amperes per square centimeter of current, and occurs only at very high temperatures. A slight biasing of the diffusion of carbon in titanium carbide in the presence of an electric current is measurable, but just barely. When you started the talk, I was anxious, but I'm relieved to hear you say that you think that is not what is going on. It is, as you suggest, very difficult to think of another plausible mechanism. Is it absolutely clear that there's no rather simple effect such as the work piece itself changing temperture as a result of what apparently was a rather heavy piece of electrical insulation between it and the lathe which might also have provided a thermal barrier allowing the work piece to get a bit warmer, which in turn would affect the tool wear? Is it clear that that is not going on?

L. J. Bredell:

We have been using very small depths of cut something like 0.1mm and also the diameter of the work piece itself was normally constant during these experiments. Also the cutting speed was kept constant.

G. Dearnaley:

Dr. Williams prompted me to suggest an alternative mechanism which
may have some connection with what I was talking about yesterday;
namely, the role of tungsten vacancies on the surface of tungsten
carbide created in our case by the injection of a lot of anions.
Could this possibly be the migration under your applied current, of
tungsten creating tungsten vacancies thereby allowing drift, or if
you prefer, an electro-migration process, which would serve the same
function, perhaps briefly, to allow cobalt migration over the surface? It seems to me there might be some connection with our model,
which I presented yesterday morning.

L. J. Bredell:

Yes. That was very interesting. This might well be something to
think about.

F. Rymas:

I presume that the tools that you used in these tests were uncoated.
My question is, have you tried these tests with CVD coated inserts?

L. J. Bredell:

These tools were uncoated. And I plan to do some with coated tools.

No Name Given:

I don't know whether in the data that you showed the inserts corresponded to the short-circuited and the unshort-circuited condition.
But if you did, I noticed that you had one or two reversals there,
where in fact you were getting the opposite effect. The considerable
variation that you get, I believe, is due to surface chemistry. The
actual EMF that you produce at that junction is dependent upon surface chemistry. For instance, in production sintering conditions,
it is not unusual to have a cobalt enrichment at the surface, and
this will not only affect the thermal EMF that you get at the surface
which would be quite different from the chemistry that you recorded
for the bulk material, but it also very much affects the wear characteristics. I wonder if we're not looking here at a simple relationship between actual composition at the surface and actual wear
of that composition. I think this method is very excellent for measuring interface temperatures, but certainly if you take a number
of inserts, you do get tremendous variations in that thermo EMF and
we have always thought that this was the explanation but it certainly
also affects the wear.

FRACTURE OF CARBIDE TOOLS IN INTERMITTENT CUTTING

H. Chandrasekaran

Swedish Institute for Metals Research
S-114 28 Stockholm, Sweden

INTRODUCTION

Resistance to wear and requisite strength against plastic deformation and fracture are the two fundamental requirements of metal
cutting tools. The progressive development of the cemented carbide
tools which has been quite impressive in the aspect of wear resistance in particular, still leaves their fracture behaviour in
intermittent cutting to possible improvement. Thus for example while
the current coated carbides have increased the tool life appreciably
their transverse reupture strength has reduced. Modern tool carbides
are far more homogeneous and free from porosity but then their
service loads have also increased. Moreover, there is a great need
for enhancing not only the available strength further, but acquire
the ability to predict tool fracture during machining. The situation
is acute specifically in the area of modern automated machining
systems. It is lack of this capability which has led to the choice
of high speed steels, with their relatively poor but predictable
high temperature performance both in terms of tool wear and fracture
in a number of multipoint intermittent cutting situations. Keeping
this in mind it is proposed to consider in the present article the
mechanism of tool failure with particular reference to fracture in
intermittent cutting so as to identify its poor predictability. It
is then possible to attempt effective tool carbide evaluation
procedures immediatley relevant to the estimation of their fracture
behaviour in machining. Such procedures while retaining the main
features of true simulation of the technological conditions
associated with a process like intermittent cutting, on the other
hand should be based upon established fundamental properties of

735

these carbide materials. Efficient exploitation of current tool
carbides and possible methods of improving their performance are
then conceivable.

BACKGROUND TO THE PROBLEM

     The intermittent cutting process may be considered as a mate-
rial testing operation in which the tool material is subjected to
a cyclic mechanical and thermal loading. In the course of the process however, unlike material testing operation, even in the absence
of tool fracture, tool wear is still present. This in turn modifies
the tool geometry (1)*, which affects the cyclic load and temperature
and consequent stress state. Moreover, the level of contact loads
(typically an average level of 2500 MPa in steel machining), high
temperatures ($850-1100^\circ C$) and the intermittent exposure to atmosphere and oxidation progressively results in structural changes of
the tool. Hence the prediction of fracture at any of these stages
becomes very difficult. On the same count it implies contradictory
constrain when the metallurgical design of such cemented carbide
tools having effective high and low temperature strength and hardness are attempted.

THE NATURE OF TOOL FRACTURE

     At this stage it is important to identify two aspects of tool
fracture.
1.   The Exogenous or Process controlled tool fracture.
2.   Indegenous or tool material controlled fracture.
The former is dependent upon the kinematics and geometry of cutting
operation (continous or intermittent, single point or multipoint)
and the mode of chip formation. Thus the often observed 'exit'
and 'entry' fracture of tool (2,3,4) and more common occurrance of
fracture of tool tip during up milling than down milling are
examples of this aspect. Similarly chip adhesion to tool surface
and consequent tool fracture in subsequent cycle is also typical
of this category (5).

     The other possible causes are the influence of the machine-
tool system stiffness and associated process damping leading to
unfavourable stress state and tool fracture (6). In other words the
ultimate influence of all the process controlled parameters is
primarily to lead to an unfavourable stress state in the tool. It
must however be kept in mind that the actual stresses present at
any instant in the tool will be a vector sum of a number of
premachining and machining stresses. These aspects are discussed
in detail by Chandrasekaran (7).

*Numbers in parentheses correspond to references at the end of
 the paper.

The indegeneous or the material controlled sources leading to tool fracture are related to the fundamental fracture behaviour of cemented carbides and its thermo-mechanical state.

It is relevant to note here that fracture and fracture-strength of the WC-Co alloys are highly variable parameters in view of the strong contrary properties of the carbide and cobalt bond phases. Even such a simple basic material parameter as compressive strength is a complex entity. According to Leuth and Hale (8), the conventionally reported compressive strength (neglecting inhomogenties) of about 4200 MPa for the WC-6Co is actually the value at which due to end effects the material fails by shear. Hence the true compressive strength when the end effect is eliminated is around 5600MPa, an increase of about 30%. The situation is probably the same in other modes of testing. Thus for example in bending, the TRS value is directly controlled by inhomogenities. According to Suzuki and Hayashi (9) the extrapolated defect free bending strength of WC-10Co alloy is about 8300 MPa, whereas the conventional TRS can be around 2500-3000 MPa.

Associated with this indefiniteness of the material behaviour we have the indefinitieness of the stress-state caused by cutting. The fluctuations of the later, often has a cumulative effect and the fracture behaviour is affected. It is this truly dynamic nature of fracture of the cemented carbides which makes its prediction difficult in metal cutting.

The fracture behaviour of WC-Co cemented carbide under standard testing conditions has been widely investigated in a number of investigations. General reviews of mechanical properties of carbides (10) as well as more specific ones related to cutting tools are available in literature (11). While no fully valid model taking into account the different metallurgical and structural aspects for a wide variety of loading conditions and temperatures is available, there are specific strength theories such as those of Kreimer (12) and Gurland (13) taking some of these factors into account. Since fracture initiation is the critical pre-requisite to tool fracture, it is worth looking into the information concerning this aspect during standard strength testing. Among other things we now know that fracture initiation in cemented carbides invariably begins at a flaw, abnormal or big carbide grain, binder pool or inclusion (14,15). In an intermittent cutting operation initiation at such sites could be from stress-state (over-load)(2) or deformation induced micropores (16), mechanical and or thermal fatigue (17). Structure change induced embrittlement can also facilitate easy fracture initiation.

Investigations devoted to deformation (16),fracture (5)and structural
changes in cutting tools (18) have indicated that in all probability
all the above causes act simultaneously in varying degrees at
different stages of the process. Consequently, when we try to base
the cause of tool fracture to a specific source we have poor agreement from cutting tests even when the more regular tool material
like high speed steel is used (19). On the other hand continuum
mechanics limit modelling of tool fracture from pure stress state
(mechanical only) controlled situation appears to be successful (20)
in the simplest plane strain form using Mohr's fracture criterion
as well as when using more sophisticated fracture criterion (3).
This is partly because it is comparitively easy to conduct limit
tests of the required type during cutting. Further, the situation
is still within the lower limit of the conventional continuum model
as applicable to cutting tools. If further refinements in results
are required (which is the case if we have to explain the inprocess
dynamic fracture of carbide tools), the first step would be to use
the modified continuum model of the type proposed by Drucker (21)
relevant to a particulate material like cemented carbide. Similar
logic is also valid when the macro plastic deformation of the
tool-tip is of interest, but the problems inmodelling this could
be far more serious if one has to explain say deformation controlled micro-pores.

Hence it is apparent from the aforesaid considerations that at
present it is rather premature to devlop a fracture mechanics
model for dynamic tool fracture, and especially the crack or discontinuity initiation.

Naturally as an alternative certain investigators have tried
to develop special non-standard testing methods for evaluating the
tool carbides for fracture (22) and wear (23). While these appear
to be successful in empirically classifying cutting tools, it is
doubtful whether these can predict or enable in understanding the
dynamic fracture behaviour of cutting tools.

The essential pre-requisites for the dynamic fracture prediction in cutting are a clear understanding of the thermo-mechanical state of the tool during an operation like intermittent
cutting and the fatigue and fracture behaviour of these carbides
under such thermo-mechanical state.

THERMO-MECHANICAL STATE AND FRACTURE MODELLING

Taking into account the highlights of the tool fracture phenomenon, let us consider the mechanical stresses associated with
intermittent cutting in a cemented carbide cutting tool.

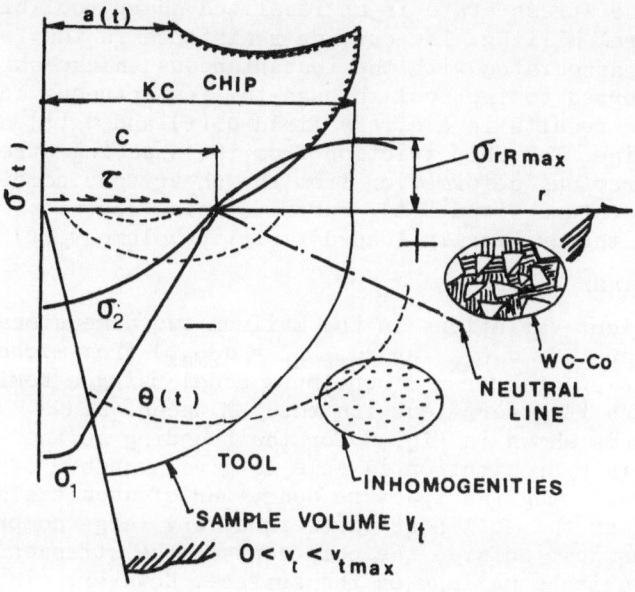

Fig. 1 The thermo-mechanical state of an intermittent cutting tool.

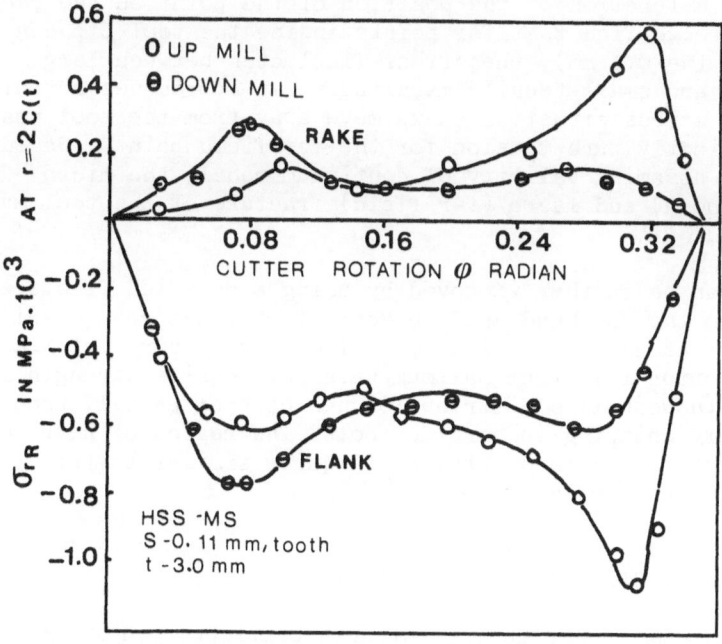

Fig. 2 Typical transient free boundary stresses σ_{rR} in a milling tool during the cutting cycle.

Fig. 1 shows the stress-state in an idealized sharp tool bit exe-
cuting peripheral milling. The cutting resistance in the form of
cutting forces associated with the instantaneous thickness of cut
a(t) is transferred to the tool through the instantneous contact
length C(t) and results in a stress field $\sigma_1(t)$ and $\sigma_2(t)$ separated
by a neutral line. The chip friction from the shearing stresses
$\tau(t)$ and the previous deformation from the shear zone combine to
produce a temperature field $\theta(t)$. Further in the course of each
milling cycle, the critically loaded tool-tip volume $\nu_T(t)$ varies
from zero to ν_{Tmax}.

   The transient variations in the maximum two dimensional elas-
tic stresses ($\sigma_{RMax} = \sigma_{1Max}$ and $\sigma_{rFMax} = \sigma_{2Max}$) from mechanical
loading only calculated from a continuum model using a semi-ana-
lytical approach (4) verified with the FEM technique (24) and· simulation model (25) is shown in Fig. 2 for the bounding surfaces of the
tool wedge. This approximation is able to give a number of useful
information concerning the fracture behaviour of such tools. Thus
we know that near the tool nose there is a very large compressive
stress, whereas just outside the chip contact the stresses are
tensile, displaying a maximum on the surface. However, since the
process is a dynamic one the tensile and compressive regions separated by a neutral line undergo cyclic changes. The implications
of this can be readily seen in Fig. 3 showing the transient σ_{rR}
stresses as a function of the position of the point on the rake
surface. Here we find that for points inside the tool-tip contact region ($r = 0.1$ mm), the stress fluctuates between large
compressive and small tensile magnitudes. It approaches a truly
alternating stress situation as we move away from the tool nose
and is distinctly pure tension for the far field points. Based
on this and assuming validity of continuum model, the micro
(near tool nose) and macro (far field) fracture of the tool may
be predicted (26).

   This can be further improved by using a modified fracture
criterion on the one hand and the concept of fracture probability
on the other as has been tried by Usai et al(3). Here triaxial
stress criterion involving maximum uniaxial tensile strength su-
proposed with Weibull's distribution has been effectivly used to
predict entry chipping in carbide tools. The region of high
fracture probability thus obtained is quite similar to those
predicted by the simpler approach (26).

MODIFICATION TO THE ELASTIC FRACTURE MODEL

   The important shortcomings of such continuum model when applied to actual carbide tools for explaining fracture in terms of
elastic stresses are the following.

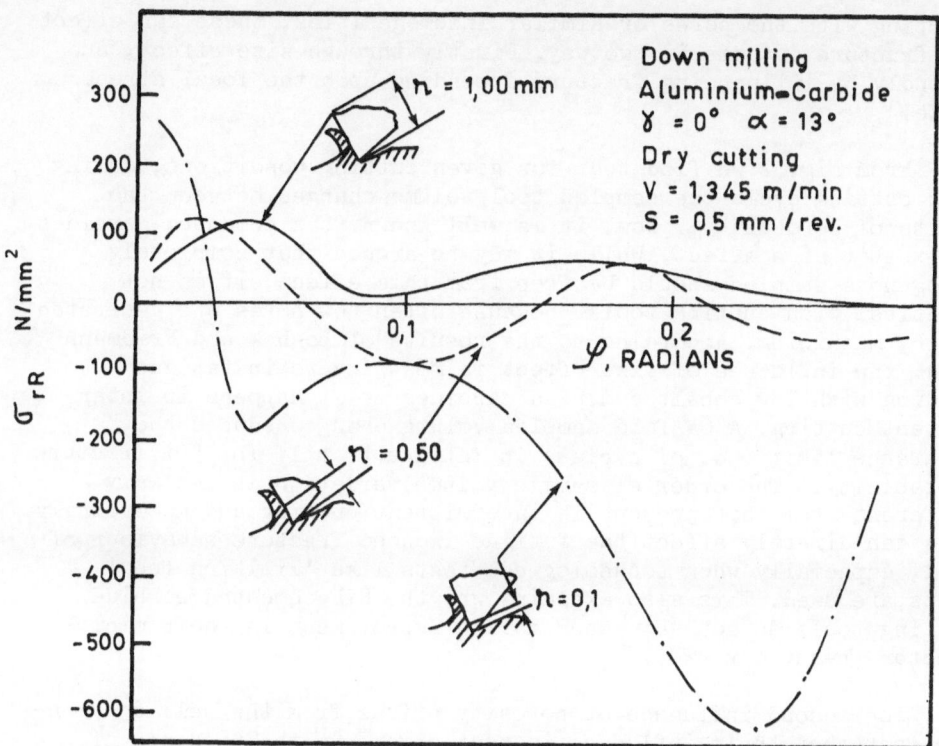

Fig. 3 Cyclic variation of the rake radial stresses as a function
        of cutter rotation and position of the point of interest.

1. The influence of inhomogenities - the discontinuities in the
   sample have not been fully taken into account.
2. The influence of large compressive stresses at the tool nose
   can lead to different consequences from those considered in the
   elastic fracture model. (Figs. 2 and 3).
3. Due to machining condition, the compressive stress loading rate
   are phenomenal and if rate influence is present then this will
   affect fracture phenomenon. Even if this is dormant at room
   temperature, at high tempratures observed during machining,
   this effect could appreciably affect fracture.
4. From pratical observations we do find plastic deformation of
   the tool during machining and accordingly plastic strain induced
   metallurgical changes and residual stresses can be expected
   to interfere with its fracture behaviour.

INFLUENCE OF INHOMOGENITIES

    Let us now consider the quantitative role of the above men-
tioned influences on the fracture behaviour of carbide tools.

Starting with the pores or similar inhomogenities, these can affect the fracture outcome in two way. Firstly through size effect, and secondly by influencing fracture depending upon the local stress state.

From Fig. 1 we find that for given cutting conditions, during each cutting cycle the sampled tool volume changes between the limits of zero and v_m. Now, it is well known that cemented carbides do exhibit size effect. While it may be argued that completely homogenous samples should be free from this effect, it is not practical with cutting tools, because often new pores are generated during machining. According to the results of Loshak and Fridmann (27), the influence of size effect is most severe in fatigue loading with low cobalt cemented carbides as it happens in inter- mittent cutting. A 64 fold sampled volume change reduces the endurance limit (no. of cycles) in fatigue to half for 50% fracture probability. The order of sample volume variation is not very different from that present in intermittent cutting and accordingly this can directly affect the fatigue induced fracture behaviour of tool, especially when technological tests like 'limiting feed' tests are used. This also explains why the HIP cemented carbide cutting tools do not show such marked improvement in their micro- fracture behaviour (28).

The second influence of porosity arises from the well documen- ted fact that their influence is most severe in the presence of tensile stresses in the fatigue mode. Accordingly as reported by Troshenko et al (29) and Romanova et al (30) in their studies devoted to the role of porosity in classical bending fatigue of WC-Co alloys, the fall in endurance limit (no. of cycles to fracture with 50% probability at a given stress level) due to porosity (0.1% volume) could be one order of magnitude or more, especially with large pores (>10 μm) and low cobalt content. The influence of such pores on other strength parameters like TRS and impact strength is less severe (15% to 30% fall). While the level of total porosity reported in a cutting tool by Jonsson (16) is comparable to this, the method used by Romanova et al (30) for inducing the pores could have led to enhanced tensile stresses in them on the basis of the work of Luyckx (14) devoted to the role of inclusions in the fracture of WC-Co alloys.

However, the total implication of the above considerations in the light of the observed transient stresses (Fig.3) is that the role of inhomogenities will not only depend upon their number and size, but in their location with respect to the local stress. This is also amply proved from the TRS test results of Suzuki and Hayashi (9) correlating the local stresses with pore sizes. Hence the presence of pores in the flank surface or chip exit point on rake will affect tool fracture behaviour differently. Thus the use of single compressive fatigue data to understand pore induced fracture initiation in intermittent cutting appears to be far fetched. Moreover the high temperatures present and the high strain loading can further affect the role of pores.

INFLUENCE OF COMPRESSIVE LOADING AND RATE SENSITIVITY

Consideration of the severe tool-tip compressive loading at room and high temperature conditions will require an elastic plastic continuum model, while flow and strain hardening aspects may extend it to visco-plastic range. In order to make it relevant to two phase cemented carbide like materials one can use the modified structural continuum model of Drucker (21).

The direct implication of severe tool nose compressive stresses at high temperature are the macro plastic deformation of the carbide tool-tip as reported by Loadze (31) or microplastic deformation as shown by Uehara et al (5). According to this plastic deformation of the subsurface cobalt layers in the vicinity WC-6Co carbide tool nose was found during face milling just after 4 cutting cycles. There was naturally no visible external deformation of the tool. In this context it is of interest to note that Lueth and Hale (8) also report micro voids in the room temperature compression fracture of cemented carbide specimens, especially in the region where calculated shear stresses were at maximum.

Under such circumstances, even when temperature levels are kept low the intermittent cutting operation could produce such compressive stress levels which facilitate plastic deformation of the cemented carbide and hence one can expect certain amount of rate influence. Information on this under room temperature fatigue is very limited.

According to Troshenko et al (32) at high stress amplitude the comparitive endurance durability (no of cycles at a given stress level and 50% probability of fracture) of low cobalt WC alloys increase marginally (about 15%) during shock fatigue when compared with conventional harmonic loading in bending fatigue (corresponding to a 30 fold reduction in loading rate) but remains practically unaffected when the stress amplitudes are low. As can be expected this picture is affected by both cobalt content and carbide grain size, but appears to be compatible with corresponding static strength changes due to loading rate. In other words, in the absence of high temperatures this is advantageous in intermittent cutting.

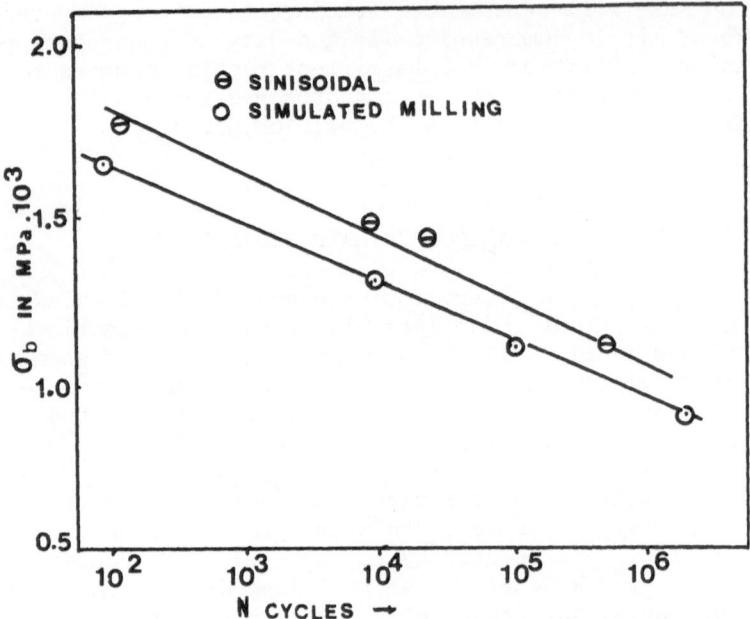

Fig. 4   Influence of pulse shape upon the bending fatigue
          endurance for the tool carbide WC+2(TiC,TaC)+6Co

         However, recent investigations (33) devoted to the
bending fatigue durability of a low cobalt tool carbide (see Fig.4)
using conventional (sinisoidal) and a special pulse shape (similar
to calculated milling stress pulse shown in Fig.2) show a fall of
almost one order in fatigue life cycles in the case of milling
simulation. The higher value of the average stress level during
milling simulation is a possible cause. Similar fall in room
temperature bending strength at higher strain rates has also been
reported (34) for WC-Co alloys. On the otherhand the high temperature strength as a function of strain rate reported by Ueda et al
(35) indicate a slight increase (about 15% for strain rate increase
of 30 times). This could very well be related with the reduced influence of microstructural defects at high temperature as reported
by Suzuki et al (36).

         In addition to the time varying loads already considered
carbide cutting tools are subjected to severe contact loads and
oxidation and it is interesting to look into these influences upon
the strength of the material.

         According to Triandafilidi and Loshak (37) the TRS of WC-15Co
alloy increased from 1000 to 2450 MPa due to surface plastic deformation and resulting residual stresses. Surface oxidation of WC

appears to enrich the same in cobalt and increase the capacity of the surface to retain greater residual compressive stresses. Results of Gillies and Lewis (38) in terms of X-ray line broadening of a WC-6Co tool subjected to machining loads, indicate a residual stress of the order of 3500 MPa. An entirely different possibility of reducing the normally present tensile stresses in the cobalt layers of WC-Co alloy is through the introduction of a thin coaating as reported by Montgomary (39). An increase of 2 orders in surface fatigue life has been reported. The average contact loads here, however, are not comparable with metal cutting.

In this context it is interesting to note the extremum type of realtionship between the TRS and the dislocation density for the WC-15Co alloy reported by Linenko-Melnikov et al.(40). Thus diamond grinding of WC-Co alloy appears to influence slip blockage more than crystal lattice distortion and hence enhances the fatigue durability.

Hence the implications of surface deformation of WC-Co alloy upon its strength are quite complex and are of direct relevance to the fracture behaviour of a cutting tool. Due to such influences and our inability to identify them individually very often we find serious contradiction in reported tool-wear-fracture data and even terminoloy at times. One way of avoiding this difficulty is through proper recognition of stable crack growth region from unstable fracture region in a cutting tool. Under material testing conditions this appears feasible from the recent fatigue crack studies of Almond and Roebuck (41).

Another essential feature of carbide tool fracture is the fracture criterion which is relevant to machining situation'. We have already seen that strength of WC-Co alloys in different testing modes appear to be different. Taking this into account and linking the micro-mechanical parameters like grain size and cobalt free path through a modified Bridgeman's necking model based upon continuum concept, Kals et al (42) have used the ultimate uniaxial strain ε_1 as the failure criterion. Here $\varepsilon_1 = (\sigma_1 - \nu(\sigma_2 + \sigma_3))/E$. Reasonable correlation between micro structural parameters and E, and the criterion strain have been established. The main shortcoming of this model appears to be the neglect of intergranular thermoelastic stresses and the possible influence of grain shape in terms of stress concentration in the cobalt layers. Hirao et al.(43) report a stress concentration factor of about 4 at the corners of triangular carbide grains based upon micro-mechanical model of WC-Co alloy using the FEM technique. This in a way explains the occurance of voids at carbide corners prior to fracture (35).

The binder composition in terms of the quantitative presence of C and W and the qualitative mode of their variation due to annealing

and cobalt domain size, also play a dominant role upon alloy strength.
However, there seem to be no data regarding their influence upon
tool fracture per se.

THERMAL STATE AND TOOL FRACTURE

Cyclic thermal influences often lead to fracture initiation in
intermittent cutting tools. While the role of some type of thermal
fatigue upon tool fracture behaviour in terms of crack initiation
and propagation has been conclusively proved, it is also often observed that thermal fatigue by itself rarely results in tool fracture.

Accurate temperature distribution in a milling cutter is not
available. Okushima et al.(44) report an average interface value
(θ_{av}) of 800-950°C during cutting and 250-400°C during the non-cutting
portion of the cycle in the speed range of 200-300 M/min. Typical
heating and cooling times are of the order of 40 and 200 milliseconds.
The actual maximum interface temperature could be much higher,
especially at higher cutting speeds. Thus according to Jonsson (16),
based upon diffusion rate and observed wear, the estimated interface
temperature is of the order of binder liquidus (1300-1450°C) during
continuous turning (V=184 M/min). Similar order of temperature
(1050 ± 40°C at V= 107 M/min) is reported by Naerheim (45) from the
estimation of W in the binderphase after turning tests. Since the
carbide strength seriousely deteriorates in the range of 800-900°C,
and the strains inducing thermal fatigue are controlled by the
difference between the instantanious maximum and average temperature
levels, it is essential to know the maximum interface temperature
(θ_{max}). There are both spatial and temporal maximums in the temperature in the cutting tool subjected to heating and cooling as in
milling.

Fig.5 shows a typical set of thermal cracks.in a milling cutter.
The characteristic features of these are the initiation of cracks on
the rake surface perpendicular to the cutting edge, above a specific
cutting speed and tooth load. Progressively a system of cracks both
parellel and perpendicular to the cutting edge on the rake and flank
faces are formed. These usually reach a steady state unless inter-
fered by tool fracture.

Numerical approximation of the transient thermal field using
simplest thermal conductivity approach (46) indicates a very steep
temperature gradient (800°C/MM) perpendicular to the rake surface.
Beyond about 3 MM from rake the temperature fluctuations are
marginal.Fig.6 shows the theoretical thermal field in a milling
cutter. Fig.7 shows the thermoelastic stresses obtained for these
temperature fluctuations using the FEM technique(46). This indicates
large tensile stresses (4000 MPa) at the end of cooling period
and larger compressive stresses (18000 MPa) at the end of

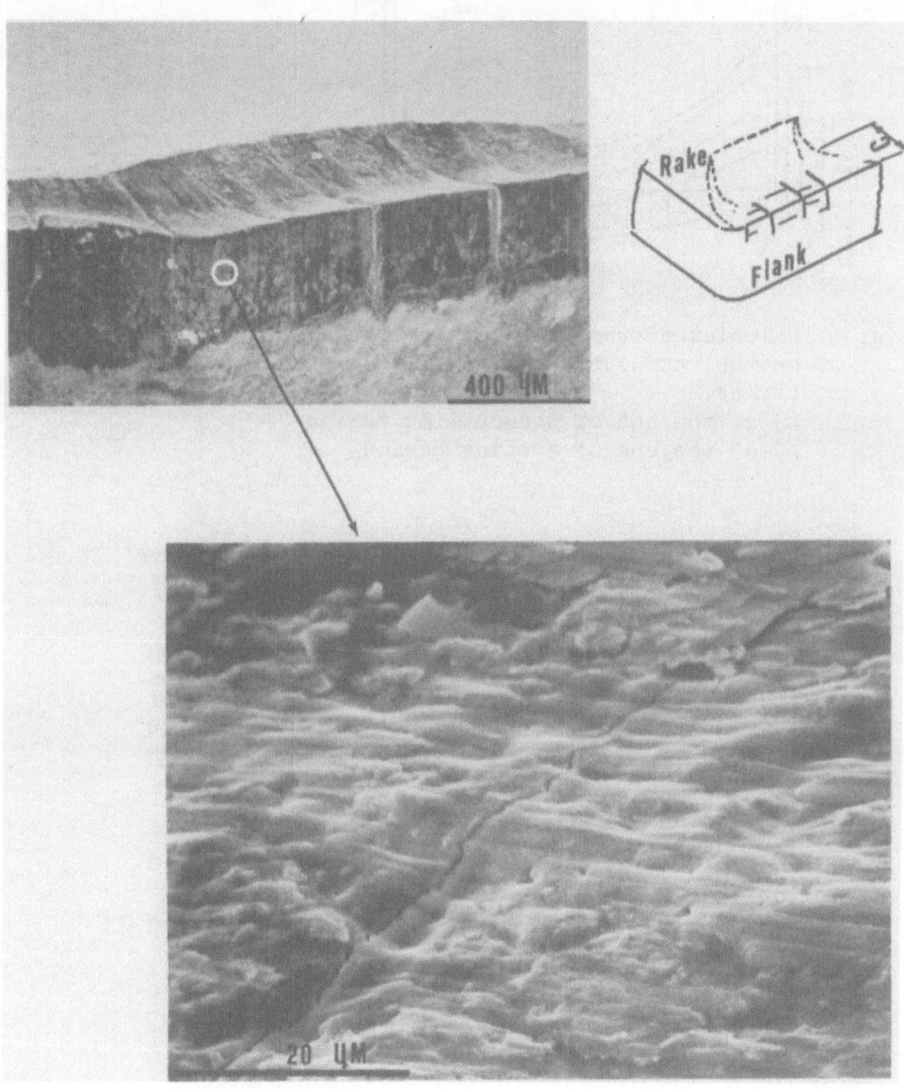

Fig. 5 Typical thermal cracks in a P40 carbide tool insert
after face milling steel. SEM picture.

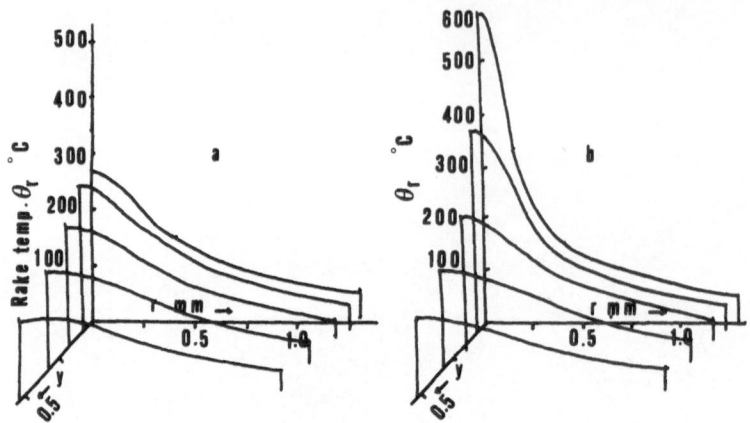

Fig. 6. Calculated transient temperature distribution
on the rake surface of a milling cutter (52) after 30
cycles.
a) at the end of non-cutting period
b) at the end of cutting period

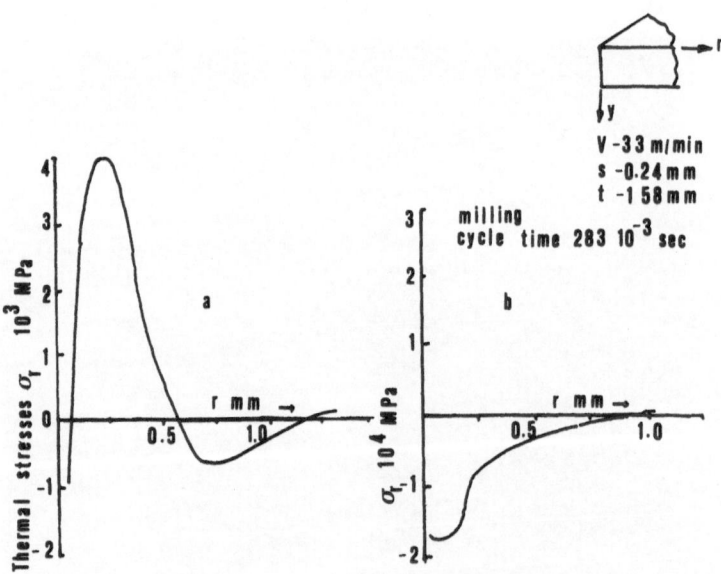

Fig. 7. Calculated transient thermal stresses (52) along the
cutting edge after 30 cycles. a) at the end of non-
cutting period. b) at the end of cutting period.

heating period. The thermal stresses thus appear to be comparable
and even more severe than the mechanical stresses (see Fig.2).
Survival of the tool indicates that the assumptions made regarding
the evaluation of the thermal and mechanical stresses as well as
the relevant fracture properties of the carbide material have been
far too simple. Further in the presence of combined thermal and
machanical loading there could be phase influences, varying strain
constraints, and interaction between macro and micro stresses in
the thermal and mechanical level.

   Moreover inview of the large compressive stresses and high
temperatures pure elastic analysis is incorrect. Under such conditions in fracture mechanism crack nucleation may not be predominant
factor (47). Different crack healing mechanisms interms of micro-
thermal stresses and their compensation with specific alloy composition (traces of TaC and NbC in a low cobalt WC) as reported by
Lardner (48) can also drastically affect fracture initiation.

THERMAL FATIGUE BEHAVIOUR OF TOOL CARBIDE

   Controlled thermal fatigue and associated fracture initiation
and propagation data for tool carbides under conditions comparable
to intermittent cutting are sparse. Thermal shock test involving
the heating of the specimen to high temperature and subsequent-
quenching is the more often used method. Either the high temperature
required to produce surface cracks after single shock or alternatively the number of shocks required to produce the first crack at a
given high temperature are both used as thermal fatigue criteria.
These tests,comparable to an impact type of test on the mechanical
analogy, are usually poorly controlled, and are usually successful
with ceramic like materials obeying Hasselman's model. Mai (49) has
successfully used this approach to evaluate the resistance of tool
carbides to thermal cracking. No comparable cutting tests were
conducted.

   On the other hand, the initiation and propagation of thermal
cracks during single tooth fly milling, as a function of binder and
WC grain size has been reported by Ekemar et al.(50), although no
temperature stress or fatigue durability estimation was made.
Attempts to use laser beams (51) to simulate the thermal cycle observed in milling has also not been successful due to the extremely
short heating time. In terms of convenience of control, capability
to simulate intense surface heating as observed in milling, repeatedly maintain a specific temperature gradient over a given distance
for given time, the high frequency induction heating is a convenient
one and the same was used by the present author.

   Fig.8 shows a typical tool carbide insert used in milling
after undergoing a thermal fatigue cycling operation. The experimental details are explained elsewhere (52). The tests indicated that

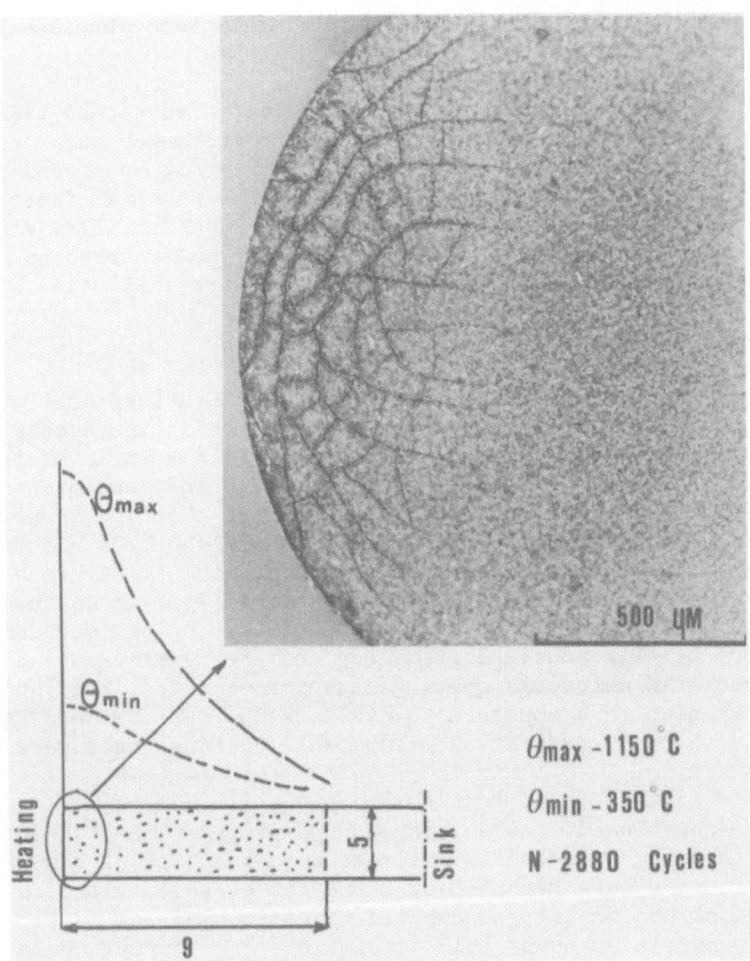

Fig. 8. Appearance of typical cracks from thermal fatigue
        cycling in WC-7.7 TiC - 1(TaC NbC)-6Co cemented
        carbide tool insert.

as in a milling operation a definite temperature amplitude (θ_{max}) has to be reached for cracks to be initiated. Cracks perpendicular to the cutting edge are first initiated at the point of highest temperature and soon develop into a system of cracks, resembling the so called "craze" cracks as shown in Fig.8. In the established stage, there appears to be a difference in the system of cracks observed in milling and the present thermal cycling. In all probability this could be due to the simultaneous action of mechanical loads during cutting and the oxidation and cleansing action of the cutting process.

CARBIDE TOOL EVALUATION FOR INTERMITTENT CUTTING

In the light of the information we have obtained on the thermo-mechanical stress state of the intermittent cutting tool and the behaviour of these tool carbides relevant to tool fracture, we find that on both these counts we have serious gaps. We are unable to correlate satisfactorily the observed fracture initiation and propagation during cutting operation interms of the stresses and temperature controlling the fracture of such materials, although a certain degree of success has been achieved in relating the micro-mechaninal properties of these carbides with their simple fracture behaviour. Similarly there is a large volume of empirical in-formation on the fracture behaviour of tool during intermittent cutting interms of the process parameters. The circle must be completed. Here it appears relevant to think interms of two additional tests with special reference to intermittent cutting.

There should be a test to simulate the mechanical loading conditions observed during milling, in terms of maximum strain magni-tude and the strain history, mode of loading the tool material under comparable volume size distribution. The other test is a thermal fatigue test interms of effective control of θ_{max}, θ_{min}, and the rate of heating and cooling. The observed parameters in both tests should include residual structural observations relevant to machining, in addition to the usual micro and macro crack initiation and propagation. The degree of disparity between the indicated tests and actual milling test results are then the true parameters of the gap that should be closed.

ACKNOWLEDGEMENT

The author wishes to thank Drs. R. Lagneborg and R. Sandström for keen interest and facilities extended in connection with the thermal fatigue tests at SIMR, Stockholm.

REFERENCES

1. H.Chandrasekaran and R.Nagarajan, Influence of Flank Wear on the Stresses in a Cutting Tool, Trans.ASME,J.of Engg. for Ind., p-566 (1977).

2. A.J.Pekelharing, The Exit Failure in Interrupted Cutting, Ann.
    of CIRP, 27/1, pp 5-10 (1978).
3. E.Usui, T.Sirakashi and T.Hara, Analytical Prediction of Entry-
    Chipping in Interrupted Turning Operation, Proc. of 4 ICPE,
    Tokyo, Aug. (1980), pp 474-479.
4. H.Chandrasekaran and R. Nagarajan, On Certain Aspects of
    Transient Stresses in Cutting Tools, Trans. of ASME, J.of Engg.
    for Ind., pp 133-141 (1980).
5. K. Uhehara and Y. Kanda, On the Chipping Phenomena of Carbide
    Cutting Tools, Ann. of CIRP, 25/1 pp 11-16, (1977).
6. M.C. Shaw, Fracture of Metal Cutting Tools, Ann.of CIRP, 28/1,
    pp 19-21 (1979).
7. H.Chandrasekaran, Fracture of Tools and Role of Stress Analysis,
    Proc. of 4 ICPE, Tokyo, Aug. (1980), pp 468-473.
8. R.C. Leuth and T.E. Hale, Compressive Strength of Cementide
    Carbide-Failure Mechanics and Testing Methods, Mat.Res. and
    Stand., Feb. p 23, (1970).
9. H. Suzuki and K. Hayashi, Strength of WC-Co Cemented Carbide in
    Relation to their Fracture Sources, Plansee Pulvermet. 23
    pp 24-36, (1975).
10. H.E. Exner and T. Gurland, A Review of Parameters Influencing
    some Mechanical Properties of WC-Co alloys. Powder Metall, 13,
    pp 13-31, (1970).
11. T.L. Chermant and F. Osterstock, Toughness, Mechanical
    Characteristics and Microstructure of Carbide Cutting Tools,
    VIth NSF Hard Materials Workshop, Brown Univ. Providence,
    June, (1977).
12. G.S. Kreimer and N.A. Aleksejva, Mechanics of Fracture in
    Sintered Tungsten Carbide-Cobalt Alloys, Phys.Metals Metallogr.
    13, No.4, pp 117-21 (1962).
13. T. Gurland, The Fracture Strength of Sintered Tungsten Carbide-
    Cobalt Alloys in Relation to Composition and Particle Spacing,
    Trans. of AIME, 227, pp 1146-50 (1963).
14. S.B. Luyckx, Role of Inclusion in the Fracture Initiation
    Process in WC-Co Alloys, Acta Met. 23, pp 109-115, (1975).
15. E.A. Almond and B. Roebuck, Defect-Initiated Fracture and the
    Bend Strength of WC-Co Hardmetals, Met.Sci. Oct. pp 451-461
    (1977).
16. H. Jonsson, Deformation in Cemented Carbide Tool Bit During
    Turning of Steel, Plansee. Pulvermet. 20, pp 39-62, (1972).
17. N.N. Zorev and H.A. Sawiaskin, Standzeit des Hartmetallwerkzeugs
    bei Unterbrochenen Schmitt mit Daurzykeln, Ann.of CIRP, 18,
    pp 555-62 (1970).
18. E. Pärnama and H. Jonsson, Influence of Binder-Phase Composit-
    ion of Cemented Carbides on Strength and Wear Resistance in
    Turning "Materials for Metal Cutting", ISI Report 126, pp 166-
    169 London (1970).
19. G. Berry and M.T. Kadhim Al-Tonnachi, Toughness and Toughness
    Behaviour of Two High-Speed Steels, Met. Technoglogy, June,
    pp 289-295 (1977).

20. T.N. Loladze, H. Chandrasekaran and A.I. Beteneli, The Stresses
     in Cutting Tools and the Meaning of Limiting Thickness of Cut,
     (in Russian), Izves.Vuzov (mashinostroenie),Dec.pp 123-130
     (1967).

21. D.C. Drucker, Engineering and Continuum Aspects of High-Strength
     Materials, "High Strength Materials", V.F. Zackary, ed. J.Wiley
     and Sons, N.Y. pp 795-833 (1965).

22. M. Hirao, R. Murata and U. Kasuya, Evaluation of Cutting Tool
     Toughness Through Non-Cutting-Test, Ann. of CIRP, 28/1, pp 29-33,
     (1979).

23. K. Uehara and H. Takashita, A Simulation Test on the Failure
     Pattern of Cutting Tools, Proc. of ICPE, Kyoto, pp 212-217 (1977).

24. T.A. Janardhan Reddy, A Study of Metal Cutting Transients in
     Peripheral milling, Ph.D.Thesis. I.I.T. Madras, (1980).

25. H. Chandrasekaran and R. Nagarajan, Incipient Cutting and
     Transient Stress in a Cutting Tool using Moire Method, Int.J.
     of Machine Tool Des. and Res. (1981).

26. H. Chandrasekaran, Fracture of Cutting Tools, Proc. of Int.Conf.
     on Fracture Mechanics and Engg. Applications, ed. G.C.Sih and
     S.R.Valluri, Sythoff and Noordhoff, Holland, pp 601-609, (1979).

27. M.G. Loshak and V.M. Fridman, The Scale Effect of the Metallo
     Ceramic Hard Alloys of the Tungsten Group, (in Russian), Prob.
     Proch. No 8, pp 41-47 (1971).

28. E.Lardner, Isostatic Hot Pressing of Cementite Carbide, Powder
     Metall. 18, No 35, pp 47-52 (1975).

29. B.T. Troshenko, L.I. Alexandrova and M.G. Loshak, Influence
     of Technological Factors on the Strength and Durability of Hard
     Alloys, Poroshk. Met. No.9, pp 40-45, (1970).

30. N.I. Romanova, G.C. Kreimer and V.I. Tumanov, Influence of
     Residual Porosity on the Fatigue Durability During Cyclic
     Bending of WC-Co Alloys, Poroshk.Met. No.10, pp 70-76 (1979).

31. T.N. Loladze, Tribology of Metal Cutting and Creation of New
     Tool Materials, Ann.of CIRP 25/1, pp 33-38 (1976).

32. V.T. Troshenko, V.N. Bakul and M.G. Loshak, On the Ralationship
     of Durability of Hard Metal Alloys in Shock and Harmonically
     Alternating loads, (in Russian), Prob.Proch. No.1, pp 41-44
     (1971).

33. K. Rajagopalan, Fatigue and Fracture of Tungsten Carbide in
     Simulated Milling Test, M.S. Thesis, I.I.T., Madras, (1980).

34. O. Knotek, E. Lugscheider and H. Stabrey, On the Deformation
     and Transverse Rupture Strength of Cementide Carbides, Proc. of
     Vth. European Symposium on Powder Metalurgy, Stockholm, June
     (1978) pp 82-86.

35. F. Ueda, H. Doi, F. Fujiwara and H. Masatomi, Bend Deformation
     and Fracture of WC-Co Alloys at Elevated Temperature, Trans.JIM,
     18, pp 247-256 (1977).

36. H. Suzuki, K. Hayashi and Y. Taniguchi, High Temperaure Trans-
     verse-Rupture Strength of WC-Co Cemented Carbides, Plansee
     Pulver. 25, p 23 (1977).

37.   I.I. Triandafilidi and M.G. Loshak, The Influence of Plastic
      Deformation on the Strength of Hard Alloys (in Russian), Fiz.
      Khim.Mekh. Mat. 11(5), pp 88-91 (1975).

38.   D.G. Gillies and D. Lewis, Lattice Strain in Tungsten Carbide,
      Powder Metall. 11, pp 400-410 (1968).

39.   R.S. Montgomery, Effect of Various Coatings on the Surface
      Fatigue and Wear Behavior of Cemented Tungsten Carbide, Trans.
      of ASLE, 13 (1), pp 47-53, (1970).

40.   Yu.L. Linenko-Melnikov, Yu.I. Sozin, M.G. Loshak, A.N.Sinchilov
      A.R. Kruchkova and L.I. Alexandrova, The Influence of Diamond
      and Other Methods of Working Upon the Structure and Mechanical
      Properties of Hard Alloys, (in Russian), Prob.Porch. No.9, pp
      100-103 (1977).

41.   E.A. Almond and B. Roebuck, Fatigue Crack Growth in WC-Co
      Hardmetals, Met. Technology, Feb. pp 83-85 (1980).

42.   H.J.J. Kals and Gielisse, The Significance of Structural
      Parameters in Failure of Cemented Carbides, Ann. of CIRP 24/1,
      pp 65-68 (1975).

43.   M. Hirao, R. Murata and H. Takeyama, Microscopic Stress
      Analysis of Sintered Carbide, Ann. of CIRP 25/1 pp 27-31 (1977).

44.   K. Okushima and T. Hoshi, Internal Temperature Distribution
      of Carbide Fly Cutting Tools, Bull. JSME, pp 566-573 (1967).

45.   Y. Naerheim, A Metallurgical method for Determining the
      Temperature in Cemented Carbide Tools, Proc. of Vth Eurepean
      Symp. on Pow.Met. Stockholm, June (1978) pp 99-104.

46.   H. Wu and J.E. Mayer, An Analysis of Thermal Cracking of
      Carbide Tools in Intermittent Cutting, Trans. of ASME, J. of
      Engg. for Ind. May, pp 159-164 (1979).

47.   J.L. Chermant, M. Coster, G. Hautier and P. Schaufelberger,
      Statistical Analysis of the Behaviour of Cemented Carbides
      under High Pressure, Powder Metall. 17, 33, pp 85-102 (1974).

48.   E. Lardner, Review of Current Hardmetal Technology, "Materials
      for Metal Cutting" ISI Report No. 126, pp 122-132, London (1970).

49.   Y.W. Mai, Therma-Shock Resistance and Fracture-Strength Behaviour
      of Two Tool Carbides, J.of Amer.Cer.Soc. 59, pp 491-494 (1976).

50.   C.S.G.Ekemar, S.A.O. Iggström and G.K.A. Hedén, Influence of
      Some Metallurgical Parameters of Cemented Carbide on the
      Sensitivity of Thermal Fatigue Cracking at Cutting Edges,
      "Materials for Metal Cutting" ISI Report No. 126, pp 133-142,
      London (1970).

51.   A.A. Darwish and S.P. Anderson, Effect of Thermal Shock by
      Laser on Carbide Tips, Proc. of IVth European Symp. on Pow.
      Met., Grenoble, France, pp 5-16-1 (1975).

52.   S. Lund, Experimental Set-up for Simulation of Thermal
      Fatigue of Cemented Carbide Tools in Machining, Swedish
      Institute for Metals Research, Stockholm, Report No. IM-1555
      (1981).

DISCUSSION

V. K. Sarin:

Referring to thermal fatigue cracking in milling, I think several
people have observed that in the first cycle or second cycle in
milling, you get cracks. Would you comment on how you would define
those as thermal fatigue cracks?

H. Chandrasekaran:

The literature survey on this aspect shows two interesting features.
The earliest work showed that within a few cycles, I'm not sure
whether it's 1 or 2, but few, say 10 or 20, cracks could be seen.
But later work from 1962 onwards have shown for whatever reason,
maybe to due to better carbide properties, that the cracks were
observed after a few hundred cycles - 200, 400, 5,000 and so on.
On the other hand, recent investigations by Hara in Japan using
scanning techniques, (because the early investigators didn't have
the scanning technique) have shown that after 4 cutting cycles in
a carbide milling tool there is residual plastic deformation in
cobalt but the tool is intact externally and fractures maybe after
200 cuts. So, it indicates that failure is by severe, low cycle-high
strain fatigue on which practically there is no data for carbides.
We have classical fatigue data, but very little on this aspect. So,
if that is the case, then, after a few cycles it can crack mechanically. On the thermal side, quenching tests have shown only a short
time for cracking. Our own investigations have shown the that you
do get cracks in a purely thermal situation starting from probably
a few hundred cycles onwards. In other words, it's not a very
large number but maybe 100 or 200 until you get the first crack.
A typical example is 100 cycles, 1100 to 300°C, 6 cracks on a 6%
cobalt insert. That's the type of thing we get.

RESPONSE OF A WC-Co ALLOY TO THERMAL SHOCK

C. W. Merten

The University of Vermont/Multi-Metals
Department of Mechanical Engineering

ABSTRACT

WC-6%Co specimens were subjected to thermal shock treatments by means of induction heating and subsequent controlled quenching to determine their resistance to fracture. Resistance to fracture is correlated to the thermal amplitude required to produce cracking. Typical thermal amplitudes ranged from 500° to 1200° C. Specimens with a thickness of 2.3 mm. required higher thermal amplitudes for failure than did specimens of twice this thickness. The fracture surfaces of thin specimens are perpendicular to the surface of largest area indicating that a plane stress situation exists. Fracture surfaces of 4.6 mm. thick specimens deviate from this condition. Scanning electron micrographs show that the fracture path has proceeded through both the Co matrix and the WC grains. The ability of thermal shock testing to discriminate between surface treatments was also investigated. Surface treatment was found to influence the thermal amplitude required for fracture. The applicability of thermal shock test results to actual cutting conditions is discussed, and, in light of these results, recommendations as to tool design are given.

INTRODUCTION

High strength and hardness are two properties which account for the popularity of WC-Co composites as materials for metal cutting and mining tools. Early studies (1,2) indicated that these properties were retained, to a large degree, at elevated temperatures, a factor which favors WC-Co as a material over others for these applications. However, tool failure inevitably occurs

757

and a number of investigators have focused their attention on its cause. In intermittent cutting, thermal cracking has long been recognized as a mode of failure (3). The location and appearance of the thermal crack has been characterized(4,5), and stress analyses have been conducted to reconcile experiment with theory(6,7). Studies of the mechanics of rock drilling also point to thermal fatigue as a failure mechanism, especially in the rotary drilling of soft, non-abrasive rock(8). Bailey and Perrott have cited thermal fatigue likely as the most important of wear processes affecting WC-Co composites in rotary mining techniques (9). Thermal effects might also be a factor contributing to failure in percussive drilling of soft, non-abrasive rock (10).

In many cases, recommendations as to optimal cutting conditions, use of coolant, and even tool composition have been made (3,4,5,8,9,10 ,11). However, the complexity of rock drilling and intermittent cutting processes does not allow sufficient separation of thermal and mechanical tool response so that one grade of carbide might be recognized as more resistant to thermal effects than another. Hasselman has proposed a test in which brittle ceramics are thermally shocked from various initial temperatures, and then tested for retained strength(12). Mai has applied this test to several cutting grade carbides(13,14), but this method, in which thermally induced sub-critical cracks are made to propagate in three point bending, tests combined thermal and mechanical resistance.

The purpose of this investigation is to study factors which influence the resistance to failure by thermal means only of a WC-Co alloy. One such method which accomplishes this is pure thermal shock.

EXPERIMENT

The specimens subjected to thermal shock were WC-6%wt.Co plates of various size, grade, heat treatment, and surface treatment. These parameters are listed in Table 1. In some cases, specimens warped during sintering, and subsequent straightening was accomplished by placing the warped plates between graphite blocks and heating to 1380° C. Approximate time above the eutectic temperature was .5 hours. The surface treatment imposed consisted of blasting the plates with SiC grit ranging from 250 to 400μm.in size. Difference in grade was due primarily to grain size, not percent Co. Physical parameters of identical grades supplied by the same manufacturer were determined by McCabe(15). These are listed in Table 2.

Table 1. Specimen Characteristics

Designation	Grade	Size	Heat Treatment	Surface Treatment
1G	1	32x76x2.3 mm.	none	grit blasted
1SG	1	32x76x2.3 mm.	straightened	grit blasted
2	2	32x76x2.3 mm.	none	none
2G	2	32x76x2.3 mm.	none	grit blasted
2NG	2	32x76x2.3 mm.	none	fresh grit
2S	2	32x76x2.3 mm.	straightened	none
2SG	2	32x76x2.3 mm.	straightened	grit blasted
2	2	51x76x2.3 mm.	none	none
2G	2	51x76x2.3 mm.	none	grit blasted
2	2	51x76x4.6 mm.	none	none
2G	2	51x76x4.6 mm.	none	grit blasted
3G	3	32x76x2.3 mm.	none	grit blasted
3SG	3	32x76x2.3 mm.	straightened	grit blasted

Heating of the plates prior to thermal shock was accomplished by means of a 375 kHz. RF generator and two turn induction coil. In all cases, the entire 76 mm. edge of the specimen was heated. Temperature was measured at the center of the heated edge with a fiber optics probe having a spot size of .75 mm. No attempt was made to correct for emmisivity. Two quenching configurations were employed. In both cases, the orientation of the specimen with respect to the quench was fixed from test to test to eliminate any effect orientation might have on fracture path. 51x76 mm. plates were held midway between two sprayers constructed so that all parts of the specimen would be quenched evenly at the same time. In the second configuration, plates were suspended in the horizontal two turn coil by means of thermometer clamps mounted in a fixture capable of vertical movement only. When the desired edge temperature was reached, the fixture was allowed to fall, passing the plate and thermometer clamps through the coil, and into a large vat of water at room temperature. Thus the heated edge of the specimen was the first to come into contact with the quenchant.

Table 2. Physical parameters by grade.

Grade	Composition(v%)		Mean WC Grain Size	Mean Co Free Path	Contiguity
	WC	Co			
1	90.2	9.8	1.2 μ	.30 μ	.63
2	90.3	9.7	2.1 μ	.51 μ	.56
3	89.9	10.1	3.0 μ	.56 μ	.40

This configuration passes a thermal shock wave across the plate as opposed to the first configuration, in which all parts of the specimen are cooled at the same time. After quenching, specimens were examined for evidence of failure.

RESULTS

Grade 2 grit blasted and unblasted 51x76 mm. plates of thickness .2.3 mm. and 4.6 mm. were heated to temperatures ranging from 625°C. to 1195°C., and subsequently quenched by the sprayer configuration. Production of failure by this method was sporadic at best, occurring in only 5 of 79 tests, and bore no relation to edge temperature prior to the spray quench. In at least one case of failure, arcing from the induction coil to the specimen occurred, producing a visible crater which was probably the fracture initiation site. Hence, the sprayer configuration was rejected as not being a method which produced useful information.

Grade 2 grit blasted and unblasted 51x76 mm. plates of thickness 2.3 mm. and 4.6 mm. were also immersion quenched. Temperatures of the heated edge prior to quenching ranged from 500° to 1185° C. The lowest prequench temperature for which a 2.3 mm. thick grit blasted specimen failed was 617° C., and the lowest temperature for an unblasted specimen of the same thickness was 607° C. 4.6 mm. grit blasted and unblasted plates always failed when quenched from temperatures above 500° C., the lower limit of temperature which could be measured with the fiber optics system. Fracture surfaces of the thin specimens were generally perpendicular to the original exterior surface of largest area. For 4.6 mm. thick specimens, this was not the case, as the orientation of the fracture surface with respect to the original exterior surfaces varied along the fracture path.

In an attempt to better characterize response of WC-6%Co to thermal shock by immersion (TSI), 32x76x2.3 mm. specimens of three different grain sizes, two different heat treatments, and two different surface treatments, were subjected to TSI. All specimens were sintered and heat treated in the same furnace load, and all plates of any one grade originated from the same powder lot. Plots of response versus edge temperature prior to immersion are given in Figure 1. Heating times for all groups ranged from 40 to 85 seconds. Explanations of response are as follows.

No failure: No evidence of cracking of the specimen when viewed by the unaided eye. In some cases, exposure of specimens to water vapor followed by drying revealed what could be construed as micro-cracking, however, examination of the same unmisted plates at 30x did not confirm the presence of cracks.

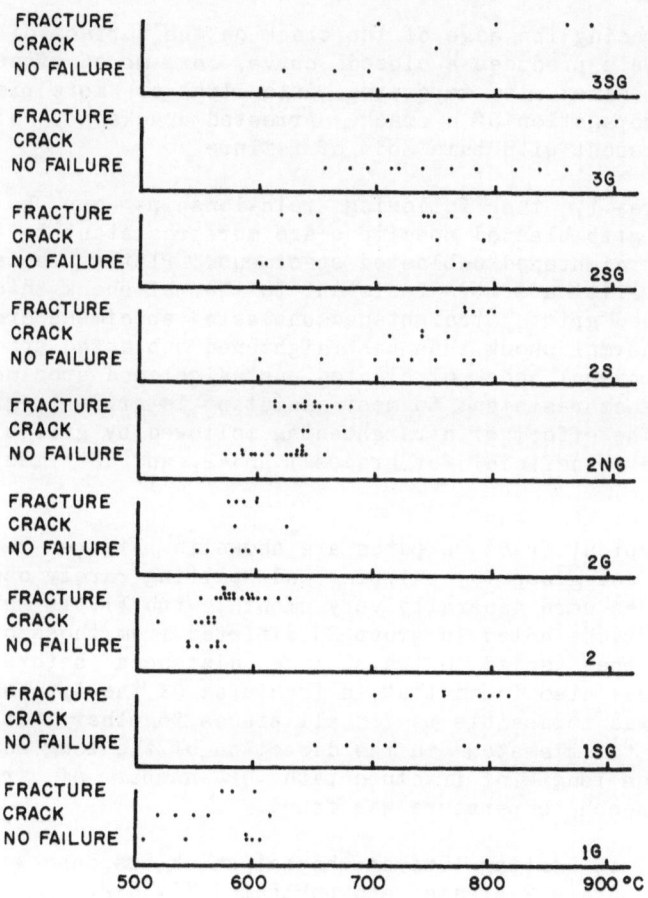

Fig. 1. Response versus thermal amplitude for 40-85 sec. heating times.

Crack: Cracks seen by the unaided eye which have not propagated
                all the way through the specimen, i.e., tracing the
                edge of the crack on the surface of the plate will not
                produce a closed curve. These cracks might also be
                referred to as arrested.

Fracture: Tracing the edge of the crack on the surface of the
                plate produces a closed curve, or more often, the
                specimen is separated into two or more pieces by
                propagation of a crack. Arrested cracks may also be
                present with this mode of failure.

From Figure 1, the following relationships may be seen. Unstraightened grit blasted specimens are more resistant to thermal shock than unstraightened unblasted specimens. Plates blasted with sharp, fresh grit are more resistant to thermal shock than those blasted with used grit. Straightened unblasted specimens are more resistant to thermal shock than unstraightened unblasted specimens. Resistance to thermal shock of blasted unstraightened specimens by grade from least resistant to most resistant is grade 1, grade 2, and grade 3. The effect of straightening followed by grit blasting appears to be beneficial for grades 1 and 2, and detrimental for grade 3.

Several typical fracture paths are shown in Figure 2. With the exception of group 1G, chipping and spalling rarely occurred. Fracture surfaces were generally very smooth, with little evidence of rippling. Cracks noted in group 2S differed from those of other groups in that they tended to travel long distances before being arrested. It was also found that in fractures of the type shown in Figure 2a, it was impossible to fit all pieces together after the fracture due to mismatch in the direction of the long edge. No relation between length of fracture path or number of fractured pieces to prequench temperature was found.

The effect of heating time on thermal shock response was also investigated. Grade 2 blasted and unblasted 32x76x2.3 mm. plates were heated to desired prequench temperatures more rapidly with heating times ranging from 16 to 24 seconds. The results appear in Figure 3. Comparison with Figure 1 shows a higher resistance to fracture by TSI for increased heating rates.

DISCUSSION

The inability of the sprayer configuration to produce failure consistently could be due to two causes. The first is that water flow is not sufficient to produce a rapid enough quench. The second and more likely possibility is that a steep gradient or even a sudden inflection in the temperature distribution across the

Fig. 2. Typical fracture paths

plate must be introduced to produce stress levels sufficient for fracture. Since quenching is even in the sprayer configuration, no change in the shape of the temperature distribution takes place during the quench. However, immersion will produce an inflection in the temperature distribution at some time on all points of the plate.

The result that 2.3 mm. thick plates were more resistant to thermal shock than plates of twice this thickness supports Mai's findings of relation of specimen size to thermal shock resistance in the Hasselman test (14). In thin plates, the facts that the fracture surface is perpendicular to the major plane of the plate and that mismatch along the heated edge occurs indicate that biaxial stress predominates. Deviation of the fracture surface from the perpendicular in thick plates shows that temperature through the thickness varied significantly upon quenching, thus giving rise to a triaxial stress state. The differing responses of thick and thin plates to thermal shock are presumed to be a result of the differing stress states, and influence of size on the distribution of Griffith flaws.

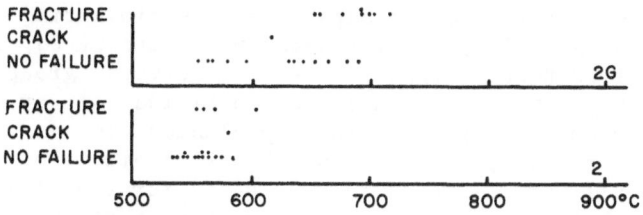

Fig. 3. Response versus thermal amplitude for 16-24 sec. heating times.

Response of WC-Co to thermal shock by immersion was found to be sensitive to heating time. The question arises as to how this result influences the response of groups plotted in Figure 1. The RF generator was calibrated at frequent intervals so that a leading edge temperature of 600 °C. was attained in approximately 45 seconds. Thus long heating times correspond to higher edge temperatures. Resistance to fracture has been shown to be greater for faster heating rates. Therefore, the effect of longer heating times within any one group is to produce fracture at lower thermal amplitudes. However, at any given temperature, this effect is present for all groups, and hence comparisons between groups are valid. It can then be concluded that thermal shock by immersion is capable of producing a response which allows for discrimination of resistance to failure between grade, surface treatment and heat treatment.

The results in Figure 1 also show significant scatter within any one group in the form of instances of no failure at higher thermal amplitudes than instances of cracking and fracture. In grit blasted groups, this scatter might be explained by variability of the blasting process in terms of the depth of carbide removed, the angle at which blasting took place, and the condition of the grit. In unblasted groups, scatter might be attributed to the statistical nature of the size and location of Griffith flaws. Mai (14) has subjected WC-8.5%Co specimens to Hasselman's test, thermal shock of specimens from various prequench temperatures followed by three point bending. It is noted that there is a marked drop in retained strength at a critical prequench temperature as predicted by Hasselman. However, for specimens quenched from 800 C. and above, retained strength increases to values nearly equal to those found for specimens quenched from below the critical temperature. Thus, Mai's findings point to a third source of scatter in the results. Groups 1SG, 2SG, and 3G exhibit behavior similar to that found by Mai. One possible explanation for this effect is that the plasticity of the Co binder increases with temperature in a non-linear fashion (16), whereas thermal stresses can only increase linearly with temperature. Thus at high prequench temperatures, the binder is capable of accommodating greater thermal strain.

As mentioned before, TSI is capable of discriminating between surface treatments. Of particular note are groups 2, 2G, and 2NG. The variation of response versus temperature between groups leads to the conclusion that fracture is initiated at or near the surface. The argument supporting this conclusion is as follows. All specimens in groups 2, 2G, and 2NG originated from the same powder lot and all were sintered in the same furnace load. The only difference between groups is their surface treatment; group 2 is as sintered, group 2G was blasted with standard production grit, and group 2NG was blasted with sharp fresh grit. Were fracture to initiate in the interior of the specimen, the crack would be well

developed and running at high speed by the time it reached the
surface. It is very unlikely that, in this case, a surface
treatment such as grit blasting would hinder the crack. As a
result, no difference in response versus temperature between groups
of the same grade and heat treatment, but different surface
treatments, would be detected. This is not the case. Therefore,
the crack must initiate in some region affected by grit blasting.
This region is at or near the surface of the plate. It is
reasonable to assume that in TSI, fracture is initiated at or near
the surface in other grades tested as well, although the same logic
cannot be applied to differences in heat treatment. The
significance of this result is that TSI may be used as a means to
evaluate the effect of coatings, and that it may give some clue as
to their strengthening mechanisms.

     In an effort to better understand the shift in response versus
temperature for the various groups, scanning electron micrographs
of the fracture surfaces were taken. These appear in Figures 4, 5,
6, 7, and 8 in order of least resistant to most resistant to TSI by
group. For each group, a series of micrographs was taken from the
juncture of the fracture surface and exterior surface of the plate
towards its central plane. A micrograph was also taken of the
fracture surface at the central plane for comparison. Three
distinct fracture modes, or combinations thereof, can be seen. In
mode 1, fracture has proceeded through the Co binder. In general,
WC grains are covered by a thick layer of Co. Mode 1 fracture is
evident in groups 2, 2G, and 2NG. Mode 2 fracture also passes
through the binder phase, but is distinct from mode 1 in that
carbide grains are covered by a very thin film of plastically
deformed Co. Mode 2 fracture can be seen at the left edge of the
group 3G series and at the central plane micrograph of group 3SG.
The third fracture mode is characterized by cleavage of carbide
grains and is evident in group 3G. Fracture of groups 1G, 2, 2G,
and 2NG is predominantly mode 1, although instances of modes 2 and
3 can be seen in micrographs taken at the central plane. In groups
2G and 2NG, some slight evidence of mode 2 fracture may be seen at
the juncture of the fracture surface and exterior surface, as
opposed to group 2 which does not display this feature. Fracture
surfaces of groups 1SG, 2SG, and 2S are similar to each other. All
display a mixture of fracture modes 2 and 3, with mode 3 being more
prevalent in group 2S than in the others. Grade 3 was the only
grade for which straightening appeared to be detrimental. Group
3SG also exhibits a fracture surface much different from those of
other groups. For a distance of 50 μm. from the juncture of
fracture and exterior surfaces, fracture is mode 1. A sharp
transition to mode 2 then occurs, although some cleavage of WC
grains has also taken place. Group 3G, the most resistant to TSI,
also shows a sharp transition, but from mode 2 to mode 3. This
occurs at 37 μm. from the juncture of fracture and exterior
surfaces.

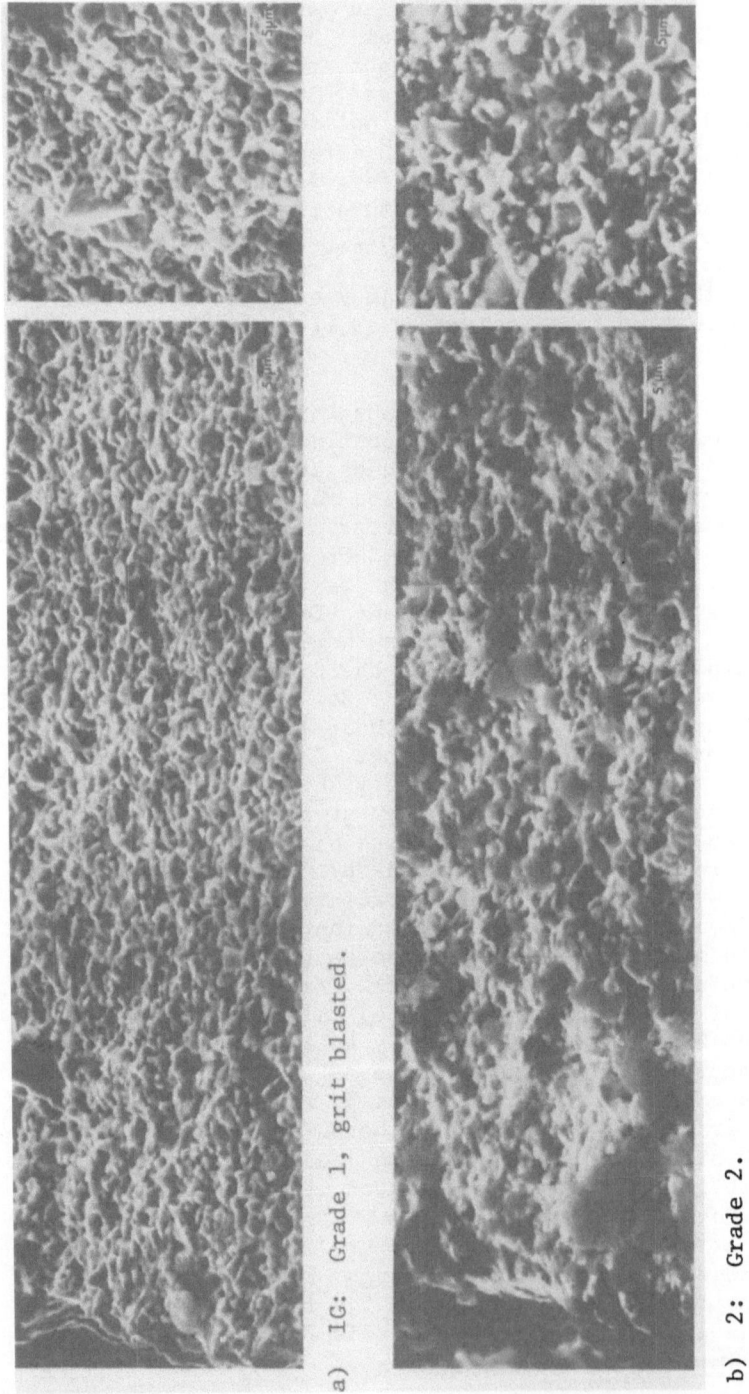

a) 1G: Grade 1, grit blasted.

b) 2: Grade 2.

Fig. 4. Left: SEM micrographs of fracture surface from juncture of fracture surface and original
         exterior surface. Right: SEM micrograph of fracture surface at central plane of the
         specimen.

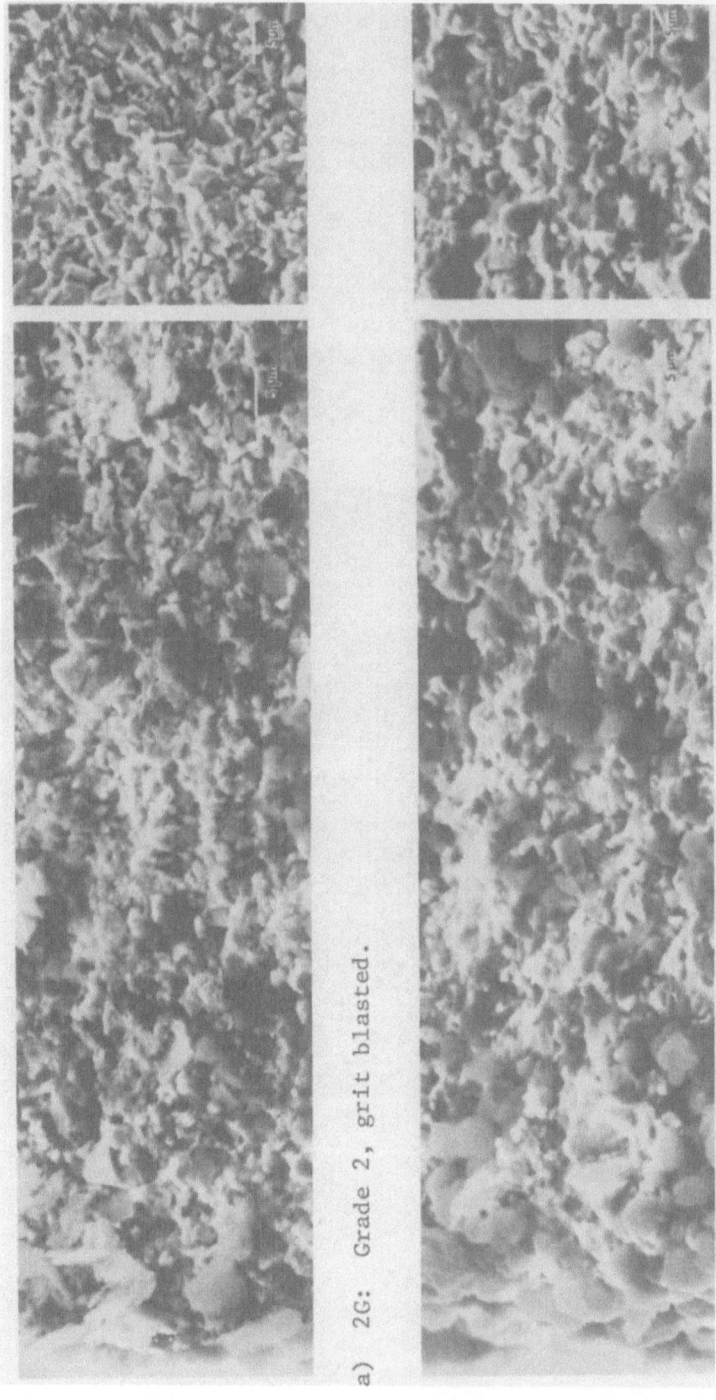

a) 2G: Grade 2, grit blasted.

b) 2NG: Grade 2, blasted with fresh grit.

Fig. 5. Left: SEM micrographs of fracture surface from juncture of fracture surface and original exterior surface. Right: SEM micrograph of fracture surface at central plane of the specimen.

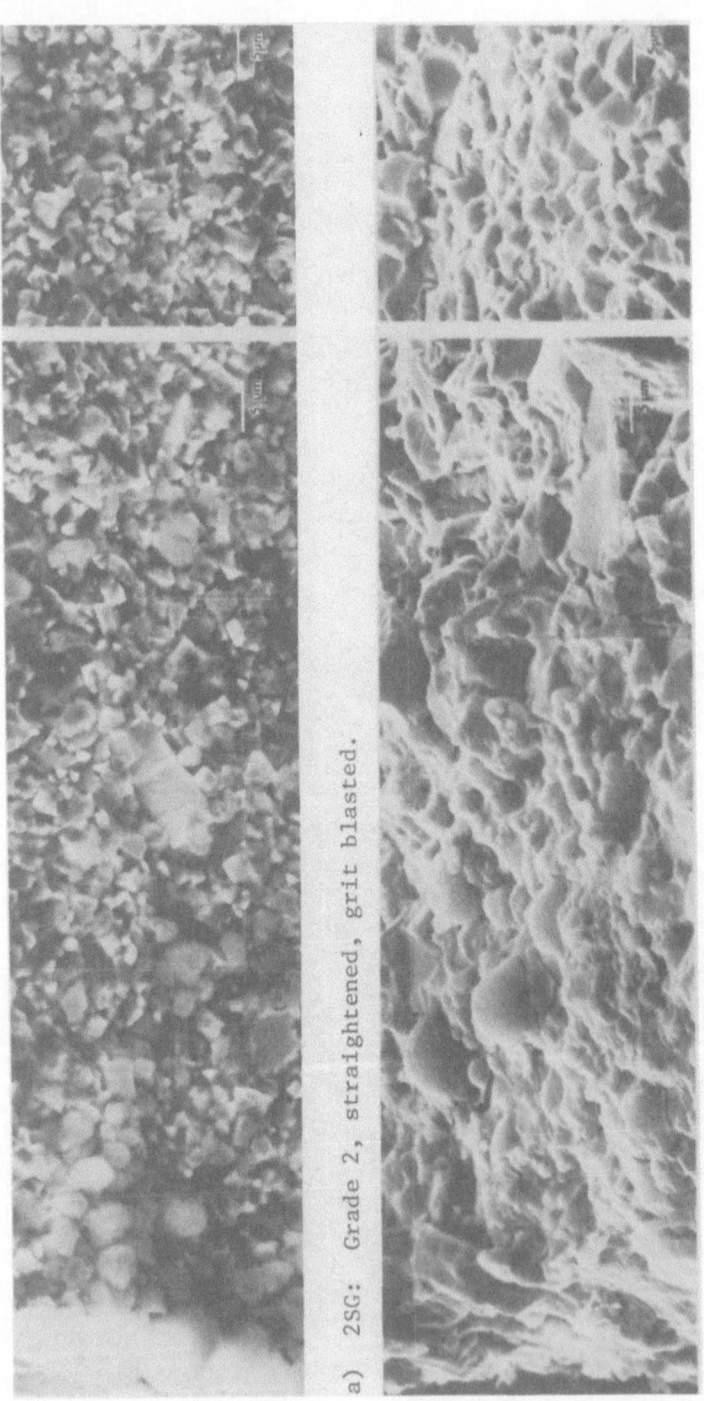

a) 2SG: Grade 2, straightened, grit blasted.

b) 3SG: Grade 3, straightened, grit blasted.

Fig. 6. Left: SEM micrographs of fracture surface from juncture of fracture surface and original
exterior surface. Right: SEM micrograph of fracture surface at central plane of the
specimen.

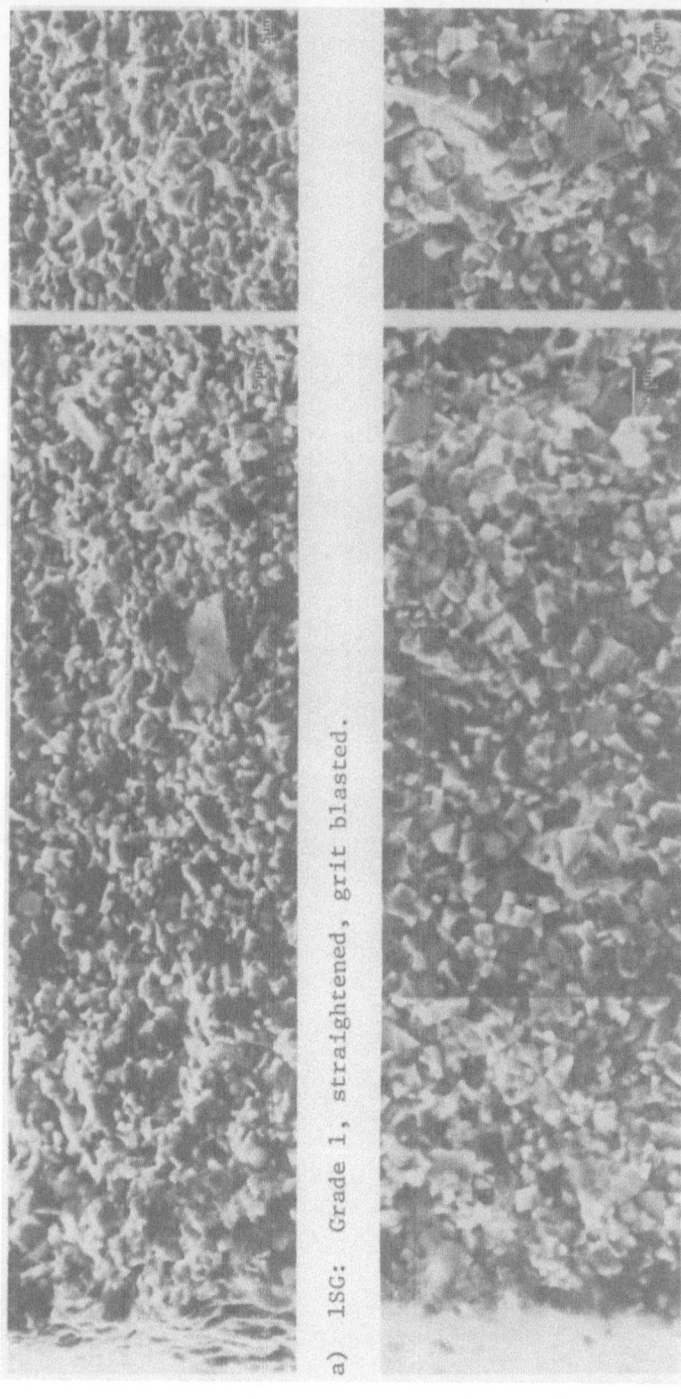

a) 1SG: Grade 1, straightened, grit blasted.

b) 2S: Grade 2, straightened.

Fig. 7. Left: SEM micrographs of fracture surface from juncture of fracture surface and original exterior surface. Right: SEM micrograph of fracture surface at central plane of the specimen.

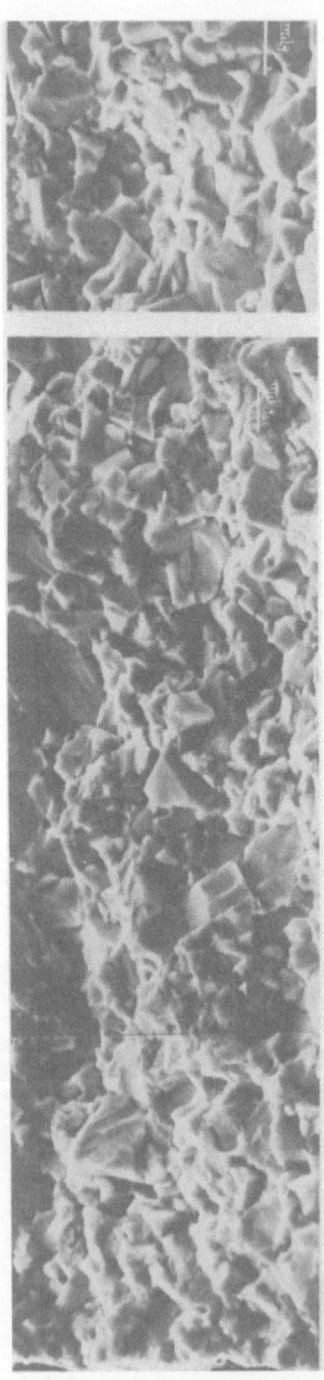

a) 3G: Grade 3, grit blasted.

Fig. 8. Left: SEM micrographs of fracture surface from juncture of fracture surface and original exterior surface. Right: SEM micrograph of fracture surface at central plane of the specimen.

Comparison by group of fracture surface morphology with resistance to thermal shock as given by Figure 1 shows a general trend. As the fracture mode progresses from one to three, there is generally an increase in resistance to failure by thermal shock. Comparison of fracture surface morphologies of groups 3G and 3SG, 2 and 2S, and 2G and 2SG, leads to the conclusion that straightening affects fracture surface morphology. Possible variables introduced by straightening are grain growth, stress relief, and any effect diffused carbon might have. The possible influence of grit blasting on fracture surface morphology has been pointed out. Although it is not clear just what effect grit blasting has on sub-surface microstructure, introduction of compressive stress is a possibility. Composite grade also influences fracture surface morphology through variables such as WC grain size, and mean Co free path. Comparison of fracture surface morphologies near the exterior surface and central plane of group 2 also reveals a difference. This could be due to variations in residual tensile stress in the binder phase by virtue of proximity to the exterior surface. Given the various fracture surface morphologies and the variables that influence them, and assuming that failure in thermal shock is due to tension, it is possible that a brazed joint model similar to that proposed by Gurland and Norton (17), might be able to predict fracture surface morphology, and hence resistance to thermal shock by immersion.

It has been shown that thermal shock by immersion is able to discriminate between WC-Co grade, heat treatment, and surface treatment and that fracture surface morphology correlates to response versus thermal amplitude. The question arises as to whether thermal shock by immersion can predict tool response in cutting applications in which thermal fatigue is deemed important. Rotary drilling in rock is one such application. Bailey and Perrott recommend the use of large grain size (3µm.) composites in weaker rock when thermal fatigue is important. Stjernberg, Fischer, and Hugoson have shown that a WC-6%Co composite with an average grain size of 5µm. outperforms a WC-6%Co composite with a grain size of 1 µm. when rotary drilling sandstone at cutting speeds above 35 m/min. Such speeds correspond to increased service temperatures. Thus it appears that there might be some correlation between response to TSI and tool life in certain mining applications. However, comparison of the response of a range of grades to rotary drilling in laboratory tests and to TSI is needed to determine the applicability of prediction of tool life by TSI.

SUMMARY

It has been shown, in accordance with Mai's results, that increased thickness of a WC-Co composite subjected to thermal shock results in decreased resistance to failure. This finding should be considered in design and mounting of tools when thermal effects are expected to play a major role. Susceptibility of WC-Co to thermal shock also warrants judicious use of coolants and dust control systems. Thermal shock by immersion has been shown to be capable of discriminating between various grades of WC-Co, heat treatments within any particular grade, and the presence or absence of grit blasting. This discrimination leads to the conclusion that fracture is initiated at or near the surface of carbide plates subjected to TSI. Scanning electron micrographs of fracture surfaces of TSI specimens show three distinct fracture surface morphologies or combinations thereof. These morphologies, in turn, show a correlation to resistance to failure by TSI. Applicability of TSI to actual cutting conditions has also been discussed.

ACKNOWLEDGMENTS

The author is indebted to the Multi-Metals Division of Vermont American Corp. for supplying the specimens and necessary support for this project. He also thanks J. D. Knox and B. F. von Turkovich for their helpful advice and many valuable discussions.

REFERENCES
1. A. A. Betser and J. Gurland, Some Effects of Temperature and Heat Treatment on the Strength of Sintered WC-Co Alloys, ASME Pap. 66-MD-17 (1966).
2. G. S. Kreimer, "Strength of Hard Alloys," Consultants Bureau, New York (1968).
3. K. Okushima and T. Hoshi, Thermal Crack in the Carbide Face-Milling Cutter-I, Bull. JSME. 6:151-160 (1962).
4. K. Okushima and T. Hoshi, Thermal Crack in the Carbide Face-Milling Cutter-II, Bull. JSME. 6:317-326 (1963).
5. S. M. Bhatia, P. C. Pandey, and H. S. Shan, Effect of Thermo-Mechanical Shocks on the Functioning of Cemented Carbide Tools in Intermittent Cutting, Int. J. Mach. Tool Des. Res. 19:195-204 (1979).
6. P. M. Braiden and D. S. Dugdale, Failure of Carbide Tools in Intermittent Cutting, in: "Materials for Metal Cutting," I.S.I., London (1970).
7. H. Wu and J. E. Mayer, Jr., An Analysis of Thermal Cracking of Carbide Tools in Intermittent Cutting, ASME. Pap.78-WA/PROD-22:1-6 (1978).
8. J. Larsen-Basse, Wear of Hard-Metals in Rock Drilling: A Survey of the Literature, Powder Metallurgy. 16:1-32 (1973).
9. S. G. Bailey and C. M. Perrott, Wear Processes Exhibited by WC-Co Rotary Cutters in Mining, Wear. 29:117-128 (1974).

10. K. G. Stjernberg, U. Fischer, and N. I. Hugoson, Wear
     Mechanisms Due to Different Rock Drilling Conditions,
     Powder Metallurgy. 18:89-106 (1975).
11. H.J. Osburn, Wear of Rock-Cutting Tools, Powder Metallurgy.
     12:471-502 (1969).
12. D. P. H. Hasselman, Unified Theory of Thermal Shock
     Fracture Initiation and Crack Propagation in Brittle
     Ceramics, J. Am. Ceram. Soc. 52:600-604 (1969).
13. Y. W. Mai and A. G. Atkins, Fracture Strength Behavior of
     Tool Carbides Subjected to Thermal Shock, Am. Ceram. Soc.
     Bull. 54:593 (1975).
14. Y. W. Mai, On the Thermal Shock Behaviour of Oxide Ceramic
     and Carbide Cutting Tools, in: "Fracture Mechanics and
     Technology," G. C. Sih and C. L. Chow, eds., Sijthoff
     and Noordhoff, Alphen aan den Rijn (1977).
15. D. D. McCabe, The Effect of Microstructure on the Erosion of
     Tungsten Carbide-Cobalt Alloys, M. S. Thesis, U. of Ky.,
     Lexington (1980).
16. R. Kamel and K. Halim, The Effect of Phase Change on the
     Mechanical Properties of Cobalt Near Its Transformation
     Temperature, Phys. Stat. Sol. 15:63-69 (1966).
17. J. Gurland and J. T. Norton, Role of the Binder Phase in
     Cemented Tungsten Carbide-Cobalt Alloys, J. of Metals.
     4:1051-1056 (1952).

DISCUSSION

R. K. Viswanadham:

The thick cobalt layer that's covering all the grains - where did
that come from?

C. W. Merten:

I am not certain. I haven't looked at the other side, and I really
have no way of accounting for it. It seems like there is a lot more
cobalt there than there should be for a 6 per cent grade.

R. K. Viswanadham:

I notice that most of your fracture surfaces look distinctly different from the fracture surfaces that one obtains if the material is
fractured at ambient temperatures.

C. W. Merten:

I believe that in failure by thermal shock, and this doesn't relate
to carbides, generally there is quite a difference. The thermal
shock surface is generally quite smooth and you don't see any of the
rippling. It's very difficult to detect a fracture initiation site,
you don't see hackling or that kind of thing which you see in a
normal fatigue fracture.

R. K. Viswanadham:

Could the combination of water and the high temperature significantly
modify the fracture surfaces that you're looking at? Is that a possibility?

C. W. Merten:

It is a possibility. This hasn't been resolved yet. The stress
state is very complex and I have't discussed the difference between
spraying the plate from both sides evenly not producing fracture and
the immersion producing fracture. But the immersion seems to pass a
thermal shock wave over the plate and an inflection in the temperature distribution takes place at the point where the plate enters the
water. The situation is generally so complex that I don't have the
answers yet. But at any rate, one of the main points is that discrimination does occur and perhaps this can be used as an inhouse
test for evaluation of powder lots and final product.

EFFECTS OF MICROSTRUCTURE ON THE EROSION OF WC-Co ALLOYS

H. Conrad[+], D. McCabe and G. A. Sargent

Metallurgical Engineering and Materials Science
Department and Institute for Mining and Minerals
Research
University of Kentucky
Lexington, Kentucky 40506

INTRODUCTION

The erosion of materials is recognized to be a serious problem in coal conversion plants, being responsible for approximately 21% of all component failures reported as of April 6, 1979, according to the National Bureau of Standards Failure Information Center summary report.[1] The components which have been especially plagued by erosion problems are the control and let-down valves[2,3] and slurry feed pumps.[3]

One group of materials frequently used for components where erosion or wear represents a problem are the cemented WC-Co alloys. However, very little information is available on the erosion benavior of such alloys. The studies by Dankin[4] and Uuémyis et al.[5] showed that the erosion rate due to the impingement of solid particles carried in an air stream increased in a hyperbolic manner with the Co content and with increase in WC grain size in the range of 1.5 to 10.6μm. Further, a higher erosion rate occurred at an impingement angle of 60° compared to 30° or 90°.[4] The particles employed by these authors were quartz sand; in one case[4] they were 100-200μm in size with a velocity of 507ms⁻¹, while in the other[5] they were 400-630μm with velocities of 75-330ms⁻¹.

[+]Presently with the Materials Engineering Department, North Carolina State University, Raleigh, North Carolina 27650.

In another study, Hansen[6] showed that the erosion rate of cemented carbides at a particle impingement angle of 90° increased
with increase in metal binder, irrespective of the binder element.
Further, erosion at 700°C was generally only slightly higher than
that at 20°C. Hanson[6] employed a sandblast-type tester using 27μm
alumina particles at a velocity of 170ms^{-1} carried in a high-purity,
dry nitrogen stream.

The present investigation was undertaken to provide additional
information on the erosion of cemented WC-Co alloys by particles
carried in a gas stream, giving special attention to the influence
of microstructure on the erosion. To this end, specimens were prepared and tested which represented a variation in such microstructural features as the volume fraction of Co(V_{Co}), the WC grain
size (d), the mean free Co path (λ) and the WC grain contiguity
(C), determined in the manner of Lee and Gurland.

Studies of the erosion of the cemented WC-Co alloys consisting
of a mixture of a hard, brittle phase and a ductile phase are also
of interest in regard to the mechanism(s) of erosion of materials.
As pointed out by Finnie,[8] the erosion of a ductile material differs from that of a brittle material in that maximum erosion occurs
at an impingement angle near 30° in the former, whereas in the latter it occurs near 90°. Very little is known regarding the erosion
behavior of materials which consist of mixtures of significant
fractions of ductile and brittle phases. A second objective of
this investigation is therefore to provide information which may be
useful in defining the erosion mechanism(s) in such two-phase
materials.

EXPERIMENTAL

Material

The sintered WC-Co alloys used in the present study were provided by Multi-Metals, a Division of Vermont American Corporation.
The composition, microstructural parameters and properties of the
alloys are listed in Table 1. The alloys represent a range in Co
content from 6 to 10.5wt.% nominal (9.7-17.1vol.%). The WC grain
size varies from 0.9 to 5.1μm and the mean free Co binder path
from 0.3 to 0.9μm; the contiguity of the WC phase ranges between
0.32 to 0.63. The microstructural parameters were determined from
scanning electron microscope (SEM) micrographs using the procedure
of Lee and Gurland.[7] Worthy of note regarding the materials is
that in the specially prepared laboratory specimen (containing the
largest WC grain size and designated LAB) the WC grains were appreciably cracked. No such cracking was observed in the other specimens. The reason for this cracking is not known.

Table 1. Properties of the WC-Co Alloys

| Spec. | Co Content | | Mean Co Free Path | Mean WC Grain Size | Contiguity | Vickers Hardness(1) | Fracture Toughness(2) |
	Nominal wt.%	Measured vol.%	μm	μm		10^4 MNm^{-2}	MNm$^{-3/2}$
XM2	6	10.2	0.27	0.9	0.63	1.529	10.4
OM2	6	9.8	0.30	1.2	0.57	1.401	11.5
1M2	6	10.1	0.36	1.5	0.53	1.343	12.1
2M2	6	9.7	0.51	2.1	0.56	1.323	12.3
3M2	6	10.1	0.56	3.0	0.40	1.245	13.2
LAB	6	9.9	0.88	5.1	0.36	1.215	14.5
2M1	9	14.8	0.46	1.8	0.32	1.294	12.6
2M12	10.5	17.1	0.76	2.5	0.32	1.225	13.4

Notes: (1) 124Kg load.
(2) Derived from correlation between hardness and fracture toughness data given by Pickens and Gurland.30

Erosion Tests

The erosion tests were performed using a sandblaster-type of tester employing 240 grit Al_2O_3 particles (30μm mean linear intercept particle size determined by SEM) carried in an air stream. The tests were carried out at room temperature and with particle velocities of 20 to $99ms^{-1}$ determined using the rotating disk method of Ruff and Ives.[9] These velocities are considered to be in the quasistatic range. The particle impingement angle was varied from 15° to 90°. The area of the specimen impacted by the particles was of the order of 1 cm^2.

The dimensions of the erosion specimens were 2.5x2.5x0.3cm except for the tests at 15° where the specimens had the dimensions 5x2.5x0.3cm. The erosion tests were performed on the as-received specimen surface, which had been lightly grit blasted with 46 to 120 grit SiC by Multi-Metals. The nature of this as-received surface is illustrated in Fig. 1. Immediately prior to an erosion test each specimen was cleaned in an ultrasonic cleaner containing acetone.

Fig. 1.   Typical surface condition of the as-received specimens. SEM micrograph of Spec. 2M12 taken at an 18° tilt.

The erosion test consisted of determining as a function of time the loss in weight experienced by the specimen (within 0.1 mg) and the weight of the abrasive which had impinged on the specimen. The latter measurement yielded the value of the flux of the abrasive, which was established to be constant throughout the test and was of the order of 0.5g abrasive per second. The duration of the erosion tests ranged from 2.5 to 40 min. and the total specimen weight loss from 7 to 128 mg, with a material removal depth of $\sim$4 to 90μm. SEM examination of the Al_2O_3 particles prior to, and following, an erosion test revealed that no change in their size or shape had occurred during the test.

RESULTS

Hardness

In general, the hardness of the WC-Co alloys (Table 1) decreased with increase in WC grain size d and with the mean free Co path λ. Lee and Gurland[7] derived the following equation for the Vickers hardness H of cemented WC-Co alloys:

$$H = H_{WC} V_{WC} C + H_{Co} (1-V_{WC} C) \qquad (1)$$

where H_{WC} and H_{Co} are the in situ hardnesses of the WC and Co phases respectively, V_{WC} is the volume fraction of the WC phase and C the contiguity of the WC phase. The in situ values of H_{WC} and H_{Co} are considered to be functions of d and λ respectively and are given by the empirical equations:

$$H_{WC} = 1382 + 23.1 \, d^{-1/2} \quad (Kg/mm^2) \qquad (2)$$

$$H_{Co} = 304 + 12.7 \, \lambda^{-1/2} \quad (Kg/mm^2) \qquad (3)$$

Lee and Gurland[7] obtained rather good agreement between measured values of the hardness of cemented WC-Co alloys and those calculated using Eqs. 1-3.

The measured hardness values obtained in the presented study (Table 1) are compared in Fig. 2 with those calculated according to Eqs. 1-3. Reasonable agreement exists; however, the fit is not as good as that found by Lee and Gurland[7] for the data they considered.

Erosion

General. Typical erosion results obtained at high particle velocities and for impingement angles $\geqslant 45^{\circ}$ are presented in Fig. 3. To be noted is that the specimen weight loss increases linearly with amount of impacting abrasive beginning with the initial data

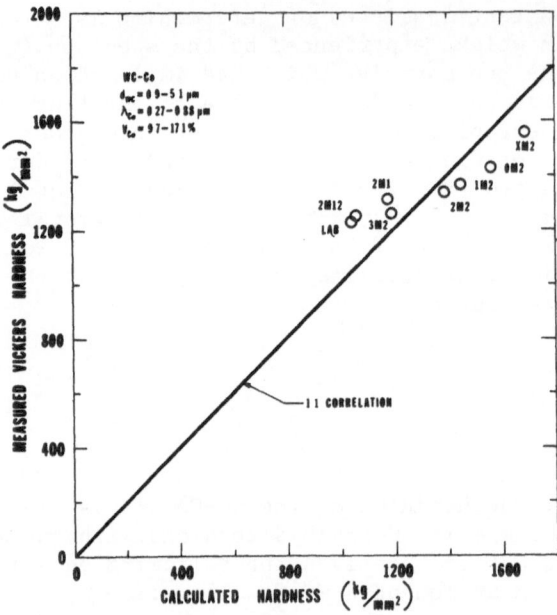

Fig. 2.   Comparison of the measured hardness values with those cal-
          culated according to Lee and Gurland.[7] The letters and
          numbers with each data point identify the specimen.

points. The reproducibility between two separate specimens of a
given alloy was found to be approximately ±10%, which is also approximately the scatter about the average straight line through the
data of Fig. 3, which represent variations in d, λ and C for the 6
wt.% Co alloy. This indicates that for the particular erosion conditions pertaining to Fig. 3, the steady-state erosion rate E_r
(slope of the straight line) is relatively independent of the microstructural parameters represented by the specimens listed on the
graph and in turn on the hardness of these specimens. However, as
will be seen below, at particle velocities at and below $42ms^{-1}$ the
erosion rate does vary with microstructural parameters.

     The data in Fig. 3 yield a small intercept on the ordinate,
suggesting that a more rapid erosion rate occurred at the very beginning of the test compared to the later steady-state value. The
size of the intercept was found to increase as the angle of impingement α decreased; see for example Fig. 4. Moreover, at the smaller
angles, the intercept showed some variation with microstructure,
tending to increase with d and λ, but not consistently. The larger
initial erosion rate suggests that a certain amount of material

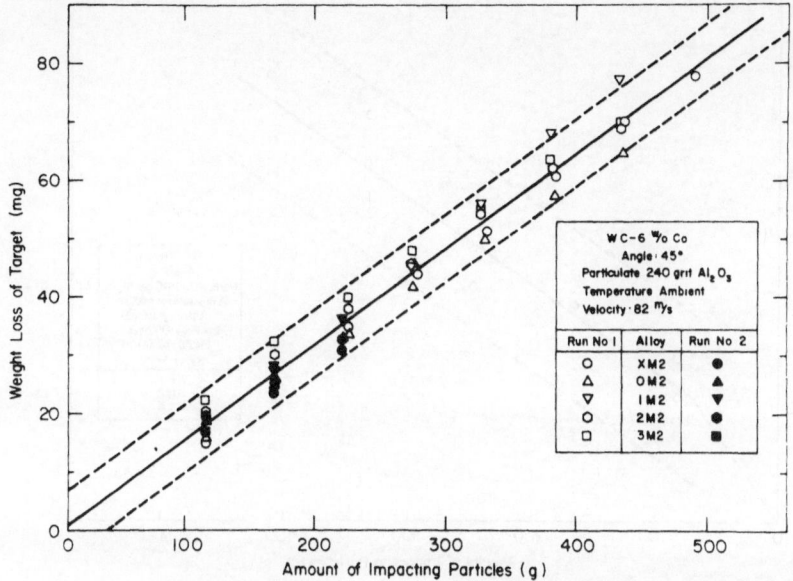

Fig. 3. Typical erosion curve for tests at high velocities and
particle impingement angles $\geq 45°$.

representing the as-ground surface has to be removed before a structure representing the steady-state is reached.

Effects of Particle Impingement Angle and Velocity. The effects of the particle impingement angle α on the steady-state erosion rate E_r as a function of particle velocity v is illustrated in
Fig. 5. The data points plotted here are the slopes of the least-
squares straight lines through all the data points for the various
specimens with a fixed Co content such as those in Figs. 3 and 4.
The angular dependence of the erosion rate depicted in Fig. 5 for
the 9 wt.% Co alloys also occurred for the 6 and 10.5 wt.% Co alloys. To be noted is that the maximum erosion rate occurs for α
near $90°$, which is typical erosion behavior for brittle materials.[8]

Also to be noted in Fig. 5 is that at all impingement angles
E_r increases with the particle velocity v. Log-log plots of E_r
versus v as a function of α for the 6 wt.% Co specimens are presented in Fig. 6. The straight lines through the data points reflect the commonly observed power relationship

$$E_r = Kv^n \tag{4}$$

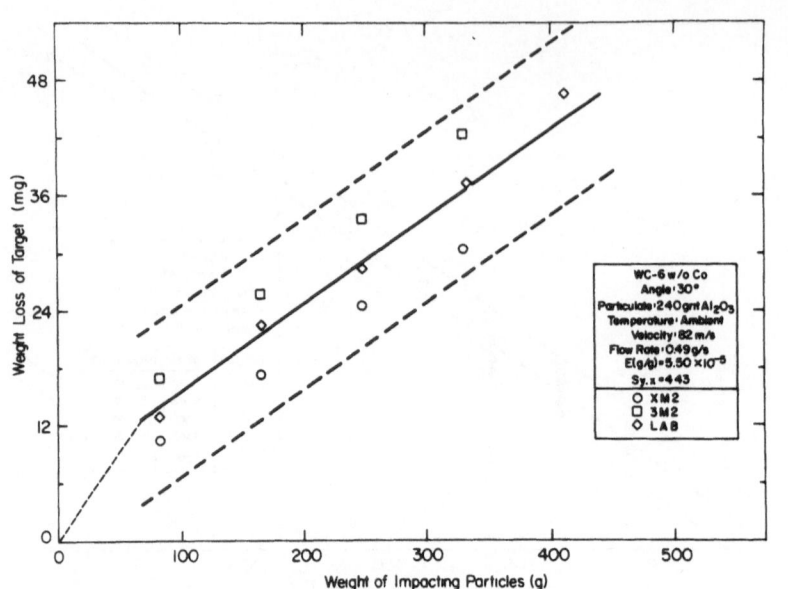

Fig. 4.   Typical erosion curve for tests at high velocities and
          particle impingement angles < 45°.

where n is approximately 2 for $\alpha \geqslant 45°$ and approximately 3 for α =
15°. Similar values of n were obtained for the higher Co alloys.
However, there was a tendency for n to increase with Co content;
see Table 2.

     The form of the angular dependence of E_r shown in Fig. 5 sug-
gests that the normal component of the particle velocity may be the
important factor. Hence, considering Eq. 4 one would expect that

Table 2.   Particle Velocity Exponent n

Co Content	Angle of Impingement		
(wt.%)	(degrees)		
	90	45	15
6	1.9	1.8	2.7
9	1.9	2.3	3.6
10.5	2.1	2.4	3.6

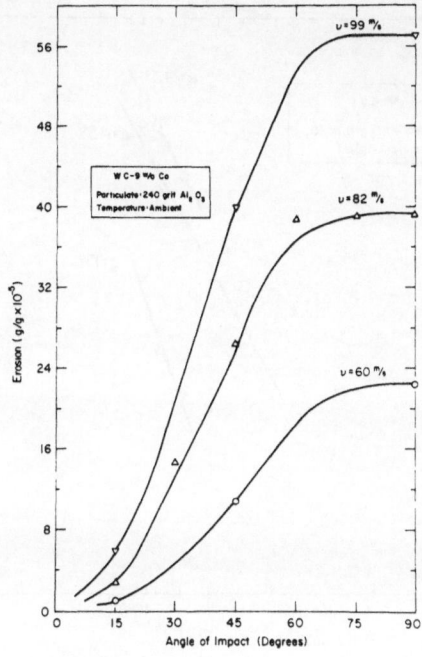

Fig. 5. The effect of the angle of particle impingement on the erosion of WC-9 wt.% Co alloy as a function of particle
velocity.

E_r should vary as $\sin^n \alpha$. This was in fact the case; see for example
Fig. 7. That the erosion of brittle solids may depend on the normal
component of the velocity had been proposed earlier by Hockey and
coworkers.[10,11]

  Effects of Microstructure. As mentioned above, the effects of
microstructure on the erosion rate varied with particle velocity.
In general, at low velocities and/or small angles of impingement E_r
was only slightly dependent on the Co volume fraction but increased
with decrease in WC grain size, decrease in Co free path and increase in WC grain contiguity, as represented by the specimens
tested. Conversely, at high velocities and/or large angles of impact, E_r was sensitively dependent on V_{Co} and relatively independent
of d, λ and C. These results are summarized in Fig. 8, which is a
log-log plot of E_r versus the normal component of the velocity v_{90}
(= $v\sin\alpha$) for all tests carried out in this investigation, including
the angular dependence. Indicated in Fig. 8 is that E_r for V_{Co} =
0.1 decreases as the WC grain size d increases (concurrently λ

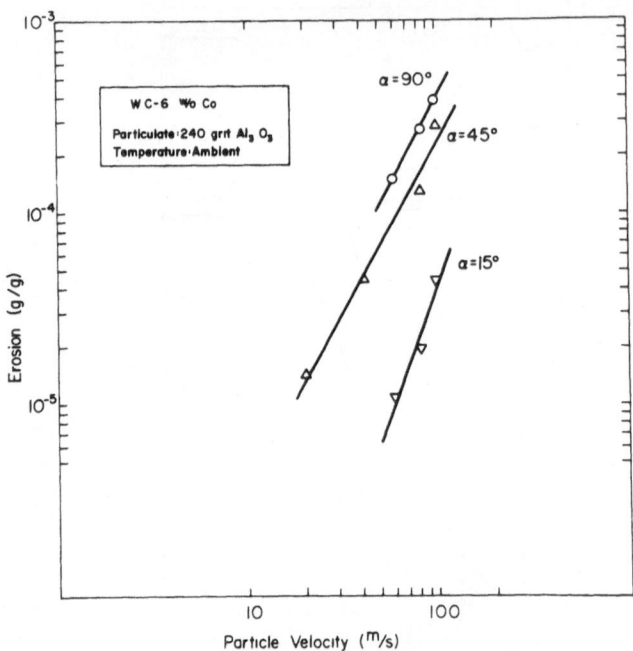

Fig. 6.   The effect of particle velocity on the erosion of the WC-6
          wt.% Co alloy as a function of impingement angle.

increases and C decreases) for $v_{90} < \sim 40ms^{-1}$ but is relatively independent of d at the higher values of v_{90}. Only one WC grain size
was investigated for V_{Co} = 0.15 and 0.17; the effect of d on the
erosion rate at these higher Co contents was therefore not
ascertained.

     The slope of the straight line drawn through the data points
in Fig. 8 for V_{Co} = 0.1 is approximately 2. To be noted is that
negligible erosion was detected for a specimen with this Co content
and d = 1.2μm tested for 480s with v = $42ms^{-1}$ and α = 15° (v_{90} =
$11ms^{-1}$), suggesting a critical velocity for erosion, as had been
proposed by others.[12-14] The data for the V_{Co} = 0.15 and 0.17 alloys in Fig. 8 can also be considered to have a slope of $\sim$ 2 for
$v_{90} \gtrsim 40ms^{-1}$. At lower values of v_{90} the erosion rate decreases
more rapidly with decrease in particle velocity than expected on
the basis of an extrapolation of the straight lines based on the
higher velocities, again suggesting the existence of a limiting velocity v_o, below which the erosion rate is essentially zero. The
existence of a critical velocity could then account for the larger

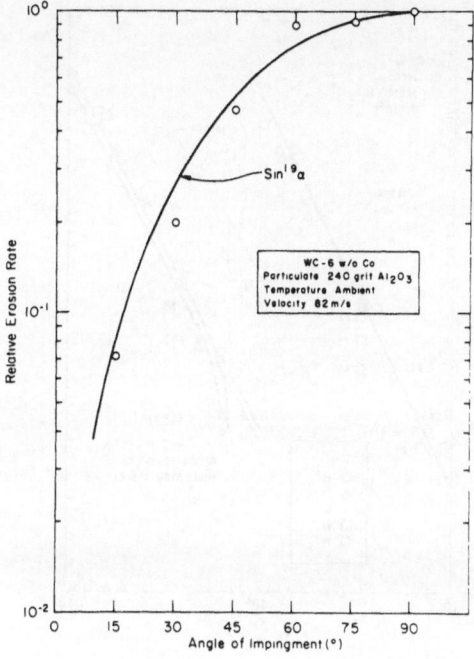

Fig. 7.   A comparison of the experimentally determined angular dependence of the erosion rate with the function $E_r \propto \sin^{1.9}\alpha$.

velocity exponent n obtained for small values of the particle velocity v and the impingement angle α.[13,14]

It should be pointed out that although the straight lines drawn
in Fig. 8 have a slope of $\sim$ 2, the data are too limited to place too
much emphasis on this specific value of the velocity exponent. Also, the data do not unambiguously establish that the combined angular
and velocity dependence of E_r is given by vsinα over the entire range
of conditions investigated. Nevertheless, the data do support a value of n at the higher velocities of the order of 2.0-2.5, which has
been reported for a number of brittle materials.[10,13-18] Also, the
angular dependence can be said to be approximated by the vsinα relation, as has been found by others.[10,16]

The results of Fig. 8 indicate that E_r increases with Co content as has been reported by others.[4-6] The present data are, however, too limited to permit any statement regarding the exact form
of the relationship between E_r and V_{Co}.

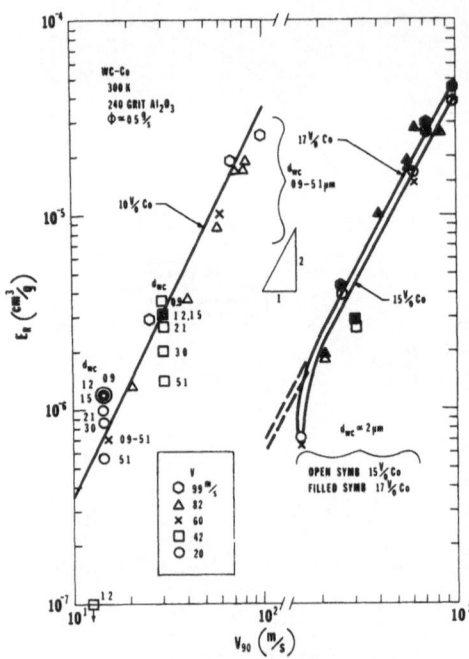

Fig. 8.   Erosion rate versus the normal component of the impinging
          particle velocity as a function of Co content. Numbers be-
          side symbols for 10 vol. % Co indicate WC grain size, d_{WC},
          in microns.   Symbols without numbers for 10 vol. % Co repre-
          sent average E_r for d_{WC} = 0.9 - 5.1μm.   All data points for
          15 and 17 vol. % Co are for $d_{WC} \simeq$ 2μm.

SEM Observations

     SEM micrographs of the surface of eroded specimens are given
in Figs. 9 and 10. To be noted in Fig. 9 is the cracking of the
carbide grains which occurred in Spec. 3M2 (V_{Co} = 0.1, d = 3μm)
following erosion at v = 42ms^{-1} and α = 45°. Similar cracking occurred for v = 82ms^{-1} and α = 45°. The cracks had more the appearance of radial or median-vent cracks, with no clear evidence of the
lateral-vent cracks observed in single-phase brittle materials.[10,13,17-19] The inability to observe lateral-vent cracks in
the WC grains may, however, simply be due to their small size. No
clear evidence of cracking of any kind was observed for Spec. 1M2
(V_{Co} = 0.1, d = 1.5μm) eroded at v = 82ms^{-1} and α = 90°, Fig. 10.
The carbide grains, however, appear to have undergone some
fragmentation.

Fig. 9. SEM micrograph of the eroded surface of Spec. 3M2 (6 wt.%
         Co; d_{WC} = 3μm) showing cracking of WC grains. v = 42 m/s;
         α = 45°.

     SEM micrographs of cross-sections of eroded specimens (which
include the eroded surface) are presented in Figs. 11 and 12. Fig.
11 shows what appears to be a protruding WC grain at the surface,
whose corners have been rounded. This suggests that an entire carbide grain may have been ejected from the microstructure during the
erosion process. Fig. 12 shows what appears to be a highly strained
or deformed layer just below the eroded surface (continuous white
band just below the surface having a thickness of about 0.5μm).

     An example of the appearance of the surface of a specimen eroded
with a very acute angle of particle impingement (15°) is given in
Fig. 13. There appears to be some "shadowing" of the Co phase by
the WC grains.

EROSION MECHANISM(S)

     The fact that maximum erosion of the cemented WC-Co alloys oc-
curred at an impact angle of ∿ 90° and that the angular dependence

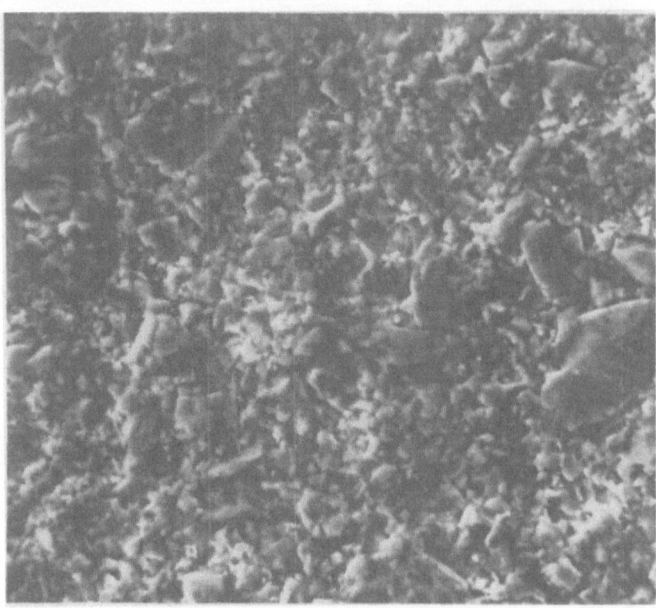

Fig. 10.   SEM micrograph of the eroded surface of Spec. 1M2 (6 wt.%
            Co; d_{WC} = 1.5μm) with no evidence of cracking of WC grains.
            v = 82 m/s; α = 90°.

of the erosion could be expressed in terms of the normal component
of the velocity leads to the conclusion that the major erosion mode
reflects the brittle fracture of the WC grains. The models proposed
for the erosion of brittle solids by impacting particles in the
quasi-static velocity range are of two general types: (a) those in
which the loss of material from the specimen surface occurs primarily by the development and linkage of Hertzian cone cracks (blunt indentors) and (b) those in which the loss occurs primarily by the
linkage of lateral cracks (sharp indentors). The former represents
elastic response with fracture initiating at preexisting surface
flaws, whereas the latter represents elastic-plastic response with
fracture initiating at a preexisting subsurface flaw (or newly created flaw) directly below the indentor, where high shear and tensile
stresses develop upon loading and unloading. As pointed out by
Evans and Wilshaw[20] the critical radius (and critical load) for the
transition from elastic Hertzian fracture to plastic indentation
and in turn elastic-plastic fracture depends on the preexisting
surface flaw size c_o and on the specimen hardness H and fracture
toughness K_c.

Fig. 11.   SEM micrograph of the etched cross-section of eroded Spec.
3M2 (6 wt.% Co; d_{WC} = 3.2μm) showing protruding WC grain
at the surface.   v = 42 m/s; α = 45°.

The various equations proposed for the two types of erosion of
brittle solids are listed in Table 3.   Using a statistical flaw distribution theory, Finnie and coworkers[21-23] developed the equation
given in Table 3 for erosion in the elastic Hertzian fracture regime.   Here the particle radius exponent m and the particle velocity
exponent n are functions of the Weibull distribution exponent,
which was found to have a value of 8 for angular particles impacting
on glass.   Following the approach of Adler,[24] Mehrotra et al.[25-27]
developed expressions for the erosion of brittle solids by spherical
particles in the Hertzian fracture regime, taking into account the
details of the Hertzian crack geometry.   A noteworthy feature of the
equations by Mehrotra et al.[25-27] is that they assume the existence
of a critical kinetic energy of the particle for the initiation of a
Hertzian crack and therefore include in the particle size and velocity relations values of the critical particle radius R_o and the critical velocity v_o.

The erosion equations in Table 3 based on lateral cracks include the dynamic hardness H and dynamic fracture toughness K_c.

Table 3. Equations for the Erosion of Brittle Materials

Authors	Equation	Comments

I. Based on Elastic Hertzian Fracture

Finnie et al.[21-23] $V=CR^m v^n$
$m=4.0-4.4$
$n=2.6-3.0$ Employs Weibull distribution flaw statistics.

Mehrotra et al.[25,26] $V=C_1(v^{0.8}-v_o^{0.8})^n$

$V=C_2R^2(R^{2/3}-R_o^{2/3})^n$

$n=1.8-2.7$ For spherical particles. Assumes critical load for Hertzian fracture.

II. Based on Elastic-Plastic Fracture

Evans and Wilshaw[20] $V=C\rho^{6/5}R^5 v^{12/5}/H^{1/2}K_c^{3/2}$ Based on lateral cracks.

Evans, Gulden, and Rosenblatt[28] $V=C\rho^{1/4}R^{11/3}v^{19/6}/H^{1/4}K_c^{4/3}$ Based on lateral cracks.

Ruff and Wiederhorn[27] $V=C\rho^{11/9}R^{11/3}v^{22/9}/H^{1/9}/K_c^{4/3}$ Based on lateral cracks.

V = volume removed   H = dynamic hardness
R = particle radius K_c = dynamic fracture toughness
v = particle velocity   C = constant
ρ = density of target

Variations in these equations result from differences in the expressions for the lateral crack size and the depth of penetration. An important consideration in these equations is the conversion of the kinetic energy of the particle into an equivalent load upon contact with the specimen.[20]

Let us compare the erosion results on the cemented WC-Co alloys with the models listed in Table 3. The velocity exponents obtained in the present tests (n = 1.9-3.6) are within the range listed in Table 3 for all of the models and therefore we cannot on this basis alone rule out any of the models listed. However, on the basis of the impacting particle size and morphology, and the velocity range investigated, it seems more reasonable to assume that erosion was due to elastic-plastic type of fracture than to pure elastic Hertzian fracture. Having made this assumption, we find that at the higher velocities the n-values are in better accord with the models of Evans and Wilshaw[20] and Ruff and Wiederhorn[27] than that of Evans et al.[28], whereas the reverse is true for the lower velocities.

Fig. 12.   SEM micrograph of the etched cross-section of eroded Spec.
OM2 (6 wt.% Co; d_{WC} = 1.2μm) showing continuous white
band next to eroded surface.   v = 82 m/s; α = 90°.

Further evaluation of the elastic-plastic fracture models is
possible by comparing the influence of hardness H and fracture
toughness K_c on erosion. K_c values were not determined on the present specimens.   However, Nakamura and Gurland[29] have shown that
there exists a direct correlation between K_c and H for cemented WC-
Co alloys.   The values of K_c listed in Table 1 were therefore taken
from a plot of K_c versus H using the data given by Pickens and
Gurland.[30]

Plots were prepared of E_r for the present tests at specific ve-
locities and angles of impingement versus $(H^{1/2}K_c^{3/2})^{-1}$, $(H^{1/4}K^{4/3})^{-1}$
and $H^{1/9}K_c^{-4/3}$, respectively.   Typical results are shown in Fig. 14,
which is a plot of E_r versus $(H^{1/4}K_c^{4/3})^{-1}$ for v = 42 and 99ms^{-1} at
α = 45°.   To be noted is that at the lower velocity some correlation
exists between E_r and $(H^{1/4}K_c^{4/3})^{-1}$; however, one may question whether there is a direct proportionality as required by the model.   A
departure from proportionality was also obtained by Evans[31] when he
compared the erosion data of a series of brittle materials with the

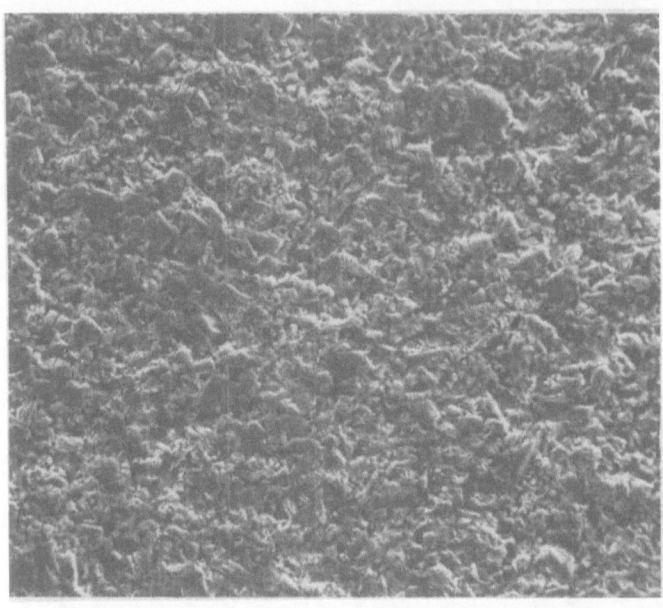

Fig. 13.   SEM micrograph of the etched surface of eroded Spec. 2M1
            (9 wt.% Co; d_{WC} = 1.8μm) showing evidence of "shadowing."
            Arrow indicates the component of particle flow parallel
            to the specimen surface.   v = 42 m/s; α = 15°.

parameter ($H^{1/4}K_c^{4/3}$). On the other hand, no correlation is evident
in Fig. 14 at the higher velocity.

   Behavior similar to that in Fig. 14 for both the low and high
velocity was obtained when E_r for the WC-Co alloys was plotted versus $(H^{1/2}K_c^{3/2})^{-1}$ and $H^{1/9}K_c^{-4/3}$, respectively, so that it was not
possible to determine which of the three models best described the
erosion data.

   The results of Figs. 8 and 13 thus indicate that for the WC-Co
alloys the erosion mechanism at low velocities ($v_{90} \leqslant 40$ ms^{-1}) may
differ from that at higher velocities. The value of the velocity
exponent and the correlation with the H-K_c parameters suggest that
the principal mechanism of material removal at the low velocities is
associated with an elastic-plastic fracture of the carbide grains,
i.e., lateral cracking. However, the present data do not permit an
identification of the specific details of the cracking as defined by
the three models listed in Table 3. It is expected that the

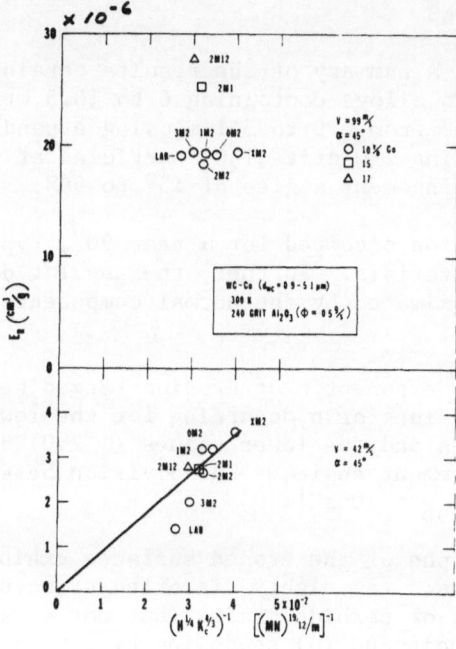

Fig. 14.   Erosion rate versus $(H^{1/4}K_c^{4/3})^{-1}$ for tests at v = 42 and 99 m/s and $\alpha = 45°$.

surrounding Co matrix will exert a modifying influence on the fracture behavior.

The erosion behavior at the higher velocities does not conform to the predictions of the elastic-plastic models, both in regard to the lower value of the velocity exponent and to the lack of correlation with hardness and fracture toughness. The results are more in accord with an elastic Hertzian fracture model; however, the operation of this type of fracture is expected to occur at lower velocities than those for elastic-plastic fracture rather than at higher velocities. Another possibility is that at the higher velocities the kinetic energy of the impacting particles is sufficiently high to eject the carbide grains in their entirety. However, if this were so, we would not expect the erosion to correlate with the normal component of the velocity as was observed here. Again, the Co matrix is expected to exert an influence on the cracking of the WC grains and hence a direct comparison of the erosion of the cemented WC-Co alloys with that for materials consisting entirely of brittle phases may not be realistic.

SUMMARY AND CONCLUSIONS

The following is a summary of the results obtained on the erosion of cemented WC-Co alloys containing 6 to 10.5 wt.% Co and with mean WC particle sizes from 0.9 to 5.1μm using a sandblaster-type erosion tester employing 240 grit Al_2O_3 particles at velocities of 20 to 99 ms^{-1} and impingement angles of 15° to 90°:

1. Maximum erosion occurred for α near 90°, typical of the erosion of brittle materials. Further, the angular dependence of the erosion was approximated by the normal component v_{90} of the particle velocity.

2. The velocity exponent n of erosion ranged between 1.9 and 3.7 with the higher values of n occurring for the lower velocities and impingement angles and the lower values ($\sim$ 2.0) at the higher velocities and impingement angles. The division between the two regimes occurred at $v_{90} \simeq 40$ms^{-1}.

3. SEM micrographs of the eroded surfaces exhibited the following damage features: (a) highly distorted structure near the surface, (b) cracking of carbide grains (but not always), (c) protrusion of carbide grains and (d) shadowing by carbide grains at shallow impingement angles.

4. The following effects of microstructure on erosion were found: (a) the erosion rate decreased with increase in WC size d at low velocities ($v_{90} \leqslant 40$ ms^{-1}) but was relatively independent of d at higher velocities and (b) the erosion rate increased with volume fraction of Co, the effect increasing with increase in velocity.

5. A comparison of the present results on WC-Co alloys with existing erosion models for brittle materials yielded the following: (a) some correlation was found with the elastic-plastic fracture models for the tests at low velocities and (b) the correlation at the higher velocities was more in accord with an elastic Hertzian fracture model, which is not expected if the elastic-plastic fracture applies at the lower velocities. It is expected that the Co exerts a modifying influence on the erosion behavior of the WC grains and hence a direct comparison with the erosion models for materials consisting entirely of brittle phases may not be realistic.

REFERENCES

1. DOE Newsletter Materials and Components in Fossil Energy Applications, No. 22, June 1, 1979.
2. Ibid, No. 19, December 1, 1978.
3. Ibid, No. 23, August 1, 1979.

4. A. A. Dañkin, Poroshkovaya Metallurgiya, No. 10 (94) 73
    (October 1970).
5. Kh. Uuémyis, I. Kleis, V. Tumanov and T. Tildemann,
    Poroshkovaya Metallurgiya, No. 3 (135) 98 (March 1974).
6. J. S. Hanson, Erosion: Prevention and Useful Applications, ASTM
    STP 664, W. F. Adler, ed. (1974),pp. 148.
7. H. C. Lee and J. Gurland, Mat. Sci. Engr. 33, 125 (1978).
8. I. Finnie, Corrosion-Erosion Behavior of Materials, K. Natesan
    (ed.), AIME, New York (1980), pp. 118.
9. A. W. Ruff and L. K. Ives, Wear 35, 195 (1975).
10. B. J. Hockey, S. M. Wiederhorn and H. Johnson, Fracture Mechan-
    ics of Ceramics Vol. 3: Flaws and Testing, R. C. Bradt,
    D. P. H. Hasselman and F. F. Lange, eds., Plenum Press, New
    York (1978), pp. 379.
11. B. J. Hockey and S. M. Wiederhorn, Proc. 5th Int. Conf. Erosion
    by Liquid and Solid Impact, Cambridge Univ., UK (1979),
    pp. 26.
12. J. G. A. Bitter, Wear 6, 5 (1963).
13. J. L. Routbort, R. O. Scattergood and E. W. Kay, J. Amer. Cer.
    Soc. 63, 635 (1980).
14. R. O. Scattergood and J. L. Routbort, Wear 67, 227 (1981).
15. J. L. Routbort, R. O. Scattergood and A. P. L. Turner, Wear
    59, 363 (1980).
16. J. L. Routbort, and R. O. Scattergood, J. Amer Cer. Soc. 63,
    593 (1980).
17. Gordon A. Sargent, P. K. Mehrotra and Hans Conrad, Ceramic Mi-
    crostructures '76, Westview Press, Boulder, Colorado (1977),
    pp. 872.
18. G. A. Sargent, P. K. Mehrotra and H. Conrad, Erosion: Preven-
    tion and Useful Applications, ASTM STP664, W. F. Adler, ed.
    (1979), pp. 77.
19. M. E. Gulden, Erosion: Prevention and Useful Applications,
    ASTM STP664, W. F. Adler, ed. (1979), pp. 101.
20. A. G. Evans and T. R. Wilshaw, Acta Met. 24, 939 (1976).
21. G. L. Sheldon and I. Finnie, J. Eng. Ind. (November 1966),
    pp. 393.
22. I. Finnie and H. Oh, Proc. 1st Congr. Int. Soc. Rock Mechanics,
    Vol. 2 (1966), pp. 99.
23. H. L. Oh, K. P. L. Oh, S. Vaidyanathan and I. Finnie, The Sci-
    ence of Ceramic Machining and Surface Finishing, NBS Spec.
    Publ. 348 (1972), pp. 119.
24. W. F. Adler, Analysis of Multiple Particle Impacts on Brittle
    Materials, Tech. Rept. AFML-TR-74-210, Air Force Materials
    Lab., Wright-Patterson Air Force Base, Ohio (September 1974).
25. G. A. Sargent, P. K. Mehrotra and H. Conrad, Proc. 5th Int.
    Conf. Erosion by Solid and Liquid Impact, Cambridge,
    England (September 3-5, 1979), pp. 28-31.
26. P. K. Mehrotra, G. A. Sargent and H. Conrad, Corosion-Erosion
    Behavior of Materials, TMS-AIME (1980), pp. 127.

27. A. W. Ruff and J. M. Wiederhorn, Treatise on Materials Science and Technology, Vol. 16 Erosion, Carolyn M. Preece, ed., Academic Press, New York (1979), p. 69.
28. A. G. Evans, M. E. Gulden and M. E. Rosenblatt, Proc. Roy. Soc. London A361, 343 (1978).
29. Morihiko Nakamura and Joseph Gurland, Met. Trans. 11A, 141 (1980).
30. J. R. Pickens and J. Gurland, Mat. Sci. Engr. 33, 135 (1978).
31. A. G. Evans, Treatise on Materials Science and Technology, Vol. 16 Erosion, Carolyn M. Preece, ed., Academic Press, New York (1979), pp. 1.

RESISTANCE OF CEMENTED CARBIDES TO SLIDING ABRASION:

ROLE OF BINDER METAL

Jorn Larsen-Basse

Department of Mechanical Engineering
University of Hawaii at Manoa
Honolulu, HI 96822

ABSTRACT

Cemented carbides of the WC-Co, WC-FeNi and TiC-MoNi families were studied. Laboratory abrasion test results for hard and soft abrasives were compared with SEM studies of removal mechanisms and with hardness and Palmquist crack resistance data.

In wear by hard abrasives material is removed from the wear surface by a mechanism involving gross plastic deformation due to yielding and extrusion of binder metal and/or by a mechanism of spalling due to crack propagation in the binder metal. In wear by soft abrasives material is removed by a series of microprocesses involving extrusion of binder followed by relaxation of compressive stresses in the carbide grains and subsequent cracking and fragmentation of these grains. In systems where the diffusion or mechanical bond between binder and carbide is weak whole grains may be pulled from the wear surface. In systems where a strong diffusion bond exists between the two phases material may be removed by brittle spalling due to cracks propagating through both phases. Under some wear conditions binder metal may be smeared back onto the wear surface. In the case of cobalt this may provide a lubricating, wear reducing effect.

INTRODUCTION

Cemented carbides are generally considered to be highly resistant to abrasive wear and this property is a major reason for the selection of these materials for a large number of applications. The straight tungsten carbide-cobalt compositions have traditionally been used in such highly abrasive situations as rock drilling, mining and excavation. However, since cobalt is a relatively soft metal and since high hardness and good abrasion resistance generally go hand in hand there is a continual search for stronger and harder binder metals which preferably also should be cheaper and more readily available. This search has, as an example, resulted in development of experimental alloys with binders of steel or iron-nickel alloys and some of these compositions can show very high hardness values and also high resistance to abrasion in conventional abrasion tests. The detailed effects of the binder on failure mechanisms is not well understood.

In practice cemented carbide wear parts usually fail due to the combined action of a number of mechanisms, such as cracking due to impact or fatigue, sliding abrasion, high temperatures at contact points, and corrosion by fluids. The present study is concerned with only one of these, namely sliding, low-speed abrasive wear. The purpose is to determine the removal rates and removal mechanisms as they are affected by the binder metal and thereby, hopefully, obtain a baseline for selection of new binder alloys for test.

EXPERIMENTAL

The abrasion testing has been described in detail previously.[1-3] Briefly, the 2-4 mm wide specimen is loaded against the lower rim of a 1.5 mm wide, 70 mm diameter low-carbon steel wheel which rotates to give a surface velocity of 8.5 cm/s. The contact zone between specimen and wheel is flooded with an abundant supply of fresh abrasives and the wear rate is determined from measured values of weight loss and density.

Some tests were performed with a diamond stylus loaded against a polished specimen surface and pulled over the surface under load in single-strokes or in multistroke reciprocating cycles.[4]

The WC-Co alloys were obtained from two sources. One batch was manufactured by Titan-Fagersta in Australia for the Division of Tribophysics of CSIRO. The other batch was manufactured by Multimetals, Inc. for tests at Brown University. The WC-FeNi alloys (0-40% Ni) and the TiC-MoNi alloys (18-22% Mo) were experimental alloys obtained from the Ford Motor Company Research Laboratories. Microstructure parameters were determined by linear analysis of a number of SEM photomicrographs of polished surfaces taken at random and counting at least 600 grains for each specimen.

Before abrasion the testing surface was hand ground on water flooded SiC abrasive papers to a depth of at least 50 μm below the original sintered surface and then rough polished by using 15 μm diamond paste.

RESULTS

Examples of wear rate-load curves are shown in Fig. 1. Straight lines are obtained in all cases, except for very high loads. The abrasion resistance defined as the reciprocal of the slope of these lines, is plotted versus specimen bulk hardness in Fig. 2 for a number of different alloy families abraded by three different abrasives. The curves show the typical transition between two regions. In one region, the "hard abrasion" region, the abrasives are considerably harder than the metal, abrasion resistance is relatively low, and minor changes in alloy hardness have little effect on abrasion resistance. In the other region, the "soft abrasion" region, the abrasives are either softer than the alloy or only slightly harder, abrasion resistance is high, and minor changes in alloy hardness can have a substantial effect on abrasion resistance. This type of behavior was described by Nathan and Jones[5] for conventional metals and alloys and has been described for cemented carbides by the author and co-workers in a number of cases.[2,3,6,7] The approximate transition hardness values estimated from Fig. 2 are listed in Table 1.

Table 1

Estimated Hardness Values (DPN) for Transition
Between Hard and Soft Abrasion

Abrasive	SiO_2	Al_2O_3	SiC
Abrasive hardness	9 - 1100	2040	2800
Alloy family			
WC-Co	1000	1800	(2050)
WC-FeNi	900	1700	2000
TiC-MoNi	1250	(>1900)	(>2200)

It is noticed from Fig. 2 that of the three alloy families tested, at any given hardness in the hard abrasion region the TiC-MoNi alloys have the lowest abrasion resistance, followed by the WC-Co alloys, while the WC-FeNi alloys have the highest resistance. It is also noticed that the rapid rise in abrasion resistance--the transition from hard abrasion to soft abrasion--takes place at a lower hardness value for the WC-FeNi alloys, followed by the WC-Co alloys, while the TiC-MoNi alloy family is a distant third. In other words, on the basis of bulk hardness values alone, at any given hardness the WC-FeNi alloys show the best overall resistance to sliding abrasion, while the TiC-MoNi alloys show the poorest resistance and the WC-Co alloys fall in between, closer to the WC-FeNi alloys.

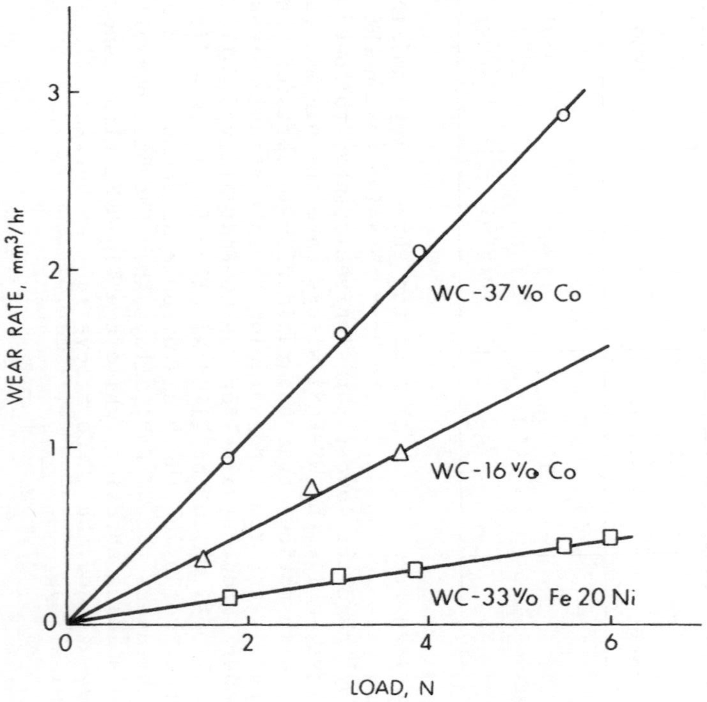

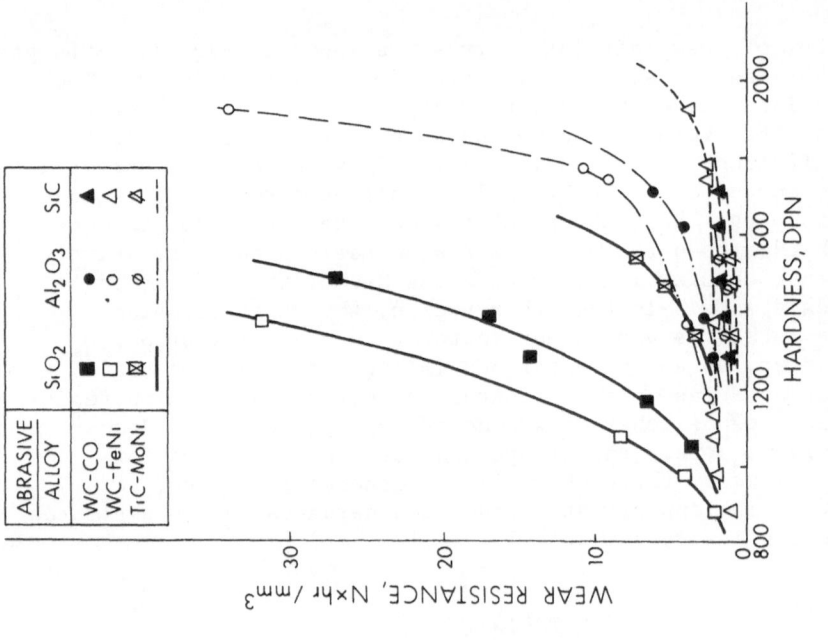

Fig. 2. Abrasion resistance as a function of alloy bulk hardness for abrasion by 120 μm, quartz, alumina and silicon carbide abrasives.

Fig. 1. Volumetric wear rate as a function of applied load for abrasion by 120 μm quartz abrasives. Alloy bulk hardness values are 850, 1040 and 1300 DPN, respectively, reading down.

Abrasion data have been compared with microstructure parameters such as carbide grain size and contiguity and binder volume and mean free path. The only parameter which shows a clear correlation with abrasion resistance is binder mean free path, as illustrated in Figs. 3 and 4. The data for the hard abrasion region are plotted in Fig. 3. It is seen that abrasion resistance decreases with increasing binder mean free path for both alloy families for which data have been obtained. Comparison of the data for WC-Co alloys abraded by different abrasives shows that the abrasion resistance decreases as the hardness of the abrasive increases from the approximately 2040 kg/mm^2 value for Al_2O_3 over the 2800 kg/mm^2 for SiC to the 3100 kg/mm^2 value for B_4C. Comparison of the data for the untreated FeNi binder alloys with the data for the same alloys treated to remove retained austenite shows that the tempering does result in some improvement in abrasion resistance in this region. Comparison of the WC-Co and the WC-FeNi alloy data for abrasion by SiC shows that in this hard abrasion region at any given mean free path the cobalt binder gives lower abrasion resistance than does the FeNi binder.

The effect of binder mean free path on abrasion resistance in the soft abrasion region is illustrated in Fig. 4. The relationships appear to be linear with a common extrapolated intercept at an abrasion resistance of around 60 for zero binder mean free path. Interestingly, in this region of soft abrasion the cobalt binders give much better resistance to abrasion than do the iron-nickel binders of the same mean free path; and the decrease in resistance with increased binder mean free path is much faster for the iron-nickel binders; the two lines follow the equations

$$R = 60 - 220 \ \lambda \tag{1}$$

for the iron-nickel binders, and

$$R = 60 - 95 \ \lambda \tag{2}$$

for the cobalt binder, where R is the abrasion resistance and λ is the mean free binder path.

SEM studies of abraded surfaces performed in this and in prior studies have resulted in a qualitative description of the material removal mechanisms for the WC-Co alloys as follows, see the sketches in Fig. 5:

- in the hard abrasion region (Fig. 5a) the abrasive can act as a cutting tool; it cuts craters or grooves in the surface which are much larger than the individual carbide grains; material is removed by gross plastic deformation which results in severe fragmentation of the carbide grains in the surface and in the extruded material; for the more brittle

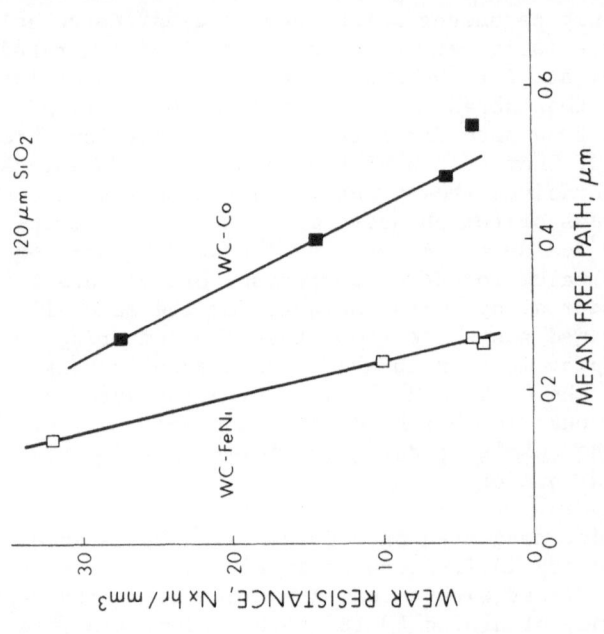

Fig. 4. Abrasion resistance vs. binder mean free path in the soft abrasion region.

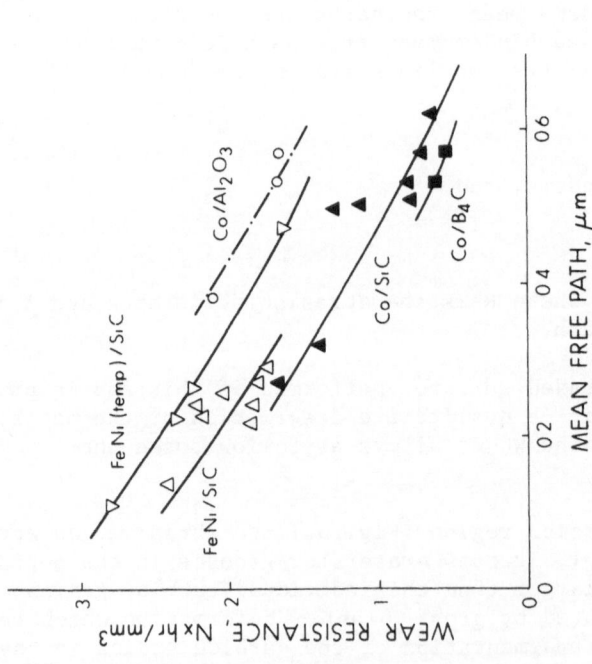

Fig. 3. Abrasion resistance vs. binder mean free path for WC alloys with cobalt, untreated FeNi and tempered FeNi binders, abraded by B₄C, SiC and Al₂O₃. All alloy-abrasive combinations are in the hard abrasion region.

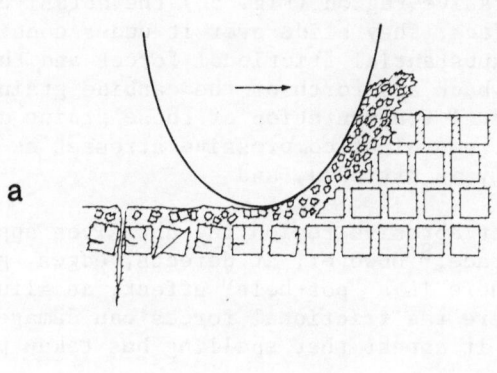

a

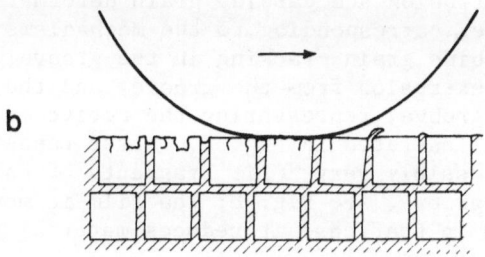

b

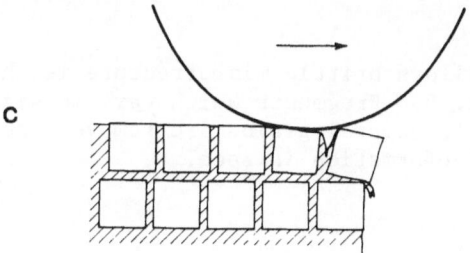

c

Fig. 5. Sketches of the proposed abrasion mechanisms
        for WC-Co alloys. a - hard abrasives; b -
        soft abrasives; c - edge effects for soft
        and very soft abrasives.

alloys some material may be lost due to brittle spall
formation,

- in the soft abrasive region (Fig. 5b) the abrasives cannot
  indent the surface; they slide over it under considerable
  load and with substantial frictional forces and this results
  in the rocking back and forth of the carbide grains on each
  cycle, and gradual fragmentation of these grains due to
  fatigue and to removal of compressive stresses as the cobalt
  binder gradually is extruded, and

- in the very soft abrasive region the abrasives appear to
  polish the surface;[8] however, at defects, edges, grinding
  marks, etc., there is a "pot-hole" effect, as illustrated
  in Fig. 5c, where the frictional forces can damage any edge
  and often make it appear that spalling has taken place.

The experiments with the diamond stylus confirm these findings,
see Fig. 6. Binder extrusion and carbide grain deformation can be
seen outside the groove, corresponding to the mechanisms active in
soft abrasion; the carbide grain cracking in the groove, the cobalt
and carbide fragment extrusion from the groove, and the spall
formation outside the groove, representing the active mechanisms
in hard abrasion are illustrated in Fig. 7. After repeated sliding
a film of binder and possibly very fine fragments of carbide forms
on the surface of the groove, see Fig. 8; the film is more evident
for cobalt binders and in that case it reduces material removal
rates.[4]

For the WC-FeNi alloys the stylus experiments showed an additional material removal mechanism--uprooting of whole carbide
grains, as found previously when these alloys were abraded by SiC
abrasive papers.[9]

For the TiC-MoNi alloys brittle microfracture is the dominant
wear mechanism, see Fig. 9. Fragments which vary in size from
much less than a grain to several grains are removed and little or
no evidence of plastic deformation is seen.

DISCUSSION

The wear rate-load curves in Fig. 1 represent data for both
the hard and the soft abrasion regions. The direct proportionality
obtained in all cases separates the cemented carbides from more
ductile conventional alloys. For these, direct proportionality is
found when using very hard abrasives,[5] while linear relations of
positive load intercept are found when softer abrasives are applied.[7]
This has been explained as due to the fact that a certain surface
strain, or a corresponding ratio of indentation depth to grit

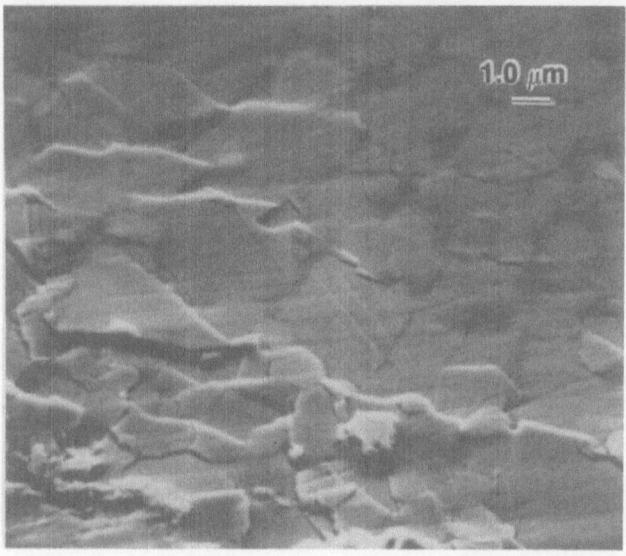

Fig. 6.   Surface of WC-22 v/o Co specimen just outside a groove
          produced by sliding a diamond stylus over the polished
          surface

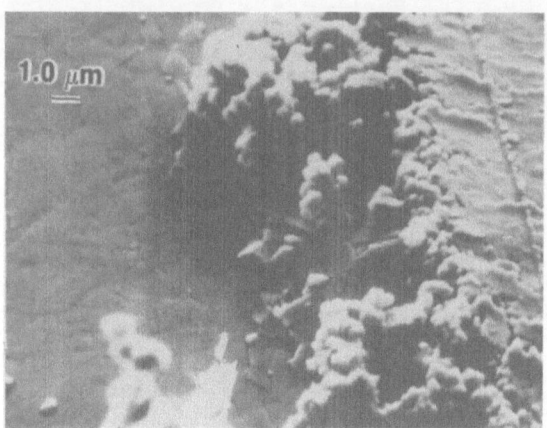

Fig. 7.   WC-13 v/o Co alloy after wear by diamond stylus.

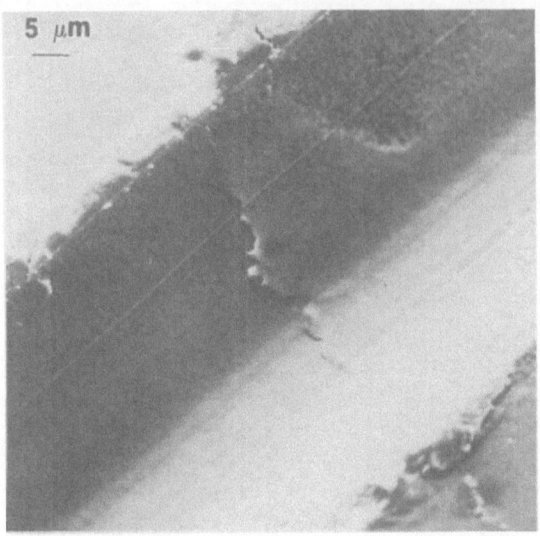

Fig. 8. WC-25% Fe10Ni after wear by diamond stylus.

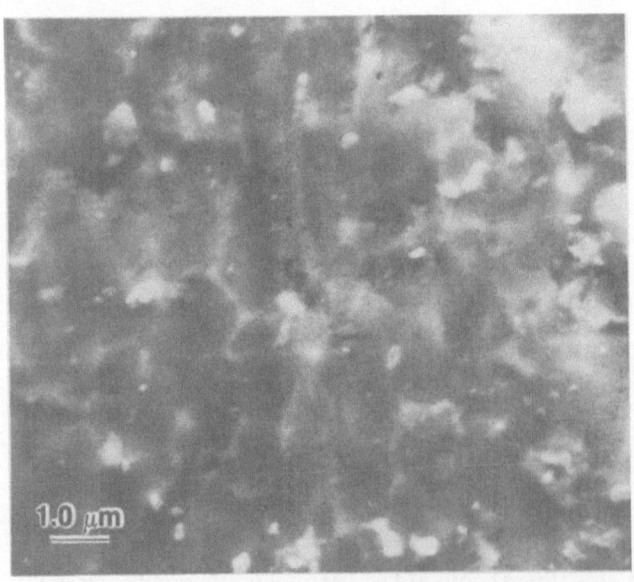

Fig. 9. TiC-26 v/o Ni21%Mo alloy after abrasion by alumina.

diameter, must be attained before plastic flow away from the surface can take place.[7] Microprocesses of a totally different nature
control the behavior of cemented carbides in soft abrasion, as
sketched in Figs. 5a and b, where no indentation is necessary for
wear to take place and wear rate is related to the number of contacts
and the forces in each contact. For this mechanism a direct pro-
proportionality between wear rate and load would be expected.

The effect of microstructure parameters on abrasion resistance
illustrated in Figs. 3 and 4 can be explained while considering
each region separately. In the hard abrasion region of Fig. 3,
where gross plastic deformation is the primary removal mechanism,
it is expected that abrasion resistance would be determined by the
material's resistance to plastic flow at a strain corresponding to
abrasion of the material in question. For ductile metals which
form distinct chips the corresponding strain has been estimated to
fall in the range 4-6.5.[10] For the cemented carbides cracking and
fragmentation take place before substantial chips can form and the
average plastic strain associated with hard abrasion may, consequently, be somewhat lower. It will, nevertheless, be substantially higher than the value found in a hardness test, for which an
average strain of approximately 0.1 has been estimated.[11]

Stress-strain curves for cemented carbides are rare and none
exist to the high strains encountered in abrasion. It is expected
that behavior at lower strains, e.g., at a hardness indentation,
may give some general indication of behavior of the same material
at higher strain values. Thus, both hardness and mean free path,
which correlates closely with hardness,[12] are expected to be good
indicators of abrasion resistance for alloys in the same family
abraded by hard abrasives. Since factors such as work hardening
and ductility control actual resistance to abrasion only a general
correlation can be expected when comparing alloys of different
families on the basis of hardness or mean free path. The data
of Fig. 3 for abrasion by SiC show that the cobalt and the austenitic and ferritic FeNi binders follow almost the same relationship with the cobalt compositions giving slightly lower abrasion
resistance, while the martensitic FeNi binders give a noticeable
improvement in abrasion resistance for the same mean free path.

In the soft abrasive region, for which data are plotted in
Fig. 4, the cobalt binder gives a great improvement in abrasion
resistance over the FeNi binders when compared on the basis of
mean free binder path. When data are compared on the basis of
bulk hardness, see Fig. 2, the position of the two alloy families
is inversed and the FeNi binders give the greater wear resistance.
Since it is evident from the mechanisms discussed previously that
no clear correlation with bulk hardness should be expected this
behavior is probably incidental.

Figure 4 appears to show an intercept at zero mean free path, which could correspond to the abrasion resistance of polycrystalline carbide for the case where large scale fracture and spalling has been suppressed. As the binder width is increased the wear resistance drops linearly because the binder areas take over from the grain boundaries of the polycrystalline carbide in controlling carbide grain microcracking and thereby wear rate. The linearity of the relations could indicate simply that a binder's resistance to removal is proportional to its width, possibly because of the number of deformation events which can take place, or it could hide a combination of factors such as decreasing residual tensile stresses in the binder and compressive stresses in the carbide and increasing binder and decreasing carbide contents as the mean free path grows plus other factors such as the deformation mechanisms, etc. At this stage it is not possible to distinguish between these. It should be pointed out also, that the compressive strength of WC-Co alloys has been found to decrease linearly with increasing mean free paths[13] and that the strain hardening coefficient increases with cobalt content[14,15] and thus generally with mean free path. The curves are expected to become horizontal at a low wear resistance, corresponding to the resistance of the binder metal itself at a point where the carbide grains are so far apart that their presence has no significant effect on abrasion resistance.

A number of factors may explain the difference in behavior between the two different alloy families:

- the cobalt binder forms a strong bond with the carbide grains; this prevents grain pull-out and makes it more difficult to remove the cobalt by plastic extrusion,

- the cobalt binder contracts more on solidification and thus is expected to set up greater compressive stresses in the carbide grains which should decrease their tendency to crack,

- the fcc-hcp transition of binder cobalt gives this material a considerable ductility which may explain its ability to extrude from the binder areas and form a lubricating film on the surface; the resulting solid lubrication may result in lower wear rates, and

- the cobalt binder usually contains very fine precipitates of WC grains; this may result in an effective mean free path for soft abrasion which is considerably smaller than the value determined by conventional means.

Additional work will be needed to further elucidate these factors.

The TiC-MoNi family is an example of a case where the interfacial bond is extremely strong. Molybdenum is added to promote wetting of the carbide grains and it diffuses into the TiC grains from the liquid. The result is a material which is more brittle and behaves more like a ceramic than the two-phase WC-Co and WC-FeNi alloy families. This can be seen, for example, from the Palmquist crack susceptibility values, which range up to 8.5 μm/kg for the cobalt binder alloys, to 9.2 μm/kg for the FeNi binder alloys and which fall between 10.0 and 16.7 μm/kg for the TiC-MoNi alloys used here. In this case, then, the interfacial bond is so strong that the binder phase has almost lost its individual properties and the composite behaves more like a single phase material of low ductility and abrasion takes place by brittle microfracture. The wear rate is higher than for WC-Co and WC-FeNi alloys of similar hardness and the transition between hard and soft abrasion takes place at higher hardness values because true indentation is not required in order to obtain substantial material removal rates.

CONCLUSIONS

It has been shown that the WC-Co, WC-FeNi and TiC-MoNi families of cemented carbides all show a transition between hard and soft abrasion. The transition takes place at higher hardness values for the brittle TiC-MoNi alloys because they wear primarily by brittle microfracture, while indentation by the abrasives into the metal surface is not required in order for material removal to take place. The material removal in hard abrasion is by plastic deformation and/or by brittle fracture and resistance to abrasion is determined primarily by bulk hardness and crack resistance. In soft abrasion different mechanisms operate, and ductile binders with a good bond to the carbide grains give the best performance. In this region hardness is only a secondary indicator of abrasion resistance.

ACKNOWLEDGMENTS

The author is grateful to Nirmal Devnani, Kam Lau Ma, Perry Tanouye, Salem Lakshmipathy and Curtis Chun for performing many of the tests. Most of the work was supported by the National Science Foundation under Grant DMR 76-17158.

REFERENCES

1. J. Larsen-Basse and P. A. Tanouye, "Abrasion of WC–Co Alloys by
    Loose Hard Abrasives," in: Proc. 1976 Int. Conf. on Hard
    Material Tool Technology, R. Kamanduri, ed., Carnegie-Mellon
    University, Pittsburg (1976), 188.

2. J. Larsen-Basse and E.T. Koyanagi, Abrasion of WC–Co Alloys
    by Quartz, J. Lub. Technol. 101:208 (1979).

3. J. Larsen-Basse, Abrasive Wear Resistance of Tungsten Carbide
    Composites with Iron-Nickel Binder, in: "Wear of Materials,"
    S.K. Rhee, A.W. Ruff and K.C. Ludema, eds., ASME, New York
    (1981), 534.

4. N.M. Devnani, Contact Fatigue in Wear of WC–Co and WC–FeNi
    Alloys, M.S. Thesis, University of Hawaii at Manoa (1979).

5. G.K. Nathan and W.J.D. Jones, Influence of Hardness of Abra-
    sives on the Abrasive Wear of Metals, in: "Lubrication and
    Wear, Fifth Convention, Proc. Inst. Engrs.," 181:215 (1966-67).

6. J. Larsen-Basse, Some Mechanisms of Abrasive Wear of Cemented
    Carbide Composites, Metaux, Corrosion-Industrie, 653:8 (1980).

7. J. Larsen-Basse and B. Premaratne, Abrasive Wear Mechanisms––
    Effect of Relative Hardness, 2nd Asian-Pacific Corrosion
    Control Conference, Kuala Lumpur, Malaysia (1981).

8. J. Larsen-Basse, Mechanisms of Wear of Sintered Carbide Dental
    Burs, J. Lub. Technol. 102:560 (1980).

9. J. Larsen-Basse, C.M. Shishido and L.K. Salem, Abrasion of
    Some Cemented Carbides by SiC Papers, in: "Proc. 1976 Int.
    Conf. on Hard Material Tool Technology," R. Komanduri, ed.,
    Carnegie-Mellon University, Pittsburgh (1976), 231.

10. J. Larsen-Basse, Abrasion Mechanisms––Delamination to Machining,
    in: "Fundamentals of Tribology," N.P. Suh and N. Saka, eds.,
    The MIT Press, Boston (1978), 679.

11. D. Tabor, The Physical Meaning of Indentation and Scratch
    Hardness, Brit. J. Appl. Phys. 7:159 (1956).

12. H. Fischmeister and H.E. Exner, Gefugeabhangigkeit der
    Eigenschaften von WC–Co Hartlegierungen, Arch. Eisenhwes
    37:499 (1966).

13. H.Y. Doi, Y. Fujiwara and Y. Oosawa, Mechanical Behavior of
     WC-Co Composite Alloys, in: "Mechanical Behavior of Materials,"
     Japan Society of Materials Science, 5:207 (1972).

14. A. Hara and T. Ikeda, Behavior in Compression Deformation of
     WC-Co Cemented Carbide, Trans. Japan Inst. Metals 13:128
     (1972).

15. R.P. Felgar and J.O. Lubahn, Mechanical Behavior of Cemented
     Carbides, Proc. ASTM, 57:770 (1957).

16. D. Moskowitz and M. Humenik, Cemented Carbide Cutting Tools,
     in: "Modern Developments in Powder Metallurgy, H.H. Hausner,
     ed., 3:83 (1966).

DISCUSSION

J. A. Peck:

The last slide you showed, where you said that the iron-nickel
binder wore considerably more than the cobalt binder in soft forma-
tions, was it martensitic, had it been quenched and then tempered,
or do you know what the condition of the binder was?

J. Larsen-Basse:

It doesn't make such a big difference. We have done some work on
both. We didn't show it in this particular case because it doesn't
make a great deal of difference. If you have an austenitic unstable
binder it will turn martensitic in the abrasion. Not always in soft
abrasion, but in hard abrasion it usually does. But here there is
not a great deal of difference and I think it mostly is a ductility
exhaustion problem rather than a straight strength problem of the
binder.

J. A. Peck:

I just wondered what the structure of the binder was.

J. Larsen-Basse:

Some of them are austenitic and some of them martensitic.

J. A. Peck:

And you got indifferent results, depending on the condition of the binder.

J. Larsen-Basse:

If there is bulk deformation it makes quite a bit of difference (20 to 30%). But when we are talking about relatively soft abrasives, they pretty much fall in the same line.

J. A. Peck:

A last question on the iron-nickel binders. Do you know how much austenite was in the binder before it was quenched?

J. Larsen-Basse:

No, I don't.

V. K. Sarin:

Your model shows that in abrasion (the way you are measuring wear) the breaking of tungsten carbide grains is critical. Yet you have not considered tungsten carbide grain size in your comparisons. Could differences from different alloys be due to different tungsten carbide grain sizes?

J. Larsen-Basse:

We have tried that, and we can't find any reasonable correlation there. We think that the real controlling factor is binder removal. Once a binder is partly removed, the carbides will break regardless of size.

V. K. Sarin:

Even in the hard abrasives like silicon carbide?

J. Larsen-Basse:

It looks like it. That once you get past the strain, remember we are talking about very, very high strains, it's all just ground up to very fine material. That must be the reason. I would say, though, that we haven't tried something with say a 5 or 10 micron grain size, but more in the 0.5 to 2 micron range.

V. K. Sarin:

There are published results which show that for 1 or 2 micron grain
size compared to 10 or 11 micron grain size you would get differences.
Not with the set-up you've got, but other types of abrasive wear, pin
on wheel, for example.

J. Larsen-Basse:

That's possible.

ABRASION RESISTANCE OF CERMETS CONTAINING Co/Ni BINDERS

W. Precht, R.K. Viswanadham* and J.D. Venables

Martin Marietta Laboratories
1450 South Rolling Road
Baltimore, Maryland 21227

ABSTRACT

Prior work in these Laboratories[1,2,3] has shown that the properties of the binder phase play a crucial role in determining the mechanical behavior of cermets. In this work we have investigated the effect of alloying the binder phase of a model VC/Ni cermet with cobalt. The results indicate that the abrasion resistance of the cermets exhibits a pronounced maximum when the binder phase composition reaches the value Co/Ni ≈ 85/15. The effect appears to be related to the presence and stability of the h.c.p. phase in the binder. For Co/Ni > 85/15, the h.c.p. and f.c.c. phases are originally intermixed, but X-ray diffraction of abraded surfaces indicates that the binder converts fully to the h.c.p. phase in the near surface regions upon deformation. For Co/Ni < 85/15, the f.c.c. phase is stabilized and does not transform during abrasion. Accordingly, we suggest that the maximum in abrasion resistance can be accounted for in the following manner: at high Co contents, the abrasion resistance is somewhat degraded because of the limited ductility of the Co – h.c.p. phase. At high nickel content, the abrasion resistance is again degraded because the binder becomes relatively soft. The maximum then appears at a binder composition where the f.c.c. phase becomes fully stabilized (allowing ductile behavior) but where the binder hardness is still high due to its large cobalt content. Similar results obtained for WC/Co with Ni additions suggest a means for conserving critical cobalt supplies while improving abrasion resistance for this important cutting tool material.

---

* Currently at Reed Rock Bit Company, P.O. Box 2119, Houston, Texas 77001.

INTRODUCTION

    The basic objective of this study is to understand how the
mechanical behavior of cemented carbides (cermets) is influenced
by the fundamental properties of their constituent phases.   In
earlier investigations it was determined that there is a definite
interrelationship between the hardness of the binder phase, the
cleavage energy of the carbide, and the properties of a cermet.[3]
For example, it was observed that for a carbide of a given cleavage energy, the failure mode of the cermet can be changed from one
where the crack propagates primarily through the binder phase to
one where the fracture propagates primarily through the carbide
phase by adjusting the hardness of the binder from a low to a high
value. The best combination of hardness and fracture toughness is
obtained when the crack propagates through the binder, although it
is most advantageous to have as high a binder hardness as possible.
The results show that, the cleavage energy of the carbide phase
plays an important, but indirect role in determining the properties
of the cermet since its value places an upper limit on the binder
hardness that may be employed without transferring the failure
mode of the system to one in which brittle fracture of the carbide
dominates.

    In this segment of the work, attention is focused on the role
played by the deformation mode of the binder in determining the
abrasion resistance of a cermet. To determine this, we have studied
the model cermet system, VC/Ni, in which Co has been systematically
substituted for Ni. The choice of Co was based upon the fact that
it: 1) forms a complete series of solid solutions with Ni, 2) is
not soluble in the carbide, 3) does not form a carbide, and 4) has
a significantly higher hardness than Ni. Upon completion of the
work with the VC/(Ni,Co) system, cermets based on the WC/(Co,Ni)
system were tested to verify the generality of the conclusions.

EXPERIMENTAL PROCEDURE

Composition Selection

    Based upon previous work,[3] it was deemed necessary in these
studies to produce cermets satisfying certain fracture mode vs.
hardness criteria. Data reported in reference (3), and reproduced
in Fig. 1, illustrate three different types of crack resistance vs.
hardness behavior labeled Types I, II, and III. (The crack resistance parameter, W, which is measured by the Palmquist method, is
related to K_{IC} as discussed in reference (1).) In general, Type I
behavior is the most desirable since for any given value of hardness, the crack resistance (or K_{IC}) is highest for a cermet system
that exhibits this type of behavior. For the present study, in
which we desired to examine the role of the binder in determining

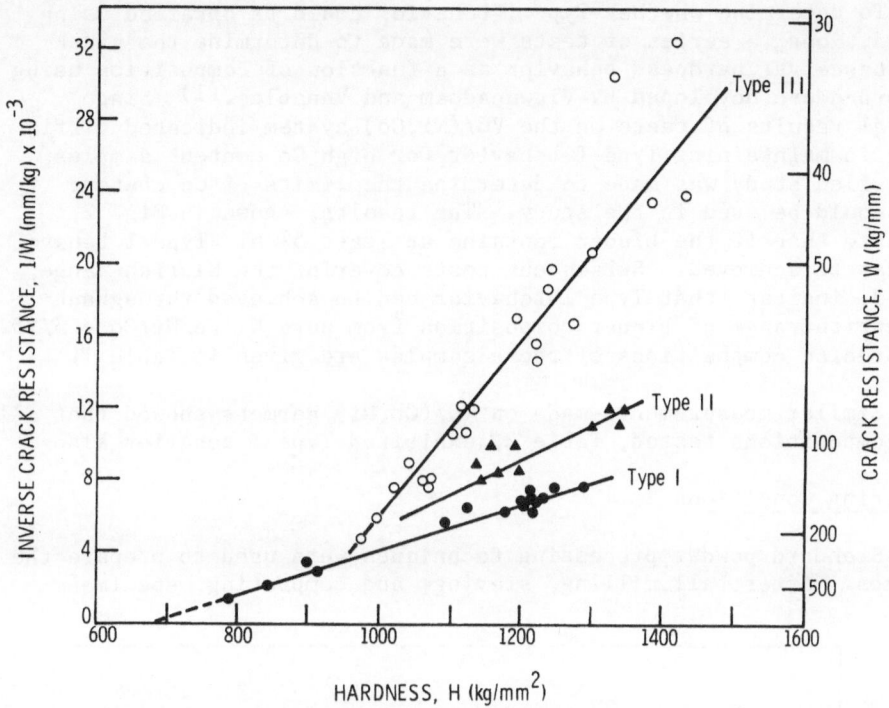

Figure 1. Inverse crack resistance vs. hardness of VC/Ni cermets.
The data fall into three groups (labeled Type I, Type II,
and Type III) each characterized by a different slope
(after Ref. 3).

cermet mechanical properties (abrasion resistance), the requirement
for obtaining Type I behavior was crucial since this is the only
type for which the effect of binder properties can be tested in a
relatively straightforward manner. Thus, using Auger electron
spectroscopy on fracture surfaces, we have observed that the dominant mode of failure in Type I cermets is fracture through the
binder; for Type III, carbide cleavage dominates; for Type II,
mixed mode fracture occurs.[3]

For this investigation we avoided circumstances that gave rise
to the Type II and Type III behavior observed in our prior work.
Specifically, we avoided using large additions of vanadium which we
employed previously to adjust the vanadium/carbon ratio of the carbide phase to a high value. Apparently, at high vanadium concentrations, the vanadium partitions into the binder forming an exces-
sively hard two-phase structure which changes the failure mode to
one in which brittle fracture of the carbide dominates giving rise
to Type II or Type III behavior.

    To determine whether Type I behavior could be obtained using
Co additions, a series of tests were made to determine the crack
resistance vs. hardness behavior as a function of composition using
the procedure developed by Viswanadham and Venables.[1] Since
initial results of tests on the VC/(Ni,Co) system indicated difficulty in maintaining Type I behavior for high Co content samples,
a detailed study was made to determine the limits of Co content
that could be used in the study. The results, shown in Fig. 2,
indicate that if the binder contains at least 5% Ni, Type I behavior can be achieved. Subsequent tests covering the Ni rich range,
Fig. 3, indicate that Type I behavior can be achieved throughout
the entire range of binder composition from pure Ni to Ni/Co = 5/95.
The precise compositions of these samples are given in Table 1.

    Similar measurements made on WC/(Co,Ni) cermets showed that all
the compositions tested, Table 2, exhibited Type I behavior also.

Sintering Conditions

    Standard powder processing techniques were used to prepare the
cermets. After ball milling, sieving, and compacting, specimens

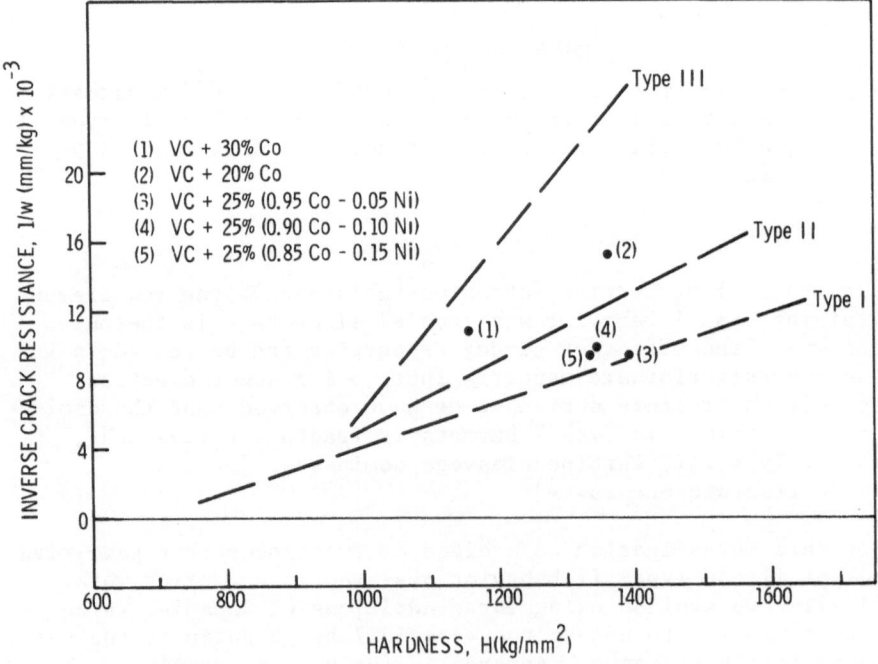

Figure 2.   Inverse crack resistance vs. hardness of VC/(Ni,Co)
             cermets. The data show that small additions of Ni
             stabilize Type I behavior.

Table 1

Starting Powder Blends For Abrasion Samples (W/O)

VC	Ni	Ni/V*	Co	Co/V*	Approximate Percent of Binder	Binder Phase
75.44	15.5	9.06	–	–	20	
70.38	20.0	9.62	–	–	25	All Ni
65.44	24.5	10.06	–	–	30	
75.44	15.5	–	–	9.06	20	
70.38	20.0	–	–	9.62	25	80 Ni-20 Co
65.44	24.5	–	–	10.06	30	
75.44	7.75	4.53	7.75	4.53	20	
70.38	10.00	4.82	10.00	4.82	25	50 Ni-50 Co
65.44	12.25	5.03	12.25	5.03	30	
75.44	–	9.06	15.50	–	20	
70.38	–	9.62	20.00	–	25	20 Ni-80 Co
65.44	–	10.06	24.50	–	30	
70.38	–	7.20	20.00	2.43	25	15 Ni-85 Co
65.44	–	8.20	24.50	1.86	30	
70.38	–,	4.76	20.00	4.86	25	10 Ni-90 Co
75.44	–	1.90	15.46	7.20	20	
70.38	–	7.20	20.00	2.43	25	5 Ni-95 Co
65.44	–	3.00	25.60	5.96	30	
70.46	–	–	19.94	9.62	25	All Co

* Prealloyed powders, containing 47.5% vanadium

Table 2

Composition, Density and Hardness of WC/(Co,Ni) Cermets

90 w/o Carbide (2.5 μ) + 10 w/o Binder

Binder Composition	Density g/cc	Hardness kg/mm^2
100% Co	14.56	1376
90% Co - 10% Ni	14.57	1306
85% Co - 15% Ni	14.58	1304
80% Co - 20% Ni	14.61	1306
70% Co - 30% Ni	14.50	1283

88 w/o Carbide (2.5 μ) + 12 w/o Binder

Binder Composition	Density g/cc	Hardness kg/mm^2
100% Co	14.29	1266
90% Co - 10% Ni	14.28	1270
85% Co - 15% Ni	14.31	1275
80% Co - 20% Ni	14.29	1225
70% Co - 30% Ni	14.38	1236

3/8" dia. and 3/8" high were sintered in a vacuum of 3×10^{-5} Torr. for approximately one hour at 1290°C for the VC/(Ni,Co) cermets, and 1430°C for WC/(Co,Ni).

Abrasion Tests

Abrasion tests were carried out using a modified Whirlimet* polishing and grinding machine with a 45 μm diamond lap and 22 psi load per sample. A constant speed of 100 rpm was maintained through-out the testing. A water flow of about 50 ml/min was used to flush the abraded material and provide cooling. The specimens were held in stainless steel mounts and indexed for repeatable alignment.

Abrasion resistance was determined by measuring the weight loss and converting it to a volume loss. The results, which were recorded in hours/cc lost, were normalized to a standard specimen of WC/Co of

* Product of the Buehler Corporation.

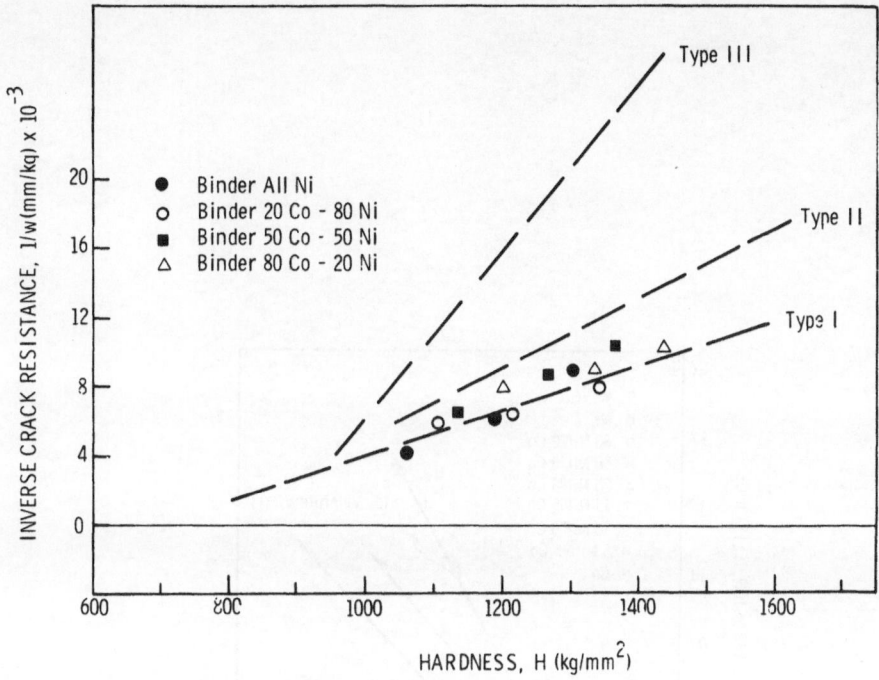

Figure 3.   Inverse crack resistance vs. hardness of VC(Ni,Co)
             cermets. The data show that Type I is exhibited
             throughout the entire composition range shown.

intermediate hardness to provide relative abrasion resistance. The
average of three 15 minute runs was used to calculate the abrasion
resistance.

RESULTS AND DISCUSSION

     The relative abrasion resistance of VC/(Ni,Co) alloys plotted as
a function of cermet hardness is shown in Fig. 4. For specific binder
contents, the relative abrasion resistance is observed to take an
interesting form as shown in Fig. 5. The figure indicates that the
relative abrasion resistance of this model cermet system exhibits a
pronounced maximum when the binder phase composition reaches the value
Co/Ni = 85/15. An explanation of this interesting observation was
obtained using X-ray diffraction to compare the structure of abraded
and non-abraded cermet surfaces.

     Table 3 summarizes the effect of abrasion on the crystallographic
structure of the binder phase in the near surface region for various

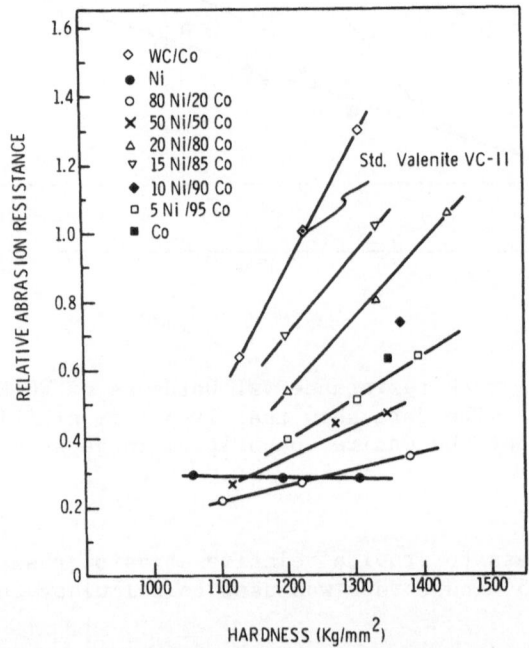

Figure 4.   Relative abrasion resistance vs. hardness of VC/(Ni,Co)
             cermets. All data were normalized to that of Valenite
             VC-11, a commercial WC/Co cutting tool material.

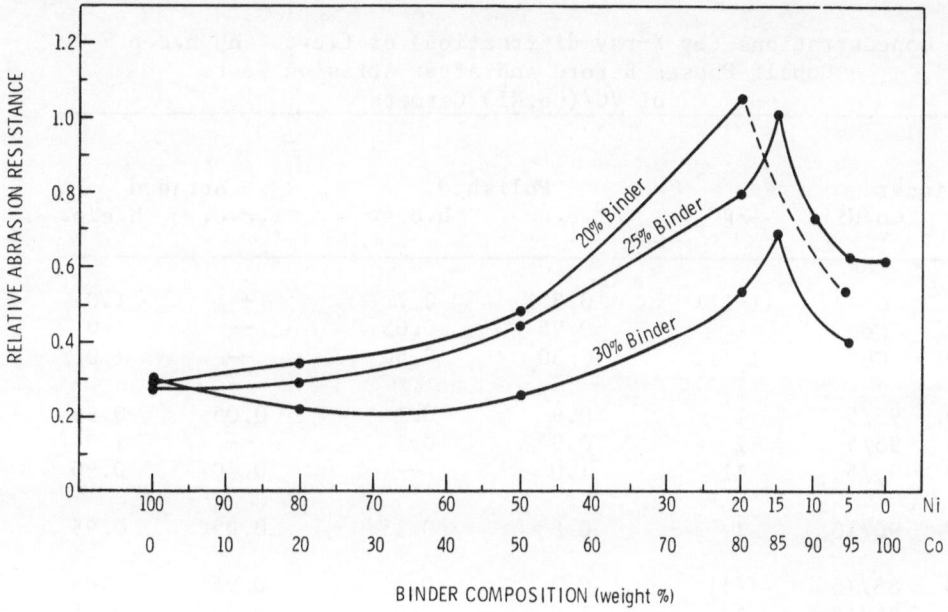

Figure 5.   Relative abrasion resistance vs. binder composition for
            VC/(Ni,Co) cermets. Data show abrasion resistance passes
            through a maximum at Co/Ni = 85/15.

cermet compositions. Two points are worth noting. First, Ni additions promote the formation of a f.c.c. binder phase in the unabraded
material. Second, although the effect of abrasion on material having
a high Co content is to transform the f.c.c. binder phase to h.c.p.,
the transformation is almost completely suppressed when the Ni/Co
ratio is greater than 15/85. These observations can be used to
account for the observed maximum in relative abrasion resistance in
the following manner: At high Co content, the abrasion resistance
is degraded because of the limited ductility of the h.c.p. cobalt
phase. This follows from the work of Kanel and Halim (4) who demonstrated that the work-hardening coefficient of the h.c.p. is four
times greater than that of f.c.c. cobalt. On the other hand, at
high Ni content the abrasion resistance is again degraded because
the binder becomes relatively soft. The maximum then appears at a
binder composition where the f.c.c. phase becomes stabilized (allowing ductile behavior) but where the binder hardness is still high
due to its large cobalt content.

   To test the generality of these results, similar experiments
were performed on the WC/Co system to which Ni additions were made
to the binder phase. The compositions studied are shown in Table 2,
and the abrasion test results are summarized in Figs. 6 and 7.

Table 3

Concentrations (by X-ray diffraction) of f.c.c. and h.c.p.
Cobalt Phases Before and After Abrasion Tests
of VC/(Co,Ni) Cermets

Binder wt%	Co/Ni	Type	Polished f.c.c.	h.c.p.	Abraded f.c.c.	h.c.p.
30	Co	11-111	0.3	0.7	--	1.0
25	Co	1	0.95	0.05	--	1.0
20	Co	11-111	0.50	0.50	--	1.0
30	95/5	1	0.9	0.1	0.05	0.95
25	95/5	1	0.9	0.1	--	1.0
20	95/5	11	1.0	--	0.70	0.30
25	90/10	1	0.85	0.15	0.05	0.95
30	85/15	1-11	0.7	0.3	0.95	0.05
25	85/15	1	1.0	--	0.25	0.75
20	85/15	1-11	1.0	--	1.0	--
30	80/20	1	0.95	0.05	0.05	0.95
25	80/20	1	1.0	--	1.0	--
20	80/20	1	1.0	--	1.0	--
30	50/50	1	1.0	--	1.0	--
25	50/50	1	1.0	--	1.0	--
20	50/50	1	1.0	--	1.0	--

The curve depicted in Fig. 7, which shows the abrasion resistance as a function of binder composition, was developed from the
data shown in Fig. 6. The procedure used was to construct curves
(not shown) having the same slope as the 100% Co binder curve and
which represented the best fit for data corresponding to each of
the different compositions. The abrasion resistance determined at
an arbitrarily chosen cermet hardness of 1300 kg/mm^2 for each composition was then used to construct Fig. 7.

When plotted in this way, it is quite evident there is a peak
in abrasion resistance at approximately the same binder composition
observed for the VC/(Co,Ni) system but the effect is not as pronounced for WC/(Co,Ni). By analogy, we suggest that the reason
for the maximum is similar in the two cases although we have not

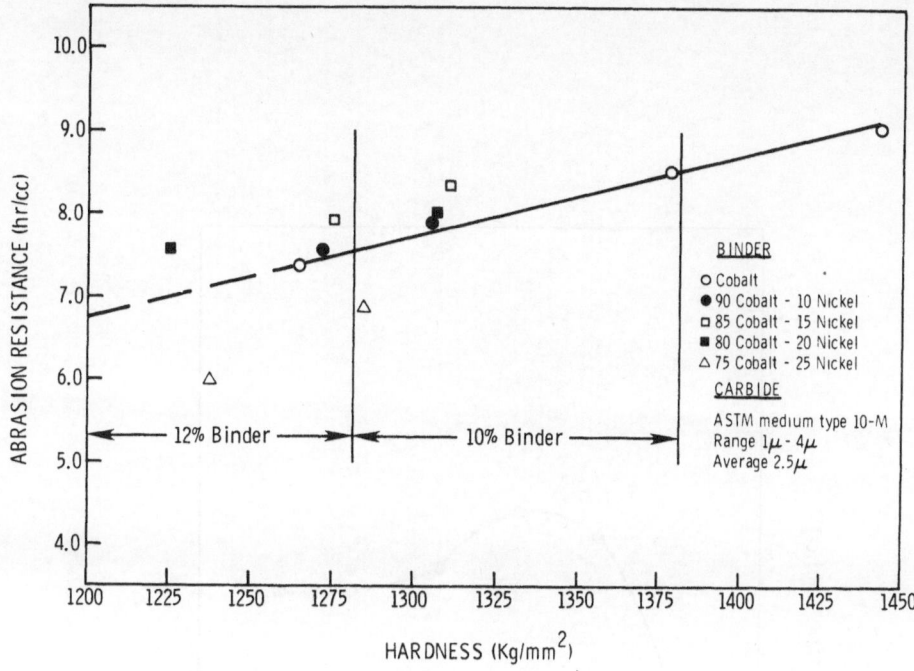

Figure 6.   Abrasion resistance vs. hardness of WC/(Co,Ni) cermets
containing binder volume fractions of 10% and 12%.

as yet performed a detailed X-ray diffraction analysis of abraded
WC/(Co,Ni).   Nonetheless, we note that Giamei et al. (5) have observed abrasion induced transformations in the binder phase of
WC/Co just as we have observed in VC/(Co,Ni) cermets.   In their
work, they determined that the cobalt in unabraded material was
100% f.c.c., whereas 90% of it converted to the h.c.p. phase in
heavily abraded material.   We suggest that the presence of Ni
suppresses this transformation, just as we have observed in the
VC/(Co,Ni) system, thus leading to the improvement in abrasion
resistance shown in Fig. 7 when Co/Ni ~ 85/15.

Finally, a comment is in order regarding our observation that
the maximum in abrasion resistance as a function of the Co/Ni
ratio is not as pronounced for WC/(Co,Ni) as for VC/(Co,Ni).   We
suggest this may be related to the fact that in the absence of Ni
additions, the f.c.c. Co phase is already stabilized to some extent
in WC/Co whereas it is not in VC/Co.   Accordingly, even though Ni
additions are apparently needed to stabilize the binder against an
allotropic phase transformation during deformation, the relative
effect is somewhat smaller for WC-based cermets than for VC-based
cermets because of the different initial conditions.   Nonetheless,

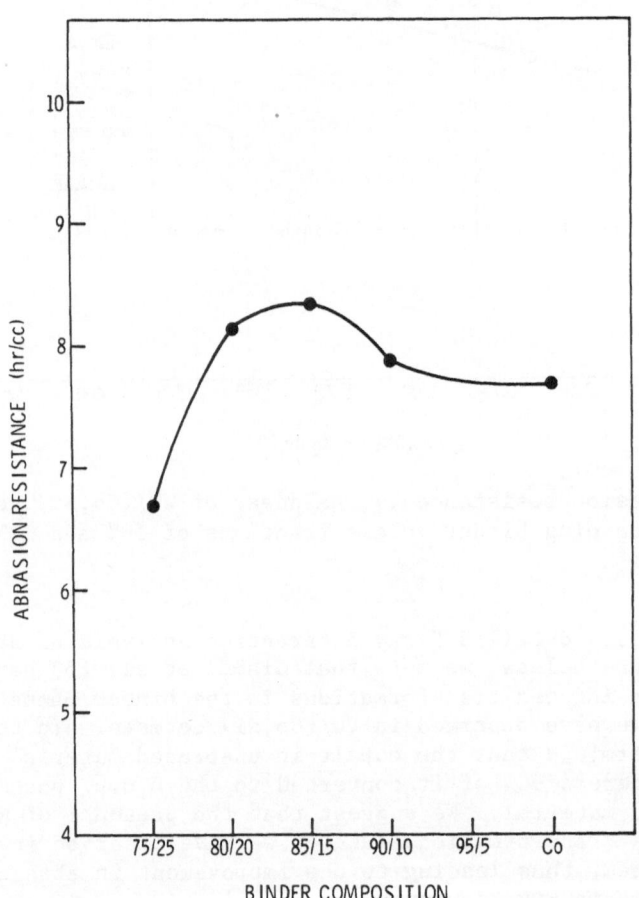

Figure 7.   Abrasion resistance vs. binder composition of WC/(Co,Ni)
            cermets at a hardness level of 1300 kg/mm^2. Data show
            maximum in abrasion resistance at Co/Ni $\approx$ 85/15.

the prospect of improving the abrasion resistance of WC/Co cermets, while at the same time conserving a critical material such as Co, appears attractive.

CONCLUSIONS

As a result of this investigation we can conclude that:

1) For VC/(Co,Ni) and WC/(Co,Ni) cermets, a significant maximum in abrasion resistance occurs when Co/Ni ~ 85/15.

2) The effect is due to stabilization of the relatively ductile f.c.c. cobalt phase by Ni, preventing it from transforming to the less ductile h.c.p. phase during deformation.

3) By replacing 15% of the cobalt binder phase in WC/Co cermets with Ni, the abrasion resistance may be improved while at the same time a critical material (cobalt) is conserved.

ACKNOWLEDGEMENTS

We wish to thank the National Science Foundation who supported this work under Grant DMR 77-08354.

REFERENCES

1. R.K.Viswanadham and J.D. Venables, "A Simple Method for Evaluating Cemented Carbides," Met. Trans. 8A:187 (1977).

2. R.K. Viswanadham, B. Sprissler, W. Precht, and J.D. Venables, "The Effect of V/Ti Ratio on the Partitioning of Mo in (V,Ti)C + (Ni,Mo) Cemented Carbies," Met. Trans. 10A:599 (1979).

3. R.K. Viswanadham and W. Precht, "Preparation and Properties of VC + Ni Cermets with Controlled Carbon-to-Metal Ratio," Met. Trans. 11A:1475 (1980).

4. R. Kanel and K. Halim, "The Effect of Phase Change on the Mechanical Properties of Cobalt Near its Transformation Temperature," Phys. Stat. Sol. 15:63 (1966).

5. A.F. Giamei, J. Burma, S. Rabin, M. Cheng, and E.J. Freise, "The Role of the Allotropic Transformation in Cobalt-Base Alloys," (Part II) Cobalt No. 40, September, 1968, 140.

DISCUSSION

E. Kimmel:

What binder levels were you working at?

W. Precht:

For the VC it was 30, 25 and 20 per cent binders by weight. For
WC-Co it was 10 and 12 per cent.

E. Kimmel:

When you were making the Co-Ni binders were you pre-alloying the
powder prior to making it, or were you admixing nickel and cobalt?

W. Precht:

Just admixing.

D. Moskowitz:

I would like to comment on the relation between abrasion resistance
and machining. I think in general this isn't a prime variable with
respect to whether a tool material is an excellent one in machining,
at least for machining of steel. It's generally recognized that
TiC-(Ni, Mo) has greater wear resistance in machining steel than
WC base alloys independent of whatever criterion you use for tool
life. However, in terms of abrasion resistance, it's generally recognized that they are not in the same class. So, apparently abrasive wear resistance, in itself, is not a good gauge for whether a
tool runs well, at least in in finish machining of steel.

W. Precht:

Yes. We also found this. We did some evaluation of WC-Co,
(V, Ti)C-(Ni, Mo) and TiC-(Ni, Mo) in steel machining. Our material
fell right in between the two. So, it was better than WC-Co for a
given hardness, but not as good as TiC-(Ni, Mo).

ADHESION AND FRICTION OF TRANSITION METALS

IN CONTACT WITH NONMETALLIC HARD MATERIALS

Kazuhisa Miyoshi and Donald H. Buckley

National Aeronautics and Space Administration
Lewis Research Center
Cleveland, Ohio

INTRODUCTION

Friction is the most commonly observed phenomenon incidental
to the interaction of two solid surfaces.

In order to gain a fundamental understanding of the surface
interactions between metals and nonmetals, it is extremely important to consider the basic material properties that determine and
influence adhesion and friction.

The authors have studied surface interactions for many years
and have been interested in fundamental experimental studies to
determine the basic material properties of metals that relate to
the adhesion and friction of metals in sliding contact with nonmetallic hard materials, such as diamond, silicon carbide, boron
nitride and manganese-zinc ferrite (refs. 1 to 4).

Diamond, the hardest known material, is generally used for
machining nonferrous alloys, abrasive materials (such as presintered carbide, borides and nitrides), graphite, fiberglass and
rubber. Diamond is also widely used in the electronics and jew-
elry industries as a wear resistant coating material and in machining tools as well.

Silicon carbide has been used and has great potential for use
in high-hardness and high temperature applications such as stable
high-temperature semiconductors, turbine ceramic seal systems, gas
turbine blades and as an abrasive in grinding.

Boron nitride would be a good solid lubricant if it's adhesion to metals could be improved. This has, however, not been achieved to date.

Manganese-zinc ferrite is very important as a typical magnetic material used for highly developed magnetic recording devices, that is, video tape recorders. It is therefore widely used.

Most of the foregoing applications of hard nonmetallic materials deeply involve surface interactions. Understanding surface interactions and reactions is therefore important.

In the 1940's Pauling recognized differences in the amount of a d-bond character associated with transition metals. The filling of d-electron band he found to be responsible for various physical and chemical properties including adhesive energy, Young's and shear moduli, tensile and shear strengths, chemical stability, and magnetic properties (ref. 5). The greater the amount or percentage of d-bond character that a metal possesses, the less active is its surface.

The effect of the chemical properties of metals on adhesion and friction should be identified. The knowledge gained in such studies will assist in achieving a better understanding of the adhesion and friction properties of metals sliding against nonmetals.

The objective of this paper is to discuss the adhesion and friction properties of various transition metals in contact with various nonmetallic hard materials. The transition metals examined include yttrium, titanium, tantalum, zirconium, vanadium, neodymium, iron, cobalt, nickel, tungsten, platinum, rhenium, ruthenium, and rhodium. The nonmetals examined were single-crystal diamond, silicon carbide, boron nitride and manganese-zinc ferrite. The investigation also examined metal transfer to the hard nonmetallic materials. All the experiments were conducted with a metal pin contacting a nonmetallic hard material flat at loads of 0.05 to 0.3 N, at a sliding velocity of 3 or $0.7x10^{-3}$ m/min, in a vacuum of 10^{-8} Pa at room temperature. The radius of the metal pin specimens was 0.79 mm.

MATERIALS

The metals were in both polycrystalline and single-crystal forms. The polycrystalline metals were used for the experiments with diamond, silicon carbide and manganese-zinc ferrite. The single-crystal metals were used for those studies with pyrolytic boron nitride. The titanium was 99.97 percent pure; the yttrium

was 99.9 percent pure; the vanadium was 99.95 percent pure; and
all the other metals were 99.99 percent in purity. For the single
crystals, the body centered cubic metals had the {110} plane on
their surface and, therefore, parallel to the sliding surface.
The face centered cubic metals had the {111} planes parallel to the
sliding interface and the hexagonal metals had the {0001}
surfaces parallel to that interface.

Natural, single-crystal diamonds were used in these experiments. The {111} plane was parallel to the sliding interface
(ref. 1).

The single-crystal silicon carbide used in these experiments
was a 99.9 percent pure compound of silicon and carbon. Silicon
carbide has a hexagonal close-packed crystal structure. The basal
plane was parallel to the interface (ref. 2).

The pyrolytic boron nitride employed in these experiments was
a 99.99 percent pure compound of boron and nitrogen (ref. 3). It
has a hexagonal crystal structure, and the basal plane was parallel to the interface.

The single-crystal manganese-zinc ferrite was 99.9 percent
pure oxide (ref. 4) The crystal is that of a spinel structure in
which the oxygen ions are in a nearly closed-packed cubic array.
The {110} plane was parallel to the sliding interface.

All the nonmetal specimens used herein were within $\pm 2^{\circ}$ of the
designated plane. All the specimens of nonmetals were in the form
of flat platelets.

APPARATUS

An Ultra-High Vacuum system was used in this investigation.
An apparatus capable of measuring adhesion, load, and friction was
mounted in the vacuum system, which also contained a tool for
surface analysis, an Auger electron spectrometer (AES). The mechanism used for measuring adhesion, load, and friction is shown
schematically in figure 1. A gimbal-mounted beam is projected
into the vacuum system. The beam contains two flats machined normal to each other with strain gages mounted on each flat. The
metal pin is mounted on the end of the beam. As a load is applied
by moving the beam normal to the disk, it is measured by the
strain gage.

The vertical sliding motion of the pin along the flat surface
is accomplished through a motorized gimbal assembly. Under an
applied load the friction force is measured during vertical trans-

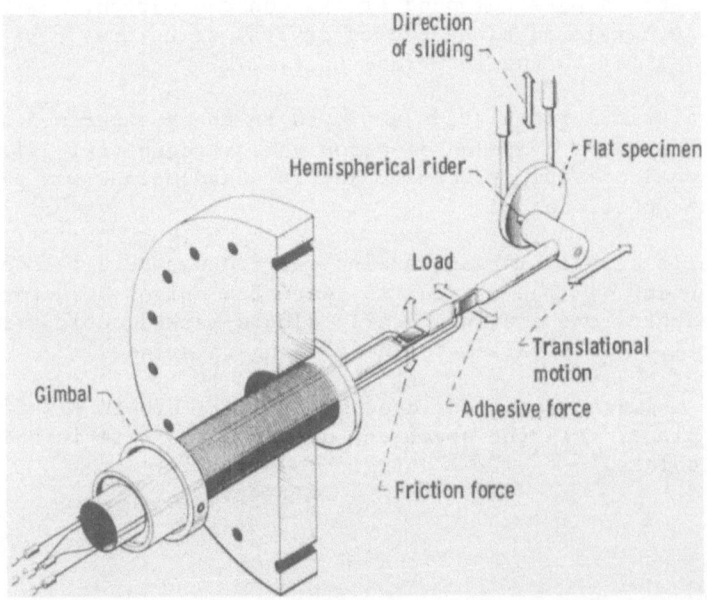

Figure 1. Ultra high vacuum friction and wear apparatus.

lation by the strain gage mounted normal to that used to measure load. This feature was used to examine the coefficient of friction at various loads.

EXPERIMENTAL PROCEDURE

The flat (nonmetallic hard material) and metal pin specimens were cleaned and polished with an aluminum oxide powder (1 μm). The radius of the pins was 0.79 mm. The flat and pin surfaces were rinsed with 200-proof ethyl alcohol.

The specimens were placed in the vacuum chamber, and the system was evacuated and baked out to a pressure of 10^{-8} Pa. Argon gas was then bled back into the vacuum chamber to a pressure of approximately 7×10^{-4} Pa in the case of diamond, and 1.3 Pa in the cases of silicon carbide, boron nitride and ferrite. Then the flat specimen was argon-ion bombarded for 30 to 60 minutes The pin specimens were argon sputter bombarded for 30 minutes at a -1000 V direct-current potential at a vacuum chamber pressure of 1.3 Pa. After the bombardment operation, the vacuum chamber was reevacuated, and AES spectra of the flat surface were obtained to determine the degree of surface cleanliness. When the desired degree of cleanliness of the flat was achieved, friction experiments were conducted.

Loads of 0.05 to 0.3 N were applied to the pin-flat contact by deflecting the beam of Fig. 1. Both the load and friction force were continuously monitored during friction experiments. Sliding velocity was 3 or 0.7×10^{-3} m/min. All friction experiments were conducted with the system evacuated to a pressure of 10^{-8} Pa.

RESULTS AND DISCUSSION

Surface Cleanliness of Nonmetals

The Diamond Surface - It is extremely difficult to expose {111} diamond faces by cleavage in the vacuum chamber for study in situ, and no entirely satisfactory cleaning procedure has yet been established for diamond. It has been suggested by Lurie and Wilson, on the basis of Auger electron spectroscopic and electron energy loss measurements, that, when diamonds are bombarded with argon ions, their surfaces become graphitized; Thomas and Evans, however, had considered that this treatment merely cleaned the surface (refs. 6 and 7). If Lurie and Wilson's conclusion is correct, surface graphitization of diamond would profoundly influence the tribological properties of the ion-bombarded diamond surface.

The main features in the vicinity of the carbon peaks of the Auger spectra from diamond are shown in figure 2. An Auger electron spectroscopy spectrum of a single-crystal diamond {111} plane obtained before argon-ion bombardment is shown in figure 2(a).

The crystal was in the as-received state after it had been baked out in the vacuum system. A carbon contamination peak is evident, and the spectrum is similar to that of amorphous-carbon. The surface was next argon-ion bombarded at a 3-kilovolt potential, under a pressure of approximately $7x10^{-4}$ Pa for 15, 30, 45, and 60 minutes.

The spectrum of the surface after 15-minutes has three peaks, which are characteristic of graphite. The spectra of the surface after 30, 45, and 60 minutes have four peaks, which are characteristic of diamond, as has been demonstrated and indicated in reference 8.

The peaks have been labelled A_0 to A_3, where A is used to denote an Auger peak. The energy of the peaks in this experiment were 267 to 269 for A_0, 252 to 254 for A_1, 240 eV for A_2, and 230 to 232 eV for A_3. The Auger spectrum of figure 2(d) is essentially the same as that obtained by Lurie for a clean surface (ref. 8). Figure 3 shows the spectrum of a diamond surface after ion bombardment for 60 minutes. A very small sulphur peak is observed in the diamond spectra.

Thus, for the adhesion and friction experiments reported herein, the surfaces of the diamond were argon-ion bombardment for 45 to 60 minutes under the pressure of approximately $7x10^{-4}$ Pa, and the Auger spectra of the surfaces were very similar to that shown in figure 2(d).

The Silicon Carbide Surfaces - An AES spectrum of the single-crystal silicon carbide surfaces obtained before sputter cleaning, but after polishing and bake out, revealed an oxygen peak in addition to the silicon and carbon. The oxygen peak and the chemically shifted silicon peaks at 78 and 89 eV indicated a layer of SiO_2 on the silicon carbide surface as well as a simple, adsorbed film of oxygen. The carbon peak was similar to that obtained for amorphous-carbon (ref. 9). Thus, the spectrum indicated a carbon containment on the silicon carbide surface as well as the SiO_2 layer.

The AES spectrum taken after the silicon carbide surface had been argon-sputter cleaned clearly reveals the silicon at 91 or 92 eV and carbon peaks at 272 eV, as shown in figure 4(a). The carbon peak is of the carbide type, which is characterized by three

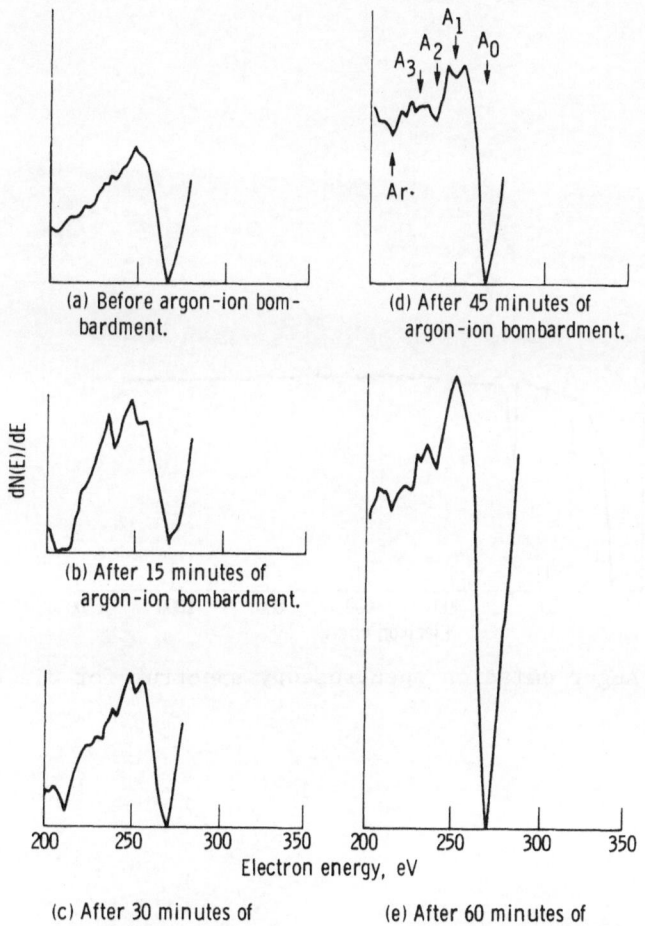

Figure 2.   Comparison of fine structure of the carbon Auger
emission spectra for diamond.

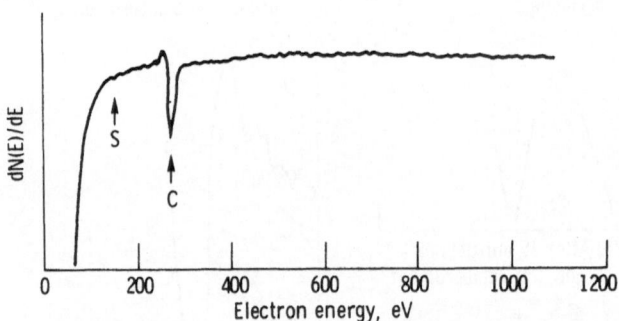

Figure 3.   Auger emission spectroscopy spectrum for diamond.

peaks labelled A_0 to A_2 in figure 4(a). A small argon peak is evident in the spectrum, but the oxygen peak is negligible.

The Boron Nitride Surface - An Auger electron spectroscopy spectrum obtained from the surface before sputter cleaning revealed that, in addition to the boron and nitrogen peaks, there was a peak adjacent to the boron peaks as well as another peak adjacent to a carbon contamination peak. Oxygen was absent from the surface. An Auger electron spectroscopy spectrum obtained after sputter cleaning is presented in figure 4(b). Sputter cleaning of the boron nitride surface resulted in elimination of contaminant peaks such as sulfur.

The Manganese-Zinc Ferrite Surfaces - An Auger spectra of the as-received single-crystal manganese-zinc ferrite surface obtained before sputter cleaning revealed that, in addition to the oxygen and iron, a carbon contamination peak was evident. An Auger spectrum for ferrite {110} surface after sputter cleaning is shown in figure 4(c). The carbon contamination peak has completely disappeared from the spectrum. In addition to oxygen and iron, the Auger peaks indicate small amounts of manganese and zinc on the surface.

Adhesion and Friction

The removal of adsorbed films (usually water vapor, carbon monoxide, carbon dioxide and oxide layers) from surfaces of metals and nonmetals results in very strong interfacial adhesion when two such solids are brought into contact. For example, when a clean titanium surface is brought into contact with a clean silicon carbide surface, the adhesive bonds formed at the solid-to-solid interface are sufficiently strong that fracture of the cohesive bonds in the metal and transfer of the metal to the silicon carbide surface results. This is indicated in the scanning electron micrograph presented in figure 5.

In figure 5 the light area of the figure, where a lot of metal transfer is evident, was the contact area before and during sliding of the rider. These are the areas where the surfaces of the metal and silicon carbide were sticking, one to the other and where strong interfacial adhesion occurred.

The friction-force traces obtained in this investigation are generally characterized by a marked stick-slip behavior. This type of friction trace clearly indicates that strong adhesion has occurred at the interface. All metals, discussed in this paper, transferred to the cohesively stronger nonmetals.

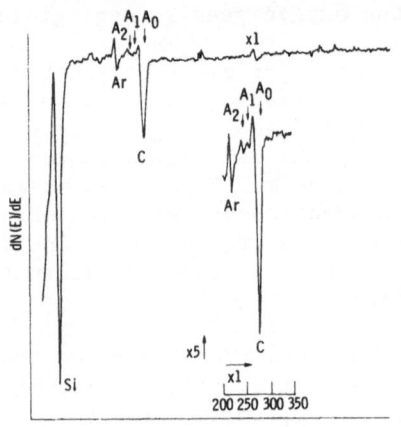

(a) Silicon carbide (0001) surface.

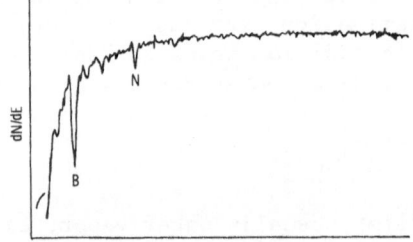

(b) Pyrolytic boron nitride (0001) surface.

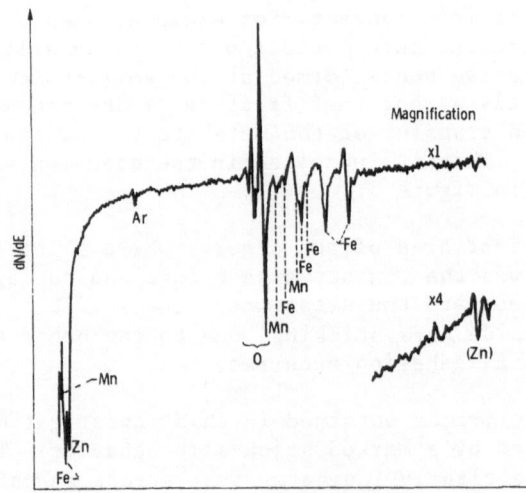

(c) Manganese-zinc ferrite (110) surface.

Figure 4.   Auger electron spectroscopy spectra for sputter cleaned nonmetals.

(a) Titanium Transfer at Commencement of Sliding

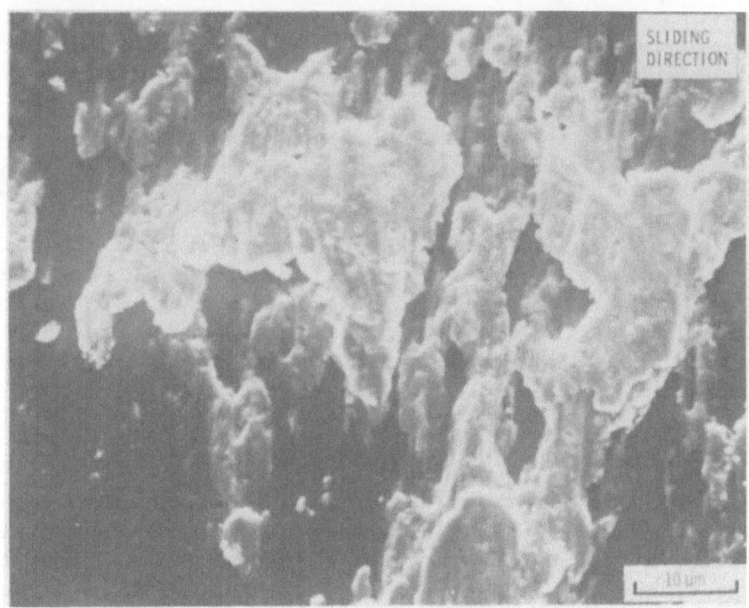

(b) Titanium Transfer during Sliding

Figure 5. Titanium transferred to single-crystal silicon carbide as
a result of single pass of rider in vacuum. Silicon
carbide (0001) surface; sliding direction <$10\bar{1}0$>; sliding
velocity, 3×10^{-3} m/min load, 0.3 N; room temperature;
vacuum pressure, 10^{-8} Pa.

Effect of Metal Activity on Friction

The data in figure 6 indicates the coefficients of friction
for some of the transition metals in contact with a single-crystal
diamond {111} surface as a function of the d-bond character of the
metal. The percentage of d-bond character can be related to the
chemical affinity of the surfaces. The greater the percentage of
d-bond character that the metal possesses, the less active its
surface should be. The data indicates a decrease in friction with
an increase in d-bond character. Titanium and zirconium, which
are chemically very active, when in contact with diamond, exhibit
very strong interfacial adhesive bonding to diamond. In contrast,
rhodium and rhenium, which have a very high percentage of d-bond
character have relatively low coefficients of friction.

Figure 6 also presents the friction data for a diamond surface
in sliding contact with a yttrium surface. Yttrium gives a higher
coefficient of friction than that estimated from data of other
metals. This may be due to the effect of oxygen. An argon-sput-
ter-cleaned yttrium surface was covered by an oxide surface layer
as shown in figure 7. It is very difficult to remove the oxide
surface layer from yttrium by argon-sputter cleaning. The effects
of oxygen in increasing the friction is related to the relative
chemical thermodynamic properties and bonding of carbon to oxygen. The greater the degree of bonding across the interface, the
higher the coefficient of friction. In the case of yttrium, oxygen on the surface tends to strongly chemically bond the yttrium
to the diamond surface (ref. 10).

Adhesion and friction properties of the transition metals
sliding on other nonmetals, such as silicon carbide, boron nitride
and manganese-zinc ferrite are the same as observed for the metals
in sliding contact with diamond, as shown in figure 8. The data
of figure 8 indicate a decrease in friction with an increase in d
character of the metallic bond. In other words, the more active
the metal, the higher the coefficient of friction. There appears
to be very good agreement between friction and chemical activity
for the transition metals.

Metal Transfer

Figure 9 presents a surface replication electron micrograph of
a wear track on the {111} diamond surface generated by a single-
pass of sliding of a titanium rider at a load of 0.2 N across the
surface. It is obvious from this photograph that a large amount
of metal transfers to the diamond surface. The transferred titanium is a thin film which is streaky and parallel to the sliding
direction.

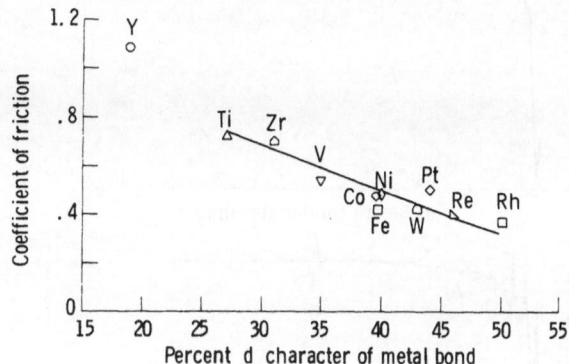

Figure 6. Coefficient of friction as function of percent of metal
d bond character for single-crystal diamond (111) surface
in sliding contact with transition metals in vacuum.
Sliding direction, <11$\bar{0}$>; sliding velocity, 3x10^{-3} m/min;
load, 0.05 to 0.3N; room temperature; vacuum pressure,
10^{-8} Pa.

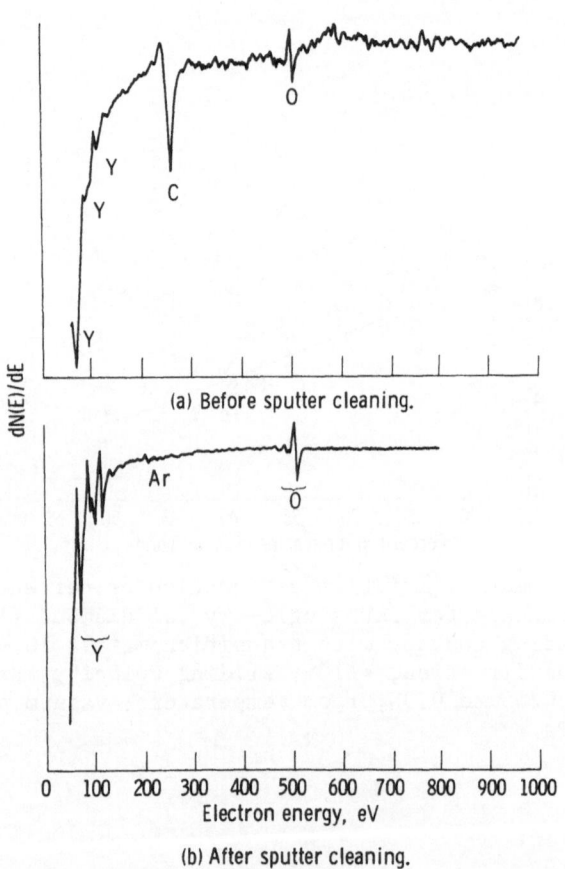

(a) Before sputter cleaning.

(b) After sputter cleaning.

Figure 7. Auger emission spectroscopy spectrum for yttrium

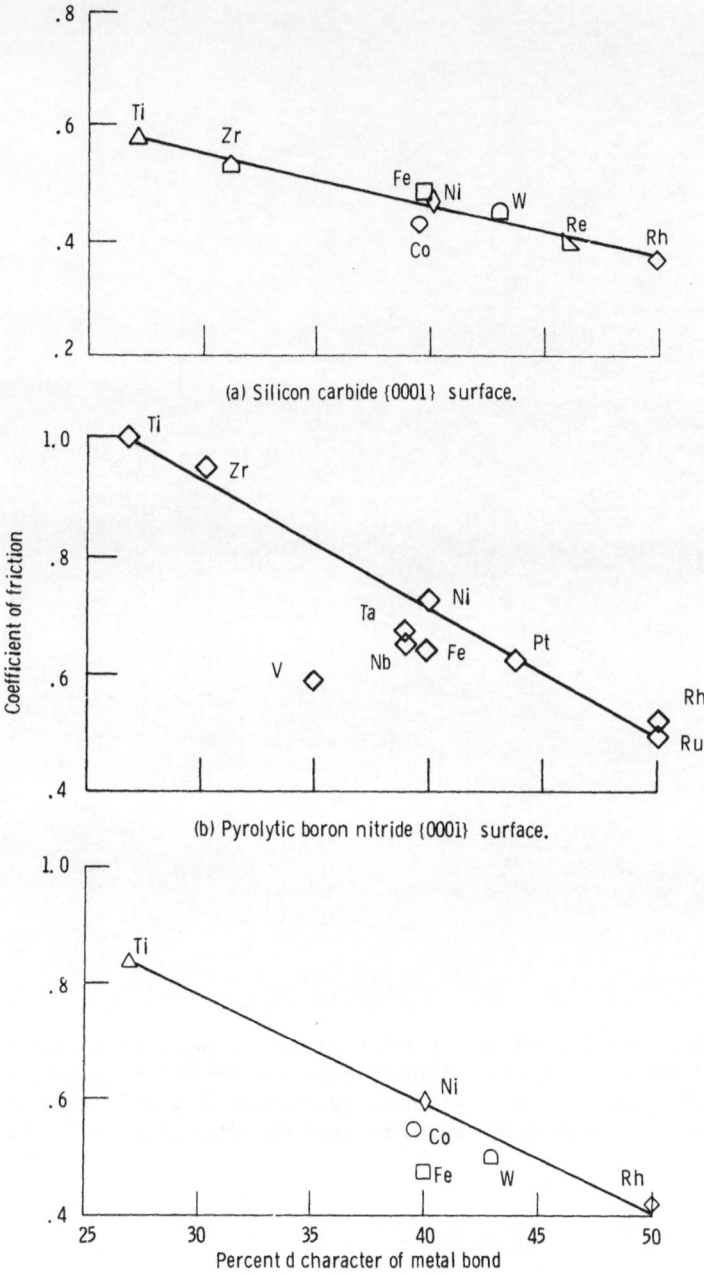

(a) Silicon carbide {0001} surface.

(b) Pyrolytic boron nitride {0001} surface.

(c) Manganese-zinc ferrite {110} surface.

Figure 8.   Coefficient of friction as function of percent of d bond character of various metals in sliding contact with non-metals in vacuum.   Single pass; sliding velocity, 3 or 0.7×10^{-3} m/min; load, 0.3 N; room temperature; vacuum pressure, 10^{-8} Pa.

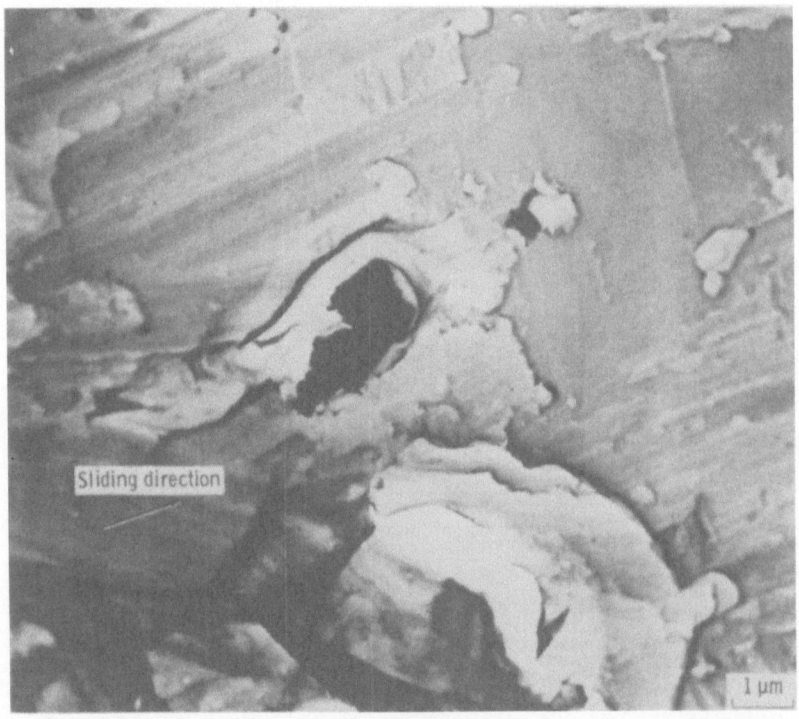

Figure 9. Replication electron micrograph of wear track on {111}
           diamond surface. Single pass of titanium rider; sliding
           direction <110>; sliding velocity, 3×10^{-3} m/min; load,
           0.2 N; room temperature; vacuum pressure, 10^{-8} Pa.

Figure 10 presents an energy dispersive X-ray profile of an area which includes the small dark (black) debris shown in the center of figure 9. The profile shows titanium and copper peaks. The copper peaks in the spectrum are associated with the element of the specimen mesh-holder. The titanium peaks reveal the element of the black debris. Thus, in figure 9 the black debris is a part of a titanium film that transferred to the diamond surfaces and that was subsequently peeled off from the surface, while the replica of the specimen surface was being prepared.

All the metals shown in figures 6 and 8 transferred to the surfaces of nonmetals with sliding except boron nitride (refs. 1 to 4). These results indicate that (1) when metals and nonmetals are brought into contact, adhesion occurs, the interfacial bond is generally stronger than the cohesive bond in the cohesively weaker metal, and (2) on separation of the metal and nonmetal in sliding, fracture occurs generally in the metal.

The chemical affinity of metal to nonmetal plays an important role in the metal transfer and the form of metal wear debris generated by fracture of cohesive bond (ref. 2). In general the less active the metal, the less transfer to the nonmetal. Titanium having much stronger chemical affinity to the elements of the nonmetals exhibited the greatest amount of transfer.

Figure 11 presents a typical scanning electron micrograph of the rider wear scar on titanium resulting from a single pass of the titanium rider over the diamond {111} surface in vacuum. The wear scar on the metal rider revealed, generally, a large number of plastically deformed grooves, and is very similar to that on a metal rider after sliding against other nonmetals.

CONCLUSIONS

As a result of sliding friction experiments conducted with various transition metals in contact with nonmetallic hard materials, such as diamond, silicon carbide, boron nitride, and manganese-zinc ferrite, the following conclusions are drawn:

(1)   The coefficient of friction for nonmetals in contact with various metals is related to the relative chemical activity (d valence bond character) of those metals in ultra high vacuum. The more active the metal, the higher the coefficient of friction.

(2)   All the metals examined transferred to the surfaces of nonmetals in sliding. The more active the metal, the greater the transfer to nonmetals.

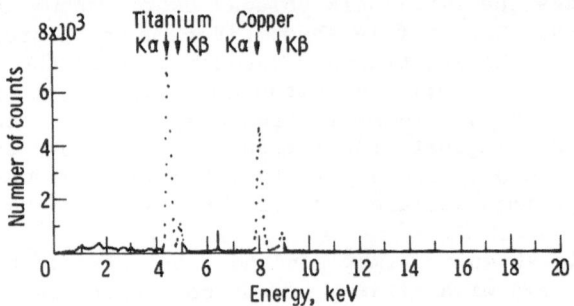

Figure 10. Energy dispersive X-ray profile of a piece of wear debris.

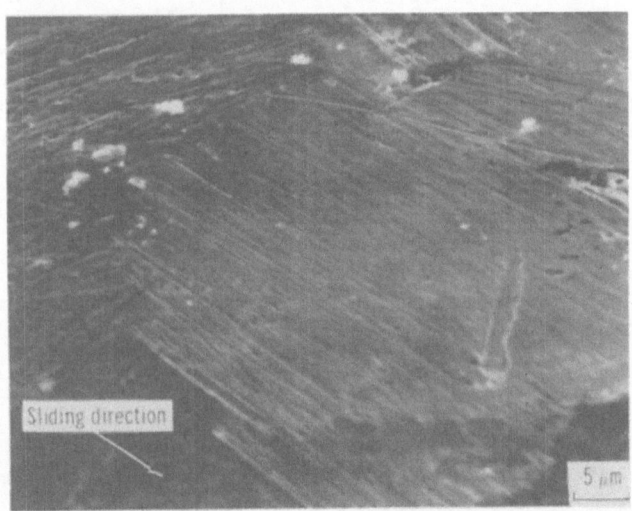

Figure 11. Wear scar on titanium rider produced during sliding
            contact with diamond. Single pass on diamond {111}
            surface; sliding direction, <110>; sliding velocity,
            3×10^{-3} m/min; load, 0.2 N; sliding distance, 2.5×10^{-3}m;
            room temperature; vacuum pressure, 10^{-8}Pa.

REFERENCES

1. K. Miyoshi and D. H. Buckley, Adhesion and Friction of Single-Crystal Diamond in Contact with Transition Metals, Appl.
   Surface Sci., 6: 161 (1980).

2. K. Miyoshi and D. H. Buckley, Friction and Wear Behavior of
   Single-Crystal Silicon Carbide in Sliding Contact with Various
   Metals, ASLE, Trans., 22: 245 (1979).

3. D. H. Buckley, Friction and Transfer Behavior of Pyrolytic
   Boron Nitride in Contact with Various Metals, ASLE, Trans.,
   21: 118 (1978).

4. K. Miyoshi and D. H. Buckley, Friction and Wear of Single-
   Crystal Manganese-Zinc Ferrite, Wear, 66: 157 (1981).

5. L. Pauling, A Resonating-Valence-Bond Theory of Metals and
   Intermetallic Compounds, Proc. R. Soc., London, A196: 343
   (1949).

6. P. G. Lurie and J. M. Wilson, A Study of the Diamond Surface
   Using Electron Spectroscopy, Diamond Res., 26 (1976).

7. J. M. Thomas and L. L. Evans, Surface Chemistry of Diamond: A
   Review, Diamond Res., 2 (1975).
8. P. G. Lurie and J. M. Wilson, The Diamond Surface, I. The
   Structure of the Clean Surface and Interaction with Gases and
   Metals, Surface Sci. 65: 453 (1977).
9. A. J. Van Bommel, J. E. Crombeen, and A. Van Tooren, LEED and
   Auger Electron Observations of the SiC (0001) Surface, Surface
   Sci., 48: 463 (1975).

10. K. Miyoshi; and D. H. Buckley; Effect of Oxygen and Nitrogen
    Interactions on Friction of Single-Crystal Silicon Carbide,
    NASA TP - 1265 (1978).

DISCUSSION

R. Reeber:

How did you prepare the surfaces of the metals and what purities did
you have of these materials?

K. Miyoshi:

Yittrium was 99.9 per cent pure and the other metals were 99.99 per cent pure. The metal surfaces were polished with 1 micron diamond powder.

W. Williams:

The model you discussed for the chemical activity of titanium and the other transition metals suggests that when the compound is formed there will be additional filling of d states by electron transfer from the non-metal species to the metal. Is that what you suggest?

K. Miyoshi:

Right.

W. Williams:

Well, that model is incorrect. According to all the bond structure calculations except one, that would be about twelve out of thirteen, and according to most of the experimental studies involving XPS and X-ray absorption and so forth, what really happens to account for this strong chemical affinity that you described, is that when the compound is made the s electrons, in the case of titanium, the 4 s electrons states are depopulated. They are pushed way up in energy and are not populated in the compound. As a result, those electrons are then able to do other things, such as filling d states. But there is a slight shift of electron density from the metal species to the non-metal, in fact. So, the picture is a little bit more subtle than just pumping electrons from an incoming particle to empty states in the other particle. There is a sort of intermediate rearrangement that occurs. So, I think that point needs to be clarified if one is to discuss the trends of yours.

CONSTITUTIONAL ASPECTS IN THE DEVELOPMENT OF NEW HARD MATERIALS

Helmut Holleck

Kernforschungszentrum Karlsruhe
Institut für Material- und Festkörperforschung
D 7500 Karlsruhe 1, Postfach 3640
(Federal Republic of Germany)

ABSTRACT

Different groups of wear resistant materials are reviewed.
Recent developments are shortly discussed and some trends for
future developments are given. The constitution can be the basis
for the optimization of material properties and for the realization of new material concepts.

INTRODUCTION

Future developments in the field of hard materials are mainly
related with: the raw materials supply, economic factors, the
general object to save energy and raw materials and with the increasing demand to have special materials with defined and optimized properties. The chemical constitution of hard materials
is one of the most important parameters determining their fabrication conditions and properties and is of high significance for
new developments. Changes in the constitution of the material
(constitution means: chemical composition, phase composition as
well as the equilibrium dependent microstructure) are mostly
accompanied by severe changes in the properties and in the application characteristics.

Typical groups of hard materials are shown in fig. 1 in respect to their hardness and transverse rupture strength. Recent
developments are added schematically by different points and
lines. With decreasing hardness one observe the fields of superhard materials, hard carbides, ceramic materials and finally the

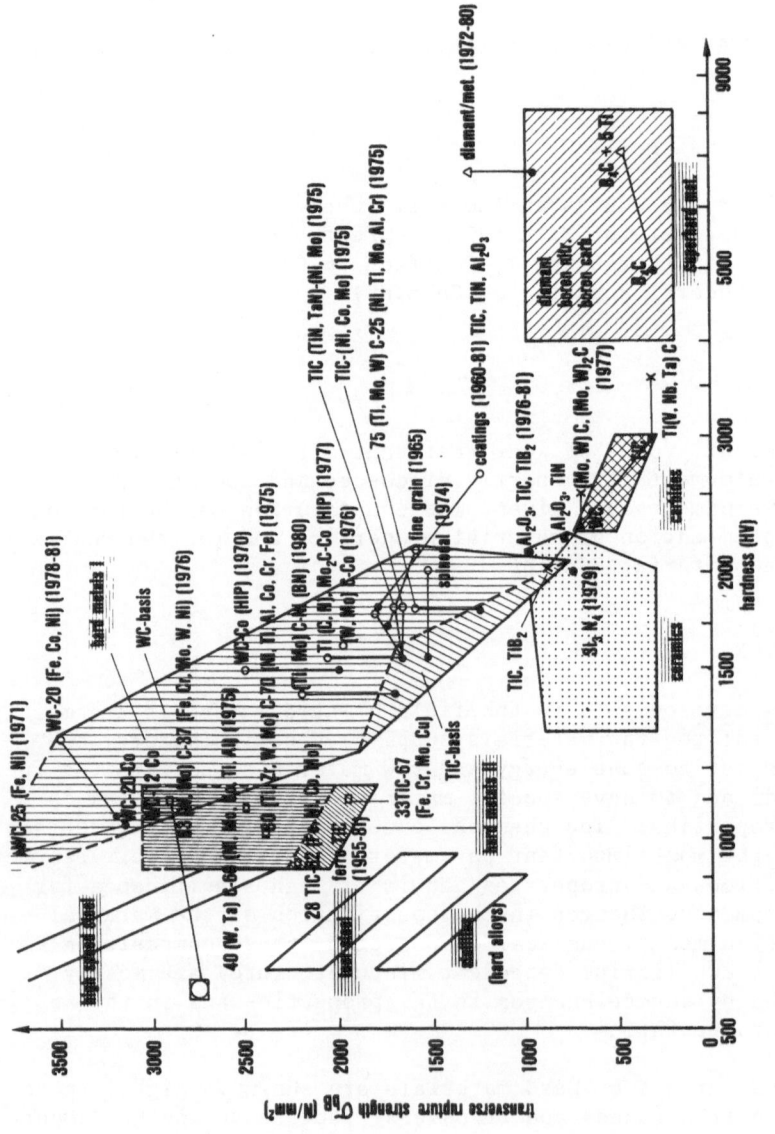

Fig. 1. Developments in the field of wear resistant materials

wide field of hard metals, subdivided in hard metals I (carbide-rich) and hard metals II (metal-rich). Both groups are again sub-divided in the TiC-based materials at lower strength and the WC-based materials at higher strength.

SUPERHARD MATERIALS

In the scope of this paper superhard materials and ceramics are not treated in detail. It should be mentioned however that recently diamond-metal composites were reported with a hardness of 11000 HV and a transverse rupture strength of 2100 N/mm^2 /1/. Also attempts to improve the properties of the cheap material B_4C should be noticed. This material has an excellent hot hardness (fig. 2). The properties can be changed sensitively for instance by metal additions as is shown in fig.3.

CARBIDES

The transition metal carbide field - bounded by TiC at the hard but low strength side in fig. 1, and by WC at the higher strength but lower hardness side - has its main significance by supplying the wear resistant constituents in steels and hard metals. The constitution of the refractory carbide systems of-fers us the opportunity of tailloring the properties of these hard phases, for instance by solid solution hardening or by the heat treatment of carbides in the region of ordered phases.

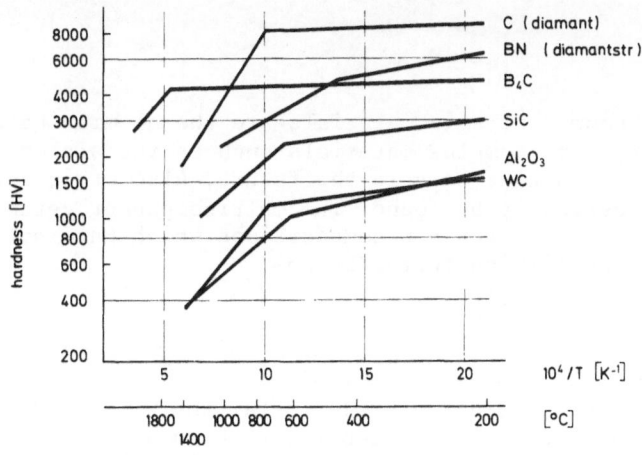

Fig. 2. Hot hardness of some non metallic hard
        materials in comparison with WC

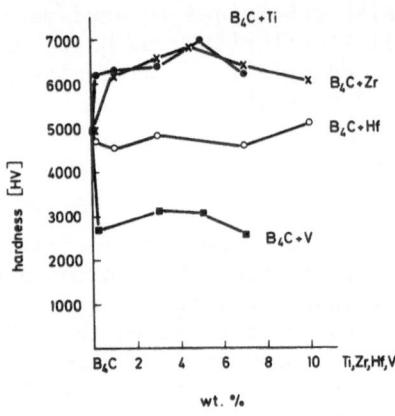

Fig. 3.   Hardness of B₄C with metal additions /2/

The pure carbides are too brittle for tool applications. Finely dispersed brittle phases can result however in materials with remarcable rupture strength, if the phase boundaries became responsible for the strength properties. An example is given in fig. 1 by the line starting from TiC. The fine grain eutectic microstructure of TiC/TiB_2 leads to a decrease of hardness but to an increase in rupture strength.

CERAMICS

The most common ceramic materials are the alumina based ceramics. Other, more complex materials such as the sialons can however play an important role in the future. Also other ceramic materials will certainly be found. The multicomponent ceramic systems offer a lot of such new compositions which have not been investigated in detail and tested by now.

HARD METALS I

Adding to the carbides a binder metal the wide hard metal field is reached (see fig. 1) extending to high strength. The phase equilibria of hard metals are characterized by a two phase region between the carbide phase and the binder metal phase (fig.4).

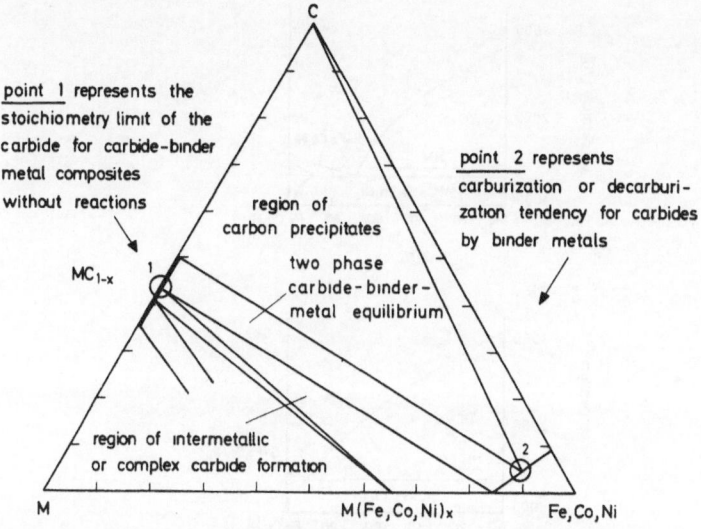

point 1 represents the stoichiometry limit of the carbide for carbide-binder metal composites without reactions

region of carbon precipitates

point 2 represents carburization or decarburization tendency for carbides by binder metals

two phase carbide-binder-metal equilibrium

MC_{1-x}

region of intermetallic or complex carbide formation

C

M   M(Fe,Co,Ni)x   Fe,Co,Ni

Fig. 4. Hard metal phase equilibria schematically

It is easy to prove, that best properties are obtained only within this two phases region and that mainly the transverse rupture strength drops down crossing the phase boundaries. The carbide-binder metal sections are eutectic systems and the special character of the WC-based materials are due partly to the constitution of this section. This is elucidated in comparison to TiC-based materials schematically in fig. 5. The systems of WC with Fe, Co or Ni show the eutectic composition more at the carbide side compared with other systems. Since the sintering temperature of the hard metals (the composition of commercial hard metals is indicated in fig. 5) lies above the eutectic temperature the high carbide content in solution leads to good densification and contributes by solution-precipitation processes to the desired microstructure. Also the highest solubilities in the solid binder metal and the best wetting behavior are observed for the sixth group transition metal carbides.

New developments in the hard metal field concern at the one side several products where the constitution is only to a minor extent involved such us: fine grain cemented carbides, hot isostatic pressing (HIP) or coatings, at the other side, those where the constitution is the base for the new materials. This second group of developments deals for instance with the replacement of W by Mo, the replacement of Co by a Fe based alloy, the attempts of extending the TiC-based cemented carbide field (see fig. 1) to higher strength or hardness and finally the metal-

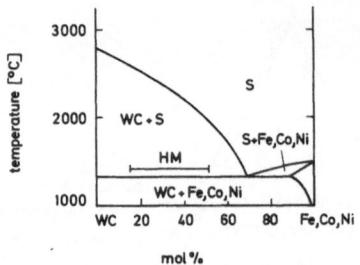

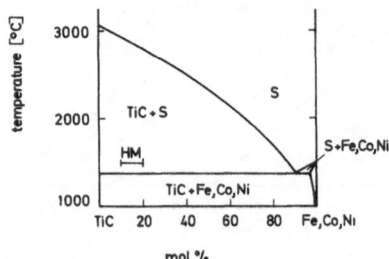

Fig. 5. Phase relations for the sections WC-Fe,Co,Ni
         and TiC-Fe,Co,Ni (schematically)

rich composites, designed as hard metals II in fig. 1.

HARD METALS II

Special phase equilibria make it possible to replace heavy
expensive transition metals by cheaper, abundant and specifically light iron- and nickel based alloys. Fig. 1 gives an
example for materials with similar properties concerning hardness
and transverse rupture strength but with different compositions
(WC-12Co and (W,Mo)C-37(Fe,Cr,Mo,W,Ni) respectively). The main
requirement for these latter materials is that the metal phase
has to take over a substantial part of the wear resistance, which
is normally provided by the higher carbide content. Temperature
dependent phase equilibria in the iron and nickel based alloys
are the basis for the hardening of the metal phase. In the case
of the iron based materials precipitation of carbides due to the
narrowing of the γ-region with decreasing temperature and the α/γ
transformation (see as example fig. 6) give the desired properties.
For the nickel based material the so called γ'-phase, precipitating from the solid solution of Ti, Mo or Al in Ni during cooling,
leads to the strengthening of the alloy (see as example fig. 7).

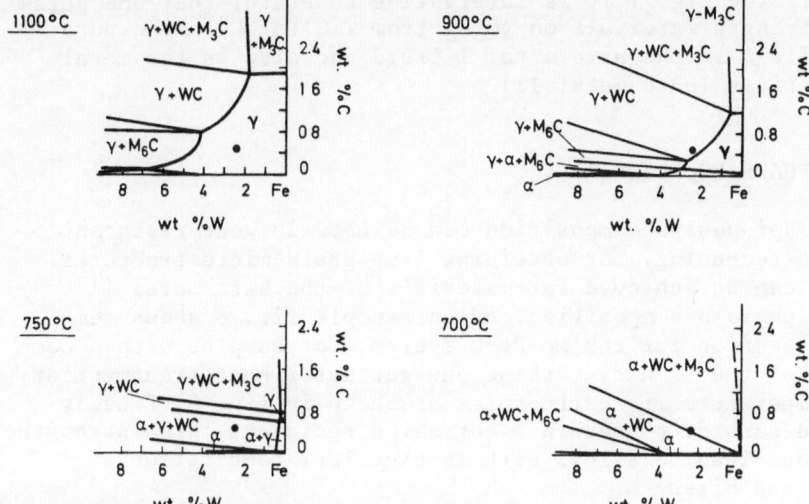

Fig. 6.   Phase equilibria in the Fe-corner of the
          Fe-W-C system (calculated /3/)

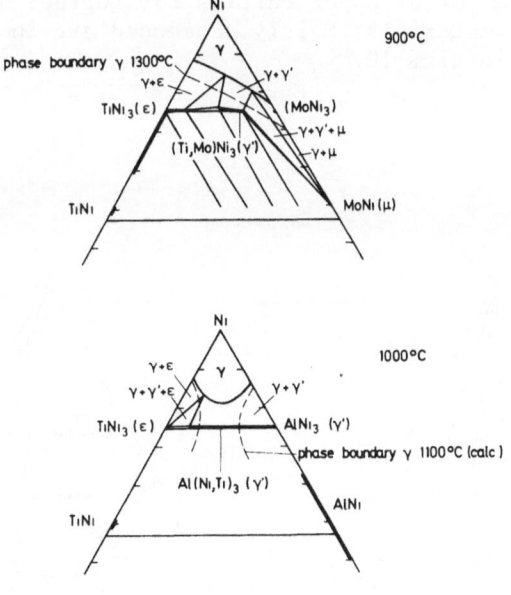

Fig. 7.   Phase relations in the Ni-corner of the
          Ti-Mo-Ni and Ti-Al-Ni systems (partly
          estimated)

Analyzing fig. 1 it is interesting to state, that one gets higher strength materials on going from TiC to WC in the pure carbide field in the hard metal I field and also in the metal rich materials (hard metal II).

SPECIAL PHASE EQUILIBRIA

The spinodal decomposition can be used in wear resistant materials technology for obtaining fine grain microstructures. The same can be achieved for materials of the hard metal II group by phase decomposition. As an example fig. 8 shows temperature section for the Mo-Pt-C system. For samples with a composition of the ternary carbide one get after heat treatment at lower temperature microstructures as shown in fig. 9. Finally dispersed carbide phases in a corrosion resistent, high strength metal phase lead to alloys with an excellent combination of hardness and strength.

FINAL OBSERVATIONS

The examples given in this paper have shown, that the constitution offers a lot of opportunities for further developments of wear resistant materials. Mainly concerned are in this respect the topics given in fig. 10.

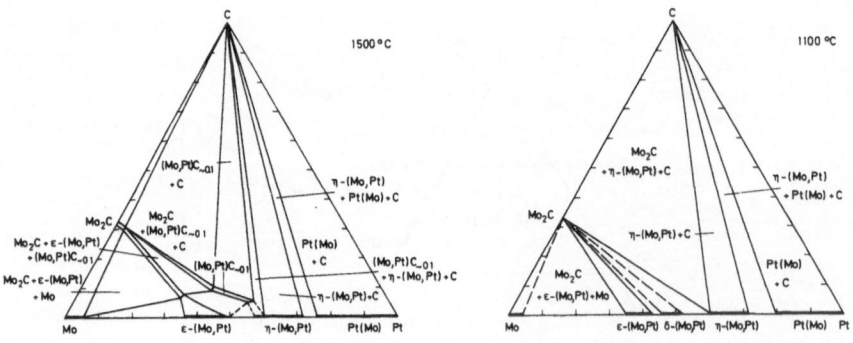

Fig. 8. Isothermal sections in the Mo-Pt-C system at
1500°C and 1100°C

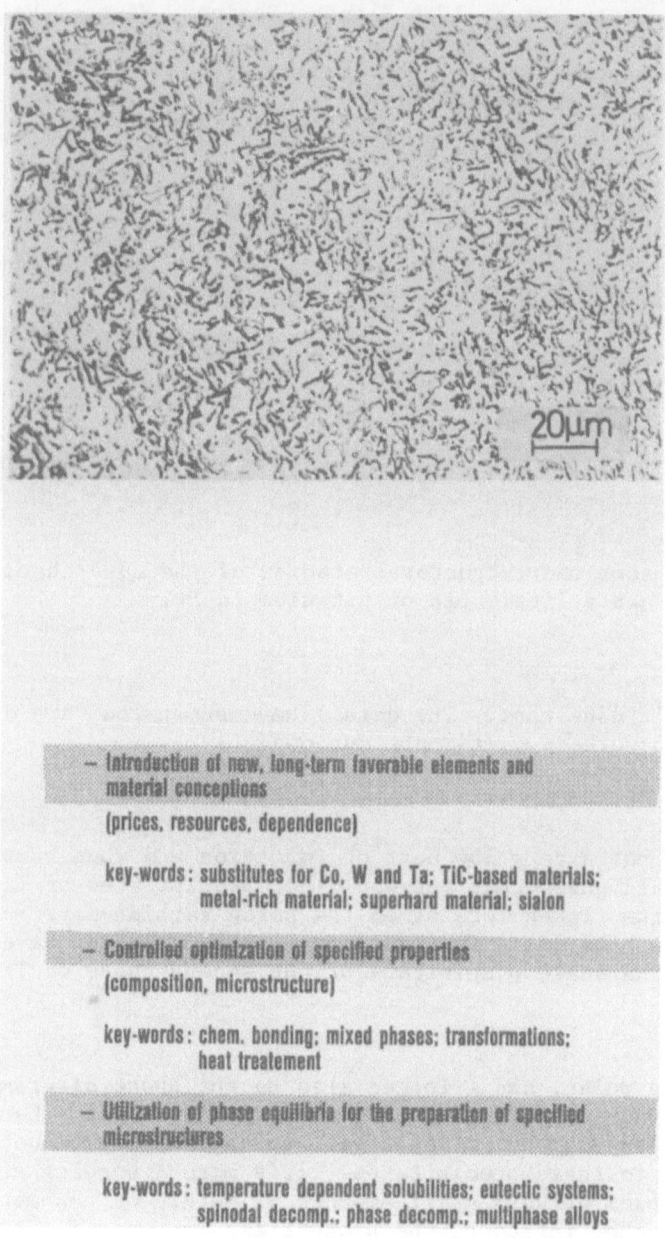

Fig. 10. Contribution of phase equilibria for the
development of wear resistant materials

REFERENCES

1. A. Hara, S. Yazu, Proc. 10th Plansee-Seminar, Reutte, Vol. I,
       581 (1981).
2. T. Ya. Kosloapova, Porosh. Met. 9:11 (1979)
3. B. Uhrenius, Calphad 4:173 (1980).

DISCUSSION

W. Williams:

Have you done some microstructural studies of the super hard boron
carbide which had a little bit of titanium in it?

H. Holleck:

No. We haven't done that. The data I have shown you, are data from
the Russian co-workers. It's not our own.

W. Williams:

Boron carbide has a very low heat of formation and when combined with
titanium I would guess that one would get precipitates of TiB_2 and
perhaps graphite flakes forming in the boron carbide matrix and perhaps if the high hardness is indeed real, that it might have to do
with a very fine scale precipitate of the second phase of TiB_2.

H. Holleck:

I look to this point, and I looked also to the phase diagram. It's
true that if you add titanium to boron carbide you get titanium diboride precipitates. But I don't believe that this pronounced effect
is really due to these precipitates. It's more a bonding effect and
this is a problem of very small amounts of metals or non-metals in a
phase and how they influence the properties.

W. Williams:

So, the understanding about the hardening effect is not yet clear.

H. Holleck:

Not yet clear.

H. E. Exner:

There is a hardness maximum in the system, chromium carbide-titanium carbide-tungsten carbide above 4,000 Vickers and we have looked into this system and they were homogeneous.

G. Dearnaley:

Taking up Williams' point, a few years ago we successfully implanted into some steel which was used for burner tips in oil fired power stations, titanium and boron in quite sizeable proportions with the intention that when this material corrodes, as it does in the combustion products within the power station, that this would form a strengthening matrix and because of bonding considerations form-ing three dimensional structures and, of course, exhibiting hard titanium boride precipitates. Both these points have just been made, and I can report that these components operated ten times more suc-cessfully than the untreated ones over a very extended period. So, this was a little bit of ceramic engineering, if you like, aimed not at strengthening the metal but in deliberately incorporating these implanted species into the scale in order to improve its resistance to erosion, which in this case is primarily by silica particles in the crude oil. So, there's another example which I think fits in with the discussion we have just had, and we were able to determine at the end of 8,000 hours of operation that the titanium was still there within the scale on the surface of the material. There had been no detectable change in dimensions of the orifice.

R. Warren:

The iron-cobalt-nickel system is known to be capable of very low thermal expansion coefficients. Have you used this consciously, or noticed any effect of this in hard metals in terms of the thermal expansion mismatch stresses?

H. Holleck:

Yes. It is a little bit too easy to take only the iron-cobalt-nickel system. You must add tungsten and carbon, and in reality you have a much more complex system.

L. Prakash:

I'd just like answer this question. The expansion coefficient of Fe-Ni-Co alloys depends on the composition. It is known that invar alloys with a very high nickel content have zero thermal expansion co-

efficients. But as Dr. Holleck said in a WC-(Fe, Ni, Co) system you
don't have the same structure of these materials. So we don't get
coefficients of expansion. We have measured the residual stresses in
WC-(Fe, Ni, Co) alloys and found that the residual stress value is
about the same. And this depends on whether you have austenite or
martensite, their amounts and type.

H. Jonsson:

The name of this session is New Horizons and maybe we should discuss
that a little. There has been a general topic, I think, in this conference that has been very interesting, to me anyway. That has been
the behavior of the binder phase. Many authors have spoken about the
importance or believed importance of many factors. But if you consider that cemented carbides usually consist of let's say 5 to 15
weight per cent of cobalt and if you take a rough figure, that means
about 10 volume per cent of cobalt. If you also roughly assume that,
for instance, a property like hardness may be taken as a weight average of the hardness of the cobalt and the tungsten carbide, we might
easily come to this very strange position that if you want to increase the hardness of the cemented carbide by about 100 in Vickers,
we have to increase the hardness of the cobalt phase with 1,000
Vickers. This is a figure that is completely impossible in normal
alloy systems. When we are talking about, for instance, martensitic
transformations that are known to create almost no change in the
mechanical properties in cobalt, have we really been exaggerating the
possibilities of changing binder phases and so on in this conference?
For instance, when we are talking about nickel-cobalt alloys as
binders. And this, in a way, also means that we've got less of the
ductile material in that type of cemented carbide. In other words,
we are making a comparison that it is not really fair because it
could be easily changed, for instance, by changing the cobalt content
of the cemented carbide. Maybe this is an old horizon. I don't know
whether I'm too pessimistic here. So, maybe we've got a lot of experts here that may probably see the horizons more positively than I
have presented here. I would be very interested to hear the opinions
of the new horizons, but also the probable limits of changing binder
phases and changing of properties.

H. E. Exner:

I do think that one of the reasons why we hear so much about the
binder phase is that we now have instruments to look into the binder
phase. I think this is part of this development.

H. Holleck:

What I would like to say is that if you can transfer some of the wear

resistance to the binder metal, then you can make materials with much
more binder metal which are much more economic. This was my point.

R. Warren:

Dr. Jonsson used the term "if we assume hardness is a simple weight
proportion between the two phases". Fortunately this is not the
case. This highlights the importance, for example, of contiguity.
By small changes in the binder phase we can drastically affect the
contiguity, maybe. And we know that contiguity has quite a profound
effect on properties.

PREPARATION OF MOLYBDENUM CARBIDE WITH WC-TYPE STRUCTURE

J.C. Schuster

Inst. of Physical Chemistry, Univ. of Vienna, Austria

H. Nowotny
Inst. of Materials Science, Univ. Connecticut, Storrs
CT. 06268, USA

ABSTRACT

Since the discovery of the complete series of solid solutions
$(Mo_{1-x}W_x)C$ (X= 0 to 1) hexagonal molybdenum carbide with WC-type
structure is considered an interesting material for WC-substitution
in several applications. With the preparation techniques used and
proposed so far no single phase MoC (Mo_2C free) can be obtained.
Thus a method to favor the formation of hex. MoC over the other
molybdenum monocarbides was developed by using additives of metal
powders to the binary alloy. To reach thermodynamic equilibrium
at the low temperatures at which the MoC with WC-type structure
is stable (below 1190°C) an auxiliary bath method is employed.
To separate the carbide from the bath metal an electrochemical
technique is used rather than an acidic-mixture in order to avoid
chemical side reactions. The final product is a dark grey, well
crystallized MoC-powder with WC-type structure.

INTRODUCTION

In the mid 70s the existence of a complete solid solution
series MoC-WC with hexagonal WC-type structure was established[1]
This opened a new horizon to materials and cost savings
especially when it was shown that the physical and chemical
properties of hex.(Mo,W)-monocarbides and hard metals are equal
or superior to conventional WC-Co hard metal; e.g. the
Mo-containing materials have about the same hardness and
wear-properties, have a better oxidation resistance and a lower
coefficient of friction[2]. Applications like dies, wear parts,
wear resistant coatings, cutting tools, mining tools etc. are
proposed. However it still is a major problem how to obtain single

phase hex.(Mo,W) carbides with WC-type structure which contain
more than 80 mol% MoC. The equilibrium diagram of the ternary
Mo-W-C system[3] (Fig.1) shows the potential for the development
of Mo-rich hex. monocarbides, but it does not reveal how to
produce them in a technologicaly simple and low cost way. It was
pointed out[4] that Mo_2C and graphite can coexist in a metastable
equilibrium indefinitely within the stability range of hex.MoC
without a trace of hex.MoC formed, since hex.MoC with WC-type
structure is a low temperature phase and decomposes at $1190°C$,
a temperature where the reactivity of molybdenum is rather low.

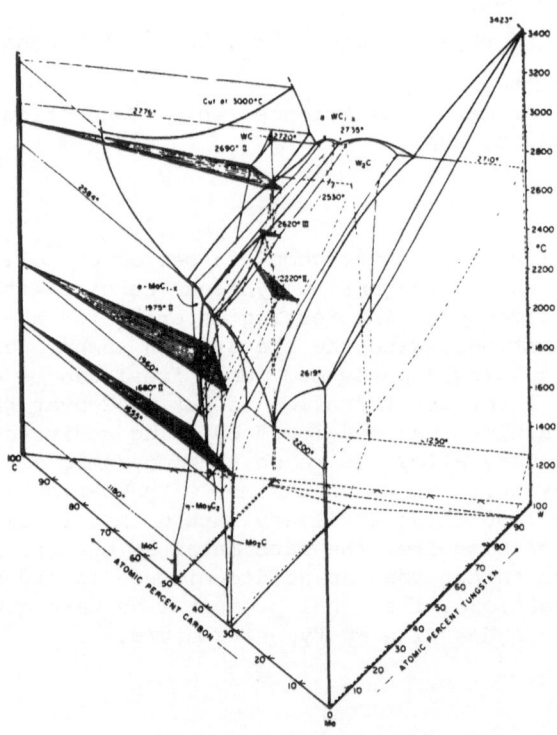

Fig. 1. Isometric view of the Mo-W-C system[3]

METHODS TO PREPARE HEX. Mo-CARBIDES WITH WC TYPE STRUCTURES

     The first reported method for synthesizing (Mo,W) carbides
with WC-structure is mixing the starting powders - Mo_2C, WC and
graphite, with a few at% of an iron metal added as diffusion aid -
and melting or heating the compacted materials up to $1700°C$[1,4].
At this temperatures the diffusion of the Mo and W atoms is fast
enough to obtain statistically ordered cub. α-$(Mo,W)C_{1-x}$ or

hex. η-(Mo,W)C$_{1-x}$ solid solutions within one hour. On cooling
these carbides decompose into very fine particles of (Mo,W)$_2$C
and graphite, which makes the diffusion path for the atoms
participating in the hex.(Mo,W)C forming transformation very short.
Thus, when reheated to temperatures within the existence range of
hex.MoC or hex.(Mo,W)C resp. the reaction proceeds within several
hundred hours to form the desired WC-type structure phase.
Another approach to synthesize hex.(Mo,W)C starts with a (Mo,W)-
alloy, which makes the high-temperature diffusion enhancing step
obsolete[3,5]. But the carburizing temperatures to ensure complete
reaction within feasable times has to be selected above the
decomposition temperatures of Mo-rich hex.(Mo,W)C. Alloys
consisting of 80 or more mol% MoC contain some subcarbide[5].

As there exists a complete solid solution series MoN-MoC[6],
stabilizing hex.(Mo,W)C alloys rich in Mo with nitrogen seemed
to be possible[1]. Experiments[5] have shown that nitriding the
(Mo,W) carbides with NH$_3$-gas at temperatures between 450°C and
800°C can produce single phase hex. alloys containing up to
0.3 wt% N. But the tendency to obtain in W-rich alloys η-phases
of composition (Mo,W)$_6$Ni$_6$(C,N) and in the Mo-rich alloys
subcarbides of composition (Mo,W)$_2$(C,N) in this experiment made
it necessary to develop a sophisticated carbon-activity-control
system to compensate the carbon loss which originated from
methane formation from NH$_3$-decomposition product H$_2$ and free
graphite in the reaction mixture. Nevertheless it is possible to
obtain single phase hex.(Mo,W)(C,N) alloys containing up to
90 mol% Mo(C,N) at 1100°C in 10 h by this method[5,7].

Using nitrogen gas for nitriding gives similar results[5],
but requires high pressures because at 1 atm only 300 to 500 ppm N
was observed to be taken up by the alloys[3]. Still control of
decarburization remains a problem and the nitrogen fills up only
part of the carbon vacancies. Still another approach is the use
of a liquid-metal bath to enhance the reaction rate. Nickel[2,5]
and cobalt[5] are proposed. But the lowest eutectic melting
temperature in the (Mo,W)C-Ni,Co systems[8] is above the
decomposition temperature of hex.MoC, and single phase hex.
carbides with high Mo-content cannot be prepared by this method.

So it was desirable to develop a method to obtain high mol%
MoC containing hex. monocarbides with WC-type-structure which
offers low cost with respect to equipment and input materials and
high efficiency with respect to high yield and quality of the
product. It seemed feasable to achieve this goal by optimizing
systematically the approaches tried and described in the
literature so far. Effort has been made in two steps by
investigating

a) the factors of favouring the stability of the hex.MoC-phase
   and by investigating
b) the factors of favouring the transformation into the
   hex.MoC phase

EXPERIMENTAL

The starting materials used were:
Mo_2C (2N8, C_{tot} 6.28 wt%, C_{free} 0.44 wt%, Treibacher Chem.Werke,
(Austria)
Mo_2C (2N8, Alpha-Ventron Corp., USA)
Wc   (C_{tot} 6.15 wt%, Metallwerk Plansee GmbH., & Co. KG. Austria)
C   (2N8, Union Carbide Corp., USA)
Mo   (2N7, Fluka KG., Switzerland)
Ti   (m2N, Alpha Ventron Corp., USA)
V   (m2N7, Alpha Ventron Corp., USA)
Cr   (m2N, Alpha Ventron Corp., USA)
Mn   (m2N, Alpha Ventron Corp., USA)
Fe   (m3N+, Alpha Ventron Corp., USA)
Co   (m2N8, Alpha Ventron Corp., USA)
Ni   (m3N, Alpha Ventron Corp., USA)
Ni   (puriss, Fluka KG., Switzerland)
Cu   (m3N, Alpha Ventron Corp., USA)
Cu   (m2N6, Metals Disintegrating Inc., USA)
Zn   (m3N, Alpha Ventron Corp., USA)
Cd   (m2N5, Alpha Ventron Corp., USA)
Al   (m2N8, Alpha Ventron Corp., USA)
Sn   (m2N5, Alpha Ventron Corp., USA)
Pb   (m2N5, Alpha Ventron Corp., USA)
Bi   (m5N, Alpha Ventron Corp., USA)
Ag   (m4N, Cominco Products Inc., USA)

To powder mixtures of the composition $(Mo_{49}Me_{02}C_{49})_{1-x}WC_x$
(x= 0.0, 0.05, 0.1; Me= stabilizing metal, see tab.1) the powdered
bath metal was added in a 1:1 ratio by weight. These alloys were
cold pressed and heat treated in a graphite crucible which was
sealed in a quartz container under vacuum. The heat treatment
conditions varied from 5 to 150 hrs. and from 800° to $1200^\circ C$. Then
the specimen were water quenched and the bath metals were removed
(e.g. Pb by electrolysis at 0.4 Amp., 15 Volts using as
electrolyte 10 vol% perchloric acid in methanol). The carbide
powder was washed, dried and examined by X-ray-powder-
diffractometry (CrK α-radiation). Lattice parameters and intensity
measurements were made with an automatic microdensitometer (KD530,
Officina Elettrotecnica di Tenno, Italy). The particle sizes were
determined by optical microscopy.
Because hex.MoC decomposes at temperatures higher than $1190^\circ C$[1]
preparation temperatures have to be relatively low. This means
very sluggish reaction rates for the solid state transformation
Mo_2C + graphite into hex. MoC with WC-type structure. The low

diffusivity of the reaction partners at this temperature makes it
necessary to use very fine powders for the MoC preparation.
Fig.2 shows how the yield of hex.MoC depends on the particle size
of the starting Mo_2C powder. The use of nowadays available
submicron carbide or metal powders should add further advantage.

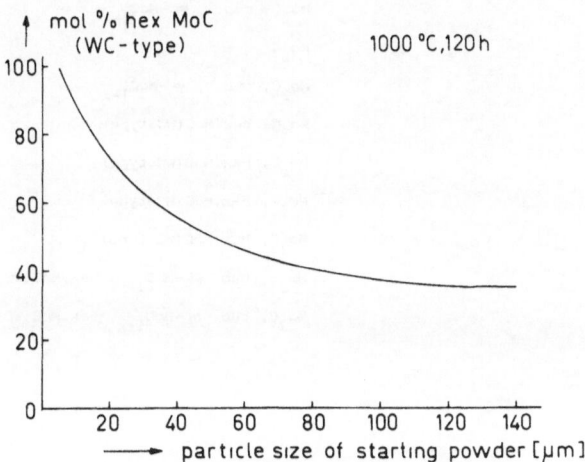

Fig. 2. Dependency of hex.MoC (WC-type structure) formation
         on particle size of starting Mo_2C powder.

RESULTS AND DISCUSSION

     First the effect of different stabilizing-metal additions to
the binary Mo-C alloys on the amount of monocarbide-phases formed
was investigated, because the small amounts of iron metals added
to the starting powder mixtures were previously shown to act not
only as a diffusion aid, but also to have a pronounced effect in
stabilizing the hex.WC-type (Mo,W)C-phase[9] .

     Among all the transition metals of the first period only the
iron group metals and the neighbouring copper show stabilization
of the hex.MoC-phase; as compared to the pure binary Mo-C alloys.
The effect of copper is not very prounounced, nickel seems to be
best (Tab.1).

     To optimize the amount of stabilizer metal additives the
dependency of the MoC-yield on the Ni-content after a reaction time
of 120 hrs. and for a reaction temperature of $1000^{\circ}C$ for various
MoC/WC-molar ratios was investigated (Fig.3). WC-containing alloys
show more hexagonal phase in compositions without stabilizing

Table 1. Effect of Transition Metal Additives on the Formation of
          Molybdenum carbides at 1000°C in Alloys of Composition
          $Mo_{.49}Me_{.02}C_{.49}$.

transition metal	molybdenum carbides formed
Ti	Mo_2C, cub. α-MoC_{1-x}
V	Mo_2C, cub. α-MoC_{1-x}
Cr	Mo_2C
Mn	Mo_2C, cub. α-MoC_{1-x}
Fe	Mo_2C, hex. MoC(WC-type)
Co	Mo_2C, hex. MoC(WC-type)
Ni	Mo_2C, hex. MoC(WC-type)
Cu	Mo_2C, hex. MoC(WC-type)
Zn	Mo_2C, cub. α-MoC_{1-x}, hex. MoC(WC-type)
Al	Mo_2C, cub. α-MoC_{1-x}, hex. MoC(WC-type)

metal; but at the temperature of 1000°C the formation of
hex.(Mo,W)C in composition with high stabilizing metal content is
slower for WC-containing alloys. The yield maximum shifts to higher
stabilizing metal content with increasing WC-content in the alloys.

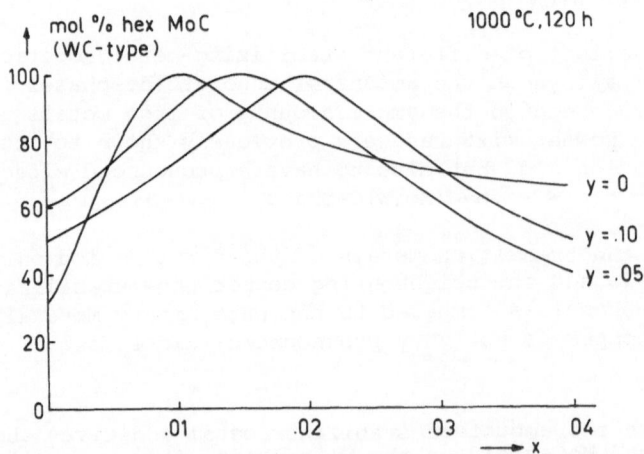

Fig. 3.   Effect of Ni-content in $(Mo_{.50-x/2}Ni_xC_{.50-x/2})_{1-y}WC_y$
          alloys on hex.MoC (WC-type structure) formation

To increase the reaction rate a bath metal had to be selected. This bath metal has naturally a stabilizing or destabilizing effect on the hex.MoC phase, too. Several low-melting metals do not display strong interactions with the interesting Mo-W-C-system that - means no tendency for the formation of binary or ternary compounds within a range around $1000^{\circ}C$ (e.g. Pb,Sn,Zn,Cd,Cu_3Ag_7). Lead was selected, because it is the only one, which showed a beneficial effect on the MoC-formation. The combination: Cu as an additive in a Pb-bath appearantly lost its stabilizing effect (Tab.2).

Table 2. Effect of Selected Bath Metals on the Formation of Hex. MoC with WC-type structure.

stabilizing metal	molybdenum phases formed		
	bath metal Pb	bath metal Sn	bath metal Zn
Fe	Mo_2C, hex.MoC(WC-type)	Mo_2C	Mo
Co	Mo_2C, hex.MoC(WC-type)	Mo_2C	Mo_2C, Mo
Ni	Mo_2C, hex.MoC(WC-type)	Mo_2C	Mo_2C, Mo
Cu	Mo_2C	Mo_2C	Mo_2C, Mo
Zn	Mo_2C	Mo_2C	Mo_2C, Mo

To remove the Pb-metal after the reaction the conventional agents acetic acid/hydrogen peroxide or nitric acid/sodium nitrite could not be used. HAc/H_2O_2 react readily with pure Pb-metal, but in the presence of carbide the lead shows remarkable passivity against this reagent. Reaction times up to 200 hrs could not dissolve it quantitatively. With $HNO_3/NaNO_2$ always insoluble Pb-nitrates were formed, which were complicated to separate from the carbide powder. But dissolving the lead electrochemically in a perchloric acid/ methanol electrolyte gave pure carbide powders.

Using a starting material with an average particle size of 5 µ or less, lead as bath metal and nickel as stabilizing additive the transformation Mo_2C + C into hexagonal MoC with WC-type structure becomes quasistationary (Fig.4). This type of kinetic behaviour is characteristic for reactions in molten alloys or salts. The reaction rate at the interface Mo_2C-particle/liquid lead is limited by the activity of carbon there:

$$v_1 = K \; a_{Mo_2C/Pb}$$

This activity, $a_{Mo_2C/Pb}$, is determined by the amount of carbon diffusing though the lead from the graphite particle/liquid lead interface

$$v_2 = J = \frac{D}{\delta} \; (a_{C/Pb} - a_{Mo_2C/Pb})$$

(J= graphite flux across the liquid Pb-layer of thickness δ ,
D= coefficient of diffusion of C in Pb,
$a_{C/Pb}$=activity of carbon at the graphite/liquid lead interface).
After a short initial reaction period this becomes a steady state system

$$v_1 = K \; a_{Mo_2C/Pb} = \frac{D}{\delta} \; (a_{C/Pb} - a_{Mo_2C/Pb}) = v_2$$

$$a_{Mo_2C/Pb} = \frac{D/\delta}{K + D/\delta} \cdot a_{C/Pb}$$

$$v_{total} = v_1 = K \; a_{Mo_2C/Pb} = \frac{a_{C/Pb}}{\delta/D + 1/K} = K^* \; a_{C/Pb}$$

For $\delta/D \gg 1/K$, which is the case for low temperatures, the reaction is diffusion controled and proceeds at a constant rate until the graphite particles are completely dissolved. Lead dissolves at 1000°C 3 at% carbon, which need to be added to the alloy in excess to the stoichiometric amounts in order to achieve complete transformation of Mo_2C into hex.MoC. On quenching the specimen this carbon is precipitated and remains with the carbide powder after electrolysis of the lead. Thus by adjusting the amount of the bath metal used (and the reaction temperature in case of strong temperature dependency of the carbon solubility in the bath metal) it is possible to control the amount of free carbon in the carbide powder. This graphite is necessary when the material is used for hard metal production, because of the graphite solubility in the binder metal, which would lead to decarbonization of the hex.(Mo,W) carbide otherwise.
It is interesting to observe that the temperature dependence of the reaction is not very pronounced. Between 800°C and 1000°C the amount of hex.(Mo,W)C found after 24 hrs. is almost the same, according to preliminary results. But a complete time-temperature-transformation diagram requires much more experimental data, because of the nonlinear interdependency of these reaction parameters.

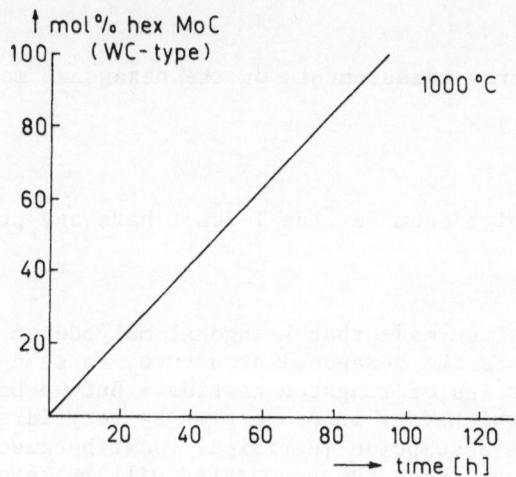

Fig. 4. Time-dependency of amount of hex.MoC (WC-type) formed

REFERENCES

1. J. C. Schuster, E. Rudy, H. Nowotny, Mh. Chem. 107:1167 (1976).
2. M. Kodama, M. Nakano, M. Miyake, A. Hara, 10th Plansee-
   seminar, Reutte (Austria), 1981
3. E. Rudy, B. E. Kieffer, E. Baroch, 9th Planseeseminar,
   Reutte (Austria), 1977
4. U. S. Pat. 4049380 (1977)
5. M. Schreiner, Thesis, Technical University Vienna 1979
6. P. Ettmayer, Mh. Chem. 101: 1720 (1970).
7. R. Kieffer, P. Ettmayer, Austrian Patent Nr. 362943 (1981)
8. W. D. Schubert, P. Ettmayer, B. Lux, W. Ohlsson, 10th Plansee-
   seminar, Reutte (Austria), 1981
9. J. C. Schuster, H. Nowotny, Mh. Chem. 110:321 (1979).

DISCUSSION

C. Politis:

What is the transformation temperature of molybdenum carbide? Do you know it exactly?

J. Schuster:

1190° C.

R. K. Viswanadham:

Are there any property measurements on the hexagonal molybdenum carbide?

J. Schuster:

Just on the pure molybdenum carbide I don't have any property data.

R. K. Viswanadham:

The assumption is often made that hexagonal molybdenum carbide, simply because it has the hexagonal structure, in some way would simulate the properties of tungsten carbide. But we have examples in cubic carbides which vary in properties by very large amounts. To what extent is this assumption justified? Just because it is hexagonal, what guarantee that the properties will be favorable?

J. Schuster:

Well, I think there are some data on the solid solution (Mo, W) C which show that the properties don't change dramatically. Hardness and mechanical properties are very much the same and abrasion resistance is improved by the Mo additions.

R. K. Viswanadham:

At what tungsten:molybdenum ratios?

J. Schuster:

I think, up to 0.7 Mo and 0.3 W.

PREPARATION AND PROPERTIES OF

WC-(Ni,Al) CEMENTED CARBIDES

R.K. Viswanadham, P.G. Lindquist and J.A. Peck

Materials Research, Reed Rock Bit Co.
P.O. Box 2119
Houston, Texas 77001

ABSTRACT

WC-Ni cermets are inferior to WC-Co when measured by their hardness-fracture toughness combination. To enhance system performance, an attempt was made to strengthen the binder in WC-Ni by alloying with Al to produce controlled dispersions of γ' (Ni$_3$Al) in the binder.

WC-(Ni,Al) cermets with 4 to 12 wt. % Al in the binder were produced by powder metallurgy techniques, liquid phase sintered and hot isostatically pressed. X-ray diffraction and transmission electron microscopy were used to characterize the amount, size and distribution of γ' in the binder. The volume fraction of γ' in the binder as measured by X-ray diffraction agreed well with values calculated using available phase diagram data. Dark field TEM studies showed that both cuboidal and heterogeneous γ' were present in the binder in the as-sintered condition. The size of γ' varied from 100 to 2000 A. The WC-(Ni,Al) cermets were then solution treated by vacuum heating and quenching in oil. They were then aged in vacuum at different temperatures and times to produce γ' of controlled size and distribution. Since all these heat treatments were carried out in the solid state, microstructural factors such as carbide fraction, size and contiguity remained unchanged. In this respect, WC-(Ni,Al) cermets provided a model system where only the binder constitution was altered.

Hardness and fracture toughness comparisons of WC-(Ni,Al) in the various heat treated conditions revealed that conventional wisdom regarding strengthening mechanisms in bulk alloys is not directly translatable to binders in cermets, and that the applicability and effectiveness of each strengthening mechanism differs considerably from bulk alloys.

INTRODUCTION

Attempts to replace Co as a binder in WC-based cermet systems have not been very successful and a variety of explanations has been offered for the failure. The most plausible of these explanations is that the yield strength and work hardening rate of Ni-W-C alloys are significantly lower than for Co-W-C alloys when the W and C contents of the two systems are adjusted to values typical of those found in the binders of WC-based systems(1). If this were the case, it should be possible to strengthen Ni binders through various strengthening mechanisms already well-known and well-established in bulk Ni-base alloys[2]. The two most widely used strengthening mechanisms are: solid solution hardening, and hardening by precipitation of Ni_3Al (γ'), an ordered, coherent intermetallic compound. To verify if a similar approach can be used to strengthen the Ni binders of cermet systems, WC-(Ni,Al) cermets containing 4 to 12 wt. % Al in the binder were fabricated using conventional powder metallurgy techniques. Solution treatment, quenching and aging were used to produce controlled dispersions of γ'. The effect of the amount, size and distribution of γ' on the mechanical behavior of the cermets was examined using hardness, crack resistance and fracture toughness measurements.

MATERIALS PREPARATION

All materials were produced by conventional powder metallurgy techniques and liquid phase sintered. The starting powders were WC, Ni and W. The Al content of the binder was varied by controlled additions of pre-alloyed NiAl to the powder batch. All cermets were sintered in a conventional vacuum furnace at temperatures ranging from 1345 to 1400°C for times up to 90 minutes. Full densification could not be achieved in cermets containing high percentages of Al in the binder. After sintering the cermets were hot isostatically pressed in argon at $1330^{\circ}C$ and 15,000 psi (103 MPa). Solution treatments were carried out in vacuum at $1204^{\circ}C$ for one hour followed by quenching in oil. Aging treatments were carried out in vacuum at temperatures of 600 and $700^{\circ}C$.

MATERIALS CHARACTERIZATION

Optical microscopy, X-ray diffraction and trans- mission electron microscopy were the principal tools used to characterize the microstructure of WC-(Ni,Al) cermets. Typical microstructure of a WC-(Ni,Al) cermet is shown in Fig. 1. The WC grains are more rounded and less angular than in WC-Co cermets. X-ray diffraction was the principal method used to detect the presence of γ' in the binder and determine its amount. Prior to X-ray analysis the carbide fraction was selectively removed by electro-etching in an aqueous solution of 10% Na_2CO_3 at room temperature for 15 minutes and a current density of 1 A/cm^2 [3]. It was observed that selective removal of the carbide resulted in

a) Considerable sharpening of all the binder peaks.
b) Incresed intensity for all the binder peaks enabling measurement of higher angle peaks and
c) Direct observation of the superlattice reflections produced by ordering in γ'.

Figure 2 shows typical X-ray diffractograms of the (111) binder peak from WC-(Ni,Al) cermets with increasing amounts of Al in the as-sintered condition. As the Al content of the binder increased, the (111) peak became a doublet -- a clear indication of the presence of γ'. The amount of γ' increased systematically with increase in the amount of Al and the measured volume fractions of γ' were in agreement with values calculated from Ni-Al binary phase diagram data at $300^{\circ}C$ [4]. The average lattice parameters for γ and γ' were 3.55 and 3.57 A respectively, resulting in a misfit of 0.6%.

Transmission electron microscopy and centered dark field techniques were used to characterize the size, morphology and distribution of γ' in the binder. Thin foils suitable for microscopy were prepared by mechani- cally polishing thin slices to a thickness of $\sim 100 \mu m$. Three mm. discs were ultrasonically machined from these slices and ion-milled at 5kV and $100 \mu A$. A typical dark field micrograph taken with a superlattice reflection from a material containing 6 wt. % Al in the as-sintered

condition is shown in Fig. 3. The γ' morphology is primarily cubic with some indication of preferential nucleation at the carbide/binder interface. The γ' particles
ranged in size from <100 Å to >1000 Å. Such a large
range in the size of the γ' particles was typical of all
the as-sintered cermets whose Al content exceeded 4 wt. %
of the binder. As the Al content and the amount of γ'
increased, large γ' precipitates at the carbide/binder
interface were more frequently observed. This large
range in the size of γ' precipitates was a result of the
slow cooling rates from the sintering and/or the hipping
temperature. It was felt that a more uniform dispersion
of γ' of controlled size could be produced if the cermets
were reheated to a high enough temperature to completely
dissolve all the γ', cooled rapidly to retain the Al in
solid solution and the γ' is reprecipitated in a subsequent controlled time-temperature heat treatment. To
accomplish this, WC-(Ni,Al) cermets with 6, 7 and 7.7
wt. % Al were reheated to 1204°C in vacuum for 1 hour
and quenched in oil.

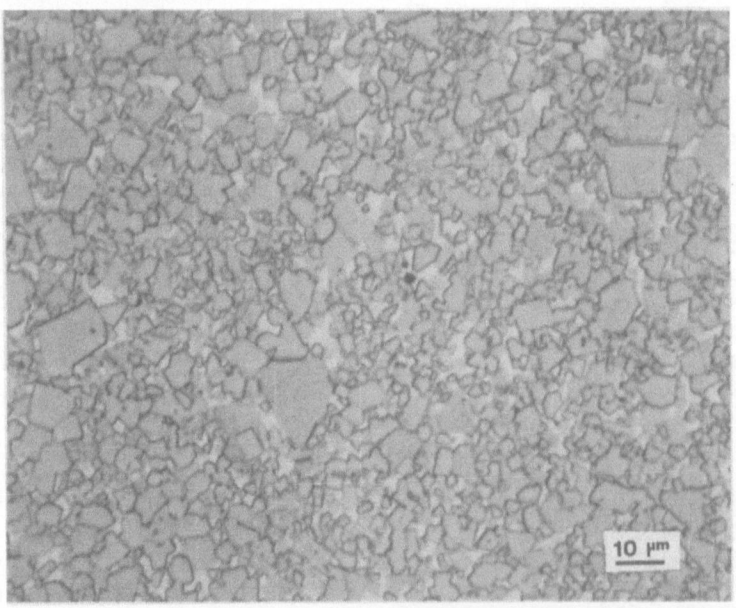

Fig. 1: Typical microstructure of a WC-(Ni,Al) cermet
         with 6 wt. % Al in the binder.

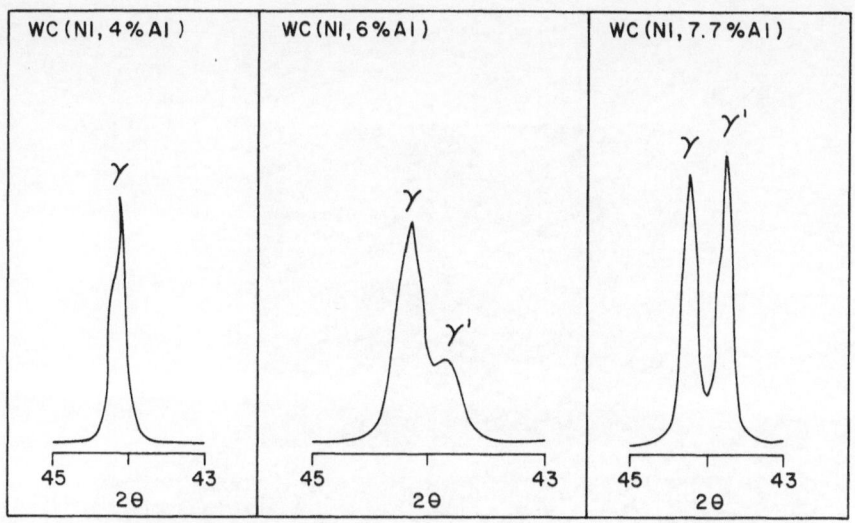

Fig. 2: X-ray diffractograms of the (111) binder peak
         of WC-(Ni,Al) cermets with 4, 6 and 7.7 wt. %
         Al in the binder.

     X-ray and TEM analysis of the WC-(Ni,Al) cermets
after the solution treatment showed that complete dissolution of γ' did not occur in all the samples. In cermets containing 7.7 wt. % Al, the solution treatment
produced only a small decrease in the amount of γ'. In
cermets containing 7 wt. % Al trace amounts of γ' remained after the solution treatment. In cermets with 6
wt. % Al the effects of solution treatment were variable.
In some cases complete dissolution of γ' was achieved,
while in others small amounts of γ' could be detected.
X-ray diffractograms from a WC-(Ni,Al) cermet with 6 wt.
% Al before and after solution treatment are shown in
Fig. 4. Diffractogram marked 4(b) was obtained when the
solution treatment was fully effective and 4(c) illustrates the case when the treatment was not completely
effective and trace amounts of γ' remained. It was
consistently noted that in the presence of γ' the (111)
peak width of γ was higher and the $K\alpha_1$-$K\alpha_2$ doublet was
not well resolved. This resulted from non-uniformity of
internal strains in the γ phase produced by the presence
of fine γ' particles. A TEM dark field micrograph from
a cermet where the solution treatment was not completely
effective is shown in Fig. 5. The presence of small
(~120Å), uniformly distributed, almost spherical γ' precipitates could be noted. No preferential nucleation at
the carbide/binder interface was observed.

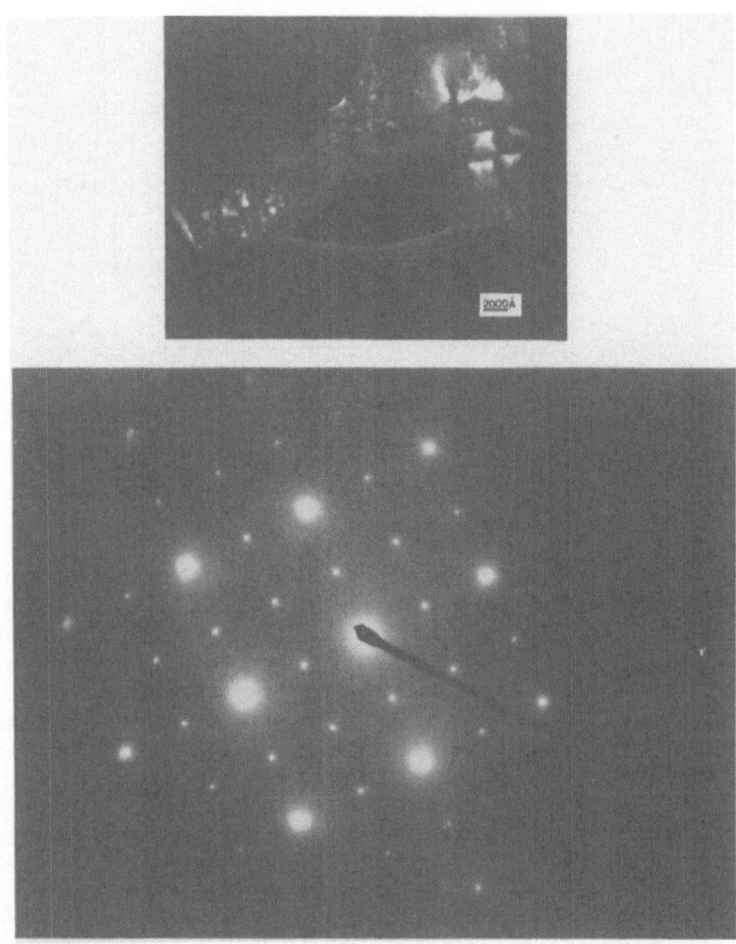

Fig. 3: (Top) Dark field transmission electron micrograph
         of a WC-(Ni;Al) cermet in the as-sintered con-
         dition showing the size and distribution of γ'
         precipitates. (Bottom) Associated Diffraction
         Pattern.

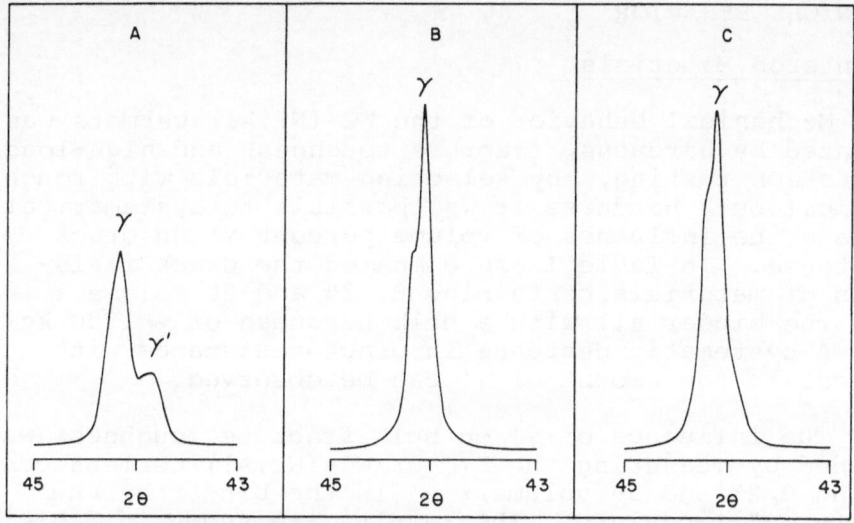

Fig. 4:   X-ray diffractograms of the (111) binder peak
          from a WC-(Ni,Al) cermet with 6 wt. % Al. (A)
          As-sintered (B) Solution treatment completely
          effective and (C) Solution treatment not effec-
          tive.

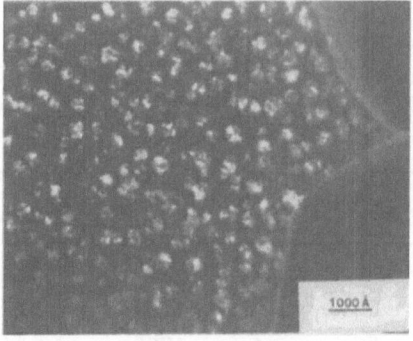

Fig. 5:   Dark field transmission electron micrograph
          showing the size and distribution of γ' in a
          cermet where the solution treatment was not
          completely effective.

MECHANICAL BEHAVIOR

As-Sintered Materials

Mechanical behavior of the WC-(Ni,Al) cermets was evaluated by hardness, fracture toughness and high-load indentation testing. By selecting materials with roughly identical bulk hardness it was possible to systematically evaluate the influence of volume percent γ' on crack resistance. In Table I are compared the crack resistances of materials containing 0, 25 and 50 volume % γ' in the binder all with a bulk hardness of ~ 1200 kg/mm^2. A systematic decrease in crack resistance with increase in the amount of γ' can be observed.

The influence of γ' on bulk fracture toughness was examined by measuring the K_{IC} of WC-(Ni,Al) cermets containing 0, 25 and 50 volume % γ' in the binder by the "short-rod" technique. The results are shown in Table II and the K_{IC} values are in general agreement with the crack resistance data presented in Table I.

Table 1. Hardness and crack resistance of WC-(Ni,Al) cerments containing 0, 25, and 50 vol. %γ' in the as-sintered condition.

Batch No.	Binder Constitution	Hardness (kg/mm^2)	Crack Resistance (kg/mm)
20 XD	100% γ with 4% Al	1215	171
20 YB	75% γ + 25% γ'	1209	150
20 ZA	50% γ + 50% γ'	1205	111

Table II. Bulk fracture toughness of WC-(Ni,Al) cerments containing 0, 25 and 50 vol. %γ' in the binder. The K_{IC} values were obtained using the "short rod" geometry.

Batch No.	Binder Constitution	Hardness (kg/mm^2)	K_{IC} (MPa$\sqrt{m}$)
20 XD	100% γ with 4% Al	1215	12.59 ± 0.36
20 YB	75% γ + 25% γ'	1209	12.12 ± 0.18
20 ZA	50% γ + 50% γ'	1205	11.49 ± 0.23

Solution Treated and Aged Materials

Since the solution treatment was not completely effective in all cases, crack resistance measurements on "solution treated" materials showed higher-than-normal scatter. Inspite of this, the crack resistance of the solution treated materials was always higher than their as-sintered counterparts. Typical results are shown in Table III. It is interesting to note that changes in bulk hardness are negligible in all cases. Bulk fracture toughness measurements on cermets containing 6 wt. % Al before and after solution treatment are compared in Table IV. A significant improvement can be noted. Even when the solution treatment was not completely effective, the amount of γ' was reduced and some of the improvement in crack resistance could be attributed to this factor alone.

Table III. A comparison of the hardness and crack re-sistance of WC-(Ni, Al) cerments in the as-sintered and solution treated conditions.

Binder Composition	As-Sintered Cermets		Solution Treated Cermets	
	Hardness (kg/mm^2)	Crack Resistance (kg/mm)	Hardness (kg/mm^2)	Crack Resistance (kg/mm)
Ni-6 wt% Al	1209	150	1212	252
Ni-7 wt% Al	1134	186	1134	228
Ni-7 wt% Al	1073	446	1070	563
Ni-7 wt% Al	1053	407	1038	1200
Ni-7.7 wt% Al	1205	111	1194	173

Table IV. Bulk fracture toughness of 6 wt% Al WC-(Ni,Al) cerments before and after solution treatment.

Binder Composition	As-Sintered $K_{IC} (MPa\sqrt{m})$	Solution Treated $K_{IC} (MPa\sqrt{m})$
Ni-6 wt% Al	12.12 ± 0.18	14.85 ± 0.43

The crack resistance vs. hardness behavior was examined through plots of inverse crack resistance vs. hardness[5]. The results are shown in Fig. 6. The influence of the amount of γ' is clearly evident. At a fixed bulk hardness, increasing amounts of γ' reduce the crack resistance. Also, the slope of the inverse crack resistance vs. hardness plot increases as the volume fraction γ' increases -- indicating the onset of non-energy dissipative fracture modes[6].

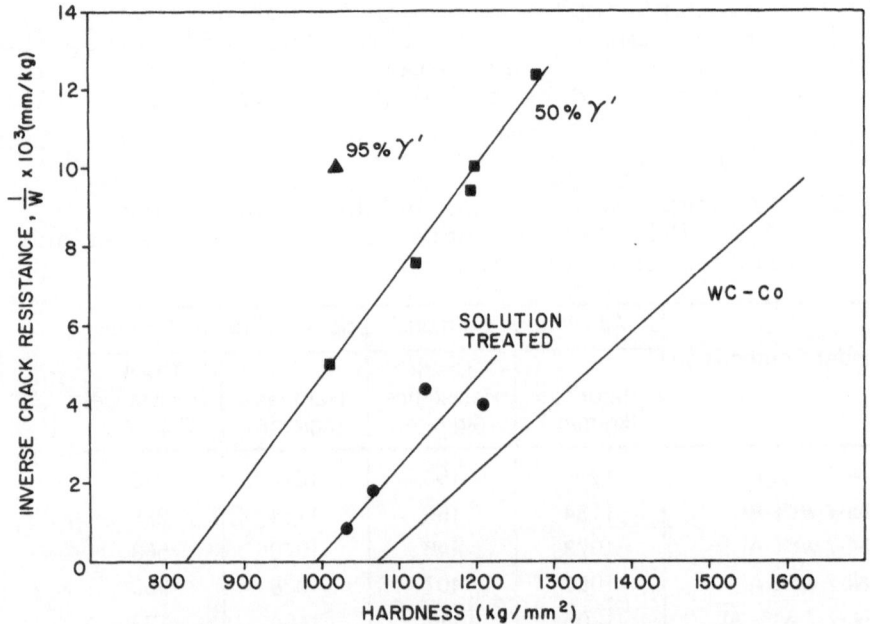

Fig. 6:   Inverse crack resistance (1/W) and hardness (H) plots of WC-(Ni,Al) cermets with increasing amounts of γ' in the binder. The plots shift to the right as the volume fraction γ' is increased. The plot for WC-Co is included for comparison.

To examine the effect of aging, solution treated and quenched cermets containing 6 and 7 wt. % Al were aged at $600°C$ for 20 hours and $700°C$ for 30 minutes. These times were selected based on an activation energy of 64 Kcal/mol[7] for the growth of γ' in bulk Ni-Al alloys. Hardness and crack resistance data obtained on the aged materials are shown in Table V. The data clearly show

Table V. Influence of aging treatments on the hardness and crack resistance of WC-(Ni,Al) cerments.

Binder Composition	Solution Treated		Aged 700°C 30 mins.		Aged 600°C 20 hrs.	
	Hardness (kg/mm²)	Crack Resistance (kg/mm)	Hardness (kg/mm²)	Crack Resistance (kg/mm)	Hardness (kg/mm²)	Crack Resistance (kg/mm)
Ni-6 wt% Al	1212	252	1216	161	1210	154
Ni-7 wt% Al	1134	228	1135	176	1125	196
Ni-7 wt% Al	1070	563	1080	440	1075	291
Ni-7 wt% Al	1038	1200	1054	730	1067	254

that crack resistance decreases after each heat treatment with little or no change in the bulk hardness.

DISCUSSION AND CONCLUSIONS

New cermet systems that combine both high hardness
and toughness can only be developed when fundamental
understanding about their flow and fracture behavior is
obtained. Such understanding is possible only through
the evaluation of the mechanical behavior of model cermet
systems where one parameter is varied at a time. In
this sense WC-(Ni,Al) is just such a system. Since all
the heat treatments -- such as solution treatment,
quenching and aging are all carried out in the solid-
state, parameters like carbide grain size, distribution,
contiguity and volume fraction remain unaltered. The
only parameter that is varied is the amount, size and
distribution of γ' in the binder. Comparison of the
mechanical behavior of WC-(Ni,Al) cermets in the various
heat treated conditions clearly shows that γ' provides
little or no strengthening to the binder irrespective of
its size and distribution. On the other hand, the
presence of γ' does have a deleterious effect on fracture
toughness. If the γ' is very fine (~ 100 Å), and uniformly distributed, its effect on fracture toughness is
neutral and at large sizes, the fracture toughness is
significantly reduced.

These observations clearly show that strengthening
mechanisms used for bulk alloys are not necessarily
effective in strengthening binders of cermet systems.
Dispersion hardening, precipitation hardening and order
strengthening fall in this category. These strengthening
mechanisms appear to be effective only when the alloys
are not under constraint. Solid solution hardening, on
the other hand, appears to be effective in bulk alloys
as well as binders under constraint. In the WC-(Ni,Al)
system, the best combination of hardness and fracture
toughness is obtained when all the Al is in solid solution. Reference to the literature on superalloys shows
Al to be a very effective solid solution hardener[2].

It is now possible to classify strengthening mechanisms into universal and restricted based on their range
of applicability. Solid solution hardening appears to
be of universal validity while many of the others are
effective only in a restricted range of microstructures.
If λ is the binder mean free path and d is the average

grain size of the carbide, then d/λ can be taken as a measure of the constraint on the binder. Strengthening mechanisms such as precipitation hardening, dispersion hardening and order strengthening are valid only when d/λ<<1. Reasons for the ineffectiveness of these strengthening mechanisms when the binder is under severe constraint are not very clear and point to a lack of understanding of flow and fracture in the binder phase of a cermet. It does appear, however, that flow and fracture in this class of materials are not necessarily interlinked. For example, the presence of γ' in the binder enhances fracture phenomena in a cermet without significantly modifying its flow characteristics. This observation might provide a clue for improving the resistance to flow of a binder without adversely affecting the fracture behavior of a cermet.

REFERENCES

1. B. Roebuck and E.A. Almond, "A Comparison of the Deformation Characteristics of Co and Ni Alloys Containing Small Amounts of W and C", Proc. 10th Plansee Seminar, 1:493 (1981).
2. C.T. Sims and W.C. Hagel, "The Superalloys," Wiley-Interscience, Newyork (1972).
3. H. Jonsson, Studies of the Binder Phase in WC-Co Cemented Carbides Heat-Treated at 650°C, Powder Metallurgy, 15(29):1 (1972).
4. A. Taylor and R.W. Floyd, The Constitution of Nickel-Rich Alloys of the Nickel-Aluminium-Titanium System, J. Inst. Metals, 81:25 (1952).
5. R.K. Viswanadham and J.D. Venables, "A Simple Method for Evaluating Cemented Carbides, Met. Trans. A, 8A:187 (1977).
6. R.K. Viswanadham and W. Precht, "Preparation and Properties of VC-Ni Cermets with Controlled Carbon-to-Metal Ratio, Met. Trans. A, 11A:1475 (1980).
7. A.J. Ardell, R.B. Nicholson and J.D. Eshelby, "On the Modulated Structure of Aged Ni-Al Alloys, Acta Met., 14:1295 (1966).

DISCUSSION

E. Almond:

Can you remind me of the grain size of this material?

R. K. Viswanadham:

It's in the range of 3 microns.

E. Almond:

We have worked on 2 micron grain size and 10 per cent binder and we do get an effect on hardness by solution treating and reprecipitation. So, I think this event you see must be very sensitive to λ/d ratio. And we can compare these effects with effects we see in binder phase alloys made up with simulated compositions.

R. K. Viswanadham:

But, in all the work we have done, we cannot measure a detectable change in the hardness.

E. Almond:

The hardness changes we saw were about 50 to 100 Vickers, at the most.

R. K. Viswanadham:

From the measurement error, I would say that if there are hardness changes, they are $\leq$ 15 kg/mm^2. We also made compositions with different binder contents and grain sizes. There were materials that were in the hardness range of about 1,000 kg/mm^2. In that case, you just begin to pick up some changes in the hardness. But in the range of about 1,200 kg/mm^2, there are no measureable changes.

A. Krawitz:

I wonder if you took the weighted average of the bulk hardness values of WC and bulk gamma-prime alloy, how much hardness change would you expect to get?

R. K. Viswanadham:

I would certainly have expected at least 75 to 100 kg/mm^2.

J. Gurland:

Well, I'd just like to mention some general ideas about the design of these alloys as far as hardness is concerned. We should not neglect the important and predominant effect of the load carried by the carbide itself, namely the stresses in the carbide phase. It seems to me if the carbide structure is continuous over long distances, then in order to plastically deform the alloy, it is necessary to plastically deform the carbide. Therefore, the controlling parameter

would probably be the yield strength, the in situ yield strength, of
the carbide phase. On the other hand, if the structure of the
carbide is dispersed, then the load transferred to each particle may
not reach the yield stress and then perhaps the binder phase could
have a very high and disproportionate effect. So, I think the effect
expected from binder phase manipulation has a lot to do with the
structure, the microstructure that is, of the cemented carbide
itself.

R. K. Viswanadham:

It definitely is. But it's interesting that this system provided us
with a model. The reason it's a model system is because with a given
microstructure you can modify the binder alone which you cannot do in
other systems without modifying something else. If you vary the
carbon content, for example, you vary the sintering kinetics, the
grain size and it becomes difficult to control. But here, because
the solution treatment and aging are carried out in the solid state,
the contiguity of the carbide, the carbide grain size and the volume
fraction carbide remain fixed. Only the gamma-prime goes in and out
of solution. I think you cannot ask for a better model system to
study the behavior of the binder. Now, irrespective of load transfer, when all these changes have taken place in the binder, one would
have expected a change in the flow behavior, albeit small, but still
some change.

J. Gurland:

Can you measure the micro-hardness, let's say in high binder alloys?

R. K. Viswanadham:

By the time you go to high binder, you have effects.

H. Fischmeister:

The alloys that you call solution treated showed a rather sizeable
volume fraction of cooling gamma-prime, in the form of very small
particles. Was that picture that you showed representative?

R. K. Viswanadham:

No, that was the case where the solution treatment was not effective.

H. Fischmeister:

I'm not sure you should look at it that way. We know about cooling
gamma-prime in superalloys, and it's a good way to produce a fine,
dispersed structure. I believe that also gave you the highest toughness, did it?

R. K. Viswanadham:

The very fine gamma-prime and where the solution treatment was completely effective were comparable.

H. Fischmeister:

Well, two things struck me. The first has really been alluded to by
previous speakers. Superalloys are rather sensitive to the Hall-
Petch relation. Decreasing the grain size, that is to say, decreasing the slip lengths, has a large effect on the flow stress. Larger
even in comparison with the volume fraction of gamma-prime. The
absence of a hardness effect may be due to the fact that hardness is
controlled by the mean free path. Also, I would like to add, you
don't ever really have a completely gamma-prime-free material
apparently. All you're changing is the size distribution of the
gamma-prime...

R. K. Viswanadham:

And the amount as well.

H. Fischmeister:

You do change the amount? O.K. Again, the amount enters only in the
square root relationship. So, if you have little, it may already be
effective. The second thought that struck me is the important improvement in toughness that you get. This means that the gamma-prime
becomes really effective and does good when the binder undergoes considerable strain. Hardness doesn't strain the binder much. So, you
wouldn't, from that point of view, expect much of an effect. I would
prefer to look at the rosy side of things and rejoice about the improvement in toughness although you will have to find a rather narrow
niche for these alloys because, in cutting, for instance, the aging
effect would wipe out the improvement.

H. Jonsson:

I notice that the aging temperatures were around 600 or 700° C. Did
you ever age at a little higher temperature?

R. K. Viswanadham:

No, we did not. The reason was because at 700° C, within less than
an hour, we were able to see effects.

H. Jonsson:

As Professor Fischmeister also indicated, the problem will probably
arise in many ways if you increase the temperature. This is a more

general aspect of heat treating cemented carbides. We all know how
large amount of grain boundaries there are in cemented carbides. If
you increase the temperature, you get precipitation within grain
boundaries. The binder properties would then vary because the width
of these precipitation zones may soon be comparable to the width of
the binder phase.

R. K. Viswanadham:

Yes. We kept the aging temperatures low intentionally to promote
homogeneous nucleation in the binder and try and avoid gamma-prime at
the carbide/binder interface.

H. Jonsson:

I also know from a five year old investigation concerning heat treating of cemented carbides (the same system as for instance Dunlop and
Wirmark spoke of) that precipitation is very sensitive to binder
layer thickness. You obtain precipitation within the very thick
cobalt layers but the thinner layers never gave homogeneous precipitation. You must always remember how enormously important the grain
boundaries are in cemented carbides. We can hardly expect to obtain
even the same type of partition that we expect from equilibrium
diagrams just because of the grain boundary effects. It was just a
more general aspect of the problem.

STRUCTURE-PROPERTY CORRELATIONS FOR TiB$_2$-BASED

CERAMICS DENSIFIED USING ACTIVE LIQUID METALS *

V. J. Tennery, C. B. Finch, C. S. Yust, and G. W. Clark

Structural Ceramics Group
Metals and Ceramics Division
Oak Ridge National Laboratory
Oak Ridge, Tennessee   37830

INTRODUCTION

The compound TiB$_2$ has numerous exceptional properties including high hardness, high melting temperature, high electrical conductivity, and nonreactivity with various liquid metals. These make it an attractive candidate for technological applications, such as cutting tools, valve trim for erosive environments, and cathodes in Hall-Heroult cells for aluminum smelting. In general, such applications require fabrication of TiB$_2$ into various shapes, and the latter requirement dictates the production of high-density polycrystalline ceramic bodies. The preparation and characterization of TiB$_2$-based materials have already been the subject of numerous previous works, exemplified by the references (1-9). These include data on the phase equilibria in TiB$_2$ containing systems,[1,3,9] the wettability of TiB2 by various metals,[1,3,4] and the results of property determinations on TiB$_2$-based specimens.

Densification of TiB$_2$ ceramics is complicated by two characteristics of this compound. In the first instance, the high melting point of 2980°C (Ref. 1) stipulates the use of sintering temperatures on the order of 2000°C to effect normal (1 atm) densification (1800°C for uniaxial hot pressing) of ceramics having greater than 95% theoretical density. These temperatures result in rapid mass diffusion and permit attainment of high densities in acceptable sintering

---

*Research sponsored by the Division of Materials Sciences, U.S. Department of Energy under contract W-7405-eng-26 with the Union Carbide Corporation.

times between minutes and hours, although maintenance of such
temperatures is costly and requires special equipment. Secondly,
the hexagonal structure of TiB_2 results in marked thermal expansion
anisotropy (TEA), i.e., expansion along the c-axis is considerably
greater than that along the a-axis.   The expansion difference is
about 42% between 25 and 930°C and this difference increases as the
temperature exceeds 930°C. The expansion anisotropy produces considerable internal stress during specimen cooling and generates
microcracking when the grain size is above a critical value.[10]
The microcracking occurs in the grains and at grain boundaries,
with a resultant degradation of macroscopic mechanical properties.
This paper describes procedures for circumventing these difficulties
to allow fabrication of dense, crack-free TiB_2 ceramics

FABRICATION STUDIES

TiB_2 Powder Characterization

     Commercially available TiB_2 powders produced by grinding the
product of carbothermic reaction between TiO_2, C, and B_4C (or B_2O_3)
were selected as the bases for this work. Several of these powders
were characterized as to particle shape, particle size distribution,
and oxygen and other impurity content. In general, the average
particle size varied between 5 and 15 μm, while the oxygen content
varied between 0.5 and 2 wt %. Impurities including carbon, nitrogen,
and other elements were also present, and the total extraneous non-
oxygen impurity concentration was 0.5 to 0.8 wt %. Scanning electron
microscope (SEM) studies indicated that the powders often contained
fragments of an extraneous material having a particle size considerably smaller than that of the predominant TiB_2 phase. This
impurity usually occurred as small particles of a lighter contrast
on the surface of the TiB_2 particles. X-ray diffraction analysis
gave no answer as to its identity, although some diffuse scattering
occurred at Bragg angles below 10° when CuK_α radiation was used.
The scattering was attributed to amorphous phases which are believed
to contain major portions of the oxygen impurity, but definite
identification of the phase was not obtained. Efforts were made
to size-classify several of the TiB_2 powders to obtain powders
having a size distribution narrower than that normally available
for the as-received powders. In addition, it was desirable to
ascertain the amount of the aforementioned extraneous phase which
could be removed by the classification procedure. The TiB_2 powders
were classified with an air classifier* into approximately 80%
coarse and 20% fine fractions. The particle size distributions of
the as-received TiB_2 powders and the classified fraction were determined by both Micromerograph[†] and Microtrac[‡] instruments and the

----

*Vortec Industries, Compton, Calif.
[†]Sharples Co., Bridgeport, Penn., Stokes law of fall principle.
[‡]Leeds and Northrop Co., Largo, Fla., Frauenhofer dispersion of
  light principle.

results of these classifications for a given powder were in good
agreement. Examples of the classification results for two commercial powders, including distribution of the coarse and fine
fractions from one of these powders, are shown in Fig. 1. Powder
"A" was a relatively fine powder, received with an average particle
size of about 3.5 μm, and containing 3 wt % oxygen. Powder "D" was
another powder received with about 1 wt % O_2 content, and size-
classified into fractions identified as "B" and "C" in Fig. 1.
Powders A, B, C, and D were used in fabricating TiB_2 ceramics by
hot-pressing, and the characteristics and properties of some of
these ceramics are presented in this paper. The appearance of the
TiB_2 particles in powders "B" and "C" is illustrated in Fig. 2.
The fine fraction (Powder "B") contains occasional TiB_2 particles
having the morphology of a hexagonal prism, while the coarse fraction
(Powder "C") contains only several particles of an irregular shape
but relatively equidimensional. The small objects visible on the
surfaces of the particles in Powder "C" consist of the extraneous
material which, as noted earlier, is believed to contain appreciable
oxygen. Powder "A" contained substantial quantities of this material,
and had an oxygen content of 3 wt. %.

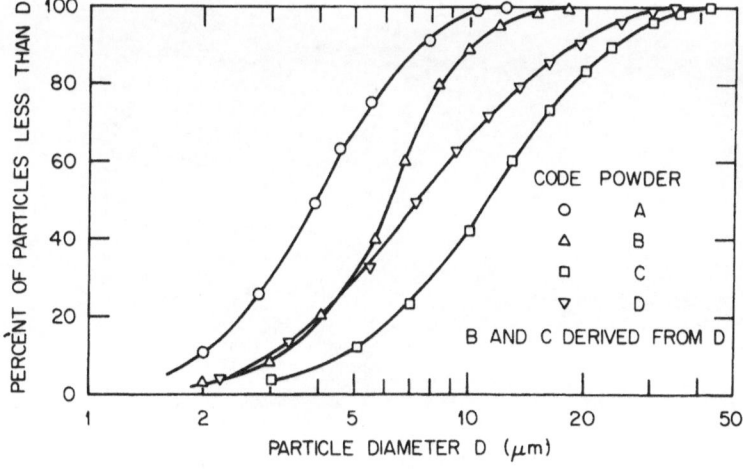

Fig. 1. Particle Size Distributions of Two Commercial TiB_2 Powders
(Powders A and D), Plus a Fine Fraction (Powder B) and
Coarse Fraction (Powder C) Derived from Powder D by an Air
Classification Technique.

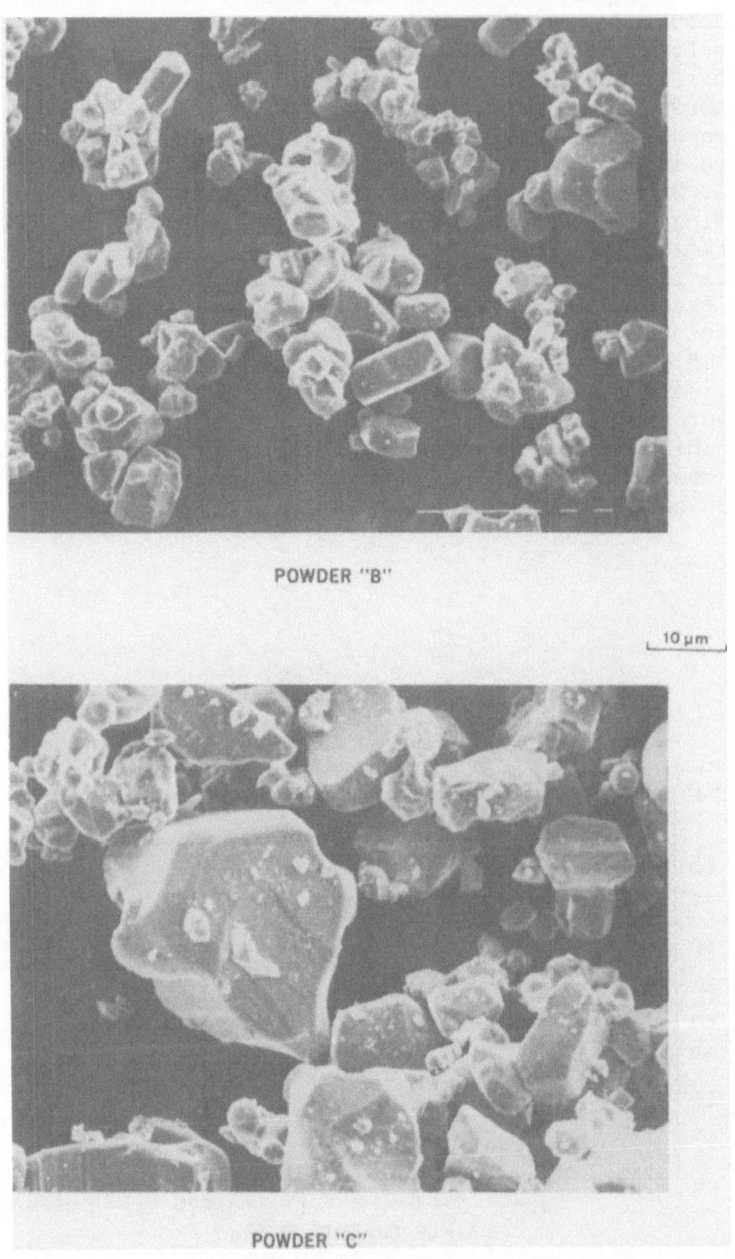

Fig. 2. Typical TiB$_2$ Particles in Powders B and C Resulting from
Air Classification of Powder D in Fig. 1.

Densification of TiB2 Ceramics

Experiments were conducted to study the densification by hot pressing of these TiB$_2$ powders, as well as to investigate the effects of liquid metals on their densification. According to earlier results at this laboratory[11] in addition to other literature data,[12-13] Ni and Fe were selected for a study of the role of transition metal content on the densification of TiB$_2$ ceramics. This paper presents the results obtained for Ni only which serves also as data for another paper in this conference.[14] It was observed that vacuum hot pressing at modest pressures (12 MPa) and temperatures (1425°C) slightly below the Ni melting point results in rapid densification of the diboride. Based on these results, a set of hot-pressing conditions was selected in which only the time at hot-pressing temperature was the variable. The hot pressing was accomplished as follows. Dry TiB$_2$ and Ni powders were weighed to ± 0.05 g and mixed 24 h by tumbling in polyethylene jars. The powder mixtures were loaded into 4.5-cm- or 6.0-cm-ID graphite die assemblies.* The cylindrical surfaces and punch faces of the dies were lined with 0.25-mm-thick grafoil† or a 0.1-mm-thick coating of BN.‡ The initial charge was selected so that the final thickness of the resulting ceramic disks was between 0.5 and 1.2 cm for the 4.5 and 6.0 cm diameters. The disks were pressed singly in a vacuum hot-press assembly§ consisting of a vertical graphite resistance heating element furnace surrounding the press assembly. The system was loaded with the dies and evacuated to 10^{-3} Pa. The furnace was heated to 1000°C and held at this temperature until outgassing was complete and the pressure reached 10^{-3} Pa. The temperature was then increased to a final hot-pressing temperature of 1425°C. In most cases, substantial outgassing occurred near 1200°C, the pressure rose to about 10^{-2} Pa and remained at this value until the die had attained its final temperature. Recovery of the pressure to about 10^{-3} Pa was usually achieved within 1.0 to 1.5 ks (0.28 to 0.42 h) at the final temperature. At this time, loading was applied to the die punches via the press rams at a rate of about 3.5 kPa/s until the total pressure was 12 MPa. The charges were maintained under pressure for times ranging from 7.2 ks to 28.8 ks (2–8 h). During application of pressure, a Ni-rich liquid was exuded at both the upper die surfaces and below the die assembly, having been ejected from the charge through the clearances of the die assembly. As will be described later, the exudation process is important in

---

*Carbonyl reduction, 0.2 wt % O$_2$, <44 μm, Cotronics Corp., N.Y.
†Union Carbide Corp., Carbon Products Division, Cleveland, Ohio.
‡Carborundum Corp., Graphite Products Division, Niagara Falls, N.Y.
§Advanced Vacuum Systems, Ayer, Mass., Model 1-VHP-4-8-2200-3-20.

achieving high densities, strengths, and other desirable properties
in the ceramics. In a typical hot press run, between 25 and 95 wt. %
of the initially charged Ni was exuded during densification. After
cooling and removal from the hot press, the densified compacts (disks
2 to 4 cm diam x 0.5 to 1.0 cm thick) were removed from their dies
using modest force applied with a hydraulic press.

Specimen Preparation

Specimens of the required geometry and dimensions were cut from
the as-pressed disks by electron discharge machining (EDM) or by an
oil and water-lubricated diamond saw. The latter operation was not
extensively used due to the rapid wear of the diamond blades and
difficulties with nonlinear cuts due to blade deformation during
cutting. Following EDM cutting, the thin oxidation layer produced
on the surface by the cutting was removed with diamond grinding
wheels of grit sizes from 60 to 6 μm. Final polishing of critical
surfaces of some specimens was performed using 0.5 μm diamond on
a nylon lap in a vibratory polisher, followed by vibratory polishing
with 0.3 μm Al_2O_3.

STRUCTURE AND PROPERTIES OF TiB_2 CERAMICS

Characterization Procedures

The densities of both as-pressed disks and specimens cut
from the disks were determined by immersion in H_2O and by direct
measurement determination of geometric density. The densities generally agreed within $\pm$ 0.02 g/cm^3. X-ray diffraction and x-ray
fluoresence were used to identify phases and to determine elemental
Ni content. The x-ray diffraction data showed that the major phase
in the compacts was TiB_2, although numerous low-intensity lines
were present, indicating the presence of secondary phases. Unambiguous identification of the secondary phase(s) could not be obtained from the x-ray diffraction data due to the complex changes
in line intensity of the secondary phase(s). Standards were prepared which allowed rapid x-ray fluorescence determination of the
elemental Ni concentration in the hot-pressed ceramics. Oxygen was
determined using a neutron activation technique. Electron diffraction and analytical electron microscopy (AEM) techniques were also
employed to identify phase(s) and optical microscopy was used to
determine the microstructures. A technique utilizing both bright
field and polarized light conditions was used to characterize the
optical microstructures.

Acoustic velocities, crack propagation and fracture, as well
as indentation behavior and thermal transport of hot-pressed specimens were used to calculate values of Young's modulus, Poisson's
ratio, fracture strength, microhardness, and the thermal diffusivity.

Fracture toughness was determined by the crack-indent method and by calculation from the measured fracture energies of double canti-lever beam specimens and measured values of the elastic moduli and Poisson's ratios. Microhardness was determined with a diamond pyramid indenter, while thermal diffusivity data were obtained by the laser flash method. Both of these properties were determined as a function of temperature. In addition, the behavior of these ceramics as cutting tools was determined in controlled metal-cutting tests. Most of the results presented are for ceramics fabricated using the TiB$_2$ powders identified "B" in Fig. 1 since these have superior microstructures and mechanical properties compared with specimens fabricated from other powders.

Microstructures

Ceramics of TiB$_2$ fabricated from powder "B" had the fine-grained microstructure shown in Fig. 3 when specimens were hot pressed at 12 MPa and 1425°C for 14.4 and 28.8 ks (4 and 8 h, respectively). The bright field micrographs of Fig. 3 display TiB$_2$ as a lighter (gray) phase in which relief polishing accentuates the grain bound-aries between TiB$_2$ grains. The dark regions delineate locations of the Ni-rich phase(s) for specimens in Fig. 3, and the micrographs illustrate the Ni-bearing secondary phase as becoming more homo-geneously distributed throughout the microstructure with prolonga-tion of pressing time. The immersion density of both TiB$_2$ specimens was 4.53 ± 0.01 g/cm^3. The microstructure in Fig. 3(b) (polarized light, crossed polarizer and analyzer) is shown in Fig. 4 for the field identical to that of Fig. 3(b). In Fig. 4, the grain structure of the TiB$_2$ is clearly visible, and careful examination shows evi-dence of more extensive secondary phase regions. The TiB$_2$ grain size in Fig. 4 varies between 4 and 12 μm. Figure 1 shows that over 95% of the particles in powder "B" had starting diameters less than about 12 μm. Comparison of these values and the grain sizes in Fig. 4 shows that only slight grain growth of TiB$_2$ occurred during densification. This is a desirable advantage of using Ni-rich liquids to activate sintering of TiB$_2$, since grain growth to the critical microcracking size of about 20 μm does not occur for hot-pressing times up to 28.8 ks (8 h). Hot-pressing for 7.2 ks (2 h) frequently resulted in disk-shaped Ti$_2$O$_3$-rich regions with a diam-eter up to about 60 μm and a thickness up to about 5 μm. These regions also appear black under bright field illumination conditions. The disk shaped geometry of these regions was determined by analysis of microscopy sections polished with their planes perpendicular and parallel to the hot-pressing direction. The plane of the disks was always oriented perpendicular to the hot-pressing direction. These regions apparently form by coalesence of liquid at the hot-pressing temperature. The disposition of this titanium oxide phase within the microstructure of these materials at hot-pressing times greater than 7.2 ks is unclear since it has not been identified in specimens hot pressed for 14.4 and 28.8 ks. This phase may play

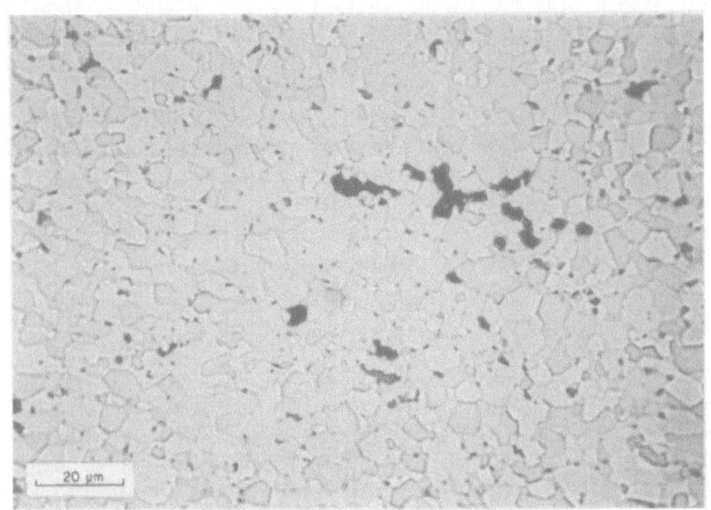

(a) Hot Pressing Time = 14.4 ks (4 h),
Final Ni Content = 0.4 wt %.

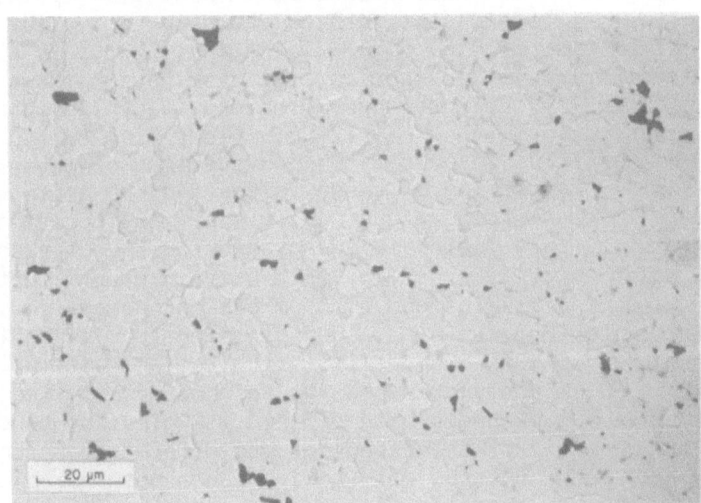

(b) Hot Pressing Time = 28.8 ks (8 h),
Final Ni Content = 0.2 wt %.

Fig. 3.   Bright-Field Optical Micrographs of Two TiB$_2$ Ceramics
Fabricated from Powder B Using Ni-rich Liquids, Pressure
= 12 MPa, Temperature = 1425°C.

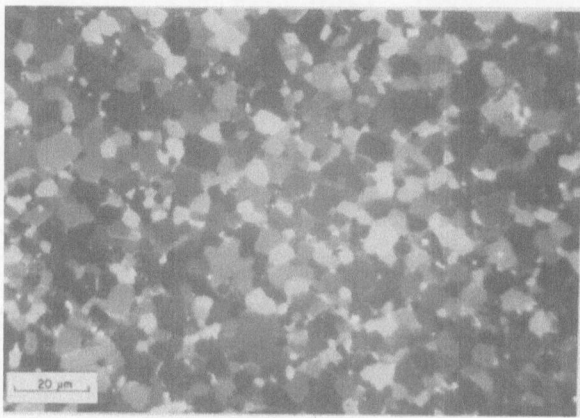

Fig. 4.   Polarized Light Micrograph of Identical Field of TiB$_2$
           Ceramic Specimen Shown in Fig. 3(b), Showing Boride
           Grain Size and Shape as Well as Structure in Ni-rich
           Second Phase Regions.

some role in affecting the distribution of the Ni-rich secondary
phase shown in Fig. 3 as black areas. An example of the disk-shaped
Ti$_2$O$_3$ rich areas is shown in Fig. 5, where the specimen was fabricated with powder "B." Under bright field conditions, the Ti$_2$O$_3$-
rich regions appear black. In this figure, the section is perpendicular to the hot-pressing direction of the original disk and the
specimen density is 4.52 g/cm^3, corresponding to the theoretical
density of TiB$_2$ (Ref. 15). The fact that the measured densities
in Figs. 3–5 are equal to or slightly exceed the theoretical density
of TiB$_2$ is apparently because the Ni-rich and Ti$_2$O$_3$-rich secondary
phases have densities near to or greater than that of pure TiB$_2$.
The actual Ni concentration of specimens fabricated from powder "B,"
hot-pressed for 7.2, 14.4, and 28.2 ks, was found by x-ray fluorescence to be 0.91, 0.39, and 0.25 wt %, respectively, although the
Ni content prior to hot pressing was 15 wt %. For hot-pressing
times of 7.2, 14.4, and 28.2 ks, the charges thus lost 93.9, 97.4,
and 98.3%, respectively, of their initial Ni content. The microstructure produced when the non-size-classified (as-received) TiB$_2$
powder "D" was used and the compact hot-pressed for 7.2 ks is shown
under polarized light conditions in Fig. 6. In this case, the
final Ni content was about 3 wt %, and the Ni-rich phase appears
as the lighter material distributed among the TiB$_2$ grains. It will

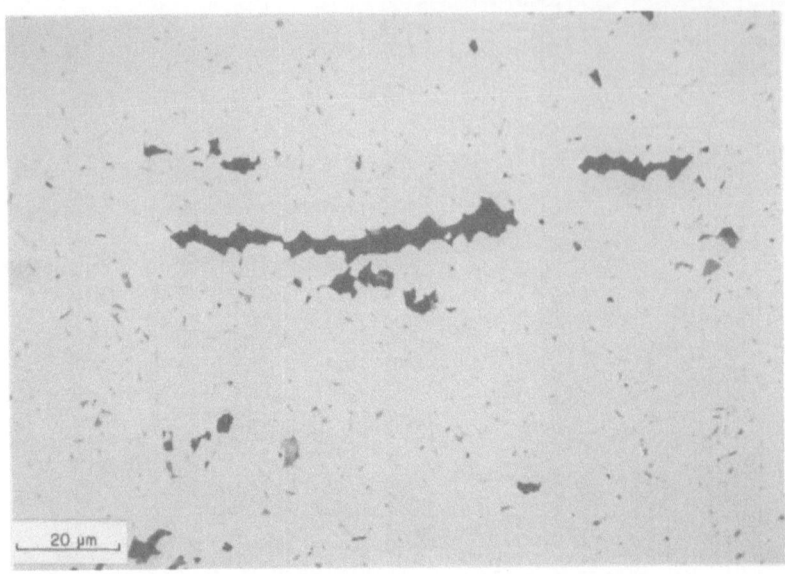

Fig. 5. Bright-Field Optical Micrograph of TiB_2 Ceramic Fabricated
   from Powder B using Ni-rich Liquid and Hot-Pressed at 12
   .MPa, 1425°C, 7.2 ks (2 h). Final Ni content = 1 wt %,
   extended Ni-rich regions appear black. Hot-pressing
   direction was vertical relative to micrograph.

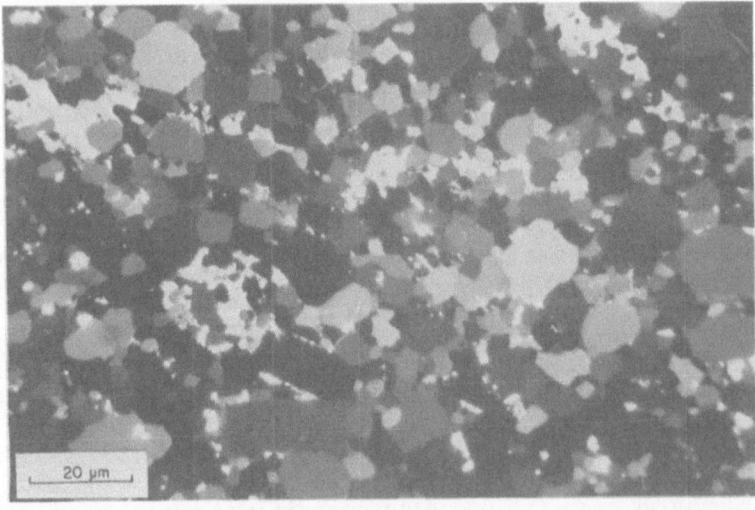

Fig. 6. Polarized Light Optical Micrograph of TiB_2 Ceramic
   Fabricated from Powder D Showing Boride Grain Morphology
   and Higher Concentration of Ni-rich Secondary Phase (Bright
   Regions) Compare to Specimens in Figs. 3, 4, and 5. Hot-
   pressed at 12 MPa, 1425°C, 7.2 ks (2 h).

be shown that this ceramic has a much lower fracture strength, fracture toughness, and Young's modulus than the ceramics illustrated in Figs. 3–5. The microstructure of the ceramics fabricated with powder "B" appears to be independent of initial Ni content for initial Ni compositions of 10 to 20 wt % Ni. As stated previously, most of the nickel is exuded from the compacts during densification, and their final Ni contents are similar. Based upon the results reported here, the enhanced densification of TiB_2 in the presence of a Ni-rich liquid appears to occur by enhancement of mass diffusion through the Ni-rich liquid forming at the hot-pressing temperature. However, the superior ceramics (having high strength, etc.) are produced only when most of the liquid is exuded from the compact when it is under the applied pressure at high temperature in the die. The exudation is strongly dependent on the particle size distribution of the TiB_2 powder and oxygen content of the powders. The liquid phase enables accelerated diffusion of both Ti and B, and results in highly dense compacts with use of lower temperatures and shorter pressing durations. Both AEM and TEM techniques were extensively employed to investigate these microstructures, including the composition of the Ni-rich secondary phase and the details of the grain structure of the TiB_2 phase itself. This work has been reported in detail elsewhere,[14] and results will be cited here only briefly. The typical TEM microstructure of these ceramics is shown in Fig. 7 for a specimen hot-pressed for 7.2 ks (2 h). The Ni-rich intergranular phase is denoted by the letter "B" in the micrograph where the TiB_2 grains are self-evident. Many of the grain triple junctions (intergranular nodes) have angles near 120°, indicating that equilibrium had been achieved during hot pressing. The Ni-rich phase lies primarily at the triple points, and electron diffraction, electron energy loss spectroscopy, and x-ray diffraction, have confirmed it to be primarily the compound Ni₃B containing a measurable quantity of Ti. Further details on the composition and structure of this phase are reported by Yust and Sklad.[14]

Elastic and Mechanical Properties

Values of Young's modulus and Poisson's ratio were determined from ultrasonic data[16] on these ceramics, while their fracture strength was determined at 25°C using four-point flexure of bars having an outer span of 3.8 cm and an inner span of 1.85 cm. These properties are given in Table 1 for selected specimens.

Ceramics fabricated from powder B and sintered using Ni-rich liquids for times from 7.2 to 28.8 ks (such as those represented by the microstructures in Figs. 3–5, and 7) have much higher flexure strengths (>700 MPa) than those fabricated from the other powders. These strength values are quite high for ceramic materials and illustrate that proper microstructural control in TiB_2 ceramics can result in significantly higher strengths than are possible in coarse grained material. The TiB_2 ceramic in Fig. 6, in which

Table 1. Properties of TiB_2 Ceramics Densified using Ni-rich Liquid

TiB_2 Powder	Hot Pressing Time (ks)	Density[a] (g/cm³)	Young's[b] Modulus (GPa)	Poisson's[b] Ratio	Flexure[c] Strength (MPa)	Fracture[d] Toughness K_{IC}, MPa m$^{1/2}$	Fracture Energy (J/m²)
D	7.2	4.62	514	0.122	386 ± 65	5.6	30
C	7.2	4.56	–	–	448 ± 28	7.8	–
B	7.2	4.52	569	0.113	731 ± 21	5.8	30
B	14.4	4.50	568	0.128	703 ± 66	6.7	40
B	28.8	4.56	574	0.112	756 ± 42	6.7	40

[a]By sample immersion in H_2O at 25°C.
[b]As derived from sonic determinations at 25°C.
[c]By four-point bend tests.
[d]From double cantilever beam data.

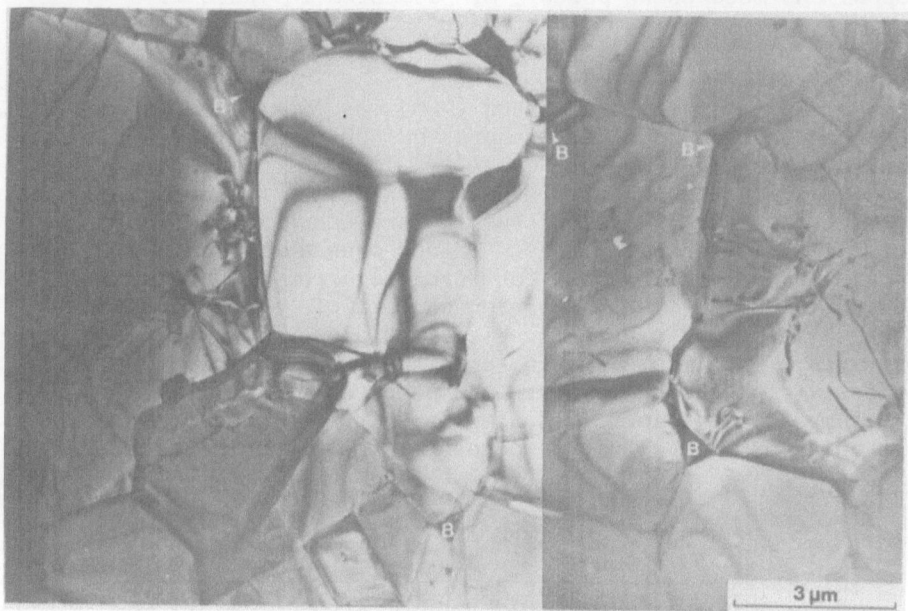

Fig. 7.   Transmission Electron Micrograph of TiB$_2$ Ceramic Fabricated
from Powder B Using Ni-rich Liquid, Hot Pressed at 12 MPa,
1424°C, 7.2 ks (2 h). Boride grain boundaries and Ni-rich
regions (B Sites) are shown.

about 15% of the initial Ni was retained, has a significantly lower
flexure strength and Young's modulus than the TiB$_2$ ceramics shown
in Figs. 3–5 and 7. For comparison, TiB$_2$ compacts containing no
Ni were hot-pressed at 1800 and 2000°C using the "B" powder. These
specimens have final densities of 4.46 and 4.43 g/cm^3, respectively,
i.e., about 98% of the theoretical value of 4.52 g/cm^3 (Ref. 15).
Appreciable grain growth occurred in specimens hot-pressed at 2000°C,
and a wide range of grain sizes (mean size ≈ 25 μm) were present,
the larger grains having dimensions greater than 100 μm.   The
specimen hot-pressed at 1800°C had a maximum grain size of only
18 μm. and the average grain size was about 12 μm. The TiB$_2$ compact pressed at 2000°C exhibited extensive microcracking due to
the large TEA of the compound discussed previously. However, the
1800°C specimen contained no obvious microcracks when examined
optically at 1000X. The Young's modulus values for the 2000 and
1800°C specimens were 503 and 563 GPa, respectively. The lower
modulus of the 2000°C specimen was due to microcracks generated
by thermal expansion stresses at the grain boundaries developed
during cooling from the hot pressing temperature. The Young's

modulus of the 1800°C specimen was approximately identical to that
of specimens made using powder "B," as noted in Table 1.

Hardness

The diamond pyramid hardness (DPH) was determined in vacuum
($\sim 10^{-3}$ Pa) between room temperature and 800°C using a diamond indenter and the procedure of Gessel.[17] The polished specimen surfaces typically showed no evidence of oxidation following indentation
at high temperature. Typical results are shown in Fig. 8 for a
specimen fabricated with powder "B" hot pressed for 28.8 ks. The
data are compared with those for pure TiB_2 from powder "B," hot
pressed at 1800°C. The microhardness data of Nakano[18-19] and of
Westbrook[20] for polycrystalline TiB_2 are also shown in Fig. 8 for
comparison. Specimen A in Fig. 8 is the pure TiB_2 sample considered
previously, while specimen B is the same as in Fig. 3(b) and Fig. 4.
The hardness of specimen B is similar to that of specimen A to
about 650°C and above this temperature, the B specimen, sintered

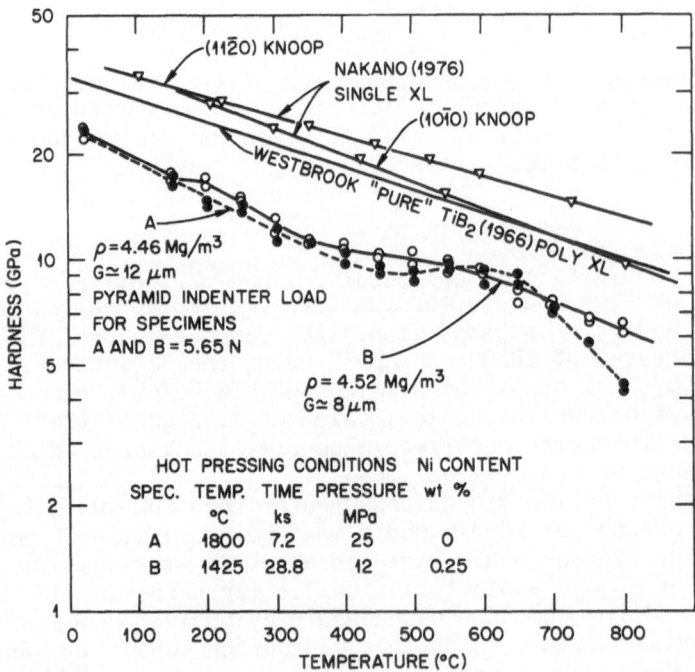

Fig. 8. Diamond Pyramid Hardness of TiB_2 Ceramics Fabricated from
Powder B without (Spec. A) and with (Spec. B) Use of Ni-
rich Liquid with Temperature. Other published values for
TiB_2 are also shown.

with the liquid Ni-bearing phase, was measurably harder than the
pure specimen. The fact that the hardness of both the A and B
specimens is less than published values may be due to two reasons.
First, Nakano's data were obtained using a Knoop indenter and a
force of 0.98 N, while Westbrook employed a pyramid indenter and
a force of either 0.98 N or 0.49 N. The present data, however, were
obtained with a diamond pyramid indenter and a force of 5.65 N.
Comparison of pyramid and Knoop hardness values is difficult due
to the different stress fields produced by the two different indenter types and this can readily explain the differences from Nakano's
results. Secondly, the hardness of some materials is often a function
of both indenter force[21] and indenter configuration; the higher
force of the present work relative to Westbrook's may have resulted
in respectively lower hardness values.

The cause of the inflections in the hardness vs temperature
curves for the A and B specimens in Fig. 8 is not presently understood. They were, however, highly reproducible for the several
specimens on which measurements were made, and hardness deviations
were well outside of the equipment limit of error (see the size of
the data points in Fig. 8). Some microcracking was observed at
indent corners of both the A and B specimens, particularly at temperatures between 25 and 450°C. Slip bands were visible in the
TiB$_2$ grains within the indent, and occasionally these extended to
one grain diameter outside the indent.

Thermal Diffusivity

Thermal transport in these ceramics was determined between 25
and 1200°C using the laser-flash technique.[22] One side of a thin
square plate specimen was flashed with a 50 J Nd-glass laser, and
the temperature behavior of the opposite side was sensed with a
liquid N$_2$-cooled infrared detector. Analysis of the thermal transient led to calculation of the thermal diffusivity. The thermal
diffusivity of two specimens fabricated using powder B is shown in
Fig. 9, including a pure TiB$_2$ specimen pressed at 1800°C and a
specimen having a final nickel content of 0.43 wt % pressed at
1425°C. The thermal diffusivities of the two specimens are similar
at a given temperature, although the 0.43 wt % Ni specimen has
a slightly lower diffusivity over most of the temperature range.
Concentration of the Ni$_3$B-type phase at grain nodes or as layers
along the grain boundaries and the presence of the Ti$_2$O$_3$-rich
regions do not significantly affect heat transport in these ceramics
up to 1200°C.

Behavior as Metal Cutting Tools

The potential of these ceramics as metal cutting tools was
evaluated. Triangular tools were fashioned from selected compacts,
and cutting experiments were conducted on a special alloy at the

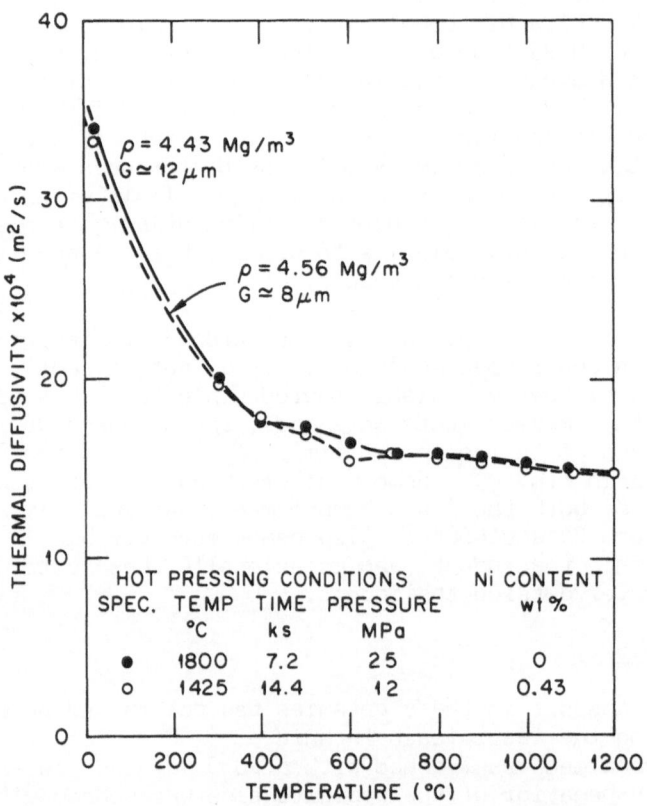

Fig. 9.   Thermal Diffusivities of TiB$_2$ Ceramics.

Y-12 facility in Oak Ridge. Tool performance was determined
according to the extent of flank wear as a function of cutting
time up to 3.6 ks (1 h). The flank wear of tool specimens, A and
B, fabricated from powder B and pressed for 7.2 ks (2 h), is shown
in Fig. 10. These specimens A and B were fabricated at 1425°C
and 12 MPa with initial Ni contents of 10 and 20 wt %, respectively, and after pressing contained approximately 1 wt % Ni. For
comparison, the flank wear for several commercial WC-Co cutting
tools is also shown in Fig. 10 for similar conditions. The TiB$_2$
ceramics thus have very promising flank wear values for use on
the special alloy.

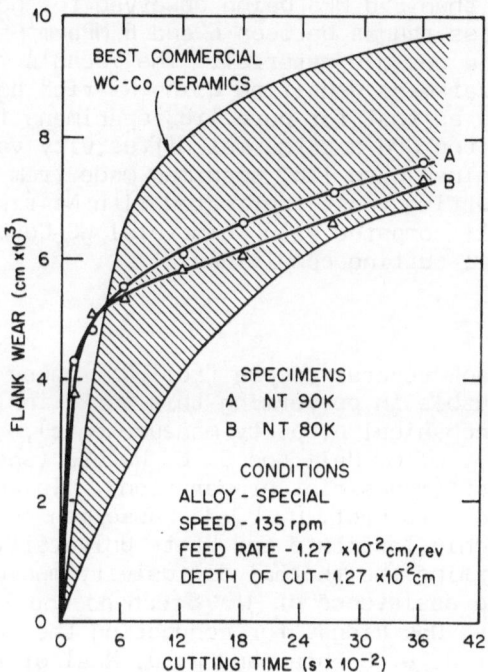

Fig. 10. Flank Wear of Two TiB$_2$ Ceramic Cutting Tools Fabricated using Ni-rich Liquid vs Cutting Time on Special U-based Alloy. Performance of best commercial WC-Co tools for this alloy is also shown.

CONCLUSIONS

It was demonstrated that TiB$_2$ powders can be sintered to near theoretical densities using modest hot-pressing procedures when prescribed diboride particle sizes are used in conjunction with liquid sintering activation agents based on metals such as Ni. An important feature of the process, which permits achievement of >99%, sintered densities, is exudation of excess Ni-rich liquid before porosity becomes prevalent. Microcompositional data obtained from electron microscopy, and the microstructural data, indicate that the nickel forms a liquid during hot pressing, and that some of the TiB$_2$ dissolves in the liquid to promote rapid diffusive transport of both Ti and B throughout the microstructure.

Under optimum conditions, the TiB_2 particles essentially maintain their original size; the grain growth and microcracking characteristic of typical TiB ceramics is therefore avoided. The flexure strength of the TiB ceramics is typified by a value of about 750 MPa, with values greater than 850 MPa being observed for some specimens. The fracture toughness ranges between 6 and 8 MPa m$^{1/2}$, which is relatively high for a ceramic material. The Young's modulus and Poisson's ratio of ceramics densified using Ni-rich agents are essentially the same as that for pure TiB_2 specimens from the same powder. Similarly, the DPH and thermal diffusivity values approximate those for a fine-grained pure TiB_2 specimen made from the same powder. Experimental TiB_2 cutting tools fabricated with Ni-rich sintering agents performed well compared with commercial WC-Co based tools under identical metal-cutting conditions.

ACKNOWLEDGEMENTS

The authors thank several people from ORNL whose contributions were especially valuable in performing this work, including W. Simpson and P. F. Becher (mechanical property measurements), W. H. Farmer (optical microscopy), L. L. Hall and S. B. Waters (specimen preparation) and F. M. Foust (manuscript preparation). In addition, we express special thanks to Prof. D. P. H. Hasselman and his colleagues at Virginia Polytechnic Institute and State University, Blacksburg, Virginia, for performing the thermal diffusivity measurements. We also acknowledge the assistance of W. Stephens and T. O. Morris of the Y-12 facility, Oak Ridge, for conducting the cutting tool measurements, and of J. O. Cochran and R. B. Neal of the Oak Ridge Gaseous Diffusion Plant for help with the particle size classification and evaluation experiments.

REFERENCES

1.   G. V. Samsonov, Boron, Its Compounds and Alloys, AEC tr 5032, part 2, Kiev (1960), pp. 520–526.
2.   B. Post, F. W. Glaser, and D. Moskowitz, "Transition Metal Diborides," Acta Metallurgica, 2 , 20–25 (1954).
3.   J. D. Schobel and H. H. Stadelmaier, Die Nickelecke in Die Dreistoffsystem Nickel-Titan-Bor, Metall 19 , 715 (1965).
4.   I. langermann, "Beitrag zum Verhalten von Schwermetallboriden gegenuber Bindemetallen," Neue Hutte 8 (6) 359–365 (1963).
5.   D. R. Prendse and P. L. Pratt, "A Study of the Physical and Mechanical Properties of Cemented Borides," in Special Ceramics 5 , P. Popper, editor, Manchester, pp. 135–145. (1972).
6.   R. C. Brewer, "Cemented Borides as Tool Materials," Engineers' Digest 20 (5), 205–208 (1959).
7.   Yu. B. Kuzma and M. V. Chepiga, "An X-Ray Diffraction Investigation of the Systems Ti-Ni-B-Mo-Ni-B, and W-Ni-B," Soviet Powder Metallurgy 10, 832–835 (1969).

8. H. J. Juretschke and R. Steinitz, "Hall Effect and Electrical
    Conductivity of Transition Metal Diborides," J. Phys. Chem.
    Solids, 4 118–127 (1958).
9. V. F. Funke and S. I. Yudkovskii, "O Bzaimodeistvii Boridov
    Perekhodnykh Metallov c Metallami Gruppy Zheleza," Issledovania
    Stalei i Splavov, Moscow, pp. 108–113 (1964).
10. E. D. Case, J. R. Smyth, and O. Hunter, "Grain Size Dependence
    of Microcrack Initiation in Brittle Materials," J. Mater. Sci.
    15, 149–153 (1980).
11. C. F. Yen, C. S. Yust, and G. W. Clark, "Enhancement of Mechani-
    cal Strength in Hot-Pressed TiB Composites by the Additional
    Fe and Ni," in New Developments and Applications in Composites,
    Transactions, AIME, Warrendale, Penn., 317–330 (1979).
12. Y. S. Touloukian, R. K. Kirby, R. E. Taylor, and T. Y. R. Lee,
    Thermal Expansion Nonmetallic Solids, Thermophysical Properties
    of Matter, Vol. 13, Plenum Press, New York (1977).
13. V. I. Matkovich, Boron and Refractory Borides, Springer-Verlag,
    New York (1977).
14. C. S. Yust and P. S. Sklad, "Characterization of TiB -Ni Ceramics
    by Transmission and Analytical Electron Microscopy," in this
    Conference.
15. P. I. B. Shaffer, Materials Index No. 1, Plenum Press, New
    York (1964).
16. W. Mason, Acoustic Properties of Solids, Section IIIF-1, Amer.
    Inst. of Physics Handbook, 74–79 (1957).
17. G. R. Gessel, "Effect of Minor Alloys on the Strength and
    Swelling Behavior of Austenitic Stainless Steel," ORNL/TM-6359,
    pp. 72–75 (1979).
18. K. Nakamo, H. Matsubara, and T. Imura, "High-Temperature Hardness
    of Titanium Diboride Single Crystal," Japan J. Appl. Phys.,
    13, (6), 1005–1010 (1974).
19. K. Nakamo, T. Imura, and S. Takenchi, "Hardness Anisotropy of
    Single Crystals of IVa-Diborides," Japan. J. Appl. Phys. 12,
    (2), (1973).
20. J. H. Westbrook, "The Temperature Dependence of Hardness of
    Some Common Oxides," Rev. Hautes Temper. et Refract, 3, 47
    (1966).
21. L. Kaufman and E. V. Clougherty, "Investigation of Boride Com-
    pounds for Very High Temperature Applications," Report
    RTD-TDR-63-4096, Part 1, ManLabs, Inc., Cambridge, Mass., (1963).
22. F. F. Lange, H. J. Slebeneck, and D. P. H. Hasselman, J. Amer.
    Ceram. Soc. 59 454–458 (1976).

CHARACTERIZATION OF TiB$_2$-Ni CERAMICS BY

TRANSMISSION AND ANALYTICAL ELECTRON MICROSCOPY*

P. S. Sklad and C. S. Yust

Metals and Ceramics Division
Oak Ridge National Laboratory
Oak Ridge, TN   37830

INTRODUCTION

Some of the physical properties of the transition metal diboride TiB$_2$, especially high hardness and chemical stability, suggest use of this compound as a wear-resistant material in valve components in coal liquefaction plants and as a cutting tool material.[1] The preparation of dense, single-phase, poly-crystalline bodies of this refractory compound (m.p. = ∿3253 K), however, requires a sintering temperature approaching 2300 K and is accompanied by extensive grain growth.[2] In addition, experience has shown that single phase TiB$_2$ polycrystals are relatively fracture sensitive.[3] A method that has been explored for improving the fracture toughness of TiB$_2$ bodies is liquid phase sintering with metals, especially iron, cobalt, and nickel.[1-4] The use of a metal-based bonding phase also offers the possibility of densification at temperatures much lower than the sintering temperature for pure TiB$_2$ through the formation of TiB$_2$-metal eutectics. In addition to promoting densification under less rigorous conditions, the use of the metal bonding phase has been found to improve the fracture strength of TiB$_2$.[1,5]

The primary research emphasis at ORNL has been on nickel as the densification aid during hot pressing. The formation of a eutectic liquid during hot pressing results in a complex multiphase microstructure. The phase diagram result of Schöbel and Stadelmaier[6] suggest that one component of the microstructure will be a ternary phase Ni$_{20.3}$Ti$_{2.7}$B$_6$, the tau phase.

---

*Research sponsored by the Div. of Materials Sciences, U.S. Dept. of Energy, under contract W-7405-eng-26 with Union Carbide Corp..

The results of a number of engineering tests to measure the erosion and wear rate of this composite material suggest that the properties may be sensitive to the volume fraction and distribution of the intergranular phases, the initial diboride particle size, and the pressing parameters. Therefore, a full characterization of the microstructure is necessary in order to establish the critical parameters in the fabrication of this dense, hard, erosion-resistant, ceramic material. The small volume fraction of the intergranular phase made positive identification of microconstituents by conventional x-ray diffraction techniques difficult; however, the structure and composition have been characterized using transmission and analytical electron microscopy (AEM) techniques.

SAMPLE PREPARATION

Hot pressed cylindrical compacts (64 mm diam × 1 mm thick) were prepared by powder metallurgical techniques using 10, 15, or 20 mol % nickel powder mixed with various TiB_2 powders. The TiB_2 starting powders were of three types as summarized in Table 1. Powder E was used as supplied by the manufacturer (Cotronics) and had an average particle size of about 8 μm. Powder G was produced by impact oscillatory milling of powder E in a WC-Co assembly. The resulting average particle size was about 5 μm. Powder B consisted of a 4-5 μm fraction with a more narrow particle size distribution separated by air classification from a batch of TiB_2 powder with an average starting particle size of about 7 μm.[5] The particle size of the nickel powder was <44 μm.

The TiB_2 powders were mixed with the nickel and hot pressed at pressures less than 10^{-3} Pa in grafoil-lined* graphite die assemblies at a temperature of about 1700 K under an applied pressure of 12 MPa for 7.2 ks (2 h). The density of compacts produced by this procedure was greater than 99% of the theoretical density. In a similar manner, tau phase powder, which was prepared by milling single crystals, was mixed with "B" type TiB_2 and pressed by the procedure described above. A more detailed description of the processing procedures can be found elsewhere in these proceedings.[5]

Samples of the cylindrical compacts were prepared for optical microscopy using standard vibratory polishing techniques. In addition, 3 mm-diam disks were cut by electrodischarge machining. These were subsequently reduced in thickness to 80-120 μm by mechanical lapping with successively finer abrasive powders. Electron transparent specimens for AEM were then obtained by standard argon ion milling techniques.

---

*Union Carbide Corporation, Carbon Products Division, Cleveland, Ohio.

Table 1. Characteristics of TiB_2, Nickel and Tau Starting Powders

	TiB_2 Powder E	TiB_2 Powder G	TiB_2 Powder B	Ni Powder	Tau Powder
Purity	>99.8% (cation)	>99.8% (cation)	>99.8% (cation)	>99.9%	--
Average Particle Size	8 μm	5.0 μm	4–5 μm	<44 μm	<44 μm
Oxygen Content	1.0	3.0	1.2	--	0.07

OPTICAL MICROSCOPY

    Optical microscopy was used to gain a preliminary assessment
of the uniformity of the microstructures produced. Typical
results are presented in Fig. 1 in which the microstructures in
compacts fabricated from TiB_2 powders E, G, and B using 15 mol %
Ni are compared. The light regions are TiB_2 grains and the dark
regions delineate the nickel-rich intergranular phase, which is
located exclusively at grain boundaries. Although the intergranular phase appears to be homogeneously distributed in both specimens, there is a striking difference in the total amount of
this phase present as well as in the size of the nickel-rich
regions. In fact, the final nickel content in the compacts
fabricated from powders E, G, and B was about 15, 4.3, and
0.5—1.0 wt %, respectively. Apparently, a large fraction of the
initial nickel was lost during hot pressing. The loss of nickel
was caused by the exudation of nickel-rich liquid during hot
pressing. The results in Fig. 1 suggest that the amount of exudation is a strong function of the particle size distribution of
the TiB_2 powder.

TRANSMISSION ELECTRON MICROSCOPY (TEM)

    The microstructures of three TiB_2-Ni pressed compacts were
examined using conventional TEM. The specimens chosen were made
using each of the three powders, E, G, and B, and correspond to
those shown in Fig. 1. Figure 2 illustrates the typical
microstructural features of these materials. All three specimens
were characterized by a matrix of TiB_2 grains incorporating an
intergranular second phase, the volume fraction of which
decreases with decrease in the mean particle size of the starting
TiB_2 powder as was observed by optical microscopy. The intergranular phase was observed primarily at grain edge intersections
(triple points), but it has not been determined whether or not a
continuous thin grain boundary film exists. Also evident in the
micrograph from the specimen fabricated using powder E (Fig. 2a)
are TiB_2 grains with a rounded morphology, one of which is
completely surrounded by the secondary phase. Similar structures
were observed in the other two specimens, Fig. 2b and 2c. These
observations suggest that reaction of TiB_2 grains with the
nickel-rich liquid during hot pressing results in formation of
secondary phases containing nickel, boron, and titanium. The
size of the intergranular regions varies considerably both in
individual specimens and from one specimen to another; however,
the sizes of these regions agreed well with those seen in the
optical micrographs.

915

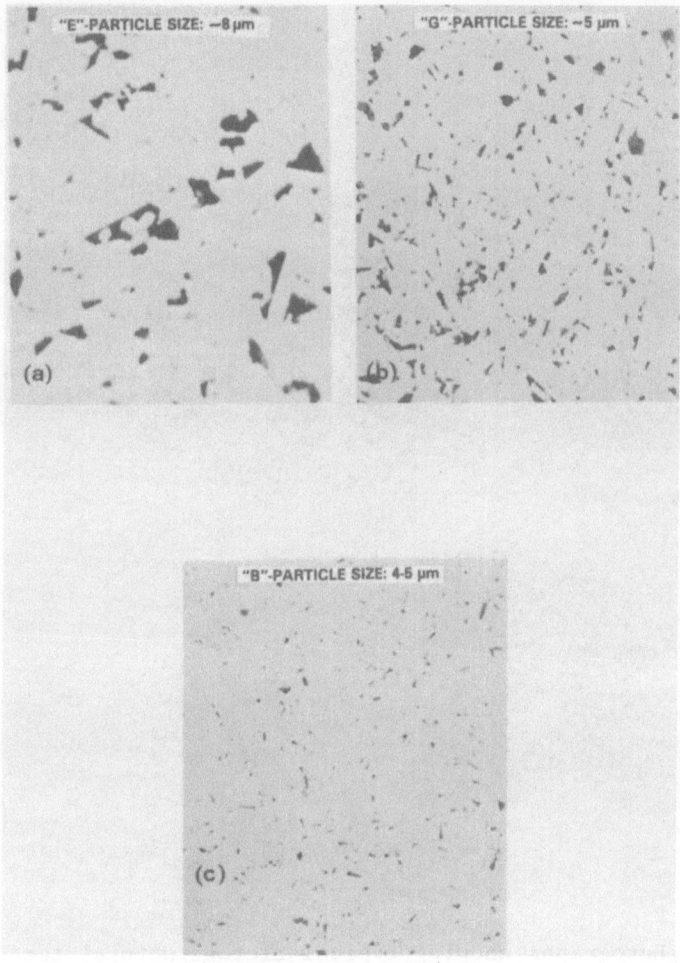

Fig. 1. Typical optical microstructures of TiB₂—15 mol % Ni ceramics. (a) Powder E, (b) Powder G; (c) Powder B. As polished condition shown in bright field.

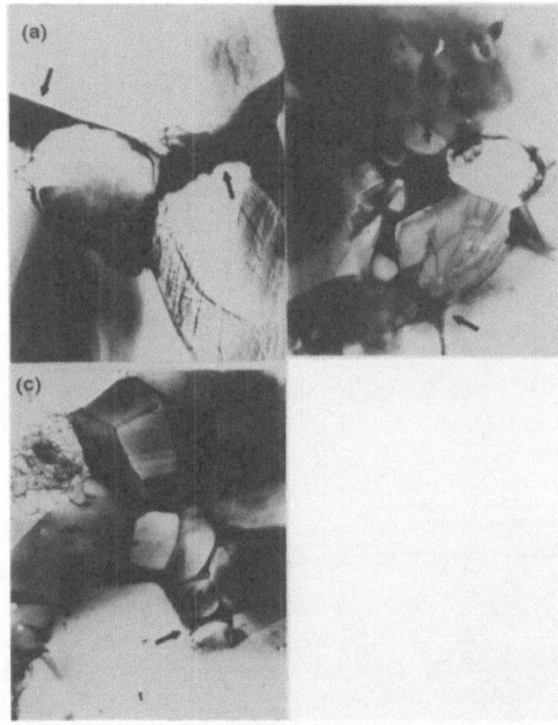

Fig. 2.   TEM photographs showing typical microstructural features
          of TiB$_2$–15 mol % Ni ceramics. (a) Powder E;
          (b) Powder G; (c) Powder B.

X-RAY ENERGY DISPERSIVE SPECTROSCOPY (EDS)

Although determination of the crystal structure and unit cell dimensions can often be the most reliable means of phase identification, a knowledge of composition can greatly simplify the identification process by reducing the number of possibilities that must be considered. A knowledge of composition is also often necessary in order to distinguish between two or more closely related compounds which have identical structures and only slightly different lattice parameters, (e.g., TiC and TiN). For these and other reasons, x-ray microanalysis was the first microanalytical technique to be used for phase identification in this work.

The EDS measurements were made using a Philips EM 400T/FEG equipped with an Edax detector and interfaced to a PDP-11/34 computer and peripherals. The microscope was optimized for x-ray microanalysis according to procedures outlined by Bentley et al.[7] In addition, beryllium specimen holders were used and areas were selected for analysis where specimen thickness did not require absorption corrections. The EDS spectra were analyzed quantitatively employing integrated peak intensities and standardless analysis routines using programs developed by Zaluzec.[8-9]

The use of probe sizes of about 10 nm and probe currents of about 5 nA allowed measurements to be made in thin regions of the specimens where complicating effects of absorption were negligible. The small probe size also allowed measurements to be made in intergranular regions sufficiently far from neighboring TiB_2 grains to minimize the possibility of exciting x rays in these regions. Figure 3 illustrates typical x-ray spectra from an intergranular region and an adjacent TiB_2 grain in a specimen fabricated from 15 mol % Ni and 85 mol % TiB_2 powder E. The results of many such analyses of the intergranular phase in the E specimen as well as in specimens prepared from powders G and B consistently indicated a composition (for elements with atomic number greater than 11) with a Ni:Ti ratio greater than 20. The variation in measured composition between specimens was within the error associated with the measurements. However, it is well established that there are a number of limitations to the accurate quantification of the titanium content in the intergranular phase when the overall composition of the whole specimen is high in titanium.[7,10] Therefore, the Ni:Ti ratio of 20 must be considered a minimum value, and the possibility exists that the intergranular phase contains less titanium than indicated.

CONVERGENT BEAM ELECTRON DIFFRACTION (CBED)

The CBED techniques allowed a determination of crystal structure since the electron probe size of about 10 nm in the Philips EM 400T/FEG made possible diffraction information from

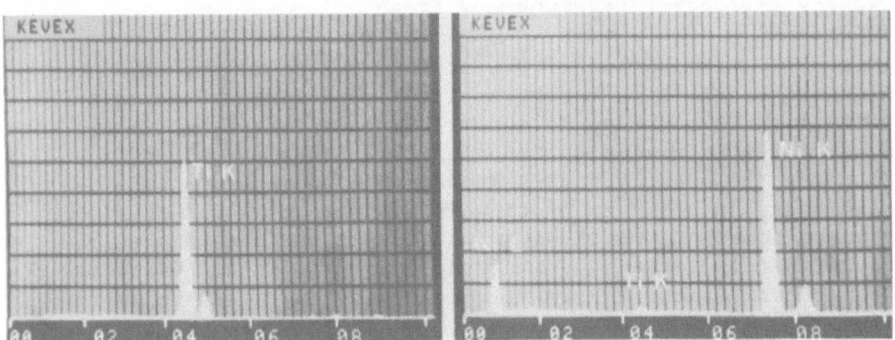

Fig. 3.   Energy dispersive x-ray spectra from a TiB_2 grain, (a),
          and an adjacent intergranular region, (b).

very small regions. The camera length of the diffraction patterns was varied over a wide range in order to examine details of
the zero and higher order Laue zones (ZOLZ and HOLZ,
respectively).

       Tilting experiments on several areas of the intergranular
phase failed to reveal any diffraction patterns with greater symmetry than 2 mm. A small camera length pattern of such a zone
axis is shown in Fig. 4 where the two mirror planes m_1 and m_2 are
indicated. An inspection of the tables of Buxton, et al.[11] which
give the relation between the symmetry of convergent beam patterns and diffraction groups, revealed that further high angle
tilting experiments were required to determine the crystal point
group. In fact, the International Tables[12] indicate that a tilt
of 90° along one of the mirror planes was required. Fortunately,
this was possible with the double tilt specimen holder ($\pm$ 60
eucentric, $\pm$ 30° secondary tilt) used in this work. These experiments showed that orthogonal zones all possessed 2 mm symmetry,
whereas all intermediate zones had only m symmetry. The point
group was thus established as orthorhombic mmm.

       The ASTM x-ray powder files contain information on three
nickel-boron phases which have orthorhombic crystal structure —
Ni_3B, Ni_4B_3, and NiB. Using CBED patterns from the TiB_2 grains
as standards, the analysis of the diffraction pattern in Fig. 5
produced values for $a_o = 0.523$ nm and $b_o = 0.663$ nm which are
close to those quoted for Ni_3B, 0.521 and 0.662 nm, respectively.
Using the diameter of the first order Laue zone (FOLZ) ring and
the procedure described by Steeds[13] a value of $c_o = 0.437$ nm was
measured which again is in good agreement with the tabulated

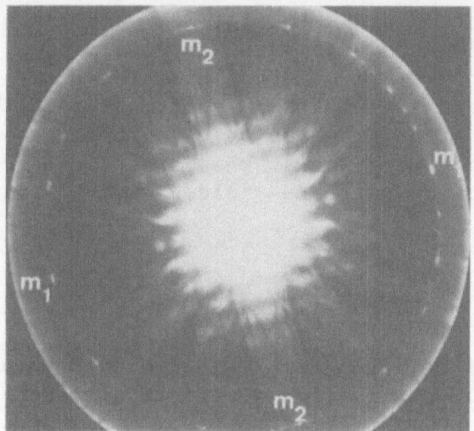

Fig. 4. Small camera length CBED pattern of the [001] zone of
         the intergranular phase. Two mirror planes, m_1 and m_2,
         are indicated.

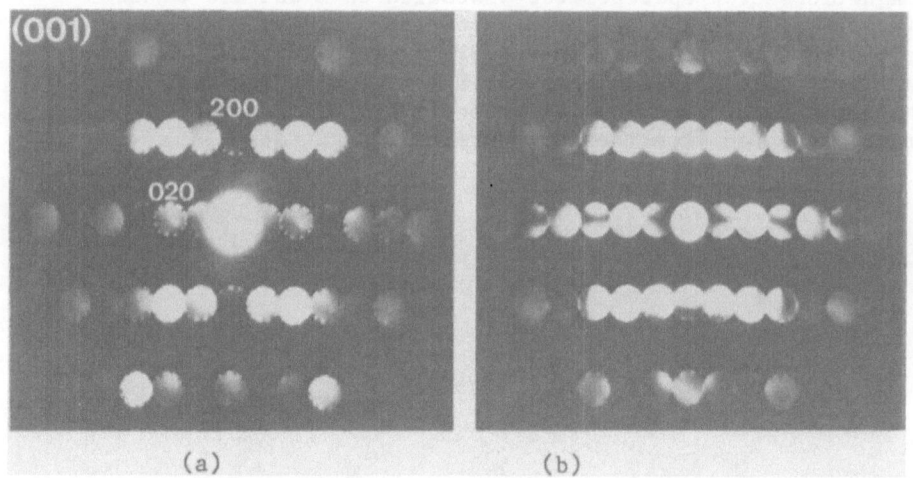

Fig. 5. Large camera length CBED pattern from the [001] zone of
         the intergranular phase. Note the presence of
         0, (2n+1), 0 reflections due to double diffraction,
         (2n+1), k, 0 reflections are absent at this orientation.
         A thin area (kinematical conditions) is shown in (a), a
         thicker area (dynamical) in (b). The indexed pattern
         corresponds to the Ni_3B structure.

value of c_0 = 0.439 nm for Ni_3B. The observed relative intensities of the diffraction disks in Fig. 5a, which is a CBED pattern from a thin region of intergranular phase, agree
qualitatively with the relative intensities for x-ray diffraction
data on Ni_3B. This provides further confirmation of the identification of the intergranular phase as one similar to Ni_3B. In
those cases where the intergranular phase was large, conventional
selected area diffraction patterns yielded similar results
(Fig. 6).

ELECTRON ENERGY LOSS SPECTROSCOPY (EELS)

The observations of TiB_2 grains with rounded morphologies
and the diffraction identification of the crystal structure of
the intergranular phase as similar to Ni_3B both suggest that the
intergranular material contains boron. Of course, no indication
of elements with atomic number Z <11 is possible with conventional EDS detectors, so in an attempt to detect and measure the
boron content of the binder phase, EELS was performed. The
instrumentation[14] consisted of a double-focusing, symmetrical,
90° magnetic sector spectrometer installed on a JEM 120 CX AEM
and interfaced with the electronics of the x-ray spectrometer.
The microscope was operated in the STEM mode at 120 kV, and specimens were cooled to less than 220 K to avoid hydrocarbon contamination. Data reduction was accomplished using computer
programs developed from the formulations of Egerton.[15-17]

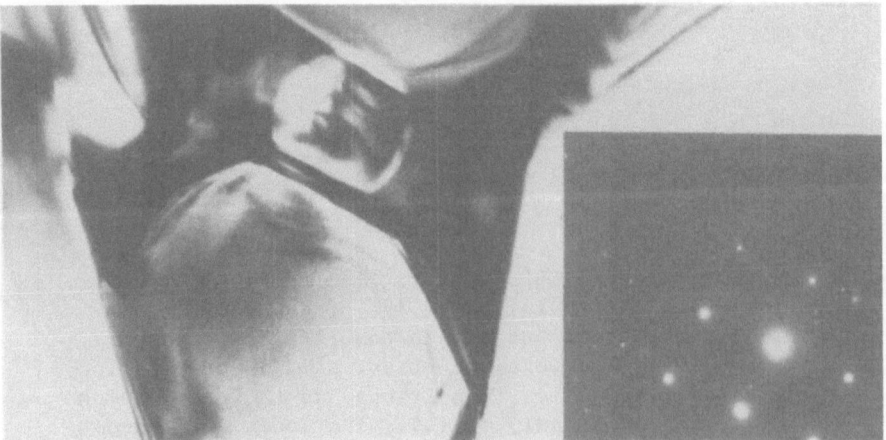

Fig. 6.   Conventional selected area diffraction pattern from
         intergranular material in a specimen prepared from
         Powder E. The selected area is illustrated in (a),
         the corresponding pattern in (b). The planar spacings
         and interplanar angles correspond to the Ni_3B structure.

Many spectra were obtained from nickel-rich intergranular regions as well as adjacent TiB$_2$ grains. Typical spectra are shown in Fig. 7. Examination of the spectra allows a qualitative comparison of the compositions of the two phases. Clearly defined boron K and titanium L$_{23}$ edges were observed in the spectra from the TiB$_2$ grains. There was no indication of any nickel edges in any of the spectra from TiB$_2$. Spectra from the intergranular regions had pronounced Ni$_{L23}$ edges with only a slight indication of boron K and titanium L$_{23}$ edges. In addition, oxygen K edges were sometimes observed.

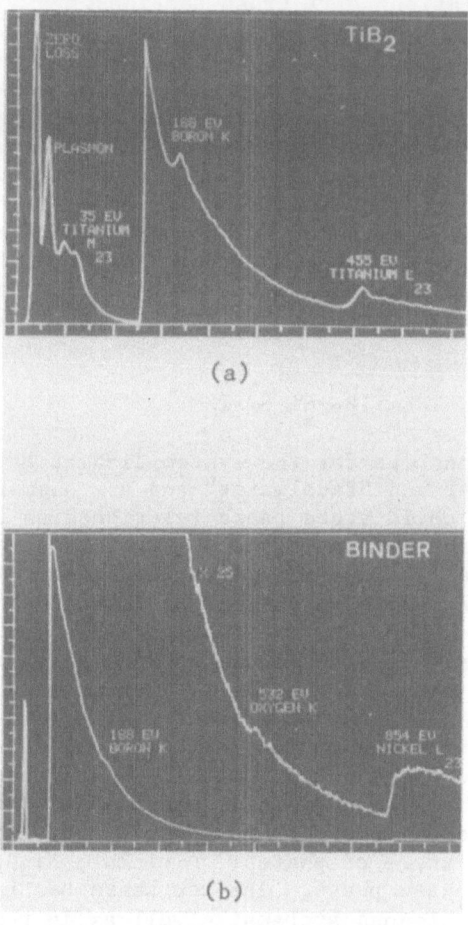

Fig. 7. Electron energy loss spectra from (a) a TiB$_2$ grain and (b) a region of nickel-rich intergranular phase. Edge energies are indicated.

To date, accurate quantification of the ELS spectra has been complicated by two major factors. First, the standard procedure of fitting a curve of the form AE^{-r} to the background was not satisfactory for a number of spectra. In this case the behavior is believed to be due to the presence of M edges from the transition metals, Ti (M_{23} = 35 eV, M_1 = 60 eV) and Ni (M_{23} = 68 eV and M_1 = 112 eV). The tails of these edges, which may superimpose on the boron K edge, do not follow a simple inverse power law form.[18] However, a modified background form utilizing a polynominal fit to the logarithms of the electron intensity and energy loss values provided a satisfactory background fit.[19] Although the background subtraction of the spectra was improved substantially, the accuracy of quantification was still unsatisfactory. It is believed that the inaccuracies are due to uncertainties in the calculations of L cross sections.[20]

In spite of these difficulties, attempts were made to obtain reasonably accurate compositional information from these spectra with a simple method utilizing ratios of effective cross sections which were obtained from measurements of materials with known compositions. Using this technique it was found that the ratio of the nickel to boron in the intergranular phase was about three. This result is in agreement with the theoretical value of Ni_3B. There were instances, however, where the size of the boron edge observed suggested that the amount of boron in the intergranular phase was less than the expected stoichiometric 25%.

PHASE RELATIONSHIPS IN THE TiB_2-Ni SYSTEM

The phase relationships for the system Ti-Ni-B have been investigated by Schobel and Stadelmaier[6] and are summarized in Fig. 8. Although the solid state phase relationships shown in Fig. 8a are for equilibrium conditions at 800°C, they are representative of the phases which may be expected on solidification. The mixture of TiB_2 and Ni representing specimens in this work are indicated by the dashed line in the figure corresponding to the Ni-TiB_2 pseudobinary diagram. During processing, a liquid eutectic phase forms, containing nickel, titanium, and boron, and such other impurities which may be present on the powder particles. Oxygen is one likely impurity in the powders used. The melt may not be homogeneous in composition and the location of the melt in the composition field is uncertain. The phase diagram, however, strongly suggests that the solidified melt should consist of a mixture of phases: α-nickel, Ni_3B, Ni_3Ti, and $Ni_{20.3}Ti_{2.7}B_6$, the tau phase. The tau phase has been observed to form in analogous systems as well as in the Ni-Ti-B system.[21,22] It has, in fact, been grown in this laboratory as a single crystal by C. Finch. The tau phase melts congruently at 1508 K, has the fcc crystal structure of chromium carbide, $Cr_{23}C_6$, and can exist at temperatures as low as 1075 K. It has a narrow homogeneity range which extends from 70.8 at % Ni,

8.5 at. % Ti, 20.7 at. % B to 69.3 at. % Ni, 10.5 at. % Ti, 20.2 at. % B. The lattice parameter for tau varies between $a_0 = 1.0507$ and 1.0538 nm. It should be noted that the nickel to titanium ratio for tau is between 6.6 and 8.3.

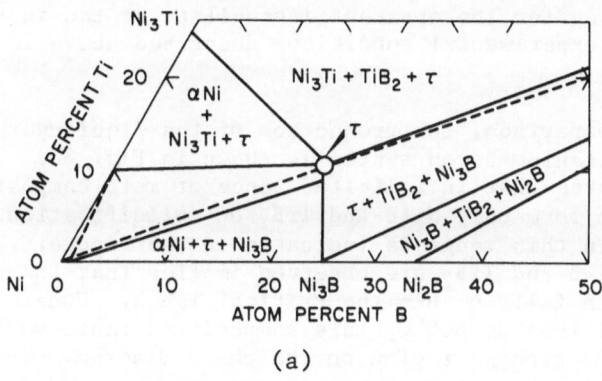

(a)

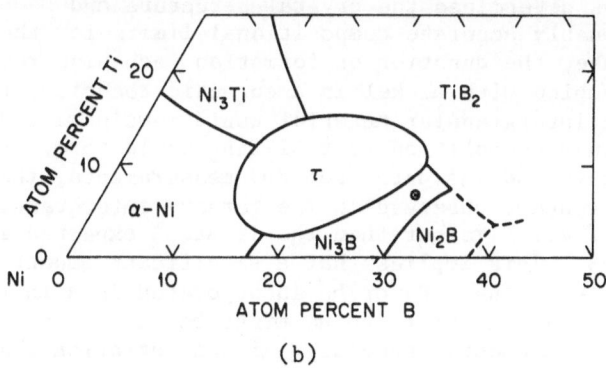

(b)

Fig. 8. (a) Phase relationships in the nickel-rich corner of the nickel-titanium-boron system. The diagram represents equilibrium conditions at 800°C. The dashed line represents the Ni-TiB₂ pseudo-binary plane, (b) The liquidus surface for this system. After Schöbel and Stadelmaier.[6]

In spite of the expectation of multiphase intergranular material, the only phase identified in any of the specimens examined was the Ni_3B phase. To further investigate bonding phase composition, a specimen was prepared in which fine particles of pure tau were used instead of nickel. Analytical electron microscopy examinations of this specimen revealed that the microstructure was almost identical to those produced from the TiB_2–Ni mixtures. In addition, EDS measurements indicate Ni:Ti ratios greater than 20, in agreement with the measurements made on the intergranular phase in specimens of TiB_2–Ni, and the results of CBED experiments are consistent with the Ni_3B structure. At present, the reason for the apparent instability of tau in this system under the experimental conditions described above is unknown.

By way of comparison, the projection of the liquidus surface for the nickel-titanium-boron system is shown in Fig. 8b. This diagram demonstrates that in a limited range of melt compositions it is possible to form only Ni_3B and TiB_2 on solidification. A composition within this range is indicated in the figure. The fact that only Ni_3B and TiB_2 are observed implies that the molten binder composition falls within the critical range. Under equilibrium conditions at 800°C, this composition range would correspond to a multiphase region on the phase diagram. The solid state reaction kinetics, however, are unknown, and formation of the equilibrium phases may require extended annealing times. Boundary melt compositions outside of the critical composition limits should lead to the formation of phases other than Ni_3B, and in particular, to the tau phase.

While AEM has determined the crystal structure and established reasonably accurate compositional limits for the intergranular phase, the question of formation mechanism remains. If TiB_2 grains combine with nickel in a eutectic reaction, the B:Ti ratio in the intergranular material would remain ≈ 2 and an M_3B structure with a composition near Ni_5TiB_2 would form. Based on the M_3B structure and the data from EDS measurements, the maximum titanium content observed in the ternary intergranular phase would be 3.7 at. % rather than the 12 at. % expected in the postulated Ni_5TiB_2. This implies that a significant amount of titanium may remain in the TiB_2 or be incorporated in a third phase. This excess amount of titanium might be undetectable by the energy loss measurements especially if concentration gradients exist. However, the possibility also exists that the intergranular phase is deficient in boron. A number of studies in the literature[21,23-25] suggest that complex interstitial alloys, in which the small metalloid atoms are positioned at sites which correspond to the centers of metal atom polyhedra, tend to have rather wide ranges of stoichiometry without large changes in lattice parameters. Some of the ELS results in the present study do indeed suggest that the intergranular phase may be nonstoichiometric.

While the available data cannot at present discriminate between the possibilities, there is one other experimental observation that may indicate that a titanium deficient phase is more likely. Some of the composites fabricated using powder B contained a third phase which was present as disk-shaped regions with diameters up to about 60 μm and thicknesses up to about 5 μm.[5] These regions are apparent in optical micrographs as shown in Fig. 9. By careful specimen preparation a TEM specimen of sufficiently high surface quality was produced to allow correlation of optical and TEM views of the same area. Such a comparison is shown in Fig. 10. The third phase grains are consistently found to be associated with cracks in the specimen. Energy dispersive x-ray analyses have shown that these regions are titanium-rich and contain no significant concentrations of other elements with $Z > 11$. Electron diffraction results have led to a tentative identification of this phase as Ti_2O_3. The observation of this phase provides a possible mechanism for accounting for the titanium although it is currently not clear whether this phase is present in other composites. Continued experiments are underway to verify the tentative identifications. Research is continuing in order to investigate the stability and composition of the intergranular phases under long-term aging conditions.

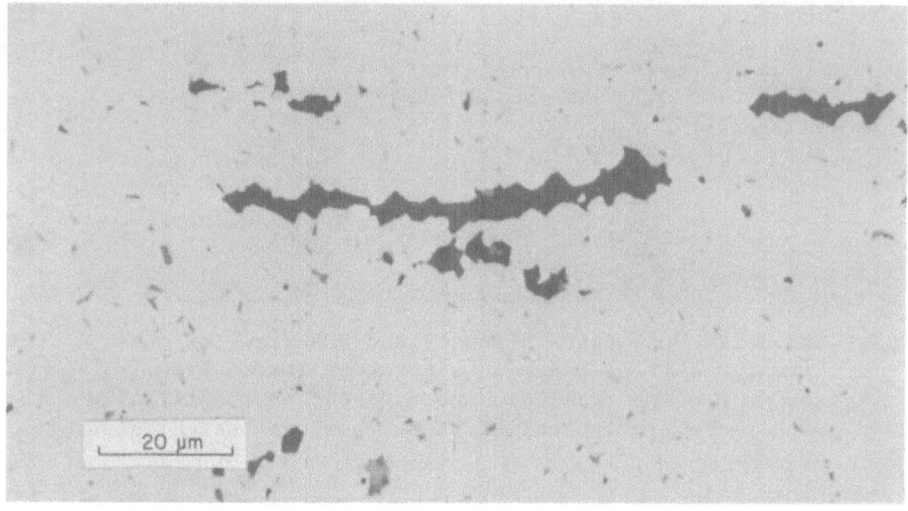

Fig. 9. Optical photomicrograph of a third phase observed in specimens formed from Powder B. These polycrystalline regions are found to be titanium-bearing by energy dispersive x-ray analysis.

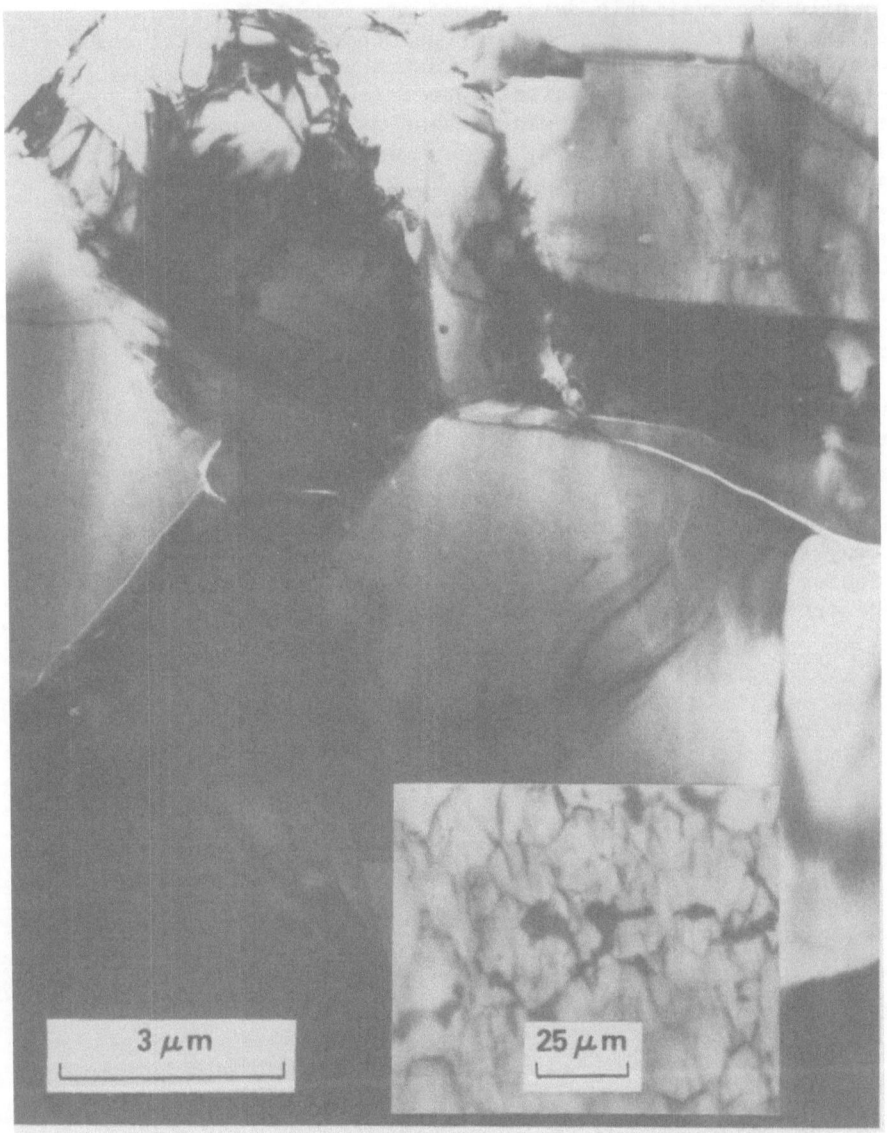

Fig. 10. Comparison of optical (inset) and TEM views of the
          titanium-bearing phase in a specimen formed from
          Powder B. The cracks in and adjacent to the titanium-
          rich phase are typical of those shown in these regions
          by TEM.

SUMMARY

     A series of TiB$_2$-Ni ceramics fabricated using different
TiB$_2$ powders have been examined. It has been determined that the
amount of intergranular phase observed decreases as the mean

particle size of the TiB_2 decreases, the excess initial nickel content being exuded from the compacts during hot pressing.

Both conventional electron microscopy and AEM techniques have been used to identify the primary intergranular phase in the hot-pressed TiB_2-Ni composites. X-ray microanalysis showed that the phase contained nickel and titanium, with Ni:Ti >20. Convergent beam electron diffraction as well as selected area diffraction established that the crystal structure was orthorhombic, point group mmm with a_o = 0.523 nm, b_o = 0.663 nm, and c_o = 0.437 nm. These observations indicate that the phase is similar to Ni_3B. While electron energy loss spectra confirmed the presence of boron, some question still exists as to whether the phase is stoichiometric. Finally, research is continuing in order to explore the stability of this phase under long-term aging conditions and to determine its relationship to observed microscopic properties.

ACKNOWLEDGEMENTS

The authors thank V. J. Tennery and C. B. Finch for helpful discussions regarding fabrication; J. Bentley for discussions of AEM results; A. P. Fisher and G. L. Lehman for technical support, S. B. Waters and J. T. Houston for specimen preparation; F. M. Foust for manuscript preparation; and M. K. Ferber and J. I. Federer for reviewing the manuscript.

REFERENCES

1. C. F. Yen, C. S. Yust, and G. W. Clark, Enhancement of Mechanical Strength in Hot Pressed TiB_2 Composites by the Addition of Fe and Ni, in: "New Developments and Applications in Composites," Transactions, AIME, Warrendale, PA, 317 (1979).
2. H. Pastor, Metallic Borides: Preparation of Solid Bodies-Sintering Methods and Properties of Solid Bodies, in: "Boron and Refractory Borides," V. I. Matkovich, ed., Springer Verlag, New York, 457 (1977)
3. V. Castaing and P. Costa, Properties and Uses of Diborides, in: "Boron and Refractory Borides," V. I. Matkovich, ed., Springer Verlag, New York 390 (1977).
4. G. V. Samsonov, Nature of the Interaction between Titanium Boride and the Iron-Group Metals, Henry Brutcher, translation HB NO. 4122, Metallovedenie i Obrabotka Metallov, 1:35 (Jan. 1958).
5. V. J. Tennery, C. B. Finch, C. S. Yust, and G. W. Clark, Property Structure Correlations for TiB_2 Based Ceramics Densified Using Active Liquid Metals, this meeting.
6. J. D. Schobel and H. H. Stadelmaier, Die Nickelecke im Dreistoffsystem Nickel-Titan-Bor, Metallwissenchaft und Technik, 19(7):715 (1965).

7. J. Bentley, N. J. Zaluzec, E. A. Kenik, and R. W. Carpenter, Optimization of an Analytical Microscope for X-Ray Microanalysis: Instrumental Problems, in: "Scanning Electron Microscopy," 2:581 (1979).

8. N. J. Zaluzec, Quantitative X-ray Microanalysis: Instrumental Considerations and Applications to Materials Science, Ch.4 in: "Introduction to Analytical Electron Microscopy", J. J. Hren, J. I. Goldstein, and D. C. Joy, eds., Plenum Press, New York, N.Y, 121 (1979).

9. N. J. Zaluzec, An Analytical Electron Microscopy Study of the Omega Transformation in a Zirconium-Niobium Alloy, Ph.D. Thesis, University of Illinois (1978); Oak Ridge National Laboratory report ORNL/TM-6705.

10. J. Bentley, Systems Background in X-ray Microanalysis, to be published in: "The Proceeding of the Third Workshop on Analytical Electron Microscopy, July 13-17, 1981" Vail, Colorado

11. B. F. Buxton, J. A. Eades, J. W. Steeds, and G. M. Rachham, The Symmetry of Electron Diffraction Zone Axis Patterns, Phil. Trans. Roy. Soc A281:171 (1976).

12. "International Tables for X-ray Crystallography," published for the Internatinal Union of Crystallography by Kynoch Press, Birmingham, England Vol. I, 25-29 (1969).

13. J. W. Steeds, Convergent Beam Electron Diffraction, Ch. 15 in: "Introduction to Analytical Electron Microscopy," J. J. Hren, J. I. Goldstein, and D. C. Joy, eds. Plenum Press, New York, N.Y 387 (1979).

14. N. J. Zaluzec, Gaussian Optics Calculations of the Parameters of a Magnetic Sector Energy Analyzer, in: "Proc. Workshop on Analytical Electron Microscopy," (Cornell University,July 24-28, 1978) P. J. Fejes, ed., Materials Science Center Cornell University, Ithaca, N.Y 40, (1978).

15. R. F. Egerton, Formulae for Light-Element Microanalysis by Electron Energy-Loss Spectrometry, Ultramicroscopy 3:243 (1978).

16. R. F. Egerton, Ultramicroscopy 4:169 (1979).

17. R. F. Egerton, SIGMAL: A Program for Calculating L-shell Ionization Cross Sections, "Proc. 39th Annu. EMSA Meeting," Atlanta, (August 1981.).

18. L. A. Grunes and R. D. Leapman, Optically Forbidden Excitations of the 3s Subshell in the 3d Transition Metals by Inelastic Scattering of Fast Electrons, Phys. Rev. B22:3778 (1980).

19. J. Bentley, G. L. Lehman, and P. S. Sklad, Background Fitting for Electron Energy Loss Spectra, to be published in: "The Proceeding of the Third Workshop on Analytical Electron Microscopy, July 13-17, 1981," Vail, Colorado.

20. P. S. Sklad, J. Bentley, and G. L. Lehman, Quantification of Energy Loss Measurements with the Use of K and L Absorption Edges, Ibid.

21. H. H. Stadelmaier, Metal-Rich Metal-Metalloid Phases, in: "Developments in the Structural Chemistry of Alloy Phases," B. C. Giessen, ed., Plenum Press, New York, N.Y. 141 (1969).

22. E. Ganglberger, H. Nowotny, and F. Benesovsky, Neue Boride Mit $Cr_{23}C_6$-Struktur, Monatsch. fur Chemie 96:1144 (1969).

23. H. J. Goldschmidt, "Interstitial Alloys," Plenum Press, New York, N.Y., 254 (1967).

24. J. M. Leitnaker, G. A. Potter, D. J. Bradley, J. C. Franklin, and W. R. Laing, The Composition of Eta Carbide in Hastelloy N after Aging 10,000 h at 815°C, Met. Trans. 9A:397 (1978).

25. J. M. Leitnaker, R. L. Klueh, and W. R. Lain, The Composition of Eta Carbide Phase in 2 1/4 Cr-1 Mo Steel, Met. Trans. 6A:1949 (1975).

DISCUSSION

J. Bhardwaj:

Do you have any speculations as to the shape of the oxide precipitate that you saw?

C. S. Yust:

I think it is evident from the optical view that it is polycrystalline, disc-like, and the disc is perpendicular to the pressing direction. I would think that hot pressing somehow causes these features to form in that manner.

O. Knotek:

Normally you see in Ni-B alloys a metastable stage in solidification between Ni_3B and Ni_2B. In many cases you can find a melting range about 150° C. Did you observe this in your binder phase?

C. S. Yust:

As far as I can tell from the diffraction data everything we have seems to be Ni_3B. I don't know if this is stoichiometric Ni_3B. The Ni_2B structure is significantly different. I did not see anything that bothered me and made me feel there was another phase there.

O. Knotek:

Did you make some DTA measurements on your binder phase or on your compounds?

C. B. Finch:

Yes we did--twelve hundred and twenty.

J. Oakes:

What was the source of the micro-cracks in the titanium diboride?

C. B. Finch and C. S. Yust:

The micro-cracks occurred in grains roughly 20 μm. The crack size
varied between one-twentieth of the grain size generally. It arises
from thermal expansion anisotropy. Only when the grain size is below
15 microns, the stress levels go down such that the body will hold
together. When you get above that, the stresses go up and cracks are
generated.

T. Hall, Jr.:

The stoichiometry of the reaction that gives you the Ni_3B should
leave you with titanium somewhere in the ratio nickel to titanium
6:1. Where has the extra titanium gone? Has it gone to the oxide,
or has it escaped with that large mass of fluid that has left during
the hot pressing?

C. S. Yust:

I think I would not be ashamed to say that I don't know. That's a
question we have asked ourselves many times and we don't have a good
resolution of it at the moment.

C. B. Finch:

Some of it is exuded and some of it goes to the oxide.

V. K. Sarin:

Did you get full density on all these when you hot pressed them?

C. B. Finch:

Yes, indeed, we did.

ATOM-PROBE MICROANALYSIS OF WC-Co BASED CEMENTED CARBIDES

M. Hellsing, A. Henjered, H. Nordén and H-O. Andrén

Department of Physics, Chalmers University of
Technology, S-412 96 Gothenburg, Sweden

INTRODUCTION

Our understanding of the microstructure and microchemistry of
metallic materials has increased dramatically during the last few
decades, as modern methods of high resolution microscopy and micro-
analysis have successively become available. The transmission electron microscope (TEM), the electron microprobe, the scanning electron microscope (SEM) and, more recently, the "analytical electron
microscope" (i.e. a scanning transmission electron microscope
equipped with an energy dispersive X-ray spectrometer, STEM/EDS)
have all produced a wealth of new knowledge and understanding.
However, the science of hard metals has probably gained less from
these methods than have other branches of metallurgy. This is no
doubt due to the difficulty to prepare, from the hard metal materials, the thin foils that are a prerequisite for the study by
the high-resolution methods, TEM and STEM/EDS. Electropolishing
of materials containing phases as chemically different as refractory carbides and a soft binder phase is very difficult[1-4], and
ion etching is very time consuming[5,6].

It is therefore encouraging that we have found it possible to
study cemented carbides with the field-ion atom-probe instrument,
an instrument which gives the very highest spatial resolution both
in microscopy and microanalysis and which has equal detectivity for
all elements. Specimens for the atom-probe have to be needle-shaped
with a tip radius less than 0.1 μm, and we have found no particular
difficulty in producing such specimens by electropolishing.

In this paper we report our first results from the study of

cemented carbides, both WC-Co containing cubic carbides and cemented
carbide specimens coated with TiC. The interpretations of data
given in this paper are preliminary and more experimental evidence
is needed before any definite conclusions can be drawn. However, we
feel that the results obtained so far can give some indications of
the usefulness of the atom-probe technique for the study of cemented
carbides.

PRINCIPLES OF ATOM-PROBE MICROANALYSIS

The two basic physical mechanisms utilized in the atom-probe
instrument, field ionization and field evaporation, both require an
electrical field strength of more than 10^{10} V/m at the specimen
surface. The specimen is therefore a sharp needle to which a high
positive potential is applied. The specimen is mounted inside a
vacuum system (Fig. 1), which is backfilled with a low pressure
of an inert gas such as neon or helium. When the required high
voltage is applied to the specimen, gas atoms are ionized close to
the surface atoms of the specimen (field ionization), and the resulting gas ions are accelerated by the field radially out from the
tip and impinge on a phosphor screen to create a highly magnified
image of the specimen surface (see e.g. Fig. 6). (For technical
reasons, the specimen is cryogenically cooled, and the phosphor
screen is preceeded by an image intensifier.)

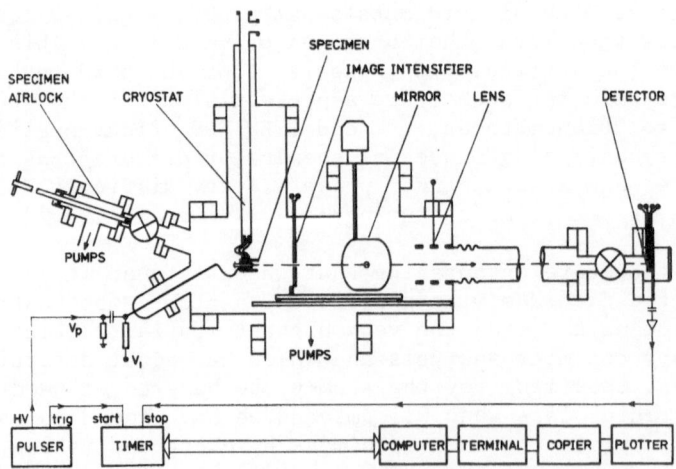

Fig. 1. Schematic diagram of the atom-probe instrument.

As the specimen voltage is increased further, the atoms of
the specimen surface may themselves become ionized and removed from
the specimen as positive ions (field evaporation). Thus, by mass
analysing these field evaporated ions, the composition of the specimen can be determined, atom by atom. This is done by time-of-flight
spectrometry; a hole in the image screen permits specimen atoms
from a small surface area to enter a flight tube and reach a detector some 1 or 2 m away from the tip. The mass-to-charge ratio
of the ion can than be computed from its flight time and the field
evaporation voltage. (In practice this is done automatically by a
computer connected on-line.) The area of analysis is selected by
tilting the specimen around its tip and by changing the specimen-
to-screen distance.

The results of atom-probe analyses are presented as mass
spectra or composition profiles. The interpretation of spectra is
in principle straightforward, since normally no sensitivity factors
need to be applied - the atom-probe detects all ions with the same
probability.

A more detailed description of the technique can be found in
the review by Müller and Tsong[7].

EXPERIMENTAL

Two sorts of cemented carbides have been investigated, both of
the type WC-Co with cubic carbide additions. The amount of cubic
carbides was large in Material 1 and small in Material 2; the compositions and grain sizes are given in Table 1. Both materials are
commercial grades, made by Sandvik Coromant, Sweden.

Table 1. Composition and grain size of the cemented carbide
materials studied.

Material	Grain size (μm)	Composition in volume percent			
		WC	(Ti,W)C	(Ta,Nb)C	Co
1	2-3	36	39	14	11
2	1-2	92	-	2	6

Pins cut from the material with a low speed diamond saw were electropolished to needles in a solution of 8% sulphuric acid in methanol. Controlled backpolishing and ion etching were sometimes used to obtain specimens with a specific feature of interest at the tip (Fig. 2). The preparation methods are described in more detail elsewhere[8].

Some of the specimens of Material 2 were coated with a thin layer of TiC in a laboratory size chemical vapour deposition (CVD) reaction chamber at a temperature of 1000°C. The CVD process can be described by the formula

$$TiCl_4 + CH_4 \xrightarrow{H_2} TiC + 4HCl.$$

TiC layers as thin as 50 nm could be prepared by using short reaction times (30 s).

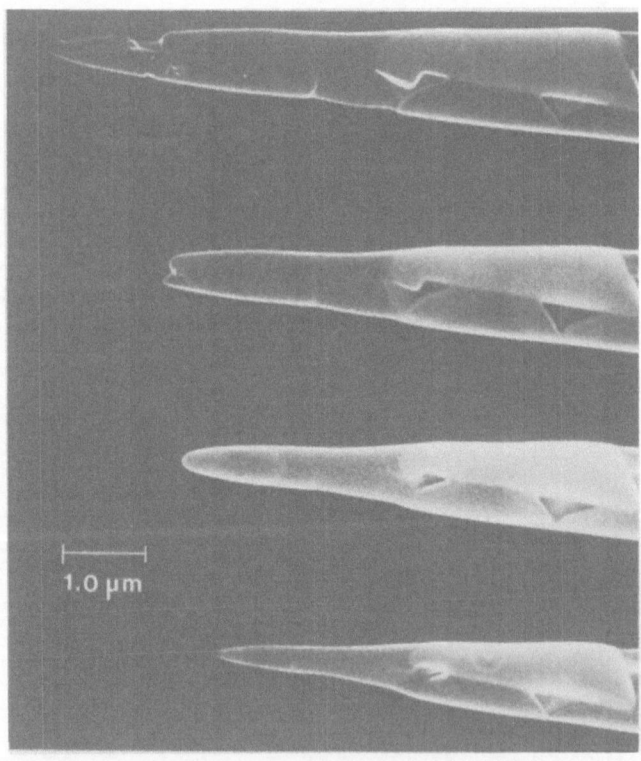

Fig. 2.   SEM micrographs showing the effect of backpolishing a cemented carbide specimen. (Electropolishing in intervals of 10 ms duration.)

All specimens prepared were inspected in the transmission electron microscope, either in the Philips EM 300 or the JEOL 200 CX. Electron diffraction or EDS analysis was used to identify the phases present at the tip. The result of backpolishing or ion etching was also determined in the electron microscope, and all specimens were examined after atom-probe analysis to determine the volume of analysed material.

The specimens were field-ion imaged at a temperature of 92 K using neon as "imaging gas". The field-ion atom-probe instrument used in this investigation has been described previously[9]. The three main phases present have very different evaporation fields, which means that they will image in the field-ion microscope with rather different intensities - WC brightest and binder-phase darkest - so that phase boundaries can easily be located in the field-ion image.

Atom-probe analyses were performed in ultra-high vacuum (less than 10^{-7} Pa) using an evaporation pulse amplitude equal to 20% of the imaging voltage. Since atom-probe spectra have virtually zero background and since the yield is mass independent, the accuracy of atom-probe analysis is normally given by counting statistics only. However, in two cases in this investigation the accuracy was somewhat lower due to the limited resolution of the present instrument. Firstly, the mass resolution was not sufficient to fully separate $_{181}Ta$ from $_{182}W$. Secondly, the dead time in the detector system prevented the second of two C^{2+} ions arriving within 10 ns from being recorded. Since only a small fraction of the C^{2+} ions

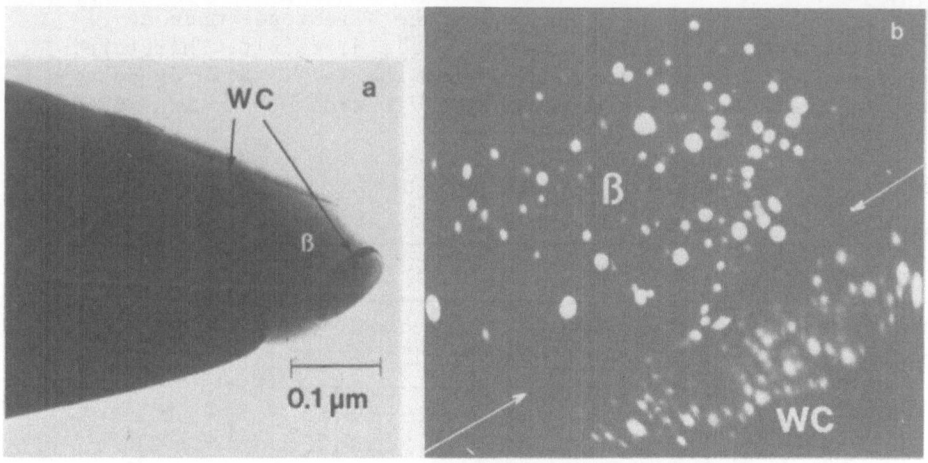

Fig. 3.   A specimen containing binder-phase and WC grains at the tip. TEM (a) and field-ion (b) micrographs.

arrived in pair this error was small and has been corrected for.
(This type of error was negligible for all other ions due to their
larger flight time and energy spread.)

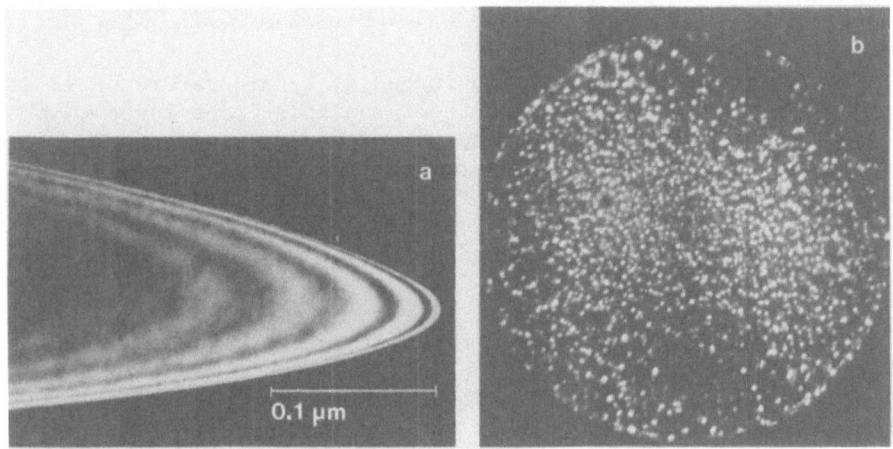

Fig. 4. Dark field TEM micrograph (a) and field-ion micrograph
         (b) of a specimen with a cubic carbide at the tip.

Table 2. Atom-probe analyses of various phases in Material 1.
The error limits given represent one standard deviation. (Ta
and Ti were only detected in the cubic carbide phase. The very
small mass difference between Ta and the W isotopes made an
accurate quantitative determination of Ta difficult. Therefore,
the Ta content given was computed from the Ta/Ti ratio as determined by chemical analysis.) Number of ions detected: WC – 6747,
β – 1058, γ – 1197.

Phase	Composition (atomic percent)							% of total	
	% of metal content								
	W	Ti	Ta	Nb	Co	Fe	Mo	C	N
WC	100.0 ±0.05							47.5 ±3.2	
β	7.3 ±0.8				89.6 ±0.9	2.6 ±0.5	0.5 ±0.2	0.2 ±0.1	
γ	27.5 ±1.7	55.1 ±2.5	11.5 ±1.0	5.9 ±0.9				44.7 ±4.5	0.9 ±0.3

RESULTS

Cobalt binder phase

A specimen containing binder-phase (β-phase) is shown in
Figure 3. Analysis of two such specimens of Material 1 gave the
composition given in Table 2. The binder-phase contained considerable amounts of W, but no Ti, Nb or Ta could be detected (detection
limit less than 0.2 at%). A very small carbon content, 0.2 at %,
was measured, and in addition some Fe and Mo was recorded.

Cubic carbides (γ-phase)

A TEM micrograph of a specimen of Material 1 containing a
cubic carbide at the tip is shown in Figure 4a. Due to the mixed
metal element composition, the field-ion image appears rather irregular (Fig. 4b). An atom-probe spectrum obtained from a cubic
carbide grain is shown in Figure 5. The results of atom-probe analyses are given in Table 2. No significant difference in composition was found between various cubic carbide grains in the material.

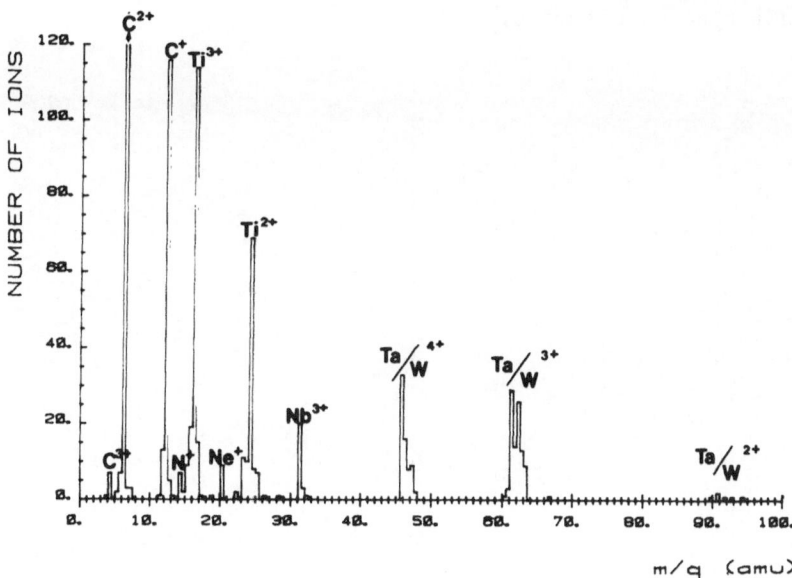

Fig. 5.   An atom-probe mass spectrum obtained from a cubic carbide
          grain.

Tungsten carbide (α-phase)

An electron micrograph of a specimen containing WC at the tip
is shown in Fig. 6a. The presence of some dislocations in the carbide can be noted. The corresponding field ion micrograph (Fig. 6b)
is rather regular, indicating a high degree of order in the structure. Within the limits of uncertainty atom-probe analysis of WC
grains gave a stoichiometric composition (Table 2), and no other
metallic atoms than tungsten were recorded (detection limit less
than 0.05 at%).

TiC-coated specimen

A specimen of Material 2, coated with a thin layer of TiC by
CVD, is shown in Figure 7. Atom-probe analysis of the layer showed
that it consisted of three distinct regions:

(i)   Coating surface. The outermost surface of the coating gave
an atom-probe spectrum containing peaks from different titanium
oxide ions, as well as from various Ti and C ions (Fig. 8). A peak
at $m/q = 28$ was interpreted as CO^+, since no iron signal could be
detected by STEM/EDS. The composition deduced from the spectrum
(Table 3) shows that the coating surface contained much oxygen and
rather little carbon.

(ii)   The major part of the coating. This was much richer in carbon
but still contained some oxygen (Table 3). The composition was constant with depth, and electron diffraction showed that the layer
had the NaCl-type structure of TiC.

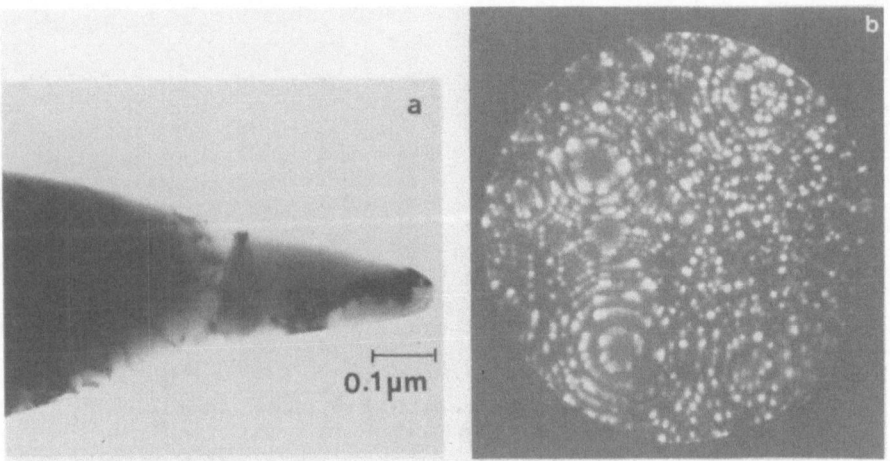

Fig. 6.   TEM (a) and field-ion (b) micrographs of a specimen containing WC at the tip. Note the dislocation contrast
in the TEM micrograph.

(iii) Coating-substrate interface. As the interface was approached, some tungsten was detected and the amount increased closer to the interface. Immediately before the coating-substrate interface was reached, the oxygen disappeared. Instead, a small amount of cobalt was recorded (Fig. 9). The substrate tip was a WC-grain, the nearest binder phase grain lying some 0.5 μm away from the tip apex.

Table 3.   Atom-probe analysis of TiC coating on WC-Co.

	Composition in atomic percent			Number of ions detected
	Ti	C	O	
Surface layer	33	15	52	600
Bulk of coating	58	30	11	2000

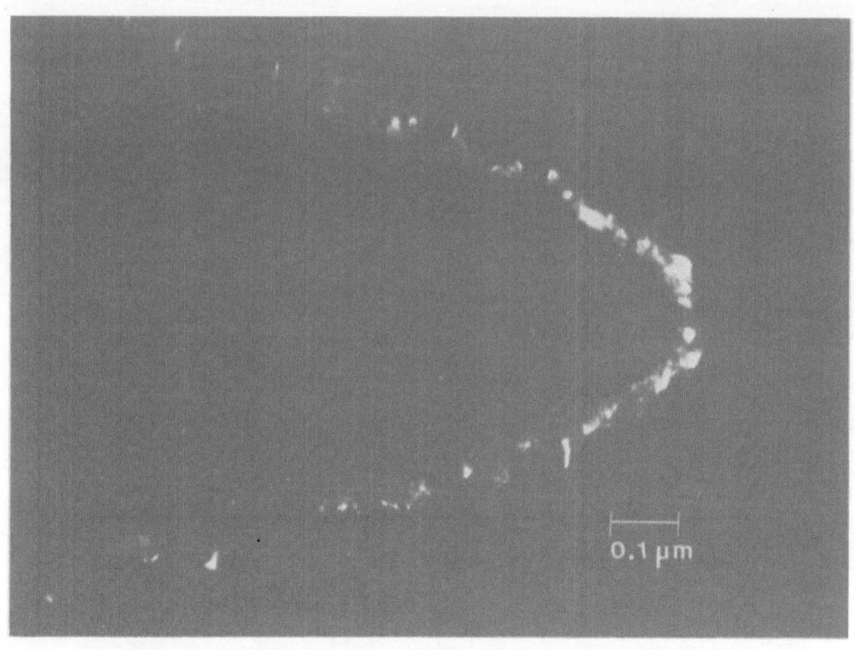

Fig. 7.   Dark field TEM micrograph showing a thin TiC layer on a WC grain. The layer was grown using chemical vapour deposition.

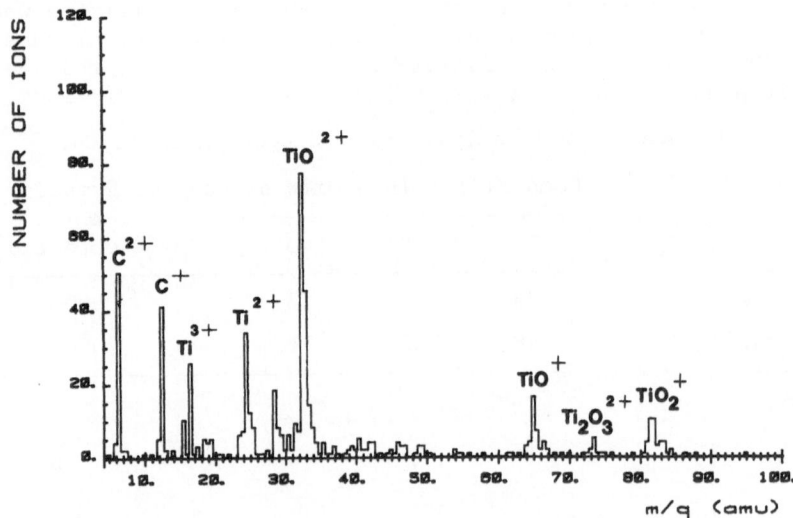

Fig. 8.   Atom-probe mass spectrum obtained from the surface of a
          TiC-coated WC grain.

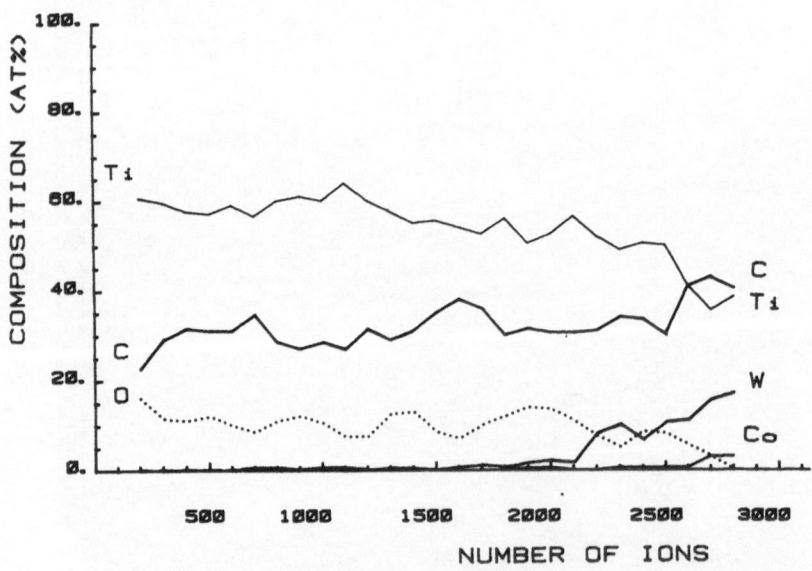

Fig. 9.   Atom-probe composition profile through a TiC coating and
          through the interface between coating and WC substrate.

DISCUSSION

Material 1

Material 1 was manufactured from a mixture of WC, (Ta,Nb)C, (Ti,W)C and Co powders of a rather small grain size, and it is interesting to note that diffusion during sintering has evened out the composition so that only one, homogeneous gamma-phase exists in the final product. Its composition can be written

$$(Ti_{.55}W_{.27}Ta_{.12}Nb_{.06})C_{0.84}$$

The maximum solubility product of WC in the binder-phase of pure WC-Co is approximately $8 \cdot 10^{-4}$ (at%)^2, according to Hoffman and Mohs[10] and Exner[11]. The influence of dissolved cubic carbides in the binder-phase on the solubility of WC has been a matter of debate[12]. Our analyses of the binder-phase gave 7.4 at% W, no detectable Ti, Nb or Ta (detection limit less than 0.2 at%) and about 0.2 at% C. Thus, we obtain a solubility product of approximately $1.5 \cdot 10^{-4}$, which is in good agreement with the results obtained by Johnson[13] who studied similar materials.

A more detailed account of our results from this material is given elsewhere[14].

The TiC coating

Electron diffraction from the coating showed the NaCl-structure and substantial amounts of oxygen were detected with the atom-probe. Both TiC and the solid solution TiC-TiO have NaCl-structure with lattice parameters differing only a few percent[15]. The majority of the coating can therefore be described by the formula

$$Ti(C_{0.74}O_{0.26})_{0.71}.$$

This TiC-TiO solid solution has a wide stability range: stoichiometric TiC dissolves oxygen at least to $Ti(C_{0.6}O_{0.4})_{1.0}$[15], and at the other extreme, TiC_x is known to be stable down to $TiC_{0.3}$[16].

The surface layer had more oxide character and can be described by an approximate formula

$$Ti(O_{0.8}C_{0.2})_2.$$

However, no electron diffraction could be obtained from this very thin layer so its structure is not known.

It is interesting to note that the large amounts of oxygen found in the TiC coating were produced by an estimated partial pressure of oxygen of only about 1 Pa, whereas the methane and titanium chloride partial pressures in the CVD reaction chamber were both 260 Pa.

The fact that the interface region between the TiC coating and the substrate WC grain was free from oxygen suggests that the coating nucleated as a carbide. The amount of cobalt found in the interface corresponds approximately to one atomic layer, and this observation supports results obtained with the Auger technique[17]. The presence of cobalt on the WC surface seems to be very important for the growth of TiC[18].

The observation of tungsten in the TiC coating is in agreement with the results of Sarin[19].

A more detailed description of the results from our investigation of the TiC coatings can be found elsewhere[20].

CONCLUSIONS

- Methods have been developed to prepare needle-shaped specimens of cemented carbides and to investigate these in the electron microscope and the field-ion atom-probe microscope.

- Various phases in WC-Co-type cemented carbide containing cubic carbides have been analysed using atom-probe field-ion microscopy.

- Cemented carbide specimen tips have been coated with TiC, using CVD, and the coating and coating-substrate interface compositions have been measured with the atom-probe.

ACKNOWLEDGEMENTS

The authors wish to thank professor B. Aronsson, of Sandvik AB, for encouragement and many stimulating discussions and L. Kjellsson for helpful discussions. Parts of this work was financially supported by the Swedish Natural Science Research Council (NFR).

REFERENCES

1.   B. Lethinen, J. Sci. Instrum. (J. Phys. E) 1:673 (1968)
2.   R. Arndt, Z. Metallkunde 63:274 (1972)
3.   G. Persson, Nature 218:159 (1968)

4. A. Hara, T. Nishikawa, and T. Nishimoto, Sumitomo Electric
    Technical Review 14:106 (1970)
5. S. Vuorinen and A. Horsewell, In "Proc. 7th Eur. Congr. Elec.
    Micr.", P. Brederoo, and G. Boom, eds., Seventh Eur. Congr.
    Elec. Micr. Foundation, Leiden 1:420 (1980)
6. N.K. Sharma, I.D. Ward, H.L. Fraser, and W.S. Williams, J.
    Am. Ceram. Soc. 63:194 (1980)
7. E.W. Müller and T.T. Tsong, Progr. Surf. Sci. 4:1 (1973)
8. A. Henjered, M. Hellsing, H. Nordén, and H-O Andrén, Field-
    ion microscopy of cemented carbides, to be published.
9. H-O Andrén and H. Nordén, Scand. J. Metallurgy 8:147 (1979)
10. A. Hoffman and R. Mohs, Metall 28:661 (1974)
11. H.E. Exner, Int. Metals Reviews 4:149 (1979)
12. H. Grewe, H.E. Exner, and P. Walther, Z. Metallkunde 64:85
    (1973)
13. H. Johnson, in "Preprints 5th Eur. symp. powder metallurgy",
    Jernkontoret, Stockholm 2:69 (1978)
14. M. Hellsing and B. Aronsson, to be published
15. H. Krainer, Arch Eisenhüttenwesen 21:119 (1950)
16. W.B. Pearson, "Lattice Spacings and Structures of Metals and
    Alloys", Pergamon Press, Oxford, p. 960 (1958)
17. N.K. Sharma and W.S. Williams, Thin Solid Films 54:75 (1978)
18. H. Gass, H. Mantle, and H.E. Hintermann, in "Proc. 5th Int.
    Conf. on Chemical Vapor Deposition", J.M. Blocher Jr.,
    H.E. Hintermann, and L.E. Hall, eds., The Electrochemical
    Society, Princeton N.J., p. 99 (1975)
19. V.K. Sarin, in "Proc. 7th Int. Conf. on Chemical Vapor
    Deposition", T.O. Sedgwich and H. Lydtin, eds., The
    Electrochemical Society, Princeton N.J., p. 476 (1979)
20. A. Henjered, L. Kjellsson, H-O. Andrén, and H. Nordén,
    Scripta Metallurgica, in print (Sept. 1981)

DISCUSSION

V. K. Sarin:

Have you observed that when you removed the layers that you get lower
dislocation densities?

H. Norden:

I think this has been one of the favorite topics, that what you see
in electronmicroscopy generally are not typical dislocations, but the
odd ones which survive whatever you do, how you thin it and how you
dump it on the floor. We actually used the dislocation structure
there as a marker to convince people that we knew how much we removed. I think this is true of many of the dislocation analyses done
on thin films. What you see is typical for surviving dislocations.

V. K. Sarin:

My second question is about the cobalt you saw in your CVD coating.
Did you heat the crystal up in a hydrogen atmosphere, and then coat
them?

H. Norden:

Yes.

V. K. Sarin:

May I suggest that you can get the formation of a very thin layer of
eta phase while you are heating the WC-Co substrate up to 1,000°C in
hydrogen. You get decarburization and some eta phase formation which,
during the coating process with carbon present, can reconvert and
give you cobalt. We have observed this and you can get cobalt at the
interface.

H. Norden:

Our coating was so thin that if you have eta phase forming, it should
be a monolayer.

R. Sivan:

Were you not a little bit surprised that you got titanium oxide
formed? Can you explain its' occurrence?

H. Norden:

This was a standard coating unit with a vacuum pump backing. So,
during cool down there could be some back streaming. It is not
typical for the coated tip layers that you have a lot of titanium
oxide. You might have a few monolayers. The whole layer is only
about 100 to 150 Å.

R. Sivan:

Did you have different layers of titanium carbide or did you get just
single layer coating?

H. Norden:

Only a single layer.

A. Horsewell:

I think it was rather nice that you could CVD coat your WC crystal
with TiC. It made me wonder, seeing that you had cobalt at that

TiC/WC boundary, whether your other result of cobalt at the WC/WC boundary could be caused by some other reason, so that the cobalt wasn't there in the first place, either.

H. Norden:

Well, it doesn't migrate in the field because then we would see it at other places on the interface. Because you are working at about 50 to 80 K, I don't think you have diffusion. It's very short times and very low temperatures and we didn't see any solubility of Co in WC at the Co/WC interface.

W. Williams:

Dr. Norden, to pursue that point, do I understand correctly that your result is consistent with the result that we showed using STEM of a very narrow layer of cobalt between the two tungsten carbide grains?

H. Norden:

Yes.

G. Dearnaley:

At one stage you said that the tungsten carbide was highly stoichiometric and I wondered to what accuracy could you make that statement?

H. Norden:

Well, on the view graph I gave about 3 per cent and the bad accuracy here is inherent and that is only for analyzing carbon in tungsten carbide. Because the field is so high you get both singly ionized and doubly ionized carbon. Which means in the same mass peak you get one or two carbon atoms. We are now calibrating at lower and lower field to see how large this contribution is. In titanium carbide, or any other carbide, we get carbon with just the count inaccuracy which for 1,000 atoms is about 0.1 per cent. This was the weak point, that we couldn't give the stoichiometric value accurately for tungsten carbide.

LOW-Z ELEMENT ANALYSIS IN HARD MATERIALS

D. T. Quinto, G. J. Wolfe, and M. N. Haller

Philip M. McKenna Laboratory
Kennametal Inc., Greensburg, PA

INTRODUCTION

The study of fundamental mechanical and chemical properties
of sintered cemented carbides and ceramics has become increasingly
dependent on analytical techniques that probe into the microstructure and microchemistry of these cermets. Examples of the uses of
the SEM, STEM, neutron diffraction, atom probe etc. are given in
other papers in this conference which contribute to our understanding of the structure/property relationships. In contrast to
pure metals and alloys (which have been the traditional research
domain of metallurgists and materials scientists) two characteristics of cermets impose special requirements on micro-analysis:
relatively fine grain sizes (e.g., several microns in carbides to
sub-micron grains in ceramics and CVD refractory coatings; thin,
intergranular binder films) which necessitate submicron resolution; and the common presence of low-Z elements such as carbon,
nitrogen, boron, oxygen that comprise a major volume fraction of
the microstructure. An adequate micro-analytical method should
therefore be capable of clearly distinguishing these microstructural features as well as providing the corresponding microchemistry of the constituents. While the state of the art in electron
microscope imaging by topographic or phase contrast is more than
adequate in delineating microstructural features of cemented carbides, it is not yet quite as satisfactory in terms of microchemical analysis. The purpose of this paper is to take an overview of
the existing technology in micro-analysis and, by comparison of
particular capabilities and limitations on light element sensitivity and microchemical resolution, gain a useful perspective in
the choice of analytical instrumentation most suited for studies
of carbide, binder or CVD coating features in tool materials.

947

Three generic types of techniques will be compared principally in which both electron microscopic imaging of microstructural morphology and micro-analysis are coupled in the same instrumentation: X-ray micro-analysis of bulk samples by scanning electron microscopy (SEM); X-ray and electron energy loss micro-
analysis of thin foil specimens by scanning transmission electron
microscopy (STEM); and Auger electron analysis of sample surfaces
by Scanning Auger Microprobe (SAM). Some practical applications
of these methods in an industrial carbide research laboratory will
be illustrated.

RESULTS AND DISCUSSION

Comparison of Sample Volumes

Signal generation in any of these techniques results from
the physics of interaction between a focused electron beam and the
excitation volume in the sample from which the characteristic signals (X-rays, Auger electrons or transmitted core loss electrons)
are collected under given instrumental parameters. We refer to
this as the sample volume which is important in considering the
spatial chemical resolution attained in each technique. A schematic of the sample volumes resulting from electron excitation is
shown in Figure 1. In relation to the electron probe beam diameter d_o, it is noted that: a) the lateral and depth resolution
(parallel and perpendicular to the sample surface) for characteristic X-ray signal detection is generally $> 1 \mu m$ in SEM/EDS;
b) with a thin foil sample in STEM/EDS the sample volume is truncated at the bottom surface thus improving both the lateral and
depth resolution; c) the detected Auger electrons are limited to
the electron escape depth (typically <2 nm) which thus defines
the depth resolution while the lateral resolution is approximately
equal to the beam diameter d_o.

The sensitivity for element detection in each technique depends on the appropriate scattering cross section for the inner
shell excitation or de-excitation process, the sample volume and
detection efficiency. For light elements ($Z \leq 11$) K-shell ionization leads to a significantly greater probability for Auger transition than X-ray transition so that SAM has an intrinsic advantage over X-ray microanalysis. Detection of light elements is
precluded in EDS with the conventional Be window in the detector
but is made possible with increased detection efficiency through
"windowless" detectors or by improved signal/noise ratio in wavelength dispersive spectrometry (WDS). Electron energy loss spectroscopy (EELS) depends on the primary ionization process as opposed to the secondary processes of Auger or X-ray transitions.
Although it has a relatively high cross section it suffers a low

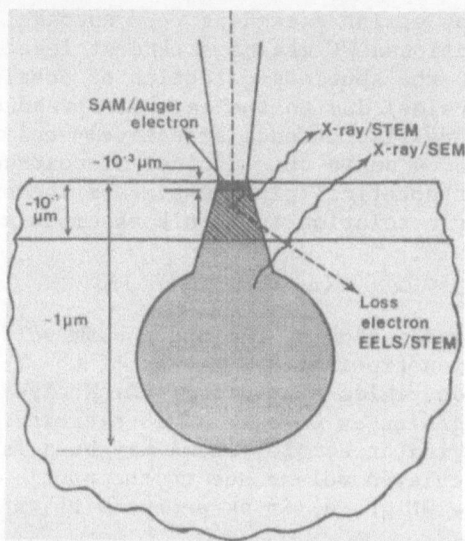

Fig. 1.   Schematic comparison of sample volumes, shaped as a
          "tear drop in bulk SEM/EDS, a conical section through the
          thin foil specimen in EDS or EELS in STEM, and a thin disc
          at the specimen surface in SAM.

signal/noise ratio with increased sample volume (thin foil thickness).

SEM/X-ray Micro-Analysis

The most common micro-analytical accessory to SEM systems is
energy dispersive spectrometry (EDS) shown schematically in
Figure 2.   EDS systems utilize a silicon crystal as a solid state
ionization device for X-ray detection, characterized by a large
signal acceptance angle which permits efficient operation with low
beam currents.   EDS has been developed to the point where quantitative micro-analysis can be accurately performed subject to the
> 1 μm resolution (lateral and depth) limits mentioned above.
However, the use of standard detectors with a beryllium window
limits the detection efficiency in conventional EDS to elements
heavier than sodium (Z=11).

To illustrate that beam-spreading is inherent in EDS, X-ray
spectra of W and Co produced by a 30 KV probe beam incident on a
WC grain were collected by point analysis from a polished WC - 11%
Co sample as seen in an SEM secondary electron image, Figure 3.
The Co/W intensity ratio obtained when the incident beam (d ~ 100
nm at 10^{-10} A current) was centered on a 1 x 2 μm^2 WC grain was
found to be 32%, which decreased to 5% for grains measuring

4 x 5 μm^2 and < 1% for grains exceeding 7 μm average diameter. Assuming that the sectioned WC grains extend at least 1-2 μm below the polished surface, the spurious detection of cobalt (presumably surrounding the WC grains) due to the lateral spread of the electron beam and secondary fluorescence effectively enlarges the sample volume size to ~ 4 μm in this typical microstructure. Reed[1] defined the theoretical minimum size of the excited volume for qualitative X-ray resolution in a bulk specimen as

$$R_{qual} = 0.077 \ (E_o^{1.5} - E_c^{1.5})/\rho \qquad -(1)$$

where ρ is the sample density, E_o the probe beam voltage, and E_c the critical excitation (ionization) energy of the X-ray transition. This expression yields R_{qual} for Co $K\alpha$ X-rays for a 30 KV incident beam on WC grains as 0.7 μm. This exercise demonstrates that the actual analyzed or sample volume may be a factor of 3-5 greater than the calculated volume due to the spurious Co X-ray detection through the WC grain, in agreement with the literature[1].

To extend micro-analysis to lower Z elements ($5 \leq Z \leq 11$), wavelength dispersive spectrometry (WDS) is used, also shown schematically in Figure 2. WDS utilizes a diffracting crystal and gas proportional counter moved in synchronization to satisfy the Bragg diffraction condition. WDS systems are characterized by higher

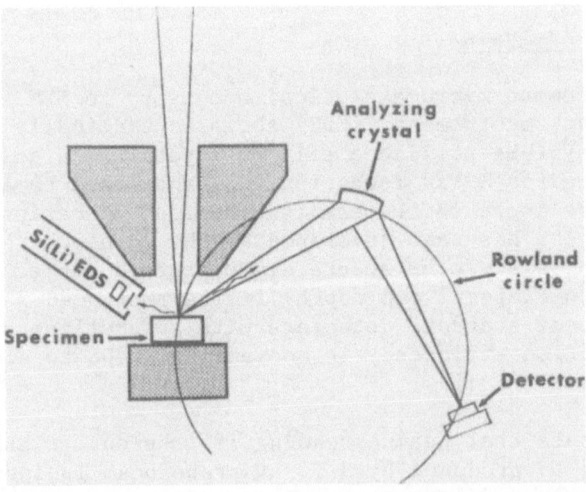

Fig. 2. Configuration for X-ray microanalysis in the SEM by EDS and WDS.

Fig. 3.   SEM micrograph of WC-11% Co polished sample. Note "carbon
          burn" left on WC grain by the electron beam during point
          analysis.

energy resolution (better element spectral separation) and higher
peak/background ratios compared to EDS systems. However, WDS
operation requires higher electron beam currents for X-ray production (due to relatively poor geometrical detection efficiency),
usually 3-4 orders of magnitude more than in EDS, which results in
an increased probe beam diameter and large sample volume.

A comparison of X-ray line scans of Ti($K\alpha$) (EDS and WDS),
$C(K\alpha)$ and $N(K\alpha)$ (WDS only) across the substrate/coating interface
of a polished cross section of a WC - Ti(Ta) C_x- Co substrate and
a TiC/TiCN/TiN CVD coating is shown in Figures 4a to 4c. CVD
coatings provide an abrupt, readily detectable change in composition. The effective lateral resolution across such a "sharp"
interface, ideally represented by a step function, may be defined
in terms of a probe function[2] Δz (see Figure 5) obtained from
the line scan across the interface. The probe beam diameter in
the EDS Ti line scan in Figure 4a is ~ 0.1 μm (at 20 KV, 6 x 10^{-11}
A absorbed current) while the measured probe function is 0.5 μm.
Figure 4b is the Ti X-ray line scan on the same specimen utilizing
a WDS system at the higher beam currents required (20 KV, 4 x 10^{-9}
A absorbed current) with a probe beam diameter of 0.3 μm. Graphi-
cal determination of Δz for Ti gives a value of 1.3 μm. Similar
WDS scans are shown for C and N, Figure 4c, obtained at 3 x 10^{-8} A
absorbed current. The Δz for C is found to 1.6 μm. These values
are in good agreement with corresponding $3R_{qual}$ values using
equation 1, which gives 1.2 μm for Ti and 1.4 μm for C. These
spatial resolution limits are adequate for micro-analysis of large

grains in WC-Co-based microstructures but would only be marginal
in analyzing fine ($\leq$ 1.5 μm) carbide grains or binder phases.

Quantitative X-ray micro-analysis has been well-developed
with the availability of computer software programs integrated
into commercial SEM instruments. These take into account
corrections to the measured integrated intensity due to matrix
effects (mean atomic number Z), absorption in the sample (A) and
secondary X-ray fluorescence (F). The concentration C of an
element is determined from its measured integrated intensity ratio

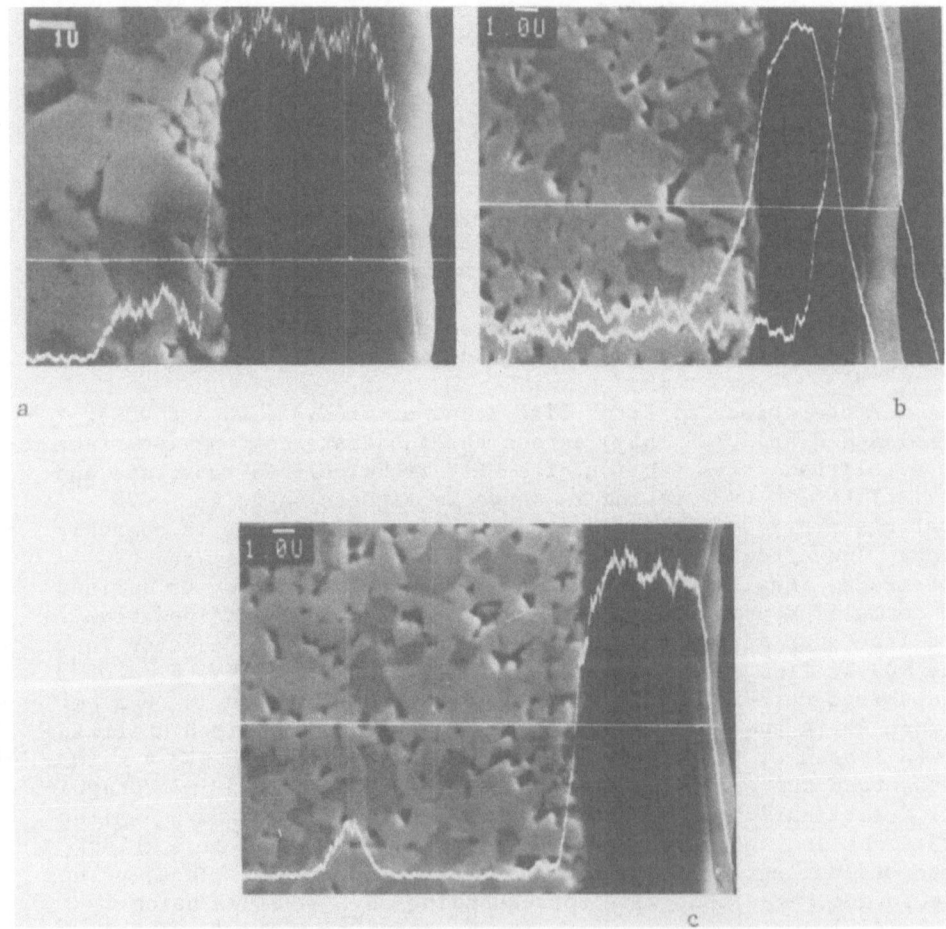

Fig. 4.   X-ray line scans across a substrate/coating interface for
          a) Ti(Kα) by EDS; b) inner scan C(Kα) and outer scan N(Kα)
          by WDS; c) Ti(Kα) by WDS.

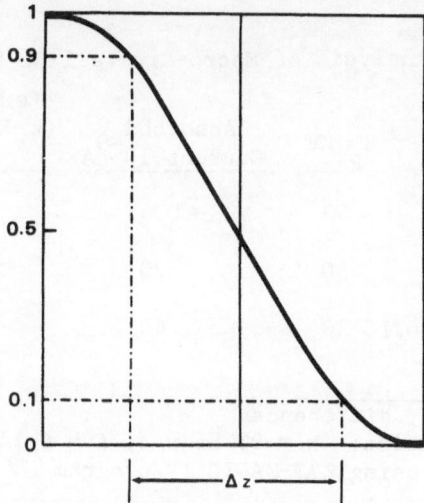

Fig. 5.   Measurement of experimental probe function, Δz, across a
          sharp interface, taken as the 10%-90% cut-off of the line
          scan signal intensity.

in the sample to its standard, K, and these calculated correction
factors K_Z, K_A, K_F:

$$C_x = K \; K_Z \; K_A \; K_F \qquad\qquad -(2)$$

        The accuracy and precision of these computer programs were
determined for standard specimens of WC at 30 KV primary beam voltage.   Large WC crystals exceeding 100 μm in size were used for
compositional determination to eliminate any binder interference.
Net integrated intensities in the EDS spectra for the W(Lα) line
were entered into the ZAF correction routine, MAGIC IV program[3]
and the C concentration was determined by difference.   For comparison, similar analyses of the same standard sample by WDS on both
W(Lα) and C(Kα) lines are given in Table 1.   The 2σ errors are
seen to be well within the predicted value, i.e. < 2 - 4%[3].   The
precision of the tungsten analysis is better in the WDS system due
to higher counting statistics than in the EDS system.

        A reference standard of boron nitride was quantitatively
analyzed by WDS to determine accuracy and precision with a
compound composed of two light elements.   Net peak intensities for
B(Kα) and N(Kα) were measured at 10 KV, 20 KV and 30 KV primary
beam voltages.   The results in Table 2 show that with the best
counting statistics and peak/background ratios, obtained at 10 KV,
the actual errors are close to the predicted error of a few
percent.   The poorer results at a higher beam voltage are due to

Table 1. Quantitative Analysis of Macro-Crystalline WC by EDS and WDS

Concentration,w/o±2σ		E_p, KV	Absorbed Current,10^{-9}A	Counting Interval (x Repeat Analysis), seconds
$W_{L\alpha}$	$C_{K\alpha}$			
EDS 94.31±0.96	5.69*	30	0.6	100(x5)
WDS 93.76±0.34	6.24*	30	20	20(x5)
WDS 93.85±0.20	6.11±0.15	30	40	20(x5)

* Carbon determined by difference
  Theoretical composition: W = 93.86 w/o, C = 6.14 w/o
  Analyses performed using ZAF-MAGIC IV program [3]

Table 2. Quantitative Analysis of BN Standard by WDS
Concentration, w/o ± 2σ

$B_{K\alpha}$	$N_{K\alpha}$	E_p, KV
43.33±0.36	56.35±0.37	10
43.62±0.83	57.21±0.83	20
44.85±2.20	57.13±1.90	30

Theoretical composition: B = 43.57w/o, N = 56.42 w/o
Counting interval: 20s, 5X repeat analysis
Analyses performed using ZAF-MAGIC IV program [3]

lower count rates and the increase in the magnitude of the
absorption correction. The absorption correction is the major
factor in the ZAF calculation and is 3 times higher at 30 KV than
that at 10 KV.

These examples indicate that, as long as there are no
spatial resolution problems, quantitative analysis by EDS or
WDS X-ray micro-analysis is quite good and in fact more accurate
than either of the two other methods described below. Care has to

be taken, however, in interpretations of quantitative composition
determinations of typical WC-Co microstructures by such X-ray
micro-analysis since the typical grain sizes are indeed comparable
to the spatial resolution for the light elements by WDS; hence,
interference from adjoining phases is likely.

STEM Microchemical Analysis

Two methods of chemical analysis have been integrated into
commercial STEM instrumentation - EDS, with the same detection
geometry as in SEM, and electron energy loss spectrometry (EELS),
as shown in Figure 6. The thin foil sample is about 100 nm in
thickness; carbide and ceramic foils are most consistently
prepared by ion thinning. The sample volume for EDS in the thin
foil (Figure 1) is significantly smaller than in SEM since the
"teardrop" excited volume occuring in a bulk specimen is truncated, which results in increased resolution. Figure 7 shows EDS
line scans for $Co(K\alpha)$ and $W(M\alpha)$ across the binder/carbide interface in the WC-11% Co thin foil sample. Adequate signal was
readily obtainable at 120 KV beam voltage and a measured beam
diameter d_o ~5 nm with a standard tungsten filament. The measured
Δz for both elements is 30 nm, which is an order of magnitude
better than in SEM/EDS.

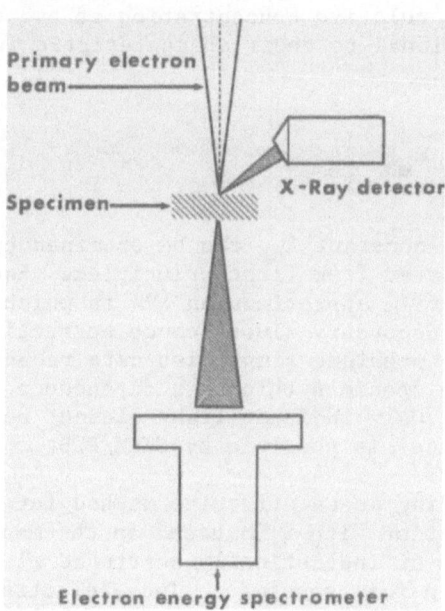

Fig. 6. Configuration for thin foil micro-analysis in the STEM by
X-ray detection for EDS and transmitted electron energy
spectrometry for EELS.

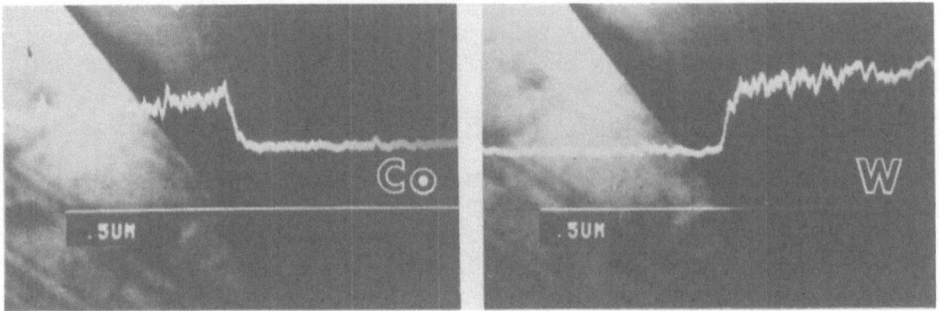

Fig. 7.   X-ray line scans by EDS in the STEM for Co(Kα) and W(Mα)
          across the binder/carbide interface in WC-11% Co thin
          foil sample.

     The same detection limitation for light element, SEM/EDS
analysis applies to STEM/EDS. Light element detection is possible
by the use of ultra-thin or "windowless" X-ray detectors but it
appears that carbon contamination on the ultra-thin window still
poses a problem in quantitation[4]. Quantitative STEM/EDS for
elements Z > 10 is performed by the Cliff-Lorimer ratio technique[5] in which the relative concentration of two elements C_A/C_B
is directly proportional to their characteristic intensity ratio
I_A/I_B:

$$\frac{C_A}{C_B} = K_{AB} \frac{I_A}{I_B} \qquad\qquad -(3)$$

The proportionality constant K_{AB} can be obtained by using suitable
standards or calculated from first principles. Equation (3) is
based on the "thin film approximation"[6] in which it is assumed
that absorption or secondary fluorescence corrections are insignificant. The ratio technique simplifies data reduction and conveniently removes the specimen thickness dependence of the measured
X-ray intensities. Only indirect light element analysis (by concentration difference) is possible by STEM/EDS.

     EELS is emerging as an effective method for light element
analysis in conjunction with STEM based on the measurement of the
energy distribution of inelastically scattered electrons transmitted through the thin foil specimen. Those electrons which have
lost a characteristic energy owing to inner shell ionization of
the atoms in the sample register as "edges" in the spectrum

superimposed over a large background signal. Since EELS reflects
the primary ionization process only, its cross section does not
decrease with lower Z compared to the fluorescence yield. In
addition, the collection efficiency is relatively high (because
the inelastic electron scattering is strongly peaked in the
forward direction), from 20 to 50% compared to about 1% for
EDS[7].

        However, a practical problem in EELS is imposed by specimen
thickness. Our limited use of EELS to date has illustrated this
difficulty. Figure 8 shows EELS spectra from a 40 nm thick amorphous carbon film and from a mixed carbide $W(Ti)C_x$ grain in a
cemented carbide 100 nm thick sample under identical instrumental
parameters. The carbon edge is seen to be riding on a much higher
background in the latter spectrum. Increasing the specimen thickness results in a linear increase in intensity of the characteristic edges and a non-linear increase in the general background
intensity, resulting in a decrease in the peak/background ratio.
This was clearly demonstrated by Zaluzec[8] for a BN specimen and
is included in Figure 9. He also showed effective use of combined
EDS and EELS in STEM studies on carbide precipitates and ceramic
samples.

        The literature in STEM/EELS indicates that analog line scans
and elemental maps of low Z elements, using loss peaks for
imaging, can be generated with a spatial resolution better than
50 nm[8,9]. Quantitation of EELS, utilizing ratio intensities of
the functional form in Equation (3), is undergoing theoretical and
experimental development. It is conceivable that this activity
will yield computer programs with accuracies approaching that of
the more established electron microprobes.

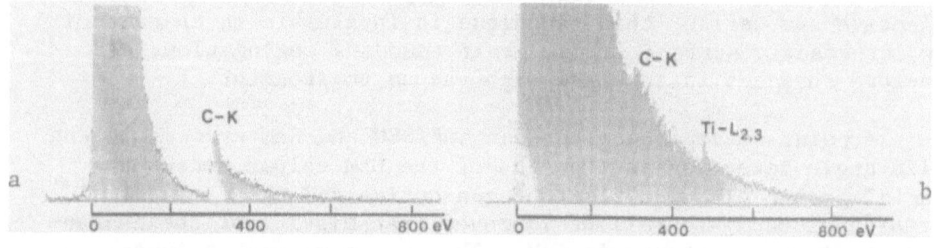

Fig. 8.   EELS spectra of a) a 40 nm thick amorphous carbon film
          b) a 100 nm thick $W(Ti)C_x$ grain in a WC-Co based cemented
          carbide sample.

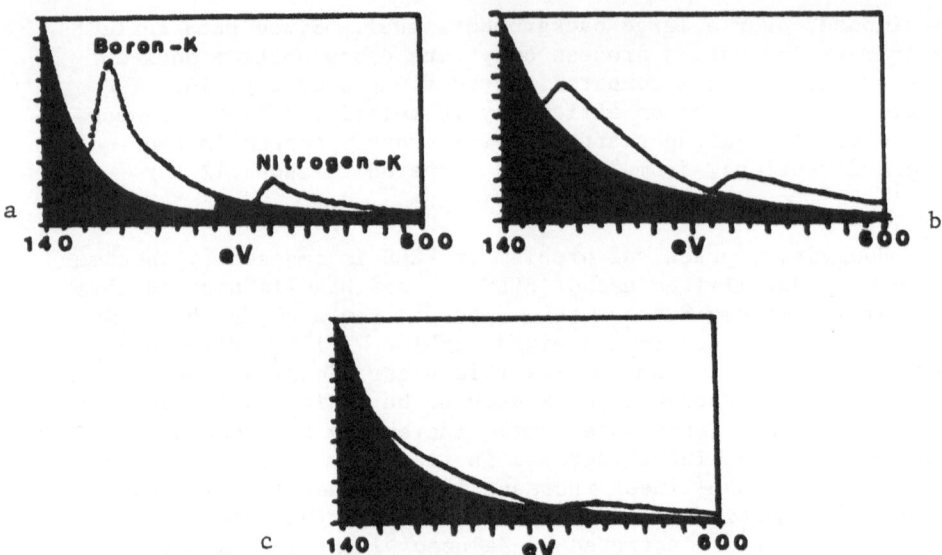

Fig. 9. Effect of specimen thickness of EELS spectra on thin foil
         sample of $BN_{(8)}$. Thickness increased from a) to b) to c).
         From Zaluzec[8].

Scanning Auger Microprobe

Auger analysis started as a macroprobe for surface chemistry
without microscopic mapping capabilities in the first commercial
instruments in the early 1970's. Rapid advances have since been
made both in electron optics and spectrometer design to the point
where the current scanning Auger microprobes can exceed the spatial chemical resolution of bulk X-ray micro-analysis. The schematic for a current scanning Auger microprobe is shown in Figure
10. Inert gas sputtering in Auger systems has traditionally been
incorporated for surface atomic layer removal with simultaneous
Auger analysis. This permits cleaning the surface of contaminants
as well as chemically depth-profiling the surface layers down to
~1 μm into the bulk. Although there may be artifacts due to
selective sputtering, this technique is invaluable in characterizing interfaces, surface films, grain boundary segregation, and
fracture surfaces in an ultra-high vacuum environment.

A point of departure between EDS/SEM and SAM systems is the
solid angle detection at the side of the SEM column versus the
coaxial geometry of the electron gun optics and the cylindrical
mirror analyzer which allows improved transmission of the characteristic Auger electrons in SAM. The sample volume in Auger
analysis (Figure 1) for a given beam diameter is limited by the

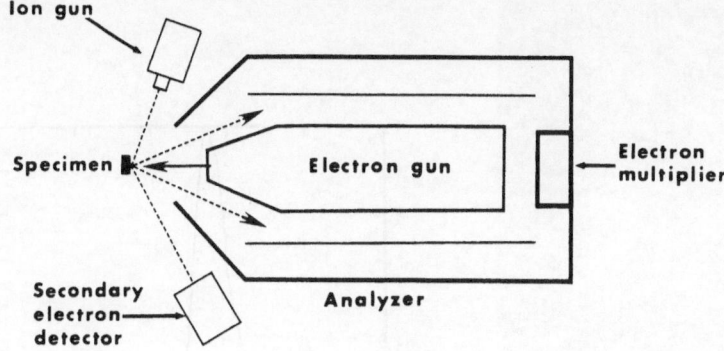

Fig. 10. Configuration for Auger electron energy analysis in SAM.

escape depth of the Auger electrons, typically 0.3 to 2.0 nm[10].
Hence, SAM sample volumes are about 3 orders of magnitude smaller
than in X-ray micro-analysis of bulk specimens. A high brightness
LaB_6 filament in the electron gun gives good current density at
focused beam diameters down to 50 nm. The important, unique
characteristic in SAM is that there is virtually no beam
spreading: the spatial resolution is roughly equal to the probe
beam size. This means that spatial resolution of 0.1 to 0.2 μm
are now routinely obtained.

     To illustrate that the probe beam size is the limiting fac-
tor in Auger spatial resolution, Auger line scans were taken on a
polished cross-section of a WC-Co substrate with a TiC CVD coating. The SAM operating parameters were 10 KV primary beam voltage
and 30×10^{-9} A beam current which give a beam diameter of
0.20 μm. The secondary electron image of this sample shows the
line of traverse in Figure 11a. The sample was sputter-cleaned to
remove surface contamination prior to analysis. The Ti_{LMM} (418
eV), W_{MNN} (1736 eV), and C_{KLL} (272 eV) Auger line scans, Figures
11b-d, were used to measure the probe function Δz at the coating/
substrate interface. The measured Δz for Ti and W are both
 0.25 μm and 0.20 μm for C--these are approximately equal to the
probe beam diameter.

     It is also of interest to examine the effective depth
resolution obtained in sputter-profiling of a CVD coating, i.e.,
simultaneous argon bombardment of the surface layers with Auger
broad beam analysis normal to the coating. Ideally, the depth
resolution is equal to the escape depth of the Auger electrons or
<2 nm. In practice the depth resolution can decrease by more than

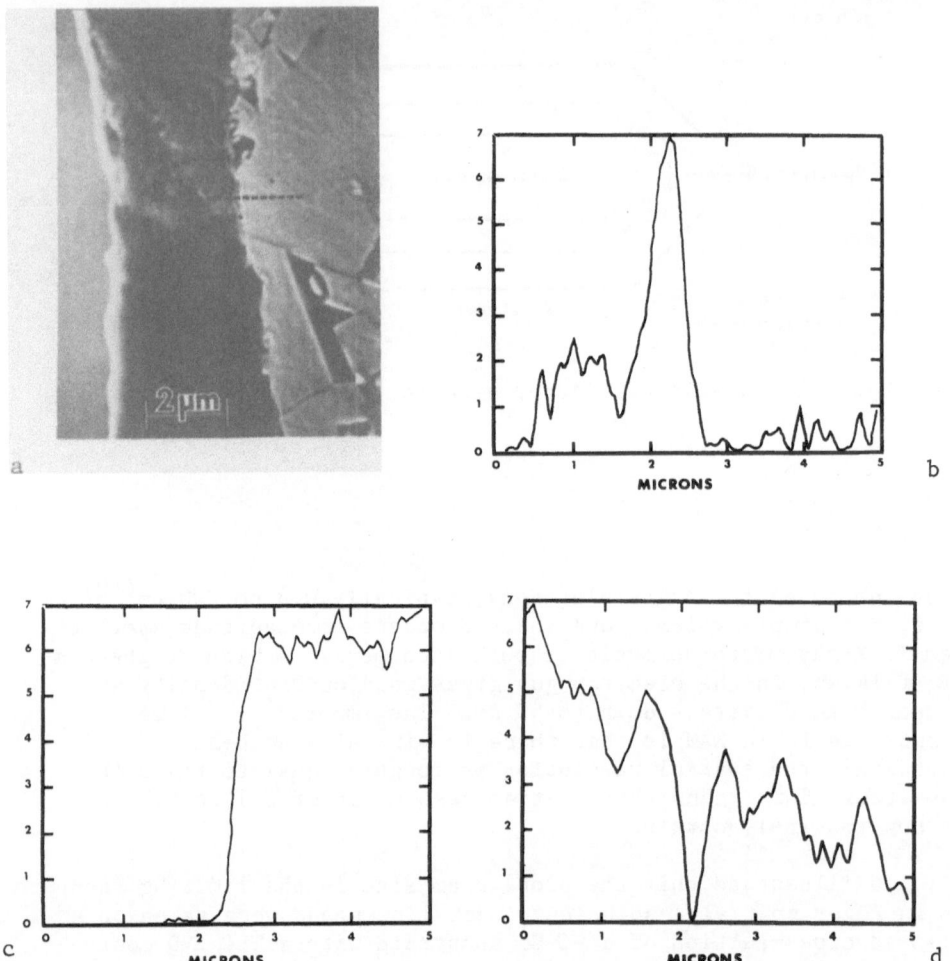

Figure 11.   Auger line scans across a substrate/coating interface in
             SAM showing a) Line of traverse in the secondary electron
             image for b) Ti_{LMM}, c) W_{MNN}, and d) C_{KLL}.

10% of the depth sputtered[11] because of non-uniform removal of
material. It has been claimed that carbide compounds such as
WC[12] or TiC[13] do not exhibit differential sputtering rates of
the constituent atoms. An implicit assumption is that the free
surface and inner interfaces considered are microscopically flat.
An Auger compositional depth profile of a TiC CVD coating on a
WC-Co substrate (similar to the previous example) is shown in
Figure 12. The sputter rate used was 20 nm/min. (calibrated with
a Ta_2O_5 thin film) giving a coating thickness of 1.80 µm from the
depth profile which agreed within 10% of the actual, measured
thickness of the TiC coating. Using the Ti_{LMM} (418 eV) profile to
determine the probe function, a value of 0.41 µm is obtained or

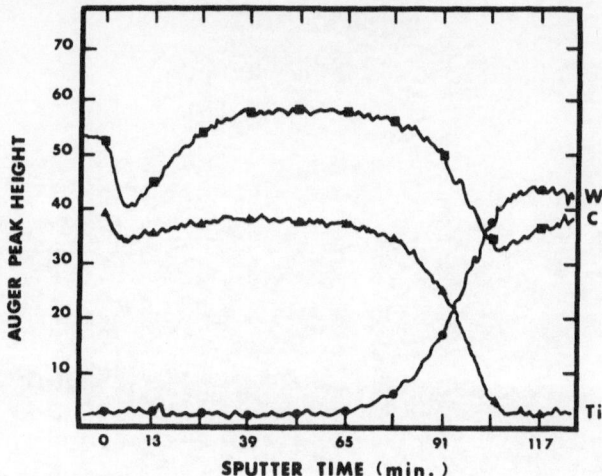

Fig. 12. Depth compositional profile of a TiC coating on WC-Co
        substrate obtained by ion sputtering and Auger analysis.

22% of the sputtered depth. This broadening of the interface is
probably a consequence of the microscopic roughness of the TiC
coating; as seen in Figure 13, the initial grain structure of the
surface is transformed to the typical cone structures, which are
an artifact of the ion sputtering process[14].

     Auger electron spectroscopy is an ideal probe for inter-
facial segregations which may only be several atomic monolayers
thick and are otherwise undetectable, e.g., by bulk X-ray micro-
analysis[15]. The thickness of such thin segregated layers can be
accurately estimated by controlled sputtering at low rates. For
thick (>1 µm) film structures such as CVD coatings, however, high
sputtering rates are required to profile through the coating/
substrate interface within a reasonable length of time (as in
Figure 12). Caution must be exercised in interpreting the apparent thickness of segregated layers revealed by rapid sputtering
because of more pronounced surface roughening and possible preferential sputtering. An alternative means of analyzing coatings of
>1 µm thickness is to use a low-angle (< 5 °) polish from the sur-
face and subsequent Auger line scans across the "extended" interfaces. Lea and Seah[16] made a comparative analysis of this technique with sputter-profiling and concluded that better spatial
resolution would be gained by low angle polishing for coatings
>1 µm thick.

     Quantitative micro-analysis by SAM is not yet as accurate as
quantitative X-ray micro-analysis, owing to the lack of a compre-

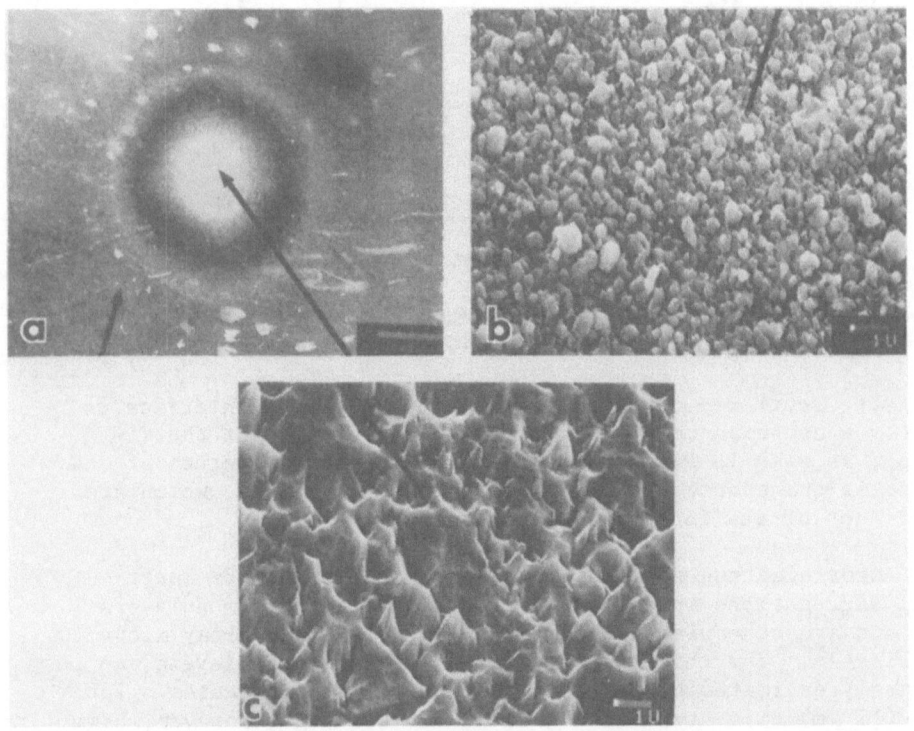

Fig. 13.   Secondary electron images of sample profiled in Fig. 12
           showing a) sputtered crater on sample surface,
           b) unsputtered coating, and c) center of crater.

hensive theoretical model analogous to the ZAF correction in X-ray
micro-analysis.   The most practical method is the use of elemental
sensitivity factors S_i obtained from reference element or compound
standards.   This entails calibration of Auger integrated intensities or derivative peak amplitudes I_i from measurements of the
unknown under identical instrumental parameters as those used for
the standard.   Atomic concentrations for an element C_x can then be
calculated from the relationship

$$C_x = \frac{I_x/S_x}{\sum\limits_i (I_i/S_i)} \qquad\qquad -(4)$$

where the summation is over all elements detected. The absolute accuracy of concentrations obtained by Equation (4) is usually no better than ± 20% although the errors in <u>relative</u> concentrations can be reduced to ±5%. This is mainly due to the neglect of matrix effects (backscattering effects and changes in the escape depth) on the Auger element signal. Of interest here is to cite what is currently available in quantitative Auger analysis with commercial instruments with no attempt to employ refinements for quantitation.

To illustrate quantitative SAM on cemented carbides, two samples from the same parent sample were examined: a polished section which was sputter-cleaned in the Auger chamber prior to analysis and a sample fractured <u>in-situ</u> for immediate examination of the fracture surface. On each sample the beam was first rastered over a 180 μm x 180 μm surface area to obtain an average composition; the beam was subsequently focused on a WC grain and a cobalt binder pool for spot composition analysis of these micro-constituents using a stationary 0.15 μm probe beam. The secondary electron images and corresponding Auger images for W, C and Co are shown in Figures 14 and 15. It is seen that there is good delineation in the Auger images of the carbide and binder phases, illustrating good spatial Auger resolution as mentioned above.

A summary of the quantitative Auger analysis results by computer* using elemental sensitivity factors S_i and integrated peak intensities for I_i in Equation (4) are given for the polished specimen in Table 3. For consistency with the previous discussion the data are presented in weight percentages after conversion from the atomic concentrations C_x. It is then assumed that the surface composition (bearing in mind that the depth of the sample volume is of the order of 1-2 nm) is equal to the bulk composition, since there is good homogeneity of microconstituents in the sample volume. It is seen that in the area analysis this agreement is within 6% for W but of the order of 30% for both C and Co. Spot analysis by SAM of the WC grain gives a deviation from the (assumed) stoichiometric composition by 20% for C and 1.3% for W. These errors indicate the deficiency in the use of elemental sensitivity factors in Equation (4). The composition of the Co binder phase obtained by SAM spot analysis is (surprisingly) within the range of binder composition expected for slowly-cooled WC-Co from sintering temperatures[17]. This probably arises from the fact that the binder phase is a Co-rich dilute solid solution where the required corrections to the elemental sensitivity factor would be minimal.

---

*MACS software program, Physical Electronics Div. of Perkin Elmer, Eden Prairie, Minnesota

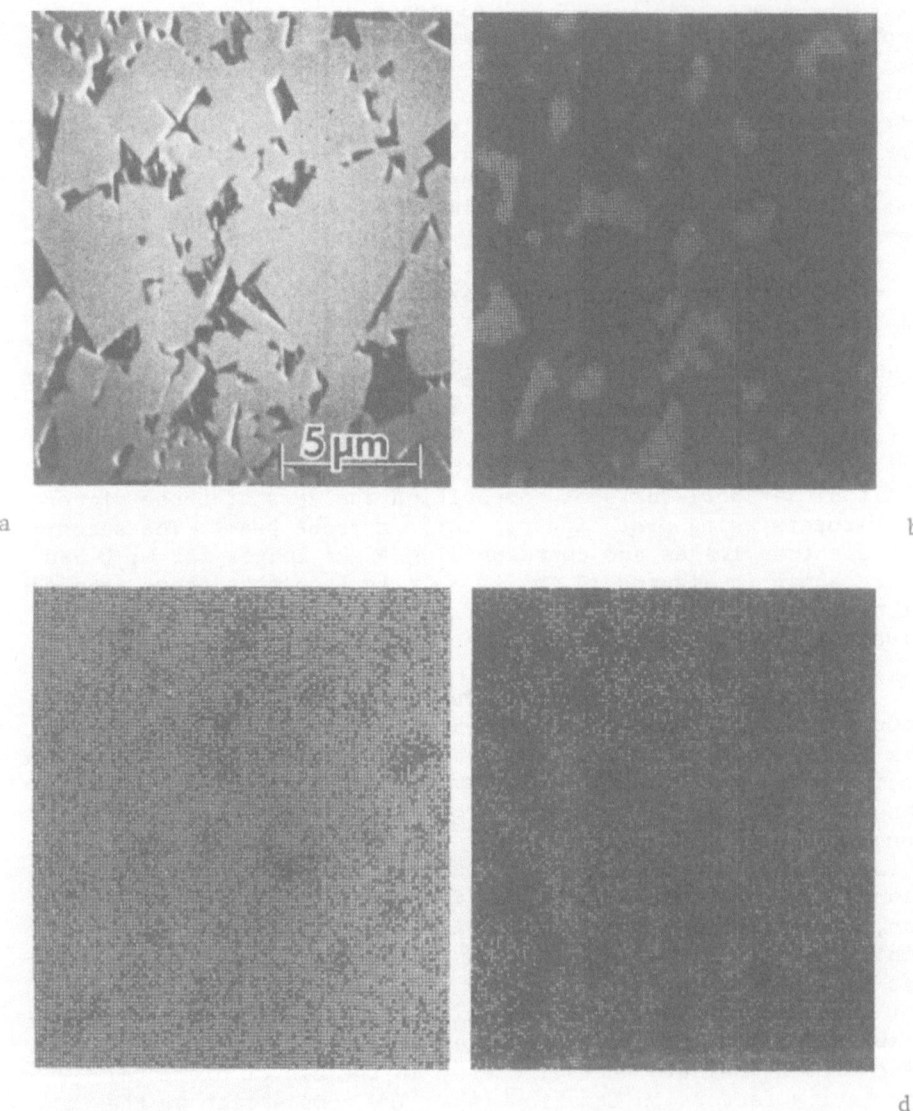

Fig. 14. SAM micrographs of WC-11% Co polished dample shown in
corresponding a) secondary electron image, b) Co Auger
image, c) W Auger image and d) C Auger image

An improvement in accuracy of quantitative AES as applied to
WC-Co has been shown by Tongson et al.[12] by calibrating the
sensitivity factors for W and C in WC relative to Co in the composite material such that the AES-determined compositions were equal
to bulk compositions determined independently by quantitative

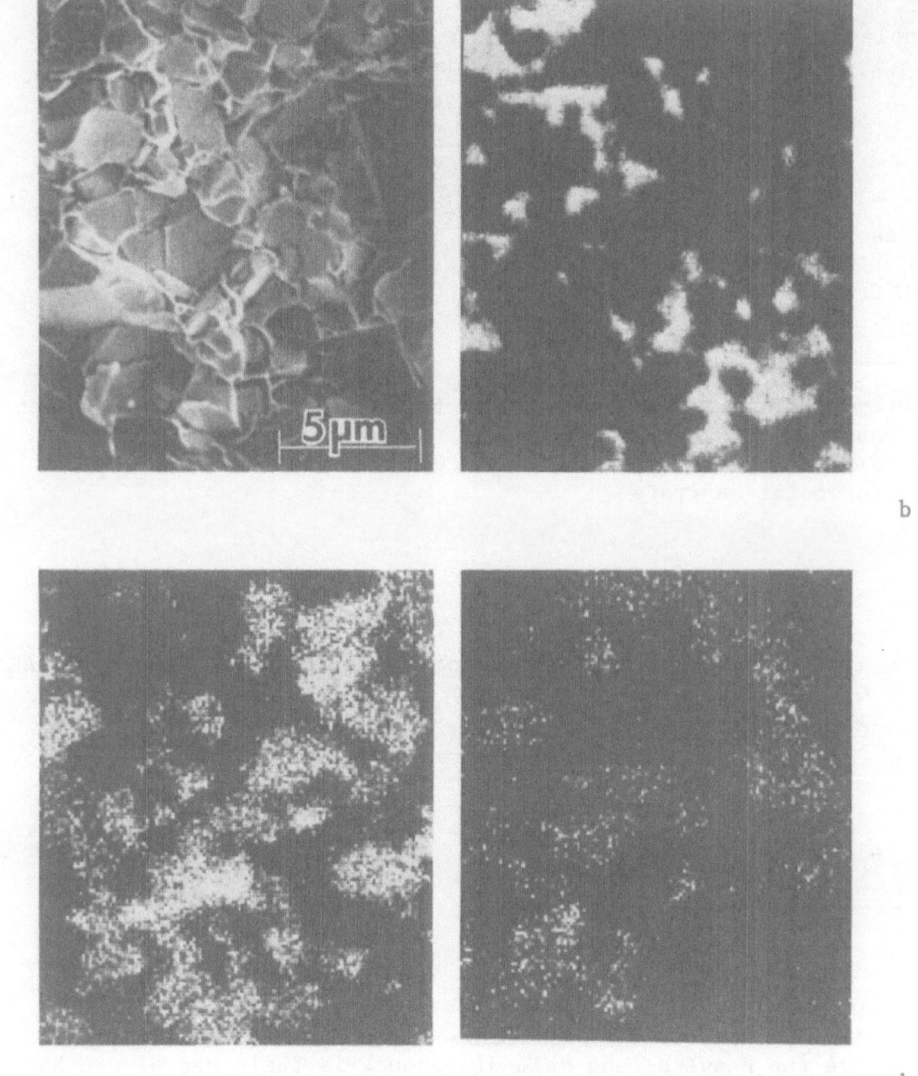

Fig. 15.   SAM micrographs of WC-11% Co fracture surface shown in
           corresponding a) secondary electron image, b) Co Auger
           image, c) W Auger image and d) C Auger image

X-ray fluorescence analysis. Use of these empirical sensitivity
factors to calculate other WC-Co compositions of powder samples
gave accuracies of ±8% of the bulk compositions. Such calibration
is obviously limited to the same instrument under identical
operating parameters. An attempt to adopt Tongson et al.'s

Table 3. Surface Composition (w/o) of Polished WC–11% Co Sample
by SAM Compared to Bulk Composition

		Surface			Bulk	
	C	W	Co	C	W	Co
Area	3.7	88.6	7.7	5.5	83.5	11.0
Co Binder	1.2	3.0	95.8	0.5[1]	3.01[1]	96.5[1]
WC Grain	4.9	95.1	–	6.1[2]	93.9[2]	–

---

[1]Equilibrium binder composition from Jonnson[17]
[2]Assumed WC Stoichiometry
  SAM Parameters: E_p=10KV, I_p=4.0x10^{-6}A for area, 20 x 10^{-9} A
  for point analysis

Table 4. Fracture Surface Composition (w/o) of WC–11% Sample by SAM

	C	W	Co
Area	5.7	67.9	26.4
Co Binder	5.7	21.2	73.1
WC Grain	13.4	81.5	5.1

SAM Parameters as in Table 3.

sensitivity factors with the SAM spectra in this work failed to
improve the results; one main difference is their use of a 3 KV
primary beam energy as opposed to 10 KV used here.

GENERAL DISCUSSION

     The state of the art in light element micro-analysis in
cermets clearly allows qualitative analysis beyond the capabilities of conventional SEM/EDS. Relative to cermet microstructures
the spatial resolution is poor in SEM/WDS but quite adequate
(~50 nm) in STEM/EELS and SAM. Accurate quantitative analysis
comparable to bulk X-ray micro-analysis, however, remains to be
developed in EELS and Auger Spectroscopy. This will probably

evolve from stringent calibration against appropriate standards close to the cermet compositions under study.

Semi-quantitative analysis with an absolute accuracy of ± 20% may sometimes not be as important as the relative accuracy of ±5% from one sample to another. As accuracies decrease with smaller concentrations the instrumentalist's concern for the minimum detectability limit - usually 1 atomic percent - becomes academic, primarily because contamination effects become significant at this concentration level and obscure interpretation. Fortunately, the elements boron, carbon, nitrogen and oxygen normally comprise 40-60 atomic percent of the cermet composition and are major constituents so that the semi-quantitative concentrations obtained are often meaningful. In the "real world" of micro-analysis aimed at solving materials problems it is often significant enough that these light elements can be identified and imaged at high magnification in relation to microstructural features.

For example, the semi-quantitative aspect of SAM coupled with the microscopic morphology characterization of the surface gives valuable information. The in-situ fracture surface of the same WC - 11% Co sample is shown in Figure 15. SAM analyses were similarly taken over a large area, a WC faceted grain surface and a Co binder pool and summarized in Table 4. The "average" fracture surface composition is seen to have a significantly higher amount of cobalt (> 3X) than on the polished surface, see Table III. This is of course a consequence of the preferential fracture path through the ductile binder rather than transgranularly through the WC grains. This corroborates the work of Viswanadham et al.[18] who made the same observation on WC-Co samples fractured outside the Auger chamber but sputter-etched to remove surface contamination. The Auger point analysis of the exposed WC grain boundary is typical of several analyses taken from various WC-Co fracture surfaces. In each case there is a detectable amount of Co, and assuming no significant topography-induced beam spreading, this tends to support the work of Sharma and Williams[19] that each WC grain is wetted by a thin film of cobalt, even at the carbide-to-carbide interfaces. The exposed binder phase on the fracture surface apparently contains more W and C than in the polished surface. A possible explanation is that the fracture path is close to the WC/Co boundary rather than through the Co phase itself; at this boundary there is higher segregation of C and W[17]. Although discussion of the significance of these analytical observations is beyond the scope of this paper it is clear that the capability of total analysis, including the light element carbon, together with morphological characterization produces insights not otherwise gained.

CONCLUSIONS

This survey of three major types of modern electron microscopes has focused on light element analysis capability and
spatial chemical resolution (Δz) which are required to fully
analyze hard materials which, typically, contain 40-60 atomic
percent of the elements B, C, N, O and fine, submicron microstructural features.   It has been shown that:

1.   SEM's have the most advanced quantitative accuracy in
     X-ray microanalysis by EDS but only for heavy ($Z \geq 11$)
     elements. With the use of WDS, light element detection
     is possible but at the sacrifice of spatial resolution
     ($\Delta z > 1.5\,\mu m$).

2.   The STEM has the smallest beam diameter which, with the
     associated decreased sample volume in the thin foil
     specimen, improves EDS spatial resolution ($\Delta z < 0.03\,\mu m$)
     for the heavy elements. Light element sensitivity is
     made possible by EELS with reportedly good resolution.
     This relatively new microanalytical method needs further
     development in quantitation particularly in terms of
     factoring out the effects of specimen thickness on the
     signal/noise ratio.

3.   SAM exhibits the highest sensitivity for the light
     element detection and adequate spatial resolution
     ($\Delta z < 0.2\,\mu m$). It also offers useful chemical depth
     profiling and in-situ fracture capabilities. Quantita-
     tive Auger analysis has only ± 20% absolute accuracy but
     is expected to improve with current theoretical develop-
     ment and use of appropriate standards.

4.   From a practical standpoint, qualitative light element
     analysis coupled with submicron spatial resolution is
     often more important than high quantitative accuracy but
     poor resolution in establishing structure/property
     relationships in hard materials.

REFERENCES

1.   S.J.B. Reed, "Proceedings IV International Congress on X-Ray
       Optics and Microanalysis, Orsay, 1965," R. Castaing, P.
       Deschamps and J. Philibert, eds., Hermann, Paris, p. 339
       (1966).
2.   S. Hoffmann, Surface and Interface Analysis, 2:149 (1980).
3.   J. W. Colby, "Proceedings of the Sixth National Conference on
       Electron Probe Analysis," p. 17 (1971).

4. L. E. Thomas, "Proceedings 38th Annual EMSA, San Francisco,"
    G. W. Bailey, ed., Claitor's Publishing Division, p. 90.
    (1980).
5. G. Cliff and G. W. Lorimer, J. Microscopy, 103:203 (1975).
6. J. Philibert and R. Tixier, J. Phys. 0 1:685 (1968).
7. D. C. Joy, "Introduction to Analytical Electron Microscopy,"
    J. Hren, J. Goldstein, and D. Joy, eds., Plenum Press,
    p. 223 (1979).
8. N. J. Zaluzec, Thin Solid Films 72:177 (1980).
9. D. Maher, "Introduction to Analytical Electron Microscopy,"
    ibid. p. 275.
10. C. J. Powell, Surf. Sci. 44:29 (1974).
11. J. F. Moulder, D. G. Jean, and W. C. Johnson, Thin Solid Films,
    64:427 (1979).
12. L. L. Tongson et al., J. Vac. Sci. Tech. 15:1133 (1978).
13. N. K. Sharma and W. S. Williams, J. Am. Cer. Soc. 62:193 (1979).
14. R. S. Berg and G. J. Kominiak, J. Vac. Sci. Tech. 13:403 (1976).
15. D. Bhattacharya and D. T. Quinto, Met. Trans. 11A:919 (1980).
16. C. Lea and M. P. Seah, Thin Solid Films 75:67 (1981).
17. H. Jonsson, "Planseeberichte Fur Pulvermetallurgie," p. 187
    (1973).
18. R. K. Viswanadham et al., J. Mat. Sci. 16:1029 (1981).
19. N. K. Sharma, W. Williams, I. D. Ward, and H. L. Fraser,
    J. Am. Cer. Soc. 63:194 (1980).

DISCUSSION

W. Williams:

The slide in which you compared the sample volume analyzed by these
various electron beam methods, did I understand you to say that the
sample volume studied by EELS is the same as that studied by EDS in
the STEM?

D. Quinto:

Yes.

W. Williams:

I don't think that's correct. The energy loss method makes use of
only forward scattered electrons. There are no other processes, just
single scattering events with a very small angle of deviation of the
electron beam; whereas the x-ray photons that are generated in those
collisions excite fluorescence in quite a large volume around the
spot. Sharma showed some time ago that can be as much as, say 200 $\overset{\circ}{A}$
even though the beam size is only about 5 $\overset{\circ}{A}$. So I think one has to
be a little cautious in making that kind of comparison. The energy

loss method does have an advantage, I think, in terms of the spatial
resolution, even for a very thin film. Also, of course, in the case
of the Auger analysis, one of the primary benefits or advantages is
the atomic scale resolution normal to the surface of the specimen.
That aspect is sometimes very helpful if one is looking at interfaces.
You can sputter off the surface and go down to the interface one
layer at a time and get very sharp concentration gradients.

E. Almond:

I may have missed it, but did you mention how you prepared your
fracture surfaces for the Auger examination?

D. Quinto:

These were fractured inside the Auger chamber.

E. Almond:

That's the same way as we've done it. We got about the same results
as you have. But I'm not sure whether different methods produce
different roughnesses, and from what I can see, yours produces a
small surface and we've done the same as you and we've produced a
smooth surface and that's easier to analyze. But I think other
methods may produce a rough surface which presents a big difficulty
for Auger examination.

E. Almond:

Yes. I see your point. The fracture surfaces generated were not as
rough as you might expect, compared to, say, ductile metals, where
you might indeed have quite a bit of roughness and might want to use
cooling before you break it to induce brittle fracture. But carbides
are inherently brittle, so we have not used any special means of
producing the fracture surfaces.

H. E. Exner:

I think this business of what is in the WC/WC grain boundaries is
discussed here a lot, and I don't want to dwell on it. I do want to
raise this point. There seem to be two types of WC/WC boundaries.
One type is high coincidence boundaries where you won't find any
cobalt. These boundaries are strong. The other type is where there
is cobalt and fracture during cracking.

W. Williams:

Yes. Thank you, Dr. Exner. I think that is a very helpful distinction. The presence of the cobalt film will help lower the surface energy in case there is a large angle of mismatch between the

two grains. For low angle boundaries, or good coincidence boundaries, that may not be necessary.

P. Sklad:

In reference to the windowless detectors, there is some speculation that they will allow us to measure quantitatively the light elements in our materials. There's still a problem there because the x-rays that are produced from those elements are being absorbed in the specimen material itself as opposed to the window in the other cases. So, it still leaves the quantification of those light elements up to some speculation.

R. K. Viswanadham:

I know that you didn't discuss the sputtering aspects. Could you tell me a little bit about how fast the cobalt decayed when you sputtered the fracture surface?

D. Quinto:

I did not include the cobalt profile in that particular case. However, I do want to make a point. If you take a look at the sputtered crater on a coating, you have a very uncharacteristic surface. In fact, this is commonly known as coning structures in ion sputtering. There are artifacts in ion sputtering that one has to be aware of. The point I was going to make was that if you try to measure thicknesses with sputtering you will effectively degrade the sharpness of the interface. What you might measure as, let's say 100 Å thickness from the probe function, may actually be half or three times less than that. So it's difficult to measure the sharpness of that deposit, unless you take into account the microstructural effects.

NEUTRON DIFFRACTION STUDIES OF CEMENTED CARBIDE COMPOSITES

A. D. Krawitz,* E. F. Drake,** R. L. DeGroot,*
C. H. Vasel* and W. B. Yelon***

*Department of Mechanical & Aerospace Engineering
 University of Missouri-Columbia
 Columbia, MO 65211

**Reed Rock Bit Co.
  Houston, TX 77001

***Research Reactor
   University of Missouri
   Columbia, MO 65211

INTRODUCTION

The appreciation of cemented carbide composite materials
(cermets) has become more sophisticated in recent years. The desire
for cermets that yield better performance and/or that contain lower
amounts of cobalt has spurred development of a variety of alloy
binders, some of which can undergo structural modification through
heat treatment. Concurrently, an increasing effort is being made
to understand at a more fundamental level the service response of
binder and carbide phases in traditional and experimental materials.
The success of such investigations depends upon analytical
approaches capable of discerning fundamental structure-property
relationships. It is the purpose of this paper to discuss one
such approach that shows promise as a new analytical tool--neutron
diffraction.

The use of neutron diffraction in the study of cermets is
predicated upon the unique properties of neutron interactions in
solids.[1] Three aspects of these interactions are of particular
importance for this application: neutrons are often highly penetrating; neutrons possess scattering cross sections which do not
decrease with increasing scattering angle; and, neutron scattering

Table 1.   Comparison of Neutron and X-Ray Scattering and
           Absorption Characteristics (X-ray f and μ values
           calculated for CuK_α at $\sin\theta/\lambda = 0.5$)

COMPARISON OF NEUTRON AND X-RAY
SCATTERING AND ABSORPTION CHARACTERISTICS

	NEUTRONS			X-RAYS		
	b	μ	†50%	f	μ	†50%
ELEMENT	$(10^{-12}cm)$	(cm^{-1})	(cm)	$(10^{-12}cm)$	(cm^{-1})	(cm)
C Gr	0.66	0.62	1.11	1.69	9.6	0.72×10^{-1}
Aℓ	0.35	0.10	7.05	5.69	131	0.53×10^{-2}
Ti	-0.34	0.45	1.55	9.12	938	0.74×10^{-3}
V	-0.05	0.56	1.25	9.63	1356	0.51×10^{-3}
Cr	0.35	0.47	1.47	10.1	1814	0.38×10^{-3}
Fe	0.96	1.12	0.62	11.5	2424	0.29×10^{-3}
Co	0.25	2.40	0.29	12.2	2980	0.23×10^{-3}
Ni	1.03	1.86	0.37	12.9	407	0.17×10^{-2}
Mo	0.69	0.48	1.44	21.6	1618	0.43×10^{-3}
W	0.47	1.05	0.66	42.3	3311	0.21×10^{-3}

cross sections do not vary systematically with atomic number. These
aspects lead to a number of possible advantages for the characterization of cermets, including:

1)   Insensitivity to surface condition and/or preparation so that
     as-produced samples may be employed.
2)   The ability to sample bulk material.
3)   An increased ability to study binder phases directly in situ.
4)   The ability to study binders with large grain sizes.
5)   The ability to examine samples before and after exposure to
     conditioning or service in the laboratory or field.

These possibilities can be better appreciated through comparison with X-rays.[2] In Table 1, scattering cross sections (b for
neutrons, f for X-rays), linear absorption coefficients (μ) and
50% absorption thicknesses for a variety of elements encountered
in cermet systems are shown. This table reveals that:

1)   X-ray scattering cross sections (amplitudes) are considerably
     greater than those for neutrons.

Table 2. Comparison of Neutron and X-Ray Absorption for
Cemented Carbide Composites (X-ray μ values
calculated for CuK_{α})

COMPARISON OF NEUTRON AND X-RAY ABSORPTION FOR CERMETS				
	NEUTRONS		X - RAYS	
MATERIAL	$\mu(cm^{-1})$	†50 % (cm)	$\mu(cm^{-1})$	†50%(cm)
W C	1.06	0 65	2542	0.27×10^{-3}
WC/10 $^W/_o$ Fe	1.07	0.65	2521	0.27×10^{-3}
WC/10 $^W/_o$ Co	1.28	0.54	2614	0.27×10^{-3}
WC/10 $^W/_o$ Ni	1.19	0.58	2192	0.32×10^{-3}

2) The scattering power for the elements shown varies, as noted
above, in an irregular fashion for neutrons while for X-rays
it increases systematically with atomic number. Thus, for
example, the ratio b_C/b_W for neutrons is 1.4 while for X-rays
f_C/f_W is 0.04. Also, the scattering power of W relative to
binder elements such as Fe, Ni and Co is considerably less
for neutrons than X-rays. In general this leads to the binder
phase being more "visible" on neutron diffraction patterns
than on X-ray patterns.

3) The values of linear absorption coefficients for neutrons are
markedly lower than for X-rays (X-ray values are sensitive to
wavelength used but the general statement remains valid).
This is reflected in the thickness of material required to
absorb 50% of an incident beam. For neutrons, the range of
thickness values varies from 3 to 70 mm while for X-rays the
range is 2 to 720 μm. The linear absorption coefficients and
50% absorption thicknesses for pure WC, WC-10 w/o Fe, WC-10
w/o Co and WC-10 w/o Ni are given in Table 2. For all these
materials it is seen that the thickness to absorb 50% of a
beam is about 6 mm for neutrons and 3 μm for CuK_{α} X-rays.
For 80% absorption these values are doubled. Since a beam
must enter and leave a sample, a beam that is 50% absorbed
would penetrate about 3 mm of a typical cermet; a beam that
is 80% absorbed would penetrate 6 mm of material. Most
samples studied to date have been 12.7 mm diameter cylinders
that are rotated. Scattering from the entire volume of such
a sample is occurring. About 85% of the intensity comes
from the outer half of the cylinder, which comprises 75% of
the volume. Another way to view this is to consider a flat
plate geometry. Neutron and X-ray beams of 1 x 1 cm cross

sections incident on flat plate specimens would sample 0.87 and 4×10^{-4} cm^3 of cermet material, respectively, i.e., the neutrons would sample 2200 times more material.

In the following sections, the potential of neutron scattering as a tool for the study of cermets is explored in a variety of applications. Binders studied to date include Co, Co-Ni, Ni, Fe-Ni and Ni-Al. Aspects investigated include phase analysis; phase transformations due to service and mechanical treatment; texture; binder stress state; and, binder deformation. The results and discussion are divided into two parts, a series of four illustrative examples and two more extensive project summaries. The illustrative examples are intended to demonstrate the advantages listed above; familiarize the reader with typical data; and, indicate the variety of systems and problems that can be addressed with the method. The more extensive cases presented deal in greater detail with the physical implications of the neutron results.

All data has been taken at the University of Missouri Research Reactor (MURR), a 10 MW light water reactor.

ILLUSTRATIVE EXAMPLES

WC-Ni

Two neutron scans of WC-15.6 w/o Ni samples[3] are presented in Fig. 1. The samples are pre-sintered and sintered, respectively. In the pre-sintered sample, which has been "dewaxed" at 700°C , the Ni powder is still heavily deformed due to milling. This is manifested in broader peaks and unresolved overlaps with WC peaks. After sintering the binder peaks sharpen, shift, and increase in integrated intensity, i.e., the area under the peaks increase.

The sharpening is due to elimination of plastic deformation during sintering. The shift is due to diffusion of W into solid solution and the establishment of differential thermal residual stresses well-known to be present in cermets.[4] Since the coefficient of thermal expansion for the binder is considerably larger than for the carbide, the binder is placed in tension and the carbide in compression. The fact that the peaks shift while remaining sharp suggests that the residual stresses are largely hydrostatic in nature. If the stresses are neglected, available lattice parameter data[5] for the Ni-W system indicate that about 9 a/o W has entered the Ni matrix. If an average tensile stress of 690 MPa (100 Ksi) is assumed to be present in the binder,[4] it would account for about half of the peak shift so that only 5 a/o W would be in solution. This is presumably closer to reality but the issue cannot be resolved with these data alone. The increase in integrated peak intensity arises from the addition of W atoms to the

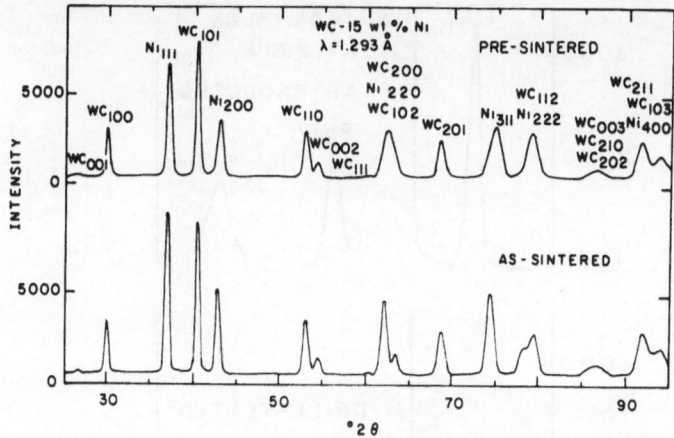

Fig. 1.   Neutron diffraction patterns of WC-15.6 Ni samples in
pre-sintered and sintered conditions.

binder without back diffusion of Ni, i.e., upon sintering a growth
in the volume of binder occurs. In this case the volume of binder
increases about 10%, from 26 to 29 v/o.

Measurements to determine if texture is present in either the
carbide or binder phase have also been made.[5] Cylindrical samples
were employed and texture was measured over a full hemisphere of
orientations centered around the cylinder axis. An as-sintered
sample as well as samples subjected to zero-compression-zero fatigue
at stresses up to -1380 MPa (-200 Ksi) were used.[3] In all cases,
the carbide phase showed no evidence of deviation from a uniform
grain orientation distribution. The binder phase showed irregular
deviations from uniformity that varied from one pole figure to the
next, and with load level, but in no discernibly systematic way.
This seems at least partly due to occasional large binder grains
but some effect due to loading may also be a contributing factor.
This is supported by corresponding integrated intensity data which
shows an increasing deviation from ideal behavior with fatigue
load level.

WC-Co

A matched set of WC-14.5 w/o Co rock-drilling inserts was
examined. One was in the as-produced, unused state while the other
had been used to drill through about 762 m (2500 ft.) of iron ore.
The results are shown in Fig. 2. The binder of the as-produced
insert is retained in the high temperature FCC (α) phase typical of
as-sintered WC-Co cermets. However, the drill-tested insert shows
that about 35% of the FCC binder has transformed during service to

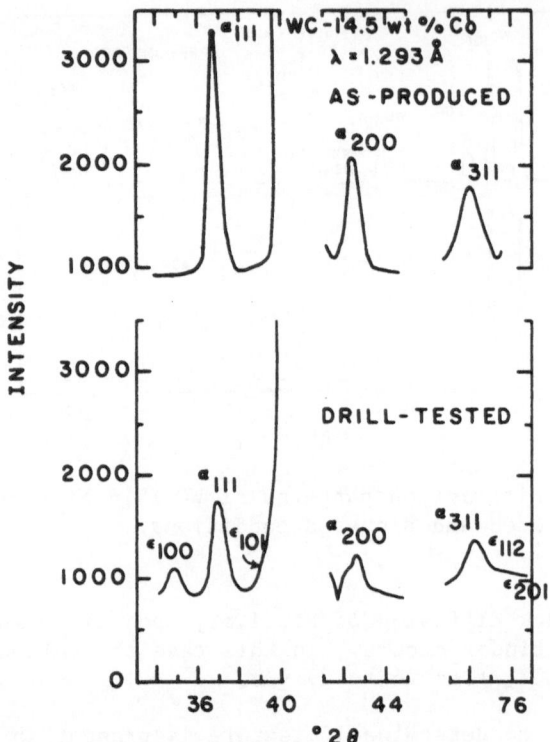

Fig. 2.   Neutron diffraction scans showing binder peaks in a
          WC-14.5 Co rock bit before and after service.

the HCP (ε) phase due to mechanical deformation. This transformation, which is martensitic, is well documented in Co[6] and Co
alloys[7-9] and has come to be appreciated more recently in Co base
cermet binders.[10-14]

The HCP is principally visible as the 100 peak, however, it
is also manifested in the low-angle shoulder of the strong WC peak
at 40° 2θ and at the FCC 311. Note also the corresponding decrease
in FCC peak intensity and the irregular shape of the FCC 200 peak,
a result, presumably, of plastic deformation.

WC/Fe-Ni

Samples containing 8.5 wt. pct. of an Fe-20 w/o Ni binder were
scanned in three states: (1) as-prepared, (2) quenched in liquid
nitrogen, and (3) quenched and tempered for 2 hrs. at 200°C. This
heat-treatable binder[15] shows property changes with thermal treatment and was the subject of an earlier X-ray study.[16] The features
of the earlier study were confirmed in the bulk using neutrons; see
Fig. 3. The binder is initially predominantly austenitic. Upon

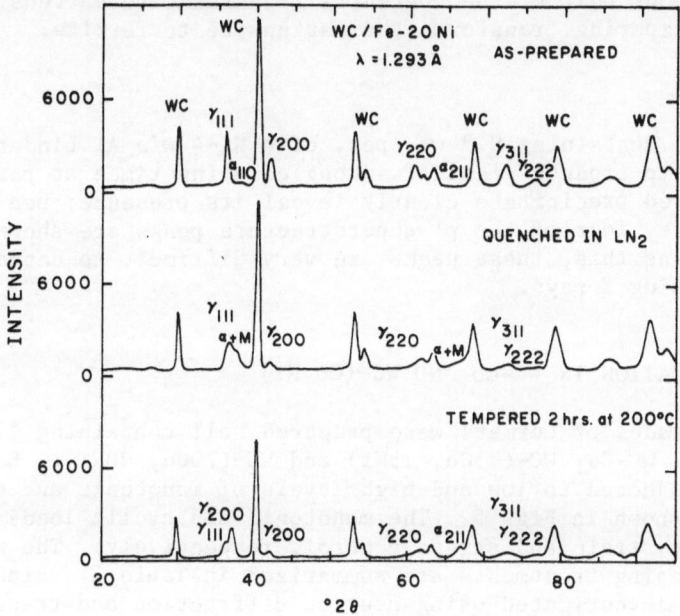

Fig. 3. Neutron diffraction patterns from WC-(80Fe,20Ni) in as-
sintered, LN$_2$ quenched and quenched and tempered states
(γ is austenite, α is ferrite and M is martensite).

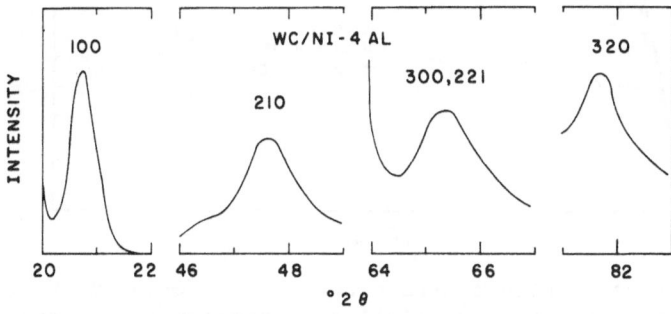

Fig. 4. Neutron diffraction scans of four superstructure peaks
from the γ'-Ni$_3$Al ordered phase in the binder of a
WC-(96Al,4Ni) cermet.

quenching, about half of the austenite transforms to martensite. Subsequent tempering transforms the martensite to ferrite.

WC/Ni-Al

A sample containing 9.3 wt. pct. of a Ni-4 w/o Al binder was heat-treated to produce γ'-Ni$_3$Al. Long counting times at positions for the ordered precipitate clearly reveal its presence; see Fig. 4. In this figure, four of the γ' superstructure peaks are shown. In cermets such as this, these peaks are very difficult to detect above background using X-rays.

BINDER DEFORMATION IN WC-Co AND WC-(Co,Ni)

Three grades of cermets were prepared, all containing 17 wt. pct. binder: WC-Co, WC-(85Co, 15Ni) and WC-(70Co, 30Ni). Each grade was subjected to low and high levels of monotonic and cyclic loading, as shown in Fig. 5. The monotonic and cyclic loads were applied under strain and stress control, respectively. The mechanical conditioning treatments are summarized in Table 3. Binder response was investigated using neutron diffraction and transmission electron microscopy (TEM).[17,18]

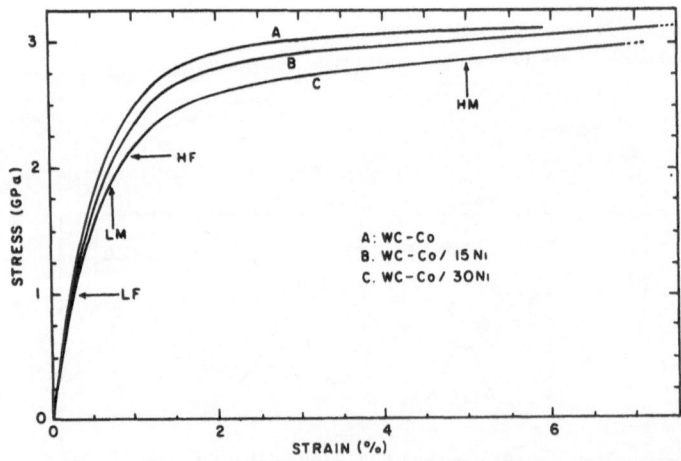

Fig. 5.   Compressive stress-strain curves for WC-Co and WC-(Co,Ni) cermets showing mechanical conditioning treatments: low fatigue (LF), low monotonic (LM), high fatigue (HF) and high monotonic (HM).

Table 3. Summary of Mechanical Conditioning Treatments
for WC-Co and WC-(Co,Ni) Cermets

MECHANICAL TREATMENT	STRESS/STRAIN LEVEL	NUMBER OF CYCLES	PLASTIC STRAIN (%)
LOW FATIGUE (LF)	1.0 GPa	5×10	0.04 - 0.09
HIGH FATIGUE (HF)	2.1 GPa	5×10^5	0.3 - 0.6
LOW MONOTONIC (LM)	0.75 %	1	0.25 - 0.3
HIGH MONOTONIC (HM)	5.0 %	1	~4.5

The amount of HCP created by strain-induced transformation of
the metastable FCC binder is shown in Fig. 6. These data were
obtained by quantitative measurement of the reduction in FCC binder
peak integrated intensity. In Fig. 7, neutron spectra examples for
as-sintered, high fatigue and high monotonic specimens of WC-Co are
shown. The results are consistent with the Co-Ni phase diagram and
studies of bulk Co-Ni alloys. It is interesting to note that, for
one fatigue cycle (nominally, the low monotonic treatment), a considerable amount of HCP is produced for the pure Co and Co-15 w/o
Ni samples; see Fig. 6. Electron microscopy revealed that, as the
FCC phase is stabilized, slip and twinning become the prevalent
modes of binder deformation.

Analysis of binder peak positions on the neutron diffraction
patterns revealed a systematic shift to lower angles. This is
interpreted as arising from a relaxation of the tensile residual
stress characteristic of cermet binders as discussed above for
WC-Ni. The data are summarized in Table 4. The origin of the
relaxation is probably plastic deformation of the binder and/or
microcracking of the carbide.

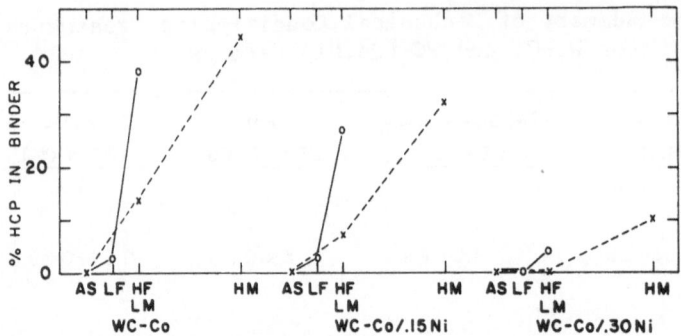

Fig. 6.   Amount of low-temperature HCP binder phase produced as a
          function of mechanical treatment for WC-Co and WC-(Co,Ni)
          cermets (AS: as-sintered).

        The widths of binder peaks generally increased with severity
of load, indicating an increasing level of plastic deformation.
The results are summarized in Table 5.   The data is presented as
change in peak breadth relative to the as-sintered samples.   The
broadening appears to decrease with increasing Ni concentration.

Table 4.   Stress Relaxation Due to Mechanical Conditioning
           in WC-Co and WC-(Co,Ni) Cermets

	STRESS RELAXATION (MPa)			
SAMPLE	LOW FATIGUE	HIGH FATIGUE	LOW MONOTONIC	HIGH MONOTONIC
WC-CO	$255 ^{+124}_{-200}$	$455 ^{+207}_{-131}$	$241 ^{+28}_{-55}$	$669 ^{+117}_{-207}$
WC-(CO,15NI)	$214 ^{+145}_{-159}$	$490 ^{+200}_{-166}$	$324 ^{+90}_{-55}$	$620 ^{+138}_{-159}$
WC-(CO,30NI)	$269 ^{+255}_{-214}$	$552 ^{+207}_{-172}$	$310 ^{+21}_{-41}$	$607 ^{+41}_{-55}$

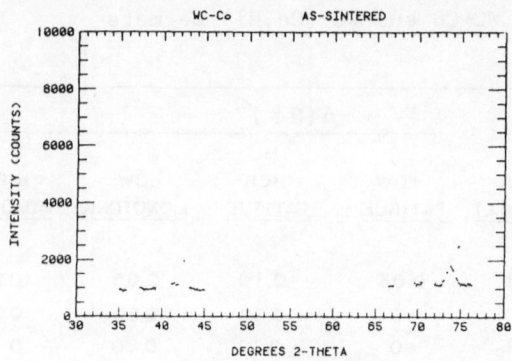

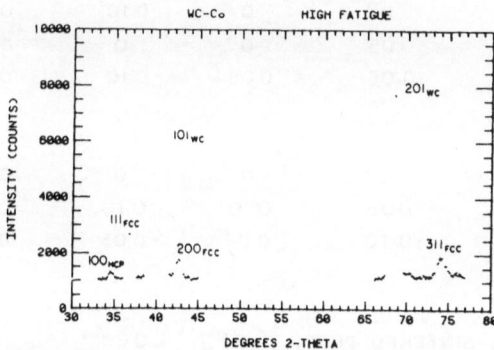

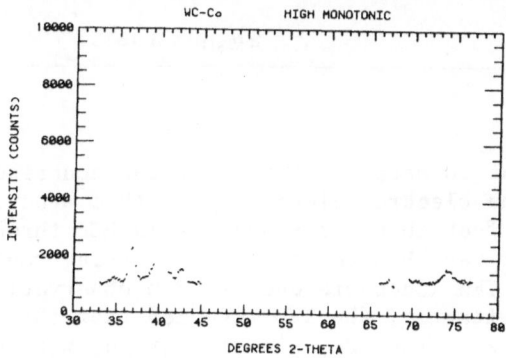

Fig. 7. Neutron diffraction scans for WC-Co in as-sintered, high
         fatigue and high monotonic states. Only regions with
         clear binder peaks are shown (λ = 1.293 Å).

Table 5. Full Width at Half Maximum (FWHM) Values of
          Binder Peaks Due to Mechanical Conditioning
          in WC-Co and WC-(Co,Ni) Cermets

SAMPLE	BINDER PEAK (HKL)	$\Delta(2\theta)°$			
		LOW FATIGUE	HIGH FATIGUE	LOW MONOTONIC	HIGH MONOTONIC
	111	0.05	0.10	0.05	0.20
WC-CO	200	0	0.10	0.20	0.30
	311	0	0.10	0.10	0.30
	111	0	0	0.10	0.15
WC-(CO,15NI)	200	-0.05	0	0	0.40
	311	-0.05	0.05	-0.10	0.25
	111	0	0	0	0.10
WC-(CO,30NI)	200	0.05	0.10	0.10	0.25
	311	-0.10	0.0	-0.05	0.05

FWHM VALUES FOR
AS-SINTERED PEAKS $2\theta_{111} = 0.60°$

                      $2\theta_{200} = 0.70°$

                      $2\theta_{311} = 1.15°$

It is important to note the nature of the contributions of the
neutron data and the electron microscopy to this study, and their
relationship. The fact that the reduction in HCP through addition
of Ni to the binder was observed in the bulk using neutrons then
corroborated using TEM lends credence to TEM observations in such
materials. Concommitantly, the TEM revealed that the reduction in
HCP, in the presence of the same plastic strain, was accomplished
through slip and twinning processes. Furthermore, only the bulk
sampling capability of the neutron measurements could enable discovery of the relaxation of bulk hydrostatic residual stresses in
the binder.

PHASE/STRESS ANALYSIS OF WC-Co ROLLS

New and used WC-16 w/o Co die rolls that had experienced premature failure by fracture were investigated.[19] It was thought that an unfavorable macroscopic residual hoop stress pattern or the presence of undesirable binder phases due to manufacture or use at elevated temperatures might be present. A schematic of the rolls is presented in Fig. 8a. Two sizes were studied in the as-received condition, 15.2 cm OD x 6.35 cm high and 20.3 cm OD x 5.7 cm high. The three rolls used are shown in Fig. 8b. A stage and an alignment device were constructed to position the (heavy) rolls on the diffractometer.

Residual stress measurements were made using the two-exposure methodology developed for X-ray diffraction.[20] Measurements were principally made on contact surfaces for both binder and carbide constituents. No consistent evidence of residual stresses was found and it was concluded that any stresses were considerably less than 690 MPa (100 Ksi). It is noted that the two-exposure method is not sensitive to bulk hydrostatic stresses. Its use in the present case was predicated on the assumption that detrimental circumferential (hoop) stresses might be present. The method employed is ideally suited for the detection of such stresses.

The neutron beam penetrates a few millimeters below the surface so that surface preparation is unnecessary and surface condition is not critical as it is for measurements with X-rays. In fact, X-ray measurements were also made. They indicated that very shallow (a few microns) stresses, presumably induced by grinding, were present on as-ground surfaces. These stresses were compressive and ranged from -690 to -1380 MPa (-100 to -200 Ksi). Grinding stresses were not found on surfaces exposed to thermal/mechanical service for extended periods or on internal fracture surfaces.

Neutron diffraction scans of several surfaces of the three rolls were made to investigate phases present. These included contact surfaces, bottom surfaces, work grooves and fracture surfaces. All patterns revealed WC and Co binder in the FCC state with the exception of the work groove of the most extensively used roll, which revealed that some HCP phase had formed in the binder. Examples of a full scan from a new roll and the scan indicating HCP are presented in Fig. 9.

The results of this study were negative in that neither macroscopic residual stresses nor detrimental phases were found to be responsible for roll failure. It did, however, establish that data can be taken on massive samples in as-received condition.

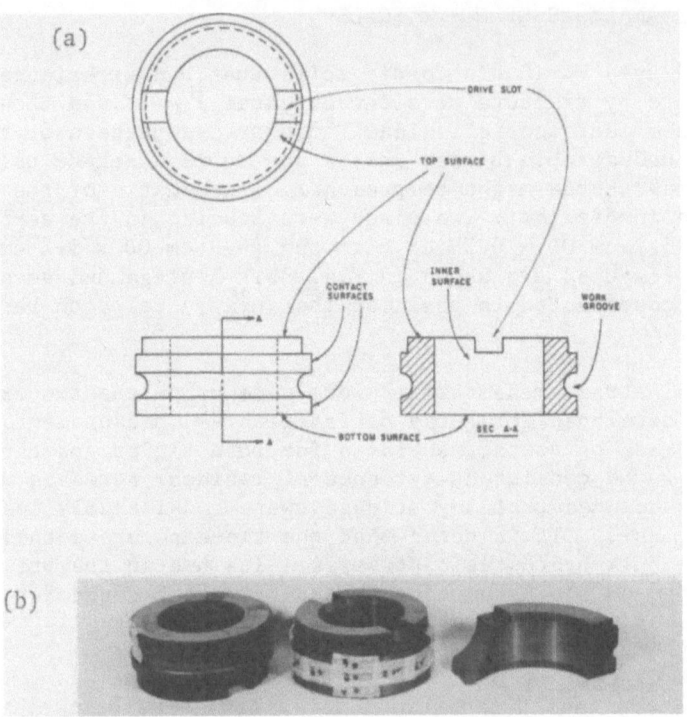

Fig. 8.   WC-16 Co die rolls. (a) Schematic of roll geometry.
(b) The three rolls studied. Roll condition left to
right: extensive service, new, and failed after brief
service.

CONCLUSIONS

Neutron diffraction has been successfully employed to study a
number of aspects, including phase relations and transformations,
texture, stress state and deformation, in a wide variety of cemented
carbide composite materials. In all cases bulk samples have been
used without any special preparation. They have been examined in
pre-sintered and sintered states; before and after thermal and
mechanical treatments; and, before and after actual service. The
technique is particularly well suited to characterization of binder
constituents.

Perhaps the most interesting aspect considered to date is that
of microscopic and macroscopic stress state and response. This is
principally due to the fact that large volumes of bulk material are

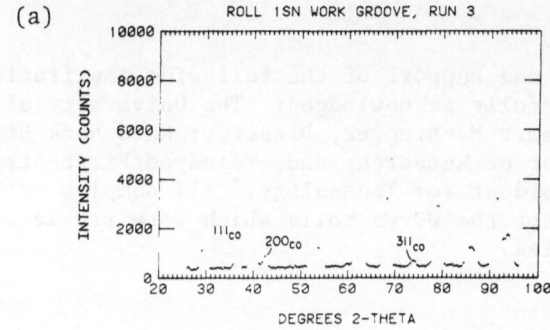

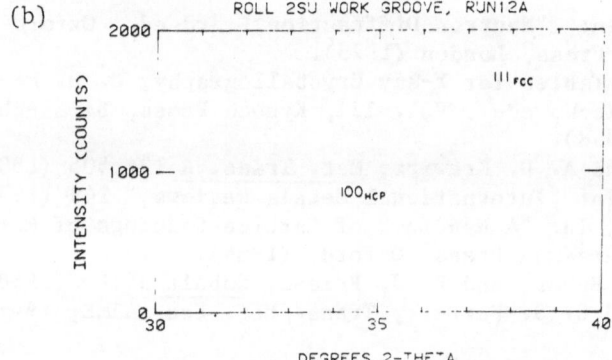

Fig. 9.   Neutron diffraction patterns from rolls (λ = 1.293 Å).
(a) Presence of HCP 100 peak next to FCC 111. From work
groove of extensively used roll. (b) Full scan from new
roll (see Fig. 1 for indexing of WC peaks).

sampled by neutrons rather than the near-surface regions to which
X-rays are confined. Observations in pre-and post-sintered WC-Ni
samples suggest that the equilibrium microscopic residual stress
state established between carbide and binder is hydrostatic in
nature when averaged over many binder regions. Thus an attempt to
apply the conventional two-exposure technique of residual stress
measurement to a volume of cermet material should reveal no net
macroscopic stress. If, however, a macroscopic stress is superimposed on the material, it would, in principle, be measurable,
even if it were triaxial in nature.[21]   To date no macro-residual
stress has been observed. Such a state was sought in the WC-Co
rolls but was not found in the bulk. It was, however, observed at
the surface due to grinding. Finally, uniaxial plastic deformation
of WC-Co and WC-(Co,Ni) cermets has been observed to alter the
microscopic residual stress state between carbide and binder. The
alteration is in the form of a progressive relaxation with increasing
monotonic or cyclic loading with, however, apparent preservation
of the hydrostatic character on the average.

ACKNOWLEDGMENTS

    The cooperation and support of the following institutions and
individuals are gratefully acknowledged: The University of Missouri
Research Reactor, Robert M. Brugger, Director; Reed Rock Bit Co.,
J. A. Peck, Supervisor of Research; and, Teledyne/Firth-Sterling,
T. Penrice, Vice-President for Technology.   All samples were provided
by Reed Rock Bit except the WC-Co rolls which were provided by
Teledyne/Firth-Sterling.

REFERENCES

1. G. E. Bacon, in: "Neutron Diffraction," 3rd ed., Oxford
      University Press, London (1975).
2. International Tables for X-Ray Crystallography, C. H. MacGilavry
      and G. D Rieck, eds., Vol. III, Kynoch Press, Birmingham,
      England (1968).
3. E. F. Drake and A. D. Krawitz, Met. Trans. A 12A:505 (1981).
4. H. E. Exner, in: "International Metals Reviews," 149 (1971).
5. W. B. Pearson, in: "A Handbook of Lattice Spacings of Metals and
      Alloys," Pergamon Press, Oxford, (1964).
6. A. Giamei, J. Burma, and E. J. Friese, Cobalt 39:88 (1968).
7. J. B. Hess and C. S. Barrett, Trans. Met. Soc. AIME, 194:645
      (1952).
8. L. Remy, Acta Met. 26:443 (1978).
9. G. Chalant and L. Remy, Acta Met. 28:75 (1980).
10. A. Krawitz, in: An Initial Study of Cemented Carbides by Neutron
      Diffraction," report to Reed Rock Bit Co., Houston, TX (1980).
11. R. Arndt, Z. Metallkunde 63:274 (1972).
12. V. K. Sarin and T. Johanneson, Metal Science 9:472.
13. H Jonsson, Plansseeber. Pulvermet 26:108 (1976).
14. S. M. Brabyn, R. Cooper, and C. T. Peters, in: "Proc. of 20th
      Plansee Seminar," 2:675 (1981).
15. D. Moskowitz, M. J. Ford, and M. Humenik, Int. J. Powd. Met. and
      Powd. Tech. 6:55 (1970).
16. A. Krawitz, E. F. Drake, and J. Peck, Int. J. Powd. Met. and
      Powd. Tech. 15:67 (1979).
17. C. H. Vasel, The Role of Binder Deformation Mechanisms on
      Mechanical Properties in WC-Co Cemented Carbide Composites,
      M.S. Thesis, University of Missouri, Columbia, MO (1981).
18. E. A. Kenik, A. D. Krawitz, and E. F. Drake, in: 39th Ann. Proc.
      Electron Microscopy Soc. Amer." 56 (1981).
19. A. Krawitz, in: A Developmental Study of Tungsten Carbide-Cobalt
      Rolls Using Neutron Diffraction," Report to Teledyne/Firth
      Sterling, Nashville, TN (1980).

20. H. P. Klug and L. E. Alexander, in: "X-Ray Diffraction Procedures for Polycrystalline and Amorphous Materials," 2nd ed., John Wiley and Sons, New York 755 (1974).
21. H. Dolle and J. B. Cohen, Met. Trans. 11A:159 (1980).

DISCUSSION

C. H. de Novion:

Were your measurements made with a single counter?

A. Krawitz:

The only thing I've had, to date, is a single counter although we have a multicounter and we hope, of course, to move into position sensitive or area detectors.

C. H. de Novion:

The linear multidetectors like the ones in Grenoble can measure simultaneously with a resolution of 0.1°2θ. And so the kinetic experiments during loading, you could have measurements nearly every minute.

A. Krawitz:

We are aware of the fact that we need to improve our detection capability. We are in the process of doing that. If nothing else, it would cut down the time for scans tremendously. We don't have the resources of Grenoble. As you may know, Grenoble uses more resources than the entire governmental support for the whole reactor program in the U.S.

C. H. de Novion:

Perhaps it's useful to mention also that high temperature experiments are much more easy with neutrons than with x-ray.

A. Krawitz:

In fact, he has brought up a very important point. Cold stages, hot stages, environmental chambers, or high pressure experiments are all considerably easier with neutrons. A very interesting and a very useful feature.

CONTINUOUS TOOL WEAR MONITORING IN TURNING

J.G. Baldoni, S.T. Buljan, and V.K. Sarin

GTE Laboratories, Inc.
40 Sylvan Road
Waltham, MA   02254

INTRODUCTION

Quantitative characterization of cutting tool wear in turning is critical to cutting tool performance evaluation. It is therefore highly desirable to have a simple, yet accurate method of measuring tool wear during such an operation.

Conventional methods of measuring tool wear usually require removal of the tool from the machine for microscopic measurement of the wear lands or crater. While in theory wear land measurement should be a relatively simple procedure, in practice such measurements are subject to operator judgement due to the shape variability and frequently encountered lack of edge definition of these lands. This can lead to large inaccuracy in this type of tool wear testing. Furthermore, repeated tool removal from and re-installation in the machine introduces unpredictable effects due to deviations on precise relocation of the working area of the tool relative to the workpiece. This technique is time consuming and cannot be automated, often resulting in limiting the number of test trials and further impairing the statistical quality of the results obtained. A tool wear monitoring system which dynamically and accurately registers physical changes in the condition of the cutting edge of a tool during the turning operation would do away with many of these problems and greatly improve tool wear measurements.

In industrial applications a common criterion of tool wear is loss of part tolerance, i.e. increase in workpiece radius to the point where the part is oversized. The increase in radius or decrease in distance between the tool post and the machined surface reflecting loss of tool material due to wear can be measured by means of pneumatic gauges, electric feeler micrometers and ultrasonic methods.

Experimental Set-Up

The tool wear monitoring system developed and in use at GTE Laboratories consists of a non-contacting differential pressure air gauge which is mounted in the tool post of a 40 h.p. American Tool CNC lathe. The gauge nozzle is positioned so that it is coincident with the cutting point of an unworn tool with respect to the center-line of the workpiece in close proximity to the point of the cutting action (see Figure 1).

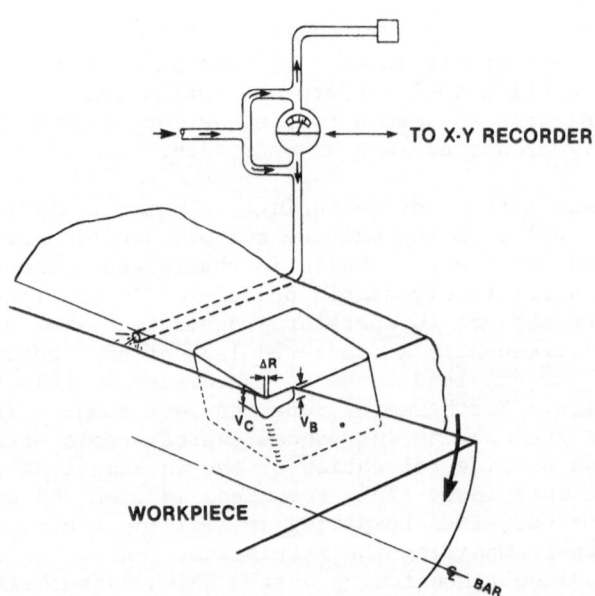

Fig. 1.   Tool Wear Monitoring System

The airguage functions in a manner similar to a Wheatstone Bridge. Regulated, pressurized air is divided into two channels. The measuring channel is balanced against the reference channel to

obtain the zero gauge reading with the gauge nozzle positioned close
to the machined surface. As more machined surface is generated the
gauge monitors the distance from the nozzle face to the machined surface. A decrease in that distance caused by tool nose wear produces
a positive deviation. The gauge output during a cut, henceforth
referred to as ΔR, is recorded continuously as a function of time
in the cut. The measured nose wear parameter, V_c, is geometrically
related to increase in workpiece radius. The flank wear parameter,
V_B, is usually measured to determine the wear rate of a cutting tool,
since the flank wear land is generally the better defined and more
easily measured wear parameter available for optical measurements.
The wear on the tool flank can contribute to catastrophic failure
of the tool but it does not directly determine the final radius of
the machined workpiece in straight turning. Therefore, ΔR is, in
general, not a measure of flank wear. With the typical air pressure
used, 100 psi, the gauge exerts virtually no force on the workpiece
and unlike mechanical contacting devices, contributes no vibration[3].

All reported results were obtained using right hand cutting
with a 15° lead angle and 5° negative rake angle.

RESULTS AND DISCUSSION

Figure 2a shows the positive change in ΔR associated with increasing tool nose wear. In this instance, a 4340 steel workpiece
was cut with a composite ceramic (Al_2O_3 + TiC) cutting tool using
the cutting conditions shown. The worn tool is shown in Figure 2b.
The low wear rates and associated small wear lands encountered when
testing ceramic cutting tools makes optical wear land measurements
very difficult and thus subject to a higher degree of error [9]. By
monitoring ΔR, the tool wear behavior is easily and continuously
observed. Initially the wear rate is high but then declines as the
steady state wear rate is attained after the tool is broken in.
Similar behavior is generally observed in microscopically monitored
tool wear experiments [4,5]. However, the break-in period may be ill-
defined since wear lands are usually measured at specific intervals
of time in the cut. Also, the procedure of removing and reinserting
the tool in the holder for such measurements may add to the extent
and variability of the break-in period. We must emphasize the Figure 2a was generated in the time interval shown and did not require
time consuming wear land measurements.

In a cutting tool development program it is at times necessary
to perform classically measured wear tests for very short total cutting times due to the large number of experimental grades which must
be tested. As shown in Figure 3, this procedure should be approached
with extreme caution. The tools used were experimental ceramic all

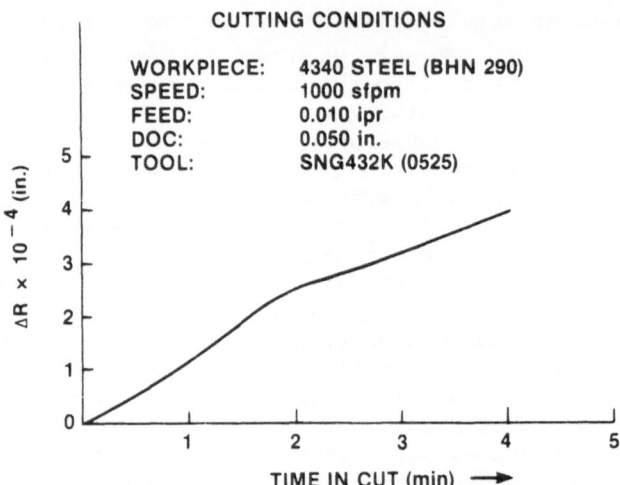

Fig. 2a.   Measured Change in Workpiece Radius Due to Tool Wear

Fig. 2b.   Appearance of Worn Ceramic Tool After Four Minutes of
           Cutting

prepared under identical conditions and ground to the proper geometry, SNG438, with identical K-lands (0.007 in. at 20°).   Edge
preparations A and B were slight modifications of the as fabricated
K-land.   The break-in of the modified tools occurred during the first
few seconds of cutting with very little change in workpiece radius
after which steady-state wear was observed.   The unmodified tools
cut for about one minute before break-in was completed and a large
ΔR was associated with this break-in period.   The steady state wear

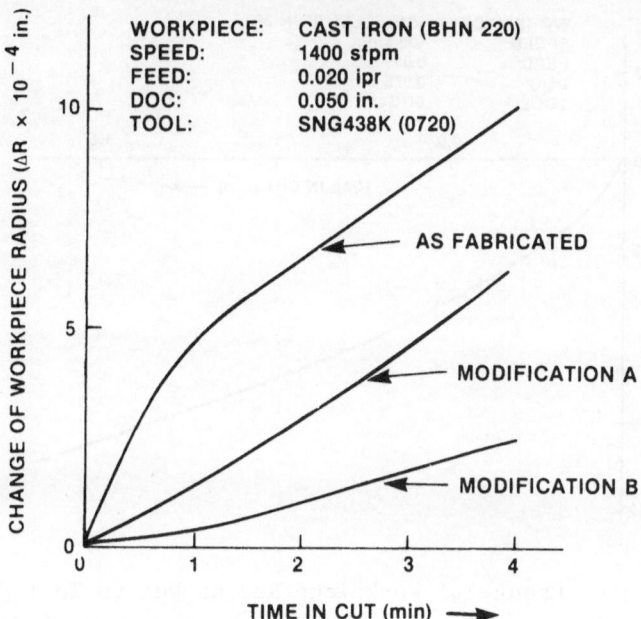

Fig. 3. Machining Performance of a Ceramic Tool with Different
Cutting Edge Preparations

rate, $d(\Delta R)/dt$, was found to be equal to that of the A modified
tools. If a machine part had a tolerance specification of plus
0.001 in., the "as fabricated" tools would reach one-half of the
limit in the first minute of cutting and, as shown in Figure 3, would
have a much shorter tool life than the modified tools. From a tool
materials designers view, cognizance of these types of effects, which
are quickly and reproducibly determined with the air gauging system,
is critical so that non-material effects are not attributed to experimental tool materials.[6]

In monitoring the behavior of cemented carbide tools during
the cutting operation phenomena other than gradual attrition of the
cutting edge have been observed dynamically with the air gauging
system. Figure 4b shows the worn corner of a commercial C-5 uncoated cemented carbide SNG432 tool as viewed from the auxiliary
cutting edge flank after 30 seconds of cutting at 400 sfpm, 0.015
ipr and 0.075 in. DOC. The tool nose has been plastically deformed.
The deformation produced a bulge in the tool nose which lowered the
cutting point and moved the cutting edge toward the workpiece. This
effectively increased the depth of cut causing an increase in the
distance from the air gauge nozzle to the newly machined surface
and an accompanying drop in gauge back pressure (negative ΔR) as

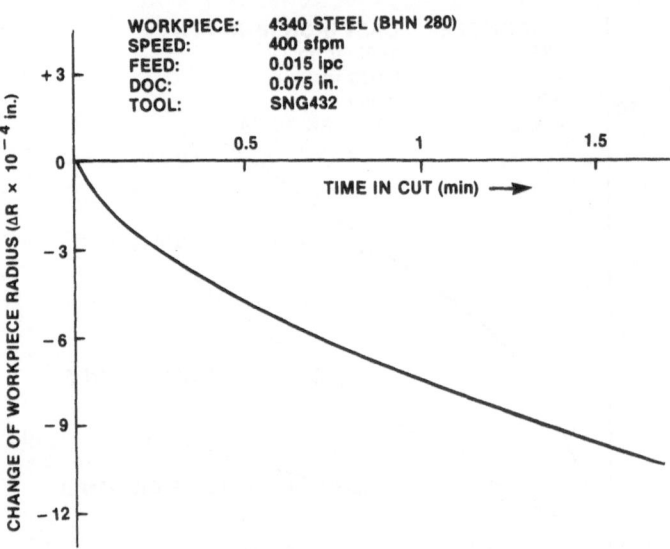

Fig. 4a.   Measured Change of Workpiece Radius Due to Tool Nose
            Deformation

shown in Figure 4a. Although this steel roughing grade is primarily
designed for toughness and not deformation resistance at elevated
temperature it was run at this upper bound of its recommended speed
range to confirm that tool nose deformation caused a corresponding
negative air gauge reading. At lower cutting speeds tool nose
plastic deformation is smaller and not easily detected at the magnification of a typical tool makers microscope, 50X. The deformation is difficult to measure quantitatively by other techniques,
i.e. SEM, due to the lack of an adequate reference surface. The
air gauge tool wear monitoring system readily identifies this contribution to wear in a quantitative manner.

     This point is further illustrated in Figure 5a. The tool shown
was a TiC coated commercial C-5 grade (8.5 w/o Co, 8 w/o TiC, 12 w/o
TaC, balance WC). If this tool had been tested in a traditional
tool wear experiment measuring wear lands, the sequence of events,
either deformation induced coating decohesion and subsequent wear
or wear induced coating decohesion and subsequent deformation,
leading to tool failure would be open to speculation. As shown in
Figure 5b, using the direct tool wear monitoring system the failure
mode was clearly determined. On cut initiation the tool nose deformed and the coating spalled at 2.3 minutes of continuous cutting
as shown by the six micron stepped increase in ΔR which corresponds
to the coating thickness. The positive increase of ΔR occurs because loss of the coating decreases the depth of cut, increases the

Fig. 4b. Worn Cemented Carbide Tool After 30 Seconds of Cutting
          Under the Conditions Shown in 4a

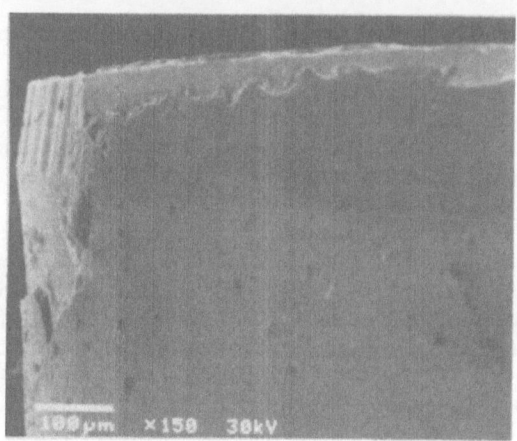

Fig. 5a. Worn Coated Cemented Carbide Tool After Nine Minutes of
          Cutting

workpiece radius, by an amount equal to the coating thickness. Subsequently, the uncoated tool corner wore in an attritious manner
until the end of the cut. The wear marks on the uncoated tool nose
can be seen in Figure 5a. In this particular case the coating was
probably poorly bonded since spalling was also observed on the tool
clearance face well below the cutting action.

     A possible wear mechanism of coated carbide cutting tools has
been reported to be thermal cracking of the coating and subsequent
coating spalling in small areas allowing rapid diffusion wear of the
substrate to occur.[7] The results of cutting tests using the air

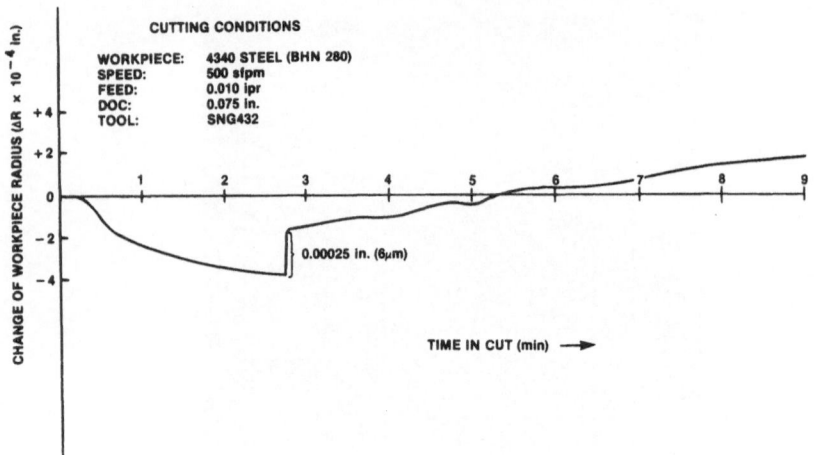

Fig. 5b.   Measured Change in Workpiece Radius Using Tool Shown in
           5a.

gauge monitoring system indicates an alternative or competitive
cause. Plastic deformation of the substrate can result in generation of microcracks in the coating. This, combined with propagation
of reported [8] microcracks in the coating due to thermal expansion
mismatch, can result in spalling. The influence of these effects
on the workpiece radius is observed in Figure 6.

     The tool used for this test was a commercial C-5 coated grade.
On cut initiation gradual plastic deformation of the tool nose occurred. During this time period the wear rate, a positive ΔR effect,
was less than the rate of deformation, a negative ΔR effect, since
the ΔR curve has a negative slope. After four minutes of cutting
the wear rate of the tool became predominant and a steadily increas-

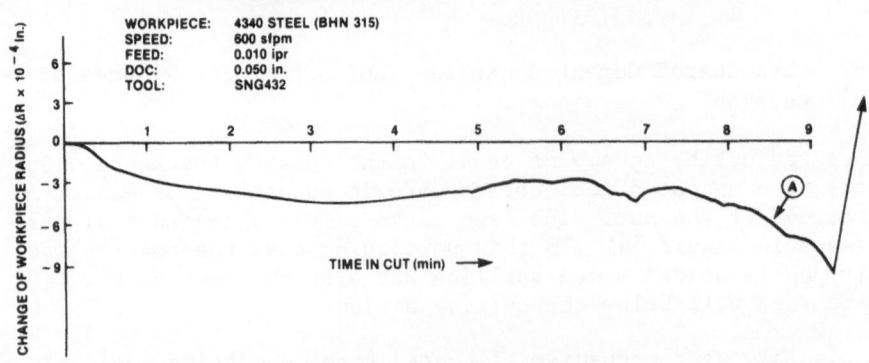

Fig. 6.   Machining Performance of a Commercial Coated Cemented
          Carbide Grade in Continuous Cutting under the Cutting
          Conditions Shown.

sing ΔR was observed. At six minutes of cutting the ΔR curve shows
a series of oscillations between negative and positive slope indicating tool nose instability which seems to be associated with a
sequential deformation-wear phenomenon. After eight minutes of
cutting the tool deformed drastically to the point where accelerated
attritious wear occurred, high positive slope ΔR. SEM photomicrographs of a tool run to failure and a tool stopped prior to accelerated wear (point A in Figure 6) are shown in Figure 7. The coating on the latter (tool stopped during the rapid deformation period
prior to accelerated wear) is beginning to break up because the substrate can no longer properly support the coating. The coating on
the tool run to failure has been completely removed and the uncoated
substrate suffered accelerated wear due to the high cutting speed
of this test.

   Built-up edge (BUE) is indicative of improper machining conditions and is avoided in machining practice since it has a deleterious effect on workpiece finish and cutting tool life under most circumstances. The formation of the BUE is a dynamic process involving
growth to a certain size until part or all of the structure is sheared
off and then the process is repeated [5]. Figure 8 demonstrates this
process. On cut initiation this tool deformed and at two minutes
of cutting began to show wear. At three and one-half minutes BUE
occurred. This caused a sharp drop in ΔR since the depth of cut
was increased by size of the BUE. After a few seconds of cutting
the BUE broke off, causing a sharp positive rise in ΔR. As the
process continued throughout the cut pieces of the tool nose were
carried away when the BUE broke off. The total wear of the tool
is the envelope formed by the positive spikes as shown.

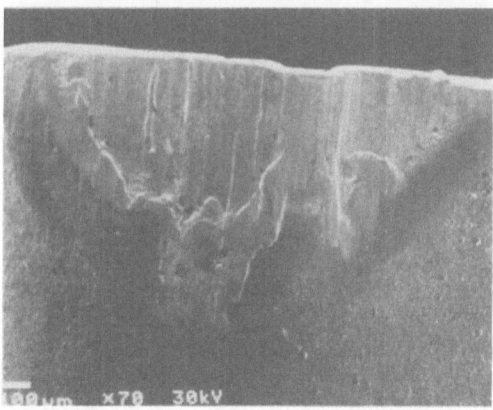

Fig. 7a. Worn Cutting Edge of a Coated Cemented Carbide Tool
          Run to Failure (9 minutes of Continuous Cutting in
          Figure 6)

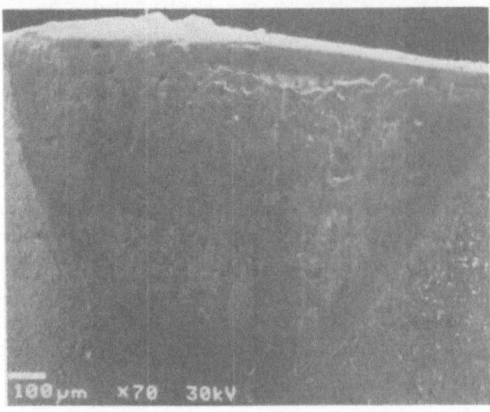

Fig. 7b. Worn Cutting Edge of a Coated Cemented Carbide Tool
          Stopped Prior to Accelerated Wear (point A in Figure 6)

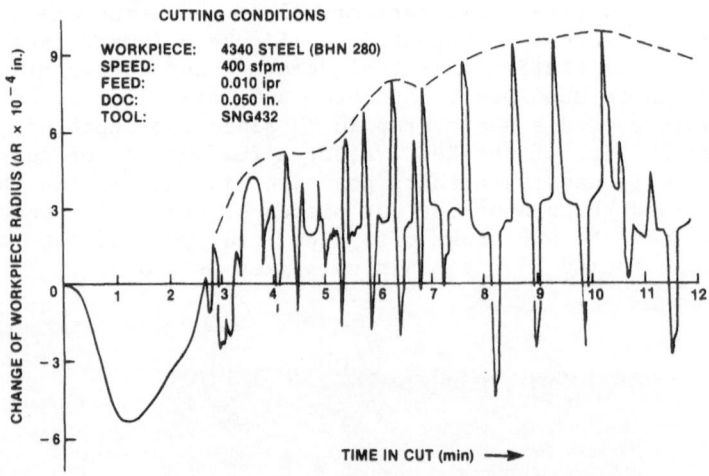

Fig. 8. Air Gauge Response Indicative of BUE

SUMMARY

    A direct monitoring system developed at GTE Laboratories which
continuously and quantitatively measures the change in workpiece
radius associated with physical changes of the cutting edge of a
cutting tool has been discussed. When used in conjunction with
microscopy, this system was shown to have the capability of differ-
entiating between various phenomena associated with the wear of
cutting tools and demonstrated the combined effects of these phenomena on cutting tool life. The direct monitoring system was demonstrated to be a substantially more powerful technique than class-

ical wear land measurements in measuring and understanding cutting tool wear since it does present a live-time representation of the tool wear process. With the addition of other techniques, i.e. measurement of tool forces and workpiece surface finish, into the system, the sensitivity to the dynamic processes involved in tool wear can be further increased and a more complete picture of this complex phenomenon can be obtained.

REFERENCES

1. G. F. Micheletti, W. Koenig, and H. R. Victor, In Process Tool Wear Sensors for Cutting Operations, Annals of CIRP 25, 2: 483 (1976).
2. N. H. Cook, Tool Wear Sensors, Wear 62:49 (1980).
3. H. R. Untenwoldt, Air Gauging Shows Greatest Progress, Microtechnic 22 (5) (1967).
4. G. Boothroyd, Fundamentals of Metal Machining and Machine Tools, Hemisphere Publishing Corporation, London (1975).
5. E. M. Trent, Metal Cutting, Butterworths, London-Boston (1977).
6. G. E. Kane and E. W. Zinners, Jr., An Investigation of the Influence of Cutting Tool Surface Finish Upon Performance Parameters, Lehigh Univ. (1967).
7. J. P. Chubb et al., Comparison of Wear Behavior of Single and Multilayer Coated Carbide Cutting Tools, Metals Technology 7 (7):293 (1980).
8. V. K. Sarin, These Proceedings.
9. S. T. Buljan and V. K. Sarin, Machining Performance of Ceramic Tools, "Cutting Tool Material, Proceedings of International Conference," ASM Publication, 335 (September 1980).

ACKNOWLEDGEMENTS

The authors are grateful to Mr. V. Samarov for his invaluable contributions to the inception and design of the air gauge system and to Mr. R. Wentzell who performed the machining. The support and suggestions of Dr. D.M. Koffman throughout the course of this work were very helpful. The efforts of Ms. K. Bucolo, who typed the manuscript, are appreciated.

APPENDIX

Changes in workpiece radius due to physical changes of the tool cutting edge.

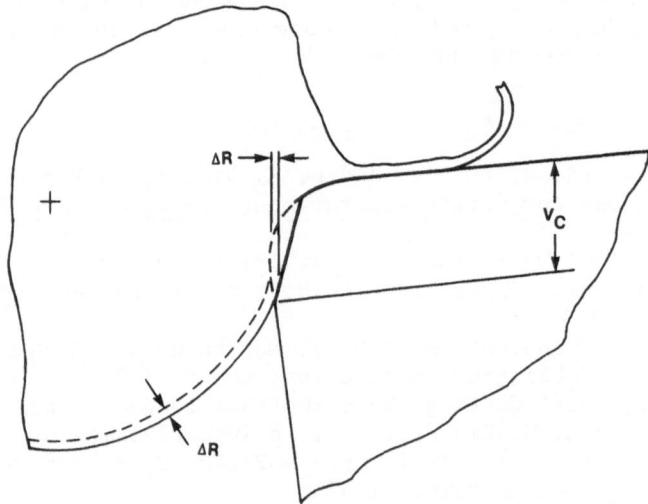

Fig. 9.   Positive ΔR (Increase in Workpiece Diameter) Caused by Tool Wear

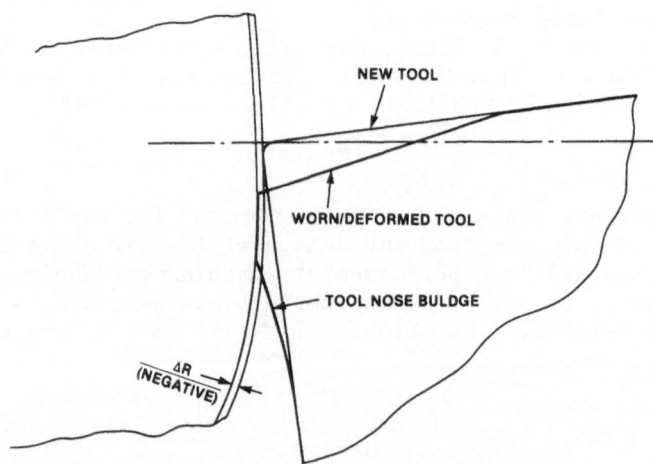

Fig. 10.   Negative ΔR (Decrease in Workpiece Diameter) Caused by Tool Nose Deformation

CLOSING REMARKS

                    H. Fischmeister

How does one sum up the Conference? I came here with about ten research ideas that have sprung from the work I pu. into preparing my
opening lecture. I'm leaving with about thirty. Part of the list is
this, and I'm not going to discuss this in detail I just want to
highlight some points. I think we need models. Models as opposed to
descriptions. If a model is good, it is predictive. If a model is
bad, it's almost as good because then it is provocative. We need
more insight into the time-load spectrum of our tools. There was one
paper at this Conference that went into that. I think we need more
of that type of work.

We need to be reminded constantly that we are really trying to
achieve hot properties. We are measuring cold properties because we
don't want to burn our fingers.

Concerning coatings, I think that we should know more about what
tries to pull the coating away. The adhesion between the work and
the coating is something that I, at least, have seen very little
about.

And another thought that came to me was, maybe the crazing that you
get because of misfit stresses could be put to use. Perhaps it
could be optimized. Crazing is a great way to relieve stresses.

Finishing up about toughness, I think we should realize that really
we are dealing with three entirely different kinds of toughness,
depending on the field in which our tools should work. Large
engineering parts, I think, we have a good grip on. Mining calls
for surface toughness, at least it seems so now. Again, it should
be hot toughness. But again I think we have a fairly good grip of
that field, or it's developing very satisfactorily. What is not
developing at all, it seems to me, is the understanding of toughness in very small stress volumes. I'm quite sure that the defects
that determine the transverse rupture strengths of large wear parts
have nothing to do with the chipping resistance of a cutting edge.

                              1003

And so, this is the final thought I would like to leave you with. A word to the researchers. The new instruments that we have now have brought in a great amount of people who would never have thought of cemented carbides ten years ago. This is very good. But, on the other hand, researchers who come from the basic side of the fence do need some bearings. I like to tell my students never to get the "so what?" criteria. When you start a piece of basic research, you should ask yourself, "When the answer is there, will the response be 'so-what?' Or will it break a log-jam?" Very often the course of basic research is set by the instruments and by the knowledge of the researcher. That's a good thing in many cases. But also it makes for research that does not really coincide with the needs of technology. And I feel that there is cause for anxiety about the divorcement of research and technology in this field as in many others.

This brings me over to the industrialists. The first point I want to make is, I believe I have learned now that there is something like a ten-year cycle in the attractiveness of ideas. And the reasons for this are many. But I think part of the reason is that it takes ten years for the dust to settle. After you have had a real tough fight with an idea, you have reached a certain level of satisfactoriness in the explanations or in the insights in the field. And it also takes ten years for the people who generated the ideas to get out of the field because now they are only doing damage and for young people to get in. And this is a very important lesson to a man who has to pay for research, because he has to pay for more than ten years in order to harvest the fruits of research. I think we've seen several examples at this Conference of ideas that one would have scrapped ten years ago and that have come to life and flowered since.

When I started a job as director of research in a small carbide producing company, a book came into my hands which said, "The first task of a research director is not to direct." And I think that summarized my sentiments at the time very well, because I thought that my first task was to provide a climate. To provide a climate of awareness; on the one hand, of the needs of technology and of the market; and on the other hand, (this is strange to the researchers that I recruited, and it is probably still to the researchers that I recruited into organizations today) to create an awareness of the state of the art in the field. Because each researcher tends to see only his particular specialty and he tries to dig into that and to avoid contact with all the other factors that bother him, or that would bother him. And I think it's the task of a research director to keep constantly rubbing these two things together: awareness of the state of the art or the state of insights in the field, and awareness of the needs. And if he keeps rubbing them together, then a spark may be kindled.

But how good is that spark? Its quality is probably going to be
proportional to the knowledge that his people have outside their own
narrow fields of work. Creative ideas are mostly generated by
entirely new elements coming in and leading to unexpected solutions
that circumvent problems that seem to block further progress. And, so
I think a Conference like this is really the best method to promote
this contact surface, this interfacing of technology and research.
This is just about the only thing, except for paying, that an organization can do to stimulate the growth of new ideas. It's really like
a farmer who sows the seeds in autumn and has to wait to see what
will grow out of them. You can't direct this or that development.
You can just create conditions that are favorable for it.

I know that modern management techniques give you directions on how
to produce inventions. But I still think that providing the necessary contact surfaces is the real important thing. In this respect,
I think the sponsors of this Conference have done themselves a
service.

From which you will probably have gathered that, to me at least, this
Conference appears as a great success. And being one of the foreign
attendees at the Conference, I trust I may speak for all of us when
I offer our very sincere thanks to Dr. Viswanadham and to the whole
of the Organizing Committee and the Staff and also to the Sponsors.
This has been a great Conference and I can only hope that the next
organizing committee will be as successful as this one was.

INDEX